6th Annual International Conference on Material Engineering and Application (ICMEA 2019)

IOP Conference Series: Materials Science and Engineering
Volume 730

Guangzhou, China
20 - 21 December 2019

ISBN: 978-1-7138-2589-0
ISSN: 1757-8981

Printed from e-media with permission by:

Curran Associates, Inc.
57 Morehouse Lane
Red Hook, NY 12571

Some format issues inherent in the e-media version may also appear in this print version.

Institute of Physics
Dirac House, Temple Back
Bristol BS1 6BE UK

Phone: 44 1 17 929 7481
Fax: 44 1 17 920 0979

techtracking@iop.org

Additional copies of this publication are available from:

Curran Associates, Inc.
57 Morehouse Lane
Red Hook, NY 12571 USA
Phone: 845-758-0400
Fax: 845-758-2633
Email: curran@proceedings.com
Web: www.proceedings.com

TABLE OF CONTENTS

MATERIAL ENGINEERING

MATERIAL ENGINEERING AND APPLICATION

ENERGY SCIENCE AND ENVIRONMENTAL ENGINEERING

IOP Conf. Series: Materials Science and Engineering **730** (2020) 011001 doi:10.1088/1757-899X/730/1/011001

Preface

The 6th Annual International Conference on Material Engineering and Application [ICMEA2019] was held successfully in Guangzhou, China during December 20th-21st, 2019. This is one of the leading international conferences for presenting novel and fundamental advances in the fields of Material Engineering and Application. The conference program covered keynote speeches, oral and poster presentations from scientists working in similar areas, and establish platforms for collaborative research projects in fundamental advances in the fields of Material Engineering and Application. Also, processing of the conference provided a platform for delegates and speakers to find potential opportunities for international cooperation.

The major goal and feature of the conference is to bring academic scientists, engineers, industry researchers together to exchange and share their experiences and research results, and discuss the practical challenges encountered and the solutions adopted. Distinguished professors and experts in the area delivered the latest knowledge in their respective expertise fields. International scientific activities, like conference, symposium and workshop are big scientific platforms for knowledge exchange among the scientists from all over the world.

The theme of this conference was the key issues associated with science and technology in this rapidly evolving field of research and to promote contact between basic researcher and technological needs for real advanced materials and industrial applications.

This event was intended as a vehicle for the dissemination of research results on latest advances made in the area of advanced materials and technological processes, which gave an excellent opportunity for the scientists and researchers around the world to have a common platform to exchange their findings and to discuss the developments and make a base for collaborations in the national and international level.

The volume consists of the selected papers presented at ICMEA2019, which were addressed to diverse topics related to recent trends mainly in advanced, functional, industrial and smart materials. I believe that this conference provided good opportunities for the delegates to exchange new ideas and application experiences face to face, to establish business or research relations and to find global partners for future collaboration.

We want to express our gratitude to all the members of conference committee, and reviewers who spared their valuable time, for their advice which have certainly helped to improve the quality, accuracy, relevance and the sincere efforts to maintain the quality of each paper selected for conference program and volume for publication. Also, we would like to thank every author and participant who has contributed to the success of the conference, and we truly hope to see you again in the next year of ICMEA!

ICMEA2019 Organizing Committee

Peer review statement

All papers published in this volume of *IOP Conference Series: Materials Science and Engineering* have been peer reviewed through processes administered by the proceedings Editors. Reviews were conducted by expert referees to the professional and scientific standards expected of a proceedings journal published by IOP Publishing.

A Study on Physical and Chemical Properties of Detonation Nanodiamond

Shengzhong Xie[1,3], Lingzhi Ma[1], Weiping Zeng[1,*], Peng Gao[1] and Xuezhang Liu [2]

1. Hunan Nonferrous Metals Vocational and Technical College, Zhuzhou 412006, China;
2. Jiangxi Normal University of Science and Technology, Nanchang 330013, China;
3. Hunan Nonferrous Metals Scientific and Technical Development Co., LTD., Zhuzhou 412006

*Corresponding author: Weiping Zeng

Abstract. Using analysis test methods, such as x-ray diffraction (XRD), x-ray photoelectron spectroscopy (XPS) and Fourier infrared spectrum (FTIR), x-ray small angle scattering (SAXS), optical correlation spectroscopy (PCS), scanning electron microscopy (SEM) and so on, the researchers systematically studied the detonation nanodiamond (DND)'s physical and chemical properties like crystal structures, surface chemical compositions, surface electrical property and particle size distribution, particle morphology and agglomeration state, etc. The results showed that the crystal structures did not be changed before or after DND disaggregation and dispersion, but the surface chemical compositions, surface electrical property, particle size distribution, particle morphology and agglomeration state all were changed obviously. The systematic data of physical and chemical properties of DND were obtained through the study in this paper, which could provide theoretical basis and technical reference for the application research of DND based on relevant technologies.

1. Introduction

With TNT, RDXX and other explosives as carbon source precursors in a high-strength steel closed container, DND is synthesized by nanodiamond through detonation reaction. Studies have shown that DND has stable chemical properties, acid-alkali resistance, corrosion resistance, hydrophilic surface, dual characteristics of super hard materials and nanomaterials, and has been applied in ultra-precision polishing, lubricant additives, composite electroplating, biomedicine, polymer material modification and other fields, and has broad market application prospects[1~7]. Characterizing the physical and chemical properties of DND, is the basis of research on the application all above. And systematically studying DND's physical and chemical properties like crystal structures, surface chemical compositions, surface electrical property and particle size distribution, particle morphology and agglomeration state, etc., which could provide theoretical basis and technical reference for the application research of DND based on relevant technologies.

2. Experiment

2.1. Sample Preparation

The raw material sample of this experiment is the DND produced by a domestic company. It is actually a nanodiamond hard aggregation, in the form of gray powder, known as DND grey powder. The preparation process of nanodiamond dispersion material sample is as follows: put the DND grey powder, grinding aid and deionized water in a ball mill for ball mill, through the ball mill-amentia's

IOP Conf. Series: Materials Science and Engineering **730** (2020) 012001 doi:10.1088/1757-899X/730/1/012001

synergy effect, the nanodiamond hard aggregation will be spread out. Then by acid treatment in removing impurity and cleaning, get the water-based DND pulp, after size grading treatment, the pulp turns into water-based DND dispersive material, in the form of dark grey liquid.

2.2. Testing and Characterization
The crystal structure of DND was studied by x-ray diffractometer (D/MAX-EAX, Rigaku, Japan). The surface chemical compositions of DND were measured by x-ray photoelectron spectroscopy (XPS) and infrared spectroscopy. The x-ray small-angle scattering method was used to measure the particle size distribution of DND original particles, and the nano-laser particle size meter (Zetasizer3000HS, Malvern Instrument co., LTD.) was used to measure the particle size distribution of DND in water phase. Anhydrous ethanol was used as dispersing medium, ultrasonic wave was used for dispersing for 15min, samples were taken, and SEM (JSM-5600LV, Japanese Electronics Company) was used to observe the morphology and agglomeration state of DND particles.

3. Results and Discussion

3.1. Crystal Structure
The test results of the crystal structure of DND (before and after dispersion) are shown in Figure 1. On the one hand, it shows that the nanodiamonds are cubic lattice structures before or after dispersion. On the diffraction pattern, {111}, {220} and {311}, these three characteristic diffraction peak positions are the same as ordinary diamonds which are synthesized by static pressure method. But since the average grain size of DND particles is about 4.5nm and their grain size does not change after dispersion, it is far less than the average grain size of ordinary diamonds which are synthesized by static pressure method. All above make the diffraction peaks of DND are wider than those of ordinary diamond. On the other hand, compared with before dispersion, the diffraction peak positions of dispersed nanodiamond does not change, and the crystal structure also does not change, the reason may be that DND is just dispersed from coarse aggregations to small aggregations and single particles in the process of ball mill-amentia's dispersion, and is just modified the surface of small aggregations and single crystals which after depolymerization, thus it does not alter the crystal structure of nanodiamond.

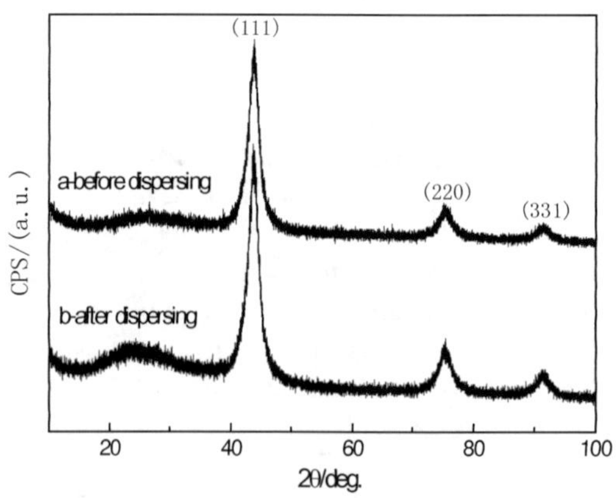

Figure 1. X-ray Diffraction Pattern of DND

3.2. Surface Chemical Compositions
The test results of DND grey powder's surface chemical elements are shown in Table 1. X-ray photoelectron spectroscopy (XPS) analysis showed that the main elements of DND surface are C, O, N, S, among them, the atomic percentage of C is 90.646%, the atomic percentages of O, N, S are 8.090%, 1.135% and 0.129% respectively. Visibly, the sum of atomic percentages of amorphous carbon is 9.354%. The reason is that while the oxygen acid was removing the carbon of non-diamond in the surface of nanodiamond, the oxygen-containing groups were grafted to the surface of

nanodiamond. It is speculated from the XPS oxygen scan, nitrogen scan and nitrogen scan of DND that the oxygen on the surface may exist in the form of carbonyl and hydroxyl, the nitrogen on the surface may exist in the form of amine, nitro and nitride, and the carbon may exist in the form of C-H, C=O and C-C.

Table 1. XPS Analysis of DND Grey Powder

Peak	O*1s*	N*1s*	C*1s*	S*2p*
Atomic Percentage	8.090%	1.135%	90.646%	0.129%

The influence of different treatments on the infrared spectra of DND is shown in Figure 2. In Figure 2, G represents the nanodiamond grey powder (DG), M represents the nanodiamond sample treated by chemical machinery, and Y and Z represent the sample treated by Y and Z steps.

Sample G has two absorption peaks at 2853.97cm^{-1} and 2922.22cm^{-1}, which can be attributed to C-H vibration, namely, symmetric and asymmetric vibration of $-CH_3$ and $-CH_2CH_3$. In O-H and N-H stretching vibration zones, there is a strong and wide peak of 3433.76cm^{-1}, which may be caused by the stretching vibration of hydroxyl groups on the surface of nanodiamond or hydroxyl groups of carboxyl groups. In addition, there may be amino groups, but their N-H characteristic stretching peak is covered by the peak of hydroxyl groups. The absorption peak at 1629.29cm^{-1} may be formed by the combination of amide peak IνC=O and II δN-H+νC-N. The absorption peak at 1764.89cm^{-1} (next to 1629.29cm^{-1}) may be the characteristic absorption of the stretching vibration of carboxyl groups' carbonyl groups. In general, the position of this peak is near 1720cm^{-1}, at which time it moves in the direction of high wave number. It may be caused that the a-carbon on the hydroxyl group is directly connected to the electron-absorbing group such as Cl, or there are acid groups on the surface of the diamond. The absorption peak at 1117.43cm^{-1} may be the superposition of the stretching vibration peak of C-OH on the surface of nanodiamond and the stretching vibration peak of C-O single bond of carboxyl group, in addition, the stretching vibration peak of amine C-N is also in this region. The absorption peak at 619.74cm^{-1} may be caused by the stretching vibration of C-Cl peak.

Sample M is the nanodiamond sample treated by ball milling. Compared with G, its carbonyl peak at 1764.89cm^{-1} disappeared, and it is probably caused that the carboxyl group on the surface of diamond generated carboxylate during the ball-milling process, and the carboxylate appeared in the spectrum as two free stretching vibration peaks (1611.74cm^{-1} and 1324.48cm^{-1}). The absorption peak at 1485.50cm^{-1} may be the bending vibration absorption peak of free hydroxyl O-H. It can also be found from Figure 2 that the hydroxyl peak of sample M was significantly weakened and moved in the direction of low wave number. The main reason is that some hydroxyl groups reacted with other ions and the number of hydroxyl groups decreased.

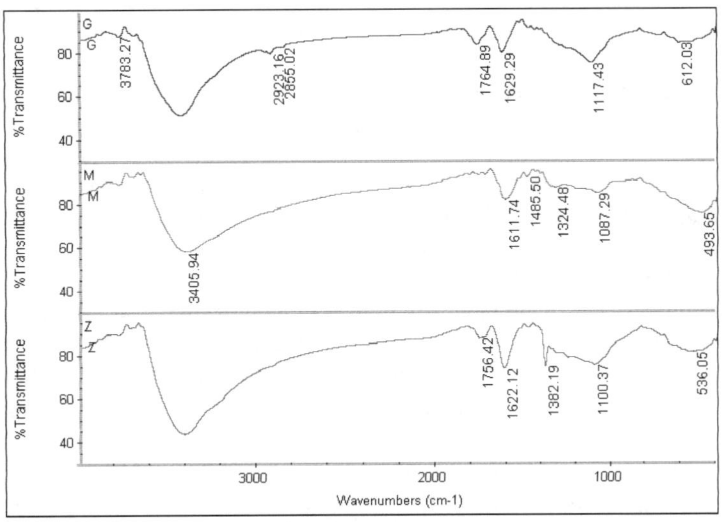

Figure 2. Effects of Different Treatments on DND Infrared Spectra

After acid treatment and impurity removal on sample M, compared with M, since the carboxylic acid anion peak on the surface of nanodiamond disappeared and the carboxyl characteristic peak reappeared, the hydroxyl characteristic absorption peak at 3405.47cm^{-1} was enhanced. In addition, a characteristic absorption peak of 1382.19cm^{-1} appeared, which may be due to the presence of -NO_2 groups on the surface of nanodiamond.

3.3. Surface Electrical Property

Generally, Zeta potential is used to measure the electrostatic repulsion force and dispersion stability of ultra-fine powder in aqueous solution. It is generally believed that when the absolute value of Zeta potential of ultra-fine particles in aqueous solution exceeds 30mV, the dispersion system has good dispersion stability. Because of ionization, ion adsorption and lattice substitution, the surface of DND in aqueous phase is often charged. As shown in Figure 3, Zeta potentials of nanodiamond change with pH values, before or after dispersion, or inorganic electrolyte sodium polyphosphate (STPP) is added after dispersion: When pH = 2.5 ~ 11.0, the Zeta potentials of DND in water medium are in 15 to 33mV, ball mill dispersion can make the Zeta potentials of DND in 22-36mV, and they all increase in the scope of pH2.5 ~ 11.0; In the range pH2.5 ~ 8.3, the Zeta potentials of dispersed nanodiamond are more than 30mV, so the system's stability is enhanced; The reason is that the nanodiamond adsorbs more positively charged functional groups on its surface after dispersion, which makes the surface potentials increase; After the addition of STPP, when pH = 8.3~11.0, the Zeta potentials of nanodiamond below -35mV, and the absolute potential values increase, thus significantly improving the electrostatic repulsion between particles. It can be found that the absolute Zeta potential values of DND can be greater than 30mV by dispersing and adding agentia STPP after dispersion. In the actual application range of pH2.5~11.0 of DND, the suspension is in a stable dispersing state.

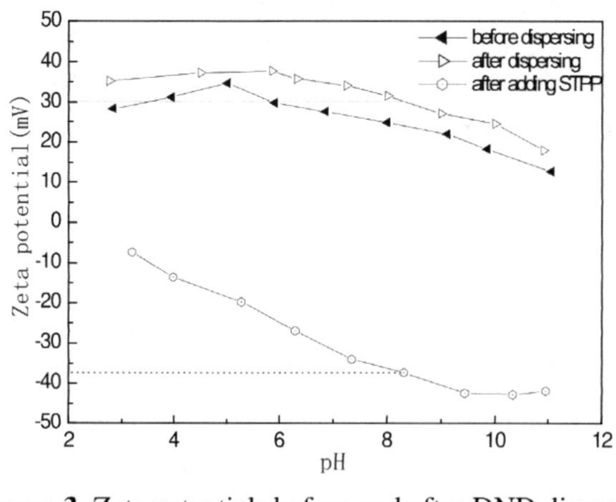

Figure 3. Zeta potentials before and after DND dispersion

3.4. Particle Size Distribution

In Figure 4, we can see the primary particle size distribution curve that SAXS measured DND grey powder: DND average particle size is about 10nm, particle size distribution are all below 60nm. But in Figure 4, we can also see the particle size distribution curve that Zetasizer3000HS measured DND in deionized water: the particle size distribution of DND in water medium is 10 to 8500nm, and most particles above 500nm, less than 20% (wt) of particles under 100nm. That is because in the process of synthesis and drying, nanodiamond is easy to form ultrasonic inseparable hard aggregates through the action of chemical bond force, Van der Waals force, coulomb force, etc. The existence of hard aggregates limits the application of nanodiamond and makes the excellent properties of nanodiamond partially lost. Therefore, it is necessary to disperse and classify nanodiamond in practical application system to meet the needs of practical application.

Figure 5 shows the change of the volumetric particle size distribution curve of nanodiamond in deionized water after ball milling dispersion and particle size classification. Zetasizer3000HS measured the average particle size of DND grey powder, ball mill dispersion and particle size classification at 364.1nm, 84.7nm and 54.9nm, respectively. From the particle size distribution curve in Figure 5, it can be seen that the particle size distribution of nanodiamond is greatly reduced after dispersion, with about 97% of the particles below 100nm and the maximum particle size at about 230nm. The dispersed nanodiamond suspension was graded to make its particle size distribution below 100nm.

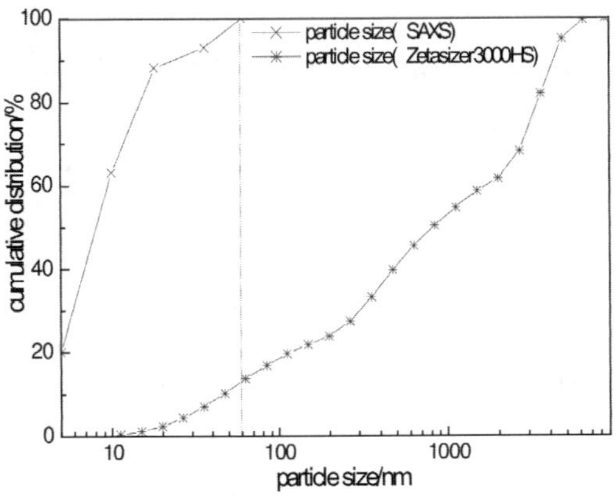

Figure 4. Particle Size Distribution of DND Grey Powder

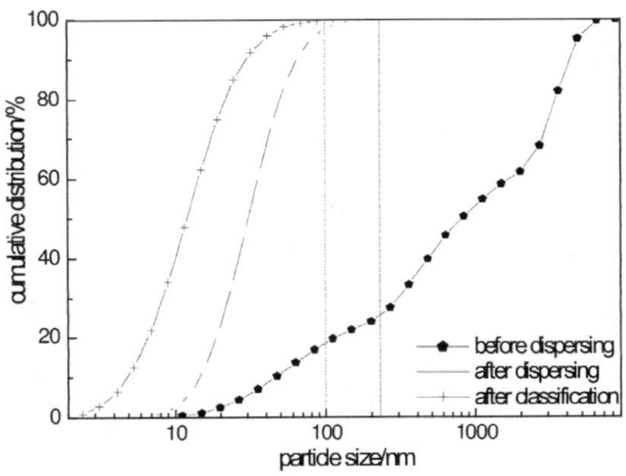

Figure 5. Particle Size Distribution before and after DND Treatment in Aqueous Phase

3.5. Particle Morphology and Agglomeration State

DND ultrasonic wave was fully dispersed in ethanol, and the dispersible solution was dipped in glass rod and glued on the copper plate to make samples, which were naturally cooled and dried , for SEM analysis, as shown in Figure 6. The picture on the left is the SEM photo of DND grey powder, and the observed DND is agglomerate state with the shape of particle snowflake. The picture on the right shows the SEM image of DND grey powder dispersed by ball milling and graded particle size and the observed DND is dispersed and the particles are spherical or quasi-spherical. SEM can directly observe the size, shape and agglomeration performance of nano powder. When the tested sample has good dispersion and no agglomeration phenomenon, SEM can measure the size of the original particle, while conversely, it can measure the size of the agglomeration. Since SEM photographs reflect the size of particles in local areas and cannot reflect the particle size distribution of the whole sample, the author believes that SEM can be used as an auxiliary method to measure the particle size of samples. In addition, SEM is not suitable for particle size detection of some samples that are unstable or even decomposed under intense electron beam bombardment. SEM images showed the particle size of nanodiamond after dispersion and grading, with the shape of spherical or quasi-spherical.

Figure 6. SEM Photos of DND

3.6. Other Physical and Chemical Properties
DND grey powder has been measured by NOVA1000 equipment: specific surface area is 293.6m^2/g, relative density is 3.05~3.30g/cm^3, conductivity is 7.7×107Ω·cm, magnetic susceptibility is 1.0×10^{-8}m^3/kg, and stomatal area is 1.314cm^2/g.

4. Conclusion
(1)The crystal structure of DND grey powder is cubic lattice structure, and dispersion does not change the crystal structure. The primary particle size distribution was below 60nm. The morphology of the particles is spherical or quasi-spherical, and they are in the state of agglomeration. The specific surface area is 293.6 m^2/g, the relative density is 3.05~3.30g/cm^3, the conductivity is 7.7×107Ω·cm, the magnetic susceptibility is 1.0×10^{-8}m^3/kg, and the stomatal area is 1.314cm^2/g.

(2)The main elements on the surface of DND grey powder are C, O, N and S, among which the atomic percentage of C is 90.646%, the atomic percentage of O, N and S is 8.090%, 1.135% and 0.129% respectively. The functional groups on the surface mainly include polar functional groups such as -OH, -NH_2, -COOH and -NO_2, and a small amount of non-polar functional groups such as -CH_3 and -CH_2CH_3. After ball-milling treatment, the hydroxyl peaks of DND grey powder were obviously weakened and the -CH_3 and -CH_2CH_3 peaks disappeared. Then, after further acid treatment, the characteristic absorption peak of hydroxyl group was enhanced.

(3)In the practical application range of pH2.5~11.0 of DND, the Zeta potential of DND in the water medium is located at 15-33mv, and the absolute value of Zeta potential of DND can be greater than 30mV through dispersion and addition of agent STPP after dispersion, and the suspension is in a stable dispersion state. However, the size distribution of DND in water medium is 10~8500nm, and most of the particles are above 500nm. The average particle size of DND grey powder, ball mill dispersion and particle size classification were 364.1nm, 84.7nm and 54.9nm, respectively. The particle size classification of dispersed nanodiamond suspension could make its particle size distribution below 100nm.

5. Acknowledgement
Fund project: Funded by Science and Education Joint Project of Natural Science Foundation of Hunan Province (2018JJ5047); Funded by Hunan Province Five 100 Major Products Innovation Projects (Zhuzhou Governmental Letter No.7, 2018); Funded by Transformation and Development Project of SME Service Center of Hunan Province (Xiang SME Office Letter No.8, 2019);

6. Reference
[1] Wen Chao, Guan Jinqing, Liu Xiaoxin, etc. Development History and Current Situation of Explosive Detonation Synthesis Nanodiamond[J]. Superhard Materials Engineering, 2009, 21(2): 46~51.

[2] Cai Yunpeng, Zeng Liming, Ye Xiaochuan etc. Research Progress on Preparation of Nanodiamond Suspension [J]. Diamond and Abrasive Tool Engineering, 2011, 31(183): 68~72.

[3] Xuezhang Liu, Xiangyang Xu. Ultra-fine Polishing of Glass-ceramics by Disaggregated and Fractionated Detonation Nanodiamond. Ceramics International[J], 2017, 43(8):6063-6068.

[4] Nicholas Nunn, Marco Torelli, Gary McGuire etc. Nanodiamond: A High Impact Nanomaterial. Current Opinion in Solid State and Materials Science[J], 2017,21(1):1~9.

[5] Caymen Novak, David Kingman, Kyle Stern etc. Tribological Properties of Paraffinic Oil with Nanodiamond Particles. Tribology Transactions [J], 2014,57:831~837.

[6] Dae Gon Lim, Racelly Ena Prim, Ki Hyun Kim etc. Combinatorial Nanodiaomond in Pharmaceutical and Biomedical Applications. International Journal of Pharmaceutics [J], 2016, 514(1):41-51.

[7] Arash Golchin, Alexis Villain, Nazanin Emami. Tribological Behaviour of Nanodiamond Reinforced UHMWPE in Water-lubricated Contacts [J].Tribology International, 2017, 110: 195-200.

IOP Conf. Series: Materials Science and Engineering **730** (2020) 012002 doi:10.1088/1757-899X/730/1/012002

Preparation of TiC based Cermet with Different Ratio of Cu and Al

Nengwei Wang[1,a,*] and Shuquan Wan[2,b]

College of Vanadium and Titanium, Panzhihua University, 617000 Panzhihua, Sichuan Province, People's Republic of China.
[a]wnw_2008@163.com, [b]48182833@qq.com

Abstract. TiC based cermet belongs to superhard tool material, which has high melting point, high hardness and excellent chemical stability. It is the preferred material for cutting tools and wear-resistant components. At the same time, it has excellent conductivity and is also the preferred material for electrode. In this experiment, TiC based cermet was prepared by atmosphere protection sintering method with TiC, Cu and Al as raw materials. The phase analysis was carried out by X-ray diffractometer(XRD), the microstructure was observed by scanning electron microscope (SEM), the element analysis was carried out by energy spectrometer, and the bending strength was tested by universal material tester. The different sintering temperature and the ratio of additive Cu and Al were systematically studied under the premise of controlling the total mass of TiC based cermet, which effect on microstructure and mechanical properties of TiC based cermet. The experimental results show that the bending strength of cermet increases obviously with the increase of copper content in a certain range, and the best mechanical properties of cermet are obtained when the ratio of Cu and Al is 6:4. With the increase of sintering temperature, the mechanical properties of TiC based cermet are significantly strengthened. When the sintering temperature reaches 1100 °C, the grain size is fine and uniform, and the comprehensive properties of the material are the best.

1. Introduction

Cermet is an heterogeneous composite composed of metal or alloy and ceramic phase, in which ceramic phase accounts for about 15% and 85%. It not only maintains the high strength, high hardness, wear resistance, high temperature resistance, chemical stability and oxidation resistance of ceramics, but also has good metal toughness and plasticity. It is a very important and excellent tool material and structural material in modern industrial production [1]. It is widely used, involving all sectors of national economic production and various fields of modern social application, and has promoted the development of industry and the improvement of productivity.

However, the fatal weakness of cermet is brittleness, which seriously restricts the application of cermet. In the 1950s, based on the shortcomings of TiC-Ni based cermet, such as high brittleness and low strength, the addition of Mo to TiC-based cermet can form an annular phase metal around it to enhance the wettability of TiC and N, and greatly improve the strength of the alloy [2]. This important technological breakthrough makes TiC based cermet become the most valuable and promising cermet material at present.

The microstructure of TiC based cermet is composed of carbide hard phase and metal bonded phase. A coating layer is formed at the edge of hard phase. It is formed by the solid solution of Mo, W, Ta, Nb and other elements in TiC. The ring structure is exactly the same as that of TiC lattice, and the lattice parameters are very close. Therefore, dislocations can pass through the ring and hard phase continuously. In fact, the ring is a structure covering the surface of TiC particles, which is formed by the diffusion of Mo, Ni and other atoms to the surface of TiC. The coating greatly improves the wettability between hard ceramic phase and metal bonded phase and holds back the aggregation and growth of hard phase. However, since the coating itself is still in the brittleness stage, its thickness must be appropriate. It is reported that the average thickness of annular phase can not exceed 0.5 μm, otherwise the bending strength of the material will decrease significantly [3]. TiC based cermet is the most widely developed material in titanium, zirconium and chromium transition metal carbides. TiC can be made into all kinds of multiphase ceramic materials with TiN, WC, A12O3 and other raw materials. TiC based cermet has been used in the cutting of high speed wire rod and carbon steel because it does not have crescent depression wear and good oxidation resistance with steel. Multiphase ceramic tools containing titanium carbide have been widely used [4].

2. Experimental Process

In the experiment, a series of TiC based cermet samples were prepared by traditional powder metallurgy method and different ratio of Cu and Al according to the optimum technological conditions. The total mass was 25g, the addition of TiC accounted for 60%, and the total addition of additive Cu and Al accounted for 40% of the total mass. The ratio of Cu and Al components was set to 5 groups: 0:10, 2:8, 4:6, 6:4, 8:2, respectively. So as to ensure that the microstructure and properties of each group were comparable, the composition ratio was shown in Table 1.

The mixture was put into nylon grinding tank and cemented carbide ball was used as grinding ball. Under the condition of rotating speed 175r/min, the continuous ball milling was carried out in the star ball mill for 24 hours. Then the ground powder and the grinding ball were poured into the powder plate, dried at 80 °C for 12 hours, the dried powder was sieved, the cemented carbide ball was taken out, and then the mixed powder was ground until there was no greater agglomeration. Polyvinyl alcohol solution was used as forming agent, and two drops of polyvinyl alcohol solution were added to the mortar for granulation every 25g dried powder for about 20 minutes, until the powder was fully mixed with the molding agent. The granulated mixture was molded by four-column hydraulic press in steel die. The samples with different ratios of Cu and Al were sintered at 900 °C, 1000 °C and 1100 °C, respectively.

Table 1. Composition design (wt.%)

Sample number	TiC	Cu	Al	Cu:Al
1	60	0	40	0:10
2	60	8	32	2:8
3	60	16	24	4:6
4	60	24	16	6:4
5	60	32	8	8:2

Then the properties of the samples were tested according to the requirements of the experiment, including porosity and bending strength. Then the phase analysis was carried out by X-ray diffractometer(XRD), the microstructure was observed by scanning electron microscope (SEM), the element analysis was carried out by energy spectrometer.

3. Analysis and Discussion

3.1. Porosity Analysis

In the process of sintering, densification and grain growth are two simultaneous processes. With the increase of sintering time and temperature, the density of TiC based cermet increases and the porosity decreases, which will improve the hardness of cermet. The porosity of cermet at different temperatures is shown in Fig. 1. Through the analysis and processing of the experimental data, sample 2 is much lower than sample 1 with the increase of Cu content at 900 °C. Through the SEM analysis of the sample, it is found that the microstructure of TiC based cermet shows a typical core-shell structure, and the addition of Cu greatly promotes the wettability of Al and makes the TiC powder reunite closely. At the same time, the electron valence compounds formed by Cu and Al are also fully filled with gaps, so the porosity of the samples decreases to a great extent. At 1000 °C and 1100 °C, with the increase of Cu content, the porosity increases to a certain extent, the Cu content is too much, the sintering temperature is not high, so the porosity of the sample continues to increase.

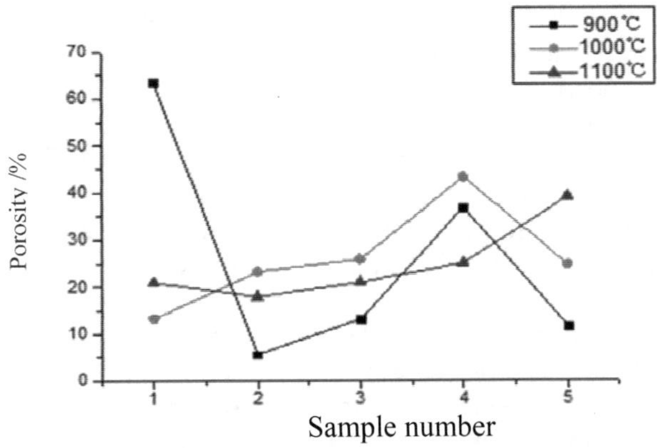

Figure 1. Porosity change at different temperatures

3.2. Analysis of Bending Strength

Bending strength refers to the ability of materials to resist bending and non-fracture, which is mainly used to reflect the strength of brittle materials such as ceramics. In order to improve the bending strength of cermet materials, it is necessary to have a good core / shell and interface bonding between bonded phase and shell. It can be found from the experimental results that the bending strength of TiC based cermet samples increases with the increase of Cu content, and the increasing trend is very obvious, the variation of bending strength is shown in figure 2.. This is because on the one hand, Cu, as a metal phase itself, has a certain strength, and the increase of Cu content is beneficial to improve the bending strength of cermet materials.On the other hand, with the increase of Cu content, the microstructure of cermet began to refine, according to the classical Hall-Petch formula, the bending strength of cermet was enhanced by fine grain strengthening, and the metal and ceramic phase dissolved in bonding phase increased with the increase of Al-Ti alloy content in TiC expansion ceramics, which resulted in solid solution strengthening.

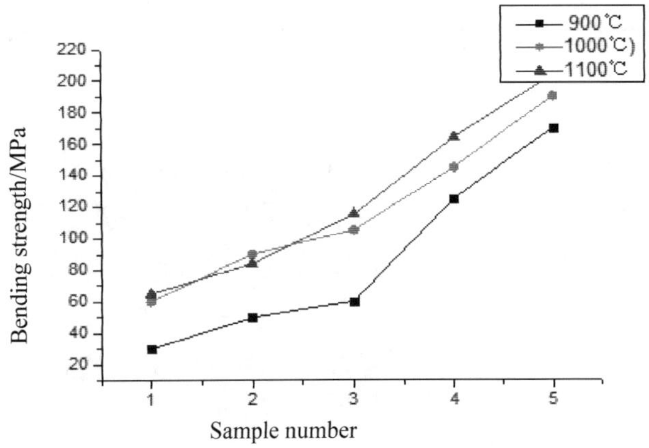

Figure 2. Variation curve of bending strength at different temperatures

3.3. Observation of Morphology of SEM and EDS Analysis

Before the addition of Cu, the cermet phase is mostly spherical. With the addition of Cu, some ceramic phase particles changed from circular to polygons (most of them were quadrilateral), and the interface between ceramic phase and bonded phase is a flat straight line. The orientation relationship of the ceramic phase / bonded phase interface is straight and straight, which indicates that after liquid phase sintering, the liquid phase of cermet is solidified and grown by heterogeneous nucleation with ceramic phase as substrate interface. At this time, the curvature size and direction (concave and convex) of the substrate interface (ceramic phase) will affect the difficulty of crystal nucleus growth, the concave crystal nucleus growth is the easiest, the plane is in the middle, and the convex crystal nucleus growth is the most difficult. There is no concave surface in ceramic phase, so the relatively straight surface part is easy to become the non-uniform nucleation part of liquid phase. Because the relationship between bonded phase nucleation and ceramic phase is beneficial to reduce the interface energy needed for nucleation, there is a certain relationship between the two. However, due to the difficulty of nucleation, there is no orientation relationship in the protruding part of ceramic phase.

The results of EDS identification of TiC based cermet elements containing 16%Cu at 900 °C, 1000 °C and 1100 °C are shown in table 2. It indicates that the burning rate of Al increases gradually, and the atomic percentage of Cu does not change obviously with the increase of sintering temperature. The melting point of Al element is about 660C, and the boiling point is about 2 500 °C, which indicates that with the increase of sintering temperature, the Al element changes from solid phase to liquid phase, the physicochemical phenomenon of dissolution and precipitation occurs in tic-based cermet. Finally, the Al vapor escapes with the protective argon gas, and the weight percentage of Al decreases.

The distribution of elements at 1000 °C for TiC-based cermet containing 16% Cu is showed in fig. 3. It can be seen from the diagram that there is no obvious boundary between the distribution of Cu and Al elements, and the distribution is very similar. The reason may be that Cu and Al mix very evenly on the one hand, and on the other hand, Cu and Al form electronic valence compounds. Copper as wetting agent wet TiC and Al,. Al is used as bonding phase to promote the agglomeration of TiC particles. At the same time, Cu and Al miscibility fills the gap between TiC particles and reduces porosity.

Table 2. The weight percent of TiC-based cermet elements containing 16% Cu at different sintering temperatures. (wt.%)

Temperature/°C	Element type					
	C	O	Al	Ti	V	Cu
900	57.33	6.67	10.02	18.21	1.22	6.77
1000	58.41	7.04	7.85	18.69	1.10	6.52
1100	59.60	7.37	6.09	19.71	1.04	6.23

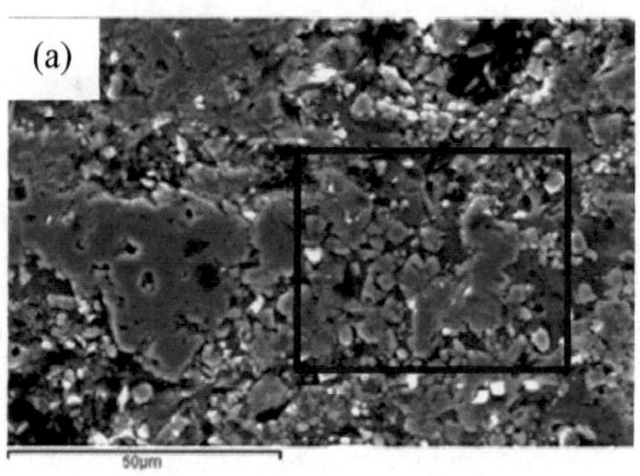

IOP Conf. Series: Materials Science and Engineering **730** (2020) 012002 doi:10.1088/1757-899X/730/1/012002

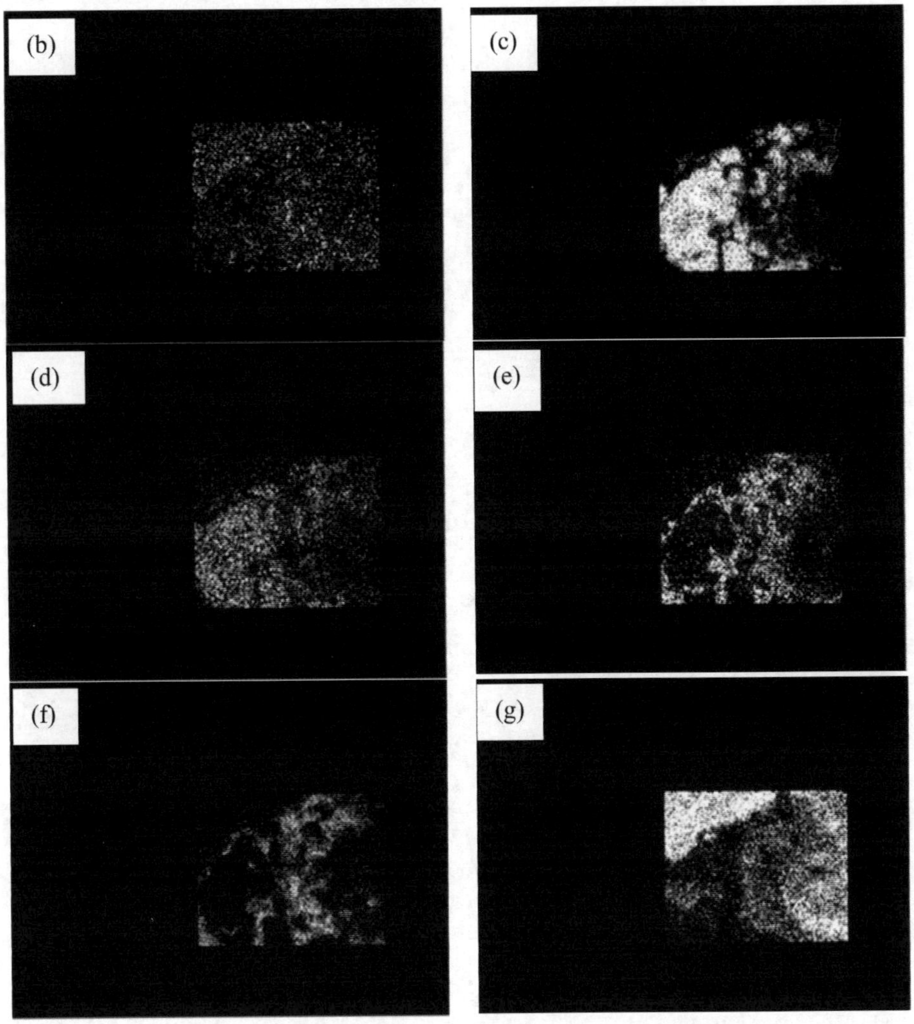

Figure 3. Distribution of Elements of the TiC-based cermet with 16% Cu at 1000 °C

(a)Selection analysis of EDS; (b) Distribution of O elements; (c) Distribution of Ti elements; (d) Distribution of V elements; (e) Distribution of Cu elements; (f) Distribution of Al elements; (g) Distribution of C elements

3.4. Phase Analysis of XRD

The XRD diffraction pattern of five different Cu and Al ratios at 1100 °C is shown in fig. 4. Through the comparative analysis of five groups of diffraction patterns, it can be seen that the diffraction peak of hard phase TiC remains unchanged, and the peak value matches well with the standard sample in PDF card with the increase of the amount of Cu metal powder. In addition to the line emission peak of TiC, there is a diffraction peak of solid solution with Cu3Al4. With the increase of copper content , the diffraction peak intensity of bonded phase Cu3Al4 is increasing. Through PDF card inspection, it is found that the diffraction peak is basically the same as that of Al. It can be considered that the continuous enhancement of the diffraction peak is the result of the superposition of the two diffraction peaks. In addition, the existence of C was found in XRD diffraction pattern. It is considered that the heating temperature of TiC based cermet sample is too high, or the residence time at high temperature is too long, decarbonization reaction occurs, which makes the C in TiC free on the surface of the sample.

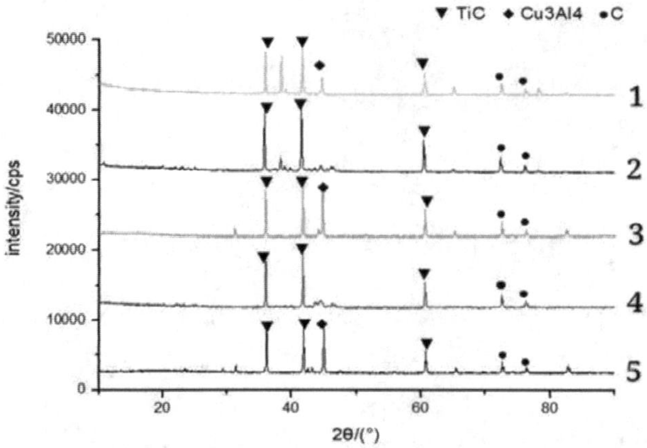

Figure 4. XRD diffraction Analysis of cermet samples
(1)32%Cu; (2)24%Cu; (3)16%Cu; (4)8%Cu; (5)0%Cu

4. Conclusions

In this study, TiC based cermet materials with different Cu and Al ratios were successfully prepared by traditional powder metallurgy method. The materials were characterized and analyzed by XRD, SEM, EDS. The main results are as follows:

(1) In a certain range, with the increase of Cu content, the microstructure of TiC-based cermet is gradually refined, and the microstructure of TiC-based cermet is the classical black-core gray shell structure.

(2) The microstructure elements of TiC based cermet containing 16%Cu were identified by EDS at 900 °C, 1000 °C and 1100 °C. The results showed that with the increase of sintering temperature, the burning rate of Al increased gradually, and the atomic percentage of Cu did not change obviously. The physicochemical phenomenon of dissolution and precipitation occurred in TiC based cermet, and finally Al exudated again to its surface, so the weight percentage of Al showed a downward trend.

(3) Cu and Al, were added to the sample, but in addition to the incident peak of TiC, the diffraction peak of the solid solution with the main composition of Cu3Al4 appeared, and with the increase of the amount of Cu alloy, the intensity of the diffraction peak of the bonded phase Cu3Al4 increased more and more.

5. Reference

[1] Ke De-qing, Pan Ying-jun, Zhang Heng, et al. Effect of Cr3C2 on Microstructure and Properties of Co-TiC Cermet [J]. Material Guide, Vol. 31, pp. 80-83, 2015.

[2] Shan Ke, Zhai Feng-rui, Li Nan, etc. Ac Impedance of Ti (C, N) Multiphase Ceramics [J]. Journal of Ceramics,Vol. 9, pp. 4- 9, 2017.

[3] Fang Min-xian, Chen Hou-sheng. Thermodynamic Principle of Preparation of Ti (C, N) by Carbothermal Reduction [J]. Powder Metallurgical Materials Science and Engineering, Vol. 11, pp. 43-49, 2006.

[4] Liang Bao-yan, Wang Yan-zhi, Zhang Wang-xi, etc. Titanium Carbide Ceramic Binder CBN Composites Prepared by Self-propagating High Temperature Sintering [J]. Diamond and Abrasive Tool Engineering, Vol. 35, pp. 25-38, 2015.

[5] Ming Lu, Li Hao-jie, Han Wen-juan, et al. 2D Titanium Carbide Electrodes with Lower-F Surface for High Performance Lithium-ion Batteries[J]. Journal of Energy, 2015, Vol. 47, pp. 86-90, 2015.

[6] Lepicka, M. Tribological Performance of Titanium Nitride Coatings: a Comparative Study on TiN-coated Stainless Steel and Titanium Alloy[J]. Wear, Vol. 95, pp. 2-4, 2019.

[7] Zhan Bin, Liu Ning. VC/Cr3C2 on Microstructure and Mechanical Properties of Ti(C,N) Based Cermets[J]. Transactions of Nonferrous Metals Society of China, Vol. 44, pp.140-149, 2012.

[8] Huang Pei-yun. Principle of Powder Metallurgy [M]. Beijing: Metallurgical Industry Press, 2012.
[9] Zhang Xiao-bo, Rong Chun-lan, Liu Ning. Effect of ZrC content on Microstructure and Properties of TiC Based Cermet [J]. Hard Metal, Vol. 32, pp. 66-72, 2007.
[10] Tang Bo. Properties and Development of TiC Based Cermets[J]. Hunan Metallurgy, Vol. 21, 45-51, 2004.
[11] Chen Geng, Sun Wen-wen, Fang Qin. Effect of TaC Content on Microstructure and Properties of TiC Based Cermet [J].Foundry Engineering, Vol. 14, pp. 25-36, 2018.
[12] Yang Tian-en, Xiong Ji, Guo Zhi-xing. Research Progress of Ti (C, N) Based Cermet Core / Ring Structure [J]. Hard Metal, Vol. 8, pp. 74-80, 2010.
[13] Liu Ning, Zhou Feng-yun, Hu Zhen-hua. Study on Microstructure of η Phase in Ti (C, N) Based Cermet [J]. Hard Metal, Vol. 1, pp. 34-45, 1996.
[14] Niknam, S A. Turning Titanium Metal Matrix Composites (Ti-MMCs) with Carbide and CBN Inserts[J]. International Journal of Advanced Manufacturing Technology, Vol. 98, pp. 126-133, 2018.
[15] Xiao Shui-qing, Liu Jie, Xiao Bai-jun. Technical Path to Strengthen and Toughen Ti (C, N) Based Cermet [J]. Material Guide, Vol. 98, pp. 74-86, 2018.
[16] Zhan Bin, Liu Ning, Cai Wei. Effect of Original Powder Particle Size of Ceramic Phase on Microstructure and Properties of Ti (C, N) Based Cermet [J]. Heat Treatment, Vol. 89, pp. 17-22, 2013.

IOP Conf. Series: Materials Science and Engineering **730** (2020) 012003 doi:10.1088/1757-899X/730/1/012003

Effect of Heat Input Parameters on Molten Pool Size during Selective Laser Melting of Ti-6Al-4V Powder Through Numerical Simulation

Liang Zhang[1], Kaiqi Wu[1], Qiyun Yang[1], Xiaoqing Ni[1], Tao Wang[1], Wenheng Wu[1, *] and Lin Lu[1, *]

[1] Shanghai Engineering Research Center of 3D Printing Materials, Shanghai Research Institute of Materials, Shanghai 200437, China

[*]E-mail: wwhwwh2004@126.com; Lulinws@163.com

Abstract. Numerical simulations were proposed to research the influence of heat input parameters on the melting pool dimension and temperature field distribution during the selective laser melting process of Ti-6Al-4V powder. Firstly, temperature field results demonstrate that in the same powder layer, the center part of the sample will experience three more obvious heating processes under the irradiation of laser. Secondly, a good exponential function relationship is found between the melting pool dimension and volume energy density. Finally, based on the melting pool dimension from our numerical simulations, two full-process energy prediction diagrams are drawn to predict hatch spacing and powder layer thickness of SLM Ti-6Al-4V powder, consistent with the experimental results.

1. Introduction

Selective laser melting (SLM) technique can directely fabricate complex parts from three-dimensional CAD data without any extra tools.[1,2] Compared with transitional subtractive manufacturing, SLM has a high level of flexibility, meaning that it can deal with various commercial metal powder materials, for example, Ti-6Al-4V.[3-5]

In SLM process, temperature filed, determined by heat input parameters such as laser power and scanning velocity, has great influence on micro-structures and mechanical properties of parts considerably. Therefore, it is crucial to study the temperature field during SLM process of Ti-6Al-4V powder. However, SLM process involves numerous complex physicochemical phenomena, which could not be revealed clearly by traditional experimental method.[6]

Finite element methods are adopted to address this problem.[7-10] In this study, a three-dimensional finite element model is firstly constructed to investigate the influence of heat input parameters on melting pool dimension and temperature field in the process of SLM Ti-6Al-4V powder. Secondly, two full-process energy prediction diagrams of SLM Ti-6Al-4V powder are plotted by the values of simulated melting pool dimension.

2. Finite Element Modeling Method

The model of SLM Ti-6Al-4V powder based on finite element method is constructed and shown in figure 1. The below block with a dimension of $1.96\times1.12\times0.20$ mm^3 is considered as the stain-steel substrate and the upper block with a dimension of $1.12\times0.56\times0.09$ mm^3 is regarded as the powder bed of Ti-6Al-4V, in which the powder bed is evenly divided into three layers of 0.03 mm each.

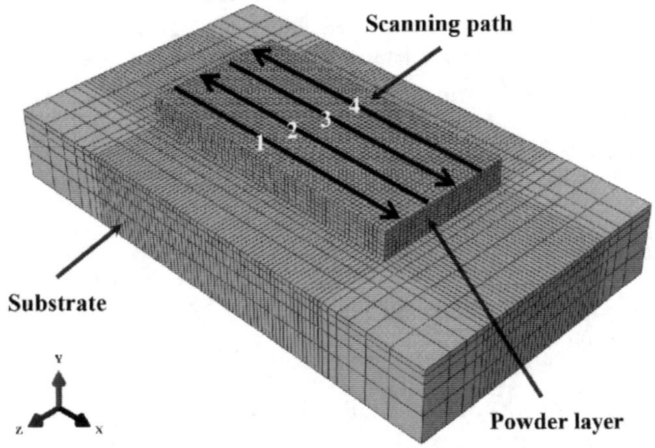

Figure 1. Three-dimensional finite element model of SLM Ti-6Al-4V powder

To save simulation resource and maintain enough precision, the Ti-6Al-4V powder bed is meshed by fine grids with dimensions of 0.0175×0.0175×0.01 mm^3 while the stainless-steel substrate is meshed by coarser grids. The coarser mesh dimension gradually increases with the distance from the Ti-6Al-4V powder bed increasing.

Considering the laser penetration depth and instantaneous heat transfer during SLM process of Ti-6Al-4V powder, a Gaussian distribution heat source is chosen and mathematically presented as, [11]

$$q(x, y, z) = \frac{\eta \cdot 6\sqrt{3}Q}{hr^2\pi\sqrt{\pi}} exp\left(-\frac{3x^2}{r^2} - \frac{3y^2}{r^2} - \frac{3z^2}{h^2}\right) \tag{1}$$

in which, the $q\ (x, y, z)$ is the heat flux density of point (x, y, z), Q is the laser absorption of powder, η is the laser power, r is radius of surface of Gaussian heat source and h is depth of Gaussian heat source.

In addition, to better simulate the thermal interaction between the powder and solidified layer, the thermophysical properties of the powder layer and solidified alloy material are temperature-dependent. The chemical compositions and thermal physical properties of Ti-6Al-4V are summarized in table 1 and table 2, respectively.

Table 1. Chemical compositions of Ti-6Al-4V

Element	Ti	Al	Fe	V	C	O
Compositions (wt.%)	89.53	6.09	0.11	4.03	0.01	0.12

Table 2. Thermal physical properties of Ti-6Al-4V at various temperatures

Temperature °C	Density g·cm^{-3}	Specific heat J/(g·K)	Conductivity W/(m·K)
25	4.425	0.548	5.845
300	4.390	0.625	11.180
600	4.346	0.720	15.934
900	4.312	1.177	21.496
1200	4.280	0.694	27.418
1500	4.220	0.766	32.230
1800	3.991	0.977	101.082

Finally, a FORTRAN program is written not only to control the magnitude and moving path of Gaussian distribution heat source, but also to realize transition of material from powder to solid. Besides that, the hatch spacing is set to 0.12 mm, the different laser powers (*P*) are 250, 280 and 310 W, and the different scanning speeds (*v*) are 1100, 1200, and 1300 mm/s.

3. Results and Discussions

3.1 Effects of Heat Input Parameters on Temperature and Cooling Rate

The rapid cooling in the tiny melting pool, one of the important characteristics of SLM process, has a great influence on the formation of micro-structure and grain refinement of the alloy. This work systemically investigates the temperature and cooling rate of the forming process under different heat input. Figure 2a and 2b show the temperature curve with time under different laser powers (*P*) and scanning velocities (*v*), respectively. It can be found that the central area of SLM part will undergo three obvious heating processes, namely, the initial heating process of powder, the melting process of powder and the reheating process of solidification state. Secondly, when the laser scanning velocity is fixed at 1200 mm/s, the peak temperature increases gradually from 2812 to 3235 °C with the laser power increasing from 250 to 310 W, shown in figure 2a. Thirdly, when the laser power is fixed at 280 W, the peak temperature decreases gradually from 3086 to 2964 °C with the laser scanning velocity increasing from 1100 to 1300 mm/s, shown in figure 2b. To sum up, laser power and scanning velocity are two important parameters during SLM process, determining the temperature field distribution in forming process.

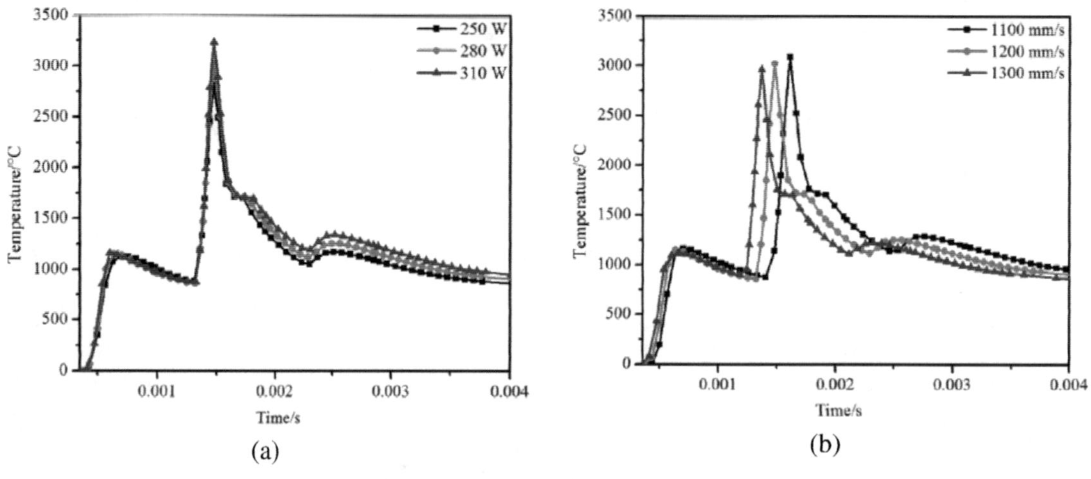

Figure 2. Temperature curve with time under different laser powers for v=1200 mm/s (a) and temperature curve with time under different scanning velocities for P=280 W (b)

3.2 Effects of Heat Input Parameters on Melting pool dimensions

As is well known, melting pool dimensions extremely affect the metallurgical bonding between the adjacent trajectories (and layers) during SLM process. For instance, if melting pool dimension is too large, it will cause balling effect; however, if melting pool dimension is too small, it will result in incomplete fusion. Therefore, the prediction of melting pool dimension is essential because it can guide the selection of the heat input parameters of SLM technology.

To better analyze the influence of various SLM processing parameters on melting pool dimension, it can be characterized by volume energy density, *E*, which is mathematically expressed as follows, [12]

$$E = \frac{P}{v \cdot s \cdot d} \tag{2}$$

in which P is laser power, v is laser scanning speed, s is laser scanning hatch spacing, and d is powder layer thickness. This section, a series of extra laser powers (100, 150, 200, 300, and 350 W) and scanning velocities (600, 900, 1500, and 1800 mm/s) are added to systematically research the influence of volume energy density on melting pool dimension.

Our simulated melting pool dimensions of SLM Ti-6Al-4V powder are showed in figure 3. It can be seen that a satisfactory exponential function relationship is found between the melting pool dimension and volume energy density, which is expressed as follows,

$$y = y_0 + A_1(1-e^{-x/t_1}) + A_2(1-e^{-x/t_2}) \tag{3}$$

where the parameters y_0, A_1, t_1, A_2, and t_2 are obtained at -79.15, 268.01, 28.23, 3.56, and 8.15 for melting pool width with R^2 values of 0.9692, and at -16.75, 42.32, 36.27, 1.13, and 3.17 for melting pool depth with R^2 values of 0.9811, respectively. In short, melting pool dimension increases with volume energy density increasing.

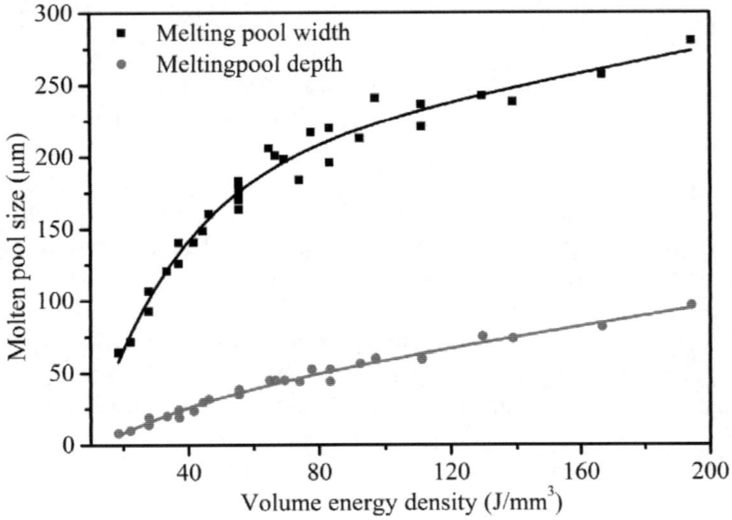

Figure 3. Effect of volume energy density on melting pool dimension

3.3 Full-Process Energy Prediction Diagrams of SLM Ti-6Al-4V Powder

It is well known that an appropriate overlap between neighbouring trajectories (and layers) is helpful to avert defect of SLM part, such as pore and crack. Presumably, the overlap between neighbouring trajectories is controlled by a combination of hatch spacing and melting pool width; the overlap between vertical trajectories in different layers is controlled by a combination of powder layer thickness and melting pool depth. Thus, based on the melting pool dimension calculated from our numerical simulations, two full-process energy prediction diagrams of SLM Ti-6Al-4V powder are drawn and demonstrated in figure 4.

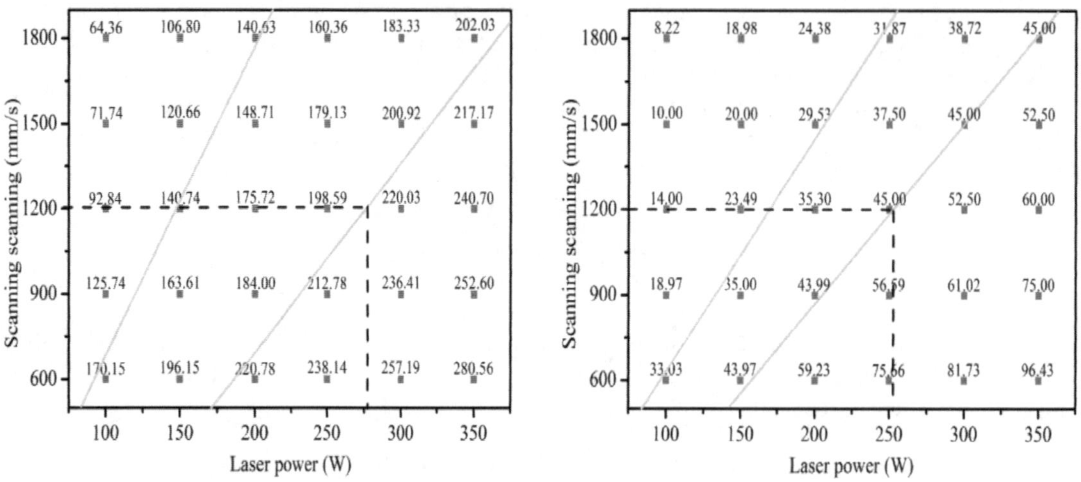

(a) (b)

Figure 4. Two full-process energy prediction diagrams of SLM Ti-6Al-4V powder based on the our simulated (a) width and (b) depth of melting pool

In figure 4, these red points are melting pool dimensions calculated from our numerical simulations under the different laser powers and scanning velocities, the two green lines in figure 4a represent the melting pool width in the same level with values of 0.14 mm and 0.21 mm, and the two green lines in figure 4b represent the melting pool depth in the same level with values of 0.03 mm and 0.045 mm.

The hatch spacing and layer thickness of SLM Ti-6Al-4V powder can be predicted via these two diagrams. For example, if the laser power, scanning velocity, and powder layer thickness are set to 1200 mm/s, 280 W, and 0.03 mm, respectively, the corresponding melting pool width will be about 0.21 mm. In this situation, to obtain a fine metallurgical bonding between neighbouring trajectories in the same layer, the best hatch spacing should be set about 0.14 mm, almost consistent with real experiment, to overlap half of melting pool. Analogously, if the laser power, scanning velocity, and hatch spacing are 255 W, 1200 mm/s, and 0.12 mm, the corresponding melting pool depth will be about 0.045 mm. In this situation, to obtain a fine metallurgical bonding between neighbouring layers, the optimal layer thickness of Ti-6Al-4V powder should be set to about 0.03 mm, almost consistent with real experiment, to overlap half of melting pool.

Not only that, the diagram can be applied to select laser power and scanning speed in the same way, as well. To sum up, the plotted full-process energy prediction diagram provides a important guidance for selecting heat input parameters of SLM process.

4. Conclusions

In this work, numerial simulations are proposed to research the process of SLM Ti-6Al-4V powder. There are three important results. Firstly, laser power and scanning velocity are two important parameters, determining the temperature field of Ti-6Al-4V alloy in the SLM process. Secondly, there is a satisfactory exponential function between volume energy density and melting pool dimension and the melting pool dimension increases with the volume energy density. Thirdly, the plotted full-process energy prediction diagram provides a important guidance for selecting heat input parameters of SLM process.

5. Acknowledgments

This work is supported by the project to strengthen industrial development at the grass-roots level (Project Number TC160A310/19), and Shanghai Rising-star program (Project Number 18QB1400600).

IOP Conf. Series: Materials Science and Engineering **730** (2020) 012003 doi:10.1088/1757-899X/730/1/012003

6. References

[1] Ali H, Ghadbeigi H and Mumtaz 2018 *Mater. Sci. Eng.* A **712** 175
[2] Song J, Li X J, Zhang Y, Yan L M, Jiang T and Peng S M 2017 *J. Electrochem. Soc.* **164** 846
[3] Gu D D, Hagedron Y C, Meiners W, Meng G B and Poprawe R 2012 *Acta Mater.* **60** 3849
[4] Zhang Y, Song J, Yan L M, Shi S P and Peng S M 2019 *Electrchim, Acta* **239** 446
[5] Song J, Gao Z B and Zhang L 2019 *Mol. Simulat.* **45** 935
[6] Sun S B, Zheng L J Liu Y Y Liu J H and Zhang H 2015 *Int. J. Adv. Manuf. Technol.* **80** 1787
[7] Parry L A, Ashcroft I A and Wildman R D 2019 *Addit. Manuf.* **25** 166
[8] Song J, Zhang L, Wu W H, He B B and Lu L 2019 *J. Mater. Res.* **34** 1395
[9] Luo C, Qiu L, Yan Y, Yang J, Uher C and Tang X 2018 *J. Mater. Process. Tech.* **261** 74
[10] Li Y and Gu D D 2014 *Mater. Design* **63** 856
[11] Song B, Dong S J, Liao H L and Coddet C 2012 *Int. J. Adv. Manuf. Technol.* **61** 967
[12] Thijs L, Verhaeghe F, Craeghs T, Humbeeck J V and Kruth J-P 2010 *Acta Mater.* **58** 3303

IOP Conf. Series: Materials Science and Engineering **730** (2020) 012004 doi:10.1088/1757-899X/730/1/012004

Mechanism of Low Cycle Fatigue Damage for Nickel-Based Powder Superalloy

Zhe Liu[1,*], Gaofeng Tian[2], Hao Fu[1], Zhen Ji[1], Shanwu Yang[1] and Chengchang Jia[1]

1 School of Materials Science and Engineering, University of Science and Technology Beijing, Beijing, 100083, China
2 Science and Technology on Advanced High Temperature Structure Materials Laboratory, Beijing Institute of Aeronautical Materials, Beijing, 100095, China
*Email: skdcllz123@126.com

Abstract. In this paper, the in-situ scanning electron microscope (SEM) was used to study crack initiation and propagation during the fatigue process of nickel-based superalloys prepared by powder metallurgy at room temperature(RT) and high temperature(HT). It could be found that the crack initiation of alloy is mainly along the <111> direction of crystallographic orientation at room temperature, cracks are parallel, perpendicular, at an angle of 53° and 60°. Cracks initiated in the crystal propagate in zigzag manner at an angle to each other, the direction of growth is perpendicular to the direction of stress. The fatigue cycle required for crack initiation at the grain boundary is much lower than that in the crystal, and the crack propagation is much faster. At initial stage, cracks expand along the grain boundary to a certain extent and then propagate in a transgranular manner, the direction of which is approximately perpendicular to the stress direction. At high temperature, the fatigue crack initiation time is much longer than the room temperature, and the cycle times required for crack initiation at the same stress ratio are 15 times higher than that at room temperature. The time and length of fatigue crack propagation along the grain boundary are larger than those at room temperature, but the length of the transgranular expansion is slightly smaller.

1. Introduction

Because of the characteristics of small grain size, uniform structure, high strength and good fatigue resistance, nickel-based power superalloy have become the material of choice for key parts such as aero-engine turbine discs [1, 2]. Fang Bin, Wang Mengya, Tian Gaofeng and other researchers [3-8] conducted in-depth research on the microstructure, mechanical properties, constitutive equation, recrystallization equation, processing map, grain boundary engineering of nickel-based powder superalloys, but further research is needed on the fatigue life and its influencing factors in service under harsh environment (high temperature, high load stress) [9]. For superalloy parts, fatigue failure is the most common failure mode. It occurs from crack initiation, propagation and fracture. It is characterized by no obvious plastic deformation before failure, so fatigue fracture usually occurrs suddenly [10-12]. Fatigue-induced fracture accounts for 70%~80% of engineering component or structure fracture accidents.

The in-situ test refers to the application of a servo fatigue testing machine with a scanning electron microscope (SEM) for fatigue testing. The advantage of this method is that the initiation and early propagation of small fatigue cracks can be observed in real time, which provides experimental supports for the study of the mechanism of the initiation and propagation of fatigue cracks. This kind of research is very important for improving material preparation technology and fatigue life of materials [13]. Liang Yan et al. [14] studied fatigue crack initiation and propagation behavior of 2524 aluminum alloy with high strength at room temperature by using in situ SEM fatigue experience. It can be found that the fatigue crack of 2524 alloy nucleates on the interface between the inclusion particles or the inclusion particles and the matrix material. The main crack propagates along the slip zone in the shear mechanism or the direction of the microcrack before the crack tip, and the propagation direction and velocity of microcrack propagation is affected by the grain boundaries. Xie Xishan et al [15] in-situ analysis of fatigue fracture caused by inclusions in nickel-based powder superalloys, the results show that fatigue cracks are easy to initiate and propagate into the matrix at the inclusion/substrate interface and inclusions.

At present, high reliability is proposed for nickel-based powder superalloys as aero-engine vortex, and low cycle fatigue performance is a key factor in its service life. In this paper, the initiation and propagation of cracks in nickel-based superalloys during low-cycle fatigue have been studied by in-situ fatigue experiments at RT and HT.

2. Experimental materials and methods

2.1. Materials

The samples of nickel-based superalloys used in this paper are prepared by powder metallurgy (PM) method. The composition is shown in Table 1.

Table 1. Composition of nickel-based powder superalloy (wt%).

Ni	Cr	Co	Mo	Ai	Ti	W	Nb	C	B	Zr
Bal.	15.78	13.04	4.33	4.14	3.88	2.26	0.82	0.03	0.01	0.03

2.2. Mechanical Properties Test

The mechanical properties of the samples at room temperature, 400°C and 650°C were measured by Universal Capacitance Testing Machine. The stress and strain values of the experimental process were converted by the extensometer and the load.

2.3. In-situ SEM Fatigue Test

In-situ fatigue test was used to characterize the fatigue properties of the samples. The alloy sample was mechanically cut into the in-situ fatigue sample shown in Figure 1. In order to ensure the purpose of the experimental observation, a gap was formed at the narrowest point in the middle of the sample, and the notch effect was obtained here, and the initiation of the crack was more easily observed. The sample was sanded to the upper and lower surfaces and side to 2000# and polished, the sample size was accurately measured for stress conversion. The stress was calculated from the thinnest part of the middle of the sample and rounded down. The stress was not adjusted during the experiment. To show the change in tissue during the in-situ stretching of the sample, the sample was subjected to mild chemical attack prior to stretching to enhance the contrast.

For the high temperature fatigue test, in order to prevent the sample from oxidizing during the high temperature experiment, the whole process was carried out in a low vacuum environment. The experimental results were observed for the first time as cracks appearing as crack initiation cycles.

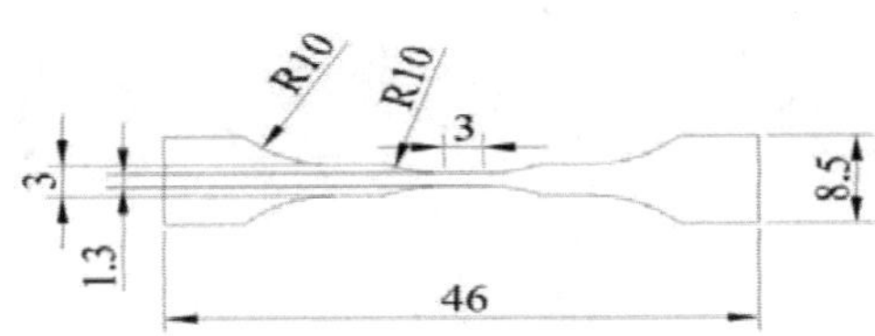

Figure 1. Samples for in situ fatigue test

IOP Conf. Series: Materials Science and Engineering **730** (2020) 012004 doi:10.1088/1757-899X/730/1/012004

2.4. Microstructure Analysis

The microscopic morphology of the sample was analyzed using a metallographic microscope and CX-200TA SEM equipped with a mass spectrometer.

3. Results and Discussion

3.1. Mechanical Properties

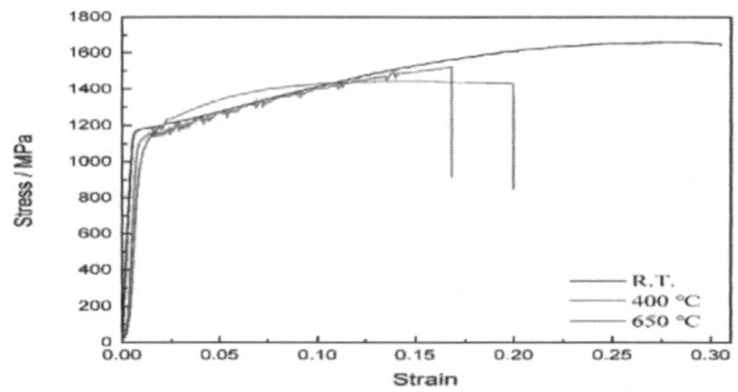

Figure 2. Stress-strain curve of nickel-based alloy

Table 2. Mechanical properties of nickel-based alloy.

T/°C	σ_s/MPa	σ_b /MPa	A/%
R.T.	1200	1650	30%
400	1100	1500	17%
650	850	1400	20%

Fig.2 and Table.2 represents the RT and elevated temperature stress-strain curves and mechanical properties of nickel-based alloys. The mechanical properties are as follows: the yield strength at R.T. is 1200 MPa, the UTS is 1650 MPa, and the elongation after fracture is about 30%.. At 400 °C, the yield stress o was reduced to 1100 MPa, the tensile strength was reduced to approximately 1500 MPa, and the elongation after fracture was approximately 17%. At 650 °C, the alloy had no obvious yield limit. Using the equivalent calculation of 0.2% deformation, it can be considered that the yield strength is 850 MPa, the tensile strength was reduced to about 1400 MPa, and the elongation after fracture was about 20%

3.2. Fatigue at RT

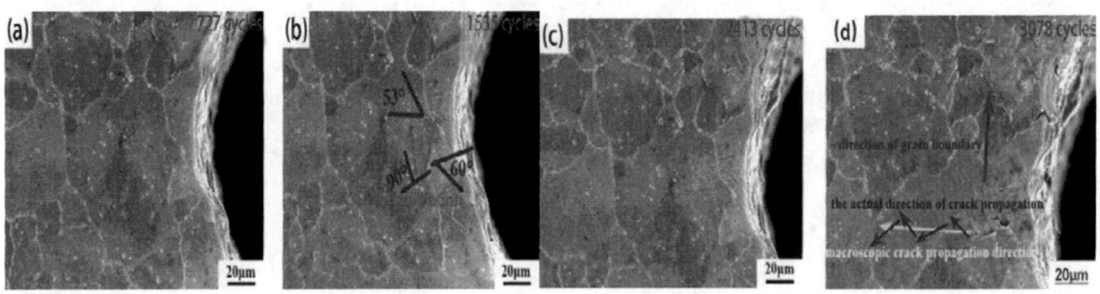

Figure 3. In-suit SEM fatigue morphology of alloy at room temperature and 1150Mpa load stress

The in-situ fatigue test in RT used a load of 800 N, 10 Hz frequency and 0.1 stress ratio; the sample thickness was 0.81 mm, the narrowest part of the width was 0.86 mm, and the load stress was about 1150 MPa. The sample eventually fractured when cycled to 6636 cycles. The process of high-stress in situ experiment at room temperature is shown in FIG. 3.

It can be found in Fig. 3(b) that crack initiation occured on the surface of the sample when it was cycled to about 727 cycles, most of the cracks were initiated in the crystal and extended along a certain crystallographic orientation, with a few cracks passed through the grain boundaries. The cracks were parallel, perpendicular or at a specific angle to each other. The geometric relationship between the cracks in the image was analyzed using ImageJ software. It can be seen from the figure 3(b) that there were mainly four orientation relationships between the cracks: parallel, vertical, 60° angle and 53° angle. Since the main component of the alloy was Ni, the crystal structure was face centered cubic (FCC). The main slip system that was activated in the FCC structure was the <111> direction, and the projection angle of the <111> direction in each direction could be calculated by crystallography, the result is shown in Fig.4. It can be seen that the projection angle of the <111> direction in the <001> direction was 90°, the projection angle in the <011> direction was 53°, and the projection angle in the <111> direction was 60°. These angles were consistent with the angles observed in Figure 3(b). Therefore, it can be judged that the crack was preferentially initiated in the crystal along the starting slip system.

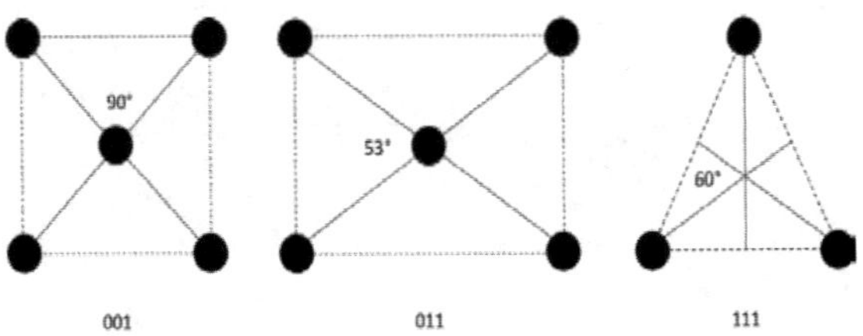

Figure 4. Projection angle of <111> direction in FCC structure

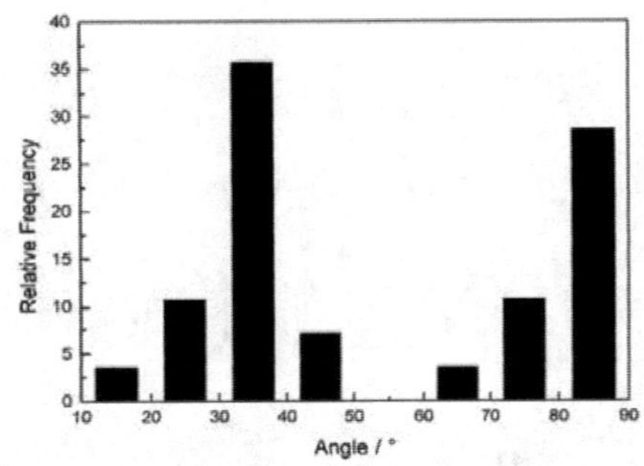

Figure 5. Angel between crack direction and stress direction

Statistical methods were used to analyze the angle between the crack initiation direction and the stress direction. Fig.5 shows that the angle between the crack and the stress direction was mainly concentrated in the range of 80°~90° and 30°~40°. This is mainly because the tensile stress was the largest when the crack was approximately 80° to 90° perpendicular to the stress direction, where the crack exhibited a tearing state and rapidly expanded under the tensile stress, which was the main cause of sample fracture.

In the direction of 30~40° with the stress direction, the direction had two characteristics at the same time: one is that the direction had a large shear stress, and the stress was along the crack initiation direction. The second is that the angle between the direction and the direction of the vertical stress was about 60°. As discussed earlier, the primary crack causing the fracture of the specimen was approximately perpendicular to the stress direction, while the intragranular crack preferentially propagated along the <111> direction. If the direction of the main crack was assumed to be along the <111> direction of a certain grain, for a crack initiated in the crystal, an angle of 60° to the direction of the fracture means that the direction was very close to the other <111> directions. It is also easy to start a slip system through which dislocation slips and dislocation moved in the opposite direction. Therefore, the above-mentioned geometric rule of crack initiation was determined by the crystal structure of the material.

From the start of the crack initiation at 1535 cycles in Fig.3(b), to the 2413 cycles shown in Fig.3(c), the crack expanded but the expansion was quite slow, and then to 3078 cycles shown in Fig.3(d), cracks began to appear at the crack crossing grain boundaries, followed by rapid crack propagation, to 3123 cycles of sample fracture.

From the process of crack propagation, when the crack initiated in the crystal, the crack propagated directly through the crystal, and the expansion process was shown in Fig.3(d). The crack was actually interconnected by two sets of cracks at an angle to each other to extend in the form of a broken line. This expansion should be related to its crystallographic orientation. However, no matter whether it is the macroscopic expansion direction of the crack or the actual expansion direction, the crack did not have a tendency to expand along the grain boundary. Even if the ultimate crack first propagated into a crack at the intersection of grain boundary, the direction of crack propagation was transgranular.

3.3. Fatigue at HT

High temperature in-situ fatigue test parameters could be found in Table 3. The sample thickness is 0.90 mm, the narrowest part of the width is 0.735 mm, the sample finally fractured at the cycle of 30879 cycles. The initiation and propagation of cracks during in-situ experiments could be found in Fig.6.

Table 3. High temperature in-situ fatigue test parameters.

Load/N	Frequency/Hz	Stress Ratio	Load Stress/MPa	T/°C
450	10	0.1	680	500

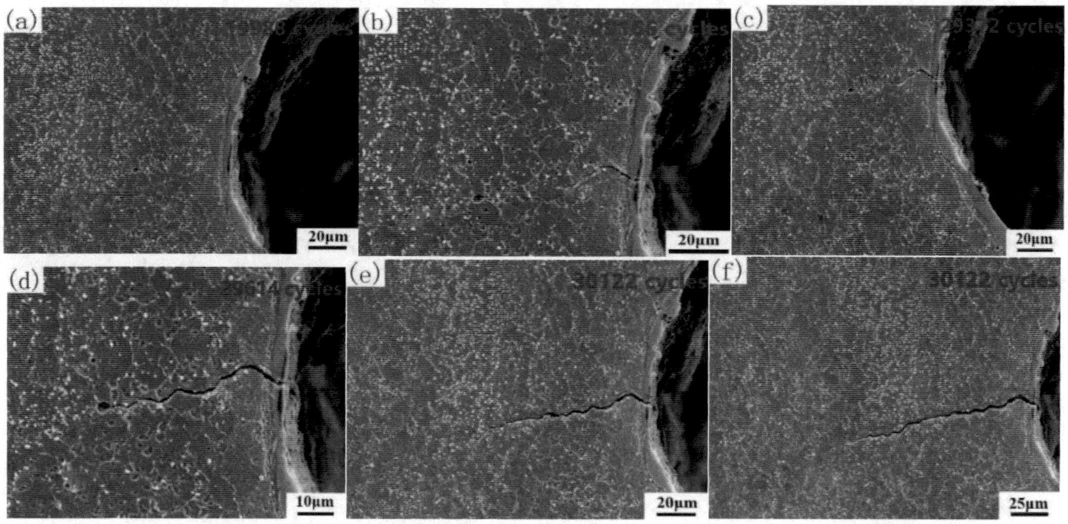

Figure 6. In-suit fatigue morphology of alloy at 500°Cand 680Mpa load stress

Fig.6 shows that crack initiation had not been observed on the surface of the sample when it was cycled to about 19988. At the cycle of 29166 cycles, crack initiation occurred on the surface of the sample, and cracks originated at the grain boundaries and propagated along the grain boundaries. At 29372 cycles, the crack propagated further, but the propagation direction was still generally along the grain boundary. By 29614 cycles, the crack continued to propagate into the interior of the sample. It should be noted that the crack did not propagate along the vertical direction of the stress, but propagated toward the intra-crystalline defect. By 30122 cycles, the crack transformed into transgranular propagation, and the propagation direction continued in a nearly straight line along the previous crack direction. By the 30359th cycle, the cracks showed a significant transgranular propagation, and the propagation direction remained in the original direction. The sample fractured when the stress was applied continuously to 30879 cycles.

Under room temperature fatigue test, when the fatigue stress was 80% of the yield strength, the alloy used in the test underwent about 11700 cycles of crack initiation and 12500 cycles of fracture, and only a few hundred cycles from crack initiation to sample fracture. At high temperatures, the same proportion of stress (80% of the yield strength of the sample) was maintained. The sample began to crack initiation when the cycle exceeded 29000 cycles. From crack initiation to sample fracture, it had undergone a cycle of about 2000 cycles. This was higher than room temperature experiments. This should be due to recovery and recrystallization when the sample deformed at high temperatures. Ni is a low-level fault metal. Due to its wide width of extended dislocations, it is difficult to dynamically recover by cross-slip and edge dislocations, and the dynamic recovery rate is low, and the subdislocation density is large. Therefore, the tendency of dynamic recrystallization occurs.

Due to the slow crack propagation rate in the high temperature, it could track the crack length change in situ. Data in Fig.7 is measured by software and its projected length in the direction of main crack propagation was taken as the effective length of crack. The result is shown in Fig.7. It can be seen that at the initial stage of crack initiation, the crack grows along grain boundary at a relatively fast speed. The crack growth rate decreases as the crack turns to transgranular growth, but increases rapidly as the crack grows larger, indicating that the stress at the crack tip is highly concentrated, which is consistent with the performance of the crack in the normal temperature fatigue experiment.

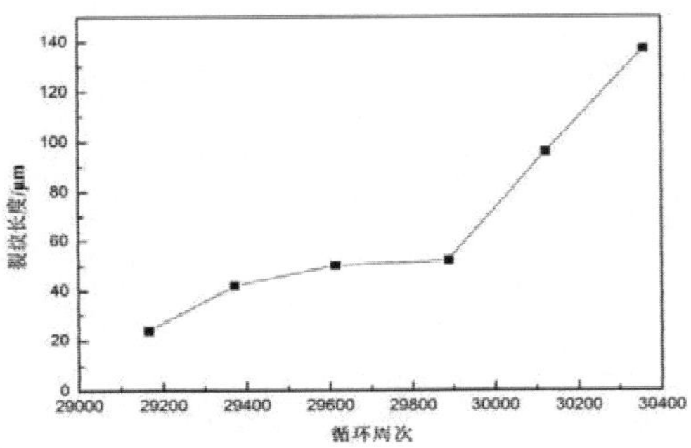

Figure 7. Relationship between crack length and cycle time at 500°C

4. Conclusions

(1) Fracture characteristics of nickel-base superalloy at RT: when the cracks start, it mainly initiation along <111> direction of crystallographic orientation, the cracks are parallel, perpendicular, and at an angle f 53° and 60° to each other. Cracks initiated in the crystal propagate in a zigzag manner at an angle to each other, expansion direction and stress direction are vertical. The fatigue cycle required for crack initiation at the grain boundary is much lower than that in the crystal, and the crack propagation is much faster. At initial stage, cracks expand along the grain boundary to a certain extent and then propagate in a transgranular manner, the direction of which is approximately perpendicular to the stress direction.

(2) Fracture characteristics at high temperatures: the fatigue crack initiation time in alloys is much longer than room temperature, the cycle times required to initiate cracks on the premise of the same stress ratio exceed 15 times of room temperature. The time and length of fatigue crack propagation along the grain boundary are larger than those at room temperature, but the length of the transgranular expansion is slightly smaller.

5. References

[1] Mao Jian, Chang Keh-Minn, Yang Wanhong et al. Materials Science and Engineering A[J], 2002, 322(1-2): 318

[2] Hu Benfu, Zhang Shouhua. Transactions of Metal Heat Treatment [J], 1997, 18(3): 28

[3] Fang B, Ji Z, Tian G F, et al. Investigation on complete solution temperature of γ' in the P/M superalloy of FGH96.Powder Metall Technol, 2013, 31(2): 89

[4] Fang Bin, Ji Zhen, Tian Gaofeng, et al. Flow behavior for FGH96 superalloy during two-pass hot. RARE METAL MATERIALS AND ENGINEERING, 2014, 43(12):3089-3094.

[5] Bin Fang, Zhen Ji, Meng Liu et al. Study on constitutive relationships and processing maps for FGH96 alloy during two-pass hot deformation. Materials Science & Engineering A 590 (2014) 255–261.

[6] Bin Fang, Zhen Ji, Meng Liu et al. Critical strain and models of dynamic recrystallization for FGH96 superalloy during two-pass hot deformation. Materials Science & Engineering A 593 (2014) 8–15.

[7] FANG Bin, JI Zhen, TIAN Gao-feng et al. Flow behavior and processing map of FGH96 superalloy during two-pass hot deformation. Chinese Journal of Engineering, 2015, 37(3): 336—344.

[8] Bin Fang, Gao-feng Tian, Zhen Ji, et al. Study on the thermal deformation behavior and microstructure of FGH96 heat extrusion alloy during two-pass hot deformation. International Journal of Minerals, Metallurgy and Materials,2019, 26(5):1-7.

[9] Hoon K J, YOON D. Evaluation of Fatigue Crack Propagation Behavior of Nickel-based Powder Metallurgy Superalloy for Aircraft at Elevated Temperature[J]. Transactions of the KSME, A, 2017, 41(8):751-758.

[10] Ma Z, Liu J, Wang G, et al. Strength gradient enhances fatigue resistance of steels[J]. Scientific Reports, 2016, 6(1).

[11] Meyer S, Brückner-Foit A, Möslang A. A stochastic simulation model for microcrack initiation in a martensitic Steel[J]. Computational Materials Science, 2003, 26: 102-I10.

[12] Ochi Y, Ishii A, Sasaki K. An experimental and statistical investigation of surface fatigue crack initiation and growth[J]. Fatigue & Fracture of Enginering Materials &Structures,1985, 8:32-39.

[13] Huang E, Barabash R I, Wang Y, et al. Plastic behavior of a nickel-based alloy under monotonic-tension and low-cycle-fatigue loading[J]. International Journal of Plasticity, 2008, 24(8):1440-1456.

[14] Liang Yan, JunKai Fan. In-situ SEM study of fatigue crack initiation and propagation behavior in 2524 aluminum alloy. Materials and Design, 2016, 110:592-601.

[15] XIE Xishan, ZHANG Lina, ZHANG Maicang, et al. Micro-mechanical behavior study of non-metallic inclusions in nickel-base P/M superalloy[J]. ACTA METALLURGICA SINICA, 2002, 38(6):635-642.

IOP Conf. Series: Materials Science and Engineering **730** (2020) 012005 doi:10.1088/1757-899X/730/1/012005

Development of a Stable Carbon Isotopes Analysis Instrument Based on Tunable Diode Laser Absorption Spectroscopy

Xiaojuan Cui*, Zhirong Zhang, Tao Pang, Hua Xia, Pengshuai Sun, Bian Wu and Runqing Yu

Anhui Provincial Key Laboratory of Photonic Devices and Materials, Anhui Institute of Optics and Fine Mechanics, Chinese Academy of Sciences, 230031 Hefei, China.

* Email: cuixiaojuan@aiofm.ac.cn

Abstract. A compact laser spectrometry instrument was developed for high precision measurements of isotope ratio of CO_2 by tunable diode laser absorption spectroscopy in the mid-infrared at 2.7 μm. The experimental spectrum of carbon isotopologues in the gas phase near 3641 cm^{-1} is very suitable for real-time analysis of these isotopologues. Simultaneous measurements of the mixing ratio and the corresponding $\delta^{13}C$ values of CO_2 in the atmosphere were performed. The achieved standard deviation (1σ) of $\delta^{13}C$ was 1.8‰. The Allan analysis of the time series of the mixing ratio of CO_2 shows a measurement precision of 0.2‰ for $\delta^{13}C$ with an optimum integration time of about 130 s. The spectrometer is capable of real-time measurements of stable carbon isotope ratios of CO_2 under ambient conditions.

1. Introduction

Carbon dioxide is one of the most important greenhouse gases and the arch-criminal of greenhouse effect. Its releasing must be controlled [1-3]. In order to search the sources and sinks of carbon dioxide, many studies on stable isotopes of atmospheric carbon dioxide have been performed [3,4]. Furthermore, ^{13}C-breath gas analysis can be used to evaluate human's health condition, such as enzyme activities, organ functions, etc. [5].

Isotope ratio mass spectrometry (IRMS) has always been the main technology used by scientific researchers to analyze the isotopic ratio with high precision from 0.01‰ to 0.1‰ [6]. But the IRMS system is bulky and the samples to be analyzed must be pretreated which make it impossible to detect the isotopic ratio in-situ or on-line [7, 8]. The tunable diode laser absorption spectroscopy (TDLAS) system is very suitable for real-time on-line detection of the trace gases isotope ratio with the advantages of simple design, low cost and small size [9]. The isotope ratios can be easily calculated with high sensitivity and precision when temperature, pressure, optical path length and absorption line strength of gases are certain. The distributed feedback (DFB) laser was selected in the current experiment. The advantages of using DFB laser are better mode stability, room temperature operation, smaller size and cheaper price than the quantum cascade laser (QCL). Particularly, the tuning range of the DFB laser is fully satisfied with the experimental requirements.

In this paper, the development of a DFB laser based instrument for high precision measurement of isotope ratio of CO_2 in the mid-infrared at 2.7 μm is reported. The derived average value of CO_2 mixing ratio is 454 ppm. The corresponding $\delta^{13}C$ value is -98.75‰ with a 1σ standard deviation of 1.8‰. The measurement precision can reach to 0.2‰ in the light of the optimum integration time 130 s obtained by the Allan variance analysis.

2. Experimental Details

2.1. Experimental Setup

The carbon isotopes analysis instrument is displayed in Figure 1. The light source is a room temperature operated DFB laser (nanoplus GmbH) with a center wave number of 3641 cm^{-1} and a single mode output power of 6 mW. The tuning rates of the DFB laser are 0.02 nm/mA and 0.22 nm/K, respectively. The slope efficiency is 0.12 mW/mA. The DFB laser is drived by a homemade current and temperature controller. A red laser beam was used to calibrate the optical path on account of the light at 2.7 μm is not visible. The absolute wave number calibration was finished by applying the nearby water vapor absorption lines from the HITRAN 2012 database. The beam from the laser was passed to a homemade multi-pass absorption cell with an optical pathlength of 107 m after several reflections.

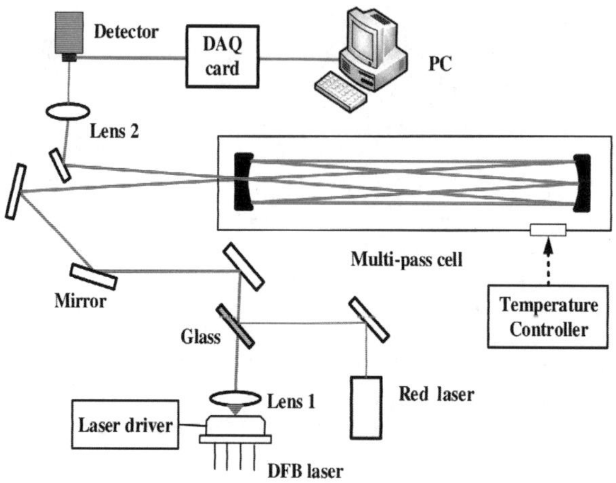

Figure 1. Experimental setup used for carbon isotope analysis

In order to improve the measurement accuracy and reduce the influence of the line strength with temperature variation, the temperature of the multi-pass absorption cell was maintained at 30 °C by using a heater band and a temperature controller. The light signal from the multi-pass cell was focused onto a lens with a focal length of 5 cm, and then directed to a thermoelectrically cooled (TEC) photovoltaic detector (VIGO, PVI-4TE-3). The information from the detector was collected by a fast data acquisition card and then transmitted to a personal computer for further data processing and analysis.

2.2. Absorption Line Selection

It is very important to choose the suitable absorption lines for the design of an isotope ratio measurement system. The selection of appropriate absorption line has obvious impact on the performance of the spectral measurement system, including sensitivity, measurement accuracy and selectivity. Isotope ratio measurements usually have some requirements for the selection of absorption lines: (1) they should have identical absorption cross sections and ground state energy; (2) they should be detected in the single scanning range of the laser; (3) there is no cross interference with other gas molecules in the atmosphere, especially water vapor [10, 11].

IOP Conf. Series: Materials Science and Engineering **730** (2020) 012005 doi:10.1088/1757-899X/730/1/012005

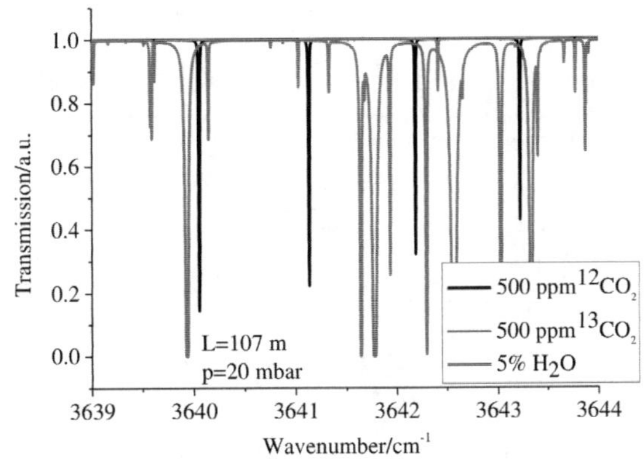

Figure 2. Signal simulation of 5% H_2O and 500 ppm CO_2 based on the HITRAN 2012 database with an optical path length of 107 m and the pressure of 20 mbar. $^{12}CO_2$, $^{13}CO_2$ and H_2O absorption lines are shown in black, red and dark yellow, respectively.

It is very simple to find appropriate spectral regions that may contain suitable absorption lines for the isotope ratio measurements using the above requirements. Signal simulation of 5% H_2O and 500 ppm CO_2 based on the HITRAN 2012 database in the spectral range of 3639-3644 cm^{-1} is shown in Figure 2. Two absorption lines are selected for isotope analysis of CO_2, one is at 3641.0311 cm^{-1} for $^{13}CO_2$ measurement, another is at 3641.1338 cm^{-1} for $^{12}CO_2$ detection. They are free of interferences of water vapor absorption lines.

3. Results and Discussion

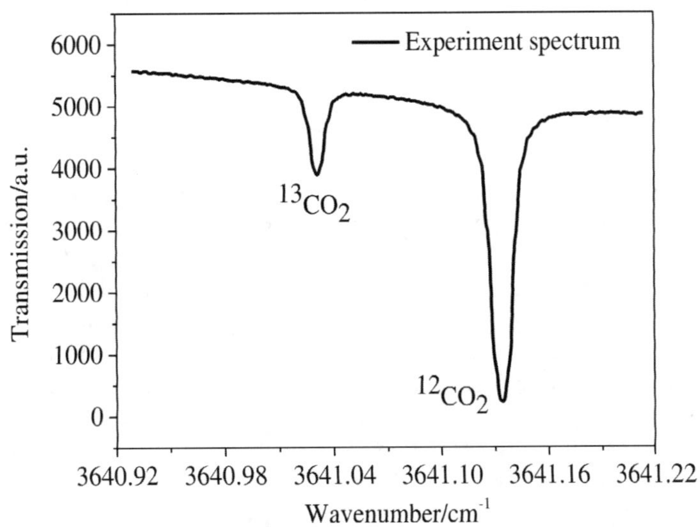

Figure 3. Experimental spectrum of $^{12}CO_2$ and $^{13}CO_2$ in the laboratory air

Figure 3 displays the direct absorption signals of $^{12}CO_2$ and $^{13}CO_2$ in the laboratory air at the pressure of 20 mbar and an optical path length of 107 m. The two absorption lines are located within a narrow scanned range of 0.1 cm^{-1}. The pressure of 20 mbar is sufficiently low to ensure a complete separation of the $^{12}CO_2$ and $^{13}CO_2$ absorption lines along with minimizing the influence of wing effects from the nearby H_2O absorption lines. The parameters of the used $^{12}CO_2$ and $^{13}CO_2$ absorption lines for carbon isotope analysis based on the HITRAN 2012 database are provided in Table 1.

Table 1. The parameters of the used absorption lines of $^{12}CO_2$ and ^{13}CO in this study

Isotopologue	Wavenumber (cm^{-1})	Line strength (10^{-21} cm^{-1} cm^2/molecule)	Ground state energy(cm^{-1})
$^{16}O^{12}C^{16}O$	3641.1338	5.64	704.30
$^{16}O^{13}C^{16}O$	3641.0311	0.64	42.92

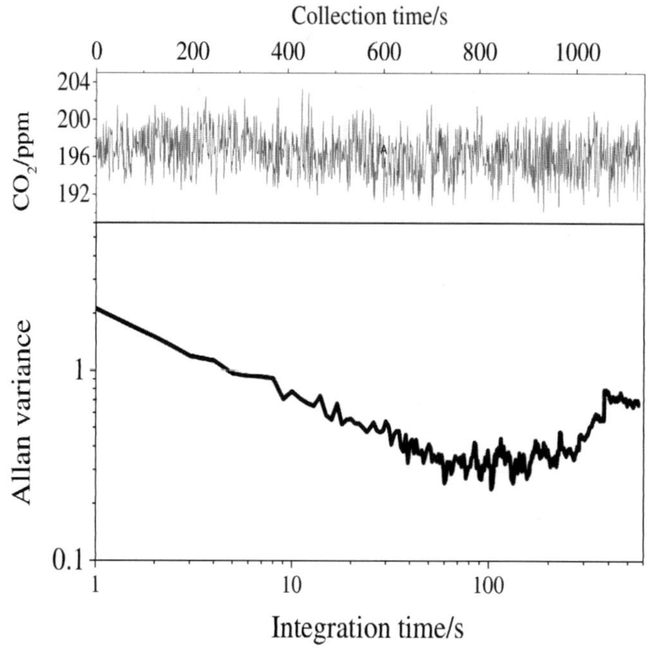

Figure 4. Allan variance analysis of 197 ppm CO_2 from a standard gas cylinder

In the absorption spectrum measurement experiment, the signal-to-noise ratio can be improved by selecting strong absorption lines or increasing the effective absorption path length (e.g using a multi-pass absorption cell or a high-precision optical cavity) in addition to reducing the internal noise of the spectrometer. In addition, for the normal white noise, it can be reduced to $N^{1/2}$ by averaging N laser scans in the stable time of the spectrometer. The optimum average time can usually be determined by Allan variance. In order to obtain the stable time of the system, the mixing ratios of CO_2 from a standard gas cylinder in the laboratory with 197 ppm CO_2 were detected. Figure 4 shows the time series and the corresponding Allan variance analysis. It can be seen that the optimum averaging time is 130 s from this figure. The room temperature was maintained at 30°C using air conditioning to dampen ambient temperatures variations during the measurement period.

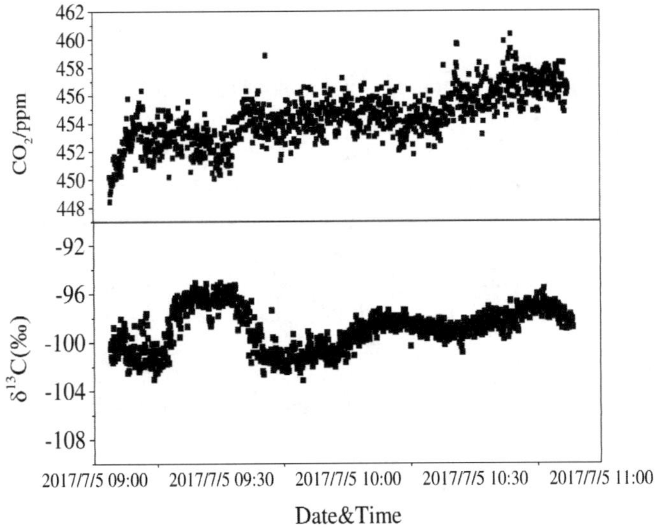

Figure 5. Time series of CO_2 in the laboratory air and the corresponding $\delta^{13}C$ values

We use this instrument to detect the isotope ratios of CO_2 in the atmosphere. The identical air in the laboratory were measured with a sample time of 1 s to obtain the reproducibility of the isotopic ratio measurements. In order to ensure the accuracy of sampling, the multi-pass cell was evacuated and purged with dry nitrogen every time before sampling and then refilled with the laboratory air. Figure 5 shows the reproducibility of the time series of CO_2 in the laboratory air profiles and the corresponding $\delta^{13}C$ values with 1 s sample time. The derived average value of CO_2 mixing ratio is 454 ppm. The corresponding $\delta^{13}C$ value is -98.75‰ with a 1σ standard deviation of 1.8‰. The measurement precision can reach to 0.2‰ on the basis of the optimum integration time 130 s derived by the Allan variance analysis. In order to apply the CO_2 isotope ratio measurement system based on TDLAS technology to the medical and other fields, we need to further improve the long term stability of the system and perform calibration to get the correct isotope ratios.

4. Conclusions

In this paper, a high precision carbon isotopic ratio analysis of CO_2 at ambient level using TDLAS technique at 2.7 um was demonstrated. The obtained results prove the feasibility to get real-time online carbon isotopic ratio at high time resolution of 1 s without any pretreatment of the samples. Experimental results explained that the system is able to measure $^{13}C/^{12}C$ isotopic ratios with a precision of 0.2‰ and is thus capable of in situ measurements of stable CO_2 isotope ratios under ambient conditions.

5. Acknowledgement

We acknowledge the financial support from the National Natural Science Foundation of China (Grant No. 41775128, 11204319), the External Cooperation Program of the Chinese Academy of Sciences (Grant No. GJHZ1726), and the Special Fund for Basic Research on Scientific Instruments of the Chinese Academy of Science (Grant No. YZ201315).

6. References

[1] G. V. Chilingar, O. G. Sorokhtin, L. Khilyuk, M. V. Gorfunkel, Greenhouse gases and greenhouse effect, Environ. Geol., 2009; 58:1207-1213.
[2] V.J. Hare, E. Loftus, A. Jeffrey, C. B. Ramsey, Atmospheric CO_2 effect on stable carbon isotope composition of terrestrial fossil archives, Nat. Commun., 2018; 9(252): 1-8.

[3] C.Yang, K. Jamison, L. Xue, Z. Dai, S.D. Hovorka, L. Fredin, R.H. Trevi, Quantitative assessment of soil CO_2 concentration and stable carbon isotope for leakage detection at geological carbon sequestration sites, Greenhouse Gas Sci. Technol. 2017; 7(4): 680–691, DOI: 10.1002/ghg

[4] C.T. Lai, J.R. Ehleringer, A.J. Schauer, P.P. Tans, D.Y. Hollinger, K.T. Pawu, J.W. Munger, S.C. Wofsy, Canopy-scale $\delta^{13}C$ of photosynthetic and respiratory $CO_{2\,m}$ fluxes: observations in forest biomes across the United States, Global Change Boil., 2005; 11: 633-643.

[5] T. N. Leea, M. M. R, C.T. Williams, Ø Tøien, B. M. Barnes, D. M. O'Brien, C. L. Buck, Stable isotope analysis of CO_2 in breath indicates metabolic fuel shifts in torpid arctic ground squirrels, Comp. Biochem. Phys. A, 2017; (209): 10-15.

[6] D. Hofstetter, J.D. Francesco, L. Hvozdara, H.P. Herzig, M. Beck, CO_2 isotope sensor using a broadband infrared source, a spectrally narrow 4.4 μm quantum cascade detector, and a Fourier spectrometer, Appl. Phys. B, 2011; 103(4): 967-970. DOI: 10.1007/s00340-011-4532-1

[7] K.Tanaka, R. Kojima, K.Takahashi, K. Tonokura, Continuous measuring of stable carbon dioxide isotopes in CO_2 with a near-IR laser absorption spectrometer, Infrared Phys. Techn., 2013; 60: 281-287.

[8] R. Aguilera,T.E.Chapman, H. Pereira, G.C.Oliveira, R.P. Illanes, T.F. Fernandes, D. A. Azevedo, F. A.Neto, Drug testing data from the 2007 Pan American Games: $\delta^{13}C$ values of urinary androsterone, etiochlanolone and androstanediols determined by GC/C/IRMS, J. Steroid Biochem., 2009; 115: 107-114.

[9] Z. Muccio, G. P. Jackson, Isotope ratio mass spectrometry, Analyst, 2014; 134: 213-222.

[10] G. Gagliardi, A. Castrillo, R.Q. Iannone, E.R.T. Kerstel, L. Gianfrani, High-precision determination of the $^{13}CO_2/^{12}CO_2$ isotope ratio using a portable 2.008 μm diode-laser spectrometer, Appl. Phys. B, 2003; 77: 119-124.

[11] B. Tuzson, M.J. Zeeman, M.S. Zahniser, L. Emmenegger, Quantum cascade laser based spectrometer for in situ stable carbon dioxide isotope measurements. Infrared Phys. Techn., 2008; 51(3):198-206. DOI: 10.1016/j.infrared.2007.05.006

IOP Conf. Series: Materials Science and Engineering **730** (2020) 012006 doi:10.1088/1757-899X/730/1/012006

The Hot Deformation Behaviourof Antibacterial Ti-10wt.%CuSintered Alloy

Liqiu Shi, Chun Xiang, Feng Yu and Zhihong Li

School of Mechanical & Automotive Engineering, Zhejiang university of Water Resources and Electric Power, Hangzhou,310017, P.R.China
Email: jmsdxshiliqiu@163.com

Abstract. Ti-10wt.%Cu sintered alloy was prepared by apowder metallurgy method to obtain strong antibacterial property. In order to optimize the deformation processing, the hot deformation behaviour of Ti-10wt.%Cu sintered alloy was investigated in this paper. The deformation temperature was set between 800°C to 1000°C and the strain rate was set in a range of0.008 s^{-1} to $5s^{-1}$. The microstructure of the deformed alloy were observed in order to reveal the effect of the hot deformation process on the microstructure. The results showed that the Ti_2Cu phase distributed homogeneously with the increasing of the deformation temperature and decreased with the increasing of the strain rate. It was concluded that the final optimized deformation process was: a deformation temperature range of 850°C to 950°C, and a strain rate of $0.1s^{-1}$ to $0.5s^{-1}$.

1. Introduction

Titanium and titanium alloys have been widely used as tissue repair materials for clinical treatment due to the good corrosion resistance, mechanical property and biocompatibility. It was reported that surgical site infection represents the most common nosocomial infection, accounting for 15% of all infections[1]. Pastides P. found periprosthetic infections in orthopaedics occur with a high frequency in primary hip and knee arthroplasties[2]. The main reasons are the colonization of bacteria on the materials surfaces and the formation of biofilm, showing a 500-5000 times more resistant to antimicrobials as compared to planktonic bacteria[3, 4]. In our previous studies, the Ti-Cu sintered alloy showed very strong antibacterial property against *S. aureus*and *E. coli*, and good biocompatibility and corrosion resistance[5-7]. Despite of those advantages, the brittleness of the sintered Ti-Cu alloy limits its uses. Hot extrusion is an effective method to improve the ductility of sintered titanium alloy by homogenizing the microstructure and eliminating the microstructure defects[8]. Recently, some research was reported on the deformation behavior of titanium alloy. For example, Bai et al[9] studied the deformation behaviour of a TLM alloy(Ti-3Zr-2Sn-3Mo-25Nb) and obtained a most peak efficiency domain at around 850°C. In Qin's research[10], the high-temperature flow behaviour of TC11/Ti-22Al-25Nb weldments was investigated in detail and the hot working processing could be carried out safely in the domain with a temperature range of 900-1060°C and a strain rate of $0.001\text{-}0.6s^{-1}$. Thus, whether the Ti-10wt.%Cu sintered alloy shows good antibacterial property after hot deformation has high research value.

The aim of this work was to optimize the deforming parameters. The hot deformation behaviour of Ti-10wt.%Cu sintered alloy was studied by an isothermal compression test in a temperature range of 800°C~1000°C with a strain rate of $0.004\sim1s^{-1}$. Based on the true stress and the true strain curves, a constitutive equation was established.

2. Methods

2.1. Preparation of Ti-10wt.%Cu Sintered Alloy

High pure Tiand Cu powder were mixed for 24h with a nominal composition of Ti-10wt.%Cu in a ball mill. The mixed powder was then cold pressed by 300MPa and high pressure sintered at 1000°C for 3h under 35MPa in a graphite moulds to obtain a cylindrical Ti-10wt.%Cu ingot with 30 mm in heightand 40 mm in diameter. Samples for the compressive test with a dimension of 12 mm in heightand 8 mm in diameter were sliced from the sintered ingot by a discharge method.

2.2. Isothermal Compressive Test

All samples before test were ground carefully by SiC papers up to 2000 grits and polished by 1μm polishing paste. The isothermal compressive strain test was conducted on a MMS-200 machine thermal simulator at temperatures of 800°C, 850°C, 900°C, 950°C and 1000°C, maximum truth strain of 0.6, and strain rates of 0.004s^{-1}, 0.02s^{-1}, 0.1s^{-1}, 0.5s^{-1} and 1s^{-1}.

2.3. Microstructure

Samples before and after compression were sliced along the longitudinal direction for the microstructure examination. Metallographic samples were ground and polished as described above, and then were etched in a mixed solution of 5% HF+15% HNO_3+ 80% alcohol. Microstructure was observed by optical microscope and scanning electron microscope (SEM).

3. Results

3.1. Stress Flow Behavior of Ti-10wt.%Cu Alloy

The intrinsic relationship between the flow stress and the thermal-dynamic behaviour can be reflected from the true stress-true strain curves, as shown in Figure 1.The flow curves were consisted of the following steps: elastic deformation, work hardening, softening and steady deformation. The flow stress showed an obvious decrease trend with the increasing of the deformation temperature and the decreasing of the strain rate. The stress increased to a peak at the initial small strain due to the effect of work hardening due to the fast increase of flow stress and generation of dislocation, then the flow stress decreased until a relatively stable stress.

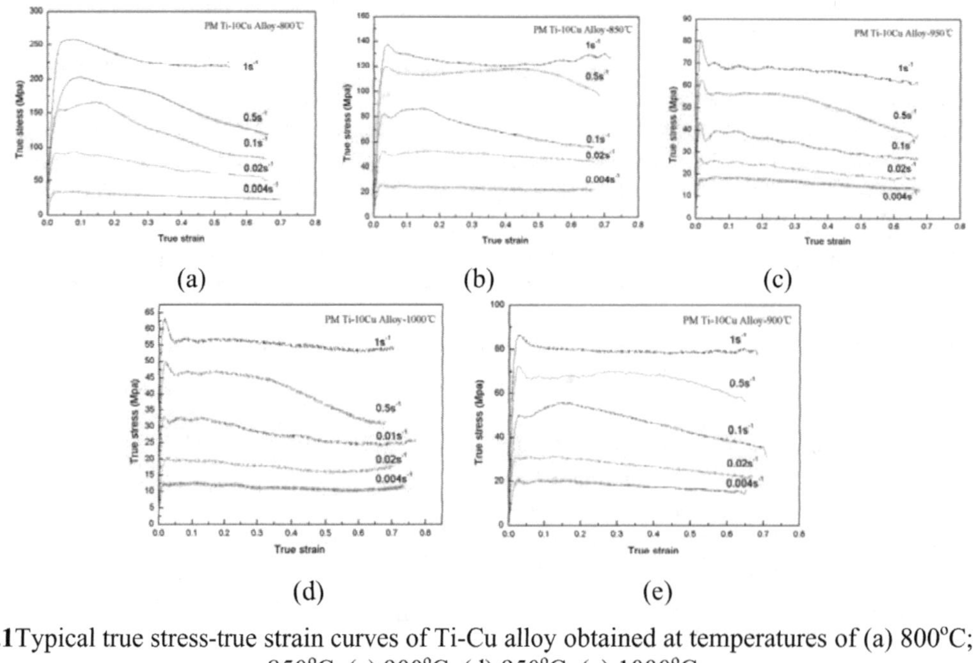

Fig.1 Typical true stress-true strain curves of Ti-Cu alloy obtained at temperatures of (a) 800°C; (b) 850°C; (c) 900°C; (d) 950°C; (e) 1000°C

3.2. Microstructure

Figure 2a shows the undeformed microstructure of sintered Ti-10wt.%Cu alloy. Figure2b~fshow the microstructure of Ti-10wt.%Cu alloys deformed at different temperatures with a strain of 0.6 and strain rate of $0.004s^{-1}$. Typical Ti_2Cu phase (lamellar-type) and Cu-rich phase(block-type) distributed discontinuously in the matrix (as shown in Figure 2a) which were systemically studied in our previous research[5]. After deformation, the grain was elongated along the vertical deformation direction and the size of secondary phase (Ti_2Cu phase and rich copper phase) were decreased. Even when the deformation temperature exceeded 950°C, the secondary phase showed fine size and distributed homogeneously in the matrix.

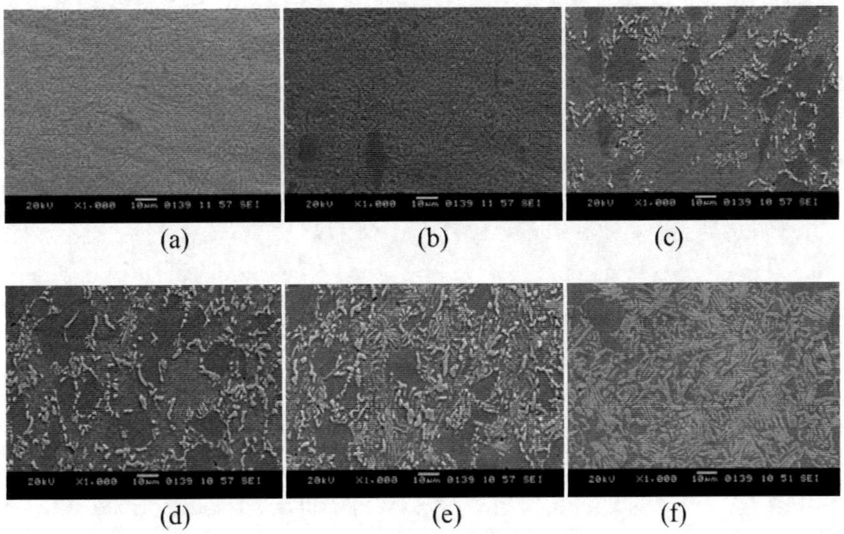

Fig. 2 The vertical microstructure of the deformed Ti-10wt.%Cu alloy at a strain rate of $0.004s^{-1}$ and different temperatures (ε= 0.6)

(a)undeformed alloy; (b)800°C; (c)850°C; (d)900°C;(e)950°C; (f)1000°C

At 800°C, original lamellar-type Ti_2Cu phase was obviously elongated and bended, obviously dynamic recrystallization was not observed. With the increasing in the deformation temperature, thermal vibration and diffusion rate of atom increase, Cu atom diffused fully to matrix Ti, which lead to the increase of α-Ti phase domain size. When the compression temperature increases to 950°C, Cu-rich phase become spherical and segregate to α-Ti phase boundaries. When temperature increase to 1000°C, recrystallization microstructure became homogeneous and grain size become larger, indicating complete recrystallization occurred in alloy. It is can assumed that α-Ti phase domain became coarse because Cu diffuse to matrix at the temperature of 900°C and no secondary Ti_2Cu phase was observed at 950°C and 1000°C.

Figure 3a~e show the microstructures of thealloys deformed at different strain rates. The deformation temperature was 900°C. The grain was also elongated along the vertical deformation direction and the size of secondary phase (Ti_2Cu phase and rich copper phase) were decreased with the decreasing of the deformation rate. When the deformation was reduced to $0.004s^{-1}$, nearly all the secondary phases were crushed and presented as tiny shape (as shown in Figure3e). At the strain rate of $0.004s^{-1}$, α-Ti phase exhibited approximately ellipse. The recrystallized Ti_2Cu phase can grow adequately due to the enough long deformation time, so recrystallization and growth were more obvious than with the strain rate of $1s^{-1}$. When strain rate increased to $1s^{-1}$, α-Ti phase became larger and grew along compressing direction, and Cu-rich phase increased. Original microstructure of the alloy showed obvious orientation, lamellar-type Ti_2Cu phase was distributed along the direction vertical to the compression direction. the abnormal growth of Cu-rich phase was observed and partial spherical Ti-Cu phase (maybe Ti_2Cu) was found along Cu-rich phase boundary. The main reason

IOP Conf. Series: Materials Science and Engineering **730** (2020) 012006 doi:10.1088/1757-899X/730/1/012006

might be due to that more deformation energy was accumulated in alloy, As reported, the presence of Ti-Cu intermetallic phase would lead to a relatively weak bond due to their brittle nature, so we supposed that the mechanical property can be improved after hot compression and the specific mechanical property need to be studied in our next research.

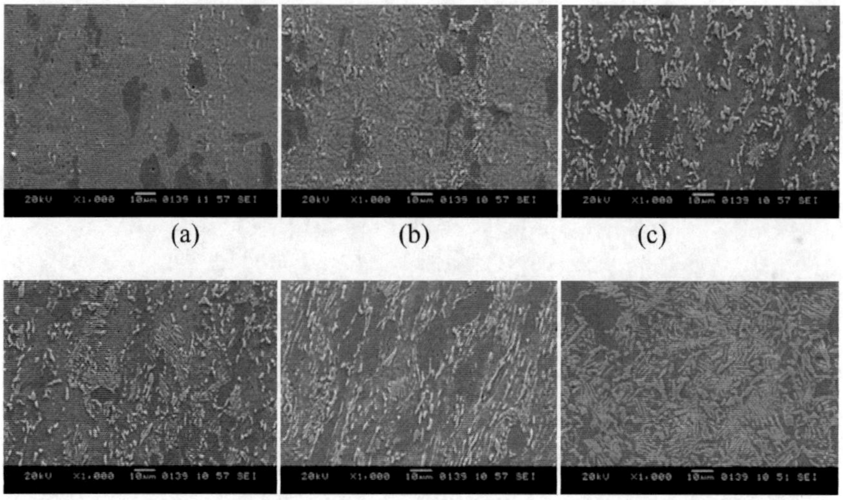

(a) (b) (c)

(d) (e) (f)

Fig.3 The vertical microstructure of the alloys deformed at 900°C and with a deformation of 0.6 and at different stain rate.

(a) $1s^{-1}$; (b) $0.5s^{-1}$; (c) $0.1s^{-1}$;(d) $0.02s^{-1}$;(e) $0.004s^{-1}$ (f) undeformed alloy

4. Conclusions

The hot deformation behaviour and the hot processing maps for antibacterial Ti-10wt.%Cu sintered alloy have been studied at a temperature range of 800°C~1000°C with a strain rate of $0.004\sim1s^{-1}$. The following conclusions were obtained from the investigation:

(1) The instability region occurred at high strain rates ($>0.5s^{-1}$) and low temperatures (<850°C). It was suggested that this region should be avoided during the hot deformation processing for the Ti-10wt.%Cu alloy. The optimal working region obtained from the processing map appears at 850-950°C/$0.1\text{-}0.5s^{-1}$ and the efficiency of power dissipation was 45-50%.

(2) Ti-10wt.%Cu alloy obtain homogenized microstructure after hot deformation.

5. Acknowledgments

This work was financially supported by Project for the Natural Science Funds of Zhejiang province (Y20E050069).

6. References

[1] Muhsin J, Wisal A, Saadia A,et al. Bacterial biofilm and associated infections. Journal of the Chinese Medical Association, 2018;81(1):7-11.

[2] Pastides P, Nathwani D. Update on the diagnosis and management of the periprosthetic knee joint infection.Orthopaedics and Trauma, 2017,31(1):53-56.

[3] Doiron K, Beaulieu L, St-Louis R, et al.Reduction of bacterial biofilm formation using marine natural antimicrobial peptides.Colloids and Surfaces B: Biointerfaces, 2018,167:524-530.

[4] Satpathy S, Kumar Sen S, Pattanaik S, et al.Review on bacterial biofilm: An universal cause of contamination.Biocatalysis and Agricultural Biotechnology, 2016,7:56-66.

[5] Liu R, Tang YL, Zeng L,et al. In vitro and in vivo studies of anti-bacterial copper-bearing titanium alloy for dental application.Dental Materials, 2018,34(8):1112-1126.

[6] Liu J, Zhang XX, Wang HY, et al. The antibacterial properties and biocompatibility of a Ti-Cu sintered alloy for biomedical application. Biomedical Materials.2014; 9(2):5013.

[7] Liu C, Zhang EL. Biocorrosion properties of antibacterial Ti-10wt.%Cu sintered alloy in several simulated biological solutions. Journal Of Materials Science-Materials In Medicine. 2015; 26: 142.

[8] Jia JB, Zhang KF, Liu LM, et al. Hot deformation behavior and processing map of a powder metallurgy Ti-22Al-25Nb alloy. Journal of Alloys and Compounds. 2014; 600: 215.

[9] Bai XF, Zhao YQ, Zeng WD, et al. Characterization of hot deformation behavior of a biomedical titanium alloy TLM. Materials Science and Engineering A. 2014; 598: 236.

[10] Qin C, Yao ZK, Ning YQ, et al. Hot deformation behavior of TC11/Ti-22Al-25Nb dual-alloy in isothermal compression. Transactions of Nonferrous Metals Society of China. 2015;25(7): 2195.

IOP Conf. Series: Materials Science and Engineering **730** (2020) 012007 doi:10.1088/1757-899X/730/1/012007

Thermal Conductivity and Sound Absorption Properties of Carbon Nanotube / Foam Aluminum Composites

Zhe Liu[1], Mingying Chen[1], Junjiao Han[1], Zhaoyang Lang[1], Xiaoqin Yan[1], Yousong Gu[1], Zhe Ji[1] and Chengchang Jia[1]

[1)] School of Materials Science and Engineering, University of Science and Technology Beijing, Beijing, 100083, China
Email: skdcllz123@126.com

Abstract. This article uses powder metallurgy technology to prepare foamed aluminum and carbon nanotubes (CNTs) / foamed aluminum composites. The laser thermal diffusion analyzer, thermal analyzer, and transfer function sound absorption coefficient test system were used to study the thermal conductivity and sound absorption properties of aluminum foam and its composites, respectively. The results show that the thermal conductivity of aluminum foam with 79% porosity is only 5W · $m^{-1}\cdot K^{-1}$, which is much smaller than the thermal conductivity of aluminum. So it can be used as an ideal thermal insulation material; The thermal conductivity of carbon nanotubes (CNTs) / foamed aluminum-based composites increases first and then decreases with the increase in the mass fraction of CNTs. As the test temperature increases, the thermal conductivity of the material gradually increases. When the mass fraction of CNTs is 0.75%, the thermal conductivity of foamed aluminum-based composites reaches the maximum. As the test frequency increases, the sound absorption performance of foamed aluminum first increases, then decreases, and then gradually increases, reaching a maximum value around 1000 Hz. The sound absorption performance of CNTs / foam aluminum matrix composites decreases at lower test frequencies as the mass fraction of CNTs increases. At higher frequencies, as the mass fraction of CNTs increases, the sound absorption coefficient of foamed aluminum decreases first and then increases.

1. Introduction

Foamed aluminum has many advantages, such as low density, high specific surface area, good heat insulation and sound insulation and noise reduction. It is widely used in automobile manufacturing, ship hull, aerospace and construction industries, and other fields[1-3]. Due to its excellent mechanical properties, thermal conductivity and sound absorption properties, carbon nanotubes (CNTs) can be used as a lightweight reinforcing phase to improve the thermal conductivity and sound absorption properties of foamed aluminum.

The heat transfer process of Foamed aluminum mainly includes four parts: thermal conduction between solid matrix; heat conduction between pores; convection heat transfer between fluid and solid matrix in the pores; Radiation heat transfer between solids and between solids and fluids[4]. Compared with traditional sound-absorbing materials such as polymer materials and asbestos, foamed aluminum metal materials have better mechanical properties, and have good processing properties, moisture resistance, high temperature resistance, flame retardancy, easy cleaning, and no release of toxic gases. So it has more advantages [5-6].

In this paper, aluminum foam and carbon nanotube (CNTs) / aluminum foam composites were prepared by powder metallurgy technology. The thermal conductivity and sound absorption properties of foamed aluminum and its composites were studied by laser thermal diffusivity analyzer, thermal analyzer and transfer function absorption coefficient test system respectively, providing experimental support for future research and application of aluminum foam composites.

2. Experiment

2.1. Sample
In this paper, pure foamed aluminum samples with porosity of 41% to 79% were prepared by powder metallurgy method, and foam aluminum composite samples with porosity of 60% and CNTs content of 0.25%~1.5% were prepared. The microstructure of the 60% foamed pure foamed aluminum and the CNTs aluminum foam composite with 0.75% added amount is shown in Fig.1.

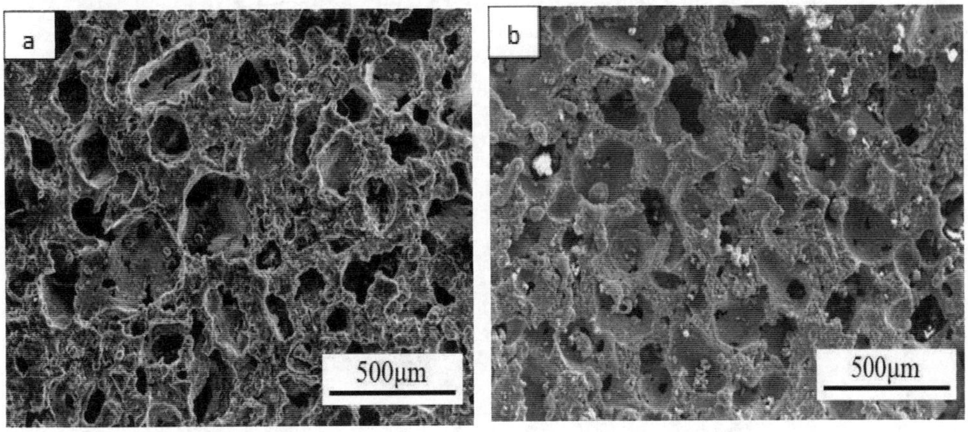

Figure 1. 60% the micro morphology of pure aluminum foam (a) and 0.75%CNTs foam aluminum composite (b) with porosity.

2.2. Thermal Conductivity Test
The aluminum foam sample was cut into discs with a diameter of 10mm and a thickness of about 1mm, which was smoothed with fine sand paper. The thermal diffusivity of foam aluminum and its composites was measured by laser thermal diffusivity analyzer, and the specific heat capacity of the material was measured on the DSC Q2000 thermal analyzer. Finally, according to formula (1), the thermal conductivity of the material was calculated.

$$\kappa=\alpha\cdot\rho\cdot C \tag{1}$$

Where α is the thermal diffusivity, unit: $cm^2\cdot S^{-1}$; κ is the thermal conductivity, unit: $W\cdot m^{-1}\cdot k^{-1}$; ρ is the actual density ,unit: $g\cdot cm^{-3}$;C is the specific heat capacity, unit:$J\cdot g^{-1}\cdot k^{-1}$.

2.3. Sound Absorption Performance Test
A cylindrical foamed aluminum sample with a diameter of 30mm and a thickness of 20mm was tested for the sound absorption coefficient of the foamed aluminum material by means of the AWA6290T transfer function sound absorption coefficient test system.

3. Results and Analysis

3.1. Thermal Conductivity of Foamed Aluminum

3.1.1. Thermal conductivity of pure foam aluminum
Fig.2 (a) is a line chart of the thermal diffusion coefficient of foamed aluminum with different porosities. It can be seen that the thermal diffusion coefficient of foamed aluminum decreases significantly with increasing porosity. Thermal diffusivity is a measure of the rate at which temperature disturbances at one point in an object are transmitted to another point. When the object is heated and heated, the heat

entering the object is continuously absorbed along the way, causing the entrapped temperature of the material to rise, and this process continues until the internal temperature of the object is all the same. Because the heat transfer rate in the matrix part of the foamed aluminum is faster than that in the pores, when the porosity is relatively high, the proportion of the matrix body is smaller, so the thermal diffusion coefficient of the material is smaller. Fig.2 (b) shows the specific heat of foamed aluminum with different porosities. It can be seen that the specific heat of foamed aluminum changes relatively little at the same porosity between 25 ° C and 100 ° C. The centralized change is between 1 ~ 1.2 $J \cdot g^{-1} \cdot K^{-1}$, and the change is relatively small. It can also be seen in the figure that the specific heat capacity of foamed aluminum first increases and then decreases with the increase of porosity. The cause of this phenomenon may be due to errors, because the size of the material sample is required when testing the specific heat of the material. Small and thin, the accidental error caused during the test will be relatively large. Fig.2 (c) is a line chart of the thermal conductivity of foamed aluminum with different porosities calculated according to formula (1). It can be seen that the thermal conductivity of foamed aluminum decreases significantly with increasing porosity. Within the range of 41% to 79% porosity, the thermal conductivity of pure foamed aluminum at room temperature is between $5W \cdot m^{-1} \cdot K^{-1}$ and $30W \cdot m^{-1} \cdot k^{-1}$, when the porosity is about 79%, the thermal conductivity of aluminum foam is 5 $W \cdot m^{-1} \cdot k^{-1}$, which is much smaller than that of solid aluminum ($237W \cdot m^{-1} \cdot k^{-1}$). Therefore, when the porosity of foam aluminum is relatively large, it can be used as an ideal insulation material.

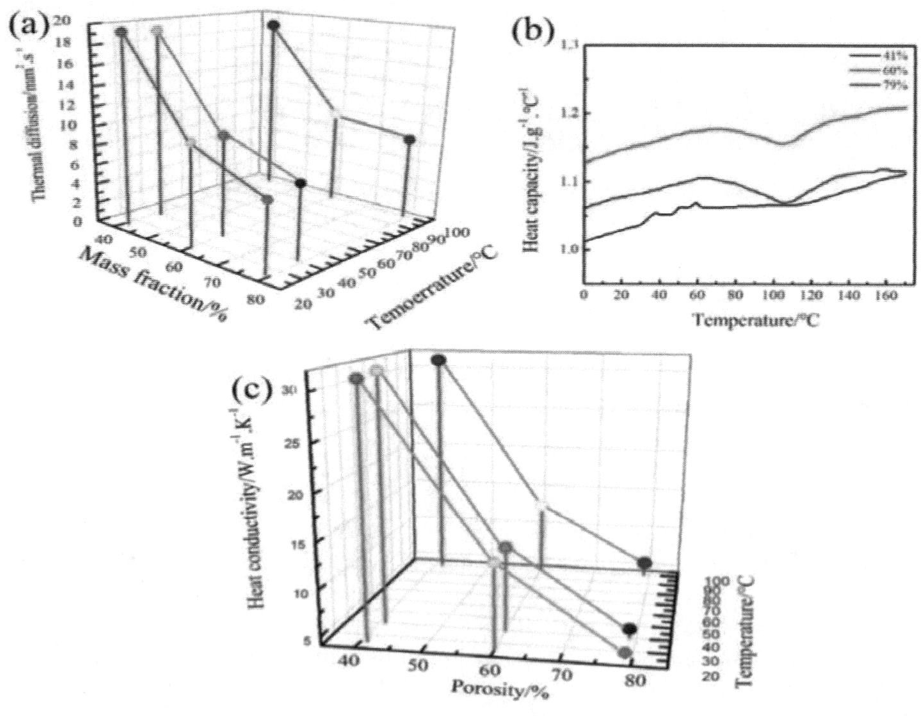

Figure 2. Effect of porosity on thermal conductivity of aluminum foam
(a) Line chart of thermal diffusion coefficient; (b) Curve of specific heat capacity; (c) Line chart of thermal conductivity

3.1.2. Thermal conductivity of CNTs / foam aluminum matrix composites

Fig.3 (a) is a line chart of the thermal diffusion coefficient of foamed aluminum matrix composites with different mass fractions of CNTs and a porosity of 60%. Fig.3 (b) is a graph of the specific heat capacity of foamed aluminum matrix composites reinforced with different mass fractions of CNTs. Fig.3 (c) is a thermal conductivity line chart of CNTs reinforced aluminum matrix composites with different mass fractions added after calculation. It can be seen that as the mass fraction of CNTs increases, the thermal conductivity of foamed aluminum-based composites also increases. This is mainly because the thermal conductivity of CNTs itself is higher than that of aluminum, so the addition of CNTs Will increase the thermal conductivity of the material. It can also be seen in the figure that the thermal conductivity of the foamed aluminum-based composite material does not change much at room temperature, but when the temperature is 100°C, the thermal conductivity of the material increases significantly, as can be seen from Fig.3 (b) that under the same curve, the specific heat capacity of the material increases with increasing temperature, which causes the thermal conductivity of foamed aluminum to increase as the temperature increases. It can also be seen in the figure that with the increase of the mass fraction of CNTs, the thermal conductivity of the material first increases and then decreases. This is mainly because when the CNTs exceed a certain mass fraction, the agglomeration phenomenon becomes more serious, which makes the foamed aluminum matrix The thermal conductivity of the composite material decreases. When the mass fraction of CNTs is 0.75%, the thermal conductivity of the foamed aluminum-based composite material reaches a maximum value, and the thermal conductivity is 19.993W·m^{-1}·K^{-1}at room temperature, which is 1.6 times of the thermal conductivity of pure foamed aluminum.

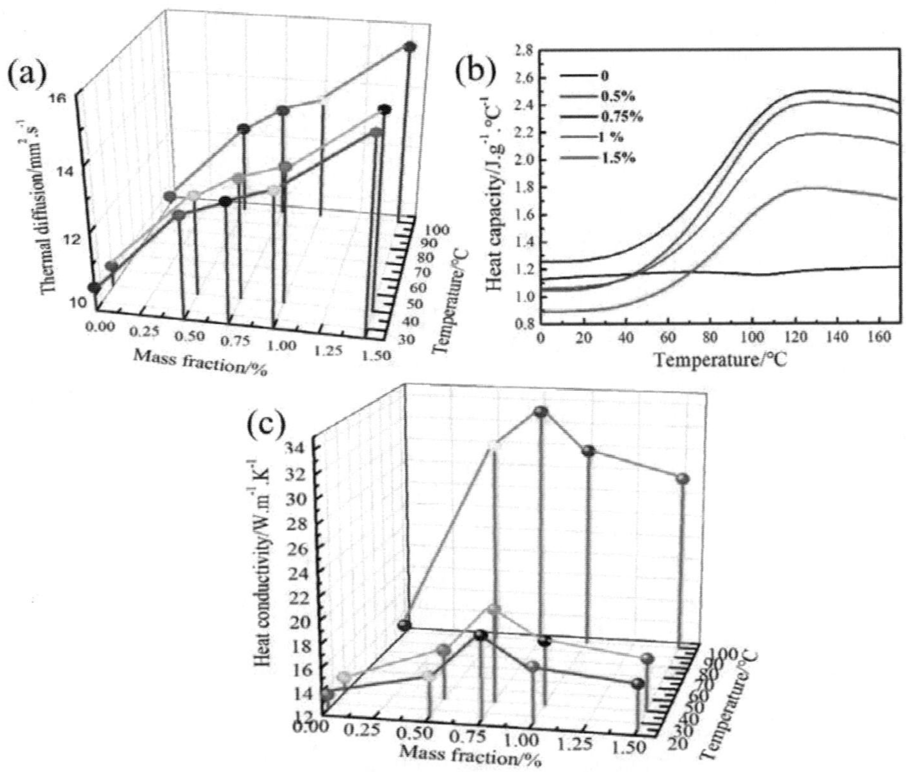

Figure 3. Line chart of the effect of CNTs on the thermal conductivity of 60% porosity aluminum foam
(a)Line chart of thermal diffusion coefficient; (b) Curve of specific heat capacity; (c) Line chart of thermal conductivity

3.2. Foamed Aluminum Sound Absorption Performance

3.2.1. Sound absorption performance of pure foam aluminum

The special pore structure inside the foamed aluminum has good sound absorption performance. When sound waves pass through the pores, the air in the pores periodically vibrates with the effect of noise, causing friction with the pore wall, and a part of the energy is generated and the heat energy is lost. In addition, the air in the hole will also undergo volume changes under the action of sound waves, thereby generating thermal energy; there will also be inelastic collisions between sound waves and the hole walls, which can also consume part of the sound energy, especially at high frequencies inelastic collisions absorb more sound.

Fig.4 is the sound absorption performance curve of foamed aluminum with different porosities. It can be seen that the sound absorption coefficient of foamed aluminum material has different behaviors at different frequencies, and all show that as the frequency increases, the curve first rises then decreases and then gradually increases, reaching a peak at around 1000Hz[7]. At lower frequencies, the sound absorption coefficient of foamed aluminum decreases as the porosity increases, but at higher frequencies, the sound absorption coefficient of foamed aluminum increases as the porosity increases, because at lower frequencies, the energy of the sound wave is lower. When the sound wave reaches the inside of the foamed aluminum, the sound wave is mainly obstructed by the foamed aluminum pore wall matrix, and the proportion of the foamed aluminum matrix with a lower porosity is larger. The obstacles are greater. When the frequency is higher, the propagation of sound waves in foamed aluminum is more complicated. When the sound is propagated internally, the air in the pores will vibrate accordingly, and the number of collisions with the pore wall will increase. Due to the retarding force, the sound energy is continuously converted into thermal energy, so that the sound absorption effect is better [8]. In addition, aluminum foam with high porosity has low strength, and the bubble wall is easy to rupture and cause through holes. And consume sound energy[9].

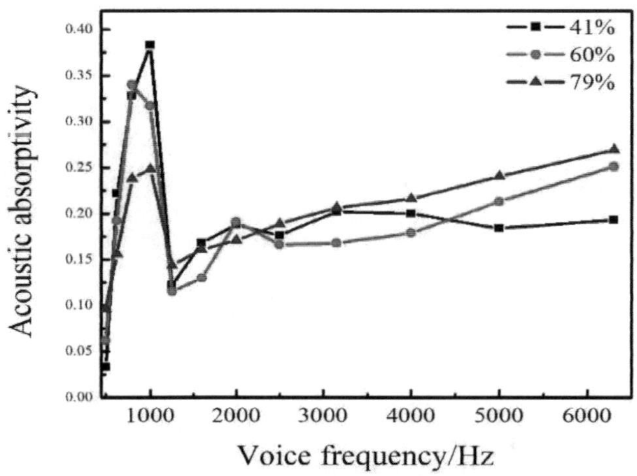

Figure 4. Curve of sound absorption coefficient of foamed aluminum with different porosities

3.2.2. CNTs enhanced sound absorption of foamed aluminum matrix composites

Fig.5 is a graph of the sound absorption coefficient of a foamed aluminum composite with different mass fractions of carbon nanotubes and a porosity of 60%. It can be seen that the sound absorption coefficient curve is similar to that of pure foamed aluminum. As the frequency increases, the material's sound absorption coefficient is first increased, then decreased, and then gradually increased, and a peak appears at about 1000 Hz. In addition, it can be seen that as the mass fraction of CNTs increases, the sound absorption coefficient curve of foamed aluminum matrix composites decreases with the increase of the mass fraction of CNTs at lower frequencies, and at higher frequencies, with the increase of CNTs mass fraction, it first decreases and then increases. This is because at lower frequencies, due to the lower energy, sound waves are mainly hindered by the foamed aluminum pore walls when propagating inside the foamed aluminum. However, as the mass fraction of CNTs increases, the relative density of foamed aluminum will be smaller. Lowering reduces this blocking effect on sound waves. At higher frequencies, due to the higher energy, the acoustic wave is more severely affected in the process of propagating inside the foamed aluminum. Related studies have shown that carbon nanotubes not only introduce additional acoustic mass resistance in this process, but also introduce Auxiliary sound absorption mechanisms such as friction and vibration at the interface between carbon nanotubes and micro-perforated plates have been used to make the sound absorption performance of this composite sound absorber better[10]. It can be seen in the figure that when the mass fraction of CNTs is 0.75%, the sound absorption coefficient curve of the foamed aluminum-based composite material is the lowest. This is because when the mass fraction of CNTs is 0.75%, the mechanical properties of the material are the strongest, resulting in when the sound wave propagates in the foamed aluminum, the vibration caused by the material is small, and the energy loss caused is also small, which results in the smallest sound absorption coefficient of the foamed aluminum-based composite material.

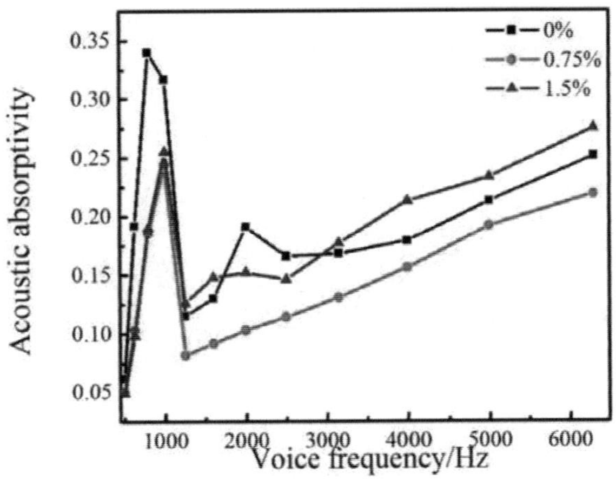

Figure 5. Effect of CNTs on sound absorption coefficient of 60% porosity aluminum foam

4. Conclusion

(1)The thermal conductivity of foamed aluminum material decreases significantly with increasing porosity. When the porosity is 79%, the thermal conductivity of foamed aluminum is only $5W \cdot m^{-1} \cdot K^{-1}$, which is much smaller than the thermal conductivity of aluminum. The ideal thermal insulation material is used; the thermal conductivity of foamed aluminum-based composites increases first and then decreases with the increase of the mass fraction of CNTs, and the thermal conductivity of the material gradually increases with the increase of the test temperature. The quality of CNTs When the fraction is 0.75%, the thermal conductivity of the foamed aluminum-based composite material reaches the maximum.

(2)As the test frequency increases, the sound absorption performance of foamed aluminum first increases, then decreases, and then gradually increases, reaching a maximum value around 1000 Hz; the sound absorption performance of CNTs / foamed aluminum-based composites at lower test frequencies, as The mass fraction increases and decreases. At higher frequencies, as the mass fraction of CNTs increases, the sound absorption coefficient of foamed aluminum decreases first and then increases.

5. References

[1] Baumeister J, Banhart J, Weber M. Aluminium foams for transport industry[J]. Materials & Design, 1997, 18(4-6): 217-220.

[2] Jiang J J, Gasik M, Laine J, et al. Electrochemical evaluation of sintered metal hydride electrodes for electric vehicle applications[J]. Journal of Alloys and Compounds, 2001, 322(1-2): 0-285.

[3] Maine E M A, Ashby M F. Applying the investment methodology for materials (IMM) to aluminium foams[J]. Materials & Design, 2002, 23(3): 307-319.

[4] XIA De-hong , GUO Shan-shan, REN Ling, DU Zheng. General calculating method of thermal conductivity for porous materials [J]. Journal of University of Science and Technology Beijing, 2012, 34(06):707-711.

[5] Chen C C, Chen W Y, Huang S H, et al. Experimental Study on the Loading Characteristics of Needlefelt Filters with Micrometer-Sized Monodisperse Aerosols[J]. Aerosol Science and Technology, 2001, 34(3): 262-273.

[6] Saleem M, Krammer G. Effect of filtration velocity and dust concentration on cake formation and filter operation in a pilot scale jet pulsed bag filter[J]. Journal of Hazardous Materials, 2007, 144(3): 677-681.
[7] Wang Lucai, Wang Fang. Sound Absorption Properties of Aluminium Foam [J]. Metallic Functional Materials, 2004, 11(5):17-18.
[8] YANG Xiao-dong, WANG Lu-cai. Research Prospects of the Aluminum Foam Sound Absorption Property [J]. FOUNDRY EQUIPMENT AND TECHNOLOGY, 2013(06):55-58.
[9] Wang Yue. Sound Absorption Characteristics of Foamed Aluminum [J].Development and Applocation of Materials, 1999(04):17-20.
[10] QIAN Yu-jie, KONG De-yi, GAO Jun. Improvement of sound absorption characteristics for micro-perforated panel absorber usingsingle-walled carbon nanotubes [J]. Technical Acoustics, 2014, 33(02):154-15.

IOP Conf. Series: Materials Science and Engineering **730** (2020) 012008 doi:10.1088/1757-899X/730/1/012008

Analysis of the Application of Laser Welding Technology on Maintenance of the Underwater Components of Nuclear Power Station

Zhiqiang Sun[1], Shaohua Yin[1], Zhongbing Chen[1], Jianlin Zhang[1], Jianlu Shang[1] and Yishi Lv[1]

1.Suzhou Nuclear Power Institution,1788 XIhuan Road, Suzhou, P.R. China
Email:13776254398@163.com, yinshaohua@cgnpc.com.cn, czbing2000@163.com, zhangjianlin@cgnpc.com.cn, shangjianlu@cgnpc.com.cn, lvyishi@cgnpc.com.cn

Abstract. This paper put its eyes on the PWR's underwater components which have the risk of early failure after long-term inservice because the material tends to deteriorate and embrittle due to irradiation and other factors, and should be maintenanced. In this paper, we contrastively analyse two main technologies used as underwater welding, named underwater wet laser welding and underwater local dry laser welding. It is difficult to obtain high quality weld seam by underwater wet laser welding for the reason of some disadvantages on laser beam underwater, such as water attenuation, bubble reflection and so on. On constract, underwater local dry laser-welding can obtain high quality weld-joints, and becomes the current hotshot. But, there are still some problems to be settled, such as the way to obtain stable underwater chamber and the protection of back of full penetration weld. Analysis of the retrieva of underwater laser welding on Web Science, we can find that underwater laser welding technology has step out laboratory to engineering application and the in-depth theoretical research is also actively carried out on the fields of picosecond, femtosecond laser.

1. Introduction

Because nuclear power has the both advantages of stable power supply and non-fossil fuel power (Ref.1), it is an indispensable way to replace the fossil power generation in China. As can be seen in Figure 1, the numbers of nuclear power station in China has entered a rapid development period since 2009, and the steps of annual growth becomes faster and faster. By the end of 2018, the installed capacity of nuclear power generation had already reached 295 billion kWh in China, which increased 18.9% more than that of previous year(Ref.2). Although China has the largest number and capacity of nuclear power including the inserve units and the under-construction units in the world, the proportion of nuclear power generation in the whole Chinese power generation is only 3.8% according to the 2017 year statistics of China which is far lower than that in the United States, France and Russia (Ref.3). So, the nuclear power construction is still rapidly developing in China. At the same time, the safety of nuclear power generation has attracted more and more public attention, and has become one of the key influnce on the development of nuclear power (Ref.4). Both the "neighborhood avoidance phenomenon"(Ref.5) and the "anti-neighborhood avoidance phenomenon"(Ref.6-7) are two completely opposite attitudes towards the nearby nuclear power facilities, which are due to the counterbalance of the economic benefits brought by nuclear power equipment and the security risks caused by nuclear power equipment. Therefore, it is significantly important for enhancing the experiencement of nuclear power profits, improving the confidence of nuclear power, and reducing anxious feeling of nuclear power risk to ensure the nuclear equipment safties and prevent the nuclear accidents (Ref.8-9).

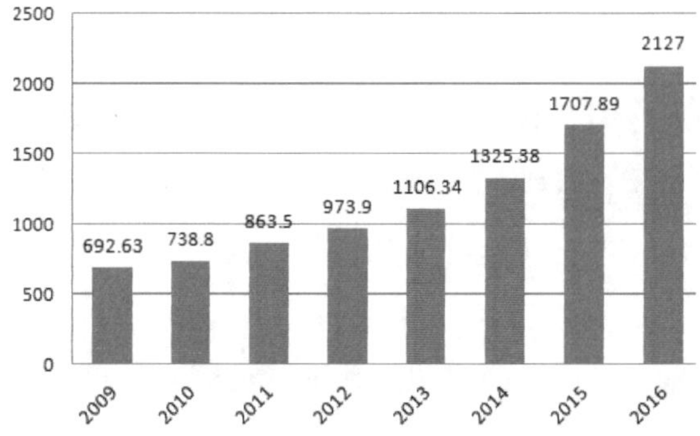

Figure 1. Statistics of Chinese nuclear power output from 2009 to 2017.2

The core structures of reactor pressure vessel are one of the most sensitive underwater equipment of pressurized water reactor nuclear power plant (Abb. PWR). The technology of maintaining these components is of great significance for PWR to win the safety awareness and support of public. So, this paper analyses the prospect of application an advanced technology which name is underwater laser welding on the maintenance of PWR's underwater components.

2. Maintenance Characteristics of PWR's Underwater Components

Boron-containing water is used as coolant and moderator in the primary loop system, spent fuel pool and auxiliary system of PWR nuclear power plant. As the results of irradiation, the materials of the components, such as core structures of reactor pressure vessel and fuel assemblies and so on, would easily come into intergranular stress corrosion cracking at the reason of hardening, swelling, and grain boundary Cr dilution due to dislocation rings and vacancies occurred in materials (Reference10) on the one hand; on the other hand, irradiation can also decompose water into H2O2 and O2, which increase the water electrode potential and promote stress corrosion cracking. The failure of connecting bolts between fencing plate and forming plate has been found on CP0 of French power plant (Reference 11). Therefore, the underwater components will cause to early failure leading by the material degradation under high temperature, high pressure and high irradiation environment of nuclear power plants.

In addition to the characteristics of high irradiation, the underwater components maintenance of nuclear power plants also has the characteristics of high quality requirements, small operating space and high cleanliness requirements. The traditional underwater arc welding technology, such as MIG/MAG and TIG welding technology, has a large space demands for complex operation of equipment. It is also difficult to avoid the generation of impurity particles such as welding slag during operation Moreover, the welding quality is hard to guarantee the weld quality to reach the qurequirements of nuclear power plants due to the negative influence of water cooling. In 1983, at the International Conference on the Application of Lasers and Optical Devices, Sepold and Teske proposed underwater laser welding technology and verified its feasibility by surfacing on the underwater carbon steel plate. From that time, a lot of researches have been carried out on underwater laser welding in developed countries. Among them, the United States and Japan are in the leading position, and have reached the practical stage of engineering maintenance. In recent years, underwater laser welding technology being used to the nuclear components maintenance become hotspot and tendency in this field (Reference13-15).

3. Analysis of the Underwater Laser Welding Characteristics

The laser is released in the duration of the transition to higher level when the electron absorbs energy and the drop back to lower level after the electron releases photons. Laser has the characteristics of high energy density, good monochrome, high brightness and good directivity, and is widely used in medical, national defense, communications, industry and other fields. At present, solid state lasers

IOP Conf. Series: Materials Science and Engineering **730** (2020) 012008 doi:10.1088/1757-899X/730/1/012008

(such as Na: YAG lasers), gas lasers (such as CO2 lasers) and semiconductor lasers are the main lasers used for metal cutting and welding.

The weld-purpose laser can be divided into continuous or pulsed laser beams. Corresponding welding principles can also be divided into conduction and keyhole deep penetration. At present, laser welding generally refers to pulse laser beam keyhole deep penetration welding, which has the characteristics of fast welding speed, deep penetration/wide penetration.

3.1. Characteristics of Laser Transmission in Liquid Medium

Laser transfers in both air and water in the form of plasma shock wave. Because the density of water is higher and the compressibility ratio is smaller than air, it is easier to observe the plasma and shock wave formed by laser transmission in water medium. Figure 2 (Reference16) displays the asymmetric laser plasma and ellipsoidal shock wave surface. We can see that the asymmetry of plasma and shock wave surface increases further following with the increase of pulse width.

There are two main mechanisms in the theory of laser underwater plasma formation, namely avalanche ionization mechanism and multiphoton ionization mechanism (Reference17-18). Free electrons collide with atoms or molecules in a strong magnetic field to absorb energy. When the inverse bremsstrahlung effect results in more than ionization energy, a new electron will be generated. When this process occurs in tandem, avalanche ionization will be formed.

In reference 19, an analytical expression of laser breakdown threshold in pure water is derived from the evolution of free electron density equation. Compared with the experimental data, it is found that avalanche ionization is dominant in nanosecond laser in water and is in good agreement with the calculated results, while multiphoton ionization is dominant in femtosecond laser, which is quite different from the calculated data (Reference19). Therefore, avalanche ionization is the main mode of laser propagation in water.

The plasma threshold formed by laser breakdown of water is closely related to the characteristics of liquid and laser. The higher the initial electron density in water is, the lower the plasma breakdown threshold is. In PWR nuclear power plants, water solution containing large amounts of initial electrons, which are not only boric acid and other substances used as moderator but also the H + and OH- particles after waterare irradiated (Reference20). So, the initial electron density in water medium is very high, which can conducive to the formation of plasma.

Ref.21 studies the experiment of laser plasma shock wave in liquid, and establishes an equivalent rotating ellipsoid wavefront propagation model of attenuated laser plasma shock wave in liquid. The results show that the power density of laser pulse and the transmission velocity of acoustic wave in medium have great influence on the initial maximum Mach number of attenuated shock wave front, which is the main factor affecting the attenuation of laser shock wave in liquid (Reference21).

From the above analysis, it can be concluded that the shape of plasma and shock wave surface formed by laser transmission under water has an important influence on the attenuation of laser beam.

IOP Conf. Series: Materials Science and Engineering **730** (2020) 012008 doi:10.1088/1757-899X/730/1/012008

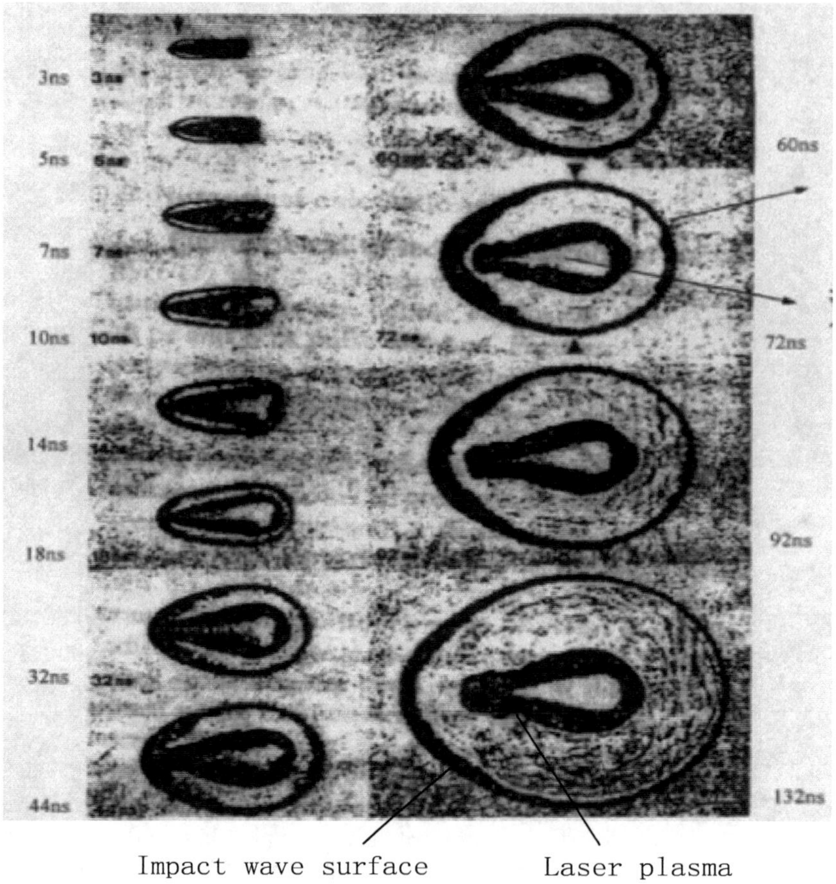

Figure 2. Diagrams of serial lasers underwater (Reference16)

3.2. Analysis of Underwater Laser Welding Technology

Underwater laser welding can be divided into underwater wet method and underwater partial dry method. Underwater wet laser-welding has not only the superiority of simple equipment and flexible operation, but also the inferior of weld defects tendency, such as shallow penetration and gas holes on the surface of weld seam. Underwater partial dry laser-welding can get higher quality welds which benefits from the local chamber, but is also limited to be only used in large operation environment by the complex equipment.

3.2.1. Underwater wet laser-welding

The underwater wet laser welding means the laser beam passes through the water directly, and lights on the surface of workpiece. The process can be divided into three stages: laser transmit during the water medium; explosive boiling occures near the workpiece surface (Reference22); laser passes through a small local chamber builded by vapour from the surface water film caused by the high temperature weld-pool (Reference23). The duration is displayed in Figure 3.

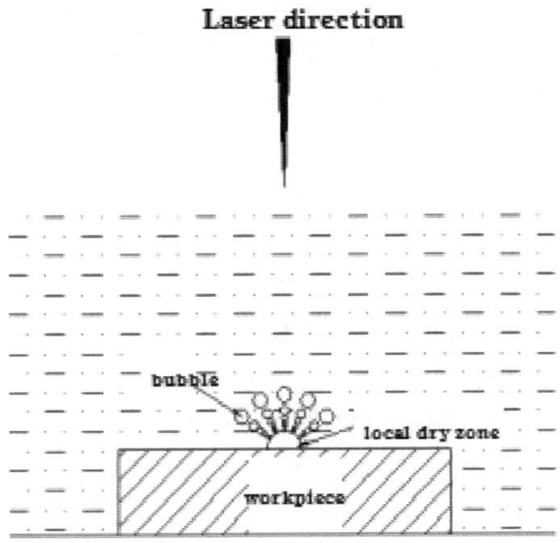

Figure 3. Illustration of the underwater wet laser welding

3.2.2. Underwater local dry laser-welding

Underwater local dry laser-welding is to transmit laser beam to the workpieces in a local dry chamber builded by some high pressure gas to discharge water, as shown in Figure 4.

There are three key problems to be solved for underwater local dry laser-welding: (1) scattering and absorption effect on laser beam from the aeroso in chamber; (2) disturbance of laser beam caused by protective airflow; (3) influence of water medium on the quality of back weld.

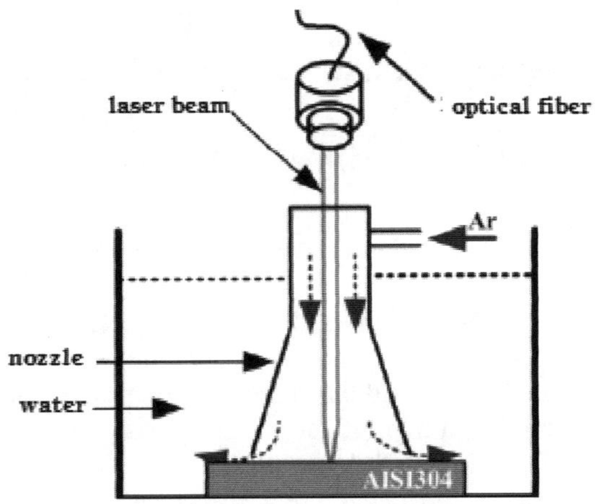

Figure 4. Illustration of the underwater local-dry laser welding

3.2.3. Contrastive analysis of underwater laser welding technology

1)Analysis of underwater wet laser-welding

Firstly, the laser beam is attenuated during the transmission in the forms of plasma ionization and shock wave through water, and the mechanism have been described in the previous section. Subsequently, the beam encounters a large number of bubbles generated by the boiling phenomenon on the surface of the workpiece. These bubbles will reflect and refract the laser beam, that will affect the shape of the spot irregularity and the expansion of the spot (Reference 23). At the same time, the liquid surface disturbance driven by bubbles will cause beam reflection. When the dip angle of the interface exceeds 30 degrees, the laser beam will be reflected enough to greatly reduce the laser quality (Reference24). Finally, after reaching the local chamber, the process is similar to that of the local dry chamber, but the beam intensity will be further attenuated due to the reflection, refraction and absorption of the aerosol in the chamber.

In addition to the above attenuation of laser beam, the water around the weld-pool will not only cool the molten metal rapidly, but also decompose into hydrogen and oxygen when the water is heated to enough temperature. The result of rapidly cooling of the weld-pool will cause the hardness increasing of the weld and heat affected zone, and lead to develop hardened microstructure in some cases. The hydrogen and oxygen will cause defects such as gas holes, oxidation and poor appearance on the weld surface, and increase the content of H and O in the weld metal (Reference 24-25).

2)Underwater local dry laser-welding

The aerosols in local chamber will scatter and absorb laser beams, then attenuate the beam intensity. The attenuation can be expressed by Bill's law, see Formula 1. The scattering effects of aerosols on beams can be described as two types of Raleigh scattering and Mie scattering. Raleigh scattering is suitable for aerosols whose size is smaller than the beam wavelength, see Formula 2. Mie scattering is suitable for aerosols whose size is idential or greater than the wavelength, see Formula 3. The two scattering modes are related to the laser wavelength and the number of aerosols, i. e. the dryness in the air chamber.

$$\mathrm{I} = \mathrm{I}_0 e^{-\mu(\lambda)l} \tag{1}$$

In the formula: I/I_0 is the transmittance. $\mu(\lambda)$ is the absorption coefficient related to wavelength lambda. L is the optical path.

$$\gamma_{\mathrm{p}}(\lambda) = \frac{8\pi^3(n^2-1)^2}{3N\lambda^4} \times \frac{6+3\delta}{6-7\delta} \tag{2}$$

In the formula, $\gamma_p(\lambda)$ is the scattering coefficient. λ is the laser wavelength. N is the number of particles. n is the refractive index of particles. δ is the radius tolerance of particles.

$$\mathrm{I}(\theta) = \frac{\lambda^2}{4\pi^2\gamma^2}(i_2 \sin^2\psi + i_2 \cos^2\psi) I_0 \tag{3}$$

In the formula, λ is the wavelength of laser. γ is the scattering coefficient. N is the number of particles. n is the refractive index of particles. Ψ is the angle between the vibration surface of incident light and the scattering surface. θ is the scattering angle.

Protecting gas flow rate has a great influence on the local chamber stability. If the flow rate of protective gas is small, water would enter into the protective zone and be turned into a large amount of vapor aerosol by the weld-pool. The number of aerosols scattering the laser beam increase to enough amount to make it is impossible to form stable metal plumes in severe cases (Reference 26). If the protective effect is poor, the elongation and section shrinkage of the weld metal would decrease greatly because of the oxygen content increases in the weld (Reference 27).

As respect to the full-penetration weld, the back weld quality is hard to guaranteed because of the lack of effective protection, which would cause defects as poor formation of welds, large numbers of metallurgical cavities and so on. Reference 26 has developed a flux for underwater backside protection (Reference 26), and achieves good results. This testifies that the use of welding protector is one of the effective ways to solve the back seam protection of underwater full-penetration laser welding. It is noteworthy that the flux on the back will produce particles such as slag. As reason of the strictly requirement of cleanliness in the reactor pressure vessel, it is necessary to evaluate the influence of these particles in PWR.

Reference 28 carried out the underwater laser-welding of Q235 steel, obtained well shaped weld, and observed the ferrite, acicular lower bainite and widmanstatten with large morphological differences in the weld and heat affected zone after metallography on the seam, which is caused by rapid cooling of surrounding water (Reference 28). Reference 25 and 29 both carried out the underwater local dry laser welding of 304 stainless steel, and obtained high quality welds with clear and uniform weld ripple and good penetration (Reference 25,29).

Based on the above comparative analysis, the underwater local dry laser-welding technology can obtain high quality weld-joints, but its operation space is limited because of the complex equipment. The underwater wet laser-welding technology still has long way to solve the problem of laser attenuation in water medium, which is not mature at present.

4. Tendency Analysis of Underwater Laser Welding Technology

Web of Science is a WEB-based product developed by ISI Company in the United States, including three citation libraries (SCI, SSCI and A&HCI) and two chemical databases (CCR, IC), covering 22,000 + journals, 3,100 million patents, 60,000 conference records, 5500 professional websites, 5,000 academic monographs and 2 million chemical structures. This paper uses underwater laser welding as a guide word to search on Web of Science. The results show that there are 303 records, the top ten of which are listed in Table 1.

Table 1. Statistics on research direction

Number	Contents	Records	%/303
1	ENGINEERING	271	89.44%
2	MATERIALS SCIENCE	172	56.77%
3	INSTRUMENTS INSTRUMENTATION	128	42.24%
4	METALLURGICAL ENGINEERING	105	34.65%
5	OPTICS	86	28.38%
6	CHEMISTRY	80	26.40%
7	PHYSICS	69	22.77%
8	COMPUTER SCIENCE	39	12.87%
9	BUSINESS ECONOMICS	33	10.89%
10	MATHEMATICS	30	9.90%

The following conclusions can be drawn from the data in the above table:

- Underwater laser welding technology is developing continuously to the level of Engineering application. This can be confirmed by the high proportion of engineering and equipment research in total records, which is 89.44% and 42.24% respectively. It is noteworthy that the records of economic analysis of underwater laser welding have 10.89% proportion, which means that the underwater laser-welding technology has been developed from experiment to application.
- Material R&D and welding metallurgy have high accounted for 56.77% and 34.65% in the whole records respectively, which means they are one of the current research hotspots. When the underwater laser welding technology is applied in different fields and different components, the material and requirements of welding quality for maintenance are different. Then, the high proportion of these two indexes also indicates that the technology has aimed at being applied in muti-fields..
- Five of the top ten items are related to theory basic research, they are optics, chemistry, physics, computer science and mathematics. While the proportion of engineering applications is in the dominant position, there is still such a high proportion of basic research. This indicates that the research of underwater laser welding technology is developing forwards to deeper stage. Because the water medium has a great attenuation of long-pulse-width lasers (Reference 19), the application of underwater laser technology is likely to be oriented to the study of ultra-short lasers, such as picoseconds, femtoseconds, etc. (Reference 31-32).

Based on the above analysis, underwater laser welding technology has developed to a relatively mature stage, and its related research directions are focus on engineering application, instrumentation, welding metallurgy, material science and economic analysis. At the same time, deeper theoretical research is also actively carried out, and the main research directions are optics, chemistry, physics, computer science and mathematics.

5. Conclusions

The PWR's underwater components have early failure risk in the influence of operational environment. Laser welding technology has the characteristics of deep penetration, high quality and pollution-free, which meet the requirements of PWR's underwater parts maintenance.

- Local dry underwater laser-welding can obtain high quality weld-joints, but there are two problems to be dealed: one is that the facilities of local chamber may limit its application in shallow spaces; the other is that although the problem of back weld could be solved by means of welding flux, but the slag formed may be the risk of nuclear power plants.
- The research of underwater wet laser-welding has enabled people to have a deep understanding of the characteristics of laser transmission in water. However, due to the attenuation of the laser-beam in water, it is impossible to obtain high quality weld-joints. So, the engineering application has not been realized at present.
- According to the results of retrieving the subject of underwater laser welding research on Web of Science, we can find that underwater laser welding technology has developed from experiment to engineering application, and the research of related technology has become a hotspots at this stage. At the same time, the direction of further theoretical research is probably about ultra-short pulse width laser, such as picosecond and femtosecond laser.

6. Reference

[1] Ou Yangyu, Wang Dasheng. International Nuclear Power Application and Its Prospect Forecast and Our Country Nuclear Electricity Development[J].Journal of North China Electric Power Univercity, 2007, 34(5):1-10.

[2] China Electric Power Enterprise Federation. China Electric Power Industry Annual Development Report 2019 [M]. Beijing, China Building Materials Industry Press,2019:165

[3] Chen yongchang. International Nuclear Power Construction and China's Nuclear Power Development [J]. Journal of Heilongjiang Vocational Institute of Ecological Engineering, 2019, 32(4):25-29+147.

[4] Du Juan, Zhu Xufeng. Literature Review on Public Acceptance of Nuclear Energy: Research Outline, Theoretical Framework, and Prospect[J]. Bulletin of Chinese Academy of Sciences, 2019, 34(6):677-692.

[5] Huang L, Zhou Y, Han Y, et al. Effect of the Fukushima Nuclear Accident on the Risk Perception of Residents Near a Nuclear Power Plant in China[J]. Proceedings of the National Academy of Sciences of the United States of America, 2013, 110(49): 19742-19747.

[6] Greenberg M R. How Much Do People Who Live Near Major Nuclear Facilities Worry About Those Facilities? Analysis of National and Site-Specific Data[J]. Journal of Environmental Planning and Management, 2009, 52(7):919-937.

[7] Jenkins-Smith H C, Silva C L, Nowlin M C, et al. Reversing Nuclear opposition: Evolving Public Acceptance of a Permanent Nuclear Waste Disposal Facility[J]. Risk Analysis, 2011, 31(4):629-644.

[8] Katsuya T. Public Response To the Tokai Nuclear Accident[J].Risk Analysis, 2001, 21(6):1039-1046.

[9] Tsujikawa N, Tsuchida S, Shiotani T. Changes in the Factors Influencing Public Acceptance of Nuclear Power Generation in Japan Since the 2011 Fukushima Daiichi Nuclear Disaster[J].Risk Analysis, 2016, 36(1):98-113.

[10] Shu Guogang, Shang guanbin, Fan liming, et al. Matal Material Handbook of Nuclear Power Station[M].Beijing: China Electric Power Press, 2017:28-31.

[11] Duan Yuangang, Zhao xiaogang, Yu Jibing. Key Issues in Life Management of PWR Reactor Internals[J].Nuclear Power Engineering,2003, 24(5): 402-404.

[12] Scpold G., Tcskc K. Results of Underwater Welding With High Power CO2 Lasers[C].International Congress on Application of Laser and Electro-Optics,1983,38:87-89.

[13] Yasuhiro Makihara, Yasuhiro Miwa, Naoya Hirose, et al. The Application of the Welding Technique at Fillet Groove by YAG-Laser Repair-Welding Robot for Underwater Environment[C].12th International Conference on Nuclear Engineering, Arlington, Virginia

USA, 2004:149-155.
[14] Takehisa HINO, Masataka TAMURA, Yoshimi TANAKA. Development of Underwater Laser Cladding and Underwater Laser Seal Welding Techniques for Reactor Components[J].Journal of Power and Energy Systems, 2009, 3(1):51-59.
[15] Chen Ying, Xu Wei, Ma Hongwei, et al. Research Status and Development Trend of Underwater Welding Technology[J]. Welded Pipe and Tube,2014,(05):29-34.
[16] A.Vogel, S.Bush, U. Parlitz. Shock Wave Emission and Cavitation Bubble Generation by Picosecond and Nanasecond Optical Breakdown in Water[J]. The Journal of the Acoustical Society of America, 1996, 100(1):148-165.
[17] Joachim Noack, Alfred Vogel. Laser-Induced Plasma Formation in Water at Nanosecond to Femtosecond Time Scales Calculation of Thresholds, Absorption Coefficients, and Energy Density[J].IEEE J Quantum Electron, 1999, 35(8):1155-1167.
[18] Docchio F, Regondi P, Capon M R C, et al. Study of the Temporal and Spatial Dynamics of Plasmas Induced in Liquids by Nanosecond Nd: YAG Laser Pluse I:Analysis of the Plasma Starting Times[J].Appl Opt,1988,27(17):3661-3668.
[19] Li Ming, Zhang Hongchao, Shen Zhonghua, et al. Simple Model for Optical Breakdown Threshold for Water Induced by Pulse Laser[J].Infrared and Laser Engineering,2005,34(6):661-663.
[20] Guangdong Nuclear Power Training Center. 900MW PWR Nuclear Power Plant System and Equipment[M]. Beijing, Atomic Energy Publishing House, 2005:122-135.
[21] Chen xiao. Studies on Mechanisms of the Interaction Between High-Power Laser and Matter in Water[M].Nanjing: Nanjing University of Science & Technology, 2004.
[22] Yin Sihua, Guo Zhongning, Chen Tieniu, Et al. The Explosive Bolling Experimental Study on Uderwater Laser Processing[J].Electromachining & Moudle,2016(3):36-41+46.
[23] Yao qi. Study on Underwater Laser Welding of Stainless Steel[M].Tianjin: Tianjing University,2014.
[24] Li Hongliang, Liu Duo, Yu Yifei, et al. Studies on Underwater Wet Welding Process of 304 Stainess Steel[C]. Papers Collection of the 12th National Welding Academic Conference of the Chinese Society of Mechanical Engineering, Lanzhou, 2015.
[25] Guo ning, Xing Xiao, Feng Jicai, et al. Design of Local Dry Underwater Laser Welding Drainage Device[C].Papers Collection of the 12th National Welding Academic Conference of the Chinese Society of Mechanical Engineering, Lanzhou, 2015.
[26] Huang zunyue. Study on Underwater Laser Welding Mechanism and Key Technologies [M].Tianjin: Tianjin University, 2018.
[27] Zhang Xudong, Chen WUzhu, Eiji AHIDA, et al. Underwater Nd: YAG Laser Beam Welding with Local Dry Cavity[J].Applied Laser, 2002,22(3):309-312.
[28] Zhu Jialei, Wang Dianshu, Jiao Xiangdong, et al. Chamber Local Dry Underwater Welding[J]. Journal of Shanghai Jiao Tong University, 2015, 49(3):329-332.
[29] Yao Qi, Luo Zhen, Li Yang, et al. Welding Forming and Mechanical Property of Stainless Steel Underwater Laser Welding[J].Journal of Shanghai Jiao Tong University,2015,49(3):333-336.
[30] Han Leigang, Zhong Qiming, Chen Guodong, et al. Development of Local Dry Underwater Welding Technology[J].Journal of Zhejiang University, 2019, 53(7):1252-1264.
[31] He Shutong, Yu Jia, Hu Mingjie. Femtosecond Laser High Precision Fabrication for Novel Applications[J].Current Nanoscience, 2016, 12(6):676-684.
[32] Yang Han, Yu Xichen, Wu Yifan, et al. Research Progress and Application of Femtosecond Laser in Microfabrication [J].Applied Laser, 2019, 39(2):346-354.

IOP Conf. Series: Materials Science and Engineering **730** (2020) 012009 doi:10.1088/1757-899X/730/1/012009

Study on Electrical Characteristics of Titanium Diboride Powder Applied in Power Field

Yu Zhu[1], Yongxing Wang[1], Enyuan Dong[1], Mingjie Li[1], Chao Sun[1] and Hongchuan Chu[2]

[1]School of Electrical Engineering, Dalian University of Technology, Dalian, China
[2]State Grid Liaoning Power Co., Ltd. Dalian Power Supply Company, Dalian, China
Corresponding author and e-mail: Enyuan Dong , dey@dlut.edu.cn

Abstract. Based on the properties of Titanium Diboride (TiB_2) combined with the knowledge of electrical engineering, a novel application of TiB_2 in the field of electric power has been proposed. The contact resistance of the particulate matter changes greatly when the external stress changes. Therefore, particulate matter can be used as a variable resistor in the field of fault current limiting. First, the pressure-resistance characteristics of TiB_2 were studied experimentally, and the law of resistance change with pressure was summarized. The explanation was given from the perspective of particle mechanics. Then, combined with the operating mechanism, a device with variable resistance was fabricated. And the curve of material resistance with time was tested. Finally, the material was passed through current, which proved that the material properties have not changed under large current. Thus the feasibility of the application of the material in the power field was verified.

1. Introduction

In the field of power, fault current limiters have always been a research hotspot. The essence of FCL is to suddenly insert a large resistor in the line to reduce the short-circuit current [1]. Therefore, the key to fault current limiting technology is to achieve an ideal variable resistor. For example, the superconducting current limiting technology utilizes the superconducting property of the material [2-3]. Under normal conditions, the material is in a superconducting state, and the resistance is substantially zero, and the load current can flow normally. When a short circuit occurs, the material enters a quench state and the resistance increases, thereby achieving the function of current limiting. However, the cost of superconducting materials is relatively high. And it requires complicated equipment to maintain its superconducting state during normal operation, and thus is bulky. There is still a long way to go to achieve widespread use of superconducting materials. Therefore, it is the main research purpose of this paper to find a easily achievable variable resistance material.

It is well known that the contact resistance is significantly affected by the contact pressure. The empirical formula concludes that the contact resistance has the following relationship with the contact pressure [4]:

$$R_j = K_c(0.102F)^{-m} \tag{1}$$

Where m is related to the contact form, which is a number greater than 0 and less than 1. K_c is the coefficient associated with the contact material. It can be seen that the contact resistance is inversely related to the pressure. Thus, a controllable change in resistance can be achieved by varying the magnitude of the contact pressure. Under normal conditions, the contact pressure is at a large value, the contact resistance is in a low resistance state, and the current is normally flowed. When current limiting is required, the contact pressure is reduced, and the contact resistance is changed to a high resistance state, thereby limiting the short circuit current.

In order to achieve the above functions, the material is required to have the following characteristics:

- The resistivity of the material should be as small as possible.
- The material itself must have sufficient hardness so that plastic deformation does not occur under large pressure. Therefore, after the pressure is reduced, the material can restore the original high resistance state.
- The material needs to flow current for a long time, so it should have a high melting point and strong antioxidant. To prevent the production of oxide film on the surface of the material to increase the contact resistance.
- In order to make the material resistance change sufficiently, the number of contact points should be sufficient. So the shape of the material should be granular.

In combination with the above requirements, TiB_2 is an ideal material. TiB_2 is a stable metal compound which is generally used as a conductive ceramic material and an electrode coating material for an electrolytic cell because of its good electrical conductivity. Its main properties are shown in Table 1 [5-6]:

Table 1. Properties of TiB_2 particle

Crystal structure	Hexagonal
Density [g/cm^3]	4.4
Melting point [°C]	2930
Hardness [GPa]	30
Young's modulus [GPa]	574
Poisson's ratio	0.11
Resistivity [μΩ·cm]	14.4
Antioxidant temperature [°C]	1100

It can be seen that TiB2 can perfectly meet all the above requirements. Its resistance will change significantly under the action of external pressure. In order to further verify the feasibility of the application of TiB_2 in the field of fault current limiting, the characteristics of TiB_2 were studied. First, the variation law of material resistance and pressure was tested by universal testing machine and analyzed theoretically. Then, a device of a variable resistance material was fabricated in combination with the operating mechanism. And the dynamic resistance curve of the material was tested. Finally, the material was subjected to a current-flow test to observe the characteristics of the material under high current impact. Through the preliminary study of material properties, the feasibility of using TiB_2 in the field of fault current limiting is verified.

2. Characteristic Test Method

2.1. Pressure-Resistance Characteristic Test

In order to verify the feasibility of TiB_2 in the power field, the pressure-resistance characteristics of TiB_2 were first studied. TiB_2 with a particle size of 30 μm, which is commonly found on the market, was selected. The test apparatus is shown in Figure 1. The material was placed in a cylindrical insulated container. Metal electrodes were placed at the upper and lower ends of the container. The lower metal electrode is fixed to the container. The upper metal electrode can be moved automatically and the material is pressed by the universal testing machine. Material resistance is measured by digital bridge. The inner diameter of the insulating container is 50 mm, and the height of the material is 40 mm. The pressure applied from 0 N to 4000 N. The data was recorded every 100 N. And a total of 10 sets of data were tested and averaged.

It should be noted that since the upper metal electrode itself has a weight of 2 kg, the material has withstood a pressure of 20 N even if no pressure is applied. Therefore, the material actually withstands a pressure range of 20 N-4020 N.

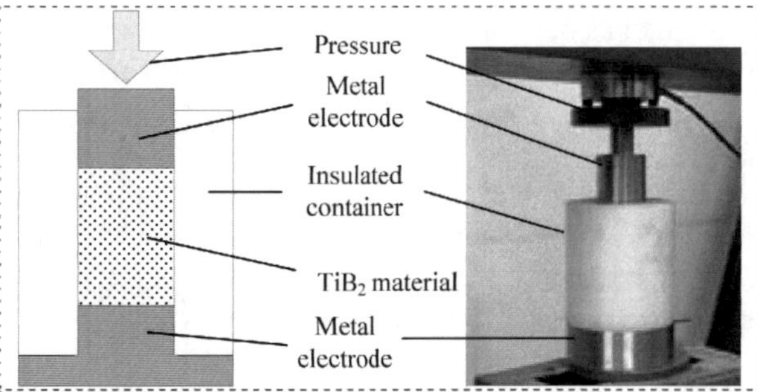

Figure 1. Test apparatus schematic

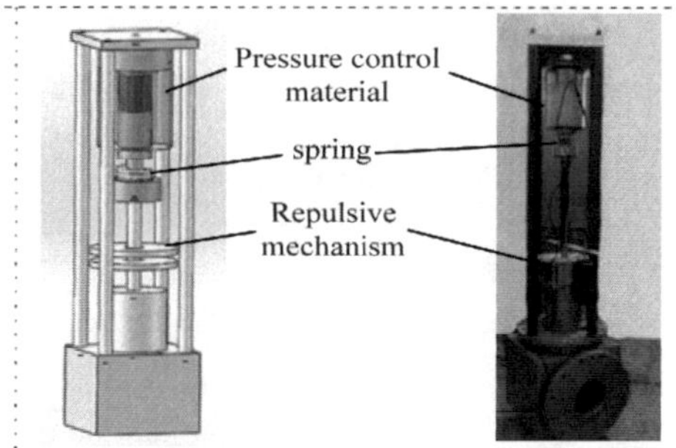

Figure 2. Variable resistance device

2.2. Dynamic Resistance Characteristic Test

Since the resistance change of TiB_2 is dependent on the change of pressure. Additional device is required to provide variable pressure. The operating mechanism of the vacuum circuit breaker can perform fast motions and can therefore be used to provide variable pressure. The overall structure is shown in Figure 2. The repulsive mechanism is connected to the metal electrode through a spring and an insulating rod. Under normal state, the mechanism is at the upper end, and the material is pressed by the spring. When the current limiting is required, the mechanism moves downward. And the pressure of the material is rapidly reduced. Material resistance will increase rapidly.

In order to measure the material resistance curve, connect it to the fixed value resistor apply a constant voltage. The voltage curve of material is the resistance curve. The changes of the material resistance under the opening and closing actions are respectively measured.

2.3. Current Test

Because the material is to be applied to the power field, the material itself has to flow a large current for a long time. Therefore, it is necessary to study the current characteristics of the material under large current. The test uses a large current generator to provide the required current, with a current range of 0-2000 A.

The test process is as follows: 1) Control the output current of the large current generator gradually increase from 0 to 2000 A, record the current and voltage values of the material every 200 A, and observe whether the material resistance changes with the increase of the current. 2) The material is subjected to a 5 minute current test. The currents are 200 A, 500 A, and 1000 A respectively. And the voltage and current data on the material are recorded every 1 min. After the current test, the pressure-resistance characteristics of the material were tested immediately. Compare with pre-flow data to observe whether the characteristics of TiB_2 have changed.

3. Results and Analysis

3.1. Pressure-Resistance Characteristic Analysis

The relationship between the pressure and the resistance is shown in Figure 3. It can be seen that the relationship between resistance and pressure is a power function, which is consistent with the description in Equation (1). And it can be clearly seen that when the pressure is greater than 500 N, the change trend of the particle resistance begins to slow down. When the pressure is less than 500 N, the particle resistance changes more severely. It can be explained by the granary effect of the particulate matter. In a cylindrical container, as the pressure increases, the pressure of the particle system on the bottom will tend to saturate, which is called the granary effect. This is because the friction between the side wall of the container and the particles counteracts the principal stress applied in the vertical direction. So after the applied pressure increases to a certain value, the pressure at the bottom of the

IOP Conf. Series: Materials Science and Engineering **730** (2020) 012009 doi:10.1088/1757-899X/730/1/012009

material reaches a saturation value. And thereafter even if the pressure is continuously increased, the principal stress of the vertical direction of the particles itself is no longer increased. Therefore, after the pressure is greater than 500 N, the resistance tends to be gentle.

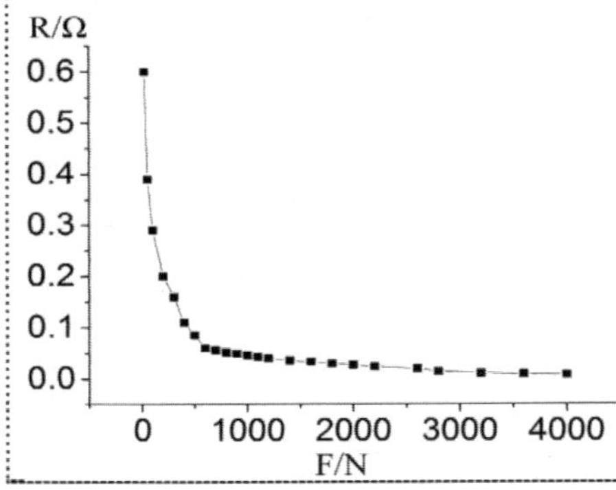

Figure 3. Relationship between the pressure and the resistance. When the pressure reaches 4020 N, the particle resistance is 0.004 Ω. When the pressure of the particles is 20 N (metal electrode gravity), the particle resistance is 0.6 Ω. From the trend of the curve, it can be reasonably estimated that when the pressure of the material is actually reduced to zero, the material resistance will become larger, which can meet the requirements of fault current limiting.

3.2. Dynamic Resistance Characteristic Analysis

The material resistance change curves measured by experiments are shown in Figure 4 and Figure 5, respectively.

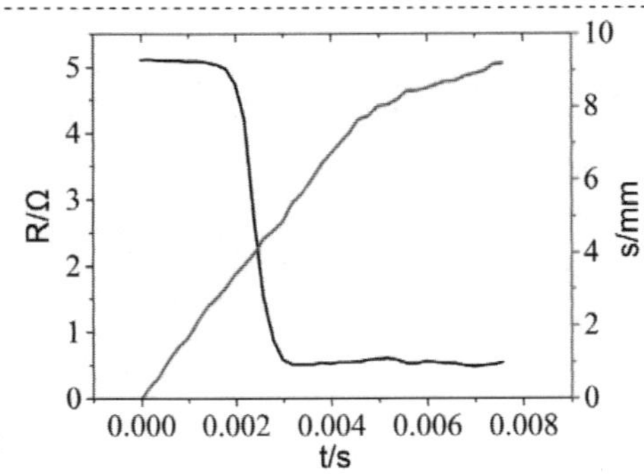

Figure 4. Resistance and displacement curve when mechanism moving up

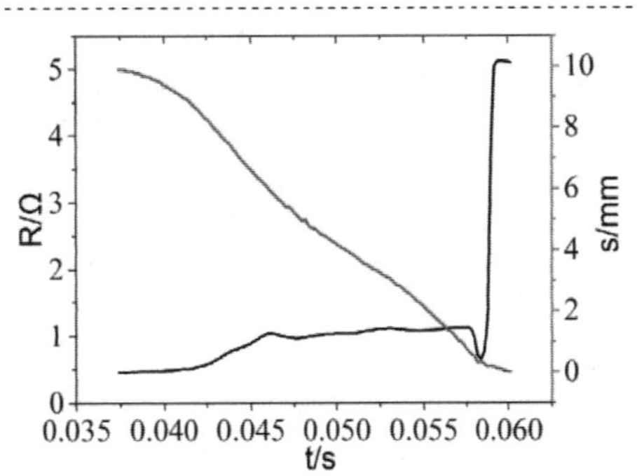

Figure 5. Resistance and displacement curve when mechanism moving down

The red line is the displacement curve of the mechanism, and the black line is the resistance curve of the material. As can be seen from Figure 4, when the mechanism moves up about 4 mm, the material resistance can be rapidly reduced and then stabilized. It can be seen from Figure 5 that when the mechanism starts to move down, the material resistance begins to have a slow rising process. But then the resistance remains stable. When the mechanism about to complete the moving, the material resistance will first become smaller and then rapidly become larger. It is because the mechanism has a bounce, which causes the pressure on the material to suddenly increase and then decrease rapidly. So the resistance curve will fluctuate at 0.058 s. However, the resistance eventually increases rapidly as the mechanism moving down.

3.3. Current Test Analysis

The results of the current-flow test of the materials are shown in Tables 2 and Tables 3. Table 2 shows the data obtained by gradually increasing the current from 0-2000 A. It can be seen that in the range of 0-1000 A, the resistance decreases as the current increases. When the current is greater than 1000 A, the resistance increases as the current increases. This is because when the current starts to increase, the internal pressure of the container increases due to the heat of the material. And when the current continues to increase, the material resistance is mainly affected by temperature. Due to the positive temperature coefficient, the material resistance is increased. However, the magnitude of the change in resistance is only 7%. And it can be considered that the resistance has not changed substantially.

Table 2. Material resistance as a function of current

Current/A	Voltage/mV	Resistance/mΩ
200	690	3.45
400	1370	3.43
600	2060	3.43
800	2660	3.33
1000	3220	3.22
1200	3970	3.31
1600	5250	3.28
2000	6700	3.35

Table 3 is the data of the materials for 5 minute current test. It can be seen that when the current is small (less than 500 A), the material resistance is mainly affected by the increase of internal pressure. As the flow time increases, the resistance is slightly reduced. When the current value is large, the material resistance is mainly affected by the temperature, and the resistance is slightly increased, but the overall variation of the resistance is still small.

Table 3. Material resistance for long-term current

Time/min	Resistance/mΩ(I=200 A)	Resistance/mΩ(I=500 A)	Resistance/mΩ(I=1000 A)
0	3.77	3.62	3.55
1	3.77	3.59	3.61
2	3.51	3.59	3.62
3	3.51	3.59	3.62
4	3.51	3.59	3.63
5	3.51	3.59	3.65

Through the above test data, it can be seen that when the material flows through a large current, its resistance value is relatively stable, indicating that the material can withstand the impact of large current. After the completion of the current-flow test, the pressure-resistance characteristics of the material were tested immediately. It was found that the overall data was not different from that before the current flows. It can be seen that the pressure-resistance characteristics of the material did not change after the current flow. So the feasibility of materials used in the power field is verified.

4. Conclusions

In this paper, a novel idea of using TiB_2 as a variable resistor in the field of electric power is proposed. The electrical properties of the TiB_2 are studied. The feasibility of using TiB_2 in fault current limiting is analyzed. The conclusion is as follows:

- Under the change of pressure, the resistance of TiB_2 will change significantly. When the pressure changes from 4020 N to 20 N, the resistance can change from 0.004 Ω to 0.6 Ω.
- The law of particle resistance and pressure is in accordance with the power function. And the change of particle resistance tends to be gentle after the pressure is greater than 500 N. This is because there is pressure saturation at the bottom of the particulate matter. So the height of the particulate material should not be too high.
- Through testing of material dynamic resistance, it is proved that the resistance can be changed quickly with the change of pressure. But in order to achieve rapid and stable change of the resistance, it is necessary to improve the operating mechanism to reduce the mechanical bouncing and action time.

- Under the impact of large current, the properties of the TiB_2 are basically unchanged, which proves the feasibility of the material applied in the electric field.

5. Acknowledgments

The authors are grateful for support from the National Natural Science Foundation of China (No. 51877026).

6. References

[1] Gromoll B, Ries G, Schmidt W 1999 Resistive fault current limiters with YBCO films 100 kVA functional model. *IEEE Transactions on Applied Superconductivity*, 9(2), 656-659.

[2] L. Kovalsky, Xing Yuan, K. Tekletsadik, A. Keri, J. Bock and F. Breuer 2005 Applications of superconducting fault current limiters in electric power transmission systems. *IEEE Transactions on Applied Superconductivity*, 15(2), 2130-2133.

[3] C. A. Baldan, C. Y. Shigue, D. S. S. Figueira, E. R. Filho and R. C. Freitas 2006 Test Results of a Superconducting FCL Using Bifilar Coil of BSCCO-2212. *IEEE Transactions on Applied Superconductivity*, 16(2), 695-698.

[4] D. Gonzalez, M. Hopfeld, F. Berger and P. Schaaf 2018 Investigation on Contact Resistance Behavior of Switching Contacts Using a Newly Developed Model Switch. *IEEE Transactions on Components, Packaging and Manufacturing Technology*, 8(6), 939-949.

[5] Guojian Cao, Erjun Guo, Liping Wang and Hongyu Xu 2011 Preparation of TiB powders by mechanical alloying. *Proceedings of 2011 6th International Forum on Strategic Technology*, Harbin, Heilongjiang, 101-103.

[6] Guoqin Chen, Ziyang Xiu, Songhe Meng, Gaohui Wu and Su Chen 2008 The thermal and electric conduction properties of high dense TiB_2P/Cu composites fabricated by squeeze casting technology. *2008 International Conference on Electronic Packaging Technology & High Density Packaging*, Shanghai, 1-3.

IOP Conf. Series: Materials Science and Engineering **730** (2020) 012010 doi:10.1088/1757-899X/730/1/012010

Synthesis and Characterization of BiOBr Microstructures Via Free Surfactant Solvothermal Route

Yuqin Mao[1], Luping Zhu[*1], Kaixiang Liu[1], Qiang Chen[1], Yongjin Zou[2], Xinying Shen[1] and Guihong Liao[3]

[1] School of Environmental and Materials Engineering, Shanghai Polytechnic University(SSPU), Shanghai 201209, China.
[2] Guangxi Key Laboratory of Information Materials, Guilin University of Electronic Technology(GUET), Guilin 541004, China.
[3]Technical Institute of Physics and Chemistry, Chinese Academy of Sciences, Beijing 100190, China.
[4]E-mail: lpzhu@sspu.edu.cn

Abstract. This work demonstrated a facilc and swift surfactant-free solvothermal synthesis of BiOBr architectures at 170°C for 6h. The structure and morphology were probed through diverse microscopy apparatus. The XRD result shows the obtained samples were well-crystallized tetragonal phase of BiOBr architecture. SEM images display these microspheres were distributed with a good polydispersity, even though most of them have inconsistent diameters. UV-vis spectrum reveals the band gap energy is approximately about 2.87 eV. The PL spectrum of the obtained BiOBr tested at room atmosphere reveals the strongest emission peaks appear at 469nm, which is in the visible region. On the basis of the above results, BiOBr samples also exhibit excellent photocatalytic performance in decomposing the organic dye RhB, which degradation efficiency reaches 99.7% after 50 min under ultraviolet light irradiation.

1. Introduction

As a layer-structured semiconductor, bismuth oxybromide (BiOBr) with tetragonal matlockite structure that $[Bi_2O_2]^{2+}$ slabs interleaved by double slabs of halogen atoms[1]. Besides this, bismuth oxybromide is interesting, not only as a good catalysts for degrading rhodamine-B (RhB)[2] or methyl orange (MO)[3], but also because of their special and excellent physical-chemical properties with other potential applications in ferroelectric materials, pigments, and catalytic fields[4-6]. Generally speaking, the properties of materials mainly rely on their size, structure, and dimensions. Considerable efforts have been devoted to the fabrication of BiOBr. For example, Xiong and co-workers prepared three-dimensional flower-like BiOBr nanoarchitectures through the direct transform of metallic Bi nanospheres[7]. Cheng's group prepared BiOBr hollow microspheres via a one-step Miniemulsion-Mediated route[8]. Zhang et al reported the synthesis of 3D flower-like BiOI nanoplates by an simple EG-assisted solvothermal method[9]. Inspired by previous work, we proposed a facile and mild surfactant-free solvothermal process to fabricate BiOBr microspheres. The optical and photocatalytical preformances were examined by UV-vis spectroscopy and fluorometer.

2. Experimental

2.1. Preparation of BiOBr Microspheres
Preparation of BiOBr Microspheres
All chemicals used in this work were purchased from the Shanghai Experiment Reagent Corporation Ltd. and were utilized as received without any further purification. In a typical experiment, 1.5 mmol $Bi(NO_3)_3 \cdot 5H_2O$ was mixed in 30 mL of ethylene glycol (EG) with vigorously stirring until $Bi(NO_3)_3 \cdot 5H_2O$ was dissolved completely. Then, 8.0 mmol NaBr was introduced into the former solution with stirring at room temperature. Subsequently, a transparent precursor was obtained and placed into a Teflon-lined stainless steel autoclave (50 mL), and then sealed and heated at 170°C for 6h at the same time. Finally, when the autoclave was cooled to room temperature naturally, the resulting white precipitate was collected, then wash with deionized (DI) water and absolute ethanol three times and dried at 80°C in air.

2.2. Photocatalysis Experiments
The photocatalytic efficiency of the obtained BiOBr samples was investigated by degrading Rhodamine-B (RhB) under UV light irradiation (λ=365nm), as described in our previous works[10-11].

2.3. Materials Characterizations
The phase purity of the products were measured by X-ray diffractometer (XRD, Rigaku D/max 2500). The overview of the sample morphology was characterized by scanning electron microscope (SEM, Hitachi S-4800), in conjunction with a system of energy-dispersive spectroscopy (EDS) analysis. Diffuse reflectance UV-vis of the powders were carried out by UV-vis spectrophotommeter (Shimadzu 2550), and photoluminescence (PL) spectrums were conducted by fluorometer (Hitachi F-4500) with a Xe lamp as the excitation light source at room temperature. Moreover, the concentrations of the RhB were measured by Shimadzu 2550 spectrophotommeter at 553nm during the photodegradation process.

3. Results and Discussion
The XRD pattern of the product is shown in Figure 1(a). The narrow broadening of peaks displayed a well-crystallized BiOBr material, and can be allocated to the tetragonal phase of BiOBr structure (JCPDS card no.73-2061). Consistent with the XRD results, the EDS pattern illustrated in Figure 1 (b) reveals the product includes Bi, O, and Br elements is approximately equal to 1:1:1.

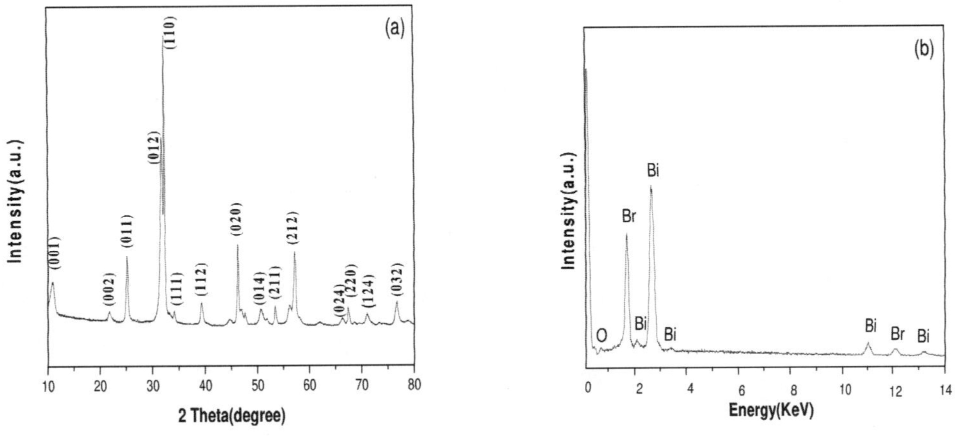

Figure. 1 Powder XRD (a) and EDS (b) patterns of BiOBr micropheres.

The morphology of the BiOBr microspheres fabricated by a surfactant-free solvothermal method are shown in Figure 2 (a-c). The SEM image with low-magnification manifests that the product composes of numerous microspheres with diameters ranging from 2 to 6 μm. An individual BiOBr microsphere is shown in Figure 2 (b)-(c) reveals the exterior surfaces of microspheres are not obviously smooth but involve many radially grown nanosheets.

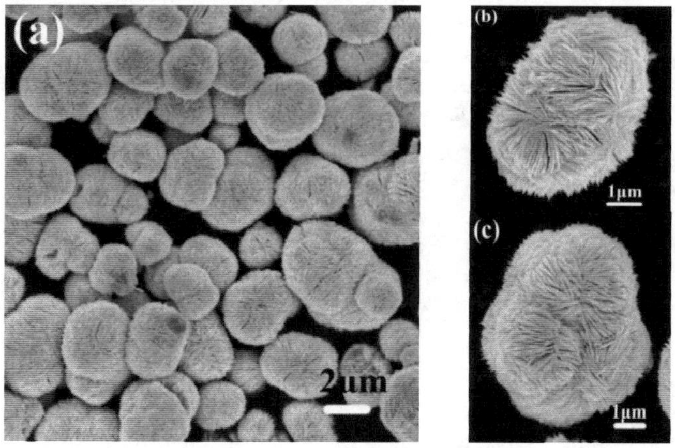

Figure. 2 SEM images in low-magnification (a), and a single BiOBr at various angle : (b) declivitous, (c) lcft-obliquc.

Figure 3 (a) depits the UV-vis diffuse reflectance spectrum of BiOBr microspheres. Assuming BiOBr is an indirect semiconductor, the $(\alpha h\nu)^{1/2}$ versus photon energy hν curve for the BiOBr samples can be shown in the inset of Figure 3 (a). The extrapolates value (the straight line to the X-axis) of band gap energy (Eg value) at α = 0 gives an absorption edge energy corresponding to Eg = 2.87 eV, which is in the visible light region. The Eg values were a slight different from the theoretic value (2.54 eV)[12], which could be ascribed not to the transition from the impurity to the conduction band but to the intrinsic transition between the valence band and the conduction band.

The photoluminescence property of the flower-like BiOBr excited at 270nm at room temperature is demonstrated in Figure 3 (b). It can be observed that the PL spectrums of the prepared samples are in the visible region (390 - 550 nm). Moreover, the strongest PL emission peaks appear at 469 nm. According to the literature[13-14], the PL intensity may be attributed to the morphology, particle size, and surface defects.

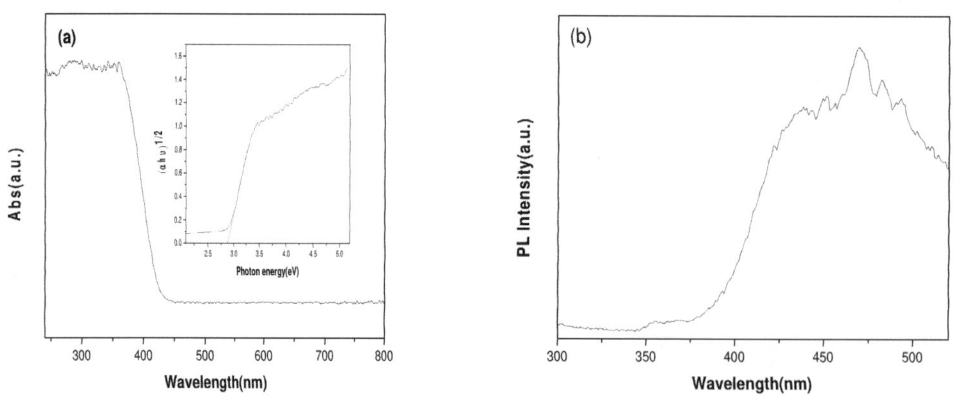

Figure. 3 UV-vis diffuse reflectance spectrum (a) and PL spectrum (b) of the flower-like BiOBr. Inset : the $(\alpha h\nu)^{1/2}$ - hν curve for the products.

The variations of RhB concentration (C/C_0) with different catalyst loading can be described in Figure 4 (b). It is note worthy that the absence of both light source and catalysts are negligible for efficient degradation. Therefore, as to photocatalytic activities, catalysts and illumination are of great importance. Generally, catalyst loading plays a critical part in the control of recombination of holes and electrons of BiOBr samples. Hence we probed with different catalyst loading ranging from 5 mg to 20 mg, while the concentration of RhB and light source remained constant. The results clearly show that the degradation efficiency reaches 89.2%, 96.1%, 99.7% for samples 5mg, 10mg, and 20mg with 50 min, respectively. Thereby, the suggested catalyst dosage is 20mg.

Figure 4 (b) demonstrates the UV-vis spectra of RhB (10mg/L) degradation on BiOBr (20mg) for time intervals of 0-50 min. The results show that the absorption peak diminishes progressively as the exposure time increases and completely vanishes after 50 min, which means that 99.7% of the RhB degradation was achieved in 50 min, which is consistent with UV and PL results.

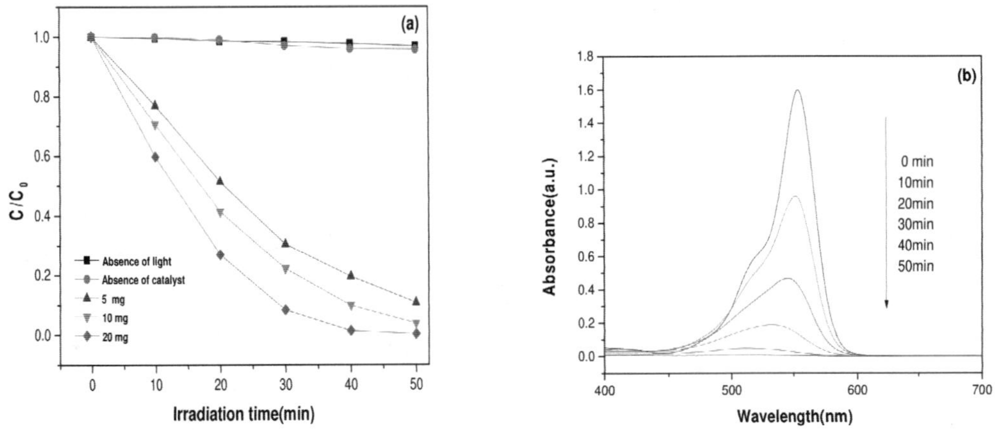

Figure. 4 (a) photocatalytic performances of different catalyst loading:5mg, 10mg, and 20mg. (b) absorption spectra of a solution of RhB in the presence of the BiOBr products (20mg) under ultraviolet light exposure.

4. Conclusions

In summary, flower-like BiOBr with high crystallinity are successfully synthesized by a simple surfactant-free solvothermal method. Flower-like BiOBr in the rang of 2- 6 μm were observed which were confirmed by SEM analysis. UV-vis demonstrates the samples prepared at 170°C show the band gap energy were about 3.07 eV. The PL spectrum indicates that the powders possess a strong and broad emission in the visible region (390-550nm), in which the strongest emission peak appears at 469nm. The suggested catalyst loading is 20mg, which the degradation efficiency reaches 99.7% in 50min. The present work suggests that these 3D BiOBr architectures may be offer new potentials in semiconductor photocatalysis, solar cells, optic and gas sensing applications.

5. Acknowledgments

This research was jointly supported from the Shanghai Municipal Natural Science Foundation (No: 18ZR1415700), the Guangxi Key Laboratory of Information Materials (GUET) (No:171011-K), the Leap Project and Postgraduate fund (SSPU) (Nos: EDG18XQD26, EGD18YJ0049), the key subject of SSPU (No. 4: Material Science and Engineering, XXKZD1601), and the Gaoyuan Discipline of Shanghai–Environmental Science and Engineering (Resource Recycling Science and Engineering).

6. References

[1] Zhang K, Liu C M and Huang F Q 2006 *Appl Catalysis B* **68** 125-9

[2] Zhang L, Cao X F and Chen X T 2011 *J Coll Interface Sci* **354** 630-6
[3] Jiang Z, Yang F and Yang G 2010 *J Photochemistry and Photobiology A* **212** 8-13
[4] Kusainova A M, Stefanovich S Y and Irvine J T 2002 *J Mater Chem* **12** 3413-8
[5] Maile F J, Pfaff G and Reynders P. 2005 *Progress in Organic Coatings* **54** 150-63
[6] Huo Y, Zhang J and Miao M 2012 *Appl Catalysis B* **111** 334-41
[7] Xiong J, Dong Q and Wang T 2014 *RSC Adv* **4** 583-6
[8] Cheng H, Baibiao Huang and Wang Z 2011 *Chemistry* **17** 8039-43
[9] Zhang B, Ji G and Gondal M A 2013 *J Nanoparticle Res* **15** 1773
[10] Mao YQ, Li YH and Zou YJ 2019 *Cera Inter* **45** 1724-9
[11] Zhu LP, Liao GH and Bing NC 2010 *CrystEngComm* **12** 3791-6
[12] Zhang J, Shi F and Lin J 2010t *Chem Mater* **39** 2937-41
[13] Zhu LP, Wang LL and Bing NC 2016 *RSC Adv* **6** 2926-34
[14] Zhu LP, Mao YQ and Chen Q 2019 *J Mater Sci* **30** 3639-46

IOP Conf. Series: Materials Science and Engineering **730** (2020) 012011 doi:10.1088/1757-899X/730/1/012011

Study on the Effect of Galacto-oligosaccharide (GOS) in Relieving Constipation and Defecating Feces Excretion

Jin Wang [1, 2], Ping-Ju Tsai [3], Pei-Hsuan Chen [3], Manxiang Ye [4], Hua Cao [5], Jiao Guo [2*] and Zhengquan Su [1*]

1 Guangdong Engineering Research Center of Natural Products and New Drugs, Guangdong Provincial University Engineering Technology Research Center of Natural Products and Drugs, Guangdong Pharmaceutical University, Guangzhou 510006, China
2 Guangdong Metabolic Diseases Research Centre of Integrated Chinese and Western Medicine, Guangdong TCM Key Laboratory for Metabolic Diseases, Key Laboratory of Modulating Liver to Treat Hyperlipemia SATCM, Level 3 Laboratory of Lipid Metabolism SATCM, Institute of Chinese Medicinal Sciences, Guangdong Pharmaceutical University, Guangzhou 510006, China
3 King-Prebiotics Biotechnology (TW) Co., Ltd. 2F.-1, No. 250, Zhongshan Rd., Linkou Dist., New Taipei City 24446, Taiwan, China
4 New Francisco (Yunfu City) Biotechnology Co., Ltd Swan-kan-chiau Ind. Dist., Kaofong Village Yunfu City, Guangdong, China
5 School of Chemistry and Chemical Engineering, Guangdong Pharmaceutical University, Zhongshan 528458, China
* Correspondence: * Authors to whom correspondence should be addressed.
Professor Jiao Guo, Guangdong Pharmaceutical University, Guangzhou 510006, China. E-mail: gyguoyz@163.com
Professor Zhengquan Su, Guangdong Pharmaceutical University, Guangzhou 510006, China. E-mail: suzhq@scnu.edu.cn. Tel.: +86-20-3935-2067; Fax: +86-20-3935-2067

Abstract. To explore the role of Galacto-oligosaccharide (GOS) in preventing constipation and promoting excretion of feces. The high, medium and low doses of GOS-1000-S (8.8g/kg, 4.4g/kg, 2.2g/kg) were administered orally to constipation mice for 7 days, and then we measured the propulsive rate of intestines, first defecation time, the number of feces particle and fecal weight in 5 hours. The results showed that GOS-1000-S can significantly promote the small bowel propulsion, shorten the first defecation time and increase the fecal weight number of fecal particle of constipation mice. These indicate that GOS has the effect of relieving constipation and defecating feces excretion. This study provided a basis for the development of GOS with constipation-relieving functions as a health supplement.

1. Introduction

The structure of Galacto-oligosaccharide (GOS) is mainly linked from the galactosyl side of lactose to 1-7 galactosyl groups by β-1,4 glycosidic bonds [1], and the chemical structure is Gal-(Gal)n-Glc/Gal (n=0-6) [2, 3]. GOS is a colorless and translucent substance with a viscosity similar to thatof high fructose syrup, and its sweetness is equivalent to 30%-60% of sucrose. It has stable physical and chemical properties [4], is easily soluble in water, has strong stability to acid and heat, and has high moisture retention.

At present, there are five main methods for the synthesis of GOS: natural raw material extraction, natural polysaccharide acid hydrolysis, chemical synthesis, fermentation and microbial enzyme synthesis. The immobilized enzymatic synthesis of GOS is a mainstream method for the synthesis of oligogalactose because it can make full use of β-galactosidase and has good production stability, high yield and few post-treatment processes [5, 6]. The separation and purification techniques of GOS mainly include the following methods: column chromatography, membrane separation, enzymatic separation and microbial fermentation [7].

As one of the most safe, highly recognized and therapeutic therapeutic effects of many functional oligosaccharides, GOS has a variety of physiological functions and has a beneficial effect on the human body[8]. GOS can be utilized by various beneficial bacteria to ferment it to produce short chain fatty acids (SCFAs) and some antibiotic substances, which can inhibit the growth of various harmful bacteria (eg Shigella, Salmonella, Staphylococcus aureus and Escherichia coli) and promote the growth of various beneficial bacteria (such as bifidobacteria and lactic acid bacteria), thereby effectively regulating the balance of intestinal flora and reducing the phenolics produced by the intestinal microbiota and improves skin health [9-11]. In addition, GOS can improve the absorption of mineral elements and improve the incidence of cardiovascular disease, obesity and type 2 diabetes [12-15]. Recent studies have also found that SCFAs produced by the use of beneficial bacteria after GOS enters the human body can also reduce colon cancer by stimulating cancer cell apoptosis [16].

Our research group has long been engaged in the research on the weight loss activity and mechanism of chitosan, chitosan oligosaccharide and GOS, and has systematically studied its weight loss activity [17-21]. In view of the fact that the GOS intestinal metabolite SCFA has a function of affecting intestinal peristalsis [16], our research group began to pay attention to and explore the role of GOS in preventing constipation and promoting excretion.

2. Materials and Methods

2.1. Materials

King-Prebiotics® GOS1000 syrup (GOS-1000-S), which contain <1% monosaccharides and disaccharides, provided by New Francisco (Yunfu City) Biotechnology Co., Ltd, and the corresponding concentration is prepared by using distilled water during the experiment.

Reagent: Activated carbon powder (Shanghai Aladdin Biochemical Technology Co., Ltd., C1607128), Acacia gum (Shanghai Aladdin Biotechnology Co., Ltd., A1619001), Loperamide hydrochloride capsule (Xi'an Yangsen Pharmaceutical Co., Ltd., 150827647), Tongbianling Capsules (Anhui Jiren Pharmaceutical Co., Ltd., 1170107).

Experimental animals: SPF Male KM mice, 6 weeks old, provided by Guangdong Medical Laboratory Animal Center. Animal production license number: SCXK (Guangdong) 2013-0002, tested after 3 days of normal feeding.

2.2 Methods

Dosage: The body recommended dosage for the GOS-1000-S is 13.3 g/d/60 kg. GOS-1000-S corresponds to three dose groups, high, medium and low doses (equivalent of 40, 20, 10 times times to the human body recommended dose), which are 8.8g/kg, 4.4g/kg, 2.2g/kg BW, respectively. The recommended dose for Tongbianlingcapsules is 1.5g/d/60kg for adults, and the dose for mice is 0.5g/kg (equivalent to 20 times the recommended amount of the human body).

Raising environment: During the experimental period, the animal room temperature was 23.0±2.0°C, the relative humidity was 40-70%, the number of air changes was >15 times/hour, 12 hours of illumination/12 hours of darkness, and the brightness and darkness alternated. Laboratory animal use license number: SYXK (GuangDong) 2017-0125.

Model establishment: After 7 days of administration of the test samples, the mice were fasted for 16 hours. Except the control group was given distilled water, the other groups were given 3 mg/kg aqueous solution of loperamide hydrochloride capsules by oral gavage, and the mice small bowel peristalsis inhibition model and mice constipation model were established respectively.

The effect of GOS on small bowel movement: 70 male KM mice were randomly divided into 6 groups according to their body weight, namely control group, model control group, Tongbianling capsule positive group, and high, medium and low dose administration groups of GOS-1000-S, with 11 rats in each group. On the 7th day, the mice were fasted for 16 hours before the administration, except for the distilled water in the blank group, and the other groups were given 3 mg/kg aqueous solution of loperamide hydrochloride capsules. After 30 min, each group of mice was given an indicator ink (5% active carbon, 10% gum arabic) containing the corresponding test sample, and the control group and the model group were administered with ink.

After 25 min, the mice were sacrificed by dislocation. The entire gastrointestinal tract of the mice was taken out, and the intestines were plated on a wet glass plate, with the pylorus of the stomach as the starting point and the ileocecal part as the end point, and the total length (cm) of the small intestine and the distance (cm) in which the ink moves in the intestine from the pylorus were measured. The percentage of ink movement distance per mice in the total length of the small intestine was calculated as the gastrointestinal propulsion rate (P) of the mice, and the data was converted (X) and then statistically analyzed $X = Sin^{-1}\sqrt{P}$.

Effect of GOS on defecation in mice: After the mice in each group were given the ink, they were kept in a single cage, and the feed and water were normally supplied. Record the time of the first black feces, the number of feces particle and weight of feces in the first 5 h after administered with indicator ink in mice

Statistical processing: The data was expressed as X±SD and statistical analysis was performed using SPSS12.0 software.

3. Results

3.1. The Effect of GOS on Small Bowel Movement

The ink propulsion rate of the model group mice was significantly lower than that of the blank group (P < 0.01), indicating that the constipation model was established. All mice had normal feces and no diarrhea. The ink propelling rates of the GOS-1000-S mice in each dose group were higher than those in the model group (P <0.01). After statistical analysis by data conversion, the difference between the conversion values of the GOS-1000-S groups and the model group was significant (P <0.01), indicating that GOS-1000-S has the effects of promoting small bowel movement in mice, and the effect shows a dose-dependent (the results shown in Table 1).

Table 1. Effects of GOS-1000-S on intestinal propulsion in mice($\bar{X}$±SD, n=11)

Group	Dose (g/kg)	Propulsion rate (P) (%)	Conversion value (X)
Control	-	0.73±0.07**	1.33±0.05**
Model	-	0.43±0.07	1.66±0.13
Positive	0.5	0.64±0.09**	1.40±0.08**
GOS-1000-S-H	8.8	0.86±0.06**	1.25±0.03**
GOS-1000-S-M	4.4	0.77±0.07**	1.30±0.04**
GOS-1000-S-L	2.2	0.65±0.12**	1.40±0.11**

Notes: Compare with model group, * P<0.05, * * P<0.01

3.2 Effect of GOS on Defecation in Mice

The first defecation time of the model group was longer than that of the control group (P<0.01), and the number of feces particle and fecal weight were lower than the blank group (P<0.01 and P<0.05), indicating that the constipation model was established. All mice had normal feces and no diarrhea. The first defecation time of each group of the GOS-1000-S was shorter than that of the model group (P<0.01). There was significant difference in the number of feces particle between the GOS-1000-S middle dose group, GOS-1000-S low dose group and the model group (P<0.01), there was no significant difference between the high dose group and the model group (P>0.05). Except for the high dose group of GOS-1000-S, the fecal weight of the middle and low dose groups was significantly higher than that of the model group (P<0.01 and P<0.05). The above results indicate that GOS-1000-S can shorten the first defecation time and increase the fecal weight and the number of feces particle of constipation mice (the results shown in Table 2).

Table 2. Effects of GOS-1000-S on defecation in mice ($\bar{X}$±SD, n=10)

Group	Dose(g/kg)	First defecation time(min)	Feces particle number(particle)	Fecal weight(g)
Control	-	78.80±13.69**	36.00±7.66**	0.44±0.09*
Model	-	153.70±65.93	18.90±9.34	0.27±0.15
Positive	0.5	60.10±25.04**	30.40±8.50**	0.52±0.12**
GOS-1000-S-H	8.8	93.90±48.34**	16.60±6.38	0.27±0.10
GOS-1000-S-M	4.4	70.00±19.19**	34.80±6.65**	0.41±0.11*
GOS-1000-S-L	2.2	91.70±29.99**	34.10±6.98**	0.46±0.08**

Notes: Compare with model group, * P<0.05, * * P<0.01

4. Conclusion

At present, our research shows that GOS-1000-S can promote small bowel propulsion in constipation mice, shorten the first defecation time of constipation mice and increase the fecal weight and the number of feces particle. According to the relevant provisions of the health food functional evaluation procedures and testing methods, these results showed that GOS has the effect of relieving constipation and defecating feces excretion.

5. Acknowledgments

This work was financially supported by the Science and Technology Planning Project of Yunfu, Guangdong, China (No. 201702-9); Guangdong Provincial University Engineering Technology Research Center of Natural Products and Drugs, China (No.2017GCZX002); and the Science and Technology Planning Project of Guangdong, China (No.201806040009, 201804010349, 201804010329).

6. Reference

[1] Osman A, Tzortzis G, Rastall RA and Charalampopoulos D 2010 A comprehensive investigation of the synthesis of prebiotic galactooligosaccharides by whole cells of Bifidobacterium bifidum NCIMB 41171. J Biotechnol. 150(1):140-148.

[2] Tomomatsu HJFT 1994 Health Effects of Oligosaccharides. 48(10):61-65.

[3] Kimura K, ., Matsumoto K, ., Ishihara C, ., Harada K, . and Miyagi A, . %J Carbohydrate Research 1995 Structure determination of galacto-oligosaccharides by pyridylamination and NMR spectroscopy. 270(1):33-42.

[4] Macfarlane G, Steed H and Macfarlane SJJoAM 2010 Bacterial metabolism and health-related effects of galacto-oligosaccharides and other prebiotics. 104(2):305-344.

[5] Neri DFM, Balcão VM, Costa RS, Rocha ICAP, Ferreira EMFC, Torres DPM, Rodrigues LRM, Jr LBC and Teixeira JAJFC 2009 Galacto-oligosaccharides production during lactose hydrolysis by free Aspergillus oryzae β-galactosidase and immobilized on magnetic polysiloxane-polyvinyl alcohol. 115(1):92-99.

[6] Krajewska BJE and Technology M 2004 Application of chitin- and chitosan-based materials for enzyme immobilizations: a review. 35(2):126-139.

[7] Martín-Ortiz A, Ruiz-Matute AI, Sanz ML, Moreno FJ and Herrero M 2019 Separation of di- and trisaccharide mixtures by comprehensive two-dimensional liquid chromatography. Application to prebiotic oligosaccharides. Anal Chim Acta. 1060:125-132.

[8] Sabater C, Prodanov M, Olano A, Corzo N and Montilla A 2016 Quantification of prebiotics in commercial infant formulas. Food Chem. 194:6-11.

[9] Bryk G, Coronel MZ, Pellegrini G, Mandalunis P, Rio ME, Portela MLPMd and Zeni SN 2015 Effect of a combination GOS/FOS ® prebiotic mixture and interaction with calcium intake on mineral absorption and bone parameters in growing rats. European Journal of Nutrition. 54(6):913.
[10] Bryk G, Coronel MZ, Lugones C, Mandalunis P, Rio ME, Gualtieri AF and Zeni SN 2016 Effect of a mixture of GOS/FOS ® on calcium absorption and retention during recovery from protein malnutrition: experimental model in growing rats. European Journal of Nutrition.1-14.
[11] Macfarlane S, ., Macfarlane GT and Cummings JH 2010 Review article: prebiotics in the gastrointestinal tract. Alimentary Pharmacology & Therapeutics. 24(5):701-714.
[12] Reynes B, Palou M, Rodriguez AM and Palou A 2018 Regulation of Adaptive Thermogenesis and Browning by Prebiotics and Postbiotics. Front Physiol. 9:1908.
[13] Conterno L, Fava F, Viola R and Tuohy KM 2011 Obesity and the gut microbiota: does up-regulating colonic fermentation protect against obesity and metabolic disease? Genes & Nutrition. 6(3):241-260.
[14] Devadder F, Kovatcheva-Datchary P, Goncalves D, Vinera J, Zitoun C, Duchampt A, Bäckhed F and Mithieux G 2014 Microbiota-Generated Metabolites Promote Metabolic Benefits via Gut-Brain Neural Circuits. Cell. 156(1-2):84-96.
[15] Cheng W, Lu J, Lin W, Wei X, Li H, Zhao X, Jiang A and Yuan J 2018 Effects of a galacto-oligosaccharide-rich diet on fecal microbiota and metabolite profiles in mice. Food & Function. 9(3):10.1039.C1037FO01720K.
[16] Sun Y and O'Riordan MXD 2013 Chapter Three–Regulation of Bacterial Pathogenesis by Intestinal Short-Chain Fatty Acids. Advances in Applied Microbiology. 85:93-118.
[17] Jian C, Gui-Dong H, Si-Rong T, Jiao G and Zheng-Quan SJIJoMS 2013 The preparation of capsaicin-chitosan microspheres (CCMS) enteric coated tablets. 14(12):24305-24319.
[18] Tao Y, Zhang HL, Hu YM, Wan S and Su ZQJIJoMS 2013 Preparation of Chitosan and Water-Soluble Chitosan Microspheres via Spray-Drying Method to Lower Blood Lipids in Rats Fed with High-Fat Diets. 14(2):4174-4184.
[19] S T, B G, Y T, J G and ZQ S 2014 Antiobese effects of capsaicin-chitosan microsphere (CCMS) in obese rats induced by high fat diet. Journal of agricultural and food chemistry. 62(8):1866-1874.
[20] Hong-Liang Z, Yi T, Jiao G, Yin-Ming H and Zheng-Quan SJII 2011 Hypolipidemic effects of chitosan nanoparticles in hyperlipidemia rats induced by high fat diet. 11(4):457-461.
[21] Yang D, Hu C, Deng X, Bai Y, Cao H, Guo J and Su Z 2019 Therapeutic Effect of Chitooligosaccharide Tablets on Lipids in High-Fat Diets Induced Hyperlipidemic Rats. Molecules. 24(3).

IOP Conf. Series: Materials Science and Engineering **730** (2020) 012012 doi:10.1088/1757-899X/730/1/012012

Preparation of PBS and $CaCO_3$ Composite Degradable Materials based on Melt Blending

Ligui Xiong[1], Zhigang Tu[2,*], Shangxian Zhang[3], Xiaoyuan Liu [4] and Xingni Peng[5]

[1,2,3,5] Guangdong Polytechnic Institute, Zhongshan, China
[4] Guangzhou Vocational College of Technology& Business, Guangzhou, China
*Corresponding author e-mail: 1049661754@qq.com

Abstract. Poly (butylene succinate)(PBS)/ Calcium carbonate ($CaCO_3$) composite were prepared by melt blending, and studied the effect of different coupling agents, PBS and $CaCO_3$mass ratio and coupling agent for composite. The results showed that aluminum acid coupling agent for the properties of composite materials improved the most obviously effect; with $CaCO_3$ content increased, the degradation properties and tensile strength of composite materials rose, the impact strength and elongation at break decreased. With aluminum acid content increased, the tensile propeties and impact strength increased and then stabilization but no influence on degradation. When the PBS and $CaCO_3$ mass ratio was 7:3, the coupling agent in an amount of 1.5%, the performance of PBS/ $CaCO_3$composite was better, meeting the basic requirements. Compared with pure PBS, it reduced the cost of nearly 30%, and application further.

1. Introduction

Polybutylene succinate (PBS) is a typical fully biodegradable polymer material that is easily decomposed and metabolized by various microorganisms or enzymes in animals and plants, and eventually decomposed into carbon dioxide and water. Compared with degradable plastics such as polylactic acid (PLA), polyhydroxybutyrate (PHB), polycaprolactone (PCL), and polyhydroxyalkanoate (PHA), PBS has low cost, excellent mechanical properties, and convenient processing. It can adapt to the advantages of conventional molding processing equipment and good heat resistance.[1-4] It can replace polyethylene and polypropylene. At the same time, it has a wide range of raw materials, which can be obtained from petroleum resources or fermented by biological resources. And industry attention, PBS is the leader in biodegradable plastic materials, but compared to traditional plastics, its price is too high, it is difficult to promote its application, so reducing its cost becomes the key to the future development of PBS plastics. Calcium carbonate is a low-cost mineral material derived from nature and has been widely used in the modification of various polymer materials. It is feasible to reduce the cost of PBS by using calcium carbonate filling.[5-6]
In this paper, PBS/$CaCO_3$ composites were prepared by blending PBS with calcium carbonate as filler. The coupling agent type, the amount of coupling agent and the mass ratio of PBS/$CaCO_3$ were studied. The effects of elongation, impact properties and degradation properties.

2. Experimental Part

2.1. Main Ingreadient

Light calcium carbonate: 2000 mesh, Zhongshan Yongfeng Chemical Co., Ltd.;
PBS: Blow molding grade, Shandong Huiying New Material Technology Co., Ltd.;

Aluminate coupling agent: DL-411, Foshan Shengyi Plastic Chemical Co., Ltd.;
Silane coupling agent: KH550, Foshan Shengyi Plastic Chemical Co., Ltd.;
Titanate coupling agent: CS-104, Nanjing Chuangshi Chemical Additive Co., Ltd.

2.2. Main Instruments and Equipment

Twin-screw extruder: TES-65, Jiangshu Nanjing Noda Extrusion Equipment Co., Ltd.;
Mixer: x(S) N-35, Jiangsu Nanjing Kaichi Machinery Co., Ltd.;
Injection molding machine: JN55E, Guangdong Dongguan Zhenxiong Machinery Co., Ltd.;
PC type intelligent electronic tensile testing machine: XLW (PC) -500N-50N, Shandong Jinan Languang Electromechanical Technology Co., Ltd.;
Cantilever beam impact testing machine: XCJ, Jilin University Science and Education Instrument Factory.

2.3. Sample Preparation

The production process is shown in Figure 1, and the production temperature is shown in Table 1.

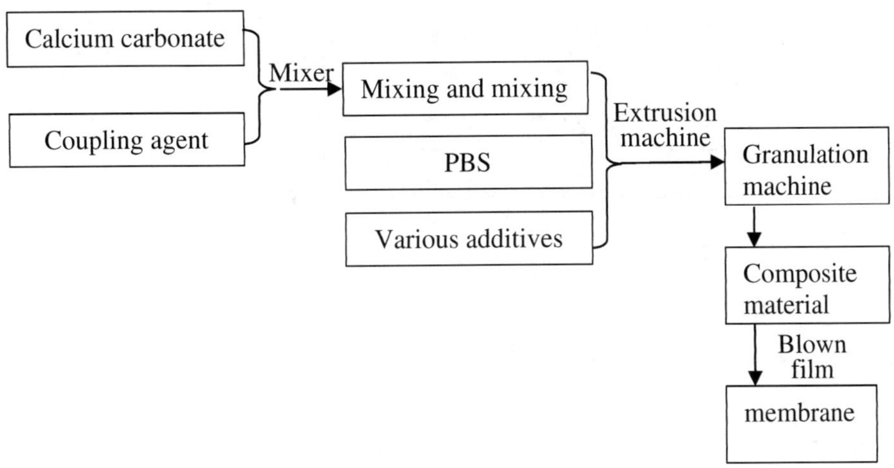

Figure 1. Production process

Table 1. Production temperature of PBS/$CaCO_3$ composite

1 area/°C	2~4 area/°C	5~6 area/°C	8 area/°C	Machine head temperature/°C	Terminal mode/°C
115~120	120~125	125~130	130~135	135~140	140

2.4. Performance Testing

The tensile properties are tested according to GB/T1040.3-2006, and the sample is type I.

For the spline, the stretching speed was 50 mm/min, and 5 samples were tested in each group. The results were averaged.

The impact performance was tested according to GB/T1843-1996.Each group was tested with 5 samples and the results were averaged.

Degradation performance was carried out by soil-buried method. The sample was cut into (25×25)mm sample. After drying and weighing, it was sandwiched between gauze and buried in the flowerbed soil about 20 cm away from the surface for a certain period of time (28d). After taking out, and calculating the mass loss rate of the sample, the calculation formula is: weight loss rate = (m1-m2) / m1 × 100%, where m1 and m2 are respectively

For the original quality of the sample and the quality after degradation, 5 samples were tested for each group, and the results were averaged.

3. Results and Discussion

3.1. Effect of Coupling Agent Type on Properties of Composites

Experimental conditions: the mass ratio of PBS and $CaCO_3$ was 7:3; the coupling agent was aluminate, titanate and silane, and the addition amount (calculated as the mass of calcium carbonate, the same below) was 2%.

It can be seen from Figure 2 that the mechanical properties of the composites are different when the coupling agents used are different.

The mechanical properties of the composites with the addition of coupling agent were significantly improved. When the coupling agent was aluminate, the effect was the most obvious. Therefore, the aluminate coupling agent was determined to be the most PBS/$CaCO_3$ composite. Good coupling agent.

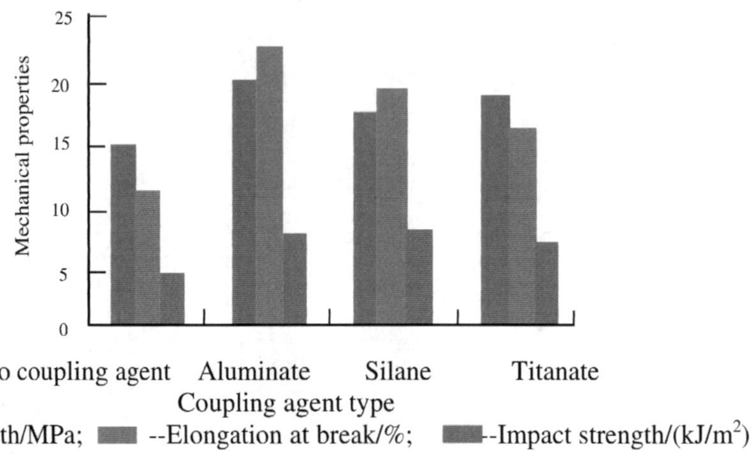

Figure 2.

3.2. Effect of PBS / $CaCO_3$ Mass Ratio on Properties of Composites

Experimental conditions: the amount of aluminate added is 2% (calculated as the mass of calcium carbonate),The mass ratio of PBS to $CaCO_3$ was set to 10:0, 9:1, 8:2, 7:3, and 6:4.

When the mass ratio of PBS to $CaCO_3$ is 6:4, the extruder is often blocked during preparation, and the required temperature is also high. If the temperature is too high, the degradation of PBS will be triggered, indicating that the amount of calcium carbonate added is too high. Therefore, the quality ratio relationship does not continue to be followed up.

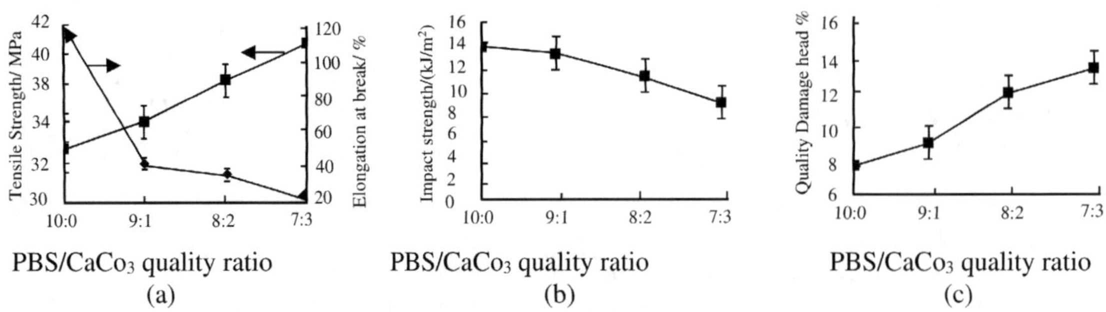

(a) tensile properties; (b) impact strength; (c) degradation properties

Figure 3. Effect of PBS/$CaCO_3$ mass ratio on properties of composites.

It can be concluded from Fig. 3(a) and (b) that as the amount of $CaCO_3$ increases, the tensile strength of the composite increases, and both the elongation at break and the impact strength decrease. This is because $CaCO_3$ acts as a filler and can act as a skeleton. When the amount is larger, the tensile strength of the composite is larger. At the same time, the presence of a large amount of powder limits the movement of the PBS molecular segment, so that $CaCO_3$ and PBS The bonding strength of the interface is weak, which leads to the decrease of impact strength and elongation at break of the composite. However, when the amount of $CaCO_3$ is as high as 30%, the elongation at break of the composite is still higher than 20%, indicating that the material still has a high Elongation, and the ultimate goal of PBS filled fillers is to minimize the cost of the product while maintaining performance. It can be concluded from Fig.3(c) that the mass loss rate of the composite increases linearly with the increase of Ca CO_3 dosage, which is much higher than that of pure PBS. This is due to the filling effect of $CaCO_3$, which makes the microorganisms more susceptible to erosion. The inside of the material [7-9], so the ratio of PBS/$CaCO_3$ was determined to be 7:3.

3.3. Effect of Coupling Agent Amount on Properties of Composites

Experimental conditions: the mass ratio of PBS to $CaCO_3$ is 7:3, The amount of the amount of the aluminate (calculated by the mass of the calcium carbonate) was 1.0%, 1. 5%, 2. 0%, 2. 5%, 3. 0%.

It can be seen from Fig. 4(a) and (b) that as the amount of aluminate increases, the tensile properties and impact strength both appear to rise first and then tend to be gentle. This is because $CaCO_3$ has poor affinity with PBS matrix resin [10-13], and the performance of the composite material is deteriorated due to uneven dispersion. The addition of coupling agent can improve the surface properties of $CaCO_3$, starting from composite materials. The effect of the binder can effectively improve the compatibility of $CaCO_3$ with the PBS matrix and increase the interfacial bonding strength of the two, but the amount of the coupling agent is not as good as possible [14-15]. For a long time, there will be an interference effect, and the excess coupling agent will not only fail to achieve the coupling effect, but will become an impurity of the system and increase the production cost of the composite material[8]. It can be seen from Fig.4(c) that when the amount of aluminate is different, the degradation performance is not much different. Considering the cost of the coupling agent and its addition amount to the mechanical properties of the composite. It can be seen from Fig. 4(c) that when the amount of aluminate is different, the degradation performance is not different. Considering the cost of the coupling agent and the amount of the addition agent on the mechanical properties and the degradation properties of the composite material, the effect is better when the amount of the aluminate is 1.5%.

(a) Tensile properties; (b) Impact strength; (c) Degradation properties

Figure 4. Effect of the amount of aluminate on the properties of composites

4. Conclusion

The aluminate coupling agent has the most obvious effect on the performance of the composite; when the amount of $CaCO_3$ is increased, the tensile strength and degradation performance of the composite increase, the elongation at break and the impact strength decrease; the effect of the amount of aluminate on the degradation performance Not large, but when the amount of aluminate is increased, the tensile properties and impact strength first rise and then tend to be gentle.

When the mass ratio of PBS and $CaCO_3$ is 7:3, and the amount of aluminate is 1.5%, the composite material has better comprehensive performance, which can meet the requirements of use, and can reduce the yield by nearly 30% compared with pure PBS. The cost of its promotion and further application.

5. References

[1] C.N.CHUI, T.J.HAUNG and Y.HE, *"Study on Aging Performance of Poly (butylene succinate)/Calcium Carbonate Composite"*. Plastic technology, Vol.41 No.11, pp.74-77.2013.

[2] Z.C.LIAO, *"Research Progress of Biodegradable Plastic PBS"*.Plastic technology, Vol.38 No.7, pp.93-98.2010.

[3] TACHIBANA Y, MASUDA T, FUNABASHI M,et al. *"Chemical synthesis of fully biomass-based poly (butylene succinate) from inedible-biomass-based furfural and evaluation of its biomass carbon ratio"*. Biomacromolecules, Vol.11 No.10, pp.2760-2765.2010.

[4] Y.C.SONG, Z.P.WANG and XCHENG. *"PreParation and PorPerties of Starch/BS Blends"*. China Plastics, vol.20, no.4, pp.65-67, 2006.

[5] J.LIU, H.J.JI, W.ZHANG, *"Development in blending modification of biodegradable Poly (butylene succinate) and natural biode gradable polymer"*. New Chemical Materials, Vol.41 No.8, pp.1-3.2013.

[6] Z.DING, Q.LING, ang X.S.PEN. *"Toughening Research Progress in Completely Biodegradable Polylactide by Blending"*. Plastic, Vol.45 No.2, pp.95-97.2016.

[7] B.YANG, Y.XIU, H.J.JI, *"Application of Double-coated Surface Modification of Calcium Carbonate in Highly Filled Composite Material"*. Plastic, Vol.43 No.1, pp.4-7.2014.

[8] L.B.GAO, *"The Study on Modification of Fully Biodegradable Poly (butylenes succinate)"*. Beijing Technology and Business University, 2006.

[9] K.K.YING, *"Study on modification of biodegradable Poly (butylenes succinate)"*. Wuhan University of Technology, 2011.

[10] X.LI, D. H.LIU and Y.H.HUANG, "Performance of Calcium Carbonate Modified by Aluminate Compound Coupling Agents". Yunnan Chemical Technology, Vol.30 No.5, pp.3-6.2003.

[11] L.X.LUN, Z.F.LUO, *"Study of the surface modification of heavy calcium carbonate"*. Journal of China University of Mining & Technology, Vol.40 No.2, pp.269-272-6.2011.

[12] Y.X.MA, P.ZHOU and H.M.LIU,*" Preparation and Properties of PLA / NPCC Composite"*, Plastic,Vol.44 No.5,pp.25-28.2015.

[13] H.YANG, L.ZHANG and X.LU, *"Study on Isothermal Crystallization Kinetics of the Calcium Sulfate Whisker/PBS Blends"*, Plastic technology, Vol.41 No.5, pp.8-12.2013.

[14] Q.WEN, *"Study on blending modification of biodegrable plastics PBS"*, Hunan University, 2011.

[15] T.LI,H.Z.LI and J.B.ZENG, "Microbial Degradability of Thermoplastic Starch/PBS Blends "Chemical Research and Application, Vol.31 No.7,pp.994-997.2019.

IOP Conf. Series: Materials Science and Engineering **730** (2020) 012013 doi:10.1088/1757-899X/730/1/012013

Analysis of T23 Steel Weld-Joint on the Stress Relaxation Crack at Medium Temperature

Zhiqiang Sun[1], Yuwei Wang[2], Zhongbing Chen[1], Yu Zhang[2], Yongsan Ding[2], Hongbin Zhu[2] and Shaohua Yin[1]

1.Suzhou Nuclear Power Institution,1788 XIhuan Road, Suzhou, P.R. China
2.China Energy Jiangsu Power Co., Ltd.,Jianbi Power Plant, Jianbi Town, Zhenjiang City, P.R.China
E-Mail:13776254398@163.com, 12065569@chnenergy.com.cn, czbing2000@163.com, zy839783@163.com, yongsan.ding@chenergy.com.cn, zhulv2012@163.com, yinshaohua@cgnpc.com.cn

Abstract. This paper issues a new type of reheat-crack of T23 steel weld which is deferent to the cracks occurred at higher temperature above 550 degree as literatures reported. The specimen to be studied taken from a unit with long service time is analyzed by means of metallographic, scanning electron microscopy and energy dispersive spectroscopy. When further in-depth analysis is carried out on the test results, we find that Ti (Al) -N compounds are precipitated at grain boundaries, promote the formation of cavities, and finally become crack. Although the temperature of the crack is only about 380 degree which is much lower than the reheat-crack sensitivite temperature limit of T23 steel, but we also regard this crack as a type of stress relaxation crack which is another form of reheat-crack in coarse grained region of T23 steel weld-joint under specific service environment.

1. Introduction

HCM2S steel, which is a kind of bainite heat-resistant stee researched and developed by Sumitomo Corporation of Japan, has been used of high temperature components such as water-wall of high parameter thermal power plants in China in recent years. This steel is developed on the basis of T22 steel with the characterise of adding tungsten (W) ,which partly replacing molybdenum(Mo), niobium(Nb) and vanadium(V) , and reducing the carbon content to 0.04%~0.10% at the same time. The ASME grade of this steel is ASTM213-T23. The chemical composition and the mechanical properties at room temperature of T23 steel are respectively indicated in table 1 and 2(Ref.1). Comparated the allowable stresses at 540~580 degree temperature of the steels of HCM2S, 12Cr1MoV and T22, the value of HCM2S is 1.3 times as much as that of 12Cr1MoV and 1.8 times as much as that of T22(Ref.2). Therefore, as a result of the high allowable stress at elevated temperatures, the use of T23 steel in water-wall can effectively reduce the tube thickness and then improve the heat transfer rate. At the same time, T23 steel has a low carbon content, which can improve the weldability and is convenient for water-wall installation (Ref.3~4).

In recent years, many failure cases of T23 water wall tube butt-joints have been reported in China(Ref.5~7). But few literatures have been published on the analysis of T23 steel water-wall fin welds. Because water-wall is the largest and most complex component in power plant, the butt-joints and the fin welds are both the key weld joint structure. So, it is necessary to analyze the performance of the fin weld of T23 steel water-wall undergoing long service time of 1000MW ultra supercritical unit. This paper does it, and aims to provide relevant experience for similar units.

Table 1. Chemical composition of T23 steel (Wt,%)

Elements	*C*	*Mn*	*P*	*S*	Si	Cr	Mo	V	Ti	W	Nb	N	Al
min.	0.04	0.10	-	-	-	1.9	0.05	0.20	n.s.	1.45	0.02	-	-
max.	0.10	0.60	0.03	0.01	0.50	2.6	0.30	0.30		1.75	0.08	0.03	0.03

Note: n.s. in this table indicates that there is no redulation.

Table 2. Mechanical properties of T23 steel

Yield strength(MPa)	Tensile strength(MPa)	Elongation(%)	Maximum hardness(HB)
≥400	≥510	≥20	220

2. Experimental Materials

The experimental materials are taken from the vertical section of water-wall of a 1000MW ultra supercritical unit which has been put into service from 2013. The materials of tube and fin are T23 steel with the tube specification of OD38.1×THK6.8(mm), and T12 steel with the 6mm thickness. The tube and fin are welded together by R307 electrode. During the maintenance period in 2019, many incomplete penetrated fin welds at the root which were welded during installation period are observed, as shown in Fig.1. We pick up sample of the this kind of weld-joint to analyse.

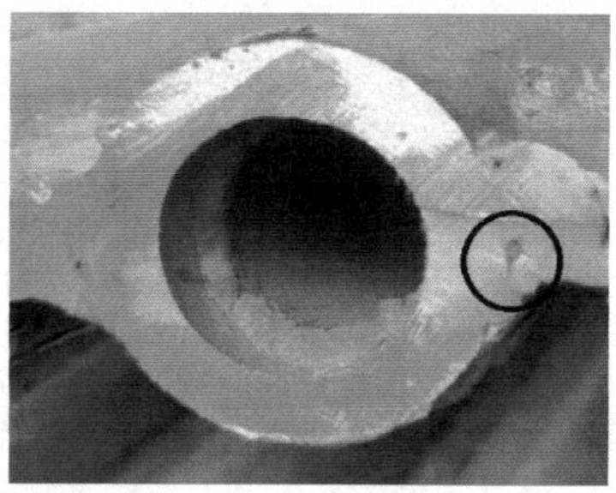

Figure 1. Incomplete penetration fin weld

3. Analysis of the Crack Characterises

3.1. Macroscopic and Microscopic Characteristics Analysis

It is observed by visual that a crack propagates from the root of the fin weld towards the inner wall of the tube after the specimen is polished and corroded by 4%H3NO3 alcohol solution, as shown in Fig.2. The same crack is also observed in another specimen with the same weld-joint structure, as shown in Fig.3.

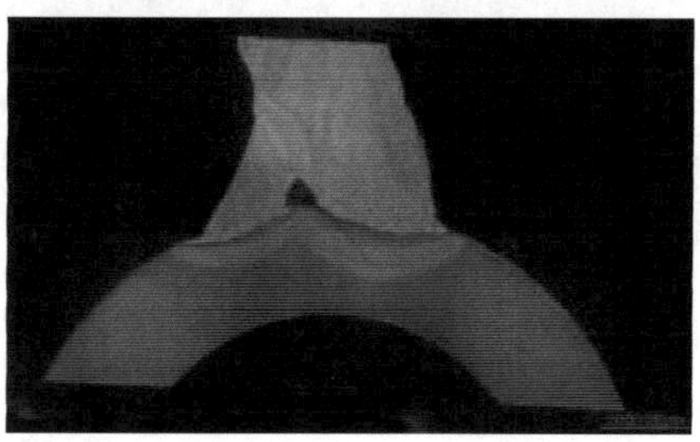

Figure 2. The crack 1 appearance of fin weld

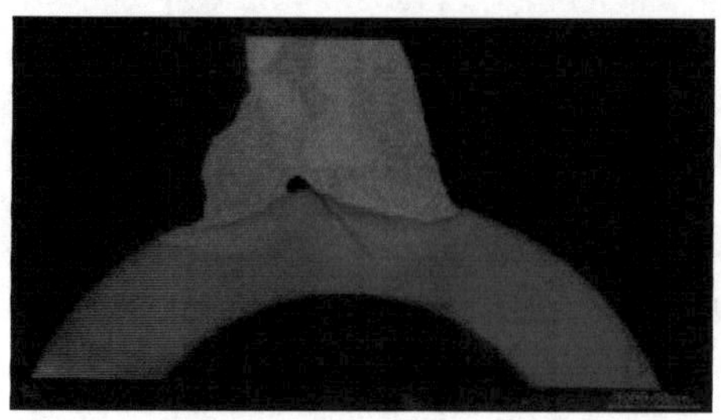

Figure 3. The crack 2 appearance of fin weld

The following characteristics of cracks are observed under the inverted universal metallographic microscope of ZEISS AXIOVERT 200 MAT.

At the starting stage, the crack initiates and extends parallel to the fusion line in the coarse grain region, and then extends towards the inner wall of the tube, finally ends in the fine grain region, as showed in Fig.4.

There are many defects are observed nearby the main crack, such as cavities, secondary cracks and so on. It is also found grey oxides in both the main crack and the secondary crack, as showed in Fig.5, 6 and7.

The crack ends are blunt, and small defects such as dense cavities are found around the ends. The cracks are also filled with gray oxide, as showing in Fig.8.

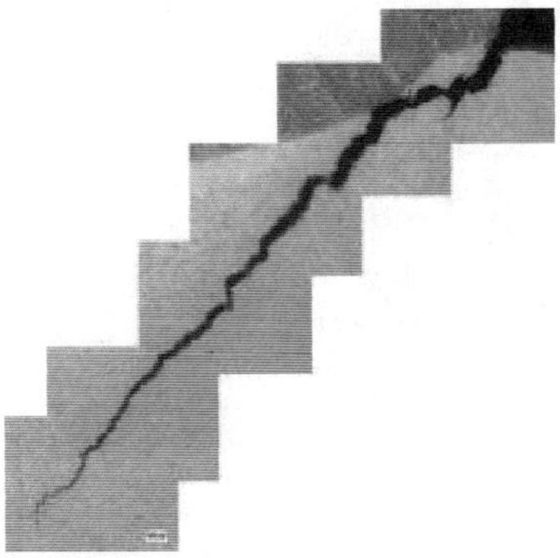

Figure 4. Appearance of the crack

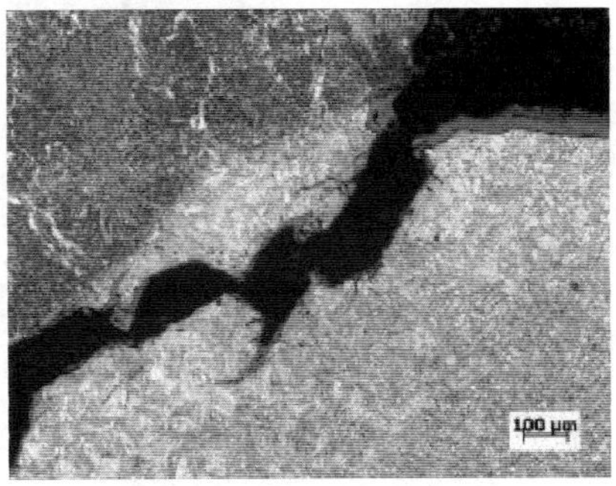

Figure 5. Appearance of the crack-start

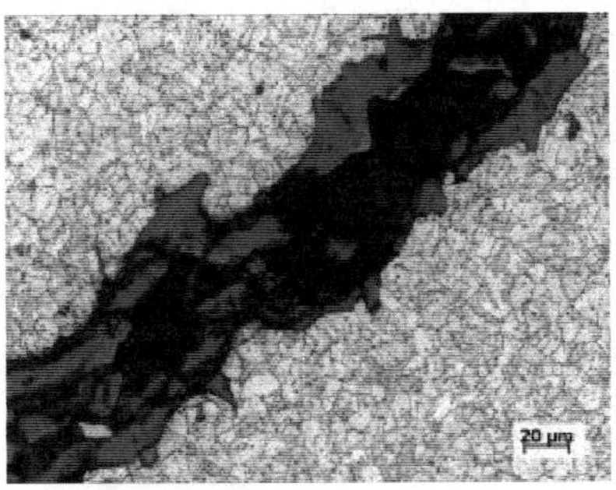

Figure 6. Appearance of the crack propagation in CGHAZ

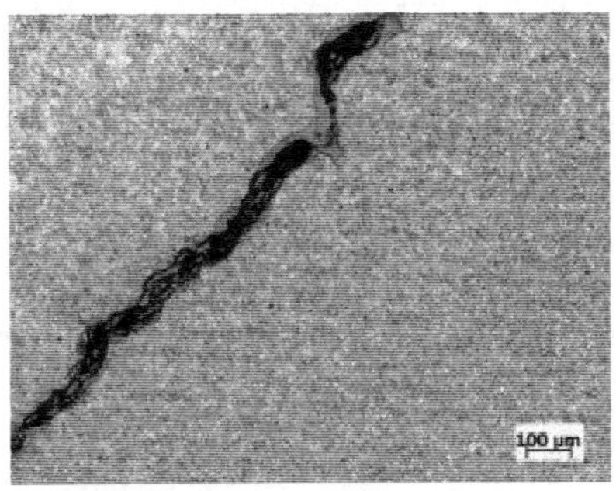

Figure 7. Appearance of the crack propagation in FGHAZ

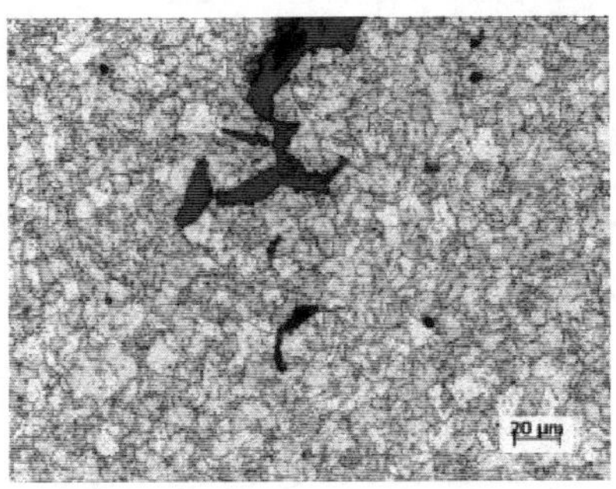

Figure 8. Appearance of the crack end

Scanning electron microscopy is used for further observation. Three kinds of defects were found around the main crack (MC), namely branching crack (CB), accompanying crack (CC) and cavity (H), see Fig.9 and 10. The crack branching (CB) and accompanying crack (CC) are all secondary cracks.

Fig. 9 displays the blunt ends of CB and CC, and the cavities morphology at the end of CB. Both the cracks are filled with gray oxide. The results of EDS displays that the grey oxide is high temperature oxide of Fe, such as Fe3O4, as showed in Fig. 10. The phenomenon indicates that the cracks are oxidized at high temperature for a long time after they are developed.

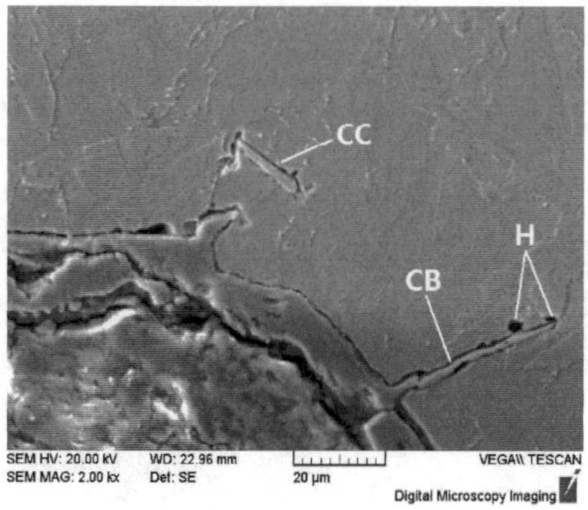

Figure 9. Appearance of secondary crack nearby the main crack

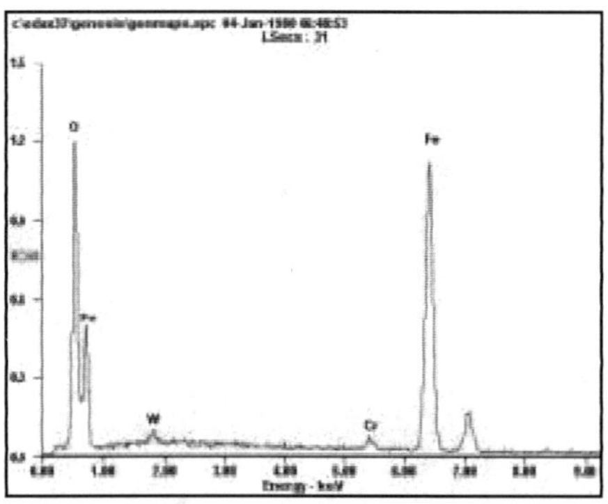

Element	*Wt%*	*At%*
OK	24.04	53.14
WM	02.99	00.57
CrK	01.66	01.13
FeK	71.32	45.16
Matrix	Correction	ZAF

Figure 10. EDS results of oxide in the secondary crack

Fig.11 reveals the appearance of two concomitant cracks (CC1,CC2) and several cavities(H1-H4) around the main crack. The concave and convex edges of CC1 show that the micro-cracks are formed after the connection of multiple cavities. But there is no grey oxide found on the surface of CC1. This is because CC1 is a newly formed crack and has not yet been oxidized. The CC2 crack has relatively straight edges, and is filled with oxide. Its characteristics are as same as those in Fig.9. The precipitates can be seen at the center of the H1-H4 cavities which grow at the triple junction of grain boundaries. EDS results of the precipitates of H4 show that it is a Ti (Al) - N compound, see Fig.12.

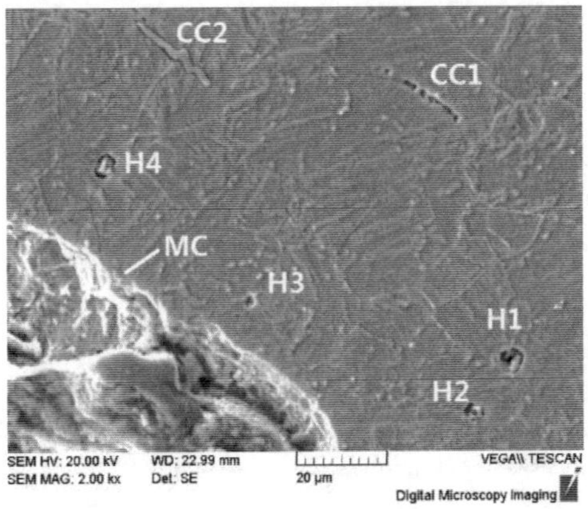

Figure 11. Appearance of concomitant cracks and cavities

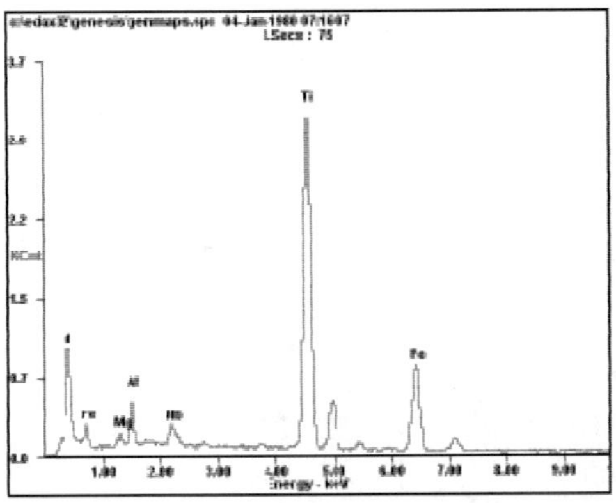

Element	*Wt%*	*At%*
NK	23.88	51.97
MgK	00.97	01.21
AlK	03.95	04.47
NbL	02.65	00.87
TiK	44.95	28.61
FeK	23.60	12.88
Matrix	Correction	ZAF

Figure 12. EDS results of the precipitate in the hole of H4

Fig.13 shows the appearance of the metal precipitate a, cavity b which is left after the precipitate shedding, and microcrack c which is formed by the cavities connection. EDS results of the precipitates c show that both the precipitate in Fig.14 and those in Fig. 11 are all Ti (Al) -N compounds.

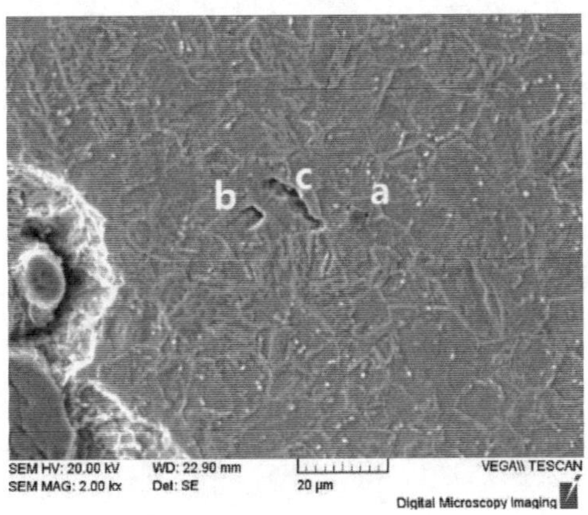

Figure 13. Appearance of the metal precipitates, voids and microcrack

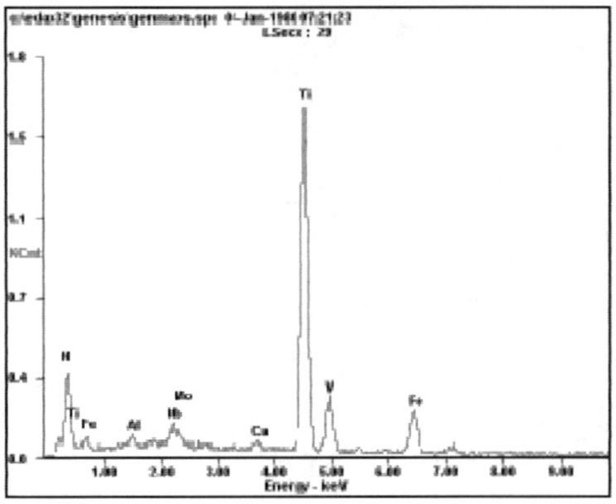

Element	*Wt%*	*At%*
NK	19.96	47.32
AlK	00.96	01.18
NbL	04.08	01.46
MoL	01.60	00.55
CaK	00.78	00.65
TiK	57.09	39.57
VK	00.67	00.44
FeK	14.84	08.82
Matrix	Correction	ZAF

Figure 14. EDS results of the metal precipitate

Fig. 15 and 17 are two wedge cracks found near the main crack. The EDS results show the elements of Ti, Al, N and O are all exist in both the cracks. Comparing the proportion of Ti and O between the two cracks(Fig.16,Fig.18) and H4 cavity(Fig.12), it is found that the proportion of Ti in the two cracks is significantly lower than that in H4 cavity. Accompany with the expansion of the wedge crack, the proportion of Ti decreases from 17.27%(Fig.16) to 9.08%(Fig.18), while the proportion of O increases from 8.06%(Fig.16) to 33.20%(Fig.18) at the same time. This indicates the process of the wedge crack that the cavities are firstly formed around the metal precipitates of Ti and other metal elements, and then partially oxidized.

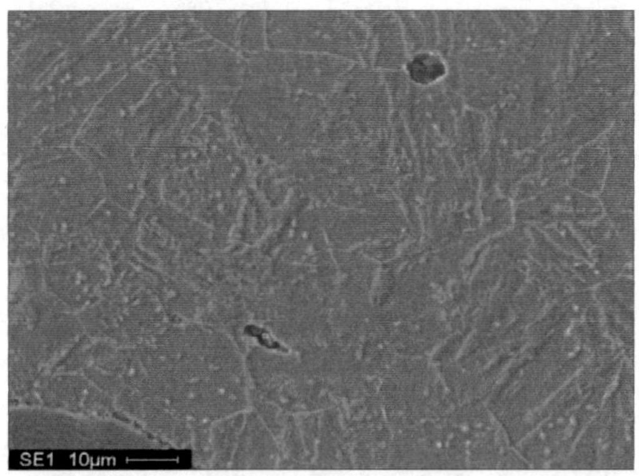

Figure 15. Wedge crack 1 near the main crack

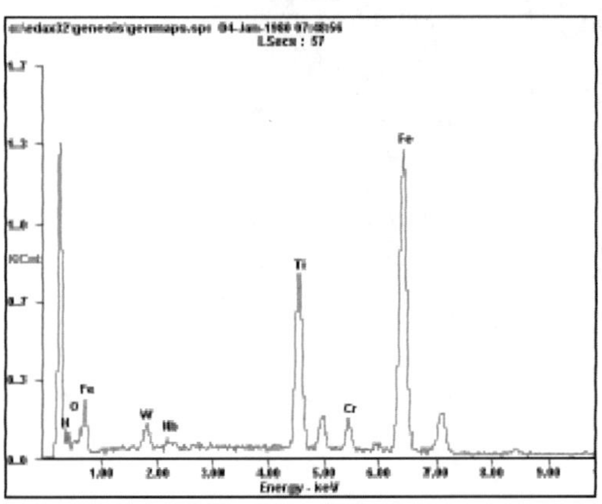

Element	*Wt%*	*At%*
NK	04.77	15.79
OK	02.78	08.06
WM	05.04	01.27
NbL	00.99	00.49
TiK	17.86	17.27
CrK	04.07	03.63
FeK	64.49	53.49
Matrix	Correction	ZAF

Figure 16. EDS results of the wedge crack1

IOP Conf. Series: Materials Science and Engineering **730** (2020) 012013 doi:10.1088/1757-899X/730/1/012013

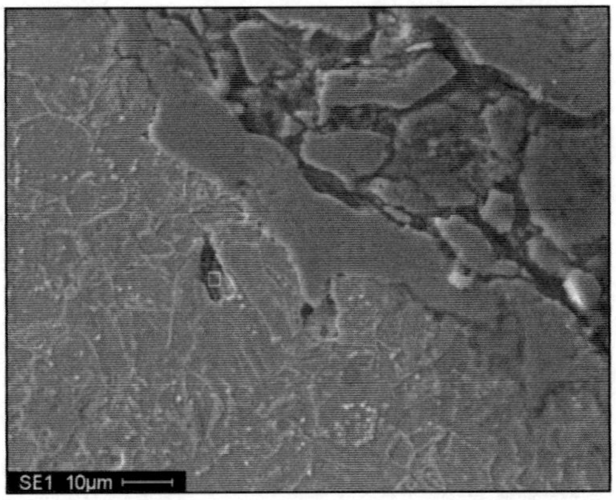

Figure 17. Wedge crack 2 near the main crack

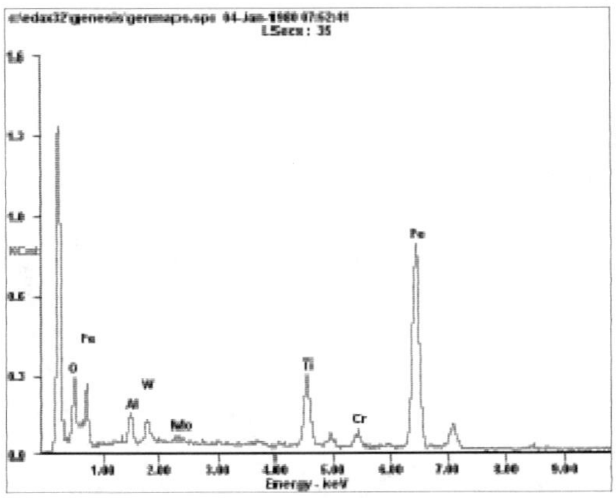

Element	*Wt%*	*At%*
OK	12.45	33.20
AlK	03.32	05.25
WM	07.22	01.68
MoL	01.35	00.60
TiK	10.19	09.08
CrK	03.23	02.65
FeK	62.24	47.55
Matrix	Correction	ZAF

Figure 18. EDS results of the wedge crack2

Fig. 19 shows the appearance of dense cavities a, wedge crack b, micro-crack c with the boundary oxidized and completely oxidized crack d. The crack end is blunt, and the crack interior is filled with gray oxide. The cavities have the exact same characteristics as those in Fig.11. The EDS results are also similar to those of both Fig.12 and Fig.21. The EDS results of the wedge crack b show that element Ti isn't found, but the proportion of O is so high to 33.97%, as showed in Fig. 22, that is probably due to the oxidation and shedding of the crack.

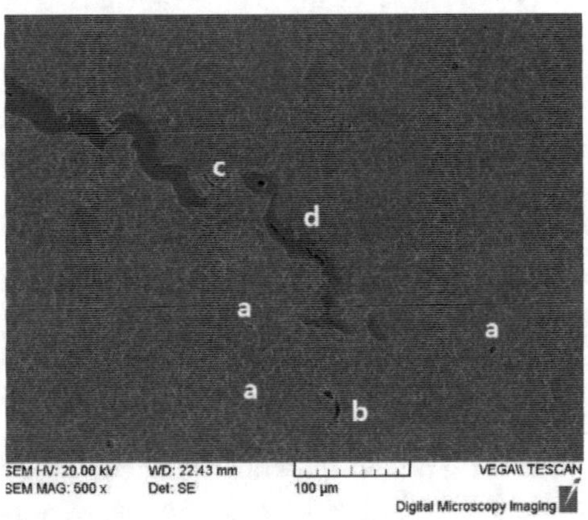

Figure 19. The whole appearance of the crak end

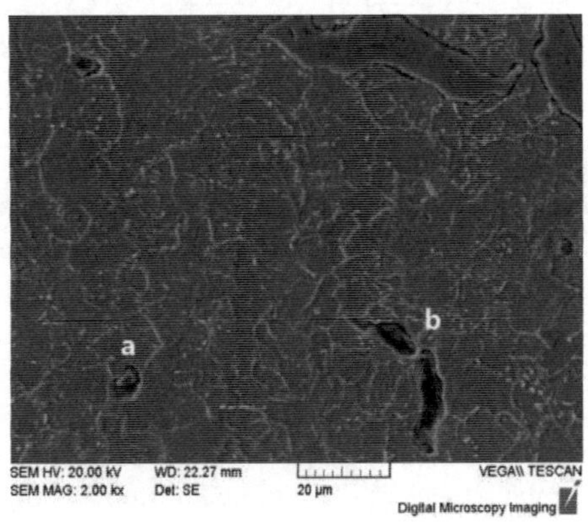

Figure 20. Appearance of the cavities and micro-crack in the crack-end

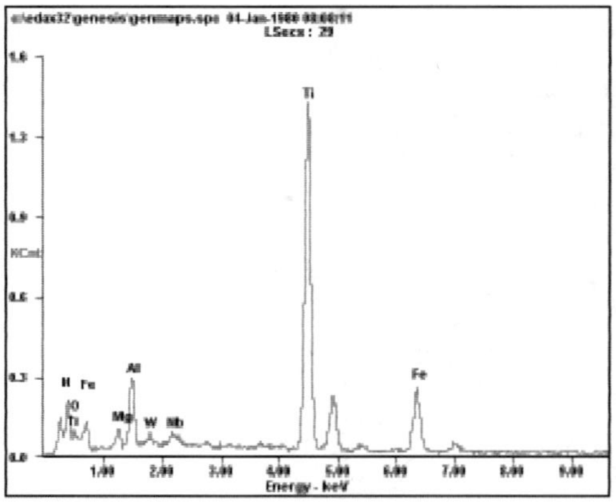

Element	*Wt%*	*At%*
NK	10.81	25.63
OK	08.43	17.50
MgK	02.15	02.93
AlK	06.21	07.64
WM	02.54	00.46
NbL	02.13	00.76
TiK	48.75	33.79
FeK	18.96	11.27
Matrix	Correction	ZAF

Figure 21. EDS results of the precipitate in the cavity

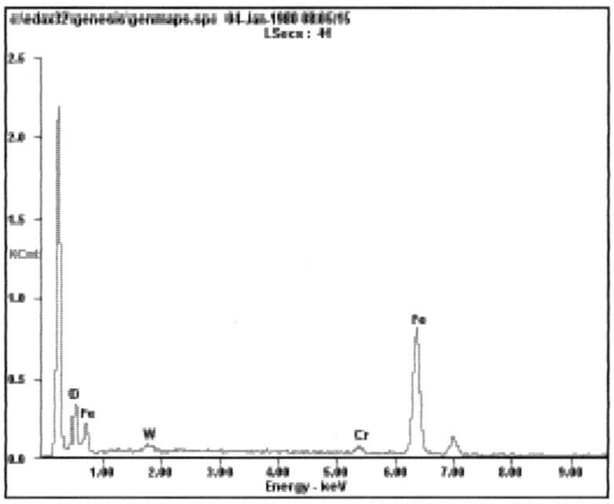

Element	*Wt%*	*At%*
OK	12.35	33.97
WM	05.76	01.38
CrK	02.54	02.15
FeK	79.34	62.50
Matrix	Correction	ZAF

Figure 22. EDS results of the wedge crack

3.2. Summary of Crack Characteristics

1)Cracks initiate in the *CGHAZ* of weld-joints. Firstly, they propagate along the fusion line in the CGHAZ, then extend towards the inner wall of the tube, finally end in the FGHAZ.

2)Many secondary cracks, such as branch cracks(CB) and accompanying cracks(CC), are found in the zone near the main crack. The main cracks and secondary cracks have blunt tips, and is filled with gray oxide, probably as Fe3O4.

3)The cavities around the crack germinate and grow at the core of metal precipitates, as Ti(Al)-N.

4)There is an inevitable relationship between the crack and the cavities from the evidence of Ti(Al)-N precipitates found both in wedge crack and cavities. The oxidation is the same step of the cracks growing, and the macrocraks are finally formed from many wedge cracks.

4. Analysis and Discuss

Based on the reheat crack theory, the crack is made as following. Firstly, the material is transformed to austenite due to primary heating, which is transformed into martensite or bainite with higher hardness during cooling stage. Secondly, in the process of reheating, the strengthening elements preferentially precipitate inside the grains, which would strengthen the grains and weaken the grain boundaries. When reheating, stress relaxation causes the slide between grain boundaries, but the grains are too strong to deformation. Under the dual action of stress relaxation and precipitate strengthening (hardening), stress will be concentrated on the boundaries of gain, so the crack is occured(Ref.8-9). The typical characteristics are that the crack initiates in the coarse grain region, propagates parallel along the fusion line and ends in the fine grain region(Ref.10-11). But the crack of this paper has its own typical characteristics in the respects of generation condition, micro-characteristics, crack evolution process.

Firstly, the service temperature of the crack in this paper is much lower than the reheat-cracke sensitive temperature of T23 steel which ranges from 575 to 750 degree(Ref.12-13). A lot of researches has been carried out in this temperature range. The elevated temperature aging tests on T23 steel weld show that new precipitates are detected neither inside nor at the boundaries of the weld grain at 500 degree temperature, while a small amount of secondary precipitates have just been detected until the temperature reaches to 550 degree(Ref.14). Reference 15 calculated the service temperature and stress diagram of water-wall of 1000MW unit. The results showed that the temperature of fin weld root(where cracks occur) was estimated to 380 degree, while the water wall tube was at 413.7 degree service temperature(Ref.15). This result is corresponding with reality according to the actual operating conditions of the unit. So, the temperature at which cracks occur is lower than not only the reheat-crack sensitive lower limit temperature, but also the lower secondary precipitation phase limit temperature. The Vickers hardness test on the heat affected zone shows that the average hardness reaches HV306, see the results of 7-11 number points in Fig.23. The hardness of T23 weld without high tempering heat treatment can reach HV300~350(Ref.16-18). Contrusting the two hardness values, we could draw a conclusion that the weld of this paper is not gone through high tempering heat treatment. So, the oxides filled in crack is gradually generated during long-term exposure in high temperature environment.

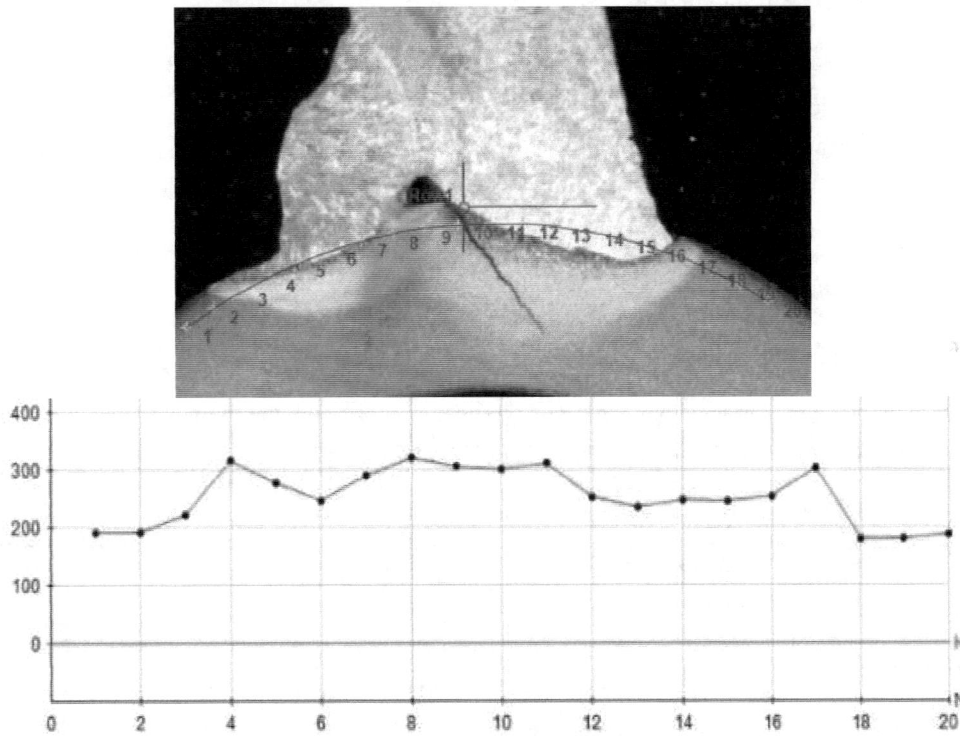

Figure 23. Hardness results of weld-joint HAZ

Secondly, according to the above analysis, it is not hard to evolute the process of cracks. In the first period, the cavities are bred with metal inclusion as the core. Then, the wedge cracks are developed at the grain boundary, especially at the trigeminal grain boundary. After a serial procedure of oxidation, propagation and connection of microcracks, the macro-cracks are finally formatted. The metal inclusion refer to Ti(Al)-N which is separated out during welding. The TiN inclusion near the trigeminal grain boundary will significantly reduce the grain boundary strength as the cause of cavities initiated by TiN, as reported in reference 19(Ref.19). The TiN appearance in this paper is exactly the same as that in reference 19, as showed in Fig.24 and Fig.13. Aluminum is ofen used as deoxidant metal during steel-making. It is also repeorted that AlN can largely reduce the creep properties of T91 steel as the core of creep cavities at grain boundaries(Ref.20). So, we can conclude that the Ti(Al)-N inclusion can cause the cavities, decrease the bonding strength at grain boundaries, and promote the formation of cracks in this paper.

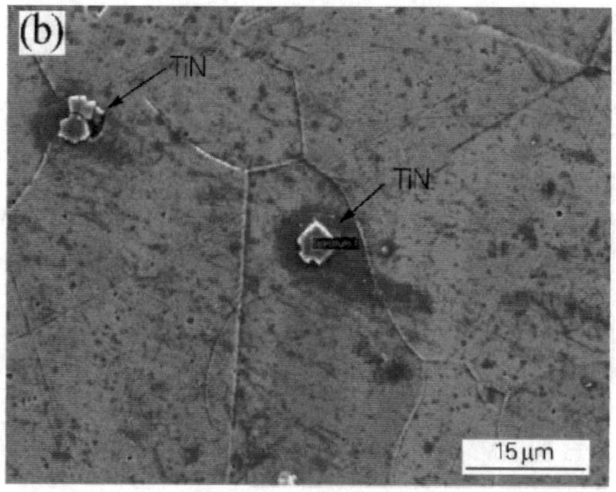

Figure 24. Morphology of TiN Compounds in Reference Document 19(Ref.19)

Finally, there are two main factors on the generation of reheat-cracks in T23. One is the material grain has so high strength that it is hard to make deformation inter grain. During stress relaxation, stress are concentrated on the grain boundaries. The other is the precipitated phases and inclusions at grain boundaries which play a pinning role to hinder the sliding between grain boundaries and cause stress concentration. In this paper, the coarse grain zone has high hardness, and the Ti(Al)-N inclusion is found at the grain boundaries at the same time. So, the two factors of reheat-crack are met. Reference 21 reports two failure modes, which are intergranular failure (IG) and plastic tearing (DT) after high temperature stress relaxation simulation test on the coarse grain zone of T23 steel weld. With the increase of temperature, the amount of Fe-rich M3C phases increases at grain boundaries, and the IG ratio relevantly increases from 40% to 60% at 575 degree to 80% to 100% at 725 degree(Ref.21). So, it is conducted that the main failure mode of crack in this paper is mainly of plastic tearing(DT) for the reason that the service temperature is much lower than 575 degree. Because the plastic tearing(DT) has much lower growthing speed than intergranular failure (IG), the crack has enough time to be oxided. It is proved by the crack blunt shape tip which is the evidence of continue oxiding.

Based on the above discussion results, the author considers that the crack in this paper is a kind of stress relaxation crack in welded joints of T23 steel, and it is also another form of reheat crack in medium temperature environment.

5. Conclusion

1)The cracks of the fin weld initiate in the coarse grain zone of the heat-affected zone, extending parallel to the fusion line at first, then developing towards the inner wall of the tube, ending at the fine grain zone at last. After a thorough study on the microcosmic characteristics and inclusion, it is concluded that the crack is stress relaxation crack caused by metal inclusion, and is another form of reheat crack of T23 steel weld-joint in medium temperature environment.

2) The Ti(Al)-N inclusion in T23 steel weld greatly reduce the grain creep strength, generate cavities at grain boundaries, promote to form wedge cracks, and finally come into being macrocracks.

6. Reference

[1] Yang fu, Zhang yinglin, Li weimin, et al. Welding of New Type Heat-Resisting Steel[M].Beijing: China Power Press,2006:69-73.

[2] Pan qiangang. Prospective forecaste of HCM2S Steel Used in Boiler of Our Country[J].Electric Welding Machine,2004,34(05):6-11.

[3] Ji xianwu, Duan peng, Li ju, et al. Application of T23 Steel to 1000MW Ultra-supercritical Units[J].East China Electric Power, 2009, 37(12):2097-2101.
[4] Zhang eying, Wang xiangbin, Tao shengzhi, et al. Application of T23 Steel in Large-Scale Conventional Utility Boilers[J].Thermal Power Generation,2005,34(3):68-73+77.
[5] Li shengwen, Lin lin, Zhang shiming. Failure Analysis and Prevention on T23 Steel Joints of Water Wall in USC Tower Boiler[J].Foundry Technology,2017,38(9):2159-2161.
[6] Li wenbo, Shen dingjie, Xie yi, et al. Failure Analysis on Weld Joint of T23 Heat-resistant Steels[J]. Power Equipment, 2015, 29(6):443-446.
[7] Sun biao. Analysis on T23 Steel Welded Joint Failure and Welding Process Based on Ultra Supercritical Unit[J].Hot Working Technology, 2016, 45(9):250-252+258.
[8] John C. Lippold. Welding Metallurgy and Weldability[M].Beijing: China Machine Press, 2016:110-114.
[9] J.G.Nawrocji, J.N. Dupont, C.V.Robino and A.R.Marder. The Stress-relief Cracking Susceptibility of a New Ferritic Steel Part 1: Single-pass Heat Affected Zone Simulations [J].Welding Journal, 2000, 35(12):355-362.
[10] Qiao yaxia, He zhenyu, Guo jun, et al. Craking Analysis of T23 Steel Weld -Joint of Reheater in a Power Plant[C]. Papers Collection of the Eleventh Power Station Welding Symposium, Jinan, 2011.
[11] Jin yujing, Zhou wei. Characteristics of reheat Cracking in CGHAZ of Modified T23 Steel[J]. Heat Treatment of Metals, 2017, 42(11):191-197.
[12] Wang tiantian, Xu mengjia, Xu jijin, et al. Influence of Second Welding Thermal Cycle on Reheat Cracking Sensitivity of CGHAZ in T23 Steel Influence of Second Welding Thermal Cycle on Reheat Cracking Sensitivity of CGHAZ in T23 Steel[J].Materials Review, 2017, 31(12):56-59.
[13] Wang xue, Xu delu, Chen yucheng, et al. Reheat-Crack Sensitivity of T23 Steel[C].Collected Papers of the 2009 Academic Conference on Modern Welding Science and Technology, Haerbin, 2009.
[14] Wang xue, Li xiqiang, Yang chao, et al. Aging Properities of T23 Weld Joint in Water Wall of USC Boilers[J]. Journal of Chinese Society of Pwer Engineering, 2015, 35(04): 325-335.
[15] Fan jin, Jia hongxiang, Chen tingkuan, et al. Analysis on the Temperature and Thermal Sress Fields of membrane Tube of Boiler in Power Plant[J].Journal of Xian Jiao Tong University, 1997, 31(05):75-80.
[16] Li xiqiang, Wang xue, Yang chao, et al. Effects of Preheating and Post-Weld Heat Treatment on the Mechanical Properities of T23 Steel Welding[J].Electric Power, 49(2):1-5+66.
[17] Liu yongchao. Study on CGHAZ and Reheat Cracking of T23 Steel[M].Shanghai: Shanghai Jiao Tong University:2013.
[18] Wang qijiang, Zou fengming, Deng yongqing, et al. Effect of T23 Steel's Microstructural Evolution on Its Properities[J].Bao Steel Technology,2006,(3):18-22.
[19] Cheng xiaonong, Chen ming, Li dongsheng, et al. High Temperature Creep Behavior of Fe-Ni Based Alloy Cr19Ni28TiN[J].Heat Treatment of Metals,2015,40(2):48-51.
[20] Zhu lihui, Zhao qinxin, Gu haicheng, et al. Effect of Aluminum on Creep Rupture Property of T91 Heat-resistant Steel[J].Iron and Steel, 2002, 37(5):50-54.
[21] C.V.Robino, A.R.Marder, J.D.Puskar, et al. The Mechanism of Stress-Relief Cracking in a Ferritic Alloy Steel[J].Welding Journal, 2003, 82(2): 25-S-35-S.

IOP Conf. Series: Materials Science and Engineering **730** (2020) 012014 doi:10.1088/1757-899X/730/1/012014

Cause Analysis of Type IV Cracking of Main Steam Pipe of Ultra Supercritical Unit

Yao Xu[1], Shaohua Yin[2], Chunpo Tang[1] and Long Zhou[1]

[1]CHN Energy Taizhou Power Generation Co., Ltd.
[2]Suzhou Nuclear Power Research institute
349655897@qq.com, jnyinshaohua@126.com, tangchunpo@163.com, 634485857@qq.com

Abstract. In order to explore the cause of cracking of the main steam pipe of an ultra-supercritical unit, A series of corresponding physical and chemical tests for crack failure pipelines. Including chemical composition analysis, room temperature tensile test, high temperature tensile test, impact test, bending test of HAZ(Heat Affected Zone)and base metal, it's also including metallographic observation and section scanning electron microscopy analysis of crack tip, overall force analysis of pipeline, etc. The results show: Failure crack has obvious characteristics of crack type IV. for example, a large number of creep holes appear in the fine-grained region of the HAZ, Creep holes accumulate, grow up, and form microcracks. In addition, the stress analysis shows that the outer arc side of the pipe is subjected to large tensile stress. In conclusion, the pipe cracking failure is caused by the combination of tensile stress and type IV crack.

1. Introduction

At present, ferritic heat-resistant steel with a mass fraction of Cr of 9%~12% is widely used in supercritical and ultra-supercritical thermal power generating units in China. Such as T/P91, T/P92, etc. The steels have high temperature strength, low thermal expansion coefficient, good thermal conductivity and low stress corrosion sensitivity, so they have been widely used [1]. This kind of steel is widely used in the process of installation and repair. Due to the influence of the welding thermal cycle, the weld heat affected zone with uneven microstructure including the coarse grain zone, the fine grain zone and the incomplete phase transition zone is formed. After the unit is operated for a period of time, the welded joint of this type of steel is prone to type IV crack defects. Tabuchi et al. [2] believe that early type IV crack failure is prone to occur in fine grain regions where the heating peak temperature is close to the Ac3 temperature. Type IV creep rupture is related to the microstructure change and performance of the base metal during welding and heat treatment. Albert et al. [3] found that the type IV crack occurred in the fine-grained heat-affected zone of the welded joint when the creep fracture of the welded joint of P122 steel was found, and the deformation of the welded joint was small when fractured, and a large number of creeping pores were observed at the fracture.

July 24, 2015, In a power plant #2 boiler, the butt welds of the straight pipe section and the elbow section of main steam pipe are cracked. Pipe design temperature 600 ° C, design pressure 25 MPa, pipe material SA335-P92, Straight pipe section specification Φ559×102mm, and elbow specification is Φ559×95mm. In order to find out the cause of the crack failure of the main steam pipeline, the author conducted a series of related tests and analysis.

2. Macroscopic Observation of Crack

The macroscopic morphology of the crack shows: The outer wall crack is parallel to the straight pipe side weld fusion line and extends in the circumferential direction, the crack is about 2mm from the fusion line. From 12 o'clock to 6 o'clock(Observing along the flow direction of the medium) in the clockwise direction, the crack length is 1000 mm or more. The surface of the crack was straight and no tearing was found. The maximum opening was at 3 o'clock. The maximum opening width of the crack is about 2mm. The crack in the inner wall extends in the circumferential direction parallel to the weld line of the root of the straight pipe. From 2 o'clock to 4 o'clock, the total length is 340 mm, and the surface of the crack is relatively straight. No opening is found.

3. Sampling and Analysis

3.1. Test Sampling

The high temperature (590 ° C) longitudinal stretching and the room temperature longitudinal tensile test specimens are respectively taken along the axial direction of the base material and the heat affected zone. The sampling positions are respectively in the vicinity of the inner wall, the middle part and the near outer wall of the main steam pipe, and a total of 24 samples. The curved specimens were taken in the near outer wall and the near inner wall respectively, including 4 in the base material area and 4 in the heat affected zone. At 3 o 'clock and 9 o 'clock three transverse impact samples from each parent metal area. Three longitudinal impact samples were taken in the heat affected zone and the parent metal zone of the near inner wall, middle section and near outer wall of the pipeline, A total of 18 longitudinal impact specimens.

The metallographic sample is taken from the crack tip of the outer wall of the pipe and the entire pipe section. Cut the crack tip sample and observe the microstructure of the different sections of the sample. The different observation surfaces are numbered as #A, #B, #C, as shown in Figure 1a. The metallographic sample number of the pipe section is #D, as shown in Figure 1b.

a b

Figure 1. Metallographic sample; a crack tip; b pipe cross section

3.2. Chemical Composition Analysis

3.2.1. Chemical composition test results

The chemical composition analysis results are as follows:

Table 1. Chemical composition and standard reference value of P92 sample

Steel grade	Composition (%)											
P92	C	Si	Mn	P	S	Ni	Cr	Mo	W	V	Nb	Al	N
HAZ	0.096	0.27	0.45	0.012	< 0.010	0.20	8.62	0.48	1.63	0.21	0.083	/	/
Base material area	0.09	0.33	0.44	0.015	< 0.010	0.28	8.59	0.39	1.8	0.21	0.079	/	/
standard value	0.07~0.13	≤0.50	0.30~0.60	≤0.02	≤0.01	≤0.4	8.5~9.5	0.3~0.6	1.5~2.0	0.15~0.25	0.04~0.09	0.04	0.03~0.07

The results of chemical composition analysis showed that the content of each element was within the normal allowable range, and no abnormalities were found in the chemical composition analysis.

3.3. Analysis of Mechanical Test Results

3.3.1. Stretching test

Performing room temperature and high temperature tensile tests on the sample taken, The test results show that the room temperature longitudinal yield strength(427.5MPa) of the sample taken only in the heat affected zone of the outer wall is slightly lower than the standard value(≥440MPa). The tensile strength of the sample is 630Mpa (standard requirement ≥ 620MPa), and the yield ratio is 0.68.The room temperature and high temperature longitudinal tensile test results of the base metal and heat affected zone in the remaining positions are higher than the standard requirements. In addition, the high temperature tensile strength, yield strength, area shrinkage and elongation of the base metal are superior to the heat affected zone. It can be seen that the weld heat affected zone has poor plasticity.

3.3.2. Bending test

The bending test results are as follows: There are 2(sampling position is on the outer wall of the pipe) complete fractures of 4 samples in the heat affected zone at the indenter position. In the other two, one crack occurred and the crack length was 10 mm. The other one showed no visible cracks. There were no visible cracks in the four specimens of the base material at the indenter position. It can be seen that the plasticity of the heat-affected zone is significantly degraded compared with the parent metal, and the plasticity of the outer wall of the pipeline is the worst, indicating that its resistance to external impact loads is poor.

3.3.3. Impact test

The average value of the impact absorption of the lateral impact sample of the base metal is 34J, which is higher than the standard value required by 27J.The average longitudinal impact result of the heat affected zone of the outer wall of the pipeline is 13.7J, which is much lower than the standard requirement of 40J. The average longitudinal impact result of the heat affected zone in the middle of the pipeline is 33.5J, which is slightly lower than the standard value. The longitudinal impact of the heat affected zone on the inner wall of the pipe is 43J, which is slightly higher than the standard value. The test results show that the toughness of the heat-affected zone of the welded joint of the pipe is getting worse from the inside to the outside, and the outer wall of the pipe is the worst and the brittleness is the highest. The longitudinal impact results of the outer wall and inner wall of the pipe are slightly lower than the standard requirements. The impact of the base material in the middle of the pipe is higher than the standard requirement, indicating that the toughness of the base metal is better than that of the heat affected zone.

From overall view, the impact value of the base metal is close to the standard requirement, and the toughness is good. The impact value of the heat affected zone of the outer wall pipe is far lower than the standard requirement, and the toughness is insufficient. The analysis may be due to the long-term operation of the pipeline, P92 steel aging in the high temperature creep process, resulting in deterioration of material properties, and the weld heat affected zone is further deteriorated due to the influence of welding thermal cycle and post-weld heat treatment.

3.4. Crack Metallographic and Fracture

3.4.1. Microscopic metallographic analysis

Take the crack tip sample #C of the outer wall of the pipe, polish the crack tip position of the #C sample, and then etch it with ferric chloride hydrochloric acid solution. After etching, washing and air drying, observe the metallographic microscope under the optical metallographic microscope. The results are as follows:

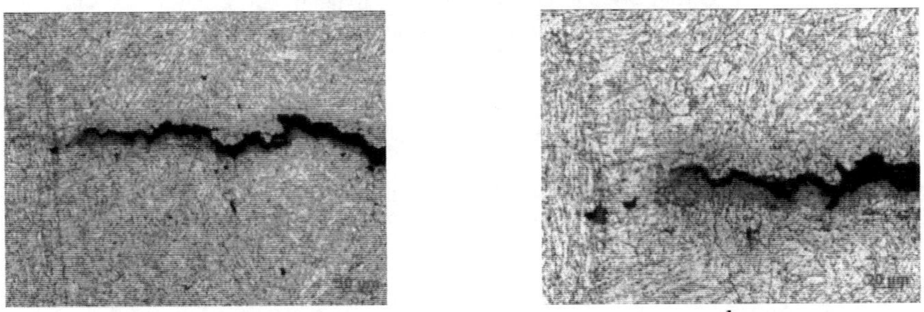

a b

Figure 2. #C surface crack tip metallographic structure - a-200X,b-500X

From the metallographic photograph of the crack tip in Fig. 2, the position is organized as tempered lath martensite, micro-cavities exist around the main crack, and micro-holes are also formed in the front of the crack tip. The crack does not expand flatly, mainly along the crystal cracking characteristics. The crack occurrence region is the recrystallization region of the heat affected zone.

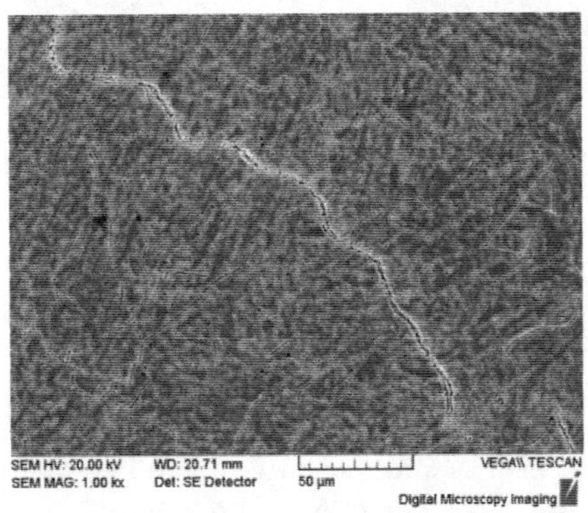

Figure 3. Scanning electron microscopy morphology of microcracks near the crack tip

In addition, during the observation of the crack tip microstructure, microcracks were found in the vicinity of the main crack, and microscopic cracks were observed by 1000X scanning electron microscopy are shown in Fig. 3.

Figure 4. #D inner wall (left side), outer wall (right side) metallographic structure (500X)

It can be seen that the inner wall and the outer wall of the sample #D are organized into tempered lath martensite, and a small amount of creep pores exist in the structure, and the creep pores are distributed along the grain boundary.

3.4.2. Observation of crack fracture morphology

In order to determine the crack initiation site, the crack needs to be opened, and the crack fracture is observed by scanning electron microscopy. Scanning electron microscopy was performed on the 3 mm outer wall fracture with the largest crack crack width. The results of scanning electron microscopy are shown in Fig. 5.

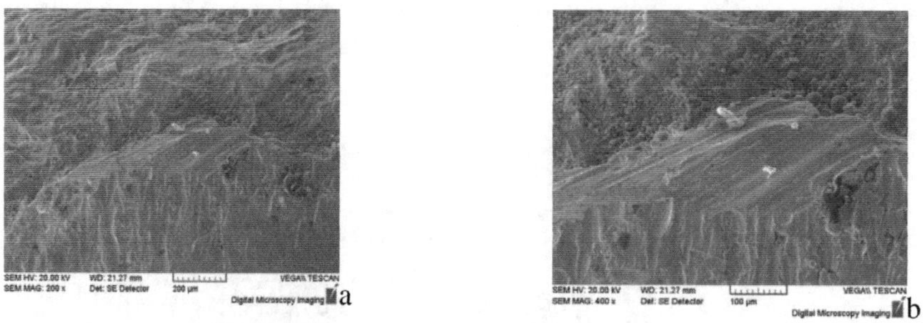

Figure 5. The outer wall crack fracture 3 o 'clock-a-200X,b-400X

Looking at the scanning electron micrograph of Fig. 5, it is found that the fracture has been oxidized in a high temperature environment, and the thickness of the oxide film is thick, which masks the true shape of the fresh crack surface to some extent. In Fig. 5a, it can be seen that there is a similar river pattern through a large angle to reverse the grain boundary morphology (in the red circle), which is a distinct brittle transgranular fracture morphology [4], and the crack cracking direction is extended from the bottom to the top, here It should be the crack initiation point. And the crack cracking direction is extended from the bottom to the top, where it should be the crack initiation. In addition, the high-magnification picture in Fig. 5b shows that the oxide film is mostly composed of oxide particles, and the fracture of the spherical particles is not flat expansion, which is characterized by the fracture along the crystal. Therefore, it can be shown that the crack is cracked by the outer wall in the long-term. During the operation, the crack gradually expands along the grain boundary and together, eventually forms a through crack.

3.4.3. #B crack tip electron microscope observation

Scanning electron microscopy observation of the #B crack tip is shown in Figure 6:

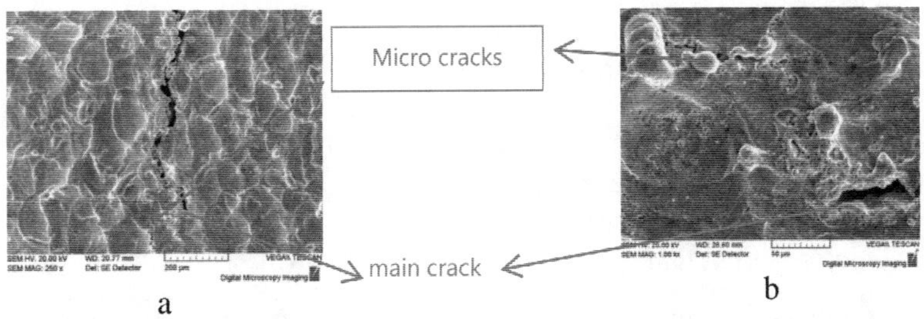

a b

Figure 6. #B side crack tip scanning electron micrograph- a-250X, b-1000X

Observing the cracked tip of the #B sample, it is found that the main crack propagates along the grain boundary, there are new microcracks in the front of the crack tip, and there are many micro creep holes between the microcrack and the main crack tip, and the holes are connected. The tendency of microcracking, the morphology has obvious IV cracking characteristics.

3.4.4. Mechanical properties test fracture scanning electron microscopy

Scanning electron microscopy observations of some samples with lower than standard values in bending test, tensile test and impact test were carried out. The results are as follows:

a b

Figure 7. Scanning electron microscopy of fracture section of fractured specimen of curved specimen—a-200X, b-500X

Observing the fracture of the curved specimen, Figure 7a, it can be seen that the river pattern in this area also has a large angle to reverse the grain boundary morphology. The crack propagation direction is from left to right, and the middle intersection line is grain boundary. Due to the large twist angle, a large step is generated in the right crystal grain, so that the crack can continue to expand, which is obvious brittle cracking. The fracture morphology of the crack initiation zone is similar to that of the fracture zone of the outer wall of the pipeline, which indicates that the load subjected to the crack initiation failure is a bending load. Observe the fracture morphology of Figure 7b. There are more diffuse creep holes in the fracture zone, and the fracture surface is flush and brittle fracture.

For the impact sample, the sample with the lower impact value was selected for scanning electron microscopy of the cracking zone. The results are as follows:

a b

Figure 8. Scanning electron microscopic observation of the cracking zone of impact specimen ZCJ3——a-500X, b-1000X

Figure 8 is a scanning electron microscope observation of the cracking zone of the longitudinal impact specimen, in which the crack propagation direction is from left to right. The crack initiation zone begins with a small amount of dimple-like morphology, followed by a river pattern, and then a distinct cleavage fracture. This is consistent with a small impact test value. In addition, there are many micro-creep holes scattered in the cracking position, and no hole connection tendency is found. Combined with the impact fracture of the sample, it is found that the fracture has no obvious macroscopic plastic deformation, and the fracture is relatively flush, which is a distinct brittle fracture.

a b

Figure 9. Room temperature tensile fracture scanning electron microscope - a-1000X, b-2000X

Figure 9 is a scanning electron microscope observation picture of the room temperature tensile test fracture. There are many dimples in the figure, and the tearing ridge is obvious. In combination with the macroscopic necking phenomenon, the fracture mode is ductile fracture.

In summary, the fractures formed by different load types are observed. It is found that the fracture morphology of the fracture surface of the main steam pipe is similar to that of the main steam pipe at 3 o'clock. The fracture morphology of the cracking zone is similar. The cracking failure was caused by the bending load of the pipeline.

4. Analysis and Discussion

Judging by the location of the crack, the direction, and the area where the creep hole is concentrated, the failure is the fracture failure caused by the type IV crack [5]. Looking for the maintenance records before the location, it was found that the pipeline was replaced at the A-level inspection in October 2012, and the welding repair was carried out at the C-level inspection in September 2014. Two welding heat treatment processes were carried out in less than two years, and the location of the organization was greatly affected. Due to the multiple thermal cycles of welding, the grain size of the fine-grained zone in the heat affected zone is reduced. The grain of the precipitated phase is coarsened, and the strengthening effect of the M23C6 carbide on the grain boundary and the subgrain boundary is weakened, and the large M23C6 carbide and there is a large stress concentration around the Laves phase particles, and the creep holes are easy to nucleate at this position first. The MX carbonitride is needle-like roughened into a spherical shape, and the dispersion strengthening effect is weakened. the original lath martensite structure becomes a sub-cell, and the dislocation density is remarkably lowered.

Due to the above changes in microstructure, the creep strength of the incomplete phase change zone of the welded heat affected zone is significantly reduced. The fine-grained zone of the weld heat-affected zone is lower in the process of welding thermal cycle due to the lower temperature, shorter time, the original carbide can not be completely dissolved, and coarsened in the subsequent cooling process, resulting in solid solution strengthening of the alloying element and second The phase strengthening effect is reduced, so it is easy to form type IV cracks [6].

In addition, there is a large residual stress in the welding process, and the stress is not completely eliminated during the heat treatment, and finally a large residual stress exists in the pipe. The relevant literature points out [7] that the residual stress of the weld is in the tensile stress state in the heat-affected zone, and the maximum value is concentrated in the fine-grained heat-affected zone of the welded joint, so more creep is likely to occur in the fine-grained zone. Holes and eventually lead to the appearance of type IV cracks.

Dispersed micro-creep holes appear in the fine-grained areas of the inner wall and outer wall heat-affected zone of sample #D. It shows that during the long-term operation of the pipeline under high temperature and low stress environment, creeping pores appear in P92 steel, and the overall performance deteriorates. In addition, during the operation of the unit, the vibration of the pipeline caused by the start and stop of the unit and the operation process may cause the support and hanger to be inclined, resulting in the outer wall of the pipeline being subjected to the bending and tensile stress load. The joint action of the two causes the crack to first open on the outer wall of the pipeline. crack. Therefore, the cracking failure is mainly caused by the bending load on the outer arc side of the pipeline, and the IV creep crack is the auxiliary interaction result.

5. Conclusion

(1) Creep holes mainly appear in the fine-grained area of P92 steel welded joints. Under high temperature and low stress environment, creep pores grow and gradually aggregate and connect, and finally form type IV micro-cracks in fine-grained areas.

(2) For the main steam pipeline, it is necessary to pay attention to the reasonable setting of the support and hanger. Control the number of weld repairs, reduce the stress applied to the pipeline and the residual stress of the weld.

(3) The main cause of crack cracking is that the outer wall of the pipeline is subjected to the bending load, and the outer arc side of the pipeline is subjected to tensile stress. The secondary cause is the long-term high temperature and low stress operation of the pipeline, which leads to the creep hole of P92 steel, and the overall performance deteriorates. Type IV microcracks appear.

6. References

[1] MURAMATSU J. Advanced heat resistant steel for power generation[M]. The university press, Cambridge, UK, 1999.
[2] Tabuchi M, Watanabe T, Kubo K, Matsui M, Abe F.J Pressure Vessel Pip, 2001.78:779
[3] Albert S K, Mstsui M, Watanabe T, Hongo H, Kubo K.J Pressure Vessel Pip, 2003.80:63
[4] Zhong Qunpeng, Zhao Zihua, etc., " Fractography ", Higher Education Press
[5] Ma Chong, Jing Hongyang, Xu Lianyong, etc. Microstructure change of type IV cracking of P92 steel, Journal of Welding, 2009; 30:8
[6] Tang Liying, Kang Yujun, Zhou Rongcan et al., Type IV cracking analysis of the main steam sampling tube of ultra-supercritical unit, Physical and Chemical Testing - Physical Volume, 2010(46)
[7] Zhang Jianqiang, Luo Chuanhong, Zhang Yinglin,P91 steel welded joint type IV crack mechanical control parameter finite element simulation, Journal of Welding, 2012(3)

IOP Conf. Series: Materials Science and Engineering **730** (2020) 012015 doi:10.1088/1757-899X/730/1/012015

Current Status and Prospects of Underwater Welding Technology for Key Sensitive Equipment in Nuclear Power Plants

Shaohua Yin[1], Zhiqiang Sun[1], Zhongbing Chen[1], Jianlin Zhang[1], Jianlu Shang[1], Yishi Lv[1] and Maotao Huang[1]

1Suzhou Nuclear Power Research institute Co., Ltd.
jnyinshaohua@126.com

Abstract. This paper analyzes the application status of underwater welding technology in nuclear power plant welding repair. Take the local dry-method underwater welding 304 austenitic stainless steel butt weld as an example. It points out the gap between the mechanical properties and corrosion resistance of underwater welded joints and the welded joints in the air. In order to adapt to the underwater welding environment with high radioactivity and high risk environment in the future, the development of local dry underwater welding automation system will be the key to underwater welding repair of key sensitive components of nuclear power plants in the future.

1. Introduction

The design life of a nuclear power plant is generally 30 to 40 years. According to the existing engineering practice experience, for nuclear power plants with a running time of more than 10 years, the workload of welding maintenance is quite large, which can be compared with the welding workload of nuclear power plant construction and installation[1]. For a variety of reasons, equipment and components in nuclear reactors operating in high-radioactivity, high-temperature environments will be partially damaged after long-term operation. In order to ensure the safe and stable operation of nuclear power plants, these equipment and components must be inspected, repaired and replaced to ensure their integrity.

At present, the underwater welding carried out in in-service nuclear reactors and spent fuel pools is mainly based on manual welding by welders, and the above areas are extremely radioactive, and manual welding in high-radiation environments adversely affects the welder's health. In addition, because the welder has greater psychological pressure during the operation, the welding quality cannot be guaranteed. Therefore, the search for feasible underwater automatic welding technology has become an urgent problem to be solved.

2. Application Status of Underwater Welding Technology in Nuclear Power Plants

Underwater welding can be roughly divided into wet welding and dry welding from the working environment, wherein dry welding includes partial dry welding, high pressure dry welding, and atmospheric dry welding. Local dry underwater welding mainly includes underwater arc local dry welding and underwater laser local dry welding[2].

Underwater arc local dry welding, the area to be welded is separated from the water to create a smaller gas phase zone for welding. Compared with wet welding, the welding quality is easier to guarantee. Compared to dry welding, the cost is lower and more flexible. However, in practical applications, the following drawbacks also exist: Due to the use of arc technology and diver operation, the arc will contract under high pressure[3], so the water depth has an important influence on the quality of the arc welding and the depth of the person's dive, and the remote operation is not high.

The basic principle of underwater laser local dry welding is to pass high pressure gas into the glass hood, and drain the small area around the part to be welded to form a partial drying space to prevent water from contacting the surface of the workpiece to be welded. Form a good gas atmosphere. At the same time, the laser beam is transmitted to the vicinity of the component to be welded through the optical fiber. The laser beam passes through the local gas phase region and acts on the surface of the component to be welded. The workpiece to be welded absorbs the energy of the laser beam and melts, thereby achieving welding. The principle of the device is shown in Fig. 1[4].

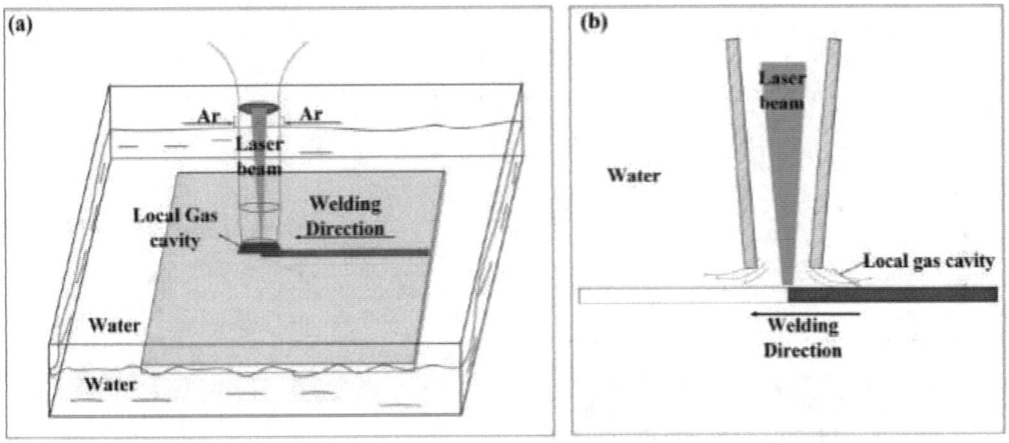

Figure 1. Schematic diagram of partial dry underwater laser welding

Compared with the arc welding method, the laser beam of the laser welding can be transmitted over long distances through the optical fiber, and the control precision is high, and it is suitable for welding and repairing the position where the welding precision is high; Laser welding has a low heat input and a fast cooling rate, so the heat affected zone of the weld is small and the residual stress level is low; The laser welding system is easy to simplify and integrate, especially for equipment maintenance where the working position is small. The local dry underwater laser welding technology combines the advantages of local dry method and laser welding, and with the continuous development of welding automation and intelligence. This technology is increasingly becoming one of the preferred technologies for remote repair of nuclear power plants, especially in the maintenance of critical sensitive equipment for reactors[2].

In foreign countries, Hitachi developed a curtain laser torch in 2001 to achieve a stable underwater dry space. The U-shaped groove underwater laser downward and horizontal transverse welding test under the pressure of 0.3MPa was carried out by means of 4KW laser and filler wire. In 2004, Japan Ishikawa Chemical Co., Ltd. carried out remote-controlled underwater laser welding operations inside the radioactive container of a nuclear power plant. The underwater laser welding system developed by Toshiba Company formed a localized dry space of laser welding by argon gas drainage, successfully welded the flat test piece, and simulated the inner circumference of the round pipe to obtain a good all-position laser welding layer. In 2009, Westinghouse Electric Company and the American Electric Power Research Institute collaborated to use underwater laser welding technology to automatic welding repair of underwater simulated cracks[5].

In China, some researchers have used the method of wire-filled thermal welding to study the influencing factors of underwater laser welding quality under the protection of local dry cavity. It was found that the gas nozzle can be used to form a local dry space, and the parameters such as the nozzle structure and the gas flow rate can be adjusted to obtain a good underwater weld quality[5].

3. Local Dry Underwater Laser Welding Joint Performance

304 austenitic stainless steel is one of the main materials for nuclear power plant application. This paper takes 304 austenitic stainless steel as an example to explore the feasibility of local dry underwater laser welding. The 304 austenitic stainless steel has a thickness of 4 mm and the chemical composition of the material is shown in Table 1[4].

Table 1. Chemical composition of 304 (unit: mass percentage)

C	Si	Mn	P	S	Cr	Ni	Cu	N	Fe
0.04	0.33	1.16	0.034	0.002	18.03	8.01	0.1023	0.0386	Bal.

The underwater laser welding parameters are shown in Table 2. For comparison, the welding parameters are shown in Table 2. The gas flow referred to herein is the gas flow rate at each inlet. The weld profiles after underwater welding and air welding are X-shaped, as shown in Figure 2.

Table 2. Underwater and in-air welding parameters

Welding method	Laser Power P (W)	Welding Speed v (mm/s)	Defocusing Dis-tance D (mm)	Gas Flow Rate F (L/min)
Underwater	1400	8	0	25
In-air	1100	8	0	15

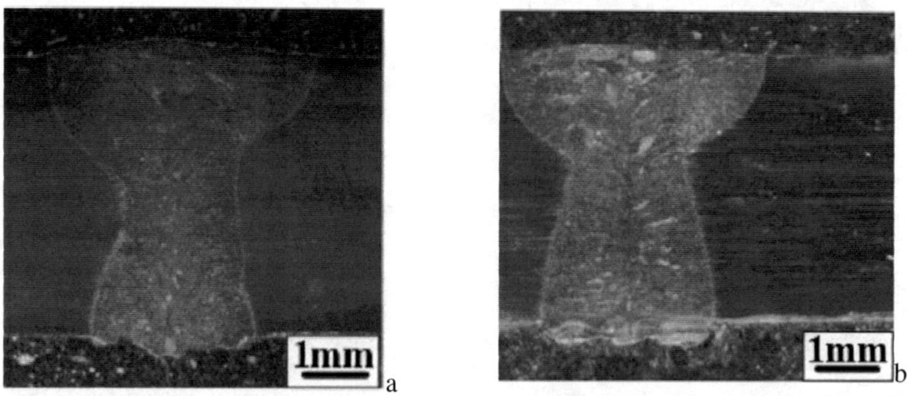

Figure 2. Weld section, a-underwater welding, b-air welding

3.1. Weld Hardness

Micro-hardness testing of underwater laser penetration welds and laser penetration welds in air, the hardness is located at the waist of the X-weld. The microhardness distribution is shown in Figure 3.

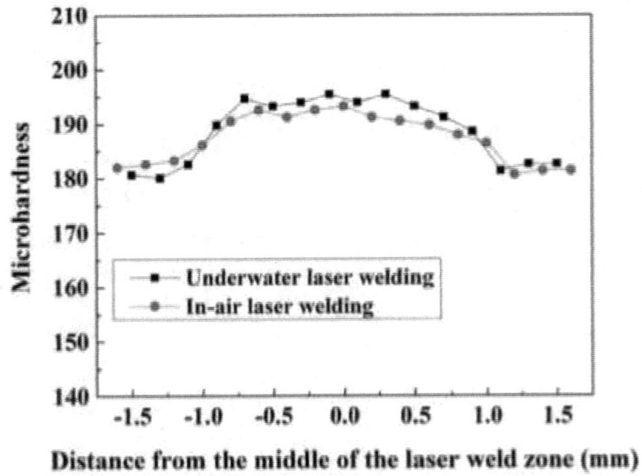

Figure 3. Weld microhardness distribution

The average microhardness of the base metal was 181 HV. The average microhardness of the underwater laser weld is 193 HV, and the average microhardness of the air laser weld is 190 HV. The microhardness of the underwater laser weld is slightly higher than the microhardness of the air laser weld. The main reasons are as follows: after the laser is attenuated by the aerosol with water vapor, the energy is irradiated to the plate member less than that of the air laser, and the cooling speed of the underwater welding is fast. The grain of the weld is fine, and the smaller the grain size, the greater the hardness value; In addition, the rapid cooling rate leads to the uncomplete transformation of ferrite into austenite structure. The residual ferrite content in the underwater laser weld increases, and a small amount of ferrite in the weld acts as a second phase strengthening effect on the weld.

3.2. Tensile Test
The transverse tensile test was carried out on the underwater weld and the air laser weld. The test results showed that the test pieces were broken at the weld. The specific results are shown in Table 3. The data listed in the table are the average of the three samples.

Table 3. Transverse tensile mechanical properties

Sample	Tensile strength (MPa)	Elongation (%)
Base metal	626	53.4
Air laser welding	613	43.3
Underwater laser welding	559	38.3

The tensile strength and elongation of underwater laser welds are reduced compared with air laser welds, which are 91.2% and 88.5% in air, respectively. This is because in the case of water immersion on the back side of the weld, the bottom weld metal is poorly formed, the alloying elements are oxidized, and at the same time, metallurgical pores are generated in the underwater laser weld, which provides a possibility for crack initiation during the stretching process.

Figure 4 shows the fracture scan of the base material, the laser weld under air, and the underwater laser weld.

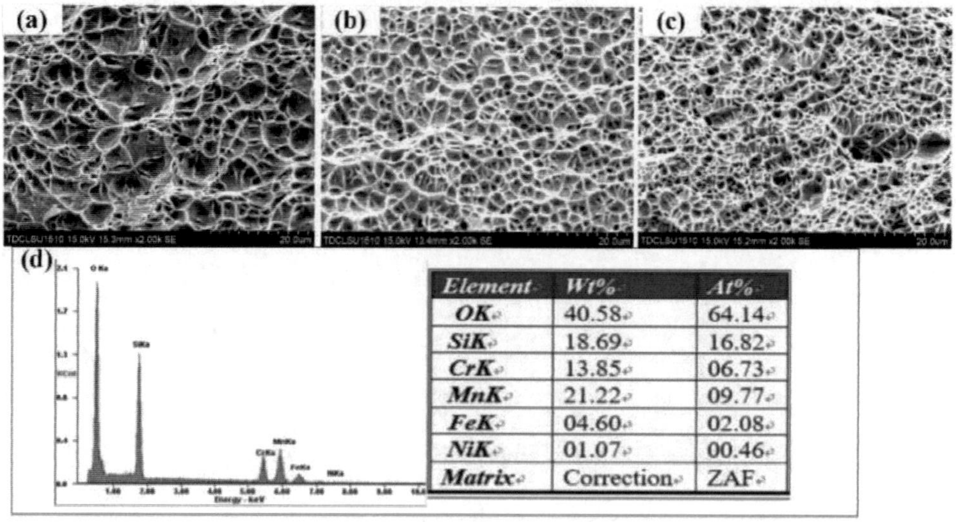

Element	Wt%	At%
OK	40.58	64.14
SiK	18.69	16.82
CrK	13.85	06.73
MnK	21.22	09.77
FeK	04.60	02.08
NiK	01.07	00.46
Matrix	Correction	ZAF

Figure 4. Typical fracture morphology: (a) base metal; (b) weld under air medium; (c) weld under water medium; (d) scanning of inclusions in underwater welds

From Fig. 4, it can be seen that the fracture of base metal, air laser welding seam and underwater laser welding seam are dimple-like, both of them are ductile fracture. The base material fracture has a large and deep dimple shape, and the weld fracture in the air presents a relatively uniform equiaxed dimple shape. The underwater weld fracture has dimples of different sizes and the overall size is small. As shown in Fig. 4 (d), there are obvious inclusions in the dimples of underwater welds, but not in figs. 4 (a) and (b). The main oxide inclusions of Si, Mn and Cr are determined by EDS. These inclusions are related to the back seam immersion during underwater welding. Oxygen element enters the weld and oxidizes the alloy element to form oxide inclusion, which makes the performance of the weld worse. In addition, the existence of pore can also be seen on the fracture surface of underwater weld. The pore of underwater laser welding reduces the effective bearing area of the weld and reduces its carrying capacity.

3.3. Corrosion Resistance of Underwater Laser Weld

Figure 5 shows the Tafel electrochemical polarization curves of underwater laser welding and air laser welding welds.

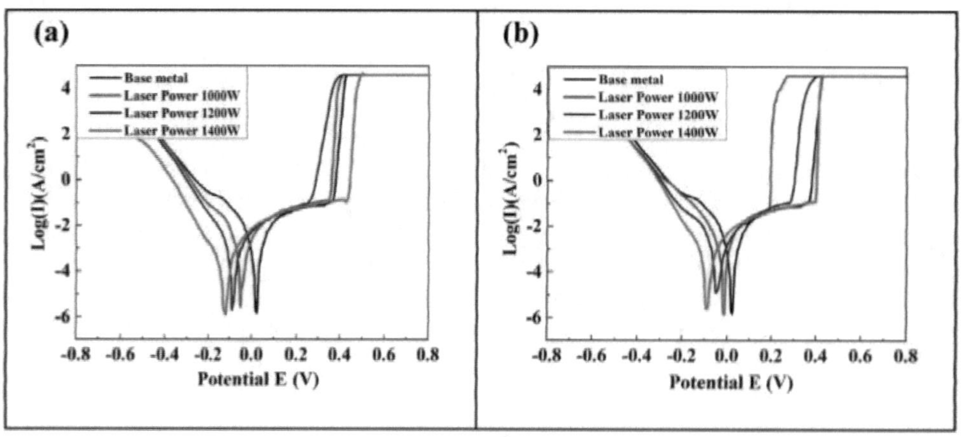

Figure 5. Tafel electrochemical polarization curve: (a) underwater laser weld; (b) in-air laser weld

In order to ensure the reliability of the test data, underwater welding samples and air welding samples of 1000W and 1200W laser power welding were added. The corrosion potential of underwater laser weld is -0.049 V, -0.087 V, -0.118 V when the laser power is 1000 W, 1200 W and 1400 W. The corrosion potential of laser weld in air is -0.015 V, - 0.05 V, - 0.092 V when laser power is 1000 W, 1200 W and 1400 W. The corrosion potential of the base metal is 0.023 V. It can be seen that the corrosion potential of the weld structure is lower than that of the base metal, indicating that the corrosion resistance of the weld is lower than that of the base metal. Under the same laser power, the corrosion potential of underwater laser welding is lower than that of air laser welding, and the corrosion resistance of underwater laser welding is lower than that of air laser welding. In addition, with the increase of laser power, the corrosion potential of underwater and air laser welds decreases, that is, with the increase of heat input, the corrosion resistance of welds decreases.

The reasons for the above phenomena are related to the process of electrochemical corrosion and the microstructure of the matrix. Pitting corrosion is the main corrosion form of austenitic stainless steel and its weld under electrochemical corrosion[6].Under the action of Cr element in austenitic stainless steel and weld, a passivation film will be formed on the surface of metal, which can effectively block the metal matrix from the medium. Once the passivation film is destroyed or weak, the metal matrix contacts with the medium, and corrosion will occur, especially the corrosive ion (Cl-) will accelerate the corrosion process. There are three main effects of Cl - on passivation film in Cl - medium, namely adsorption, penetration and local passivation film destruction. Cl-and hydroxyl groups compete to adsorb on the surface of passive film. While Cl-and dissolved metal cations combine to form soluble chloride, which have low lattice binding energy, low activation energy for migration, and accelerate the passivation film. Dissolved, Cl- is further adsorbed on the surface of weak areas, causing increased corrosion [7,8]. When the electric field at the interface between solution and film is very strong, Cl - adsorbed on the passive film will soak and pollute the passive film, and enter into the metal matrix to dissolve the metal. This Cl-contaminated passivation film is more conductive than the original passivation film, speeding up the transport speed and transfer speed of the particles. The metal ions continuously diffuse from the interface between the metal and the passivation film to the interface between the passivation film and the solution, which may cause vacancies on the interface between the passivation film and the metal. When the size of the vacancy reaches a certain level, the passivation film will be cracked. When the passivation film is damaged locally, Cl - causes the dissolution rate of the exposed matrix metal to be greater than that of its re-passivation, which leads to the corrosion of the matrix. Summarize the process of electrochemical corrosion of the matrix containing Cl-media as shown in Figure 6.

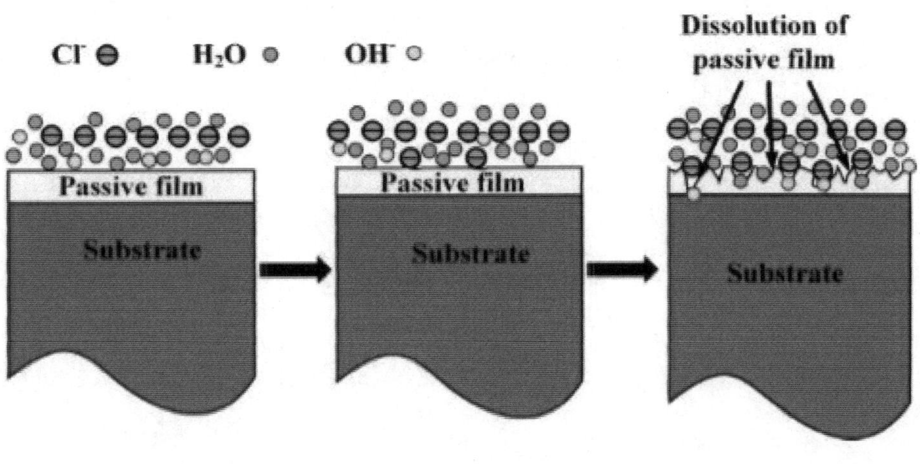

Figure 6. Schematic of matrix electrochemical corrosion in Cl- medium

In summary, after analyzing the hardness, tensile strength, elongation, fracture scanning electron microscopy and corrosion resistance of the local dry underwater laser welded joints, it is found that the performance of underwater laser welded joints is still far from that of air welded joints. Mainly due to the water immersion in the back weld, oxygen enters the weld and oxidizes the alloying elements to form oxide inclusions, which deteriorates the mechanical properties and corrosion resistance of the weld. Zunyue Huang[4] of Tianjin University and others used the method of applying activator on the back to prevent the back seam from being immersed in water, and the effect was good.

4. Development Trend of Underwater Welding Automatic Welding Technology in Nuclear Power Station

At present, the local dry-based underwater welding technology is one of the main methods for underwater welding of nuclear reactors and spent fuel pool components. However, this method still has the limitations of necessary welder diving repair. Therefore, the development of high-radiation working environment welding robots for nuclear reactors and spent fuel pools, using robots instead of manual operations in high-risk areas, is the future development trend of underwater welding technology.

The Key Laboratory of Underwater Robotics of Harbin Engineering University has developed a six-degree-of-freedom isomorphic master-slave underwater manipulator for underwater operation of nuclear power plant reactors and spent fuel pools[9]. Furthermore, through the hydrodynamic modeling of the underwater manipulator, a set of master-slave nonlinear fuzzy PID closed-loop control scheme was designed according to the working environment conditions, and the reliability of the control system was verified by setting up the experimental platform. This research provides a new idea and possibility for the next step to realize the local dry underwater welding robot.

5. Conclusion

At present, although there are still many problems in the performance of underwater welded joints, with the accumulation of underwater welding experience and the systematic cooperation of various scientific research institutions[10], the quality of joints will eventually break through. On this basis, the development of local dry underwater welding automation system will be the key to underwater welding repair of key sensitive components of nuclear power plants in the future.

6. Reference

[1] Jialei Zhu. Research on Local Dry Automatic Underwater Welding Technology for Nuclear Power Plant Maintenance[D] Beijing University of Chemical Technology.2010
[2] Qi Yao. Research on underwater laser welding of stainless steel [D]. Tianjin University,2014
[3] Leigang Han, et al.: Development of local dry underwater welding technology [J]. Journal of Zhejiang University: Engineering Edition, 2019, 53(7): 1252–1264.
[4] Zunyue Huang. Research on Laser Welding Mechanism and Key Technologies in Water Medium [D]. Tianjin University, 2018
[5] Xiangdong Jiao, Jialei Zhu. Application Status and Prospect of Underwater Welding Automation Technology in Ocean Engineering [J], Metal Processing Hot Processing, 2013, 2, 24-26
[6] Baotian Song, Lin Wang. Study on the gas hole problem in semi-automatic CO2 welding with partial drainage under water [J]. Journal of Welding, 1984, 3:15-22+71-74
[7] Yingjun Ai. Electrochemical study on pitting corrosion of AI Yingjun.304 stainless steel in 3.5% NaCl solution [D]. Nanchang: Nanchang Aeronautical University, 2016
[8] Marcus P, Maurice V, Strehblow HH. Localized corrosion(pitting): A model of passivity breakdown including the role of the oxide layer nanostructure[J]Corrosion Science, 2008, 50(90):2098-2704
[9] Gang Wang, Facheng Wang, et al. [J]. Research on Master-Slave Underwater Operating Manipulator for Nuclear Power Station Maintenance [J]. High-tech Communication, 2017, 27 (5) 442-449
[10] Yongchang Chen, International Nuclear Power Construction and China's Nuclear Power Development Trend [J] Journal of Heilongjiang Vocational Institute of Ecological Engineering. 2019, 32(4)

IOP Conf. Series: Materials Science and Engineering **730** (2020) 012016 doi:10.1088/1757-899X/730/1/012016

First-Principles Investigation of *β*-Titanium Doping with Trace Transition Metal Elements

Jia Song[1], Luyu Wang[2], Kaiqi Wu[1], Wenheng Wu[1, *] and Liang Zhang[1, *]

[1] Shanghai Engineering Research Center of 3D Printing Materials, Shanghai Research Institute of Materials, Shanghai 200437, China

[2] Institute of Fiber-Based New Energy Materials, The Key Laboratory of Advanced Textile Materials and Manufacturing Technology of Ministry of Education, School of Materials Science and Engineering, Zhengjia Sci-Tech University, Hangzhou 310018, China

[*]E-mail: wwhwwh2004@126.com; lzhang0126@hotmail.com

Abstract. The lattice parameters, formation enthalpy, and electronic structures of *β*-Ti doping with trace transition metal element X (X=Cr, In, Mo, Nb, Pd, Rh, Ru, Sc, Tc, and Y) were systematically calculated by first-principles simulations to explore the effect of these metal elements on the stability of *β*-Ti at 0 K. Based on our calculations, a satisfactory linear function is found between atomic radius of dopant X and corresponding volume difference. The formation enthalpy results show that the addition of trace transition metal X plays a positive role in the thermodynamical stability of *β*-Ti at 0 K. This intriguing behavior is well explained by the density of state (DOS).

1. Introduction

Due to titanium and its alloys possessing good strength-to-weight ratio, high heat resistance, and outstanding biocompatibility, they are widely used in numerous engineering fields, especially biomedicine applications.[1, 2] However, on the one hand, traditional *α* type and *α*+*β* type titanium alloys have a higher elastic modulus, which can result in stress shielding effect; On the other hand, they also contain elements such as Al and V, which may take poisonous effect on biological cells.[3] Therefore, development of non-toxic and low-modulus *β*-Ti alloy has become a hot and attractive research topic in recent years.

Many traditional experiments have been done and showed that the stability of *β*-Ti alloy is closely related to the type and content of alloying element.[3-6] So, in order to improve properties of *β*-Ti alloy, a better understanding of the influence mechanism of alloying element on *β*-Ti alloy is crucial. However, the influence mechanism is difficult to be obtained by the traditional experiments. Thus, some scholars attempt to investigate the influence mechanism by theoretical calculations.[2]

In this work, first-principles simulations are performed to systematically investigate the *β*-Ti doping with trace transition metal elements X (X=Cr, In, Mo, Nb, Pd, Rh, Ru, Sc, Tc, and Y) with a chemical formula of $Ti_{0.98}X_{0.02}$. The equilibrium volume and formation enthalpy are firstly calculated to study the effect of dopant X on the structure and thermodynamic properties of *β*-Ti. Secondly, the total density of states (TDOS) and the partial density of states (PDOS) of $Ti_{0.98}X_{0.02}$ systems are investigated to explore the structural stability mechanism from the electronic level.

2. First-Principles Methods

Vienna Ab-Inition Simulation Package (VASP) is used to conduct all first-principles calculations.[7]

The generalized gradient approximation (GGA) of Perdew-Burke-Ernzerhof (PBE) functional is used to calculate the exchange-correction energy of electron.[8] The ion-electron interaction is described by the projector augmented wave (PAW) approach. For consistency, the electronic wave functions are expanded in a plane-wave basis with a cutoff energy of 500 eV for all calculations, which is roughly 1.3 times higher than the default value. The pseudopotentials describe the valence states of metal elements are as follows: Ti($3p^6 3d^2 4s^2$), Cr($3d^5 4s^1$), In($5s^2 5p^1$), Mo($4s^2 4p^6 4d^5 5s^1$), Nb($4s^2 4p^6 4d^4 5s^1$), Pd($4d^8 5p^2$), Rh($4d^7 5s^2$), Ru($4s^2 4p^6 4d^6 5s^2$), Sc($3d^1 4s^2$), Tc($4s^2 4p^6 4d^5 5s^2$), Y($4s^2 4p^6 4d^1 5s^2$), respectively. The Methfessel-Paxton method with smearing width of 0.05 is applied to relax structure and atom positions of $Ti_{0.98}Xo_{0.02}$ system, while tetrahedron method with Blöchl correction is used to get accurate energy and stress during the static calculations. To ensure a good precision in the calculation, the energy convergence criterion of the electronic self-consistency is set to 10^{-6} eV per atom, and the force convergence criterion is 10^{-2} eV/Å per atom. Spin-orbit coupling is included in all simulations, as implemented within the PAW method in the VASP code.

To research the effect of transition metal element on physical and chemical properties of β-Ti, we construct the a series of supercells (containing 27 Ti unit cells) with one of the titanium atoms substituted by a metal element X atom (X=Cr, In, Mo, Nb, Pd, Rh, Ru, Sc, Tc, and Y), shown in figure 1. The Brillouin zone is sampled by a 3×3×3 Monkhorst-Pack **k**-point grid for $Ti_{0.98}X_{0.02}$ systems and by a 15×15×15 Monkhorst-Pack **k**-point grid for pure Ti and X cell, respectively.

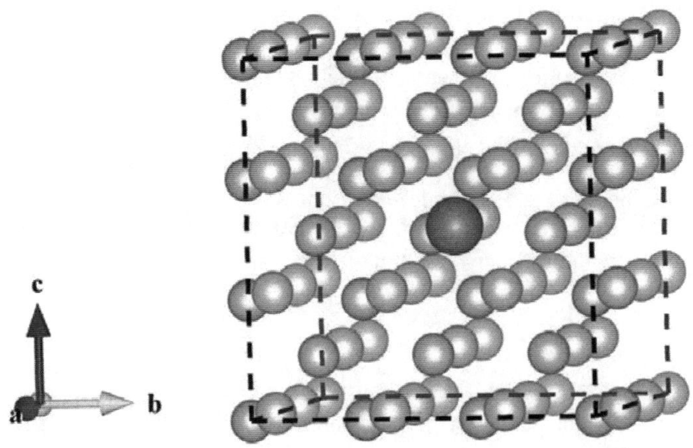

Figure 1. The atomic structure diagram of $Ti_{0.98}X_{0.02}$. The red ball represents the metal X atom and the rest blued balls represent Ti atoms.

3. Results and Discussions

3.1. Structural Optimization

The lattice parameter of pure β-Ti calculated from first-principles simulation is 3.252 Å, which is similar to experimental value of 3.283 Å with an absolute deviation of 0.9%.[9] Hence, we have reason to consider that the zero-pressure geometry optimization method in this study is reliable. Using the same method, the equilibrium lattice parameters of $Ti_{0.98}X_{0.02}$ systems are obtained and summarized in table 1. It can be found that $Ti_{0.98}X_{0.02}$ systems are still bcc structures after β-Ti is doped with a low concentration of X. What's more, it is obvious that different metal elements have different effects on the equilibrium volume of β-Ti. This intriguing behavior can be explained from atomic radius. As shown in figure 2, the relationship between atomic radius of X and volume change can be fitted satisfactorily to a linear function with an R^2 value of 0.95,

$$y = 42.6x - 58.1 \tag{1}$$

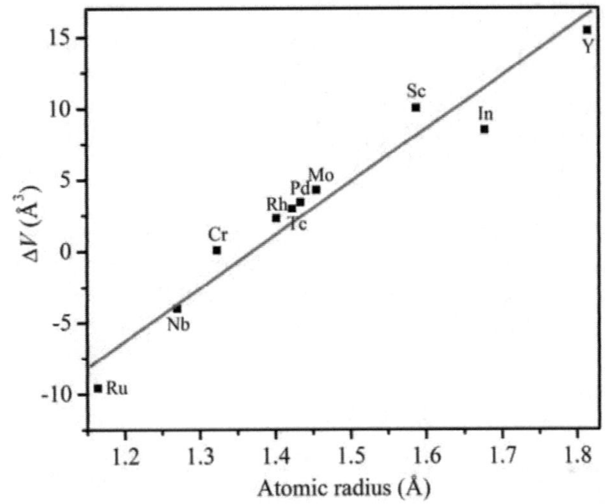

Figure 2. Effect of atomic radius of dopants X on the volume change.

Table 1. The DFT calculated lattice parameters (a, b, and c), equilibrium volume (V_0), volume change (ΔV), zero-point energy (ΔZPE), and formation enthalpy (ΔH_f) of $Ti_{0.98}X_{0.02}$ systems.

System	a (Å)	b (Å)	c (Å)	V_0 (Å^3)	ΔZPE (kJ/mol)	ΔH_f (kJ/mol)
$Ti_{0.98}Cr_{0.02}$	9.751	9.750	9.750	926.95	1.44	-8.99 -9.77[a]
$Ti_{0.98}In_{0.02}$	9.781	9.779	9.779	935.35	1.49	-5.32
$Ti_{0.98}Mo_{0.02}$	9.736	9.736	9.736	922.88	1.43	-10.36 -10.33[a]
$Ti_{0.98}Nb_{0.02}$	9.731	9.739	9.739	922.97	1.34	-9.30 -10.95[a]
$Ti_{0.98}Pd_{0.02}$	9.762	9.762	9.762	930.29	1.36	-9.77
$Ti_{0.98}Rh_{0.02}$	9.756	9.759	9.759	929.14	1.24	-11.20
$Ti_{0.98}Ru_{0.02}$	9.717	9.716	9.716	917.29	2.75	-11.40
$Ti_{0.98}Sc_{0.02}$	9.783	9.786	9.786	936.88	1.85	-8.02
$Ti_{0.98}Tc_{0.02}$	9.759	9.761	9.761	929.81	2.42	-9.82
$Ti_{0.98}Y_{0.02}$	9.801	9.805	9.805	942.25	1.86	-8.56

[a] Zou et al., first-principles calculation with PAW-GGA-PBE.[2]

3.2. *Thermodynamic Property of $Ti_{0.98}X_{0.02}$ System*

After the $Ti_{0.98}X_{0.02}$ systems and pure transition metal (Ti, Cr, In, Mo, Nb, Pd, Rh, Ru, Sc, Tc, and Y) cells are successfully optimized, the ground state total energies of these systems are obtained by static calculations. The formation enthalpy of $Ti_{0.98}X_{0.02}$ can be obtained from these total energies. In this study, to make simulation results more reliable compared with experimental results, the zero-point energy correction is added to the formation enthalpy, defined as follows,[10]

$$\Delta H_f(Ti_{0.98}X_{0.02}) = E_{total}(Ti_{0.98}X_{0.02}) - 0.98E(Ti) - 0.02E(X) - \Delta ZPE \quad (2)$$

where $E_{total}(Ti_{0.98}X_{0.02})$ is the total energy per atoms of $Ti_{0.98}X_{0.02}$ system, E(Ti) and E(X) are the energy per atom of Ti and transition metal X in their ground states, ΔZPE is the zero-point energy correction originated from lattice vibration.

The formation enthalpy of $Ti_{0.98}X_{0.02}$ system is calculated and listed in table 1. Previous DFT simulating results are also introduced in table 1 for comparison to verify the accuracy of present calculations. We can find that the formation enthalpy values of $Ti_{0.98}Cr_{0.02}$, $Ti_{0.98}Mo_{0.02}$ and $Ti_{0.98}Nb_{0.02}$ are -8.99, -10.36 and -9.30 kJ/mol, respectively. These values are in good agreement with previously reported values of -9.77 kJ/mol for $Ti_{0.95}Cr_{0.05}$, -10.33 kJ/mol for $Ti_{0.95}Mo_{0.05}$ and -10.95 kJ/mol for $Ti_{0.95}Nb_{0.05}$.[2] Thus, it can be concluded that the method of our formation enthalpy is reasonable.

Based on the known reports, the formation enthalpy can be used to preliminarily judge the thermodynamic stability of structure.[11] From the table 1, it can be found that the formation enthalpy of $Ti_{0.98}X_{0.02}$ system is slightly less than 0, about -10 kJ/mol. The negative formation enthalpy indicates that formation of $Ti_{0.98}X_{0.02}$ from pure titanium and X is exothermic, that is to say, adding a small amount of transition metal X tends to enhance the thermodynamic stability of β-Ti at 0 K. And the enhancement ability of these metallic elements X on the β-Ti is in the order of Ru > Rh > Mo > Tc > Pd > Nb > Cr > Y > Sc > In.

3.3. Electronic Structure of $Ti_{0.98}X_{0.02}$ System

The TDOS and PDOS of these systems are calculated to explore the strengthening mechanism of trace metallic element X on β-Ti from the electronic level. Figure 3 describes the TDOS of $Ti_{0.98}X_{0.02}$ systems and β-Ti and the corresponding site projected PDOS of Ti and transition metal X, respectively. Firstly, it is evident that the TDOS curves of $Ti_{0.98}X_{0.02}$ systems and β-Ti located in a range from -6 to 2.5 eV mainly originate from the contribution of Ti-d orbit. The results are in good agreement with Kuroda et al. proposed theory that the d electronic orbit plays an important role in mechanical properties and phase stability of β-Ti.[12] Secondly, the Fermi level marked by a vertical pink dash line is set to zero energy in figure 3. The left side of the Fermi level is the valance band and the right side of the Fermi level is the conduction band. Obviously, we find that the value at the Fermi level of β-Ti is about 60 and the values at the Fermi level of $Ti_{0.98}X_{0.02}$ systems are in the range from 15 to 30. The value at Fermi level is related to the phase stability, the lower value at Fermi level, the more structure stable. So, the greatly reduced the value at Fermi level indicates that all dopants X in this study help to strengthen the structural stability of β-Ti at 0 K. This can be attributed that the transition metal elements X can provide additional d-type electrons required to improve stability of β-Ti. From these discussions, we explain that the addition of trace transition metal X plays a positive role in the stability of β-Ti from the atomic and electronic level.

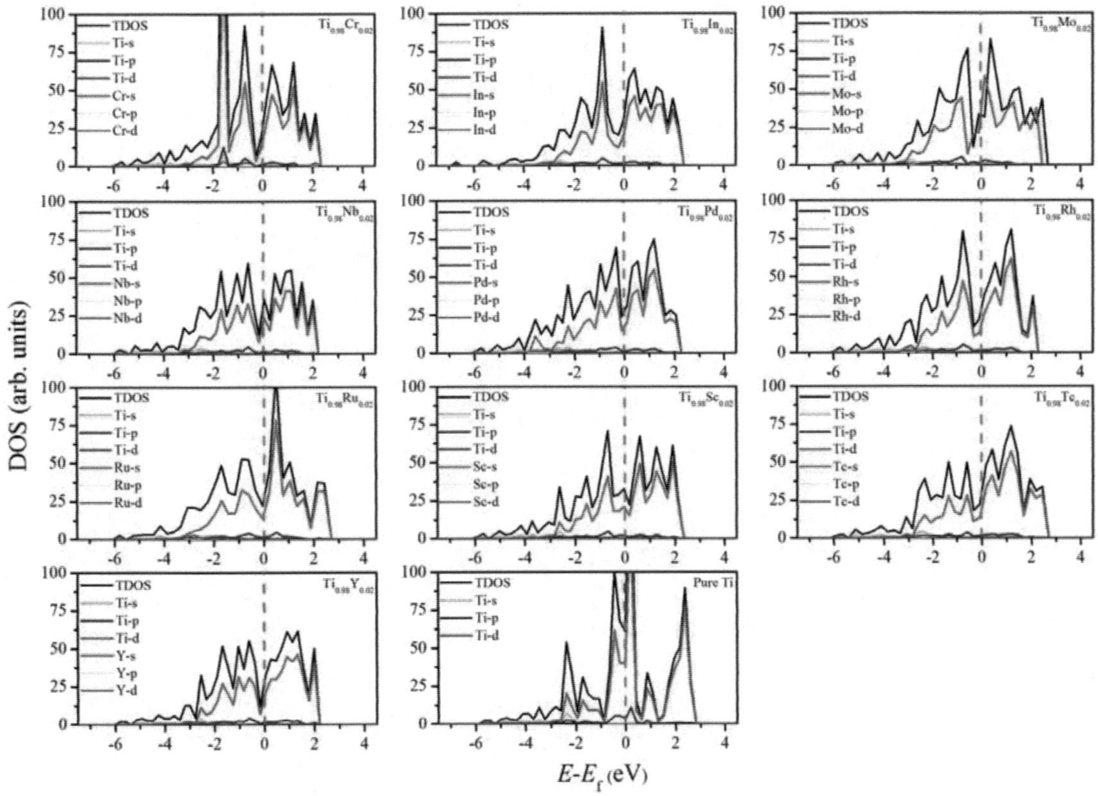

Figure 3. The calculated TDOS and PDOS of $Ti_{0.98}X_{0.02}$ systems and β-Ti. A vertical dotted line denotes the Fermi energy.

4. Conclusions

To improve our fundamental understanding of the effect of trace metallic elements on β-Ti, first-principles simulations are proposed to systematically investigate the lattice constant, formation enthalpy, electronic structure of $Ti_{0.98}X_{0.02}$ systems (X=Cr, In, Mo, Nb, Pd, Rh, Ru, Sc, Tc, and Y). We disclose a favorable linear function between the atomic radius and corresponding volumetric change. Furthermore, the negative formation enthalpy suggests that adding trace transition metal has a tendency to improve thermodynamic stability of β-Ti in the ground state, which is well explained by the displacement of TDOS at the Fermi level.

5. Acknowledgments

This work is supported by the project to strengthen industrial development at the grass-roots level (Project Number TC160A310/19), Young science and technology talents upgrading scheme (Project Number 18SG-15), Technical innovation project of Shanghai research institute of materials (Project Number 19SG-04) and Shanghai Rising-star program (Project Number 18QB1400600) and Shanghai Industry Technology Institute Innovation Pioneer Program (Project Number 16CXXF006).

6. References

[1] Hao Y L, Li S J, Sun B B, Sui M L and Yang R 2007 *Phys. Rev. Lett.* **98** 216405
[2] Zou C X, Li J S, Wang W Y, Zhang Y and Tang B 2018 *Comp. Mater. Sci.* **152** 169
[3] Al-Zain Y, Kim H Y, Hosoda H, Nam T H and Miyazaki S 2010 *Acta Mater.* **58** 4212
[4] Kim H Y, Ikehara Y, Kim J I, Hosoda H and Miyazaki S 2006 *Acta Mater.* **54** 2419
[5] Tang X, Ahmed T and Rack H J 2000 *J. Mater. Sci.* **35** 1805
[6] Zhou Y L, Niinomi M and Akahori T 2004 *Mater. Sci. Eng.* A **371** 283
[7] Kress G and Hafner 1993 *Phys. Rev.* B **47** 558

[8] Perdew J P and Burke K 1996 *Int. J. Quant. Chem.* **57** 309
[9] Levinger B W 1953 JOM **5** 195
[10] Song J, Gao Z B, Zhang L, Wu W H, He B B and Lu L 2019 *Mol. Simulat.* **45** 935
[11] Ma Z N, Wang X, Yan T T, Li Q, Xu Q C, Tian J L and Wang L 2017 *J. Alloys Compd.* **708** 29
[12] Kuroda D, Niinomi M, Morinaga M, Kato Y and Yashiro T 1998 *Mater. Sci. Eng.* A **243** 244

IOP Conf. Series: Materials Science and Engineering **730** (2020) 012017 doi:10.1088/1757-899X/730/1/012017

A Genetic Algorithm for Economic Order Quantity with a Short Product Life Cycle

Bing Chang Ouyang

Department of Business Administration Vanung University, No. 1, Wanneng Rd., Zhongli Dist., Taoyuan City 32061, Taiwan, ROC
Email: ouyang@vnu.edu.tw

Abstract. Considering a finite time horizon crossing over multiple stages of a product life cycle, this study presents a genetic algorithm to deal with an economic order quantity model with multiple demand rates under a non-periodic policy. In real, the demand of the product life cycle is a non-linear function but we assumes it as four-segment linear or constant approximations in this work. In addition, a multiple-segment to be combined with linear or constant functions can be approximated a nonlinear function. This study does not focus on this, but it provides a genetic algorithm to deal with this proposed inventory problems. The particular of this research is that we develop a proposed replenishment scheme by the differentiating equation of the total cost with respect to replenishment time. Then, calculate the total cost of the proposed scheme as the fitness function to evaluate the populations. In this paper, an explicit procedure to obtain an approximating solution is provided and numerical examples to illustrate the proposed model are shown as well.

1. Introduction

The collaborative planning, forecasting and replenishment (CPFR) is an aim to assist and support joint practice among entities for supply chain integration. Under scenario of the CPFR, a suppler would request a buyer to build up a medium- to long-term proposal replenishment schedule for their collaborative forecast and inventory policy. Therefore, the planner of the supplier has to map out the interval and demand of each stage and set up an intermediate-range distribution requirement plan according to the buyer's forecast. This planning horizon is perhaps across over multiple stages on product life cycle. Due to technology advancement and innovation, product life cycle are becoming shorter and shorter especially electronic consumer products recently. For this phenomenon of product life cycle, most previous studies have assumed that the demand pattern is one linear function on production or inventory problems. Nevertheless, the previous researches consider only one linear function over the planning horizon, which is not able to deal with the above situation with different demand trend and multiple stages. One primitive solution following previous one linear function algorithm is to sum each separated stage with the single-piece linear model; but it is neither practical nor effective. To acquire a better solution and release a strict assumption that is no inventory to be held at the end of each stage, this study proposes a genetic algorithm for the economic order quantity problem when the trend of demand is a piecewise linear function.

To consider an increasing demand on the product life cycle or bloom season, Resh et al. [1] were the first to introduce the classical lot-size model with deterministic and time-proportional demand rate. Donaldson [2] first proposed an analytic approach for replenishment problem with a linear (increasing) trend in demand. Based on the above contribution, Henery [3] put forward a recursive procedure for determining the optimal replenishment schedule under the condition of a specified replenishment lots.

Hariga [4] developed an iterative algorithm to derive replenishment schedule for both increasing and decreasing trend in demand. Rau and Ouyang [5] investigated how to model an economic order quantity with a piecewise linear trend in demand. Hill [6] first introduced a general, time-varying, continuous, deterministic demand pattern through a complete product life cycle that broke the demand pattern up into stages and approximated the pattern for each stage by a cubic polynomial.

Genetic algorithms (GAs) are one of most common search methodologies to mimic the process of natural selection and natural genetics for optimization and search problems. The genetic algorithm was first introduced by Holland [7]. Many contributions have widely applied to solve issues for operations and supply chain management, such as inventory control, facility layout, line balancing, production scheduling, and logistics distribution etc.. For characteristics of the chromosome in genetic algorithms, the formulation is ideally suited for using GAs, so Khouja et al. [8] proposed a genetic algorithm to handle the economic lot size problem (ELSP) with discrete demand. Genetic algorithms only need a computable objective function with no requirements of mathematical theory proof such as convexity. Consequently, Gaafar [9] selected genetic algorithms for the deterministic time-varying lot sizing problem with batch ordering and backorders. Recently, Bera et al [10] studied GAs applying to a realistic inventory model with continuous demand under finite horizon. Considering deteriorating, inflation, budget constraints, shortages and finite planning time horizon, Ouyang [11] and Jana et al [12] also selected GAs for inventory models with continuous demand.

2. Assumptions and Notation

2.1. Assumptions

According to characteristics of product life cycle stages, the demand starts to increase gradually in the introduction stage. Next, the demand increase rapidly during the growth stage. Then, the demand becomes to steady over the maturity stage. Finally, the demand ends to decrease fast under the decline stage. In real, the demand of the product life cycle is a non-linear function but we treat it as four-segment linear or constant approximations in here. The demand pattern of this proposed product life cycle is shown in Figure 1. Except the above assumptions, we still have some relative inventory restrictions in our study as follows:

- A finite foreseeable time horizon is considered.
- The demand of each stage is a constant, linearly increasing or decreasing function.
- The ending point of the first stage is the starting point of the second stage and time is continuous.
- A single item is considered.
- Lead-time or shortage is not considered.
- Demand is always greater than zero at the end of the time horizon.
- No stock is held at the beginning and the end of the time horizon.

2.2. Notation

n	number of replenishment cycles.
n^*	optimal replenishment cycles.
H_1	end time of the introduction stage.
H_2	end time of the growth stage.
H_3	end time of the maturity stage.
H_4	end time of the decline stage.
H	planning horizon.
W	total cost, including set up and holding cost.
$W^*(n)$	optimal total cost for n replenishment cycles.
c_1	set up cost per order.
c_2	holding cost per unit per year.
$f_i(t)$	demand function during the ith stage.
a_1	demand at $t = 0$
b_1	rate of demand change per unit of time during stage i and $b_i = 0$ during maturity stage.

t_i Terminating time of ith replenishment cycles time or starting time of i+1th replenishment cycles period i = 0, 1,…, n-1.

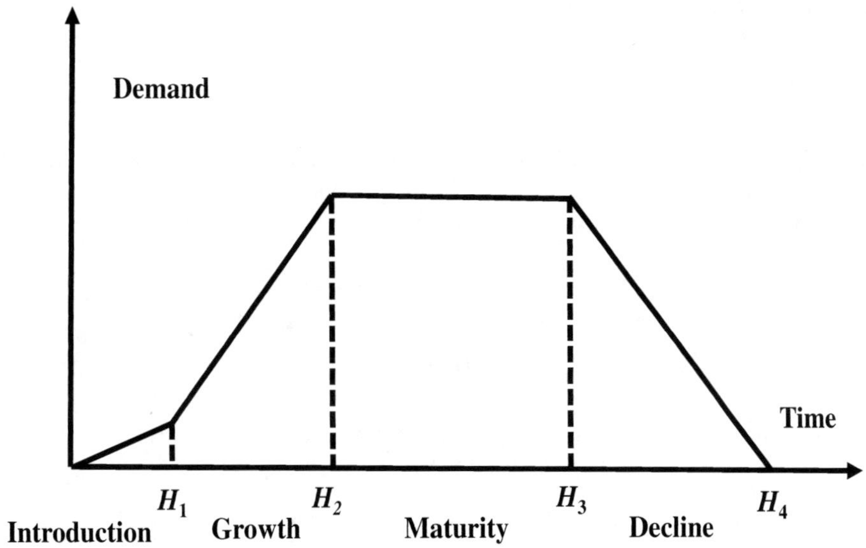

Figure 1. Demand pattern for this proposed product life cycle

3. The Mathematical Development

3.1. Single-Stage Model

The single-stage is a case that only one demand rate over the planning horizon. Suppose a fixed number of n replenishments and the demand function is $f(t) = a_1 + b_1 t$ under the planning horizon H. Therefore, we have the total relevant cost, including the replenishment cost and holding cost as follows:

$$W = nc_1 + c_2 \sum_{i=1}^{n} \int_{t_{i-1}}^{t_i} \int_{t}^{t_i} f(u)du dt \tag{1}$$

From equation (1), it is easy to understand that this proposed inventory model is a nonlinear problem, where we have to resolve the replenishment schedule t_i and number of production cycles n over a planning horizon H. Under a s deterministic n, the derivative of equation (1) with respect to t_i, i =1, 2….n-1, is

$$\frac{\partial W}{\partial t_i} = c_2 \left[\left(t_i - t_{i-1}\right) f(t_i) - \int_{t_i}^{t_{i+1}} f(t)dt \right] \tag{2}$$

Hence, W has the stationary point at t_i if equation (2) = 0, which is a necessary condition for obtaining the optimal solution, that is

$$\int_{t_i}^{t_{i+1}} f(t)dt = (t_i - t_{i-1}) f(t_i) \qquad i = 1,2,........n-1. \tag{3}$$

Solve equation (3) and derive

$$t_{i+1} = (-a + \sqrt{(-2b^2 t_{i-1} t_i + 3b^2 t_i^2 - 2abt_{i-1} + 4abt_i + a^2)}) / b \tag{4}$$

3.2. Two-stage model

The first stage demand function is $f_1(t) = a_1 + b_1 t$ with planning time from 0 to H_1 and the second demand function is $f_2(t) = a_1 + b_1 H_1 + b_2(t - H_1)$ with planning time from H_1 to H, which is the end of the whole planning horizon, as shown in equation (4).

$$f(t)=\begin{cases} a_1+b_1 t & t \le H_1 \\ a_1+b_1 H_1+b_2(t-H_1) & H_1 < t \le H \end{cases} \quad (5)$$

Suppose the kth replenishment period is across stages one and two and there are two linear demand functions $f_1(t)$ and $f_2(t)$ at the kth replenishment period. Then, the total cost, including the replenishment and inventory holding cost, is given by

$$W = nc_1 + c_2\left(\sum_{i=0}^{k-2}\int_{t_i}^{t_{i+1}}\int_{t}^{t_i+1} f_1(u)dudt + \int_{t_{k-1}}^{H_1}\int_{t}^{H_1} f_1(u)dudt + (H_1 - t_{k-1})\int_{H_1}^{t_k} f_1(t)dt \right.$$
$$\left. + \int_{H_1}^{t_{k1}}\int_{t}^{t_{k1}} f_2(u)dudt + \sum_{i=k1}^{n-1}\int_{t_i}^{t_{i+1}}\int_{t}^{t_{i+1}} f_2(u)dudt\right) \quad (6)$$

Similarly, under a deterministic n value, the derivative of equation (6) with respect to t_i, i =1,2….n-1 and then we have a stationary point at t_i as follows:

$$\begin{aligned} &(t_i - t_{i-1}) f_1(t_i) = \int_{t_i}^{t_{i+1}} f_1(u)du && i = 1,..,k-1 \\ &(t_i - t_{i-1}) f_1(t_i) = \int_{t_i}^{H_1} f_1(t)dt + \int_{H_1}^{t_{i+1}} f_2(t)dt && i = k-1 \\ &(t_i - t_{i-1}) f(t_i) = \int_{t_i}^{t_{i+1}} f_2(u)du && i = k+1,..,n-1 \end{aligned} \quad (7)$$

We solve equation (7) and obtain the following equation

$$t_{i+1} = (-a + \sqrt{(-2b_1^2 t_{i-1} t_i + 3b_1^2 t_i^2 - 2ab_1 t_{i-1} + 4ab_1 t_i + a^2)})/b_1$$
when $t_{i+1} \le H_1$ and $i = 1,..,k-1$

$$t_{i+1} = \frac{-a + \sqrt{H_1^2 b_1 b_2 + H_1^2 b_2^2 + 2b_1 b_2 t_{i-1} t_i - 3b_1 b_2 t_i^2 + 4H_1 ab_2 + 2ab_2 t_{i-1} - 4ab_2 t_i + a^2}}{b_2}$$
when $t_i \le H_1$ and $t_{i+1} \ge H_1$ and $i = k-1$

$$t_{i+1} = (-a + \sqrt{(-2b_2^2 t_{i-1} t_i + 3b_2^2 t_i^2 - 2ab_2 t_{i-1} + 4ab_2 t_i + a^2)})/b_2$$
when $t_{i+1} \ge H_1$ and $i = k+1,..,n-1$ (8)

3.3. Multiple-Stage Model

Actually, the two-stage model can be easily extended into multiple-stage models because of deterministic the period of each stage. We are able to follow the procedure for handling two-stage model when we encounter piecewise linear demand.

4. Genetic Algorithm

In this proposed model, the number of replenishment cycles n and replenishment schedule t_i have to be solved, which is a complicated nonlinear problem. Thus, we provide this genetic algorithm with the chromosome of real number type for seeking to an approximately optimal solution. Based on Bellman's principle of optimization [13], once t_1 is found correctly and the remaining replenishment schedule t_2 to t_n can be determined by equation (4) for single-stage and equation (8) for multiple-stage. Thus, we only have to search t_1 and neglect n. Our proposed genetic algorithm to derive an approximate solution is shown in Figure 2 and illustrated as follows:

Chromosome representation:
Due to short life product cycle, we assume a finite planning horizon that is always less than one year. Consequently, randomly select a chromosome consisting of 20-bits for the first replenishment cycle time t_1 under a reasonable range. For example, the terminating time of the first replenishment cycles $t_1 = 0.036$ is near bit string 00001001001101110101 by binary expression.

Fitness function:
With a trial t_1, we can obtain t_2 by equation (4) for single-stage and equation (8) for multiple-stage. Repeatedly solve t_3, t_4,….., t_{n-1}, t_n until $t_{n-1} < H$ and $t_n \geq H$. Then, let $t_n = H$ and compute total cost $W(n, \{t_i\})$ and let $t_{n-1} = H$ and compute total cost $W(n\text{-}1, \{t_i\})$ from equation (3). Finally, compare with the costs of these two scheme and select the lower one. Therefore, the fitness function can be expressed as follows:

$$\begin{cases} W(n,\{t_i\}) & \text{if} \quad W(n,\{t_i\}) < W(n-1,\{t_i\}) \\ W(n-1,\{t_i\}) & \text{if} \quad W(n,\{t_i\}) > W(n-1,\{t_i\}) \end{cases} \tag{9}$$

GA operators:
GA operators create a new generation from parent chromosomes. There are three standard GA operators are used, namely, reproduction, mutation and crossover.

Searching direction:
This GA searching direction is controlled by the self-adjustment rate of operators based on survivor off-springs' rates that are determined after each generation in order to the next search.

Input parameters:
Parameters of genetic algorithm, including population size, probability of mutation, probability of crossover, generation size, initial rate of crossover operation, initial rate of mutation operation, initial rate of reproduction operation and stop condition, are reasonably provided.

Output: The local optimal replenishment schedule and the total cost.

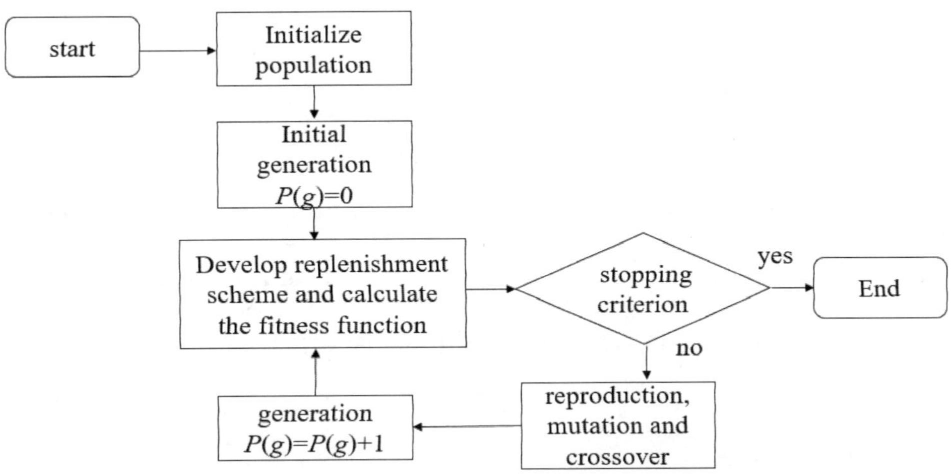

Figure 2. Genetic algorithm for this proposed model

5. Numerical Examples
A VB6.0 program was written to solve the following examples. In order to illustrate our model, we offer two examples as follows:

Under c_1 = 4.5 and c_2=1, suppose a short product life cycle item that $H_1 = 0.2$, $H_2 = 0.5$, $H_3 = 0.8$, and H_4 =1.0. The demand functions are $f_1(t) = 6000$, $f_2(t) = 1200+9000t$, $f_3(t) = 3900$ and $f_4(t) = 3900-19500t$.

Example 1: the planning horizon is from 0 to 0.4 and Example 2: the planning horizon is from 0.6 to 1.0. Applied to the proposed solution procedures, replenishment schedule for these two examples are shown in Table I. The total costs for these two examples are 67.1572 and 89.6796.

Table 1. Replenishment schedules for numerical examples

Example / *i*	1	2	3	4
1	0.1423	0.2444	0.3273	0.4
2	0.6779	0.755	0.8372	1.0

6. Conclusions

For technology advancement, the product life cycle for 3C items are becoming very short. Many industries may encounter that their planning horizon pass through different stages of the product life cycle. In real, this inventory model is a very complex nonlinear problem that is difficult to provide mathematical theory proof such as convexity. Therefore, this study develop a genetic algorithm for relaxing the previous researches with a single-stage linear trend in demand. The particular of our proposed algorithm is that we select the total cost's differentiate equation to build up a proposed replenishment scheme. Then, compute the total cost of this replenishment schedule as the fitness function to evaluate the populations for solving this inventory problem. In terms of the future research, people are able to deliberate our proposed method applying to genetic algorithm for other unconstrained nonlinear problems especially in more complex inventory model with relative issues under a production system.

7. References

[1] Resh M, Friedman M and Barbosa L C 1976 On a general solution of the deterministic lot size problem with time proportional demand, *Oper. Res.* 24, 718–725.
[2] Donaldson W A 1977 Inventory replenishment policy for a linear trend in demand - an analytical solution, *Operl. Res. Quar.*, 28, 3, 663-670.
[3] Henery R J 1979 Inventory replenishment policy for increasing demand, *J. Operl. Res. Soc.*, 30, 7, 611-617.
[4] Hariga M 1993 The inventory replenishment problem with a linear trend in demand, *Comput. & Ind. Eng.*, 24, 2, 143-50.
[5] Rau S and Ouyang B C 2011 Modelling of the inventory replenishment problem with a two-segment piecewise linear demand, *Int. J. Syst. Sci.*, vol. 42,10, 1613-1624
[6] Hill R M 1996 Batching policies for a product life cycle, *Int. J. Prod. Eco.*, 45, 1-3, 421-427.
[7] Holland J H 1975 Adaptation in Natural and Artificial Systems, University of Michigan Press, Ann Arbor, MI.
[8] Khouja M, Mickalewicz Z and Wilmot M 1998, The use of genetic algorithms to solve the economic lot size scheduling problem, *Eur. J. Oper. Res.* 110, 2, 509-524.
[9] Gaafar . 200, Applying genetic algorithms to dynamic lot sizing with batch ordering, *Comput. & Ind. Eng.* 51, 3, 433–444.
[10] Bera U K, Maiti M K, and Maiti M, 2012 Inventory model with fuzzy lead-time and dynamic demand over finite time horizon using a multi-objective genetic algorithm, *Comput. & Math. Appl.* 64, 6, 1822-1838.
[11] Ouyang B C 2013 A genetic algorithm for the production policy in a supply chain under discrete demand and time-vary cost, *Appl. Mech. & Mater*, 321-324, 2137-2140.
[12] Jana D K, Das B and Maiti M 2014, Multi-item partial backlogging inventory models over random planning horizon in random fuzzy environment, *Appl. Soft Comput.*, 21, 12-27.
[13] Bellman R E 2003 Dynamic Programming. *Princeton University Press*, Princeton, NJ. Republished.

IOP Conf. Series: Materials Science and Engineering **730** (2020) 012018 doi:10.1088/1757-899X/730/1/012018

Structural Response Analysis of Nonlinear Pavement Based on Granular Material

Longwei Zhang [1], Chunlei Wang [1], Xupeng Dong [2] and Xiaochun Zhang [3]

Taizhou Traffic Design Institute Taizhou of Zhejiang Province 317600
Tianjin Municipal Engineering Design Institute Tianjin 300000
ITS Research Center of Southeast University, Nanjing of Jiangsu province, 210096
zxc01@263.net

Abstract. For containing gravel layer structure of flexible asphalt concrete pavement structure, the response laws of gravel pavement structure under elastic and nonlinear were comparatively analyzed through the UMAT subroutine programming and FEM. The results show that when considering the nonlinearity of gravel base, the tensile stress and vertical stress of base layer in pavement structure are quite different from linearity condition. With the increase of the load, the tensile stress of the base layer increases nonlinearly, and is less than that of the elastic gravel material under heavy load. Considering the non-linear condition of gravel, the vertical stress of pavement structure increases and mainly occurs in the gravel base.

1. Introduction

Semi-rigid base asphalt pavement is widely used in China's high grade highway because of its high bearing capacity, strength and performance. However, there are still many shortcomings in this kind of pavement structure, Such as high stiffness lower in permeability and easy to produce reflection cracks, etch [1].This phenomenon has aroused the high attention of road workers, which has gradually broken the single form of semi-rigid basement pavement in China, appearing flexible asphalt pavement that take the grading macadam. Recently, Asphalt gravel, especially graded grading macadam, has been proven to have good performance in stress releasing, drainage and preventing reflective cracks that has been used in high-grade pavement design in some areas. [2] However, grading macadam aggregate have a low modulus and exhibit a large nonlinearity due to the looseness of the aggregate. These two aspects have become the main reasons restricting its application and become the focus of domestic and foreign research [3].

At present, regarding the nonlinear constitutive relationship of the granular base, various models have been proposed at home and abroad, and the regression analysis of the model parameters has been carried out through a large number of indoor triaxial tests. The most important thing of application of granular base is how to apply these results into pavement design. The finite element method has been widely used in pavement structure design. Finite element software, such as ABAQUS, is favored by designers because of its powerful functions. For example, designers can use UMAT subroutine to import nonlinear model of road material into software for analysis. However, the nonlinearity of most material is based on the constitutive relationships of elastoplastic and viscoelastic of materials, The elastic modulus of the granular materials, like gravel, changes with the stress state whose constitutive relation exhibits elastic nonlinearity and that cannot be ignored. This paper is aimed at the nonlinear characteristics of granular materials, Though a summary analysis of its nonlinear constitutive model, we could analysis the change between the linear and nonlinear constitutive structure of granular base structure and the respond of pavement structure, providing a theoretical basis for the design and maintenance of the pavement structure, providing a theoretical basis for the design and maintenance of the pavement structure.

2. Nonlinear Constitutive Model of Granular Base

After the road surface is subjected to long-term load, the permanent deformation of the primary load is very small, so under the repeated action of traffic load, most of the road deformation is considered to be recoverable elastic deformation. But due to the looseness of the aggregate, the elastic modulus will change with the stress [4]. The nonlinear properties of the granule base material have been well known whose research also has gone through a long period of time. Earlier, Lekarp[5] try to set a systematically description to this aspect, pointing out that the stiffness of the base and subbase materials is mainly affected by the water content and the stress state of the material. The most important influence is the equivalent stress, followed by the shear stress in stress. Subsequently, many scholars experimented with the granular base and proposed a series of stress-dependent models characterized by elastic modulus. Such as formula(1)~(7).

$$M_R = k_1 \sigma_3^{k_2} \quad \text{(Seed et al. 1967)} \tag{1}$$

$$M_R = k_1 \theta^{k_2} \quad \text{(Hicks and Monismith 1971)} \tag{2}$$

$$M_R = k_1 \theta^{k_2} \sigma_d^{k_3} \quad \text{(Uzan 1985)} \tag{3}$$

$$M_R = k_1 \theta^{k_2} \tau_{oct}^{k_3} \quad \text{(Witczak and Uzan 1988)} \tag{4}$$

$$M_R = k_1 P_a \left(\frac{\theta}{P_a}\right)^{k_2} \left(\frac{\tau_{oct}}{P_a}\right)^{k_3} \quad \text{(Witczak and Uzan 1988)} \tag{5}$$

$$M_R = k_1 P_a \left(\frac{q\text{-}3k_6}{P_a}\right)^{k_2} \left(\frac{t_{oct}}{P_a} + k_7\right)^{k_3} \quad \text{(NCHRP 2004)} \tag{6}$$

$$M_R = k_1 P_a \left(\frac{\theta}{P_a}\right)^{k_2} \left(\frac{\tau_{oct}}{P_a} + 1\right)^{k_3} \quad \text{(ARA, Inc. 2004)} \tag{7}$$

Among them

$\theta = \sigma_1 + \sigma_2 + \sigma_3$, First stress invariant;

$\sigma_d = \sigma_1 - \sigma_3$, deviatory stress;

$\tau_{oct} = \dfrac{\sqrt{(\sigma_1 - \sigma_2)^2 + (\sigma_1 - \sigma_3)^2 + (\sigma_2 - \sigma_3)^2}}{3}$, octahedral shear stress;

P_a——atmospheric pressure, $p_a = 1.03 \times 10^5$;

k_1, k_2, k_3, k_6, k_7—— multiple linear regression parameter;

At present, the ARA model in equation (7) is used extensively in the stress-dependent model of aggregate materials. Dieter [6] repeated experiments on different grades and water content of gravel by indoor dynamic triaxial test. Using this model for regression analysis, the average value of material model parameters was obtained.

$k_1 = 128.5MPa$ (average), $k_2 = 0.520$ (average), $k_3 = -0.1$ (average)

k_2 Mainly represents the sensitivity of aggregate modulus to stress changes. The more sensitive the material, the larger the K; The size of k_3, ranging from -0.4 to 0.6, is mainly affected by water content and k_2.

For the nonlinear model of grain material, domestic scholars Yingjun Mei and Naixing Liang [7] combined with the actual situation, referenced the Wellner model, and introduced the LH nonlinear model:

$$E=(a+b\sigma_1)\sigma_3+c \tag{8}$$

Tianqing Ling [8] established the regression model relationship between the rebound modulus of the granular base and the thickness of it and the modulus of underlying stratum in 2004:

$$E_2=(ah_2^b)E_0 \tag{9}$$

h_2 is structural course thickness of naturally graded aggregate; E_0 is Rebound modulus of the cushion of the corresponding measuring point; a, b are coefficients related to the type of material.

Through regression test and overseas relevant information, we figure out a=0.92 and b=0.40, meanwhile it is found out that the model can be directly used for nonlinear analysis of flexible pavement structures.

3. Nonlinear Algorithm of Finite Element

3.1. Fundamental Principle of nonlinear Algorithm of Finite Element

The finite element analysis firstly discretizes the structure into finite elements, and selects the element as the research object. By introducing the shape function, the relationship between the displacement of the element node and any displacement of point inside the unit is established.

$$\{d\}=[N]\{q\}^e$$

$\{d\}$ is displacement of point inside the unit; $[N]$is shape function of a certain point, which only related to the location of the point; $\{q\}^e$ is shift of element node.

After obtaining the displacement expression of a certain point inside the element, the relationship between the displacement of the element node and strain of any point inside the unit can be established by the geometric equation:

$$\{\xi\}=[L]\{d\}=[L][N]\{q\}^e=[B]\{q\}^e$$

Then, the relationship between the displacement of the element node and the internal stress of the element can be obtained by the physical equation:

$$\{\sigma\}=[D]\{\xi\}=[D][B]\{q\}^e \tag{10}$$

$[B]$ is a constant which is related to the position of the element node; $[D]$Is stiffness matrix.

Using the principle of virtual work, the work done by the node on the virtual displacement is equal to the virtual strain energy of the element, which can be obtained:

$$\delta W=\delta U \Rightarrow \{\delta q\}^{eT}\{F\}^e=\int_V \{\delta\xi\}^T\{\sigma\}dV$$

Substituting the formula (2) into the above equation can get to the link between the external force and the displacement of the unit node. the unit characteristic equation is as follows:

$$\{F\}^e=[k]^e\{q\}^e \tag{11}$$

Among them, $[k]^e=[B]^{\mathrm{T}}[D][B]^{\mathrm{T}}$, is element stiffness matrix

However, the unit characteristic equation cannot be used to find the displacement of the element node. It is needed to establish the relationship between the external force of the structure and the

IOP Conf. Series: Materials Science and Engineering **730** (2020) 012018 doi:10.1088/1757-899X/730/1/012018

displacement of all the element nodes in the structure. It is necessary to add the element characteristic equation to eliminate the influence of the internal force. We apply balance equations to all nodes:

$$\sum_{i=1}^{n}\sum_{e}\sum_{s=i,j...}[k_{is}]\{q_s\}^e=\sum_{i=1}^{n}R_i \quad \text{Or} \quad [K]\{q\}=\{R\}$$

where n is the number of structural element, s is the number of nodes in a element, $[k]$ is element stiffness matrix, $[K]$ is total stiffness matrix.

The essence of finite element calculation is to obtain the total stiffness matrix. The total stiffness matrix can be used to get the displacement of each element node according to the external force, and then the internal displacement, stress and strain of the element are acquired by the shape function, geometric and physical equation. The whole process is shown in Figure 1.

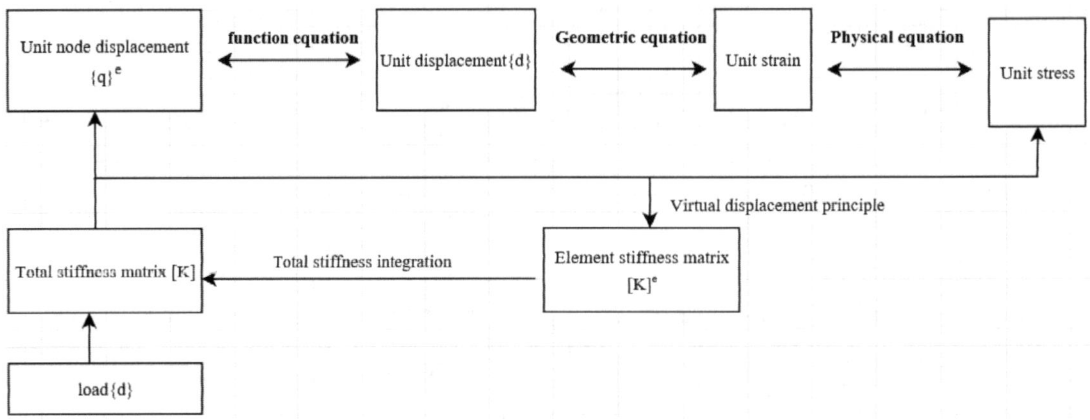

Figure 1. Fundamental principle of finite element solution

It can be seen from the above process that the total stiffness matrix is synthesized by the element stiffness matrix, the element stiffness matrix is related to the stiffness matrixd D, which is composed of the material's elastic modulus E and Poisson's ratio μ.

When the material exhibits nonlinearity and the elastic modulus E changes with the change of stress, the structural load is applied step by step by incremental method and tends to be finalized. In each incremental step we need to solve multiple equations to find the structural balance, and seek an acceptable solution for the incremental step. In each incremental step, the condition of structural balance is that the force generated by the external load on the node is approximately equal to the internal force at the same.

$$\boldsymbol{P - I = 0}$$

In an incremental step, to acquire an approximate solution, there is loop iteration on the above equation. The iteration formula is as follows:

$$\{\Delta\delta^e\}=[{}^tK]^{-1}\{\Delta F\}$$
$$\{{}^{t+1}F^e\}=\{{}^tF^e\}+[{}^tK^e]\{\Delta\delta^e\}$$

Among them, $[{}^tK^e]=\int B^{\mathrm{T}}DBdV=\int B^{\mathrm{T}}f({}^t\sigma^e,{}^t\xi^e,\Delta\xi^e)$, is Element stiffness matrix when the iterating step arrive at t; $[{}^tK]=\sum[{}^tK^e]$, total stiffness matrix for the iterating step t.

The calculation process of analysis of nonlinear finite element is shown in Figure 2.

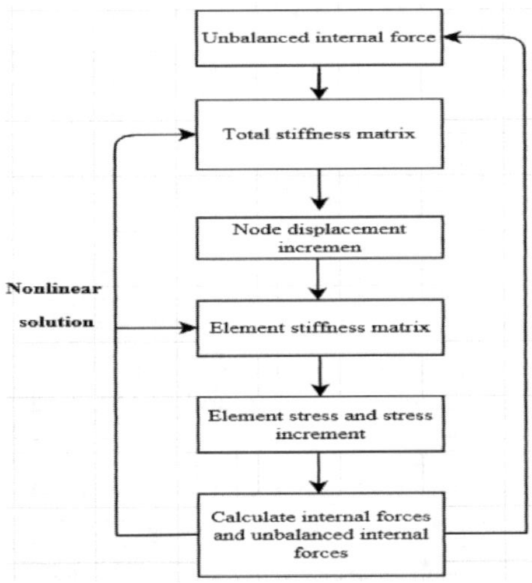

Figure 2. Process of nonlinear solution

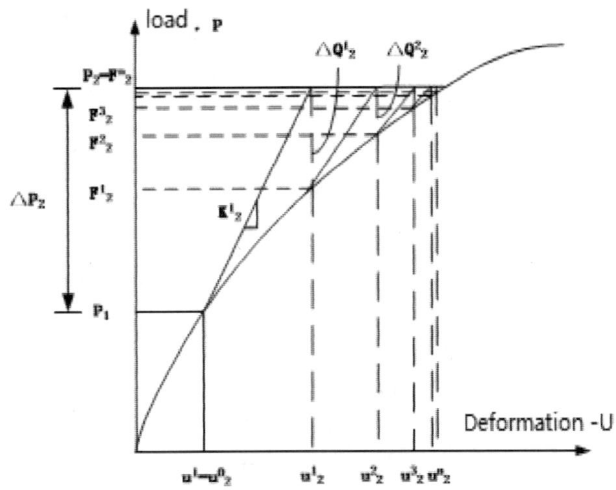

Figure 3. Tangent stiffness algorithm

3.2. Tangent Stiffness Method

The nonlinear finite element mostly uses the incremental iterative method, but in order to ensure the iterative convergence in each incremental step, which need to adopt different algorithms for different constitutive models. The elastic nonlinear model of the gravel subbase and soil foundation is relatively simple. The secant, tangent and constant stiffness method can be used to calculate. Elastic-plastic materials, when their stress can be expressed as a continuous and ductile function of strain, generally adopt the tangent stiffness method, which is also applicable to gravel materials.

A stress softening material refers to a material whose material's elastic modulus has a decreasing tendency as the stress increases, such as some special soil foundation. For this material, in order to ensure iterative convergence, we should take the tangent stiffness method as shown in Figure 3. In the figure,P_1 is the load increment; F^1_2 is the load increment in the 2nd incremental step after the 1st iteration; u'_2 is the total deformation in the 2nd incremental step after the 1st iteration; K^1_2 is The tangent stiffness at the beginning of the 2nd increment step that corresponds to the overall stiffness matrix of the model; $\Delta Q^1_2 = P_2 - F^1_2$, which is the unbalanced internal force after the first iteration; N is the number of iterations.

Combined with the general process of finite element analysis, the iterative algorithm process of stress softening material in the analysis step is as follows:

Step 1: Calculate the displacement increment of node from the load increment,

$$\Delta u_2^1 = \left[K_2^1 \right]^{-1} \Delta P_2$$

Step 2: Update node displacement and calculate element strain:

$$u_2^1 = u_2^0 + \Delta u_2^1, \varepsilon_2^1 = B u_2^1$$

Step 3: Solving element stress by element strain:

$$\sigma_2^1 = D \varepsilon_2^1$$

In this process, the elastic matrix D is not a constant, but a function of the stress state, which is expressed by the formula $D = f(\sigma)$, and the result of the elastic matrix is approximated in term of the stress state at the beginning of the iterative step.

Step 4: Calculate the elastic matrix of the current material and update the overall stiffness matrix according to the stress magnitude and material of nonlinear model obtained in step 2.

$$D(M_R) = f(\sigma_2^1), S_2^2 = \int_V B^T DBdV$$

Step 5: Calculating internal node stress

$$F_2^1 = \int B^T \sigma_2^1 dV$$

Step 6: Calculate unbalanced internal forces

$$\Delta Q_2^1 = P_2 - F_2^1$$

Return to step 1, we substitute ΔQ_2^1 with ΔP_2 to continue the loop, until Δu_2^n and ΔQ_2^n are less than the allowable value, then enter to the next incremental step.

After each iteration is complete, it needs to be updated stress. For gravel stress hardening materials, the stress update after iteration is shown in Figure 4.

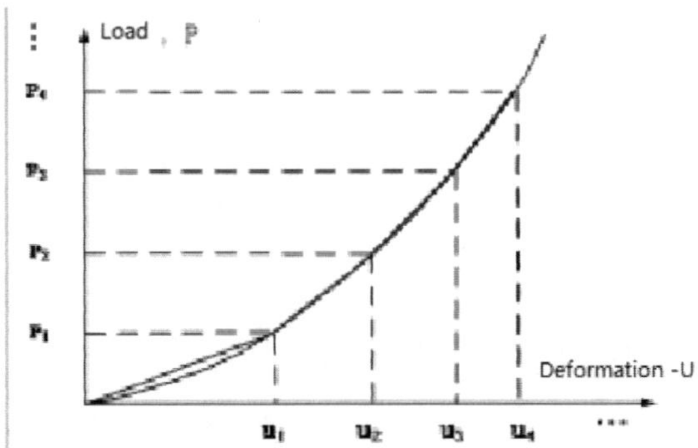

Figure 4. Stress update after iteration

4. Calculation and Analysis of Pavement Structure

4.1. Pavement Analysis Model

The pavement result model usually adopts a half-space infinite extension model. Therefore, to accurately predict the mechanical response of the pavement result, it is necessary to select an appropriate size of the pavement finite element model. Studies have shown that, when the horizontal dimension of the model is 20 times as big as the equivalent circle radius and the vertical dimension is 140 times, the road response of the uniaxial load can be better predicted. Therefore, the selection level of model size: 20R=20*0.213=4.26m; Vertical: 140R: 70*0.213=14.91m.

To ensure that the structure is easy to converge and facilitate meshing, the double rounds uniform load in the specification is converted into the following shape of load distribution. The final model of the pavement structure is shown in Figure 6.

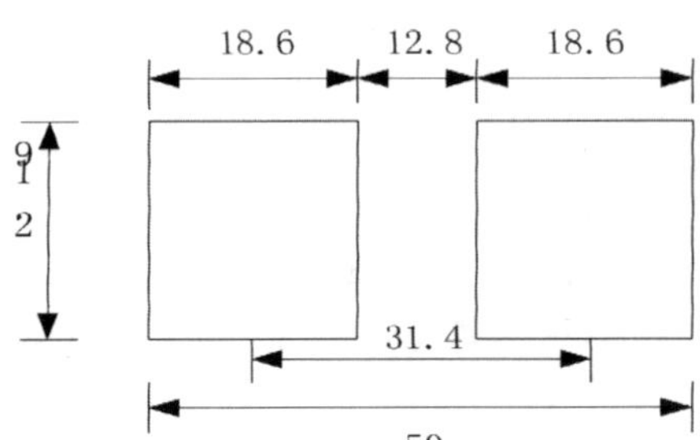

Figure 5. Rectangular load size (unit: cm)

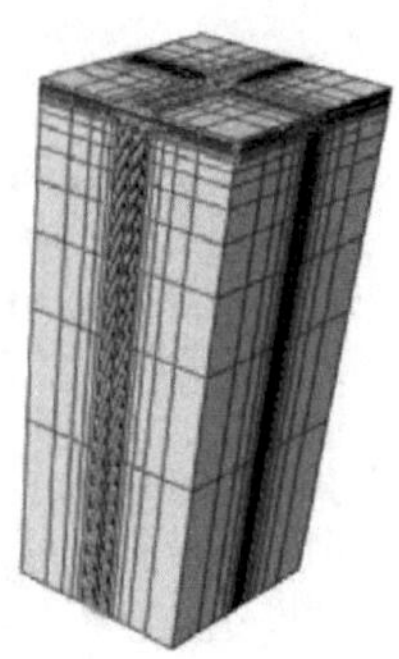

Figure 6. Finite model of pavement structure

4.2. Selection of Pavement Structure Size and Material Parameters

In China, graded gravel is mainly used for the transition layer, base course, and sub base of the road structure [10] .To study the difference of the pavement structure between the linear elastic and nonlinear conditions, the graded gravel is selected as the sub base, As shown in Table 1, the sand cushion in the typical pavement structure was removed for ease of calculation.

Table 1. Asphalt pavement structure of gravel sub base

layer	thickness	modulus	Poisson ratio
Asphalt surface	10	1200	0.3
Asphalt macadam base	10	800	0.35
crushed stone subbase	20	200	0.25
natural gravel cushion	-	-	-
soil	(14.36m)	100	0.3

In this calculation, the structural layers of the pavement are regarded as linear elastic structures except the graded gravel. For the nonlinear structural model of graded gravel, the widely used ARA model is selected for calculation. There are three different crushed stone whose nonlinear model parameters are shown in Table 2 [11].

Table 2. Selection of gravel material parameters

Gradation type	model	K_1 (KPa)	K_2	K_3	Initial elastic modulus
Continuous grading(SH1)	ARA	969.32	0.71	-0.04	200MPa
Gab grading(SH2)	ARA	1125.05	0.657	-0.012	200MPa
Design grading(SH3)	ARA	1147.78	0.623	0.073	200KPa

4.3. Calculation Results and Analysis

After finite element calculation, the results show that the nonlinearity of crushed stone sub base material has a significant impact on the tensile stress of the bottom of the base course and the longitudinal stress of the pavement structure.

(1) Influence on the tensile stress of the base course

Figure 7 shows the tensile stress-strain diagram under elastic and nonlinear conditions of the gravel base when the load is 0.7MPa. As can be seen from the chart, when considering the bottom base's nonlinearity, the tensile stress-strain diagram is approximately the same as that under elastic conditions, but under nonlinear terms, the tensile stress has appeared on both sides of the road surface load and the top of the bottom subbase.

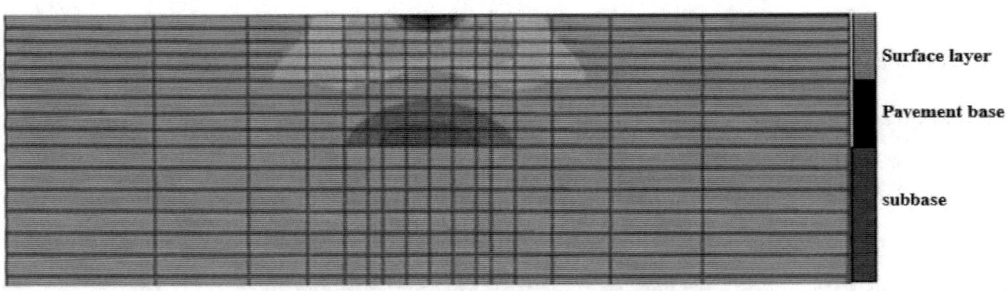

(a)Tensile stress-strain diagram under elastic

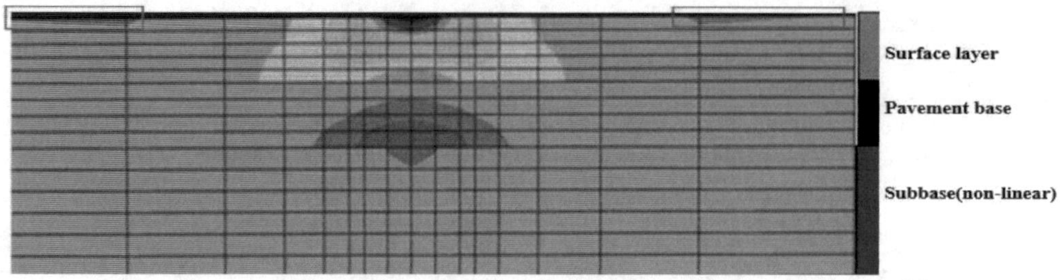

(b)Tensile stress-strain diagram under nonlinear conditions of the gravel base

Figure 7. Tensile stress-strain diagram under elastic and nonlinear conditions of the gravel base

The maximum tensile stress at the bottom of the base course under elastic and nonlinear conditions varies with the load, as shown in figure 8. It can show from the picture that, as the load increases, the tensile stress at the bottom of the base under elastic conditions increases linearly, while the base course under the nonlinear condition, the growth of the tensile stress of the base appears to be nonlinear. When the base is elastic, its modulus of resilience is 200MPa, but under nonlinear conditions, its variation range is 150MPa~350MPa. In terms of the nonlinear model of macadam, the elastic modulus of materials is proportional to the stress. Therefore, when the road load is small, the elastic modulus and stiffness of the base course are low, which makes the maximum tensile stress of the base under the nonlinear condition larger than that under the elastic condition. However, under the heavy load (such as 1.5MPa), the gravel subbase becomes stiffer due to the increased force. So that the maximum tensile stress at the bottom is less than the elastic condition under nonlinear conditions. It indicates that under heavy load conditions, the nonlinearity of the subbase has a favorable influence on the pressure of base.

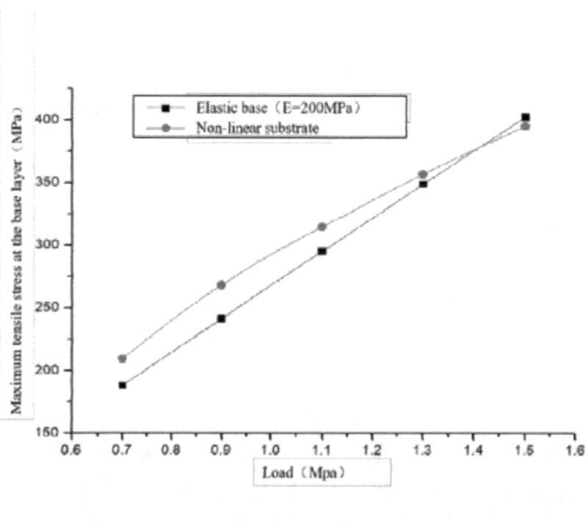

Figure 8. The maximum tensile stress at the bottom of the base course changes with the load

(2) Influence on longitudinal force of pavement structure

Under the conditions of elasticity and nonlinear subbase, the vertical stress diagram of the road structure with a heavy load of 1.5mpa is shown in figure 9. It can be found from the value that the longitudinal stress under the two conditions is different, and the load has an increasing influence on the stress of the base course under the nonlinear terms of the subbase. Figure 10 reflects the variation of vertical pressure along the depth direction under elastic and nonlinear conditions. The red line in the right picture is the selected longitudinal node connection. As shown in the figure, the vertical stress under the nonlinear terms of the subbase is more magnificent than under the elastic condition. Among them, the increase of vertical pressure under the nonlinear condition varies with depth, as shown in Fig. 11. Because the rise of vertical stress mainly occurs on the subbase, it also has a significant influence on the base stress but has a small impact on the soil foundation and surface course.

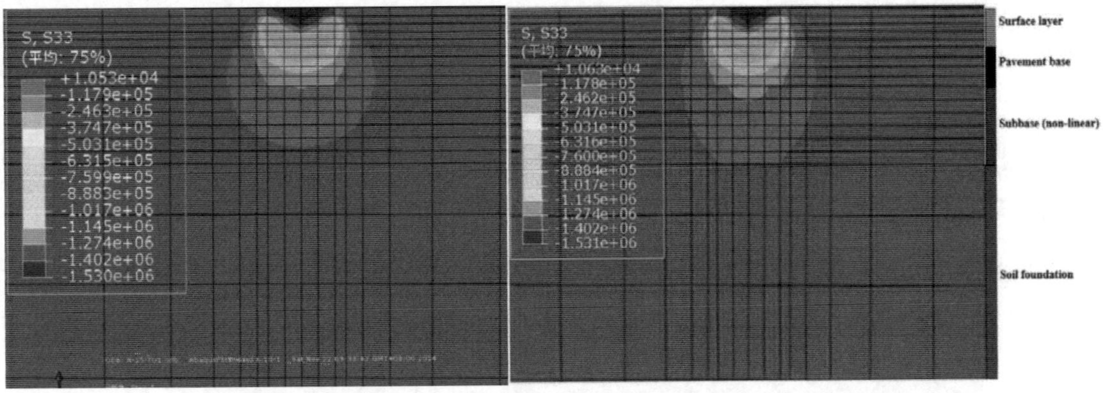

Figure 9. Vertical stress cloud diagram of pavement structure (elastic on the left and non-linear on the right)

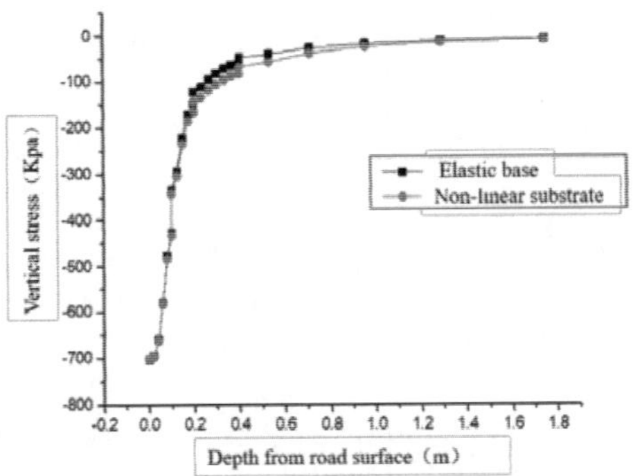

Figure 10. Variation of vertical stress along the depth direction under heavy load (1.5 MPa) elastic and nonlinear conditions

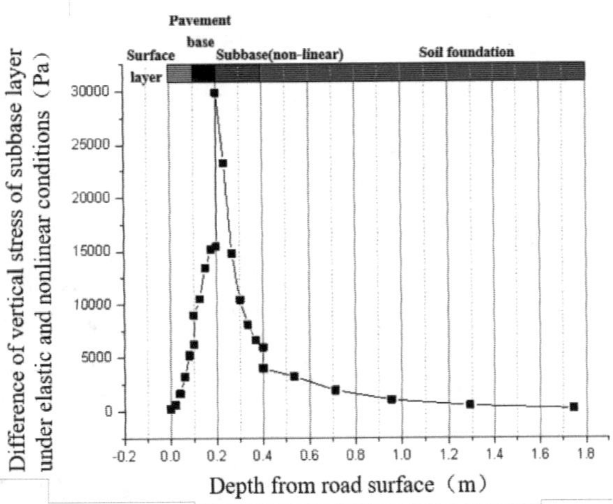

Figure 11. Change of vertical stress between elastic and nonlinear conditions along the depth direction

5. Conclusion

In this paper, the nonlinear finite element analysis of the flexible asphalt pavement structure with gravel structure is carried out by writing the nonlinear constitutive relation subroutine of gravel material. What is important is the difference of pavement structure response under the elastic and nonlinear conditions of gravel material. The results show that the longitudinal stress of the pavement is more significant when the nonlinearity of the gravel material is considered. Under the heavy load condition, the base tensile stress under the nonlinearity of the gravel material is smaller than that under the elastic condition. The research results can provide a reference for the design and research of flexible asphalt pavement.

6. References

[1] X. Xiao, N.X. Liang, Z.J. GAO, Nonlinear analysis of granular asphalt pavement structure [J]. Fujian Building Materials. 2010.

[2] X.S. Wang, Study on Material and Structural Characteristics of Graded Crushed Stone Base Asphalt Pavement [D]. Chang'an University. 2010.

[3] G. Jing, Triaxial test study of graded gravel [D]. Dalian University of Technology. 2007.

[4] Kim M, Tutumluer E, Kwon J. Nonlinear pavement foundation modeling for three-dimensional finite-element analysis of flexible pavements[J]. International Journal of Geomechanics, 2009, 9(5): 195-208.

[5] Lekarp F, Isacsson U, Dawson A. State of the art. I: Resilient response of unbound aggregates [J]. Journal of Transportation Engineering, 2000, 126(1): 66-75.

[6] Stolle D., Guo P., Liu Y. Resilient modulus properties of granular highway material [J]. Canadian Journal of Civil Engineering. 2009(36):639-654.

[7] Y.J. Mei, N.X. Liang. Nonlinear analysis of sandy and gravel base mixture [J]. Journal of Chongqing Jiaotong University, 2004, 23(3): 57-60.

[8] T.Q. Ling, S.J. Wang, Z.H. XU. Some problems in nonlinear analysis of flexible pavement structure. Journal of Chongqing Jiaotong University, 1994, 13(5):30-35.

[9] Kim M, Tutumluer E. Modeling nonlinear, stress-dependent pavement foundation behavior using a general-purpose finite element program [J]. GEOTECHNICAL SPECIAL PUBLICATION, 2006, 154: 29-36.

[10] J. Liu. Analysis and Application of Graded Crushed Stone Flexible Base in Urban Road Asphalt Pavement Engineering [J]. Sichuan Building Materials, 2010 (4): 147-148.

[11] X.P. Dong. Structural response analysis of asphalt pavement under multiple axial loads based on material nonlinearity [D]. Southeast University, 2015.

IOP Conf. Series: Materials Science and Engineering **730** (2020) 012019 doi:10.1088/1757-899X/730/1/012019

Research on Damping Characteristics of Photo-controlled Electrorheological Damper

Xiang Liu, Yusong Chen and Xinjie Wang

School of Mechanical Engineering, Nanjing University of Science and Technology, Nanjing, China.
Email: xjwang@njust.edu.cn

Abstract. Firstly, the expression of PLZT ceramic driving voltage is obtained with external load. Furthermore, the hydrodynamic modelling of the electrorheological fluid in the circular damping channel is carried out, and the relationship between the pressure difference between the two ends of the damping channel and the PLZT ceramic driving voltage is derived, and the hydrodynamic simulation of the electrorheological fluid in the circular damping channel was carried out in the fluid simulation software Fluent. The simulation result shows the correctness of the mathematic model and the feasibility of photo-controlled electrorheological damper.

1. Introduction

Electrorheological fluid is a kind of intelligent soft material with potential application. It has the electrorheological effect, that is, it exhibits the characteristics of Newtonian fluid when no electric field is applied. When the electric field is applied, its viscosity changes and it changes from liquid to solid. The change will quickly return to a liquid form after the electric field is removed. Electrorheological fluids are widely used in dampers, shock absorbers, and automotive suspension systems. However, the conventional ER damper requires an external boost circuit to provide a high-value electric field, which causes the entire damper device to have a complicated structure, be susceptible to electromagnetic interference, and not easily miniaturized.

Another smart material is called yttrium-modified zirconate titanate ceramic (PLZT), which generates anomalous photovoltaic effects under the excitation of a specific wavelength of ultraviolet light source, and produces a high-value photo-generated voltage of several kV/cm in the polarization direction. Therefore, combining the above two smart materials, the high-value photo-generated voltage generated by the PLZT ceramic under the high-energy ultraviolet light source is applied to the electrorheological damping device, such that the electrorheological damping device has a simple structure, no electromagnetic interference, and can work in a vacuum environment.

The anomalous photovoltaic effect of PLZT ceramics was first discovered by the phenomenon that the photo-generated voltage in BaTiO3 observed by Volk[1] et al. in 1973, which is much larger than the solid forbidden band width. In 1975, Brody[2] pointed out that the photo-generated voltage of PLZT ceramics depends on factors such as material, grain size, and material bulk temperature. Later scholars tried to explain the mechanism of this phenomenon with an electrical model. In 1983, Brody[3] used the RC equivalent electrical model to describe this phenomenon. In 1988, Poosanaas[4] studied the anomalous photovoltaic effect under different components, and pointed out that the maximum photo-generated current and photo-generated voltage appeared under the components of 4/48/52 and 5/54/46, respectively. Sun and Tong[5] et al. studied the effect of temperature on the photo-induced voltage of PLZT ceramics and proposed a corresponding equation. Using the anomalous photovoltaic effect of PLZT ceramics, Morikawa[6] proposed a new type of photoelectric motor.

In 1949, Winslow[7] prepared an electrorheological fluid by dispersing flour and lime in silicone oil and found the electrorheological effect. Later scholars proposed many related models. In 1983, Stangroom[8] and others proposed a water bridge model, and the flow of electrorheological fluid must overcome the "water bridge force" between the particles. Klass[9] proposed an electric double layer model, which better explains the effects of external temperature and voltage and the concentration of electrorheological fluid on the electrorheological effect. Davis[10] and Atten[11] proposed a conductivity model for the anhydrous electrorheological fluid. The model considers that the main cause of the electrorheological effect is the ratio of the conductivity of the suspended particles to the base fluid. The electrorheological fluid has the advantage of fast response speed, convenient regulation and easy control, so it is widely used in engineering fields such as damping systems, vibration and noise reduction systems, brake systems, electromechanical coupling control and liquid valves.

In this paper, the expression of PLZT ceramic driving voltage is obtained with external load. Secondly, the hydrodynamic modeling of the electrorheological fluid in the circular damping channel is carried out, and the relationship between the pressure difference between the two ends of the damping channel and the PLZT ceramic driving voltage is derived, and the hydrodynamic simulation of the electrorheological fluid in the circular damping channel was carried out in the fluid simulation software Fluent. Finally, the comparison between theoretical and simulation result is made to verify the correctness of the mathematical model.

2. Mathematical Modelling of Photo-controlled Electrorheological Damping

2.1. Photo-controlled Electrorheological Damping Device Structrue

The schematic diagram of the structure of the photo-controlled electrorheological damping device is shown in figure 1.

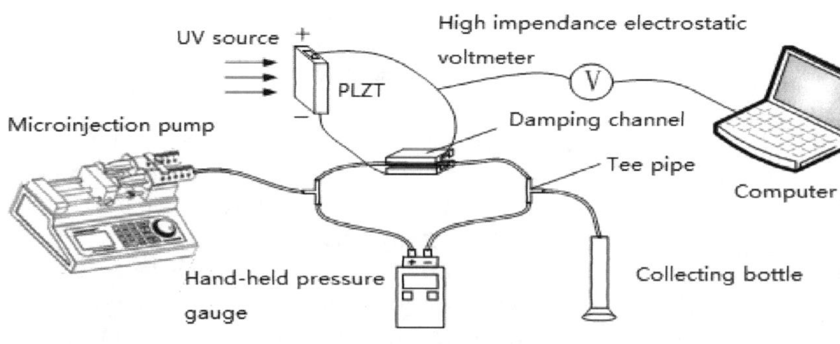

Figure 1. Schematic diagram of experimental platform for light-controlled electrorheological damping device.

When PLZT ceramic is irradiated vertically by the ultraviolet light with a wavelength near 365 nm, the photo-generated carriers flow along the polarized direction, so that the positive electrode and the negative electrode of the PLZT ceramic collect the positive and negative charges respectively to generate a high photovoltaic voltage of several kV/cm. The silver wire is used to connect the electrodes of the PLZT ceramic to the two pieces of copper foil which are attached to the top and the bottom side of the damping channel respectively with the insulating adhesive. Therefore, the electric field required for the electrorheological fluid in the damping channel is provided.

2.2. Mathematical Modelling of Photo-generated Driving Voltage

When the two ends of the PLZT ceramic are respectively connected to the parallel copper foil plate load through the silver wire, the equivalent electrical model of the external load is shown in figure 2.

R_d and C_d are the resistance and capacitance of the parallel plate generated by two parallel copper foils respectively.

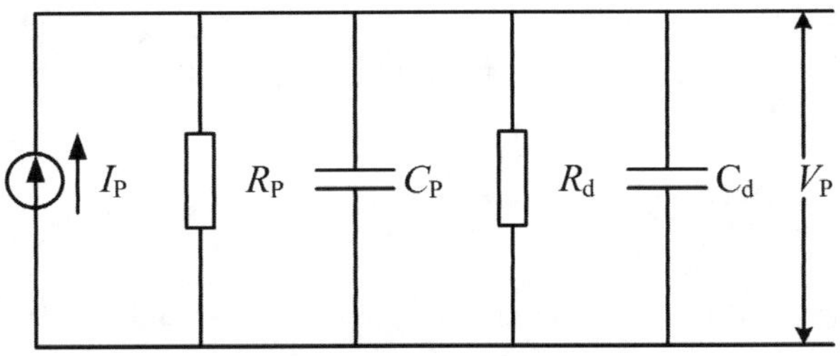

Figure 2. Equivalent electrical model of PLZT ceramics with external load.

According to the new equivalent circuit model above, the driving voltage produced by PLZT ceramic for electrostatic parallel plate is:

$$V_P = I_P \frac{R_P \cdot R_d}{R_P + R_d}(1-e^{-\frac{t}{\tau}}) = V_s(1-e^{-\frac{t}{\tau}}) \tag{1}$$

where τ is the time constant, the expression is:

$$\tau' = \frac{R_P \cdot R_d}{R_P + R_d}(C_P + C_d) \tag{2}$$

2.3. Hydrodynamic Model of Electrorheological Fluid in Damping Channel

When the electrorheological fluid flows in a steady state in the circular damper channel, the force is shown in figure 3.

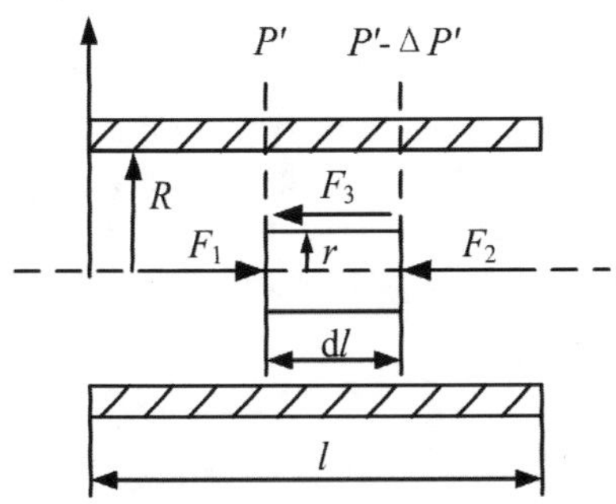

Figure 3. The force of electrorheological fluid in a circular damping channel.

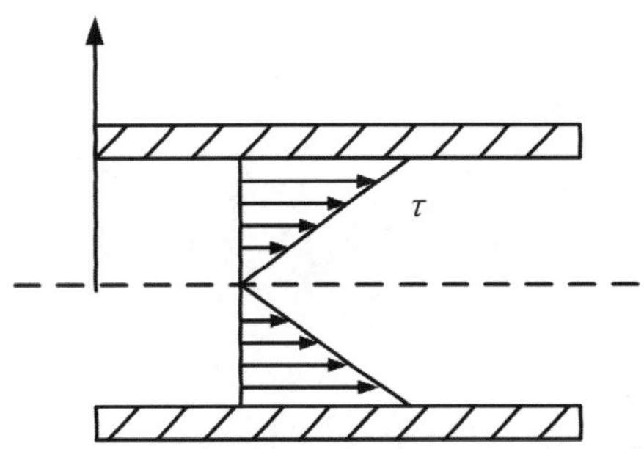

Figure 4. Electrorheological fluid shear stress distribution map.

From the liquid balance equation: $\Sigma F=F_1+F_2+F_3=0$, we can get:

$$\pi r^2 P' - \pi r^2 (P' - \Delta P') - 2\pi r \tau(r) \mathrm{d}l = 0 \tag{3}$$

Simplify formula (4):

$$\tau(r) = \frac{r\Delta P'}{2\mathrm{d}l} = \frac{r\Delta P}{2L} \tag{4}$$

Where ΔP is the pressure difference across the circular damper channel; L is the length of the circular damper channel.

It can be seen from equation (4) that the electrorheological fluid is proportional to the radius in both the Newtonian fluid and the Bingham fluid in the circular damping channel, as shown in figure 4.

2.3.1. Dynamic Model of Electrorheological Fluid without Electric Field. When there is no electric field, the electrorheological fluid exhibits the characteristics of Newtonian fluid. In a circular damper channel, the constitutive model is:

$$\tau(r) = -\eta \frac{\mathrm{d}u(r)}{\mathrm{d}r} \tag{5}$$

Where τ(r) is the shear stress of the electrorheological fluid; du(r)/dr is the shear rate; u(r) is the horizontal flow velocity of the electrorheological fluid; η is the plastic viscosity or dynamic viscosity independent of the electric field strength.

Bringing equation (5) into equation (4) and integrating it. Note that u(r)|r=R=0, from which the Poiseuille equation for the radial flow velocity distribution of the Electrorheological fluid along the damper channel is obtained:

$$u(r) = \frac{\Delta P}{4\eta L}(R^2 - r^2) \tag{6}$$

From the flow formula of the liquid ,we can obtain that:

$$Q = 2\pi \int_0^R u(r) r \mathrm{d}r = \frac{\pi R^4}{8\eta L}\Delta P \tag{7}$$

When there is no electric field, the relationship between the pressure drop and the flow when the electrorheological fluid exhibits Newtonian fluid properties is:

$$\Delta P = (P_{in} - P_{out}) = \frac{8\eta L}{\pi R^4} Q \tag{8}$$

2.3.2. Dynamic Model of Electrorheological Fluid when loading Electric Field. When the photo-generated voltage generated by the PLZT ceramic excited by the high-energy ultraviolet light source is applied to the copper foil plate on the upper and lower sides of the damping channel, the electrorheological fluid exhibits Bingham fluid characteristics under the action of the electric field. The velocity distribution of Bingham fluid in a circular damper channel is shown in figure 5.

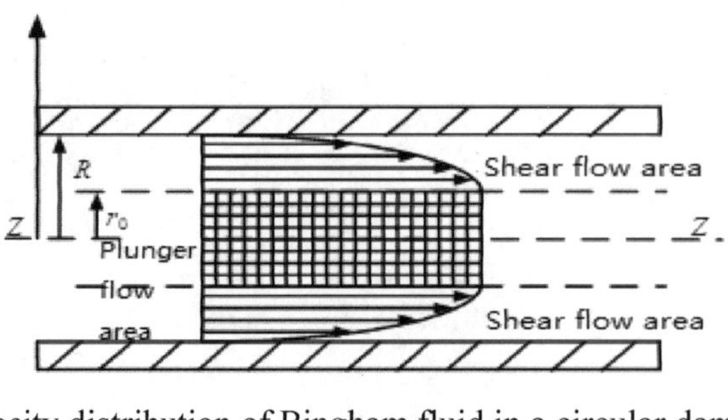

Figure 5. Velocity distribution of Bingham fluid in a circular damping channel

As can be seen from Figure 4, the flow rate of the electrorheological fluid is divided into two parts, when $r < r_0$, it is the plunger flow area; when $r > r_0$, it is the shear flow area. Where r_0 is the radius of the "cylinder plug".

Within the region I ($r_0 < r < R$):

$$\tau(r)=\tau_y(E)-\eta\frac{\mathrm{d}u(r)}{\mathrm{d}r} \tag{9}$$

Note that $u(r)|_{r=R}=0$, you can get:

$$u(r)=\frac{\Delta P}{4\eta L}(R^2-r^2)-\frac{\tau_y}{\eta}(R-r) \tag{10}$$

In the region II ($0<r<r_0$), since $\mathrm{d}u(r)/\mathrm{d}r=0$, there is $\tau(r_0)=\tau_y$, and substituting into the equation (4) can obtain the "cylinder plug" radius r_0

$$r_0=\frac{2\tau_y L}{\Delta P} \tag{11}$$

Since $u(r)=u(r_0)=C$, there are:

$$u(r)=\frac{\Delta P}{4\eta L}(R^2-r_0^2)-\frac{\tau_y}{\eta}(R-r_0) \tag{12}$$

According to the flow formula of the liquid and integrate it, you can get:

$$\Delta P^4-(\frac{8\tau_y L}{3R}+\frac{8\eta L}{\pi R^4}Q)\Delta P^3+\frac{1}{3}(\frac{2\tau_y L}{R})^4=0 \tag{13}$$

Equation (13) can be simplified when the flow rate Q is large:

$$\Delta P=(P_{in}-P_{out})=\frac{8\eta L}{\pi R^4}Q+\frac{8\tau_y L}{3R} \tag{14}$$

The relationship between the dynamic yield stress τ_y and the electric field strength E can be expressed as:

$$\tau_y=kE^\alpha \tag{15}$$

Therefore, when the high-value electric field generated by the PLZT ceramic excited by the high-energy ultraviolet light source is used as the driving electric field of the electrorheological fluid, the expression of the pressure difference ΔP at both ends of the circular damping channel is obtained:

$$\Delta P=\frac{8\eta L}{\pi R^4}Q+\frac{8L}{3R}k\left(\frac{V_{\mathrm{mod}}}{d}\right)^\alpha=\frac{8\eta L}{\pi R^4}Q+\frac{8L}{3R}k\left(\beta I_{\mathrm{p}}(\frac{R_{\mathrm{p}}R_{\mathrm{d}}}{R_{\mathrm{p}}+R_{\mathrm{d}}})(1-e^{-\frac{t}{(C_{\mathrm{p}}+C_{\mathrm{d}})R_{\mathrm{p}}R_{\mathrm{d}}/(R_{\mathrm{p}}+R_{\mathrm{d}})}})/d\right)^\alpha \tag{16}$$

2.4. Influence of Light Intensity I on the Pressure difference across the Damping Channel

The relationship between the pressure difference across the damper channel and the light intensity I can be obtained by the relationship between the driving voltage V_{mod} and the light intensity I:

$$\Delta P=\frac{8\eta L}{\pi R^4}Q+\frac{8L}{3R}k\left(\beta A_{\mathrm{cur}}A_{\mathrm{r}}I^{\alpha_{\mathrm{cur}}-\alpha_{\mathrm{r}}}\frac{R_{\mathrm{d}}}{A_{\mathrm{r}}I^{-\alpha_{\mathrm{r}}}+R_{\mathrm{d}}}(1-e^{-\frac{t}{\tau'}})/d\right)^\alpha \tag{17}$$

The fitting values of the PLZT ceramic photo-generated voltage characteristics index obtained through experiments are shown in table 1:

Table 1. PLZT ceramic photo-generated voltage characteristics index fitting value.

Photocurrent coefficient A_{cur}	Photo-resistance coefficient A_r	Photocurrent varies with light intensity α_{cur}	Photo-resistance varies with light intensity α_r
7.368×10^{-14}	1.052×10^{16}	1.512	1.293

Considering the damping channel upper and lower copper foil plate geometry is 20mm (l) × 5mm (w), the gap(d) is 3mm, according to the parameter fitting value in table 1, the variation curve of the pressure difference ΔP saturation value across the corresponding damping channel when the light intensity is gradually increased from 0 to 400 mW/cm^2 is shown in figure 6:

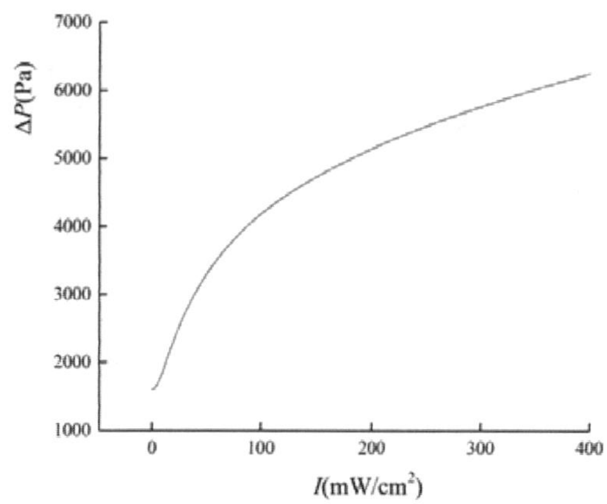

Figure 6. Curve of pressure difference across the damper channel as a function of light intensity

It can be seen from figure 6 that as the light intensity I gradually increases from 0 to 400 mW/cm^2, the pressure difference ΔP across the damping channel gradually increases, and when the light intensity is 0, the pressure difference is not zero since the electrorheological fluid still has viscosity at this time. When the light intensity changes in the range of 0-100 mW/cm^2, the pressure difference ΔP grows faster, and then the pressure difference increases gradually with the increase of light intensity, which is caused by the characteristics of PLZT ceramics: the PLZT ceramic driving voltage saturation value does not increase endlessly with the increase of light intensity.

3. Simulation of Light-controlled Pressure Difference Response Characteristics based on Fluent

Firstly, the model is established and meshed for the circular damping channel. The two ends of the damping channel are defined as velocity input and pressure output respectively. Secondly, the appropriate fluid model and solver are selected in Fluent. Finally, the calculation results in Fluent are imported. In the processing software Tecplot, the pressure and velocity clouds can be obtained, as well as the pressure difference ΔP across the damping channel, as shown in figure 7.

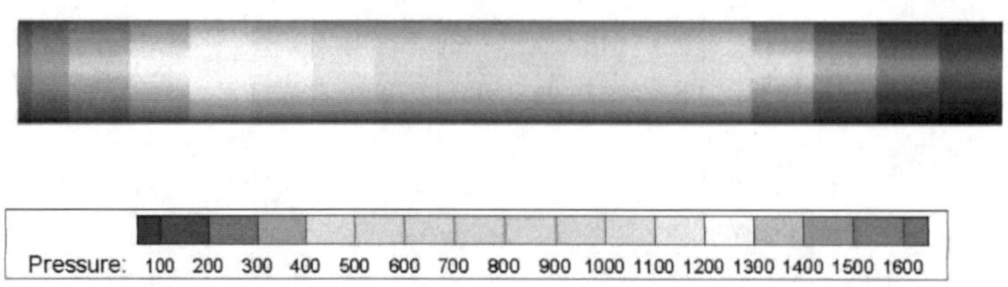

Figure 7. Damping channel pressure cloud map ($\tau_y = 0$)

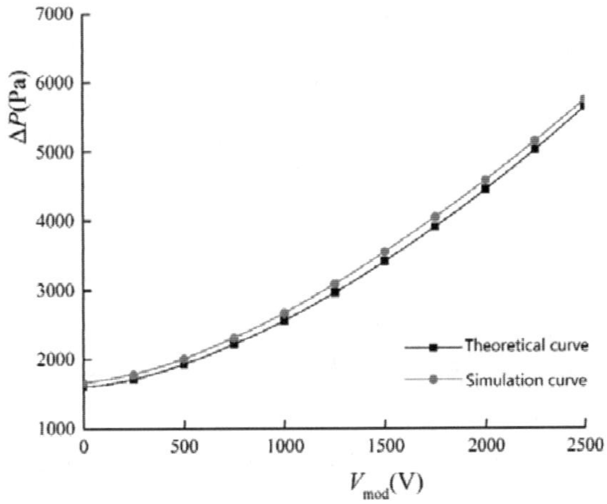

Figure 8. Curve of pressure difference across the damping channel as a function of driving voltage

According to the different driving voltage V_{mod} of PLZT ceramics, the dynamic shear yield stress τ_y of the electrorheological fluid can be calculated, and the corresponding dynamic shear yield stress can be set in the material property setting interface. After the flow field is initialized, the simulation can be obtained. The comparison curve between the pressure difference simulation results and the theoretical results is shown in figure 8.

It can be seen from figure 8 that the theoretical curve and the simulation curve are almost coincident, and the error between the theoretical pressure difference and the simulated pressure difference is within 5%, which proves the correctness of the theoretical mathematical model.

4. Conclusions

In this paper, a new way of photo-controlled electrorheological damper based on PLZT ceramics is proposed. By modifying the expression of the static voltage of PLZT ceramic without load, the equivalent electrical model of PLZT ceramic with external load of the electrorheological channel is obtained. The hydrodynamic modelling of the electrorheological fluid in the circular damping channel is deduced, and the relationship between the pressure difference between the two ends of the damping channel and the PLZT ceramic driving voltage is derived.

The pressure difference ΔP across the damping channel increases with the increase of light intensity under the premise of other conditions remain unchanged. The correctness of the mathematic model is verified by comparing the theoretical result and the simulation result which is obtained through Fluent.

5. Acknowledgements

The authors gratefully acknowledge the funding support from the National Natural Science Foundation of China (No.51675282)

6. References

[1] Volk T R, Grekov A A, Kosonogov N A, et al. (1973). Influence of illumination on the domain structure and Curie temperature of BaTiO3. *Sov. Phys.-Solid State*, **14**, 2740-2743.

[2] Brody P S. (1975). High voltage photovoltaic effect in barium titanate and lead titanate-lead zirconate ceramics. *Journal of Solid State Chemistry*, **12(3)**, 193-200.

[3] Brody P S. (1983). Optomechanical bimorph actuator. *Ferroelectrics*, **50(1)**, 27-32.

[4] Poosanaas P, Uchino K. (1999). Photostrictive effect in lanthanum-modified lead zirconate titanate ceramics near the morphotropic phase boundary. *Materials Chemistry & Physics*, **61(1)**, 36-41.

[5] Sun D, Tong L. (2007). Modeling of wireless remote shape control for beams using nonlinear photostrictive actuators. *International Journal of Solids and Structures*, **44(2)**, 672-684.

[6] Morikawa Y, Ichiki M, Nakada T. (2006). Electrostatic Optical Motor Using PLZT Elements. *The Japan Society of Mechanical Engineers*, **72(718)**, 1807-1811.
[7] Winslow W M. (1949). Induced Fibration of Suspensions. *Journal of Applied Physics*, **20(12)**, 1137-1140.
[8] Stangroom J E. (1983). Electrorheological Fluids. *Physical Technology,* 1983, **14(6)**, 290–296.
[9] Klass D L, Martinek T W. (1967). Electroviscous Fluids. I. Rheological Properties. *Journal of Applied Physics*, **38(1)**, 67-74.
[10] Davis L C. (1992). Ground state of an electrorheological fluid. *Physical Review A*, **46(2)**, 719-721.
[11] Atten P. (1994). Conduction model of the electrorheological effect. *International Journal of Modern Physics B*, **8(20-21)**, 2731-2745.

Fracture Analysis of the Chain Pin Shaft of an Escalator Ladder

Chenghong Peng[1,*], Weiheng Zhu, Ling Chen, Xiaoke Zhu and Zhiling Chen

School of Materials Science and Engineering, South China University of Technology, Guangzhou 510641, China

*mehpeng@scut.edu.cn

Abstract. This study analyzed the cause of fracture for the chain pin shaft of an escalator ladder, and found that the pin material satisfied the chemical composition content requirements of the GB 20CrMo steel standard and similar to ASTM A29/A29M-2012 grade 4120. Abnormal metallurgical quality defects were not found in the pin. The pin shaft was carburized, and internal oxidation defects were not observed on the surface as a result of carburization. The internal structure of the material was uneven, there existed a mixed grain structure, and the martensite structure was coarse at particular regions, to seven grade. Additionally, the martensite structure in the carburized layer was slightly coarser, four to five grade, which had an adverse effect on the pin strength. The pin break was a fatigue fracture caused by multiple cumulative damages. The fracture originated on the surface of the pin, and may have been caused by indentation on the pin's surface.

1. Introduction

Escalator accidents occasionally occur; minor accidents cause property damage, while serious accidents result in human casualties. Determining the cause of such accidents is very important for avoiding escalator accidents. By investigating a particular escalator accident, it was found that the chain pin shaft had broken. The appearance of the broken pin shaft is shown in figure 1. As can be seen in the figure, the pin break occurred on a smooth surface at a distance of approximately 4 mm from the R position of the variable cross-sections. The fracture surface was relatively flat and approximately perpendicular to the axial direction. There existed significant bruising at the edge of the fracture near the end of the larger diameter, as indicated by the arrow in figure 1(b).

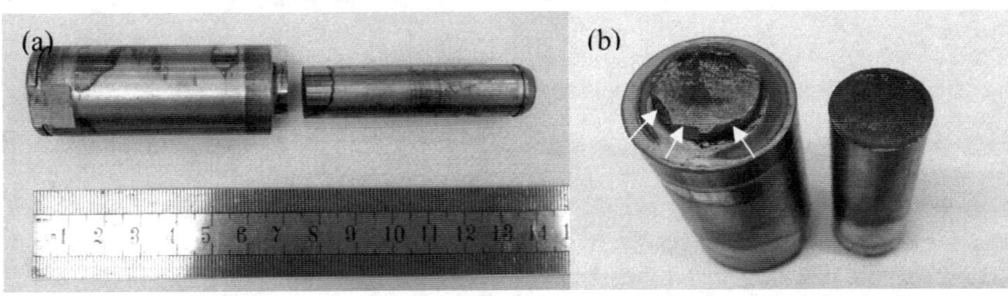

Figure 1. Appearance of stepped pin shaft fracture: (a) overall appearance of broken part of step chain pin; (b) macroscopic morphology of uncleaned fracture.

The total length of the step chain was approximately 135 mm; the larger end diameter was approximately Φ25×66 mm, and the smaller one was approximately Φ15×69 mm. The pin was made

IOP Conf. Series: Materials Science and Engineering **730** (2020) 012020 doi:10.1088/1757-899X/730/1/012020

of 20CrMo steel and used in the automatic walking elevator step chain; the surface hardness requirement was 56-60 HRC. In this study, fracture analysis, material chemical composition analysis, material metallographic analysis, and mechanical property testing were carried out to determine the main cause of the pin shaft fracture.

2. Experimental Procedure

The failure shaft was visually and macroscopically inspected by optical microscopy and chemical analysis. The hardness was also measured to complete the characterization of the materials. The chemical analysis was carried out using FOUNDARY MASTER desktop vacuum spark emission spectroscopy. The fractured surfaces were ultrasonically cleaned and observed using a scanning electron microscope (SEM) equipped with Energy Dispersive X-Ray Spectrometer (EDX).

Metallographic observation was carried out using the Leica DMI5000M microscope. The sample was etched using 4% nitric acid alcohol solution.

3. Fracture Macro- And Micro-Analysis

3.1. Macro Analysis

The macroscopic morphology of the broken shaft fracture after cleaning is shown in figure 2, the main cracking source of the broken shaft was located on the surface of the shaft, and a secondary cracking source was observed near the shaft surface in the fracture. The expansion direction of the main crack is indicated by the red arrows in the figures. Moreover, there existed a faint beach-like pattern on the fracture, which indicated that the pin fracture may have been fatigue fracture.

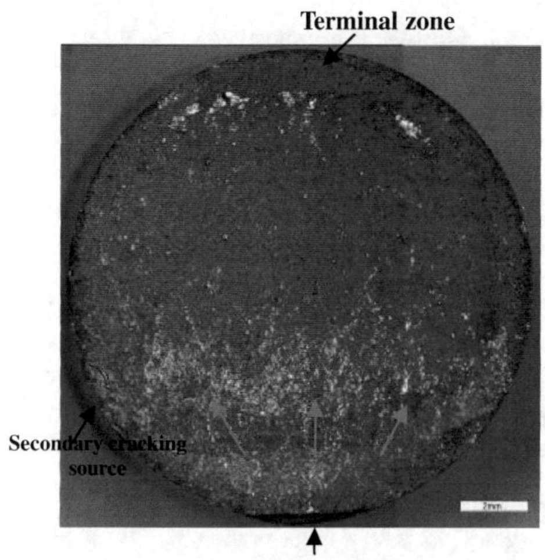

Figure 2. Low optical magnification fracture image of broken shaft fracture after cleaning.

3.2. Microscopic Analysis

The microtopography of the broken shaft fracture is shown in figure 3-1(a)-(b), there is a trace of collision in the main cracking source, which may have been formed after the pin broke. Therefore, it is impossible to confirm whether the cracking source cracked owing to metallurgical quality defects or processing defects. This fracture near the surface of the shaft was mainly intergranular brittle fracture, which indicates that the material of the shaft surface was brittle.

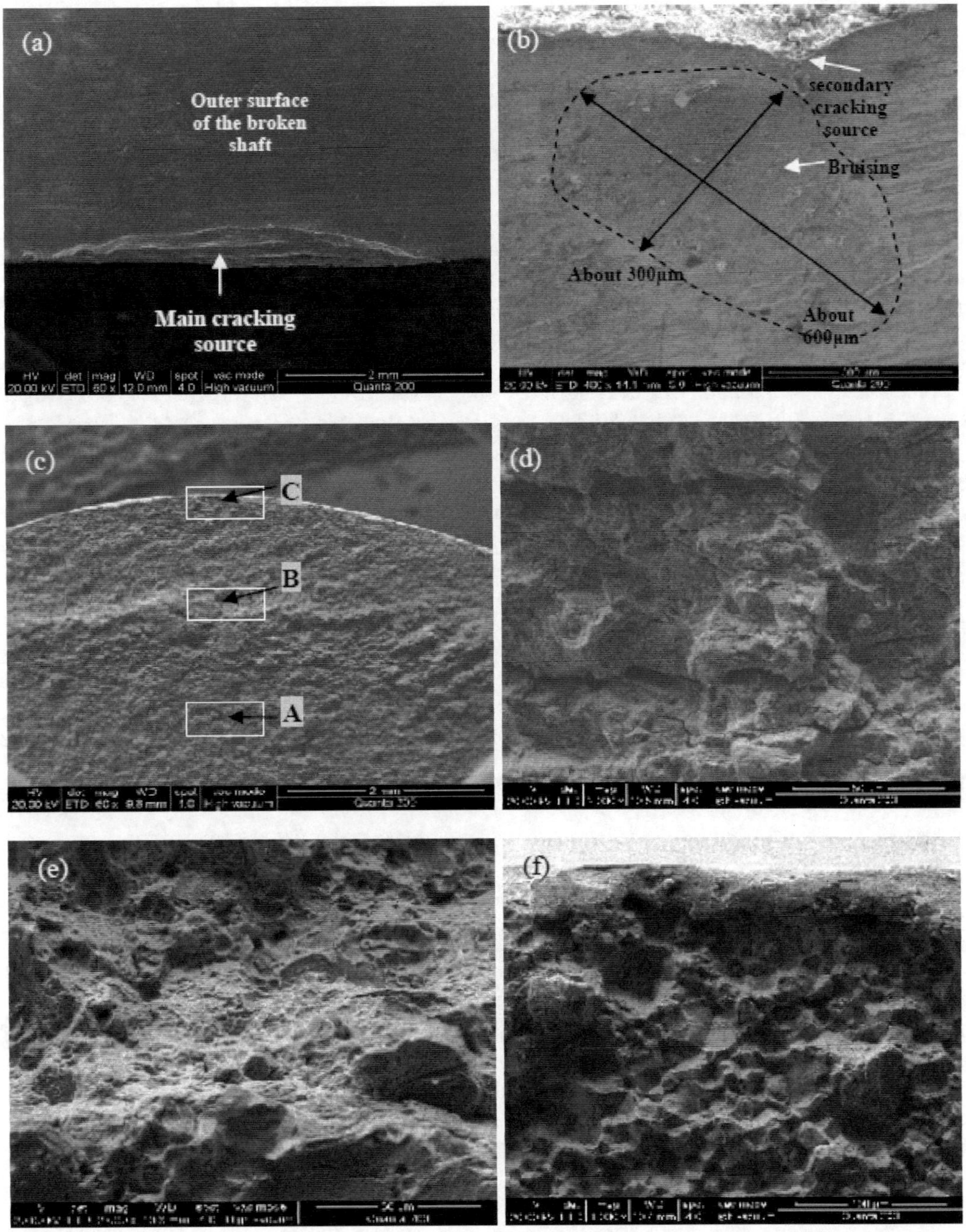

Figure 3-1. Microscopic morphology of broken shaft fracture. (a) Main crack source area shown in figure 2; (b) Outer surface close to secondary crack source; (c) Near termination zone shown in figure 2; (d) Area A in (c) (fatigue striation); (e) Area B in (c) (dimple); (f) Area C in (c) (brittle intercrystalline fracture in rock candy form).

In figure 3-1(c)-(d), in addition to the main cracking source, a secondary cracking source can be observed in the vicinity of the main cracking source. Abnormal metallurgical quality defects were not observed in the secondary cracking source region of the fracture, but a bruise mark was observed on the pin surface near the secondary cracking source region. The size of the crushed part was approximately 600×300 μm. Considering the small probability of only forming such a small crush at the secondary cracking source after the pin broke, it is believed that the crush mark formed before the fracture and caused the creation of the cracking source.

In figure 3-1(e)-(f), there existed a fatigue striation on the fracture, which indicated fatigue fracture. Thus, the fracture of the pin was actually fatigue fracture in the termination zone belonging to the instantaneous fracture zone, and the microscopic features of the fracture near the core were mainly dimples. Additionally, the microscopic features of the fracture near the surface area mainly indicated a brittle intercrystalline fracture in the form of rock candy. This indicates that the pin was subjected to surface strengthening treatment, and thus the core material had good toughness.

In summary, the pin fracture was fatigue fracture, but the main fatigue source was damage by crushing after the pin broke. Therefore, it was impossible to confirm the crack source defect that caused the pin to undergo fatigue fracture by the secondary crack source; namely, the pin surface pressure injury defects. Hence, it was assumed that the main cracking source may also had similar crushing defects resulting in fatigue cracking.

3.3. Material Chemical Composition Testing

The chemical composition of the sample was detected by desktop vacuum spark emission spectroscopy according to ASTM E415-2017. The test results of the chemical composition of the investigated broken shaft material are presented in table 1.

Table 1. Results of broken shaft material's chemical composition analysis.

Material	chemical composition (wt. %)								
	C	Si	Mn	P	S	Cr	Mo	Cu	Ni
Measured value	0.20	0.26	0.62	0.018	0.009	1.08	0.20	0.041	0.026
GB 20CrMo	0.17~ 0.24	0.17~ 0.37	0.40~ 0.70	*	*	0.80~ 1.10	0.15~ 0.25	*	≤ 0.30
ASTM A29/A29M 4120	0.18~ 0.23	0.15~ 0.35	0.90~ 1.20	<0.035	<0.040	0.40~ 0.60	0.13~ 0.20		

According to the chemical composition analysis results, the chemical composition of the broken shaft material satisfied the requirements regarding the standard component content of the "China GB standard 20CrMo steel", and indicated special-grade high quality steel. The chemical compositions are similar to those of the ASTM A29/A29M-2016, grade 4120 also.

3.4. Metallographic Analysis of Materials

According to ASTM E3-11 (2017), the axial and radial metallographic analysis specimens were collected near the cracking source area at the end of the smaller diameter of the broken shaft to carry out metallurgical quality analysis and microstructural analysis.

The microstructure of the non-etched metallographic grinding surface of the axial specimen was observed. The non-metallic inclusion grade assessment according to ISO 4967-2013 results are presented in Table 2.

Table 2. Non-metallic inclusion rating.

Assessment project	Group A (sulfide type)		Group B (aluminate type)		Group C (silicate type)		Group D (globular oxide type)		Group DS (single globular type)
	Fine series	Thick series	Fine series	Thick series	Fine series	Thick series	Fine series	Thick series	
Grade	0	0	1.0	0	0	0	1.0	0	0

The non-metallic inclusions of the inspection materials were mainly "point type + chain type" oxide-based non-metallic inclusions. The degree of non-metallic inclusions in the material was not severe, and abnormal metallurgical quality defects were not observed in the material. At the same time, internal oxidation defects resulting from carburization were not observed on the surface of the pin shaft.

The metallographic structure of the material was evaluated according to "GB/T 25744-2010 Metallographic examination for carburizing quenching and tempering of steel parts".

The microstructure of the axial surface of the axial specimen after etching is shown in figure 3-2, the microstructure image of the radial surface after etching the specimen is shown in figure 3-3. The microstructure of the material indicates that the surface of the broken shaft was carburized. The tempered martensite structure of the carburized layer was slightly coarser, four to five grade, which led to an increase in the brittleness of the shaft surface material. Moreover, the grain of the pin matrix material was uneven, and there existed an abnormally coarse tempered martensite microstructure in particular regions. The hardness of the coarse grain zone determined using the Micro Vickers hardness method was 430-460 HV, and the hardness of the fine grain zone was 360-380 HV. The mixed grain microstructure had a negative effect on the strength of the material[1, 2, 3, 4]. Because the sample to be inspected was very small, tensile testing and impact testing could not be performed.

Figure 3-2. Microstructural morphology of axial matrix specimen near outer surface in cracking source area: (a)Low-magnification morphology (arrow on matrix indicates coarse martensite area); (b) High-magnification morphology of coarse-grained matrix region, light block area in (a).

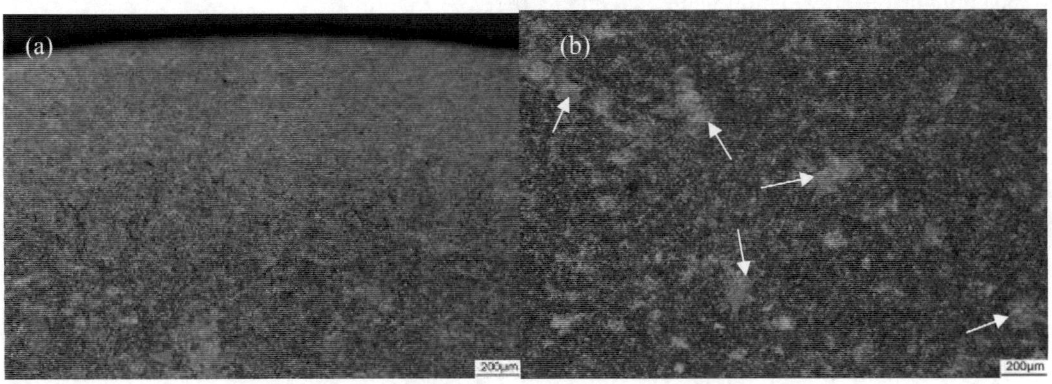

Figure 3-3. Microstructural morphology of radial matrix specimen: (a) near surface; (b) close to core (arrow indicates coarse martensite area).

According to the Chinese machinery industry standard "JB/T 9211-2008 Martensite grade medium-carbon steel medium-carbon alloy structural steel", the tempered martensite structure grade to base in the coarse grain zone was assessed as 7 grade, and the tempered martensite grade in the fine grain zone was 4 grade. A mixed grain structure appeared inside the material, maybe owing to the abnormal growth of local grain resulting from the short duration of over-temperature during the heat treatment of the material. Alternatively, it may have occurred owing to improper forging also, such as the forging ratio may have been insufficient , or the final forging temperature may have been too high, or uneven temperature, thus resulting in an uneven material structure and partial coarse grain regions[5,6,7]. Additionally, the mixed grain structure was more likely to occur under normal heat treatment conditions; grain refining by micro-alloying was effective in suppressing the formation of mixed grain[8].

The mixed grain of carburized steel and quenched and tempered steel was investigated under normal heat treatment conditions. A simulation test was carried out using cylindrical samples with a size of approximately 12 mm×12 mm, whose original microstructure was a normal annealed state. Thus, it was observed that in the process of heat treatment, low-carbon steels, such as 20Cr, 20CrMo, and 20CrMnTi, were prone to the occurrence of abnormally coarse grain in local areas. Some hold that mixed grain is produced after heat treatment. However, it is not a heat treatment defect, but rather a defect in the steel itself, such as uneven deformation or uneven composition.[7]

Energy spectrum analysis confirmed that there existed a coating with a thickness of approximately 1 μm as a galvanized layer on the shaft surface (figure 3-4). From the appearance of the coating distribution in the fracture source region and the deformation of the structure, it was assessed that the region formed by collision after the pin broke.

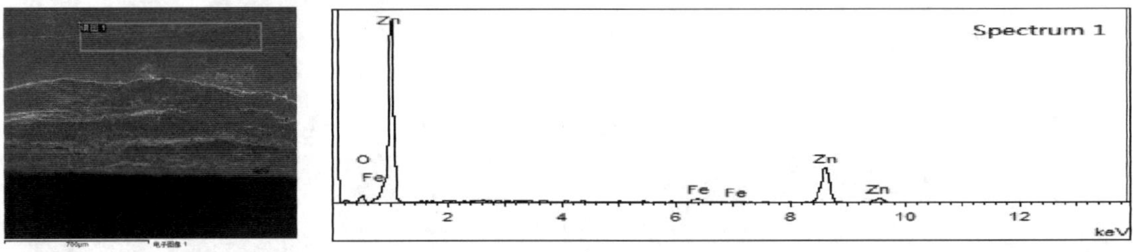

Figure 3-4. Energy spectrum analysis results for coating on pin surface.

4. Micro-Hardness Test

Because the sample was too small, impact and tensile tests could not be performed. According to the ASTM E18-2019, Surface and core Rockwell hardness test was performed on the broken shaft surface and the core, their average hardness in 56.6 and 35.0HRC.

According to the test results, the surface hardness of the broken shaft achieved the required hardness value (56-60 HRC) provided by the client.

The hardness distribution of the hardened layer on the surface of the pin shaft was tested according to "ASTM E92–17 Standard Test Methods for Vickers Hardness and Knoop Hardness of Metallic Materials" and "ISO 2639-2002 steels–determination and verification of the depth of carburized and hardened cases". The detection location of the broken shaft was on the cross-section of the pin shaft with a thin end and a distance of approximately 10 mm from the fracture surface. The test results are presented in figure 4.

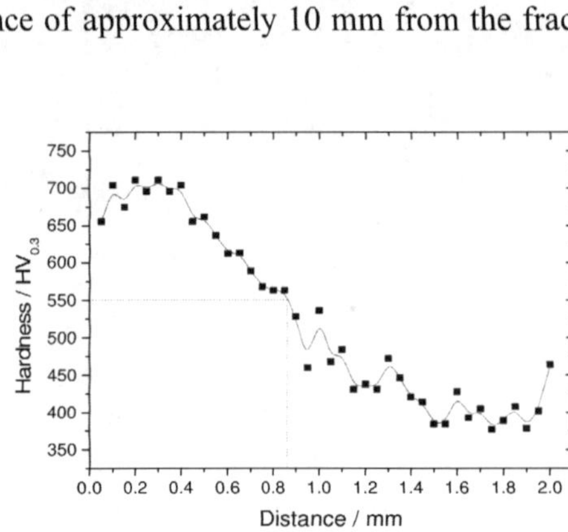

Figure 4. Hardness distribution curve of hardened layer on broken shaft.

The surface hardness change of the broken shaft conforms to the general rule of hardness change after carburizing treatment, as can be seen, the effectively hardened layer depth of the broken shaft was approximately 0.86 mm, according to ISO 2639-2002. And the hardness of the matrix fluctuates greatly, further indicating that the microstructure of the matrix is uneven.

5. Conclusion

(1) The chemical composition of the pin material satisfies the chemical composition content requirements of 20CrMo steel according to the "GB/T 3077-2015 alloy structural steels" standard, which similar to "ASTM A29/A29M-2012 Standard Specification for Steel Bars, Carbon and Alloy, Hot-Wrought, General Requirements for grade 4120". Abnormal metallurgical quality defects were not observed in the pin.

(2) The internal structure of the material was uneven, a mixed grain structure was observed, and the martensite structure was coarse in particular regions (7 grade), while the martensite structure in the layer was slightly coarser (four to five grade), which had an adverse effect on the pin strength. Hence, the material's strength decreased.

(3) The pin break was fatigue fracture caused by cumulative damage. The fracture originated on the surface of the pin, and may have been caused by indentation on the pin's surface.

6. Acknowledgment

We thank Experimental Center of Materials Science and Technology, SCUT, for composition analyses and SEM observation on the chain pin shaft.

7. References

[1] Wang Kai, Dang Shu'e, He Yan, Liu Zhilong 2013 Methods of Eliminating Mixed Grains in 30Cr2Ni4MoV Steel, *Heavy Casting and Forging* **6** 9
[2] Feng Shihuan Liang Huidong 2011 Research on Formation Causes of Mixed Austenite Grain in 20Cr Steel, *Metallurgical Collections* **8** 1
[3] Tian Xiugang, Feng Xianyong, Song Xiaojuan 2018 Analysis and Control of Surface Mixed Crystal in Ultra- Low Carbon Hot Strip Steel, *Hebei Metallurg* **6** 47
[4] Anders From and Rolf Sandström 1999 Influence of Mixed Grain Size Distributions on the Toughness in High and Extra High Strength Steels, *Materials Characterization* **42** 111
[5] SHEN Qunfang, WU Shaobin 2011 20CrMnTi"MixGrain"Analysis, Modern machinery **3** 90
[6] Zhinan Yang, Fucheng Zhang, Chunlei Zheng , Ming Zhang , Bo Lv, Lin Qu 2015 Study on hot deformation behaviour and processing maps of low carbon bainitic steel, *Materials and Design* **66** 258
[7] Song Haigang 2018 Reasons for mixing crystals of F91 steel and elimination methods, *China Metal Bulletin* **8** 71
[8] Zhu Hua 2017 Analysis of Reasons for Formation of Mixed Grain in Carburized Gear Steel 18CrMnB with Different Heat Treatment Processes, *Special Steel Technology* **2** 40

IOP Conf. Series: Materials Science and Engineering **730** (2020) 012021 doi:10.1088/1757-899X/730/1/012021

Construction and Application of Height Control Model of Urban Building based on GIS: Taking Guanyun County as an Example

Lei Zhang[1], Lin Yao[1,2] and Xiaohua Chen[1]

1.School of Architecture & Planning ,Anhui Jianzhu University , 856th Jinzhai South Road, Baohe District, Hefei, Anhui, China.
2.Anhui Urban & Rural Planning and Design Institute, 363 [th] Tongcheng South Road, Baohe District, Hefei, Anhui, China.
Email: zl12597@126.com

Abstract. It has become an urgent demand of urban residents to shape the appropriate urban space style and strengthen the sharing of landscape resources through the height control of urban buildings. Starting from the visual corridor of urban landscape, taking Guanyun County , Jiangsu Province as an example , a scientific and efficient height control model and planning technical framework for urban buildings were constructed based on GIS. Based on the demonstration of the surrounding area of Dayi Mountain in Guanyun County, scientific viewpoint selection and multi-dimensional line-of-sight control analysis were carried out to construct the three-level urban building height control model: reference model of urban height control, modified model of urban height control, status quo correction model of urban height control. Finally, the height control zones of urban buildings were divided and planning guidelines were proposed. It can provide reference for the research on height control of urban buildings and related planning and construction.

1. Introduction

In 2013, the central urbanization work conference proposed that "relying on the existing landscape context and other unique scenery, let the city blend in with nature, so that residents can see mountains, water and remember homesickness ". As a concrete expression of the vertical dimension of urban space, the height of urban buildings is an important window to display overall landscape pattern and overall spatial form of a city. In the process of urban overall planning and design, both the urban spatial form design at the overall level and landmark system, visual corridor system and skyline landscape planning at the sector level are closely related to the urban height form. Therefore, the overall planning and design for the height of the city is very necessary.

Building height control on the overall level of a city was traced back to Urban Beautification Movement and Zoning regulations[1] that emerged successively in western countries in the 19th century .Its core idea is to strengthen the control ability of the height of urban buildings and realize overall order of urban space through targeted planning, design and control strategy. At present, relevant researches at home and abroad are classified into three types:(1) Traditional methods of urban vertical control based on intuitive judgment. In order to realize the overall control of urban vertical form, the corresponding height limits for different types of blocks or land are set qualitatively. This method is suitable for macro and overall control, but it relies more on the intuitive judgment of the planner. Typical applications included the height control of the old city of Paris in France and the height control zoning of the ancient city of Matsumoto in Japan[2]. At the beginning of this century,

Beijing also applied this height control method to protect the historical features of the ancient capital[3]. With expansion of urban scale and increasing complexity of urban structure, it was difficult to accurately grasp the overall urban pattern by relying solely on the subjective judgment of planners. (2) Quantitative instrumental assistance such as multi-factor evaluation method based on GIS database. This method was widely practiced in China. For example, based on the conditions of urban spatial development and layout of Nanjing City, we the impact evaluation factors were determined, including historical features, land price, accessibility, culture, construction potential and urban landscape and proposed specific optimization plan[4]. The method was also been applied in Hangzhou, Qingdao and other cities[5, 8]. (3) Visual analysis based on aesthetic perception, that is, the vertical form control method through human visual perception. It was mainly applied in the specific key areas of city. It can obtain the precise limit of vertical form control through the field landscape feedback and become a supplement to overall vertical form control.

In London, "strategic overlooking control" was put forward. In China, Weihai, Jianou ancient city and Jiangxin island of Wenzhou discussed the application of visual analysis method in the control of urban vertical form[9,11]. The above methods and practices provided valuable references for the height control of urban buildings, but it was difficult to deal with the overall and comprehensive requirements in the practice of overall urban planning and design due to different degrees of subjectivity and one-sidedness.

Aiming at the limitation of height control in the current urban planning and construction, based on the current working framework of overall urban design compilation, we constructed an vertical form control method that included rigid pattern, evaluation model, multiple scenarios, aesthetic correction, and other technical processes in order to explore a relatively rational and objective technical method for the overall urban planning and design practice .

2. Research Objects and Methods

2.1. Research Objects

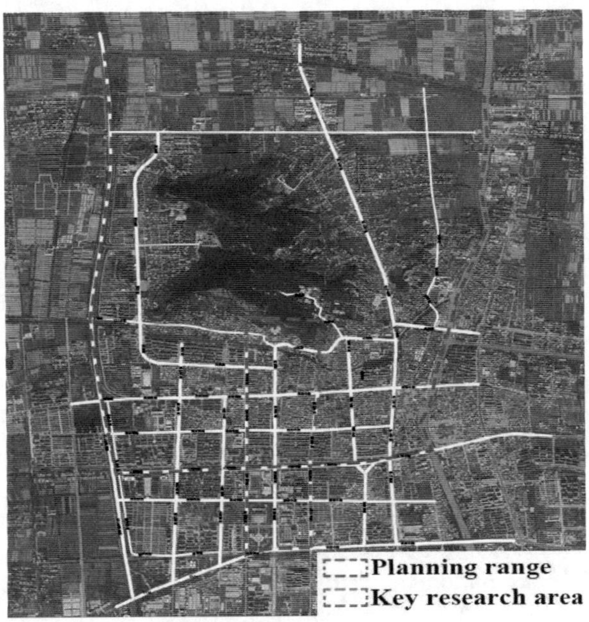

Figure 1. Study object range

Guanyun County which is under the jurisdiction of Lianyungang City is located in northeast Jiangsu Province, bordering the Yellow Sea on the east(Fig.1). The overall urban plan of Guanyun County (2017-2030) proposed to build an ecological leisure city with the integration of mountains and waters. Dayi Mountain Scenic Area which was a buddhist culture tourist area with Shifosi temple as the center was located in the central city of Guanyun County. Building height control around Dayi Mountain Scenic Area was a key factor to build Guanyun County's urban features. Building a height control system around Dayi Mountain was conducive to improving the quality of Guanyun City, guiding the healthy and orderly development of the city, and shaping a more beautiful three-dimensional spatial pattern of the city. The planning area was 18.3km2, north to Xinxinggou, west to Changshen highway, south to Renmin Road, east to Yanhe, among which the key control research area was 1.92 km^2.

2.2. Research Methods

Based on ArcGIS platform, the inverse skyline interface method was established to guide the process of viewpoint selection through 3D modeling. A comprehensive technical framework was constructed for height control of urban buildings based on "subtraction thinking"(Fig.2). Firstly, the space of possible viewpoints was collected extensively through primary selection. Secondly, some viewpoints were eliminated through modifying models and correcting models, so as to enhance the rationality of selected viewpoints and urban height control process.

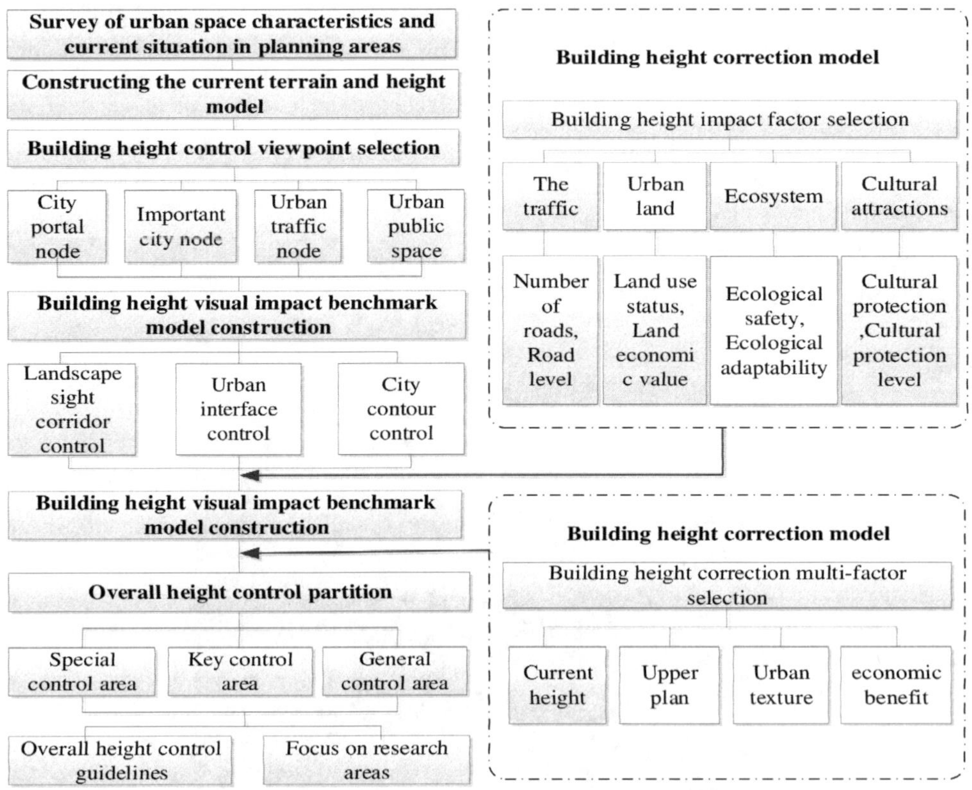

Figure 2. Urban building height control technology route

3. Construction and Modification of Urban Building Control Model in Guanyun County

3.1. Reference Model of Urban Building Height Control with Visual Impact

*3.1.1 Viewpoint selection and visual corridor analysis control.*With Dayi Mountain as the main visual landscape, the vision wasn't blocked by buildings as far as possible to create a good landscape corridor when looking at Dayi mountain from each important node within the planning scope. According to the domestic general planning standards and guidelines, the distance between each viewpoint and the mountain meets the requirement of D/H>=3, so as to relize to see 20% of the mountain scenery below the mountain contour and ensure the continuity and integrity of the city contour. The observation points must be directly facing Dayi Mountain because of the actual situation of the central urban area in Guanyun County. Considering the above factors as well as urban gateway, traffic nodes and urban public space, 33 viewpoints were initially selected.We selected the main travel data of people in four different periods including working days, weekends at 9:00am, 2:00pm, 5:00pm and 9:00pm (Fig. 3)based on Baidu thermal data, big data technology , analyzed the vitality of people in central urban areas, further guided the selection of viewpoints. Finally, 14 viewpoints were determined to construct height control reference model.

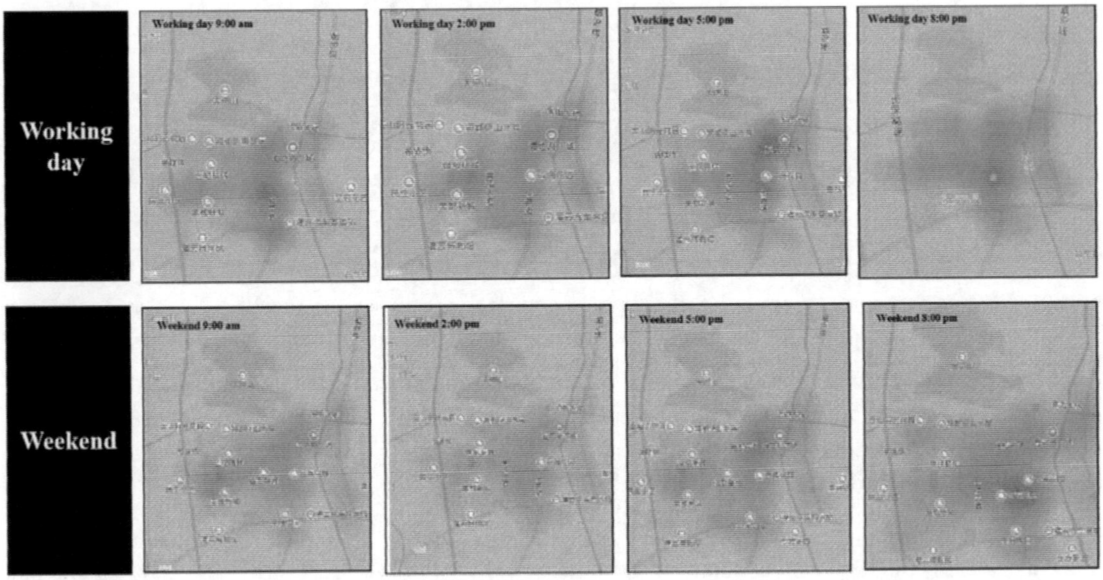

Figure 3. Population distribution of central urban population in different periods

By connecting important viewpoints in the central urban area of Guanyun County to form visual corridor, the height of buildings within the visual corridor can be controlled to form a good visual Line of landscape. Taking Dayi Mountain and buddha statue as viewing objects, combined with the overall planning and related planning of Guanyun County, 6 important viewing corridors were selected to form viewing corridor space by controlling surrounding buildings.

Firstly, applying visual psychology theories such as horizontal/vertical perspective theory and dynamic vision theory, viewing corridor was controlled comprehensively from maximization of visual sharing and the integration of citizens' dynamic visual points. Taking Dayi Mountain and buddha statue as viewing objects, the street interface requiring analysis of dynamic visual rhythm was determined by controlling dynamic visual rhythm sense of pedestrians. Secondly, according to the height of visual objects and street interface, the number of dynamic visual levels were divided to form hierarchical and rhythmic dynamic visual experience. On this basis, the view space was divided into three levels: near scenic area, middle scenic area and prospect area(Fig.4).

IOP Conf. Series: Materials Science and Engineering **730** (2020) 012021 doi:10.1088/1757-899X/730/1/012021

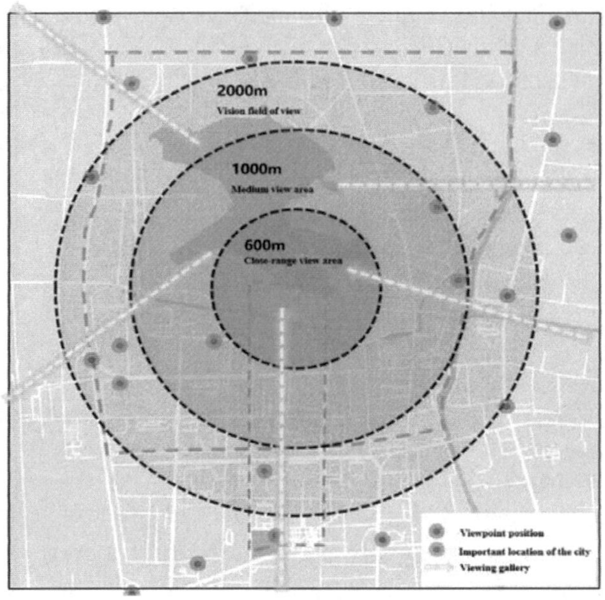

Figure 4. Distribution of important corridors and multi-dimensional control of the field of view

*3.1.2 Construction of building height control reference model.*The building height control grid of each view point was obtained by dealing with the obstacle surface of the skyline . The grid graph of terrain was converted to a point, and then to a raster. By subtracting terrain grid graph with ArcGIS grid calculator tool, the building height control grid of viewpoint 1 was obtained by using height as attribute, and the building height control grid of other viewpoints was generated similarly. Each block height limit was divided into 5 levels: 0-24m,24-40m,40-60m,60-80m,>80m. Building height control grid analyzed by sight comprehensive model was constructed through superposition calculation of multiple grid (Fig.5).

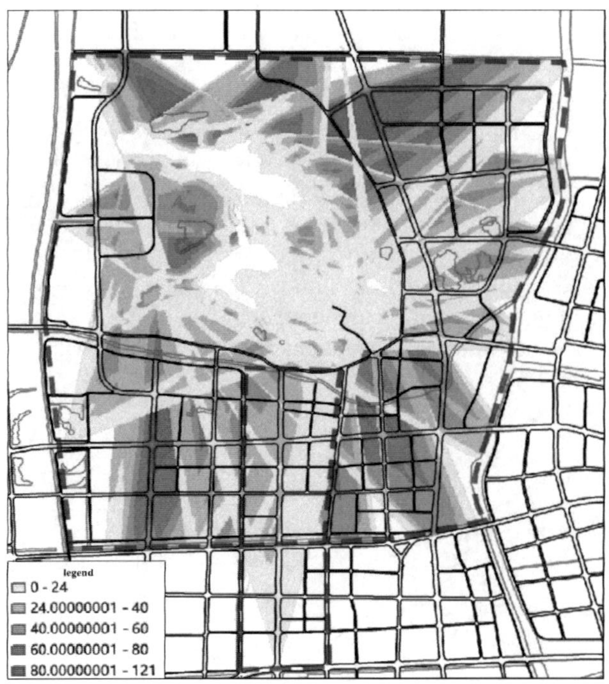

Figure 5. Building height control benchmark model

3.2. Construction of Modified Model of Building Height Control

The reference model only considering the visual corridor was insufficient for the overall control of urban building height, multi-factor analysis should be added to further supplement and modify the reference model.

*3.2.1 Comprehensive evaluation of impact factors based on analytic hierarchy process.*On the basis of relevant case studies and expert consultation and demonstration, we preliminarily constructed the framework of building height "growth force" evaluation system of 6 first-grade factors, including land location factor, traffic capacity factor, public service facilities factor, ecological environment factor ,and cultural landscape factor,and 13 sub-factors .

After classifying each factor by Delphi method, the weights of each factor were determined by analytic hierarchy process. Firstly, each factor was numbered, and the evaluation index system matrix was constructed. During multiple comparison, according to the "nine-bit" evaluation standard of the importance of the two factors, the importance of the factors was graded for several times, and finally the weight of each factor was obtained (Tab.1) . In order to avoid the error caused by the difference between before and after standards caused by human factors, the consistency test was carried out on the above evaluation results.

Table 1. Building height impact factor weight

principal factor	Weights	Secondary factor	Weights	principal factor	Weights	Secondary factor	Weights
Land location	0.28	Urban structure	0.19	Open space	0.07	Green space accessibility	0.34
		Distance from the center	0.49			Park Plaza Accessibility	0.66
		Land use	0.32				
Traffic capacity	0.16	Road convenience	0.56	ecosystem	0.16	Water system protection	0.39
		Road nature	0.44			Near mountain sensitive area	0.61
Public service facility	0.18	Educational research service	0.18	history and culture	0.15	Historical heritage protection	0.15
		Commercial financial services	0.65				
		Medical service	0.17				

Six factors were assigned to weights by applying YAAHP. The grade distribution of comprehensive evaluation of building height within the planning scope was obtained by ArcGIS multi-factor weighted overlap analysis(Fig.6). Since result of factor evaluation was the grade value, and result of sight influence model was the value of building restricted height, the original data of evaluation index was processed dimensionless and converted into specific evaluation value. Calculation results showed that the higher the evaluation coefficient, the higher the allowable height of the building; On the contrary, the smaller the evaluation coefficient, the stricter the control of the building height. Its calculation formula was as follows :

$$Z_i = \frac{X_i}{X_{il}}$$

Where, Z_i : evaluation value of X_i, X_i : actual original value,X_{il} : standard value.

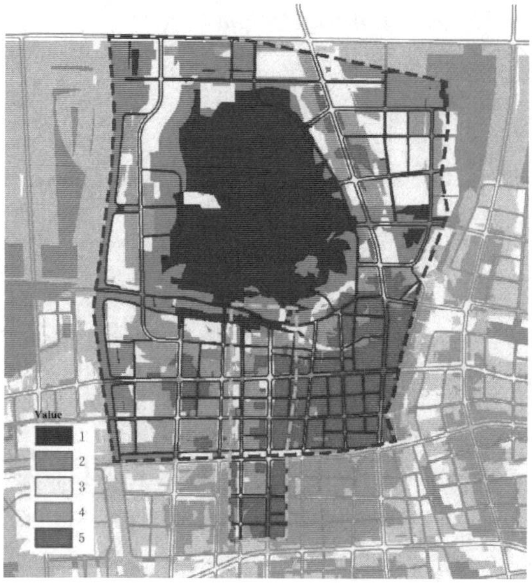

Figure 6. Multi-factor integrated overlay results

3.2.2 Construction of modified model of building height control. Applying GIS platform, the results of visual impact model and multi-factor comprehensive evaluation were given reasonable weights. After the weighted overlap analysis, the building height control distribution within the planning area was calculated .Based on the approved restricted-height of building and the upper limit of building height (100m) determined by the upper planning, the building height of each block was divided into 7 levels: 0m~15m, 15m~24m, 24m~36m, 36m~45m, 45m~60m, 60m~80m, 80m~100m.

3.3. Construction of Status Quo Correction Model of Building Height Control.

According to the correction model of current obstacle surface of each viewpoint, the height of some current buildings exceeded the obstacle surface of viewpoint(Fig.7).It mainly included high-rise residential buildings and public service buildings such as hospitals under construction on the southwest of the study area.

Figure 7. Evaluation of viewpoint obstacle surface correction model

After one round of multi-factor influence correction and two rounds of status quo correction, the results of building height control were obtained(Fig.8) .According to the building height control regulations approved by Lianyungang and Guanyun County, the building height control of each block was divided into 7 levels: 0m-15m,15m-24m,24m-36m,36m-45m,45m-60m,60m-80m,80m-100m.

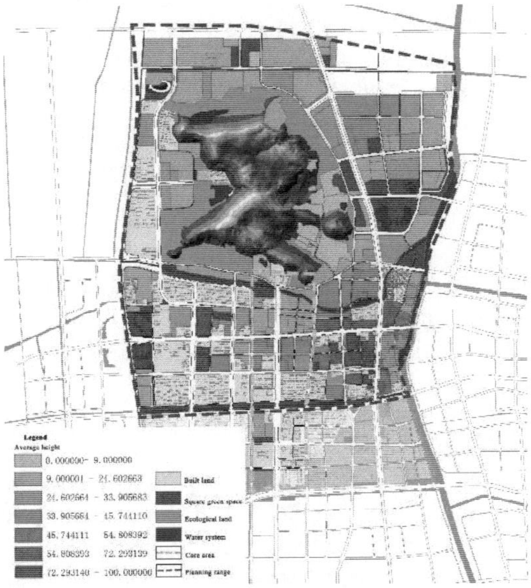

Figure 8. Urban Building Height Correction Model

4. Zoning and Control Guidelines of Overall Height Control

4.1 Zoning of Overall height control

According to construction of reference model, modified model and correction model of urban height control in planning area, the overall height control of Guanyun County was zoned to control and manage of planning.

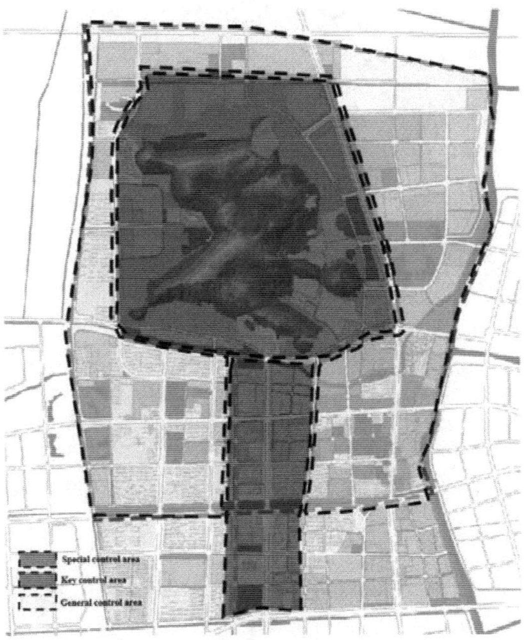

Figure 9. Overall height control partition map

The special control area included range of urban blue line and urban green line, the surrounding 100-200m was determined as ecological landscape coordination area, the 30-50m area near the Dayi

Mountain Scenic Area was designated as characteristic landscape protection area. The special control area adopted rigid control which could not exceed the upper limit.

The key control area of Guanyun County was delimited according to current development situation and relevant planning. The building height was controlled according to the functions and construction features of blocks in the key area, and it was connected with viewing the Buddha statue in Yinshan. The building height of the visual corridor in the key control area adopted rigid control and could not exceed the maximum limit. The building height outside the visual corridor adopted flexib control and the control strength could be appropriately relaxed in combination with other factors.

Other area of planning area was general control area .The building height which wasn't regulated accord with urban function and relative standard requirement. The height of most of buildings were middle-high level. Under the condition of reasonable capacity of road and municipal facilities, the high-rise buildings can be developed moderately.

4.2 Indexing of Guidelines for building height control

The control method of "point, line and surface" was adopted in the height control guidelines of block buildings in order to control urban building height scientifically and effectively.

Point control: The high-rise construction area was determined through comprehensive analysis, the height control of other regions was lower than that of planning analysis value.

Line control: To ensure a good view, the building height within the main visual corridor area was strictly controlled.

Surface control: The building height in the characteristic control area was strictly controlled. The area was dominated by multiple and low-rise buildings with no high-rise buildings to protect integrity of city style.

5. Conclusion

Based on ArcGIS technology platform, we constructed a three-level control model: reference model, modified model and correction model of urban height control . Applying the big data method, POI data of urban public services, commercial facilities, etc. and population thermal data were selected to provide quantitative support for the selection of viewpoints and visual corridor of the control model. The building height control zones were carried out by urban building control model, and the control guidelines were implemented at the level of "point, line and surface". However, this study has a certain subjectivity, such as: the viewpoints were determined subjectively, supplemented by population thermal data screening. In addition, after multi-factor evaluating, the previous round of control model needed to be re-examined to realize dynamic two-way feedback. With the development of spatial analysis technology and big data, etc. it will provide beneficial assistance and reference for the study of urban building height control planning.

6. Acknowledgments

This research was supported by the National Natural Science Foundation of China(51778002) and the Humanities and Social Sciences Project of Anhui Jianzhu University (SK2019JD11).

7. References

[1] Bagheri, H., Schmitt, M., & Zhu, X. 2019. Fusion of Multi-Sensor-Derived Heights and OSM-Derived Building Footprints for Urban 3D Reconstruction. ISPRS International Journal of Geo-Information, 8(4), 193. https://doi.org/10.3390/ijgi8040193

[2] Bender, A., Freitas, E. D., & Machado, L. A. T. 2019. The impact of future urban scenarios on a severe weather case in the metropolitan area of São Paulo. Climatic Change, 156(4), 471–488. https://doi.org/10.1007/s10584-019-02527-1

[3] Karimi, M., Nazari, R., Vant-Hull, B., & Khanbilvardi, R. 2015. Urban Heat Island Assessment with Temperature Maps Using High Resolution Datasets Measured at Street Level. International Journal of the Constructed Environment, 6(4), 17–28.

[4] Hazzard, V. M., Borton, K. A., Bauer, K. W., & Sonneville, K. R. 2018. Cross-sectional associations between gender-linked personality traits and use of weight-loss and

muscle-building products among U.S. young adults. Eating Disorders, 26(5), 418–429.

[5] Sattar, A. M. A., Elhakeem, M., Gerges, B. N., Gharabaghi, B., & Gultepe, I. 2018. Wind-Induced Air-Flow Patterns in an Urban Setting: Observations and Numerical Modeling. Pure & Applied Geophysics, 175(8), 3051–3068. https://doi.org/10.1007/s00024-018-1846-5

[6] Hongyan Cai, & Xinliang Xu. 2017. Impacts of Built-Up Area Expansion in 2D and 3D on Regional Surface Temperature. Sustainability (2071-1050), 9(10), 1862.

[7] Yang, X., & Li, Y. 2015. The impact of building density and building height heterogeneity on average urban albedo and street surface temperature. Building & Environment, 90, 146–156. https://doi.org/10.1016/j.buildenv.2015.03.037

[8] Bagheri, H., Schmitt, M., & Zhu, X. 2019. Fusion of Multi-Sensor-Derived Heights and OSM-Derived Building Footprints for Urban 3D Reconstruction. ISPRS International Journal of Geo-Information, 8(4), 193. https://doi.org/10.3390/ijgi8040193

[9] Feifei Peng, Jianya Gong, Le Wang, Huayi Wu, & Jiansi Yang. 2016. Impact of Building Heights on 3D Urban Density Estimation from Spaceborne Stereo Imagery. International Archives of the Photogrammetry, Remote Sensing & Spatial Information Sciences, 41(B3), 677–683. https://doi.org/10.5194/isprsarchives-XLI-B3-677-2016

[10] Borck, R. 2016. Will Skyscrapers Save the Planet? Building Height Limits and Urban Greenhouse Gas Emissions. Regional Science and Urban Economics, 58, 13–25. https://doi.org/http://www.sciencedirect.com/science/journal/01660462

[11] Xuebin Wei, & Xiaobai Yao. 2014. A Hybrid Gwr-Based Height Estimation Method for Building Detection in Urban Environments. ISPRS Annals of Photogrammetry, Remote Sensing & Spatial Information Sciences, 2(2), 23–29. https://doi.org/10.5194/isprsannals-II-2-23-2014

IOP Conf. Series: Materials Science and Engineering **730** (2020) 012022 doi:10.1088/1757-899X/730/1/012022

Study on Temperature Recognition of Thermal Paint Based on Color Camera

Ziqiang Hu and Ningfang Liao

School of Optoelectronics, Beijing Institute of Technology, Beijing 100081, China.
Email: liaonf@bit.edu.cn

Abstract. As a temperature-sensing material, the thermal paint will show different colors at different temperatures, so the temperature of the thermal paint can be judged by observing the color. It is an efficient way to identify the color of the thermal paint by the color image, so the calibration of the color camera is very important. The main work of this paper is to calibrate a color industrial camera and compare our proposed method with several commonly used methods. Finally, it is concluded that the proposed method combining BP neural network and maximum entropy estimation works best.

1. Introduction

In the experimental study of aero-engine, it is very important to measure the wall temperature of hot-end components (including stationary parts and rotating parts), such as turbine disk, turbine blade, casing, flame tube, power cylinder heat insulation screen and so on. Thermal paint is an essential tool for accurate analysis of high temperature materials and maximum heat resistance [1].

Thermal paint is a kind of special temperature sensing coating used to measure the surface temperature and distribution of the attached object by using the color change of the coating [2].

The principle is that when the coating is heated to a certain temperature, some physical or chemical changes occur in the heat-sensitive pigment in the coating, resulting in changes in molecular structure and molecular morphology, and even decomposition into other substances. As a result, the color of the coating film is changed to indicate the temperature.

The commonly used manual interpretation method [3] to judge the temperature by the color of thermal paint has some problems such as large analysis error, low efficiency, difficulty in quantitative analysis and so on. However, it is an efficient and convenient way to identify the color of thermal paint to judge its temperature based on color camera. Therefore, it is required that the color calibration of the color camera used to identify the color of thermal paint should be very accurate.

In this paper, a color industrial camera is calibrated by the combination of BP neural network and maximum entropy estimation (CBPNNMEE). Comparisons between the proposed method and several common calibration methods are also done. The results show that the combination of BP neural network and maximum entropy estimation is the best.

2. Color Camera Color Calibration Method

In this paper, the imaging measurement is used in the experiment. Color images of samples are taken with a color camera, and the RGB values of the samples are extracted. In fact, the RGB values measured by color cameras produced by different manufacturers are not the same, and even the products produced by the same manufacturer may not be consistent. If the RGB values measured by color cameras are not accurate, it is difficult to ensure accuracy in the process of color measurement and evaluation. Therefore, it is necessary to establish the mapping relationship between the RGB values measured by color cameras and the standard color tristimulus values (XYZ), which is the calibration process of color cameras. There are three common calibration methods, PAL algorithm, three-dimension Look-Up-Table (3D-LUT) algorithm and BP neural network algorithm.

2.1. PAL Algorithm

In this method, the mapping between the RGB color space of color camera and the standard CIE XYZ color space is processed in a linear way, that is, there exists the following matrix conversion relationship between them.

$$\begin{bmatrix} X \\ Y \\ Z \end{bmatrix} = M * \begin{bmatrix} R_0 \\ G_0 \\ B_0 \end{bmatrix} \quad (1)$$

M is PAL color conversion matrix which is known. There is a specific conversion relationship between (R,G,B) and (R_0,G_0,B_0). The specific steps are as follows:

First, the RGB value of the camera response of the whiteboard is measured, and its camera tristimulus value X_t, Y_t, Z_t is calculated according to the PAL color conversion formula, and then the correction coefficient of the camera system can be obtained, as follows:

$$k_x = \frac{X_r}{X_t}, k_y = \frac{Y_r}{Y_t}, k_z = \frac{Z_r}{Z_t} \quad (2)$$

X_r, Y_r, Z_r is the CIE standard chromaticity value of the whiteboard.

When measuring the actual samples, the camera tristimulus values of the samples can be obtained following the same steps above, and then according to the correction coefficient, the actual tristimulus values of the samples can also be obtained.

2.2. Three-Dimension Look-Up-Table Algorithm

This algorithm realizes the spatial mapping of data by establishing a point-to-point comparison table. It uses several points in 3D space to estimate a new point. In general, the steps to realize the 3D-LUT are as follows: first, measure a large number of data pairs as data nodes to construct the lookup table, then look for the points participating in interpolation, and finally realize the data interpolation of non-nodes. The emphasis of the method is the establishment of the lookup table and the interpolation method.

When the data nodes in the three-dimensional space are uniformly distributed, we can divide the three-dimensional space into several cubes through the uniformly distributed data. At this time, the interpolation algorithms include cube interpolation, tetrahedral interpolation and so on. However, in general, the obtained data nodes are not uniformly distributed, and the Euclidean distance between the interpolation point and the node is usually adopted in interpolation method. The lookup table algorithm mentioned below is this method. Now that the tristimulus values of various color samples can be determined in PAL algorithm, maybe they have a large error. But, on this basis, we can establish the mapping relationship between these tristimulus values and the standard measurement tristimulus values, and establish the database. The method is as follows:

$$\mathrm{A} = \sum_{i=1}^{n} W_i * B_i \quad (3)$$

$$W_i = \frac{\frac{1}{r_i^2}}{\sum_{i=1}^{n} \frac{1}{r_i^2}} \quad (4)$$

r_i is the distance between the interpolation point and the neighboring nodes. B_i is the corresponding point of the neighboring nodes in the target space.

2.2.1. Maximum entropy estimation [4]. For each input point, the maximum entropy estimation can reconstruct the output points linearly in the target space by calculating the weights of the sampled points within its local neighborhood. In the color space, each input point is assumed to be in a convex hull composed of sampled points, which means that the weight sum is one. Therefore, the problem of solving weights is equivalent to determining the most possible allocation problem under the constraint of known average expectation. Under the constraint of known expectation, the most probable distribution of the maximum entropy is the distribution of the maximum probability. Therefore, the maximum entropy estimation weight is the maximum possible weight. The maximum entropy estimation algorithm is as follows:

Let the sampled points be the pairs$\{x, f(x)\}$. Given a new point y, $f(y)$ is estimated by using the maximum entropy estimation.

Step 1 Taking the input point y as the center, the adjacent k points are selected to form a convex hull.

Step 2 Calculate the weights w_i:

$$\max(-\sum_{i\in k} w_i * ln w_i) \quad (5)$$

subject to the constraints:

$$\sum_{i\in k} w_i * x_i = y \quad (6)$$

$$\sum_{i\in k} w_i = 1 \quad (7)$$

$$w_i \geq 0 \quad (8)$$

Step 3 Form the estimate $f(y)$ as:

$$f(y) = \sum_{i\in k} w_i * f(x_i) \quad (9)$$

2.3. BP Neural Network Algorithm

The error back propagation neural network has the ability to realize nonlinear mapping. By measuring an appropriate amount of samples as a training set and adjusting the training parameters of the network, the mapping relationship between input and output can be obtained. Typical diagram is as follows:

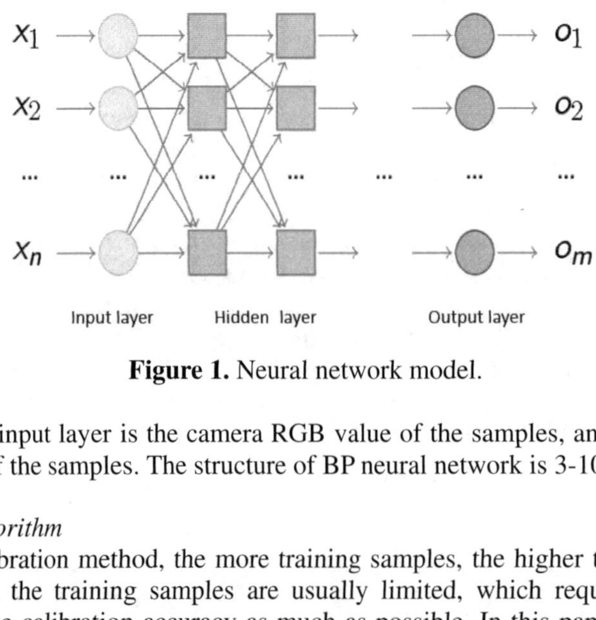

Figure 1. Neural network model.

For this paper, the input layer is the camera RGB value of the samples, and the output layer is the standard XYZ value of the samples. The structure of BP neural network is 3-10-10-10-3.

2.4. CBPNNMEE Algorithm

Regardless of the calibration method, the more training samples, the higher the calibration accuracy. However, in practice, the training samples are usually limited, which requires the use of limited samples to improve the calibration accuracy as much as possible. In this paper, a method combining BP neural network with maximum entropy estimation is proposed. Firstly, a certain number of training sample sets are measured, and the corresponding mapping relationship between samples is determined by BP neural network. Then the RGB color space of the camera is divided into 27*27*27 data nodes to expand the samples. The trained network is used to determine the values of the target space corresponding to 27*27*27 data nodes. Finally, the maximum entropy estimation method is used as the interpolation method to obtain the values of the target space corresponding to the unknown points.

3. Calibration Experiment

3.1. Experimental Instrument
Spectral radiometer (PR715), SG color card, NCS color card, light box, camera, whiteboard

3.2. Experimental Method and Process
- Place the SG color card in the light box and turn on the D65 light source.
- Place the camera at the distance of the light box 50cm, and the geometric conditions of lighting and measurement are 45°/ 0°.
- Turn on the computer and connect to the PR715.
- Turn on the camera imaging software, adjust the camera aperture, exposure time and other parameters to make the image clear, the RGB value of the whiteboard part is around 230.
- Shoot the color card with the camera and store pictures.
- Replace SG color card with NCS color card, use the camera to shoot the color card, store the picture.
- Each color block of the two color cards is measured by PR715 in the same measurement environment as the camera, and the XYZ data obtained is reserved.

Figure 2. Shooting of SG color card.

3.3. Data Processing and Analysis
The obtained images are processed by the image processing software, and the RGB values of each color block are extracted. And then, the correspondence table based on the SG color card is established.

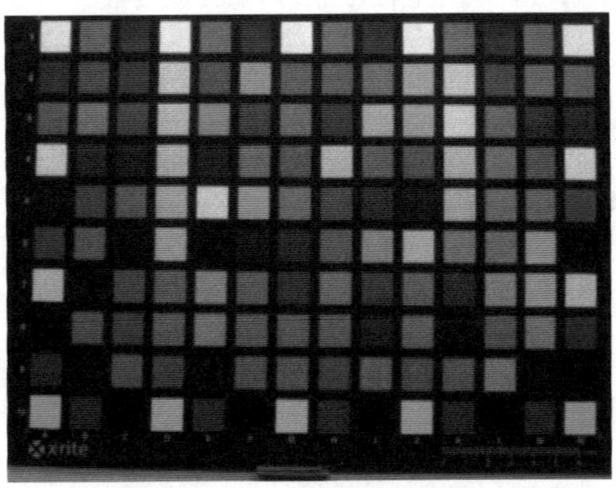

Figure 3. Picture of SG color card.

Based on the calibration methods mentioned above, the corresponding XYZ values of 900 NCS blocks selected from NCS color card are calculated. The color differences between the XYZ values measured by PR715 and the XYZ values calculated are calculated in CIE1976 L*a*b* uniform color space. The results can be seen in Table 1.

Table 1. Comparison of color differences between calibration methods.

Error	Mean	Maximum
PAL	18.06	29.31
LUT	4.85	11.13
BP neural network	2.68	6.91
Maximum Entropy Estimation	3.85	6.54
CBPNNMEE	1.86	5.93

As can be seen from the table, PAL calculation method has the worst calibration accuracy, which indicates that the linearity between the RGB value of this camera and the standard XYZ value is not high. The performance of the lookup table based on Euclidean distance is not bad, and this method is adequate for the camera with low precision requirements. The BP neural network method and the maximum entropy estimation perform well, indicating that these two methods have good ability for nonlinear mapping. The method of combining BP neural network with the maximum entropy estimation has the best performance, which shows that the strong nonlinear mapping ability and sufficient samples play a key role in improving the calibration accuracy.

4. Temperature Recognition of Thermal Paint [5]

4.1. Fit the Color Temperature Characteristic Curve Function of Thermal Paint

Color-temperature database is established by photographing the thermal paint samples with the calibrated color camera. The color is represented by *L, a, b*. Regardless of the correlation between the *L, a* and *b* components, and assuming that the values of the respective color components are functions of temperature *t*, denoted as *L(t), a(t), b(t)*. The color temperature characteristic curve function of thermal paint is fitted with a cubic spline interpolation method.

4.2. Temperature Recognition Algorithm

For the temperature of the unknown point, the Euclidean distance between it and any point on the color temperature characteristic curve of thermal paint is calculated in the CIE1976 L*a*b* color space. Denote the Euclidean distance by ΔE.

$$\Delta E = [(L(t) - L)^2 + (a(t) - a)^2 + (b(t) - b)^2]^{1/2} \tag{10}$$

The above formula is solved in the temperature measuring range of t thermal paint to find the *t* value that minimizes the ΔE. The *t* value is the corresponding temperature value on the thermal paint picture.

5. Conclusion

There is a corresponding relationship between the temperature of thermal paint and the color. The color of the thermal paint can be obtained through imaging, and then the temperature can be known. Therefore, this paper focuses on the camera color calibration, and the proposed method combining BP neural network with maximum entropy estimation has satisfactory results. In the future work, it is necessary to study new techniques and methods to improve the accuracy of camera color calibration and temperature identification of thermal paint.

6. Acknowledgments

This research work has been supported by the project of National Natural Science Foundation of China. The project is construction of the BRDF of representative water surface and its image color appearance prediction (61975012). I would like to express my thanks again.

7. References

[1] Lin MS and Chen NN 2006 Research on Color Quantization and Image Segmentation of Thermal Paint *J. Journal of Southwest University of Science and Technology* **21(2)** 48-53.

[2] Liu Z K, Ge J F and Zhang Y F 2015 Study on the automatic temperature recognition technology of thermal paint *J. China Measurement & Test* **41(9)** 20-23.

[3] Zhang Z L and Cao C T 2001 Thermopaint Color-Temperature Property Analysis and Temperature-Recognition System *J. Computer Measurement & Control* **9(3)** 20-21.

[4] Gupta M R and Gray R M 2001 Color conversions using maximum entropy estimation *C. 2001 International Conference on. IEEE* p 119.

[5] Ma C W, Jiang B and Chen F Y 2007 Automatic Temperature Interpretation and Digital Image Processing Systems of Temperature-Sensitive Paints *J. Aeroengine* **(02)** 15-18.

IOP Conf. Series: Materials Science and Engineering **730** (2020) 012023 doi:10.1088/1757-899X/730/1/012023

A Preliminary Study on Hakka Characteristic Soup Plant Resources in Eastern Guangdong

Qihe Yang, Lubing Li, Hesheng Yang, Lihui Mou and Wei Kuang

Jiaying University, Meizhou Guangdong 514015, China
Email: yangqh@jyu.edu.cn

Abstract. In this paper, we investigated and analyzed the Hakka characteristic soup plants in eastern Guangdong. The results showed that there were 79 soup plants , belonging to 74 genera of 44 families, 37 species are wild and have not yet been cultivated artificially, 11 species are cultivated and there is no wild distribution, 31 species are both wild and planted, accounting for 39.2% of the total species. The species used to cook soup with the whole plant were the most, 38 species, accounting for 48.1% of the total soup plants, and those cooked only with root ranked second, 24 species, accounting for 30.4%. 29 species can be harvested throughout the year. Based on the analysis of growth types, there were 37 herbs, 28 shrubs, 7 vines and 7 trees. 68 species are naturally distributed, indicating that the local Hakkas mainly use soup plant materials locally. There are about 20 plants which are widely distributed and seen easily in the field. These genera belong to 11 distribution types. Among them, there are 52 genera of tropical distribution types in total. Among these plants, 53 species were only distributed in the south of the Yangtze River, accounting for 67.1%, 46 species were shade-tolerant, accounting for 58.2%. 37 species have the effect of clearing heat, accounting for 46.8%, 25 species can promote blood circulation and remove blood stasis, accounting for 31.6%.

1. Introduction

As an important part of local residents in Guangdong, Fujian, Jiangxi and Taiwan Province, Hakka is a Han ethnic group with Hakka language as the native language. As an ancient Han immigrants group in South China, Hakka is one of the ethnic groups with wide distribution and far-reaching influence in the world. It is estimated that there are about 80 million Hakkas in the world, of whom about 50 million are distributed in more than 180 counties (including county-level cities) in 19 provinces, such as Guangdong, Jiangxi, Fujian, Guangxi, Sichuan, Hunan and Hainan, etc.

In old times, the Hakka area was surrounded by mountains on all sides, and the transportation was inconvenient. In terms of diet, only local materials were obtained, and the local resources in the mountains were fully excavated for domestic livestock and wild animals and plants in the mountains which grew in high mountains and deep valleys, and wild vegetables in the field often became the original materials of food[1-2]. The local, natural and convenient food has created the "wild, vegetarian, coarse, miscellaneous" style of Hakka dishes, and also endows them with the green effect of health care. Hakkas ardently love soup, and almost believe that "soup material is the soul of soup", "it is better to eat without meat than to eat without soup". No matter rich and poor, soup is essential for them in daily life, and they pay attention to it. Whether or not to cook delicious and nutritious soup is also a criterion for evaluating whether women at home are virtuous or not. Soup is not only an important supplement to daily diet, making good use of medicinal materials to make soup, but also can strengthen the body and prevent and cure all kinds of diseases [3-4]. Soup is not only an important supplement to daily diet, but making good use of medicinal materials to make soup also can strengthen the body and prevent and cure all kinds of diseases [3-4].

In the 1930s, there was already a saying about "homology of medicine and food" in China. With the rise of health-preserving and health-care, in the 1980's, some domestic nutritionists began to put forward some views on "the same origin of medicine and food". In fact, Hakka ancestors had been applied in the life practice for hundreds of years ago [5-6], and they added some herbs with certain health-care functions and cooked the medicinal and dietary soup. The adverse effects of climate on the health of the human body had been conditioned in their daily diet. With the improvement of living standard, the traditional health-keeping method is more and more important to the present day, and the meal soup based on folk herbs folk herbal medicine is more and more popular [7]. Hakkas attach great importance to disease prevention and believe the so-called "way to maintain health is not ahead of diet." So usually, they eat different medicinal soup according to the season.

In this paper, soup plants refer to plants whose roots, stems, leaves, flowers, fruits or whole plants are often used to cook soup with animal meat and bones. Therefore, tea, honeysuckle and others cooked only with water, without adding animal meat or bone, are beverage plants, not soup plants. The Hakka characteristic soup plant refers to the soup plant used in cooking by Hakkas, which are wildly distributed or small-area cultivated in Hakka area, but do not include peanut, soybean, urad, sweet potato, corn, radish, Chinese cabbage, white gourd, towel gourd, pumpkin, balsam pear, tomatoes, Chinese yam, lotus, kelp, laver, lily, day lily, etc., which are widely used as main crops or soup materials in most parts of China, even the whole of China and the world. The *Cassia* and *Cinnamon* species and the like which are more frequently used by other nationalities and Han groups as the auxiliary materials, and rarely used for the soup are not included.

The Hakka area in eastern Guangdong mainly refers to Meizhou City, which is one of the most important places of origin and concentration of Hakkas, known as "the Capital of Hakkas". Meixian, Fengshun, Jiaoling and Dapu County under the jurisdiction of Meizhou City have been selected as "the Hometown of Longevity in the World", accounting for half of the selected counties and cities in Guangdong Province [8]. Understanding the species and characteristics of soup plants in this area is not only conducive to their development and utilization, but also brings inspiration to people's health care in other areas.

2. General Situation of the Study Area and the Ethnobotanical Survey

Meizhou City is a prefecture-level city under the jurisdiction of Guangdong Province, located in Northeast Guangdong Province, and at the junction of Fujian, Guangdong and Jiangxi provinces. The geographical coordinates are between 115°18'-116°56' E and 23°23'-24°56' N. It belongs to the subtropical monsoon climate zone and is the transitional zone of the subtropical and tropical climatic zone. The sunshine is sufficient, the average annual temperature is 20.6°C ~21.4°C, the frost-free period has exceeded over 300d, and annual rainfall is 1483~1798mm.

From 2016 to 2018, this study was carried out using ethnobotanical methods, on the basis of literature review, combined with field and questionnaire survey, folk visits, specimen collection and identification and post-collation and analysis, and so on. The interviewees included vegetable farmers, medicinal herb farmers, folk grass doctors, drugstore owners, Chinese herbal medicine operators, restaurant chefs, TCM veteran practioner, ordinary residents, and so on. In the course of the interview, the "5W+1H" method was used to ask questions, that is, what, when, where, who, why and how much [9]. Such as "What's the name of the soup plant? How to collect it? When to harvest? Which parts to pick up? What are the medicinal or health effects? " The survey was recorded, and the plant specimens were taken digital photos and preserved as wax leaf specimens. By consulting the relevant data and the relevant plant classification experts, the family and species name, growth type, material position and medicinal or health care effect were sorted out.

3. Result

3.1. Specific Composition

The investigation showed that there were 79 Hakka characteristic soup plants in eastern Guangdong, belonging to 74 genera, 44 families. Among them, there were 8 Leguminosae species, 7 Compositae species, 4 species for every family of Labiataceae, Moraceae and Rubiaceae, 3 species for Cactaceae,

Euphorbiaceae and Rosaceae, 2 species for Liliaceae, Portulacaceae, Campanulaceae, Acanthaceae, Myrtaceae and Orchidaceae, only 1 species for Ericaceae, Lindeaceae, Annonaceae, Menispermaceae, Poaceae, Elaeagnaceae, Cucurbitaceae, Apocynaceae, Solanaceae, Thymelaeaceae, Saururaceae, Apiaceae, Symplocaceae, Pteridaceae, Rhamnaceae, Clusiaceae, Sapindaceae, Araliaceae, Amaranaceae, Scrpophulariaceae, Convolvulaceae, Melastomataceae, Polygalaceae, Myrsinaceae, Magnoliaceae, Polygonaceae, Bombacaceae, Aquifoliaceae, Valerianaceae and Plantaginaceae (Table 1) .

Among these 79 plants, 37 species are wild and have not yet been planted artificially, 11 species are cultivated and there is no natural or wild distribution in eastern Guangdong, 31 species are both naturally distributed and cultivated, accounting for 39.2% of the total soup plants (Table 1). Of these soup plants, 68 species are naturally distributed in eastern Guangdong, the proportion is as high as 86.1%, indicating that the local Hakkas mainly use soup plant materials locally, and 45.6% of these natural distribution species have been cultivated artificially.

In 11 cultivated species (only artificially cultivated, there is little natural distribution in the wild) are *Hylocereus undatus*, *Epiphyllum oxypetalum*, *Opuntia stricta*, *Psidium guajava*, *Bombax malabaricum*, *Eriobotrya japonica*, *Vigna umbellata*, *Uraria crinita*, *Dimocarpus longan*, *Ixora chinensis* and *Ocimum basilicum*.

6 soup plants, such as *Epiphyllum oxypetalum*, *Opuntia stricta*, *Psidium guajava*, *Cassia mimosoides*, *Talinum paniculatum*, *Hylocereus undatus*, accounting for 6.3% of the total soup plants counted in this paper, are alien species and native to South America and India, but after domestication and breeding, they have become naturalized plants in eastern Guangdong. Once, Hakkas were used to using *Alsophila spinulosa* and *Dendrobium sp.* as soup materials, but after these plants were listed as protective plants, they almost stopped eating them and preferred to use common and easy-to-grow soup plants.

3.2. Edible Parts and Harvesting Period

Of these 79 soup plants, the species cooked soup with whole grass (or whole plant) were the most, 38 species, accounting for 48.1% of the total number of soup plants, such as *Pholidota chinensis*, *Plantago depressa*, *Artemisia lactiflora*, *Andrographis paniculata*, *Portulaca oleracea*, *Houttuynia cordata, Centella asiatica*, *Lobelia chinensis*, *Artemisia capillaris*, *Gynura divaricata*, *Hypericum japonicum*, *Gnaphalium affine*, *Euphorbia thymifolia*, *Inula cappa*, *Elephantopus scaber*, *Rabdosia serra*, *Emilia sonchifolia*, *Leonurus artemisia*, *Mimosa cassia*, *Lycianthes biflora*, etc. Those cooked only with root ranked second, 24 species, accounting for 30.4%, such as *Fissistigma oldhamii*, *Ixora chinensis*, *Flueggea virosa*, *Ilex asprella*, *Ardisia gigantifolia*, *Millettia speciosa*, *Berchemia lineata*, *Imperata cylindrica*, etc. There were 4 species cooked only with flower, accounting for 5.1%, including *Bombax malabaricum*, *Hylocereus undatus*, *Epiphyllum oxypetalum* and *Eriobotrya japonica*, 2 species only with fruit, including *Rosa laevigata* and *Dimocarpus longan*, 2 species only with stem, including *Nephrolepis auriculata* and *Opuntia stricta*. *Vigna umbellata* is a soup plant cooked with only its seed. In addition, some plants have multiple parts that can be used as soup materials, such as root and fruit of *Gardenia jasminoides*, stem and leaf of *Tadehagi triquetrum* and *Agrimonia pilosa*, stem and leaf of *Schisandra lancifolia* and *Tinospora sinensis*, fruit and leaf of *Ficus pumila*, and leaf and fruit of *Psidium guajava* (Table 1).

From the harvest time, there were 29 species that can be harvested throughout the year, such as *Aralia chinensis*, *Ardisia gigantifolia*, *Fissistigma oldhamii*, *Flueggea virosa*, *Ilex asprella*, *Ixora chinensis*, gardenia, cactus, tuber fern, *Dicliptera chinensis*, *Abrus cantoniensis*, stone olive, *Gaultheria yunnanensis*, *Alyxia fascicularis*, etc. 7 species, *Artemisia argyi*, *Hedyotis diffusa*, *Artemisia lactiflora*, *Artemisia capillaris*, *Plantago depressa*, *Dichondra repens* and *Gnaphalium affine*, were harvested in spring and summer. 2 soup plants, *Andrographis paniculata* and *Flemingia philippinensis*, in autumn and winter. 4 soup plants, *Glochidion puberum*, *Gynostemma pentaphyllum*, *Corchorus acutangulus* and *Vigna umbellata*, were harvested only in autumn. 14 plants were harvested in summer and autumn, including *Hypericum japonicum*, *Ardisia gigantifolia*, *Ficus pumila*, *Hylocereus undatus*, *Agrimonia pilosa*, *Tadehagi triquetrum*, *Houttuynia cordata*, *Melastoma dodecandrum*, *Cassia mimosoides*, *Desmodium caudatum*, *Gynura divaricata*, *Lobelia chinensis*,

IOP Conf. Series: Materials Science and Engineering **730** (2020) 012023 doi:10.1088/1757-899X/730/1/012023

Lycianthes biflora and *Rabdosia serra*. 9 soup plants were harvested in spring, summer and autumn, they were *Patrinia villosa, Leonurus artemisia, Talinum paniculatum, Polygala fallax, Amaranthus spinosus, Emilia sonchifolia, Ocimum basilicum, Portulaca oleracea* and *Elephantopus scaber* (Table 1). However, the harvest period is not completely fixed, even in eastern Guangdong, if some soup plants are in a special habitat, their harvest period will be different. For example, in the humid gullies of Wuhua and Fengshun County in Meizhou, *Patrinia villosa* and *Elephantopus scaber* can be harvested almost all year round, but in higher and drier areas, they are often impossible to collect in winter.

3.3. Analysis on Utilization and Reserve

According to the utilization frequency, dosage and cultivation scale of residents, these soup plants are divided into four categories: A (most commonly used), B (more commonly used), C (generally used) and D (occasionally used). A refers to 90% of the Hakka residents who have used and have used at least three times a year, and men, women and children like to use them as raw materials to cook soup, and there are 21 species, such as *Patrinia villosa, Leonurus artemisia, Talinum paniculatum, Amaranthus spinosus, Emilia sonchifolia, Ocimum basilicum, Artemisia argyi, Eriobotrya japonica, Gynostemma pentaphyllum, Vigna umbellata, Wikstroemia indica, Rhodomyrtus tomentosa, Ficus simplicissima, Smilax china, Smilax glabra, Agrimonia pilosa, Cassia mimosoides, Desmodium caudatum, Lycianthes biflora, Rabdosia serra*, and *Paederia scandens*. B refers to 70% of the Hakka residents have used and at least once a year, there are 19 species, such as *Elephantopus scaber, Portulaca olerac*ea, *Rosa laevigata, Hedyotis diffusa, Artemisia lactiflora, Plantago depressa, Dichondra repens, Psidium guajava, Andrographis paniculata, Flemingia philippinensis, Millettia specisoa, Ficus hispida, Ficus stenophylla, Berchemia lineata, Symplocos* laurina, *Imperata* cylindrica, Uraria *crinita, Tinospora sinensis, Abrus cantoniensis, Pholidota chinensis, Alyxia fascicularis, Gardenia jasminoides, Dimocarpus longan, Bombax malabaricum, Hylocereus undatus, Houttuynia cordata, Goodyera hachijoensis, Centella asiatica*(Table 1). C refers to 50% of the residents who have used, but on average less than once a year, there are 15 species, such as *Epiphyllum oxypetalum, Inula cappa, Artemisia capillaris, Gnaphalium affine, Glochidion puberum, Elaeagnus pungens, Fallopia multiflora, Nephrolepis auriculata, Campanumoea javanica, Hypericum japonicum, Tadehagi triquetrum, Melastoma dodecandrum, Gynura divaricata, Lobelia chinensis, Euphorbia thymifolia*. D refers to less than half of the residents have used it, most of them have only occasionally drank it, such as *Polygala fallax, Corchorus acutangulus, Fissistigma oldhami, Flueggea virosa, Ilex asprella, Aralia chinensis, Ixora chinensis, Opuntia stricta, Schisandra lancifolia, Dicliptera chinensis, Gaultheria yunnanensis, Ardisia gigantifolia, Ficus pumila* and *Scutellaria barbata* (Table 1).

According to the survey and visit, the reserves of these soup plants can be divided into at least four grades from more to less: A (a great many), B (many), C (few) and D (very few). The so-called Grade A refers to those are widely distributed and can be seen easily in the field, such as *Paederia scandens, Rhodomyrtus tomentosa, Ficus simplicissima, Portulaca oleracea, Artemisia argyi, Plantago depressa, Fallopia multiflora, Emilia sonchifolia, Nephrolepis auriculata, Glochidion puberum, Melastoma dodecandrum*, etc. There are about 20 species of this grade, accounting for 25%. Grade B refers to those are widely distributed, their quantity is also large, but they often has certain habitat limitations, such as *Houttuynia cordata* is often distributed in wetter paddy fields and ditches, and is mostly planted in the shade and wet places around the house, *Leonurus artemisia* is often planted in the wet and fertile areas around of the house or in the vegetable field, although the number of planting for each family is not small, but almost all Hakka families plant, so its quantity is also very large. *Artemisia lactiflora, Gynura divaricata, Patrinia villosa, Amaranthus spinosus, Ficus pumila, Millettia specisoa, Lycianthes biflora, Hypericum japonicum, Artemisia capillaris, Gardenia jasminoides, Smilax glabra*, Ilex asprella, *Fissistigma oldhami*, Eriobotrya japonica, Opuntia stricta, Dichondra repens, etc., also belong to this grade, and there are 26 species, accounting for 23.0%. Grade C refers to those are planted by some residents, although their application amount is not very large, the overall quantity is still not very small. *Lobelia chinensis, Scutellaria barbata, Abrus cantoniensis, Cassia mimosoides, Elaeagnus pungens, Berchemia lineata, Rabdosia serra, Agrimonia pilosa, Desmodium caudatum, Inula cappa, Gaultheria yunnanensis, Tadehagi triquetrum, Symplocos laurina, Flemingia philippinensis*, *Tinospora sinensis*, *Vigna umbellata*, *Psidium guajava*, *Vigna umbellata*, etc, 31 species, all belong to this grade, accounting for 27.4%. Grade D refers to those are less abundant and less common, for many reasons, either because their natural population distribution is less and they has not yet been cultivated artificially, or the serious artificial extraction results in a shortage of resources, or because the local residents rarely cultivate, so their cultivation amount is less. The soup plants belonging to this category include are *Campanumoea javanica*, *Talinum paniculatum*, *Pholidota chinensis*, *Polygala fallax*, *Wikstroemia indica*, *Flueggea virosa*, *Ardisia gigantifolia*, *Striga asiatica*, *Ficus hispida*, *Dicliptera chinensis*, *Corchorus acutangulus*, *Alyxia fascicularis*, *Schisandra lancifolia*, *Aralia chinensis*, *Ficus stenophylla*, *Hylocereus undatus*, *Uraria crinita*, *Epiphyllum oxypetalum*, Goodyera hachijoensis(Table 1).

3.4. Ecological Habit

Based on the analysis of the growth types of 79 soup plants, there were 37 herbs (21 perennial herbs, 16 species of annual and biennial herbs), accounting for 46.8%, 28 shrubs, 7 vines and 7 trees (Table 1). There are about 53 species only distributed in the south of the Yangtze River, including Guangdong, Guangxi, Qiong, Fujian, Zhejiang, Jiangxi, Hunan, Yunnan, Anhui and other provinces, accounting for 67.1% of the total soup plants, such as *Wikstroemia indica*, *Flueggea virosa*, *Gaultheria yunnanensis*, *Ardisia gigantifolia*, *Symplocos laurina*, Ixora *chinensis*, *Desmodium caudatum*, *Fissistigma oldhami*, *Millettia specisoa*, *Berchemia lineata*, *Dimocarpus longan*, *Eriobotrya japonica*, *Elaeagnus pungens*, *Inula cappa*, *Gynura divaricata*, *Lobelia chinensis*, *Elephantopus scaber*, etc.

Among which 6 species, *Dimocarpus longan*, *Gynura divaricata*, *Dicliptera chinensis*, *Abrus cantoniensis*, *Hypericum japonicum*, *Euphorbia thymifolia* are only distributed in Lingnan area and Southern Yunnan Province. 21 species, accounting for 26.6% of the total species, were distributied in the north and south of China, such as *Artemisia lactiflora*, *Imperata cylindrica*, *Patrinia villosa*, *Scutellaria* barbata, *Amaranthus spinosus*, *Rabdosia serra*, *Agrimonia pilosa*, *Artemisia capillaris*, *Fallopia multiflora*, *Tinospora* sinensis, *Smilax china*, etc. There are 6 alien plants, 5 species from South America, which are *Opuntia stricta*, *Cassia mimosoides*, *Hylocereus undatus*, *Epiphyllum oxypetalum*, *Talinum paniculatum* and 1 species from tropical Asia, which is *Vigna umbellata*.

According to the adaptability to light intensity, these plants can be divided into 3 categories: positive plants, negative plants and shade-tolerant. Among them, 46 species were shade-tolerant, accounting for 58.2%, such as *Gynura divaricata*, *Dicliptera chinensis*, *Hypericum japonicum*, *Euphorbia thymifolia*, *Wikstroemia indica*, *Flueggea virosa*, *Gaultheria yunnanensis*, *Ixora chinensis*, *Desmodium caudatum*, *Fissistigma oldhami*, *Berchemia lineata*, *Elaeagnus pungens*, *Inula cappa*, *Lobelia chinensis*, *Centella asiatica*, *Nephrolepis auriculata*, *Houttuynia cordata*, *Paederia scandens*, *Dichondra repens*, *Glochidion puberum*, *Polygala fallax*, *Lycianthes biflora*, *Ilex asprella*, *Schisandra lancifolia*, etc.

Followed by positive plants, 27 species, accounting for 34.2%, such as *Corchorus acutangulus*, *Dimocarpus longan*, *Abrus cantoniensis*, *Rhodomyrtus tomentosa*, *Gardenia jasminoides*, *Bombax malabaricum*, *Smilax glabra*, *Psidium guajava*, *Artemisia argyi*, *Imperata cylindrica*, *Amaranthus spinosus*, *Rosa laevigata*, *Portulaca oleracea*, etc. However, there were only 6 negative plants, accounting for 7.6%, which were *Vigna umbellata*, *Ardisia gigantifolia*, *Goodyera hachijoensis*, *Pholidota chinensis*, *Campanumoea javanica*, *Artemisia lactiflora* and *Talinum paniculatum*.

Table 1. List of Hakka characteristic soup plants and their edible parts, source, usage status, health function, harvesting period in Eastern Guangdong

NO.	Latin name	Family	Edible parts	Growth type	Source	Usage status	Health function	Harvesting period
1	*Nephrolepis auriculata*	Nephrolepidaceae	Tuberous stem	Perennial herb	Both	C	Clearing heat, dampness-removing, moistening lung and relieving cough, softening hard masses and eliminating accumulation	All year round
2	*Schisandra lancifolia*	Magnoliaceae	Stem root	Deciduous woody vine	Wild	D	Astringent hemostasis, removing stasis and detumescence	All year round
3	*Fissistigma oldhami*	Annonaceae	Root	Evergreen woody vine	Wild	D	Dispelling wind and dampness, promoting menstruation, activating blood circulation and hemostasis	All year round
4	*Tinospora sinensis*	Menispermaceae	Stem root	Perennial trailing shrub	Wild	B	Relaxing muscles and activating collaterals, dispelling wind and relieving pain	All year round
5	*Houttuynia cordata*	Saururaceae	Whole grass	Perennial herb	Both	B	Clearing heatand detoxifying, diuresis and drenching and anti-inflammatory	Summer-autumn
6	*Polygala fallax*	Polygalaceae	Root, stem and leaf	Shrub or small arbor	Wild	D	Tonifying spleen,removing dampness, promoting blood circulation and regulating meridians, tonifying Qi and blood	Apr.-Oct.
7	*Talinum paniculatum*	Portulacaceae	Root	Annual herb	Both	A	Invigorating spleen and tonifying Qi, tonifying lung and relieving cough, promoting lactation and relieving asthma	Apr.-Oct.
8	*Portulaca oleracea*	Portulacaceae	Whole grass	Annual herb	Both	B	Clearing heatand detoxifying, diuresis and dysentery, moistening intestines and eliminating stagnation, cooling blood and hemostasis	Apr.-Sept.
9	*Fallopia multiflora*	Polygonaceae	Tuberous root	Perennially entangled vine	Wild	C	Tonifying kidney and essence, strengthening muscles and bones, resolving turbidity and reducing dampness	All year round
10	*Amaranthus spinosus*	Amarantaceae	Whole grass	Perennial herb	Both	A	cooling blood and hemostasis;dampness-removing;removing toxicity and eliminating carbuncles	Apr.-Oct.r
11	*Wikstroemia indica*	Thymelaeaceae	Root	Evergreen shrub	Wild	A	Antiinflammatory and detoxifying, dispersing stasis and expelling water	All year round
12	*Gynostemma pentaphyllum*	Cucurbitaceae	Whole grass	Perennial climbing herb	Both	A	Relieving cough and dispelling phlegm, clearing heat and detoxifying, strengthening with tonics	Autumn
13	*Hylocereus undatus*	Cactaceae	Flower	Epiphytic fleshy shrub	Cultivated	B	Relaxing muscles and activating collaterals, detoxification and detumescence	Summer-autumn
14	*Epiphyllum oxypetalum*	Cactaceae	Flower	Epiphytic fleshy shrub	Cultivated	C	Clearing heat and moistening lung, relieving cough and eliminating phlegm, nourishing yin and tonifying deficiency	June-Oct.
15	*Opuntia stricta*	Cactaceae	Stem	Tufted fleshy shrub	Cultivated	D	Activating qi and activating blood circulation, cooling blood and hemostasis, detoxification and detumescence	All year round
16	*Psidium guajava*	Myrtaceae	Leaf, fruit	Arbor	Cultivated	B	astringing to stop diarrhea, anti-inflammatory and hemostasis	Spring-autumn
17	*Rhodomyrtus tomentosa*	Myrtaceae	Root	Shrub	Wild	A	Expelling wind and activating collaterals, astringent and relieving diarrhea, tonifying deficiency and hemostasis	All year round
18	*Melastoma dodecandrum*	Melastomataceae	Whole grass	Creeping shrub	Both	C	Clearing heatand detoxifying, dispelling wind, dampness-removing, enriching blood and hemostasis	Summer-autumn
19	*Hypericum japonicum*	Clusiaceae	Whole grass	Annual herb	Both	C	Clearing heatand detoxifying, dampness-removing, removing jaundice, , detumescence and dispersing stasis	Summer-autumn

IOP Conf. Series: Materials Science and Engineering **730** (2020) 012023 doi:10.1088/1757-899X/730/1/012023

Continued Table 1. List of Hakka characteristic soup plants and their edible parts, source, usage status, health function, harvesting period in Eastern Guangdong

NO.	Latin name	Family	Edible parts	Growth type	Source	Usage status	Health function	Harvesting period
20	*Corchorus acutangulus*	Tiliaceae	Whole grass	Annual herb	Wild	D	Clearing heatand detoxifying	
21	*Bombax malabaricum*	Bombacaceae	Flower	Deciduous arbor	Cultivated	B	Heat-clearingand detoxifying, diuresis, dispelling summer heat and hemostasis	
22	*Flueggea virosa*	Euphorbiaceae	Root	Evergreen shrub	Wild	D	Clearing heat and detoxifying, relieving swelling and pain, stop itching and hemostasis	
23	*Euphorbiathymifolia*	Euphorbiaceae	Whole grass	Annual herb	Wild	C	Clearing heatand detoxifying, dampness-removing, detumescence and relieving itching	Summer-autumn
24	*Glochidion puberum*	Euphorbiaceae	Root	Shrub	Wild	C	Clearing heat and dehumidification, detoxification, benefiting pharynx, activating Qi and activating blood	Autumn
25	*Rosa laevigata*	Rosaceae	Fruit	Shrub	Both	B	Fixing essence and reducing urine, curing metrorrhagia and leukorrhagia, relieving diarrhea by astringency	Sept.-Nov.
26	*Eriobotrya japonica*	Rosaceae	Flower	Evergreen arbor	Cultivated	A	Moistening the lung and relieving cough, promoting body fluid to quench thirst, regulating stomach and promoting digestion	Winter-spring
27	*Agrimonia pilosa*	Rosaceae	Stem and leaf	Perennial herb	Both	A	Astringent hemostasis, stoping malaria and dysentery, detoxification and tonic deficiency	Summer-autumn
28	*Vigna umbellata*	Leguminosae	Seed	Annual herb	Cultivated	A	Eliminating edema, detoxifying and discharging pus, dredging intestines and moistening stools	Autumn
29	*Abrus cantoniensis*	Leguminosae	Whole grass	Annual herb	Both	B	Removing dampness and jaundice, clearing heat and detoxifying, dispersing the depressed liver-Qi and alleviating pain	All year round
30	*Cassia mimosoides*	Leguminosae	Whole grass	Perennial herb	Both	A	Clearing heat and detoxifying, dispelling inflammation and relieving heat, eliminating dampness and diuresis	Summer-autumn
31	*Millettia specisoa*	Leguminosae	Root	Evergreen woody vine	Wild	B	Tonifying deficiency, moistening lung, strengthening muscle and activating collaterals	All year round
32	*Flemingia philippinensis*	Leguminosae	Whole grass	Perennial trailing herb	Wild	B	Dispelling wind and removing dampness, eliminating blood stasis and detoxification	Autumn-winter
33	*Uraria crinita*	Leguminosae	Root	Perennial herb	Cultivated	B	Clearing heatand detoxifying, stopping bleeding and relieving pain	All year round
34	*Desmodium caudatum*	Leguminosae	Whole grass	Evergreen shrub	Both	A	Clearing heat and detoxifying, dispelling wind and removing dampness	Summer-autumn
35	*Tadehagi triquetrum*	Leguminosae	Stem and leaf	Deciduous shrub	Wild	C	Clearing heat and detoxifying, removing dampness and jaundice, dispersing accumulations and killing worms	Summer-autumn
36	*Ficus simplicissima*	Moraceae	Root	Arbor	Wild	A	Tonifying deficiency, eliminating phlegm, relieving asthma, relaxing muscles and activating collaterals	All year round
37	*Ficus hispida*	Moraceae	Root	Arbor	Wild	B	Heat-clearing, dispelling wind and dampness eliminating accumulation and dissipating phlegm	All year round

IOP Conf. Series: Materials Science and Engineering **730** (2020) 012023 doi:10.1088/1757-899X/730/1/012023

Continued Table 1. List of Hakka characteristic soup plants and their edible parts, source, usage status, health function, harvesting period in Eastern Guangdong

NO.	Latin name	Family	Edible parts	Growth type	Source	Usage status	Health function	Harvesting period
38	*Ficus pumila*	Moraceae	Fruit and leaf	Climbing or creeping shrub	Both	D	Removing dampmess and diuresis, Tonifying kidney, strengthening essence, promoting blood circulation and dredging collaterals	Summer-autumn
39	*Ficus stenophylla*	Moraceae	Root	Arbor	Wild	B	Heat-clearing, dispelling wind and dampness, eliminating accumulation and reducing phlegm	All year round
40	*Ilex asprella*	Aquifoliaceae	Root	Deciduous shrub	Wild	D	Clearing heat and detoxifying, promoting body fluid, promoting pharynx, removing blood stasis and relieving pain	All year round
41	*Berchemia lineata*	Rhamnaceae	Root	Evergreen climbing shrub	Wild	B	Removing blood stasis and hemostasis, expelling wind and eliminating dampness, detumescence and detoxification	All year round
42	*Elaeagnus pungens*	Elaeagnaceae	Root	Evergreen shrub	Both	C	Dispelling wind and removing dampness, promoting blood stasis and hemostasis, strengthening spleen and eliminating food, relieving cough and asthma	All year round
43	*Dimocarpus longan*	Sapindaceae	Fruit	Evergreen arbor	Cultivated	B	Calming heart and tranquilizing mind, nourishing the blood and Qi, benefiting mind and brain	Summer
44	*Aralia chinensis*	Araliaceae	Root	Shrub	Wild	D	Dispelling wind and removing dampness, promoting diuresis and dampness, activating blood circulation and dispersing blood stasis, analgesia and anti-inflammation	All year round
45	*Centella asiatica*	Umbelliferae	Whole grass	Perennial herb	Wild	B	Heat-clearing dampness-removing, detoxification and detumescence	Summer-autumn
46	*Gaultheria yunnanensis*	Ericaceae	Whole plant	Evergreen shrub	Wild	D	Expelling wind and dampness, relaxing muscles and activating collaterals, promoting blood circulation and relieving pain	All year round
47	*Ardisia gigantifolia*	Myrsinaceae	Root	Evergreen shrub	Wild	D	Expelling wind and promoting blood circulation, removing blood stasis and swelling, strengthening muscles and bones	Summer-autumn
48	*Symplocos laurina*	Symplocaceae	Root	Evergreen shrub	Wild	B	Clearing heat, nourishing Yin and purging fire, invigorating spleen and appetizing stomach, dispelling dampness and removing stagnation, clearing liver and moistening lung	All year round
49	*Alyxia fascicularis*	Apocynaceae	Whole grass	Rattan shrub	Wild	B	Dispelling wind and removing dampness, activating blood circulation and relieving pain	All year round
50	*Hedyotis diffusa*	Rubiaceae	Whole grass	Annual herb	Both	B	Clearing heat and detoxifying, relieving pain and eliminating stagnation, promoting urination and removing dampness	Spring-summer
51	*Paederia scandens*	Rubiaceae	Whole grass	Perennial herbaceous vein	Wild	A	Dispelling wind and removing dampness, dispersing blood stasis and relieving pain, invigorating spleen and removing accumulation, detoxification and detumescence	Summer-autumn
52	*Ixora chinensis*	Rubiaceae	Root	Evergreen shrub	Cultivated	D	Clearing heat, cooling blood, removing blood stasis and relieving pain	All year round
53	*Gardenia jasminoides*	Rubiaceae	Root, fruit	Shrub	Wild	B	Clearing heat and removing dampness, detumescence and relieving pain	All year round

IOP Conf. Series: Materials Science and Engineering **730** (2020) 012023 doi:10.1088/1757-899X/730/1/012023

Continued Table 1. List of Hakka characteristic soup plants and their edible parts, source, usage status, health function, harvesting period in Eastern Guangdong

NO.	Latin name	Family	Edible parts	Growth type	Source	Usage status	Health function	Harvesting period
54	*Patrinia villosa*	Valerianaceae	Whole grass	Perennial herb	Wild	A	Clearing heat and detoxifying, removing blood stasis and apocenosis	Mar.-Nov.
55	*Artemisia argyi*	Asteraceae	Root	Perennial herb	Both	A	Warming meridians to stop bleeding, dehumidifying and expelling cold, relieving asthma and cough, eliminating phlegm	Spring-summer
56	*Artemisia lactiflora*	Asteraceae	Whole grass	Perennial herb	Both	B	Promoting blood circulation, removing blood stasis, regulating Qi and removing dampness	Spring-summer
57	*Inula cappa*	Asteraceae	Whole grass	Perennial herb	Wild	C	Expelling wind, promoting Qi, detumescence, relieving pain and detoxification	Spring-autumn
58	*Elephantopus scaber*	Asteraceae	Whole grass	Perennial herb	Wild	B	Clearing heat and detoxifying, promoting diuresis and apocatastasis, dispelling wind, dispersing cold and relieving pain	Apr.-Nov.
59	*Artemisiacapillaris*	Asteraceae	Whole grass	Perennial herb	Both	C	Clearing heat, dispelling dampness, promoting diuresis and removing jaundice	Spring-summer
60	*Gynura divaricata*	Asteraceae	Whole grass	Perennial herb	Both	C	Clearing heat and detoxifying, relaxing tendons and setting bones, cooling blood and hemostasis	Summer-autumn
61	*Gnaphalium affine*	Asteraceae	Whole grass	Annual herb	Wild	C	Preventing cough, eliminating phlegm, expelling wind, dehumidification and detoxification	Spring-summer
62	*Plantago depressa*	Plantaginaceae	Whole grass	Annual or biennial herb	Both	B	Promoting diuresis, clearing heat, improving eyesight and eliminating phlegm	Spring-summer
63	*Lobelia chinensis*	Campanulaceae	Whole grass	Perennial herb	Both	C	Clearing heat and detoxifying, diuresis and apocatastasis	Summer-autumn
64	*Campanumoea javanica*	Campanulaceae	Root	Perennial herbaceous vine	Both	C	Invigorating spleen and replenishing Qi, tonifying lung and relieving cough, relieving asthma and stopping asthma	Summer
65	*Lycianthes biflora*	Solanaceae	Whole grass	Shrub	Both	A	Clearing heat and detoxifying, relieving cough and tonifying deficiency	Summer-autumn
66	*Dichondra repens*	Convolvulaceae	Whole grass	Perennial creeping herb	Both	B	Clearing heat and detoxifying, clearing damp and promoting diuresis, promoting blood circulation and detumescence	Spring-summer
67	*Striga asiatica*	Scrpophulariaceae	Whole grass	Annual herb	Wild	D	Clearing heat and detoxifying, eliminating accumulation and stagnation	Summer-autumn
68	*Dicliptera chinensis*	Acanthaceae	Whole grass	Annual or biennial herb	Wild	D	Clearing heat and cooling blood, diuresis and detoxification	All year round
69	*Andrographis paniculata*	Acanthaceae	Whole grass	Annual herb	Both	B	Clearing heat, Anti-inflammatory and anti-bacterial, liver protection and cholagogue	Autumn-winter
70	*Scutellaria barbata*	Lamiaceae	Whole grass	Perennial herb	Wild	D	Clearing heat and detoxifying, activating blood circulation to dissipate blood stasis, diuresis and apocatastasis	Summer-autumn
71	*Ocimum basilicum*	Lamiaceae	Whole grass	Annual herb	Cultivated	A	Relieving exterior syndrome, clearing heat, activating blood circulation to dissipate blood stasis	Apr.-Oct.
72	*Rabdosia serra*	Lamiaceae	Whole grass	Perennial herb	Both	A	Heat-clearing dampness-removing, removing jaundice and dampness, cooling blood and removing blood stasis	Summer-autumn
73	*Emilia sonchifolia*	Lamiaceae	Whole grass	Annual herb	Both	A	Clearing heatand detoxifying, activating blood circulation to dissipate blood stasis	Apr.-Oct.

Continued Table 1. List of Hakka characteristic soup plants and their edible parts, source, usage status, health function, harvesting period in Eastern Guangdong

NO.	Latin name	Family	Edible parts	Growth type	Source	Usage status	Health function	Harvesting period
74	*Leonurus artemisia*	Lamiaceae	Whole grass	Annual or biennial herb	Both	A	Promoting blood flow to regulate menstruation, diuresis and apocatastasis, clearing heatand detoxifying	Mar.-Sept.
75	*Smilax china*	Liliaceae	Root	Perennial woody vine	Wild	A	dampness-removing, removing turbidity, wind and arthralgia	All year round
76	*Smilax glabra*	Liliaceae	Root	Woody vine	Wild	A	Detoxification, analgesia and diuresis	All year round
77	*Pholidota chinensis*	Orchidaceae	Whole grass	Perennially attached herb	Both	B	Heat-clearingand nourishing Yin, preventing phlegm from forming and stopping coughing, dampness-removing, detoxification and dispersing blood stasis	All year round
78	*Goodyera hachijoensis*	Orchidaceae	Whole grass	Perennial herb	Both	B	Clearing heat and detoxifying,]promoting blood circulation and relieving pain	Summer-autumn
79	*Imperata cylindrica*	Poaceae	Root	Perennial herb	Wild	B	cooling blood and hemostasi, relieving pain	All year round

Note: For descriptions of the columns where use status is located, A (most commonly used), B (more commonly used), C (generally used) and D (collectively used), see the text for details. The efficacy of soup plants is based on the combination of interview materials and literature. Mainly refer to the Chinese Pharmacopoeia (National Pharmacopoeia Commission, 2010), the National Chinese Herbal Medicine Repertory (National Chinese Herbal Medicine compilation Group, 1996), the Traditional Chinese Medicine Enquiry Network (http://www.zhongyoo.com/gx/), Chinese Herbal Medicine Classified by Efficacy Category Inquiry Network (http://www.yongyao.net/new/zycgxfl.aspx?fl=1), Chinese herbal medicine query-traditional Chinese medicine center, etc.).

These 79 soup plants belong to 74 genus, and these genus belong to 11 distribution types. Among them, there are 52 genus of tropical distribution types (including pan-tropical, intermittent in tropical Asia and tropical America, tropical in the old world, tropical Asia to tropical Oceania, tropical Asia to tropical Africa, tropical Asia) in total, accounting for 70.3% of the total genus, and 22 genus of world distribution, north temperate, East Asia and North America intermittent, old world temperate distribution types, accounted for 29.7% (Table 2).

Table 2. The statistic of the areal-types of seed plants of this community

Distribution patterns	Number of genera	Percentage%
1. Cosmopolitan	8	10.8
2. Pantropic	26	35.1
3. Tropical Asia and Tropical America disjunct	5	6.8
4. Old World Tropic	6	8.1
5. Tropical Asia and Tropical Australasia	5	6.8
6. Tropical Asia to TropicalAfrica	3	4.1
7. Trop. Asia (Indo-Malesia)	7	9.5
8. North Temperate	4	5.4
9. East Asia to North American disjuncted	4	5.4
10.Old World Temperate	2	2.7
14.East Asia	4	5.4
Total	74	100

3.5. Analysis of Health Care and Medicinal Efficacy

Almost all of these soup plants have been recorded in the list of Chinese herbal medicines, which can be divided into: relieving exterior syndrom drug, clearing heat drug, purgating drug, antirheumatic drug, diuretic dampness excreting drug, regulating QI flow drug, hemostatic drug, blood activating and stasis removing drug, antitussive and antiasthmatic drug, tonifying drug, warming interior drug, astringent drugs, digestant drug, anthelmintic drug, tranquilizer drug, liver pacifying and wind subduing drug, resuscitation drug, insecticidal and antipruritic drug, removing poison and promoting muscle drug (Liu and Zhang, 2014). Through logging into Chinese Herbal Medicine Enquiry-Traditional Chinese Medicine Center (http://yc.xjlz365.com/), Traditional Chinese Medicine Inquiry Network (http://www.zhongyoo.com/gx/), Inquiry Network of Chinese Herbal Medicine Classified by Efficacy Category (http://www.yongyao.net/ new/zycgxfl.aspx?fl=1), etc., combined with investigation and interview, it is found that Hakka ordinary people are familiar with the efficacy of these soup plants. Different soup cooked with different plants has different effects. Many of the valuable life experience of Hakka ancestors handed down from generation to generation has been constantly summarized and has developed into a very rich Hakka medicinal meal culture. The survey found that almost all of the elderly over the age of 50 know the efficacy of these soup plants. For example, *Ficus simplicissima* and *Ficus hispida* can be used for invigorating spleen and removing dampness, *Mimosa cassia* can reduce inflammation and diuresis, *Rabdosia serra*can protect and nourish the liver, *Artemisia argyi* be used for warming and tonifying, *Lycianthes biflora*, *Elephantopus scaber* and *Wikstroemia indica* can be used to reduce inflammation, *Gynostemma pentaphyllum* can moisturize the lungs, etc.

The efficacy of these 79 soup plants was analyzed. 37 species have the effect of Clearing heat, accounting for 46.8%, including *Symplocos laurina*, *Glochidion puberum*, *Corchorus acutangulus*, *Scutellaria barbata*, *Emilia sonchifolia*, *Goodyera hachijoensis*, *Houttuynia cordata*, *Lobelia chinensis*, *Portulaca oleracea*, *Tadehagi triquetrum*, *Hypericum japonicum*, *Euphorbia thymifolia*, *Dichondra repens*, *Elephantopus scaber*, *Desmodium caudatum*, *Melastoma dodecandrum*, *Patrinia villosa*, *Ilex asprella*, etc. At least 12 species have the effect of reinforcing deficiency, accounting for 15.2%, including *Talinum paniculatum*, *Millettia specisoa*, *Campanumoea javanica*, *Dimocarpus longan*, *Asparagus cochinchinensis*, *Polygonatum sibiricum*, *Ficus simplicissima*, *Fallopia multiflora*, *Ficus hispida*, *Gynostemma pentaphyllum*, *Lycianthes biflora*, *Epiphyllum oxypetalum* and *Agrimonia pilosa*. 16 species have the effect of eliminating phlegm, relieving cough and relieving asthma, accounting for 20.3%, including *Elaeagnus pungens*, *Gnaphalium affine*, *Pholidota chinensis*, *Hylocereus undatus*, *Eriobotrya japonica*, *Ficus simplicissima*, *Gynostemma pentaphyllum*, *Lycianthes biflora*, *Epiphyllum oxypetalum*, *Artemisia argyi*, *Paederia scandens*, *Talinum paniculatum*, *Campanumoea javanica*, *Polygonatum sibiricum*, *Ficus hispida* and *Plantago depressa*. 25 species can promote blood circulation and remove blood stasis, accounting for 31.6%, such as *Ardisia gigantifolia*, *Ilex asprella*, *Fissistigma oldhami*, *Polygala fallax*, *Alyxia fascicularis*, *Ficus pumila*, *Wikstroemia indica*, *Ixora chinensis*, *Ocimum basilicum*, *Schisandra lancifolia*, *Pholidota chinensis*, *Hypericum japonicum* and *Berchemia lineata*, etc. 20 species have the effect of clearing damp and promoting diuresis, accounting for 25.3%, such as *Leonurus artemisia*, *Hypericum japonicum*, *Inula cappa*, *Plantago depressa*, *Smilax china*, *Nephrolepis auriculata*, *Abrus cantoniensis*, *Artemisia capillaris*, *Vigna umbellata*, *Leonurus artemisia*, *Rhodomyrtus tomentosa*, *Elephantopus scaber*, *Dicliptera chinensis* and Portu*l*aca *oleracea*, etc. 16 species have the effect of dispelling wind and dampness, accounting for 20.3%, such as *Alyxia fascicularis*, *Ficus* pumila, *Fissistigma oldhami*, *Gaultheria yunnanensis*, *Ardisia gigantifolia*, *Flemingia philippinensis*, *Tinospora sinensis*, *Millettia specisoa*, *Elephantopus scaber*, *Inula* cappa and *Fallopia multiflora*, etc. 16 species have the effect of hemostasis, such as *Agrimonia pilosa*, *Amaranthus spinosus*, *Imperata cylindrica Opuntia stricta*, *Blechnum orientale*, *Flueggea virosa*, *Uraria crinita*, *Elaeagnus pungens*, *Rabdosia serra*, *Fissistigma oldhami* and *Dicliptera chinensis*, etc. 5 species have the effect of suppressing hyperactive liver and calming endogenous wind, accounting for only 6.3%, including *Imperata cylindrica*, *Ixora chinensis*, *Abrus cantoniensis*, *Andrographis paniculata*, *Cassia mimosoides* and *Symplocos laurina*. 9 species can promote digestion, accounting for 11.4%, they are *Eriobotrya japonica*, *Paederia scandens*, *Tadehagi triquetrum*, *Striga asiatica*, *Eriobotrya japonica*, *Striga asiatica*, *Eriobotrya japonica*, *Ficus stenophylla* and *Elaeagnus pungens*. In addition, among these soup plants, *Schisandra lancifolia* and *Rosa laevigata* have astringent effect, *Ocimum basilicum* and *Berchemia lineata* have the effect of relieving exterior syndrome, *Fallopia multiflora* and *Wikstroemia indica* have diarrhea effect.

It can be seen from Table 1 that some soup plants have many effects, such as *Artemisia argyi*, which can warm and stop bleeding, remove dampness and eliminate cold, relieve asthma and relieve cough and expectoration, *Smilax china* can relieve dampness and turbid, remove wind and remove arthralgia, detoxify and disperse blood stasis. These soup plants have also attracted the attention of physicians and pharmacists. The study found that they tend to have high medical value. For example, *Rhodomyrtus tomentosa* contains phenols, tannin, and the like in its root, and can be used for treating chronic dysentery, rheumatism, hepatitis and blood fat reducing, etc. *Hedyotis diffusa* can activate blood circulation and detumescence, disperse knot to relieve pain, can treat all kinds of inflammation and prevent cancer. *Rabdosia serra* can be used for treating acute and chronic hepatitis, acute cholecystitis, dysentery, enteritis, and traumatic injury, etc. *Uraria crinita* can treat kidney-yang deficiency, waist and knee tenderness, wind-fire toothache, neurasthenia, dyspepsia, cancer prevention and detoxification, etc. *Paederia scandens* has effects in promoting blood circulation and relieving pain, and can be used for treating hepatitis, dyspepsia, acute conjunctivitis, and traumatic injury. *Elephantopus scaber* can relieve cough and diuresis, prevent influenza, etc., and has auxiliary therapeutic effect on colds and upper respiratory tract inflammation. *Rosa laevigata* contains acid substances, saponins and other medicinal ingredients in its fruit, which can be used for strengthening essence and astringent intestine, but also can treat diarrhea caused by spleen deficiency. *Abrus cantoniensis* can be used to treat jaundice, hepatitis, mastitis, stomachache, breast abscess, scrofula and traumatic injury.

In the Notice of Further Standardizing the Management of the Raw Materials for Health-care Food published by the Ministry of Health of China in 2002, specific provisions have been made on the articles of the same origin of medicine, the articles that can be used in health food and health food. In the regulations, *Talinum paniculatum*, *Portulaca oleracea*, *Dimocarpus longan*, *Vigna umbellata*, *Houttuynia cordata*, *Gardenia jasminoides*, *Imperata cylindricaare* included in the list of items that can be used as both food and medicine, *Talinum paniculatum*, *Smilax glabra*, *Plantago depressa*, *Rosa laevigata* and *Gynostemma pentaphyllum* are included in the list of items that can be used for health food.

IOP Conf. Series: Materials Science and Engineering **730** (2020) 012023 doi:10.1088/1757-899X/730/1/012023

4. Discussion and Conclusion

The investigation showed that there were 79 species of traditional Hakka characteristic soup plants in eastern Guangdong, belonging to 44 families and 74 genus. Of these 79 soup plants, 37 species are wild and have not yet been planted artificially, 11 species are cultivated and there is no natural or wild distribution in eastern Guangdong, 31 species are both wild and planted, accounting for 39.2% of the total number of soup plants.

68 species (more than 85%) are naturally distributed in Eastern Guangdong, indicating that their choice of soup plants tends to be locally available. The visit learned that the elderly prefer to go to the wild to collect, and rarely go to the market or drugstore to buy. They stress that each place has its streams in from all over the country, believe in authentic dining materials, health care and therapeutic value of local soup plants. Soup plants are also medicinal plants, like other medicinal plants, they are not only medicine, but also food, which has the function of health care, disease prevention and treatment, and prolonging life [10]. When Hakka ancestors took wild food, they found that some species could not only satisfy hunger, but also their condition would be alleviated or even recovered, which was the medicine and food homology. Soup cooked with these plants, as a kind of diet therapy, can be used with other treatments, give play to the effect of health care and strengthen the immunity of the immune body [11]. There are 119 county-level administrative regions and 8 longevity townships (counties) in Guangdong Province. Meixian, Dapu, Jiaoling and Fengshun County in Meizhou, eastern Guangdong, are all included, accounting for half of them, which suggests that there is an inevitable relationship between Hakka residents' living habits and health and longevity. The scientific and reasonable elements of their living habits will be revealed to give people inspiration and be conducive to improving people's quality of life.

Through the investigation and diagnosis of the body constitution of Lingnan population, it was found that among the healthy population, Qi-deficiency constitution was the most, followed by Yang-deficiency, and the third is damp-heat constitution. According to the theory of traditional Chinese medicine, the climate in eastern Guangdong is hot and humid, damp evil is the first of six pathogenic factors. Spleen likes dryness and dislikes dampness, Lingnan people are in the environment of heavy moisture for a long time, and the spleen Qi is easy to be damaged. Cold, fried, greasy diet should damage the spleen Qi. Thirdly, there are more mountains and less land in Eastern Guangdong, the pressure of survival is great. Long-term overwork leads to excessive Qi consumption, which mainly manifests in fatigue, shortness of breath, dizziness, palpitation and other symptoms. Qi-deficiency and weakness leads to poor metabolism of water and fluid, accumulation in body and forming endogenous dampness. If the endogenous dampness evil coincides with exogenous dampness, the depression turns into heat after a long time, the damp-heat constitution can be formed. Therefore, the constitution of Qi-deficiency, Yang-deficiency and damp-heat are more in the population in East Guangdong. There are many Hakka soup plants with the effect of invigorating spleen, tonifying deficiency, clearing heat and dispelling dampness [12-13].

Compared with Hakkas in Southern Fujian and Southern Zhangxi[6,13], those in Eastern Guangdong select more soup plants. For the same soup plant, such as *Houttuynia cordata*, *Portulaca oleracea*, in Hakka area of Southern Fujian and Southern Zhangxiare, they are usually be used as fried food. *Dicliptera chinensis*, *Hedyotis diffusa* and *Gardenia jasminoides* are usually used to soak in boiling water for drinks, but in Eastern Guangdong, they are usually used to cook soup, probably because the climate in eastern Guangdong is hotter, in addition to health care, especially in summer, the water and salt loss in body is too much, which needs to be supplemented with soup.

The proportion of herbs in these soup plants is nearly half. There are 11 species of genus and 55 genus of tropical distribution, accounting for more than 75% of the total genus, which coincides with the geographical location of eastern Guangdong. East Guangdong is located on the edge of subtropical zone, and its flora is very tropical, but the proportion of tropical distribution pattern of the genus is slightly lower than that of natural forests or plantations, which may be due to the fact that more of these forests are high bud plants (mainly trees and shrubs), and nearly half of these soup plants are herbaceous. Hakkas are relatively emancipated and not confined to the present situation. Many foreign medicinal or health plants have also entered their soup recipes one after another. Foreign returning plants such as *Ocimum basilicum*, *Epiphyllum oxypetalum*, *Talinum paniculatum*, *Hylocereus undatus*,

Opuntia stricta, *Psidium guajava* and *Cassia mimosoides* have been widely cultivated in Hakka in eastern Guangdong, which not only enriches the soup recipe, but also reduces the consumption pressure of local soup plant resources.

In this study, the variety, resources, habitat and folk utilization of soup plants used (mainly commonly used) in Hakka area of Guangdong Province have been relatively thoroughly investigated to determine the list of special soup plants in this area, so a Hakka soup plant resource bank can be established. A botanical garden was established to carry out the introduction, domestication and collection preservation of these plants, and the biological and ecological characteristics and cultivation and propagation techniques of these plants (especially those with few resources, such as *Pholidota chinensis* and *Talinum paniculatum*) were studied. In addition, some soup plants have a lot of local resources, but for Chinese medicinal materials, their related species, especially the related species of the same genus, often have similar or the same effect. If the pharmacological components of these related species are analyzed, whether they can be used as an alternative product remains to be further studied. Such as *Wikstroemia indica*, its related species have *Wikstroemia monnula* and *W. nutans*, do they have similar effects?

At present, there are still some risks in the identification of soup plants, and word of mouth is the main way of using and inheriting soup materials. There are some cases of poisoning caused by accidental collection of toxic plants to make soup, which indicates that there is a certain risk for ordinary residents to collect wild plant resources by themselves because of the lack of professional knowledge, so publicity and education should be strengthened to prevent the risk. Because people's experience and understanding are different, there is chaos between soup plant species, and herbal medicine market is often filled with second best, such as silver lotus pretending to be golden lotus, so it is necessary to standardize the management of the market.

Because people's experience and understanding are different, there is confusion among different kinds of soup plants. In the herbal medicine market, people often pass off the species of relatively poor quality as the good one, such as *Goodyera hachijoensis* as *Anoectochilus roxburghii*, so it is necessary to standardize the management of the market. The phenomenon of synonym and homonym is common, such as *Campanumoea javanica*, is also named as Native Ginseng and Native Dangsen in Eastern Guangdong, but for the plants named native Ginseng, there is also *Lilium paniculatum* in addition to *Goodyera hachijoensis*. For mixed and substitute medicinal materials, the folk often think its efficacy is the same as that of the genuine product, whether the mixed product and the genuine efficacy are consistent, whether or not the other species can be used as the genuine replacement product, and still need to be analyzed and verified from the effect component and the efficacy, and the verification[14].

At present, Hakka diet and soup products are famous and popular in southern China, and the demand is becoming more and more exuberant. Therefore, it is necessary to carry out industrial production research on these soup plants, establish wild medicinal plant cultivation bases in appropriate places, standardize cultivation and reasonable processing of commonly used species, or establish scientific and standardized chain stores to meet the needs of all kinds of markets [15]. It is urgent to strengthen the propaganda of popular science, make use of a variety of propaganda means and hold some exhibitions of soup plants, let people understand more species uses, methods, cultivation methods and awareness of environmental protection, and correctly guide the rational use of the masses, such as the use of spare land or balcony in front or back of the house in order to meet the needs of daily conditioning and avoid exploiting without limitation regardless of the consequences. The Hakka area in eastern Guangdong is the underdeveloped region of Guangdong Province and the back garden of the Pearl River Delta. Strengthening the research on these soup plant resources and the marketing of these soup products can not only improve people's quality of life, but also contribute to the development of local economy.

5. Acknowledgements

This work was supported by the Project of Guangdong Natural Science Foundation (2016A030307011), the "Striving for First-class, Making up for Shortcomings and Enhancing Characteristics" Promotion Plan of Higher Education in Guangdong Province (Project of Subject Construction of Agricultural Resources and Environment); the Fourth National Survey of Traditional

Chinese Medicine Resources Project (2017/152/013); the Special Fund for Science and Technology of Guangdong Province (2019A103009).

6. References

[1] Liu CX. The migration of Hakkas and the formation of their Cultural characteristics[J]. Guangxi Ethnic Studies, 1998(1):74-76.
[2] Au DT, Wu J, Jiang Z, et al. Ethnobotanical study of medicinal plants used by Hakka in Guangdong, China[J]. Journal of Ethnopharmacology, 2008, 117(1):0-50.
[3] Lin H. A survey of the research on the traditional medicinal diet materials in Guangdong Province[J]. Pharmacy Today, 2010, 20(11):8-9.
[4] Yuan TH, Xian SX, Yang ZQ, et al. Lingnan traditional Chinese medicine culture and health care[J]. Journal of Traditional Chinese Medicine , 2013, 54(3):266-268.
[5] Shan F, Huang LQ, Guo J, et al. History and development of "One Root of Medicine and Food"[J]. Chinese Bulletin of Life Sciences, 2015, 27(8): 1061-1069.
[6] Huang XZ, Zou XH. The investigation on utilization of wild medicinal plants in dietary culture of Quanzhou,Southern Fujian[J]. Plant Diversity and Resources, 2015, 37(6):891-902.
[7] You ML, Wen KL. The Hierarchy Analysis of Hakka Food via Grey Relational Analysis[J]. Journal of Grey System, 2013, 16(4):9.
[8] Gong CZ, Qin Y. Count the things in the hometown of longevity[J]. China statistics, 2018, (10), 56-58.
[9] Zheng XL, Sun W, Li RT. Ethnobotanical study on wild vegetable resources of Li Nationality [J]. Hubei Agricultural Science, 2013, 52(16):3856-3862.
[10] Zou P. Traditional Chinese Medicine, Food Therapy, and Hypertension Control: A Narrative Review of Chinese Literature[J]. The American Journal of Chinese Medicine, 2016, 44(8):1-16.
[11] Liu Y, Qi L, Ping L, et al. Plants traditionally used to make Cantonese slow-cooked soup in China[J]. Journal of Ethnobiology & Ethnomedicine, 2018, 14(1):4.
[12] Lu L, Huang Q, Chen Z, et al. Knowledge, attitudes and practices of food-borne diseases and surveillance among physicians in Guangdong, China[J]. Food Control, 2012, 28(1):69-73.
[13] Wand D, Wang XL, Chen BG. Brief discussion on the specific ways to classify herbal diet ingredients according to constitution [J]. Tianjin Traditional Chinese Medicine, 2011, 28(1):54-55.
[14] Cao LM, Zhu XY, Liu ZQ, et al. A Preliminary Study on Plants Used as Tea in Hakkas Area—A Case Study from Gannan Hakkas Area, Jiangxi Province[J]. Journal of Gannan Normal University, 2015, 36(3):72-76.
[15] Song SL. Four discussions on the development of Chinese medicinal diet and dietetic therapy--Development of medicinal diet therapy and protection of wildlife resources. [J]. Oriental Medicated Diet, 2008(1):4-7.

IOP Conf. Series: Materials Science and Engineering **730** (2020) 012024 doi:10.1088/1757-899X/730/1/012024

Preparation of MnO_2/Attapulgite Composites for adsorption of Ni(II)

Jianli Zhang, Yuan Yu and Weidong Liang*

School of Petrochemical Engineering, Lanzhou University of Technology, Lanzhou, GanSu, China.

Email: wdliangh@lut.edu.cn

Abstract. The emission of nickel ion has caused serious environmental pollution. Compared other techniques, adsorption has been used effectively to remove heavy metal ions. Attapulgite in Linze, a county of China, has low purity grade and poor adsorption performance. In order to improve the performance, a composite manganese dioxide/Attapulgite was prepared using simple stir. The effect of ATP-M on the adsorption researched of Ni^{2+} was investigated. The experimental results showed that adsorption capacity was related to the initial concentration pH and the dosage of adsorbents. ATP-M can reach max adsorption capacity to 63.69 mg/g under neutral conditions. The adsorption process was in line with pseudo-second-order kinetic model and the adsorption process was monolayer adsorption, which was more agreed with the Langmuir adsorption isotherm equation. Adsorption process was spontaneous with heat generation.

1. Introduction

In recent years, with the rapid development of industry, heavy metals have caused serious environmental pollution [1]. Due to the difficulty of remove, heavy metals could enter the human body through the digestive tract and pose a threat to human health [2]. Current methods for the treatment of heavy metals include physics, chemistry and biology [3]. Manganese oxide (MnO_2) is a kind of transition metal oxide, which has many advantages, such as large specific surface area, negative charge and many adsorption sites. Attapulgite(ATP) is a clay mineral with crystalline hydrated magnesia-aluminosilicate, which has the characteristics of specific surface area, ion exchange, adsorption capacity and so on. In this experiment, ATP-M composites were prepared by potassium permanganate and ethanol. The effects of the amount of adsorbent, the initial concentration, pH, and temperature on the adsorption of nickel ions were discussed.

2. Materials and Methods

2.1. Materials

The Attapulgite(ATP) from Gansu Linze, Hydrochloric acid (analytical purity, 37%), Potassium permanganate ($KMnO_4$), Ethanol, Deionized water.

2.2. Method

2.2.1. Preparation of ATP-M composites. ATP and sodium hexametaphosphate were dispersed at the ratio of 20:1, stirred for 12 h, dried at 50 °C, grinded and screened. Then, the dispersed ATP was transferred to a flake after using ultrasonic irradiation for 30min, and added a certain amount of $KMnO_4$ while keeping magnetic stirring for 1 h then 12 ml ethanol was added for 4h to make it come into full contact. Finally, the composites were prepared by filtration, drying and grinding (named ATP-M).

2.2.2. Adsorption experiment. Firstly, the Ni^{2+} was determined by spectrophotometer and the standard curve was drawn. In a certain concentration of Ni^{2+} solution, adding a certain amount of ATP-M stirring for a period of time and standing for 2 minutes, the supernatant was filtered, and the concentration of Ni^{2+} was measured by spectrophotometer (V-1800, Shanghai, China). Each experiment was repeated twice to ensure the accuracy. The adsorbed amounts of nickel q_e, q_t and the removal rate R (%) were calculated.

$$qe = \frac{(C0-Ce)\times V}{M} \quad (1)$$

$$R(\%) = \frac{C0-Ce}{C0} \times 100 \quad (2)$$

Where q_e(mg/g) stands for equilibrium adsorption capacity of adsorbed Ni^{2+} ions, respectively. C_0 (mg/L) and C_e(mg/L) stands for the initial and equilibrium concentration. V(mL) stands for the volume of the Ni^{2+} solution, and m (g) stands for the amount of the adsorbent.

3. Result and Discussion

3.1. SEM Analysis.
Figure 1 shows the SEM of ATP and ATP-M. ATP rods can be watched in the SEM of ATP. ATP rods were relatively less and the aggregates were relatively serious, most of them were flake-like and plate-like [4], which indicated that there are other clay minerals such as montmorillonite, mica and feldspar. Figure b shows SEM of ATP-M. A mass of particles were observed on the surface of ATP-M, which indicated that part of the MnO_2 nano-particles were loaded on ATP.

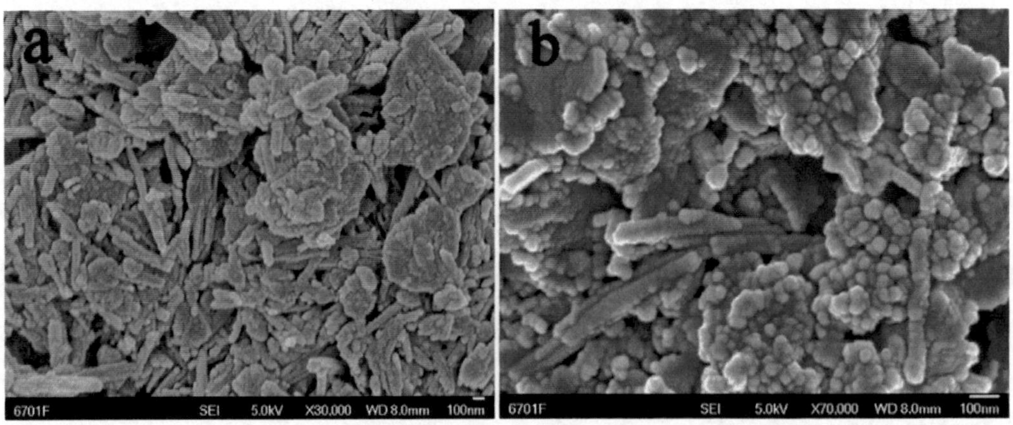

Figure 1. SEM images of (a)ATP and (b)ATP-M.

3.2. XRD Analysis.
Figure 2 shows the XRD diagram of ATP and ATP-M. The results show that ATP (2θ=8.27, 19.59, 27.91, 34.88) [5]. Muscovite(2θ=17.29, 29.89), chlorite(2θ=6.22, 12.27). quartz(2θ=20.75, 26.39, 50.02, 59.85). Feldspar(2θ=27.91). Montmorillonite(2θ=19.98, 35.06) (Standard Card: JCPDS NO.29-1498). For the ATP-M, the characteristic peak of MnO_2 were not very strong but there were characteristic peaks in 2θ=12.76 and 36.77 (110, 211) [6], which indicate that MnO_2 was successfully combined with ATP.

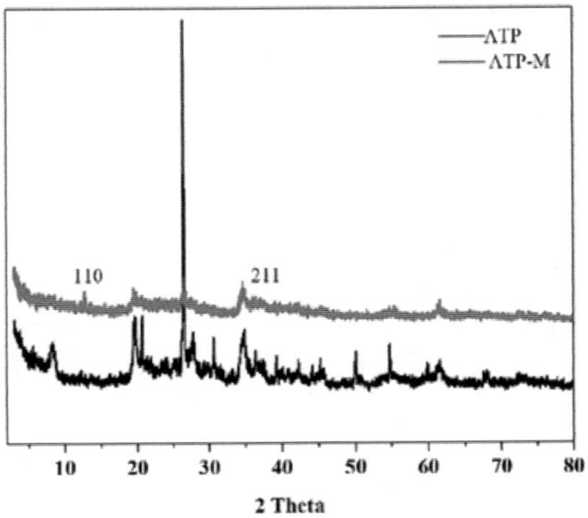

Figure 2. XRD of ATP and ATP-M.

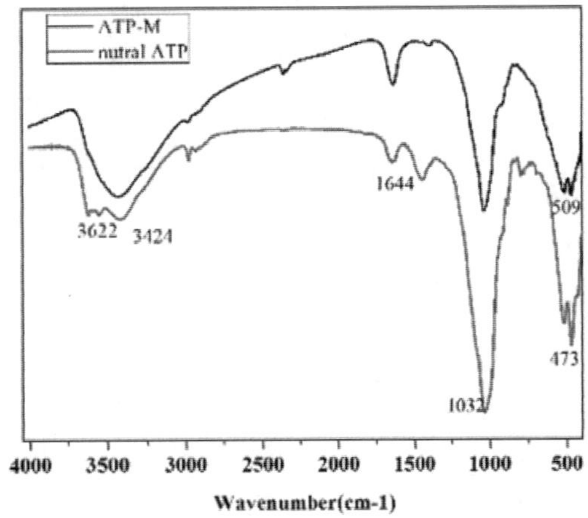

Figure 3. FTIR spectra of ATP and ATP-M.

3.3. FTIR Analysis.
Figure 3 shows the FT-IR of ATP and ATP-M. The band at 3622 cm^{-1} was caused by the stretching vibration of (Mg-OH, Al-OH, Fe-OH), the band at 3424 cm^{-1} was caused by the stretching vibration -OH of zeolite water and surface water [7].The weak band at 1644 cm^{-1} was caused by bending vibration H-O-H of coordination water and zeolite water. The poignant band at 1032 cm^{-1} was caused by stretching vibration of Si-O-Si. For the ATP-M, the characteristic peak appeared at 509 cm^{-1}, but it was not obvious. Through the comparison of literature and XRD analysis, this characteristic peak is the characteristic structure of Mn-O. The results show that MnO_2 was successfully combined with ATP.

4. Adsorption Isotherms and Parameters

4.1. Effect of Dosage of ATP-M, pH, Initial Concentration.
Figure 4 shows the effect of dosage of ATP-M, pH, initial concentration. From figure (4a), it can be seen that with the extension of the reaction time, the adsorption efficiency increases at first and then tends to equilibrium. The adsorption efficiency reached more than 90% in 60 min, which indicate that adsorption process was balanced in 60 min. The figure (4b) shows that the adsorption efficiency increases with the increase of the dosage of adsorbent. When the amount of adsorbent was 0.1g, the adsorption efficiency was more than 95%. However, the change of adsorption capacity was just opposite. These results might because the active site increases with the increase of the dosage of adsorbent. Figure (4c) shows that the removal rate increased with the increase of pH. When pH>8, Ni^{2+} forms precipitation, so the removal rate was more than 95%. Under the neutral condition, the removal rate reached about 90%, so the neutral condition was used in this experiment. Figure (4d) shows that the adsorption capacity increased with the increased of the C_0 and Adsorption capacity reached the maximum at 100 mg/L. However, the removal rate decreased. It might because of the number of ions increased with the increase of C_0, but 0.1 g ATP-M has reached the saturated and could not adsorb the excess ions, so the adsorption capacity decreased. Considering the comprehensive factors, 50 mg/L was used to be the initial concentration in this experiment.

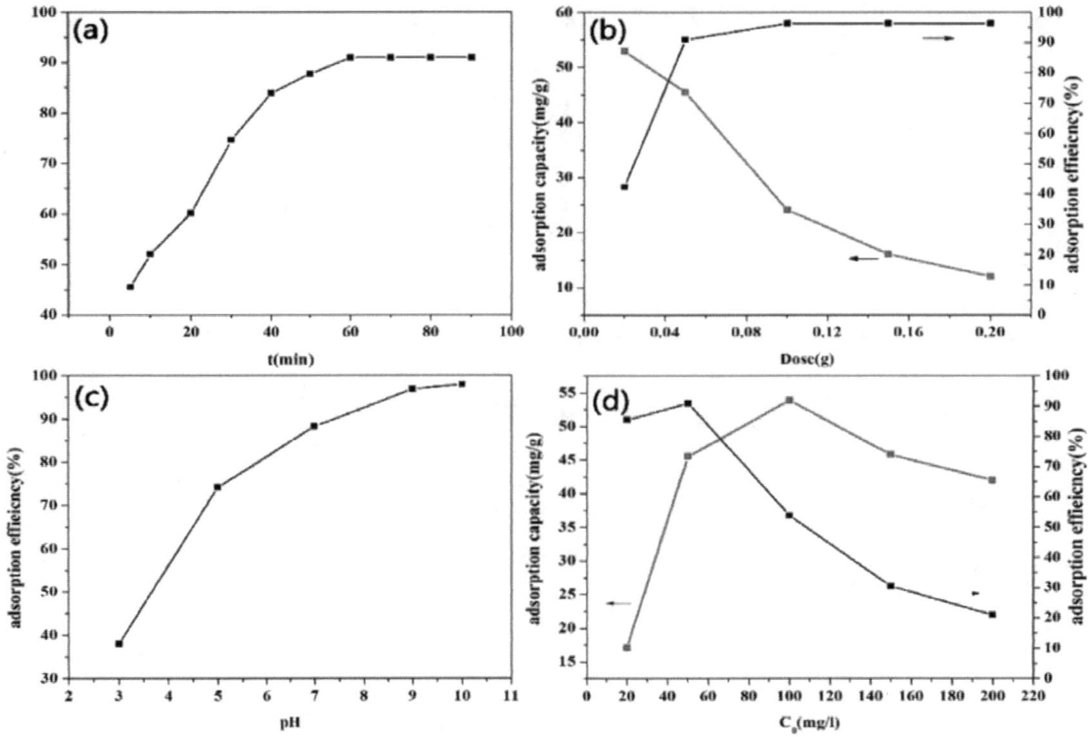

Figure 4. (a) Effect of contact time on Ni^{2+} adsorption (b) Effect of dosage of ATP-M on Ni^{2+} adsorption (c) Effect of pH on Ni^{2+} adsorption in solution (d) Effect of initial concentration of solution on Ni^{2+} adsorption.

4.2. Adsorption Kinetic Studies.

It can be seen from Figure 5 that concentrations of different of Ni^{2+} have the same kinetic curve. With the increase of time, the adsorption capacity increased at first and tended to adsorption equilibrium. Figure 6 shows the linear fitting of Ni^{2+} onto ATP-M, The pseudo-first order kinetic shown as (6a), The pseudo-second order kinetic shown as (6b), Results from linear Fitting of different dynamic equations shown in Table 1. It can be seen from Table 1 that the adsorption of Ni^{2+} by ATP-M accords with the pseudo-second order kinetics.

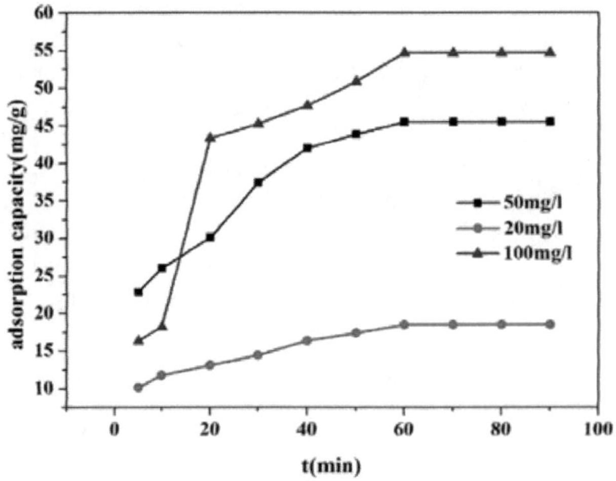

Figure 5. Adsorption kinetic curve of Ni^{2+} onto ATP-M

IOP Conf. Series: Materials Science and Engineering **730** (2020) 012024 doi:10.1088/1757-899X/730/1/012024

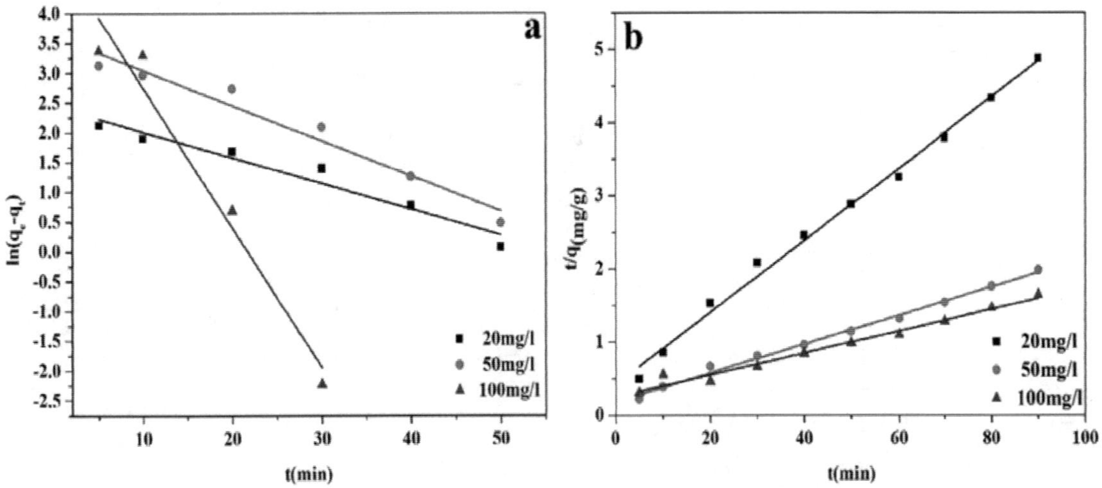

Figure 6. Linear fitting of Ni^{2+} onto ATP-M (a) The pseudo-first order kinetic (b) The pseudo-second-order kinetic.

Table 1. Results from linear Fitting of different dynamic equations.

C_0(mg/l)	pseudo-first order kinetic		pseudo-second order kinetic	
	k_1/min^{-1}	R^2	k_2/(g/mg/min)	R^2
20	0.100	0.941	0.00574	0.994
50	0.136	0.940	0.0021	0.994
100	0.540	0.945	0.0009	0.976

4.3. Thermodynamic Study.

Adsorption isotherm models for Ni^{2+}on ATP-M shown as Figure 7, Table 2 shows the two kinds of adsorption isotherm fitting parameters. The parameters of thermodynamic for the Ni^{2+} adsorption on ATP-M are shown in Table 3. According to the R^2, the adsorption process is more consistent with the Langmuir model.$\Delta G<0$ indicates the adsorption process was a spontaneous process. $\Delta H<0$ illustrates that the adsorption process was an exothermic reaction.

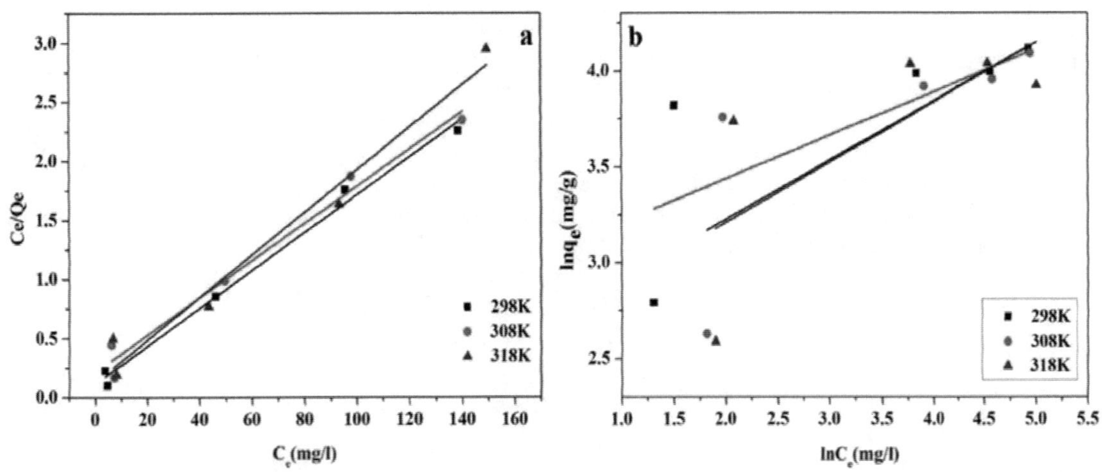

Figure 7. Adsorption isotherm models for Ni^{2+} on ATP-M.

IOP Conf. Series: Materials Science and Engineering **730** (2020) 012024 doi:10.1088/1757-899X/730/1/012024

Table 2. Two kinds of adsorption isotherm fitting parameters.

		Temperature(K)		
		298	308	318
Langumir	R^2	0.98915	0.9756	0.96333
	q_m	62.893	63.694	55.866
	d_L	0.1366	0.0726	0.1385
Freundlich	R^2	0.37507	0.43602	0.35961
	1/n	0.22576	0.30834	0.31332
	d_F	19.7964	13.5765	13.21893

Table 3. The parameters of thermodynamic for the Ni^{2+} adsorption on ATP-M.

T(K)	ΔG^0(KJ/mol)	ΔH^0(KJ/mol)	ΔS^0(J/mol/K)
298	-3.671		
308	-2.087	-31.37	-93.62
318	-1.827		

5. Conclusion

ATP-M is a cut-price and effective adsorbent for heavy metal ions. This simple prepared method makes the ATP of low purity plays a great application. The prepared composites have higher removal rate of Ni^{2+} and the adsorption process was spontaneous. The material of ATP-M could be used in treatment of heavy metal wastewater, which provides certain potential of application.

6. References

[1] Gupta S S , Bhattacharyya K G . Immobilization of Pb(II), Cd(II) and Ni(II) ions on kaolinite and montmorillonite surfaces from aqueous medium[J]. Journal of Environmental Management, 2008, 87(1):46-58.

[2] Jiang L, Liu L. Adsorption of Cobalt Ions and Nickel Ions from Wastewater by Modified Orange Peel[J]. Hubei Agricultural Sciences, 2016, 55(18):4675-4678.

[3] Saravanan R, Khan M M, Gupta V K, et al. ZnO/Ag/Mn_2O_3 nanocomposite for visible light-induced industrial textile effluent degradation, uric acid and ascorbic acid sensing and antimicrobial activity[J]. RSC advances, 2015, 5(44): 34645-34651.

[4] Zhang Z, Wang W, Tian G, et al. Solvothermal evolution of red palygorskite in dimethyl sulfoxide/water[J]. Applied Clay Science, 2018, 159: 16-24.

[5] Bradley W F. The structural scheme of attapulgite[J]. American Mineralogist: Journal of Earth and Planetary Materials, 1940, 25(6): 405-410.

[6] Huang C, Wang Y, Gong M, et al. α-MnO_2/Palygorskite composite as an effective catalyst for heterogeneous activation of peroxymonosulfate (PMS) for the degradation of Rhodamine B[J]. Separation and Purification Technology, 2020, 230: 115877.

[7] Li J J. Research of Flotation Process of Attapulgite Clay Mineral[D]. Beijing University of Chemical Technology, 2013.

Preparation of CNTs / Aluminum Foam Composites

Zhaoyang Lang[1], Mingying Chen[1], Junjiao Han[1], Zhe Liu[1], Xiaoqin Yan[1], Yousong Gu[1], Zhen Ji[1] and Chengchang Jia[1]

[1)] School of Materials Science and Engineering, University of Science and Technology Beijing, Beijing, 100083, China
Corresponding author's: lang_zy5992@163.com

Abstract. This article uses powder metallurgy to prepare aluminum foam and carbon nanotubes (CNTs) / foam aluminum composites. The effects of the surface treatment of carbon nanotubes, the amount of pore-forming agent, and the amount of carbon nanotubes on the microstructure of aluminum foam and its composites were studied by means of infrared spectroscopy, Raman spectroscopy, and scanning electron microscopy. The results show that the mixed acid treatment of CNTs can effectively disperse CNTs and introduce new functional groups on their surfaces. Using NH_4HCO_3 as a pore-forming agent and using powder metallurgy, a porosity of 41% to 79%, Cellular pure foam aluminum material and CNTs / foam aluminum matrix composite with 60% porosity.

1. Introduction

Aluminum foam has many advantages such as low density, high specific surface area and corrosion resistance. Therefore, it is widely used in automobile manufacturing, ship hull, aerospace materials, construction and chemical industry [1-3]. Due to its excellent mechanical properties, carbon nanotubes (CNTs) can be used as a lightweight reinforcing phase to improve the mechanical properties of aluminum foam.

The preparation methods of aluminum foam include liquid phase method, solid phase method and electrolytic deposition method [4]. However, it is not easy to obtain the microstructure of aluminum foam with uniform pore distribution and fewer defects. Due to the low mechanical properties of pure aluminum foam, it is necessary to add the reinforcing phase to make aluminum foam composite material, which makes the preparation process more complicated.

This article uses powder metallurgy to prepare aluminum foam and carbon nanotube / aluminum foam composites. The effects of surface treatment, pore-making agent and addition of carbon nanotubes on microstructure of aluminum foam and its composites were studied by infrared spectroscopy, Raman spectroscopy, and scanning electron microscopy and other characterization methods, providing experimental support for future studies on aluminum foam.

2. Experiment

2.1. Experimental Materials

Aluminum powder was purchased from Beijing Xingrongyuan Co., Ltd. The particle size was 600 mesh, the average diameter was about 23 μm, and the purity was> 99.7%, as shown in Figure 1. CNTs were purchased from Deco Island Gold Co., Ltd., with a length of 20-30 μm and a purity of 97%. The pore-forming agent is NH_4HCO_3.

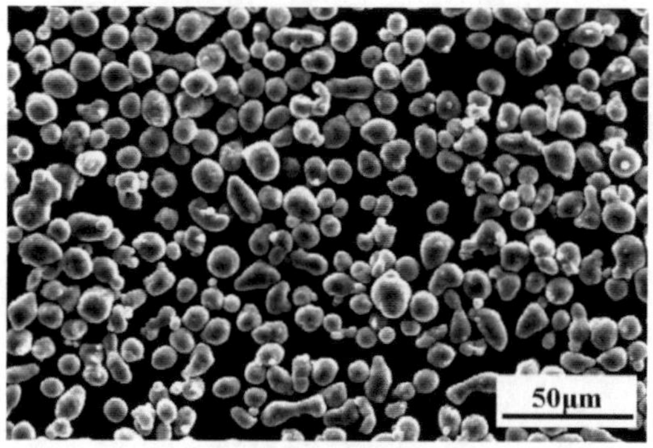

Figure 1. SEM photo of aluminum powder

2.2. Experimental Method

Pure aluminum foam was prepared by powder metallurgy. First weigh a certain mass of aluminum powder and pore-forming agent NH_4HCO_3, where the mass fraction of NH_4HCO_3 is 30% ~ 70%, then mix the two evenly, put it into an alloy steel mold with a diameter of 30mm, and press-mold at 400MPa pressure. The pore-forming agent is dried and decomposed at low temperature, and the preform body is finally sintered in a tube furnace with argon gas to obtain a small-pore foam aluminum material with certain pores.

Carbon nanotubes / foam aluminum composites were prepared by powder metallurgy method. A certain quality aluminum powder and a CNTs surface treated with concentrated sulfuric acid and concentrated nitric acid and 0.25%~1.5% of mass fraction were first mixed. The mixture was mixed in the mixer and added to NH_4HCO_3. Then the CNTs reinforced small pore aluminum foam material with porosity of 60% was prepared according to the above method.

2.3. Characterization Method

Raman spectroscopy and infrared spectroscopy were used to analyze the surface structure of the surface modified carbon nanotubes. The microstructure, CNTs distribution and pore size of foam aluminum composites were observed by JSM-6510A scanning electron microscope and NeXView three-dimensional white light interferometer.

3. Experimental Results and Analysis

3.1. Surface Analysis of Modified CNTs

Figure 2 is a SEM photograph of CNTs before and after acid treatment. It can be seen that the CNTs after acid treatment are no longer entangled with each other like the original CNTs, and the dispersion is effectively improved.

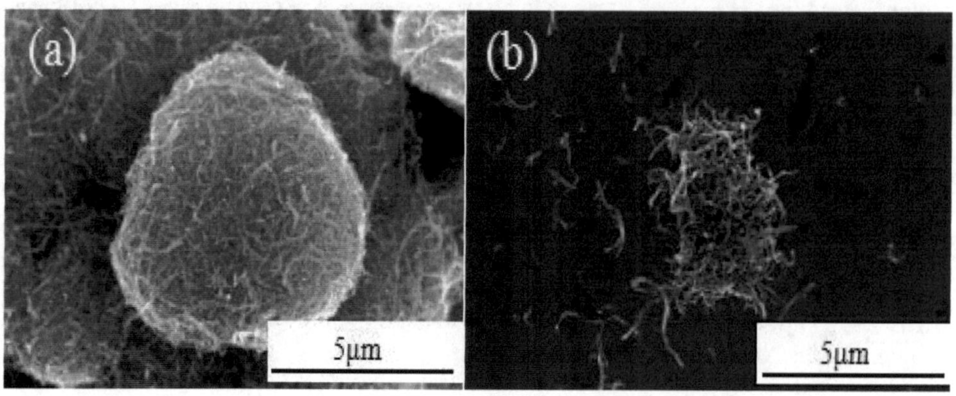

Figure 2. SEM photos of CNTs before (a) and after (b) mixed acid treatment

IOP Conf. Series: Materials Science and Engineering **730** (2020) 012025 doi:10.1088/1757-899X/730/1/012025

Figure 3 is the photo of alcohol suspension before and after CNTs surface treatment. It can be found that the original CNTs suspension has completely precipitated after standing for 1h, while the treated CNTs suspension has not been completely deposited after standing for 10h, which may be due to the introduction of some hydrophilic functional groups on the surface of CNTs after treatment, so that CNTs can be more completely dispersed.

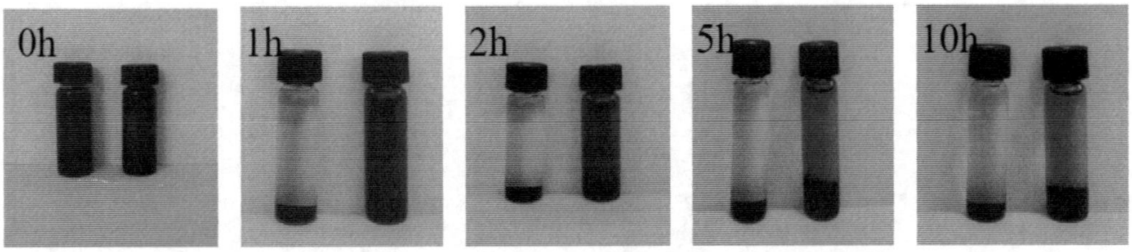

Figure 3. Photos of alcohol suspension before and after mixed acid treatment

Infrared spectrum test was carried out on CNTs before and after treatment. From the curve in Figure 4, it can be seen that the untreated CNTs have the stretching vibration peak of carbon oxygen double bond in the carboxyl group at 1622cm^{-1}, the absorption peak of hydroxyl group at 3427cm^{-1}, and the peak at 2921cm^{-1} is - CH_2 symmetric and antisymmetric vibration peak, indicating that there will be a small amount of hydroxyl and carboxyl containing functions on the surface of CNTs prepared by CVD Group [5]. Compared with the original CNTs, the intensity of the treated CNTs is large and broad, and a new carbon-oxygen double bond stretching vibration peak appears at 1067 cm^{-1} [6], indicating that the treated CNTs are not only purified, but also new oxygen-containing functional groups were introduced on its surface, which made the distance between CNTs larger, and it was easier to achieve the dispersion effect [7-9].

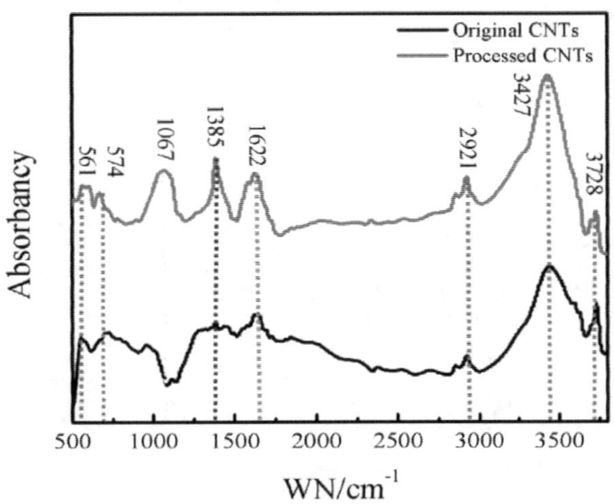

Figure 4. Infrared spectrum of CNTs before and after mixed acid treatment

Figure 5 shows the Raman spectrum of CNTs at 1000 cm^{-1} to 3000 cm^{-1} before and after mixed acid treatment. Three strong peaks appear in the figure. The peak at 2680 cm^{-1} corresponds to the double wave number of the D peak, and is also called the G'peak. The Raman frequency shift of the G' peak is related to the excitation wavelength [10]. The peaks at 1346cm^{-1} and 1571cm^{-1} represent the D peak of the carbon atom lattice defect in the CNTs and the G peak of the graphitized structure. The ratio of the two peaks, i.e. ID / Ig, represents the ratio of amorphous to crystalline carbon content in CNTs, which can be used to characterize the structural changes of CNTs [11]. The Id / Ig of CNTs increased from 0.69 to 0.79 before and after mixed acid treatment, indicating that defects were generated on the surface of CNTs, the reactivity was enhanced, it was easier to introduce oxygen-containing functional groups, and the dispersibility of CNTs was better.

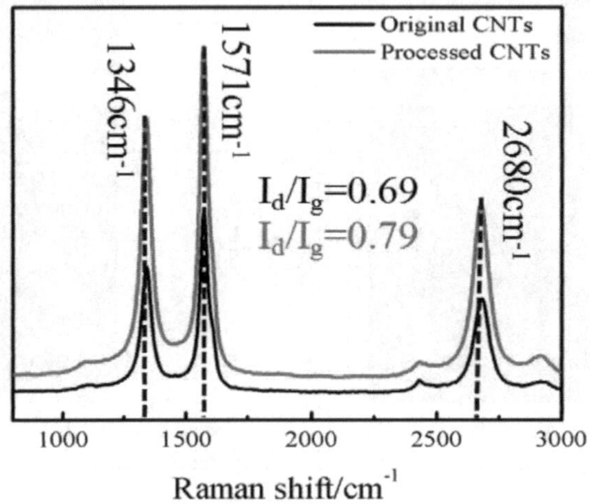

Figure 5. Raman diagrams of CNTs before and after mixed acid treatment

3.2. Effect of Pore Former Mass Fraction on Microstructure of Aluminum Foam

Figure 6 is a histogram of the porosity of aluminum foam prepared by adding different mass fractions of NH_4HCO_3. It can be seen that as the mass fraction of NH_4HCO_3 increases, foamed aluminum materials with porosity from 41% to 79% can be obtained.

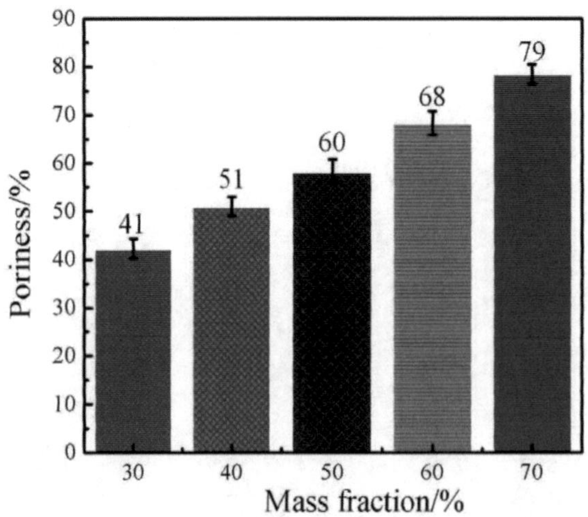

Figure 6. Porosity histogram of foamed aluminum with different mass fraction of NH_4HCO_3

Figure 7 shows the SEM photographs of the different porosity aluminum foams and their corresponding three dimensional photographs. It can be seen from Fig. 7a that the distribution of pores is even more uneven with the increase of the porosity of the foam aluminum, and even the structure of the pores is destroyed. In Fig. 7b, different interference colors represent different depths of the surface, and the size of the aperture can be calculated by the relative value of the corresponding color. It can be seen that the pore size does not change significantly with the increase of porosity, and the maximum pore size is about 400 μm.

IOP Conf. Series: Materials Science and Engineering **730** (2020) 012025 doi:10.1088/1757-899X/730/1/012025

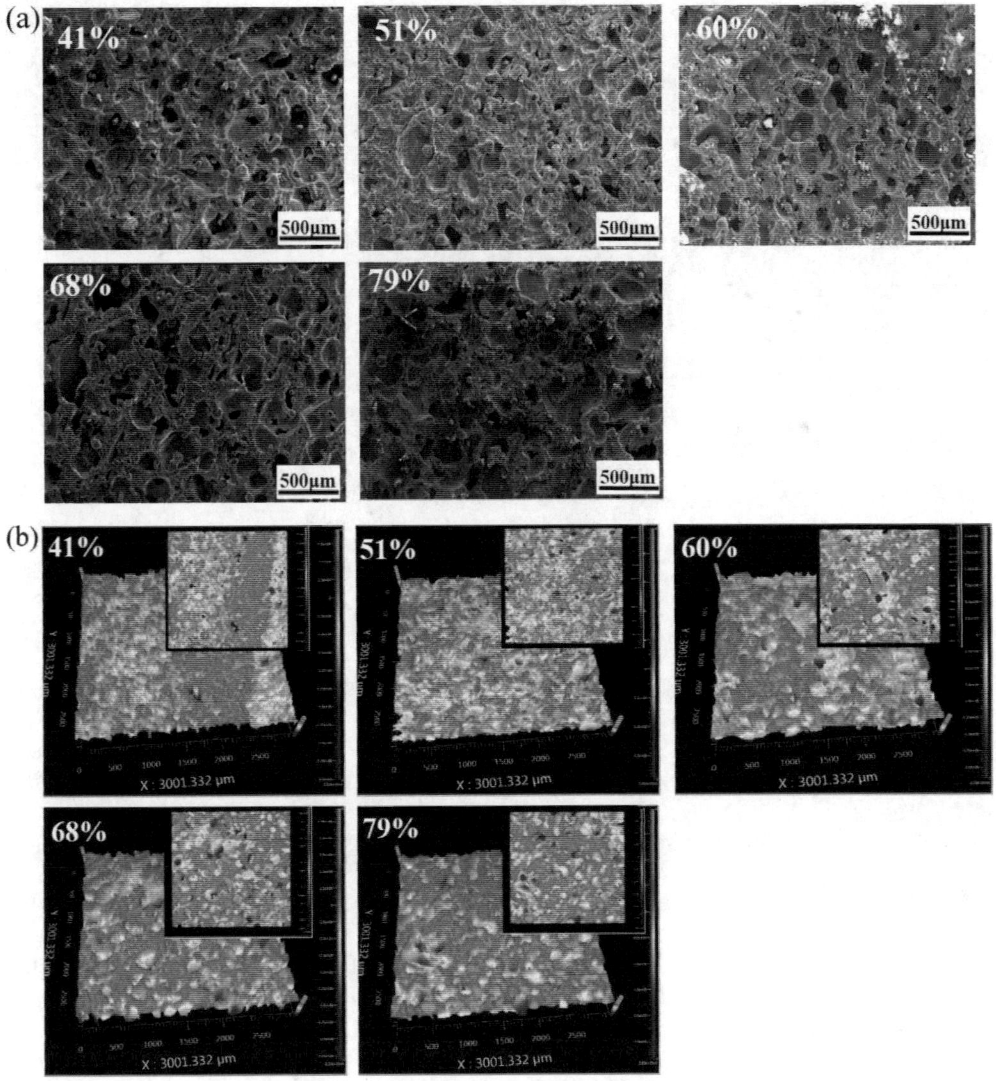

Figure 7. SEM photos (a) and three dimensional photographs of foamed aluminum with different porosity (b)

3.3. *Effect of CNTs Mass Fraction on Microstructure of Composites*

Figure 8 is a SEM photograph of fractures of composite materials with different mass fractions of CNTs and 60% porosity. It can be seen that the structure and distribution of pores in foamed aluminum and the increase in mass fraction of CNTs have no obvious changes, and can be seen in the fracture morphology It can be seen that there is a connection between the pores and the pores of the aluminum foam, which belongs to the open-cell small-pore foam aluminum.

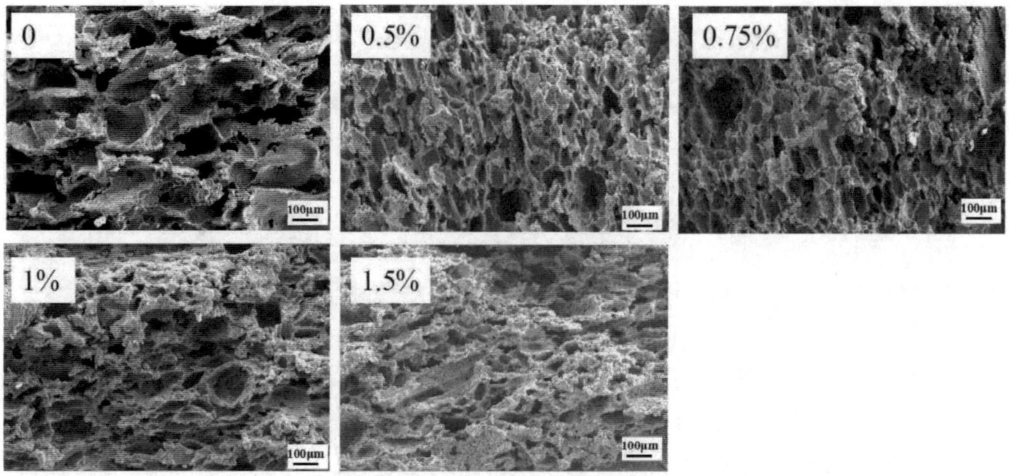

Figure 8. Cross-section SEM photos of aluminum foam with different mass fractions of CNTs

4. Conclusions

(1) CNTs can be effectively dispersed by mixed acid treatment. Compared with the original CNTs, the treated CNTs can maintain one-dimensional nanostructure despite some damage, and new functional groups are introduced into the treated CNTs.

(2) Using NH_4HCO_3 as a pore-forming agent, powder metallurgy can be used to prepare pure aluminum foam with open porosity of 41% ~ 79% and CNTs reinforced aluminum foam composite with 60% porosity.

5. References

[1] Baumeister J, Banhart J, Weber M. Aluminium foams for transport industry[J]. *Materials & Design*, 1997, 18(4-6): 217-220.

[2] Jiang J J, Gasik M, Laine J, et al. Electrochemical evaluation of sintered metal hydride electrodes for electric vehicle applications[J]. *Journal of Alloys and Compounds*, 2001, 322(1-2): 0-285.

[3] Maine E M A, Ashby M F. Applying the investment methodology for materials (IMM) to aluminium foams[J]. *Materials & Design*, 2002, 23(3): 307-319.

[4] Zhang Yong. Vibrant new functional material foam metal [J]. *Journal of Shandong Institute of engineering*, 1994, 8 (3): 1-6

[5] Li-hua, TENG, Tian-di, et al. IR study on surface chemical properties of catalytic grown carbon nanotubes and nanofibers[J]. *Journal of Zhejiang University-Science A(Applied Physics & Engineering)*, 2008, 9(5): 720-726.

[6] Bu Luxia, Wang Chunjie, Yin Lihui, et al. Surface modification and dispersion of multi-walled carbon nanotubes by nitric acid oxidation [J]. *Plating & Finishing*, 2018, 40 (6).

[7] Qu romeng, Li Yanping, Wang Mei, et al. IR and Raman analysis of acidified MWCNTs [J]. *Chemical engineer*, 2013, 27 (11): 61-63.

[8] Ma Jianchun, Zhang Jun, Du Peng, et al. Preparation of functionalized modified multi-walled carbon nanotubes [J]. *Shanxi Chemical Industry*, 2015, 35 (2): 12-14.

[9] Zheng Zheng, Chen Rong, Hu Huating, et al. Research on characterization methods of chemically functionalized modified carbon nanotubes [J]. *Colloid and Polymer*, 2009, 27 (3): 37-40.

[10] Sood A K, Gupta R, Asher S A. Origin of the unusual dependence of Raman D band on excitation wavelength in graphite-like materials[J]. *Journal of Applied Physics*, 2001, 90(9): 4494.

[11] Gao Yun, Li Lingyun, Tan Pingheng, et al. Application of Raman spectroscopy in carbon nanotube polymer composites [J]. *Chinese Science Bulletin*, 2010, 55 (22): 2165-2176.

IOP Conf. Series: Materials Science and Engineering **730** (2020) 012026 doi:10.1088/1757-899X/730/1/012026

The Study of Microstructure and Mechanical Properties of Two High Strength Steels

Shihan Li, Xiaolin Wang, Wei Zhao and Zunzhao Li

SINOPEC Dalian Research Institute of Petroleum and Petrochemicals, Liaoning, Dalian, 116000

E-mail address: lishihan.fshy@sinopec.com

Abstract. The microstructure and mechanical properties of two high strength steels with different carbon content were studied by means of metallographic microscope, scanning electron microscope, transmission electron microscope and mechanical testing machine. The results showed that the microstructure of the 0.2C (wt.%) steel and 0.4C (wt.%) steel was tempered sorbite, a large number of granular carbides dispersed on the ferritic matrix. The carbides contained Fe_3C and M_6C (Fe_3Mo_3C) phase. Compared with 0.2C steel, the content of the carbides of 0.4C steel was larger, while the particle size was smaller. The tensile and yield strength of 0.4C steel were higher, but the elongation was lower and the toughness was worse. In addition, the hardness values of the two materials were basically the same.

1. Introduction

With the increase of oil and gas exploitation depth, the terrain conditions become increasingly severe, which requires higher strength on the oil casing. Meanwhile, the use of high strength pipeline steel has also gradually highlighted its necessity and importance, due to the increase of oil and gas transmission pressure. High strength steel is playing an increasingly important role in oil and gas exploitation and transportation. From the earliest three carbon steel grades to the successful promotion and application of X65 and X80 and now the successful development of X100 and X120, the research on higher strength grades has never stopped and the development and application of high strength steel have always been one of the key directions in the field of oil and gas development and transportation [1]. This article selects two kinds of self-developed high strength steel as the research object. Through the analysis of scanning electron microscope, transmission electron microscopy and tests of tensile strength and hardness, the characterization of structure and mechanical properties of the two steels under different carbon content was compared.

2. Materials and Experiments

The carbon content of the two materials prepared in this experiment are 0.2% (wt.%) and 0.4% (wt.%) (0.2C steel and 0.4C steel for short). The specific chemical composition is shown as table 1. It was melted in a vacuum induction melting furnace and then cast into ingots. After forging, the ingots were kept in the air furnace at 1150 °C for 2.5 hours for homogenizing treatment. Then, they were hot rolled into 25 mm thick plates and quickly cooled in water. The rolling process was started at 1080 °C and finished at 820 °C. Finally, the samples were tempered at 600 °C for 1 h.

Table 1. The chemical composition of the experimental steels (wt.%)

Samples	C	Si	Mn	Cr	Ni	Ti	Mo
0.2C	0.2	0.2	1	1.2	0.02	0.02	1.2
0.4C	0.4	0.2	1	1.2	0.05	0.05	1.5

The samples were eroded by 4% nital (volume fraction) for 20 - 80 seconds after grinding and polishing. The erosion time of SEM samples is slightly longer than that of metallographic samples. OLYMPUS BX53M metallographic microscope and Quanta 200F scanning electron microscope were used for the observation. The microstructure was also analyzed by JEOL JEM-2100 transmission electron microscopy. The samples were first mechanically polished below 80 microns and punched into 3 mm wafer by Gatan 659. The specimens were then further thinned by Gatan 691. cs with ions reduction at 3° incidence angle and 5 KeV energy.

The tensile testing was conducted with Cortest® machine at room temperature. According to GB/T 228.1-2010 standard, samples were prepared with an intermediate standard distance of 25.4 mm and a diameter of 3.8 mm. In addition, three groups of parallel samples were selected for each material. The tensile fracture was observed by Quanta 200F scanning electron microscope. In addition, the hardness was measured by DHV-1000Z micro Vickers tester. 3*3 points were evenly selected on each sample.

3. Results

3.1. Microstructure

Fig. 1 shows the metallographic diagram of the two steels. Fig. 1(a) shows that the microstructure of 0.2C steel is tempered sorbite, which is a tempered structure of martensite. It is a mixture of ferrite and granular carbide, a composite structure with carbide (including cementite) pellets distributed in the ferrite matrix [2]. Fig. 1(b) shows that the 0.4C steel is also tempered sorbite. No discernible difference of metallographic structure of the two steels was observed.

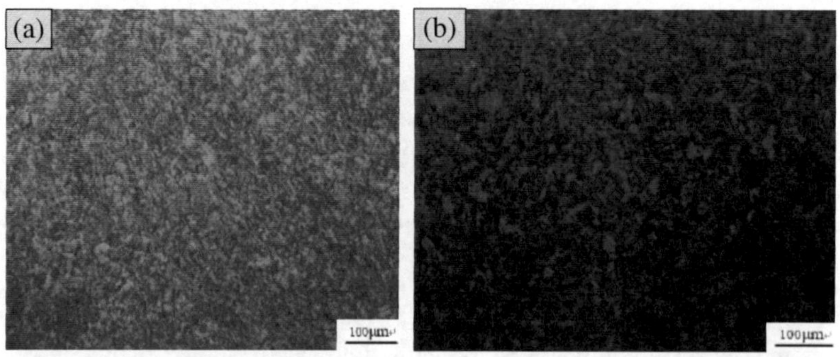

Figure 1. Metallographic diagram of the two steels. (a) 0.2C steel, (b)0.4C steel.

Fig. 2 shows the SEM examination of the two steels. Fig. 2(a) shows that large numbers of carbides are dispersed in 0.2C steel and no grain boundaries can be clearly distinguished. Fig. 2(b) shows details of the carbides in Fig. 2(a) at higher magnifications. It can be seen that the sizes of the granular carbides are below 1 μm. Fig. 2(c) shows the large amounts of fine carbides uniformly distribute in 0.4C steel. The enlarged view shown as Fig. 2(d) reveals the particle sizes of carbides are below 0.5 μm. The comparing of Fig. 2(a) and Fig. 2(c) shows that the carbide content of 0.4C steel is higher than that of 0.2C steel. In addition, the particle sizes of the carbides in 0.4C steel are smaller from the analyzation of Fig. 2(b) and Fig. 2 (d). It can be attributed to the higher content of the alloying elements like Mo and Ti, which could inhibit the growth of carbides and lead to the smaller of precipitates [3]. In addition, the larger amounts of carbides itself can also prevent the growth of carbides.

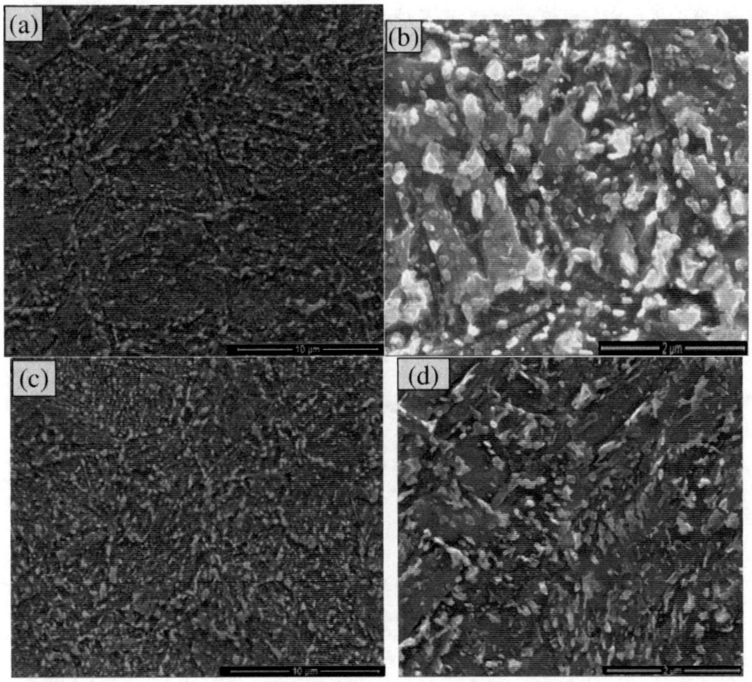

Figure 2. SEM images of the two steels. (a) 0.2C steel, (b) was higher magnification of (a), (c) 0.4C steel, (d) was higher magnification of (c).

Fig. 3 shows the TEM images of the two steels. Fig. 3(a) shows that the average equiaxed grain size of the 0.2C steel was about 0.4 μm. The submicron carbide precipitation is mostly distributed along the grain boundary. Fig. 3(b) shows that the TEM morphological feature of 0.4C steel is similar to 0.2C steel. Fig. 3(c) reveals that intensive granular carbides about 100 nm in size precipitate in 0.4C steel. Fig. 3(d) shows details of carbides marked in Fig. 3(c) at higher magnification. Through the analysis of selected area electron diffraction, the carbides are identified as Fe_3Mo_3C phase, which is an M_6C structure. It indicates that the carbides contain M_6C phase apart from the Fe_3C.

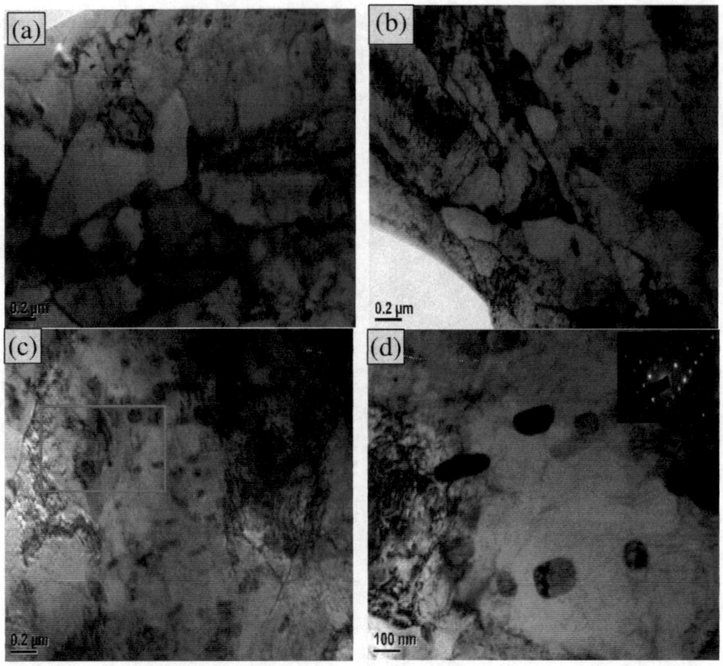

Figure 3. TEM images of the two steels. (a) 0.2C steel, (b) and (c) 0.4C steel, (d) was higher magnification of (c) and the corresponding selected area electron diffraction.

3.2. Tensile Property

Fig. 4 shows the stress-strain curves of the two steels. Fig. 4(a) shows that an obvious yield stage exists in the tensile curve of the three parallel samples of 0.2C steel. Fig. 4(b) shows that no obvious yield platform forms in 0.4C steel. Table 2 shows the tensile properties of the two steels gathering from the curves. The 0.2C steel showed a rupture strength of ~ 800 MPa, yield strength of ~ 740 MPa and an elongation and a shrinkage of ~ 25% and 65%. In comparison, the 0.4C steel showed higher strength, ruptured at ~ 1150 MPa with yield strength of ~ 1060 MPa but a decrease elongation to ~ 18% and shrinkage to ~ 63%. The result shows that the tensile and yield strength of 0.4C steel are higher. While, the elongation and shrinkage are worse. This is because of the strengthening effect of more carbides precipitation in 0.4C steel.

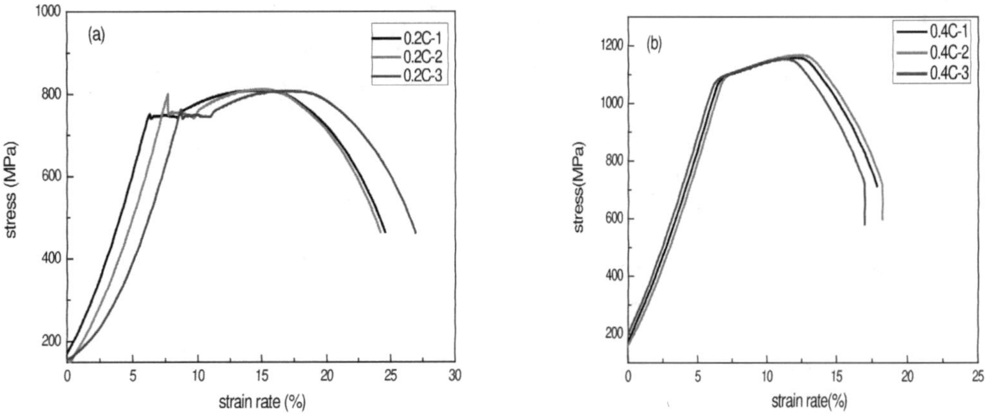

Figure 4. Tensile curves of the two steels. (a) 0.2C steel, (b) 0.4C steel.

Table 2. Tensile properties of the two steels

Average	0.2C	0.4C
Tensile Strength (MPa)	810	1160
Yield Strength (MPa)	749	1069
Elongation (%)	25.2%	18%
Shrinkage (%)	65.5%	63.4%

3.3. Fracture Analysis

Fig. 5 shows the fracture morphology of the two steels. Fig. 5(a) shows that necking occurs in 0.2C steel. It is seen that several cracks extend along the edge to the centre and a 45° shear plane appears. Fig. 5(b) shows that large number of deep dimples exist in the fibre region with no parabolic shear dimples, which indicates the good toughness of 0.2C steel. Fig. 5(c) shows that a deep penetrating crack appears in 0.4C steel and no obvious boundaries differentiate fibre zone and shear zone. Fig. 5(d) shows that the dimples of 0.4C steel are small and shallow and some sections have shown the tendency of approaching quasi-cleavage, which indicates that the toughness of 0.4C steel is low and brittleness is large. It can be concluded that the toughness of 0.4C steel is worse compared with 0.2C steel, which is consistent with the tensile elongation result before.

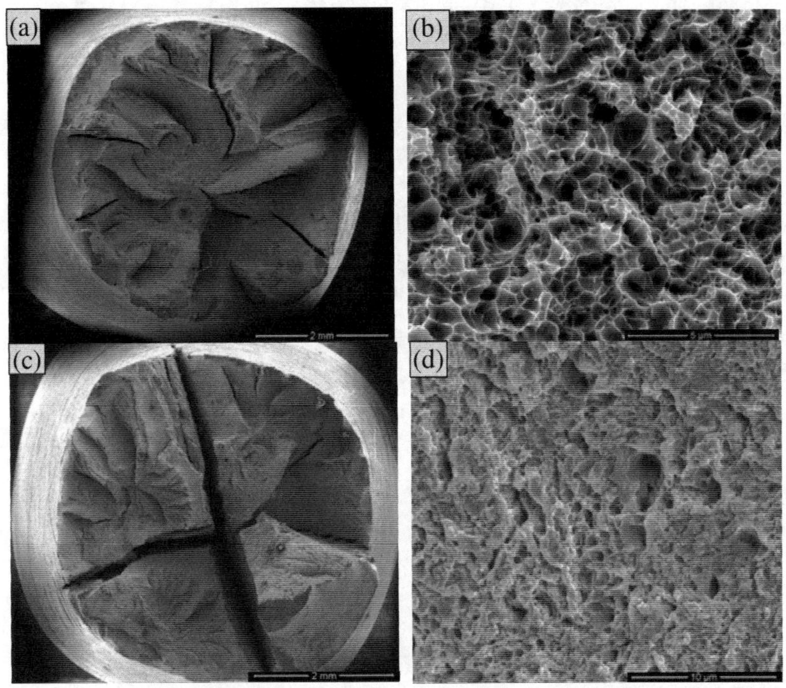

Figure 5. Fracture diagrams of the two steels. (a) 0.2C steel, (b) was higher magnification of (a); (c) 0.4C steel, (d) was higher magnification of (c).

3.4. Hardness

Fig. 6 and Fig. 7 show the Micro Vickers hardness of the two steels. Fig. 6 shows that the hardness values of the 9 points in 0.2C steel at different positions are little difference, which indicates the uniform distribution of the microstructure. The average hardness value was 268. Fig. 7 shows that the harness values of 0.4C steel range from 251 - 272 with a relatively uniform microstructure. The average value was 266. Different from the tensile properties, the hardness difference between the two steels was little. It verifies that the microstructure of the two steels is similar, which is consistent with the microstructure analysis before.

IOP Conf. Series: Materials Science and Engineering **730** (2020) 012026 doi:10.1088/1757-899X/730/1/012026

Figure 6. Micro Vickers hardness at different locations of 0.2C steel.

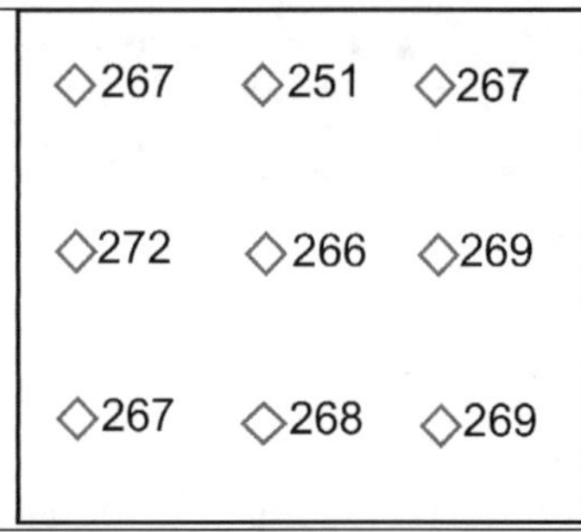

Figure 7. Micro Vickers hardness at different locations of 0.4C steel.

4. Conclusions

(1) The microstructure of the 0.2C and 0.4C steels were all tempered sorbate, which was a composite structure with carbide (including cementite) pellets distributed in the ferrite matrix.

(2) The carbides contain Fe_3Mo_3C phase, which is a M_6C structure. Compared to 0.2C steel, the carbide content of 0.4C steel was larger and the size was smaller.

(3) The tensile and yield strength of 0.4C steel were higher but the toughness was worse. While, the hardness values of the two steels were basically the same.

5. References

[1] Helin Li, Lingkang Ji, Lihua Xie, The prospect of the development of China petroleum steel pipe [J]. Journal of Hebei University of Science and Technology, 2006, 27 (2), 97-102.

[2] Xuefeng Li, Chunfen Wang, Jiamin Wang, Analysis of tempered martensite and tempered sorbonite [J]. Heat Treatment, 2012, 27 (4), 12-16.

[3] D. Leea, J.K. Kima, S. Lee, K. Lee, B.C. De Cooman, Microstructures and mechanical properties of Ti and Mo micro-alloyed medium Mn steel, Mater. Sci. Eng. A 706 (2017) 1-14.

Analysis of Chemical Constituents in *Ficus Hirta* Vahl. by LCMS-IT-TOF and GC-MS.

Wenjing Tang[1], Jianping Chen[2*], Mengjiao Du[1], Lishi Chen[1], Yi Yang[1], Shufen Wu[1], Biting Zhang[1] and Chuqin Yu[1*]

1 Guangdong Engineering & Technology Research Center of topical precise drug delivery system, Guangdong Pharmaceutical University; Centre for Drug Research and Development, Guangdong Pharmaceutical University; Guangdong Provincial Key Laboratory of Advanced Drug Delivery Systems, Guangdong Pharmaceutical University, East Waihuan Road 280, Guangzhou, Guangdong, China.
2 The First Affiliated Hospital, Sun Yat-sen University, Guangzhou, Guangdong, China.
* Corresponding authors: (Jianping Chen), E-mail: cjp3935@163.com
(Chuqin Yu), E-mail: pn333@163.com

Abstract. Objective: Analysis the chemical constituents from the roots of *Ficus hirta*. **Methods:** The *Ficus hirta* Vahl. were extracted with 75% ethanol. Qualitative analysis of ethanol extracts was carried out by using high performance liquid chromatography-tandem mass spectrometry (LCMS-IT-TOF) and gas chromatography- mass spectrometry (GC-MS). **Result:** Twenty compounds were identified by LCMS-IT-TOF. Nine compounds were identified by GC-MS. Psoralens and bergapten were identified by LCMS-IT-TOF and GC-MS. **Conclusion:** LC-MS-TOF and GC-MS have identified 27 compounds, including 9 flavonoids, 6 coumarins, 3 organic acids, 3 organic alcohols, 2 organic esters, 1 terpene, 1 polyphenol, 1 alkaloid and 1 anthraquinone, which laid a foundation for further study of *Ficus hirta* Vahl. and its compound preparations.

1. Introduction

Ficus hirta Vahl. is the dry root of *Ficus hirta* of Moraceae plants, sweet flavor, neutral in nature, has the effect of promoting dampness and relaxing tendons, strengthening the spleen and invigorating the lung. It is a traditional Chinese herbal medicine used in the south of the five ridges and mainly used in the treatment of spleen deficiency and edema, hepatitis, rheumatism and arthralgia, tuberculosis and cough. Studies have shown that *Ficus hirta* Vahl. has the functions of immune regulation[1], antibacterial[2], anti-inflammatory, analgesic, liver protection[3], antioxidant, apoptosis induction[4]and anti radiation[5]. In addition, Cheng[6] et al. found that *Ficus hirta* Vahl. can inhibit the production of NO induced by lipopolysaccharide in rats. However, the active components of the ethanol extracts of *Ficus hirta* Vahl. are not clear. In order to ensure the safety and effectiveness of the product, the determination of its active ingredients is particularly important. In this study, LCMS-IT-TOF and GC-MS were used to separate and analyze the chemical components in the extract of *Ficus hirta* Vahl., and to identify the chromatographic peak qualitatively, so as to provide theoretical basis for the development and utilization of *Ficus hirta* Vahl..

2. Instruments, Samples and Reagents

2.1. Instruments

LCMS-IT-TOF mass spectrometer(Shimadzu, Kyoto, Japan), equipped with standard electrospray ionization (ESI), Mass Hunter, Qualitative Analysis and photo-diode array(PDA); TRACE DSQ GC-MS(ThermoFinnigan, USA); BS423S One thousandth electronic balance(Sartorius); CP225D One in 100,000 electronic balance(Sartorius); HWS26 Electro-Thermostatic Water Bath(Shanghai yiheng Scientific Instruments Co., Ltd); crusher DFT-200(Wenling Linda Machinery Co., Ltd, Zhejiang); IKA -RV8 V Rotary Evaporator(IKA, Germany).

2.2. Materials and Reagents

Ficus hirta Vahl. provided by Heyuan Jinyuan green life Co., Ltd, identified by Professor Liu Jizhu, teaching and esearch Office of traditional Chinese medicine, Guangdong Pharmaceutical University as the dry root of *Ficus hirta* of *Ficus* of Moraceae; Psoralen standard(purity≥98%, Guangzhou Qiyun Biotechnology Co., Ltd); Apigenin standard(purity≥98%, Guangzhou Qiyun Biotechnology Co., Ltd); ultrapure water(A.S. Watson Group Ltd); Formic acid(Tianjin kemio Chemical Reagent Co., Ltd); methanol(Thermo Fisher Scientific); Acetonitrile(Thermo Fisher Scientific);ethyl alcohol absolute(AR, Tianjin Damao Chemical Reagent Factory), CO_2 (Guangzhou Zhenyu weiou Trading Co., Ltd).

3. Methods and Results

3.1. Preparation of Test Solution

Thc 200g of *Ficus hirta* Vahl. powdcr was passed 16 mesh sieve and refluxed two times with 6 times 75% ethanol, 3h per time. The filtrate was combined and concentrated. Ethanol was recoveried under reduced pressure. Then the filtrate was diluted with methanol to 200 mL, and filtered again to get the subsequent filtrate. The subsequent filtrate was filtered by 0.22 μm membrane and stored in a small liquid phase bottle for LCMS-IT-TOF and GC-MS analysis as test solution.

3.2. LCMS-IT-TOF Chromatography and Mass Spectrometry Conditions

3.2.1. LCMS-IT-TOF chromatographic condition. Chromatographic column: YMC-Pack ODS-A C18 column(250 ×4.6 mm, 5 μm); mobile phase: 0.1% formic acid-water solution(A)/acetonitrile(B); gradient: 10 min, 90:10(V(A) V(B)); 30 min, 80:20(V(A) V(B));60 min, 60:40(V(A) V(B)); 80 min, 30:70(V(A) V(B)), 90 min, 20:80(V(A) V(B)); 100 min, 0:100(V(A) V(B));100 min, 95:5(V(A) V(B)); flow rate: 1.00 mL· min^{-1}; column temperature: 35 °C ± 5 °C; detection wavelength:300 nm and 350 nm.

3.2.2. LCMS-IT-TOF mass spectrometry conditions. Mass spectrometry conditions: positive and negative ion modes; scanning range: *m/z* 50~600; ESI source spray voltage:1.50 KV; the ionization temperature: 200°C; detector voltage: 1.65KV; Separator voltage: 8.5 KV; Pressure in TOF zone: 1.4×10^{-4} Pa; TOF temperature: 40 °C; Ion accumulation time: 10 ms.

3.3. GC-MS Chromatography and Mass Spectrometry Conditions

3.3.1. GC-MS gas phase conditions. Chromatographic column:30.00 m×0.25 mm×0.25 μm DM-5MS capillary column; injector temperature: 250°C, Interface temperature: 230°C, Helium was used as carrier gas, flow rate: 1.0 mL·min^{-1}, Column pressure:80 kp, splitless injection mode, Injection volume: 1.0 μL; Temperature program: column temperature: 40 °C, constant temperature for 2 min, The temperature is raised to 70°C by the program of 10°C·min^{-1}, and then to 280 °C by the program of 10°C and constant temperature for 10 min.

3.3.2. GC-MS mass spectrometry conditions. Electron impact ion source(EI), the ionization temperature:200 °C, Electron multiplier voltage:520, mass range:15-380 amu, total ion scanning,

IOP Conf. Series: Materials Science and Engineering **730** (2020) 012027 doi:10.1088/1757-899X/730/1/012027

Scanning gap:1.0 s, Scanning rate: 1000 amu · s^{-1}.

3.4. Results

3.4.1. LCMS-IT-TOF analysis of ethanol-extracts from Ficus hirta Vahl.. LCMS-IT-TOF was used to detect the chemical components of *Ficus hirta* Vahl., The total ion chromatogram is shown in Figure 1, 2. The obtained mass spectrogram was retrieved by Metlin mass spectrometry database and Chemsipider database, and the analysis results are shown in Table1.

Table 1. Identification results of *Ficus hirta* Vahl.

peak	T_R	Formula	m/z	MS/MS fragments	Compound
1	19.128	$C_{15}H_{14}O_6$	291.0806[M+H]$^+$	165,139,123	Epicatechin
2	19.433	$C_{16}H_{18}O_9$	355.1010[M+H]$^+$	289,163,145	Chlrorgenic acid
3	24.850	$C_{15}H_{14}O_5$	275.0887[M+H]$^+$	149,139,207	(-)(2R,3R)Epiafzelechin
4	26.573	$C_{17}H_{18}O_9$	389.0823[M+Na]$^+$	227,185,153	Psoralenoside
5	27.355	$C_{18}H_{20}O_{10}$	419.0943[M+Na]$^+$	257,213,185	Methoxypsoralenoside
6	29.220	$C_{21}H_{21}NO_6$	384.1450[M+H]$^+$	178,164,136	Hydrasine
7	32.152	$C_{17}H_{20}O_9$	391.1000[M+Na]$^+$	229,185,149	4,5-dihydrogenpsoralenoside
8	33.333	$C_{21}H_{20}O_{10}$	431.0946[M-H]$^-$	283,311,341	Pelargonidin 7-glucoside
9	33.592	$C_{21}H_{22}O_9$	417.1171[M-H]$^-$	255,256	Aloin A
10	33.692	$C_{15}H_{12}O_4$	257.0791[M+H]$^+$	239,147,137	Isoliquiritigenin
11	34.367	$C_{21}H_{20}O_{10}$	433.1133[M+H]$^+$	364,337,283	Vitexin
12	35.103	$C_{18}H_{22}O$	399.1304[M+H]$^+$	319,263,237	Picraquassioside A
13	44.430	$C_{11}H_{10}O_4$	205.0525[M-H]$^-$	161,187	Isoeugenitol
14	49.938	$C_{15}H_{10}O_6$	287.0570[M+H]$^+$	269,153,135	Kaempferol
15	52.560	$C_{11}H_6O_3$	187.0402[M+H]$^+$	159,143,131	Psoralen
16	54.975	$C_{15}H_{12}O_5$	273.0757[M+H]$^+$	189,153,147	(±)-Naringenin
17	56.062	$C_{15}H_{10}O_5$	271.0660[M+H]$^+$	243, 187,163	Apigenin
18	60.505	$C_{12}H_8O_4$	217.0501[M+H]$^+$	202,174,115	Bergapten
19	70.257	$C_{14}H_{12}O_3$	229.0861[M+H]$^+$	211,187,145	Resveratrol
20	80.110	$C_{18}H_{30}O_2$	279.2333[M+H]$^+$	261,173,145	Pinolenic Acid

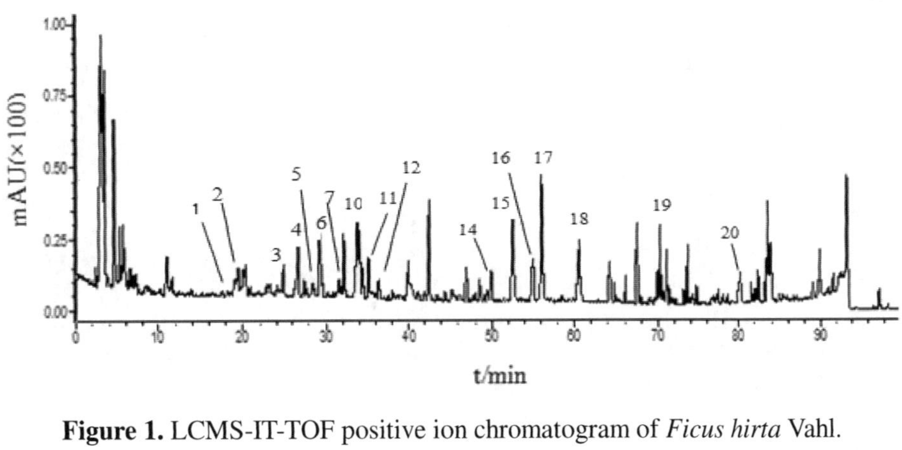

Figure 1. LCMS-IT-TOF positive ion chromatogram of *Ficus hirta* Vahl.

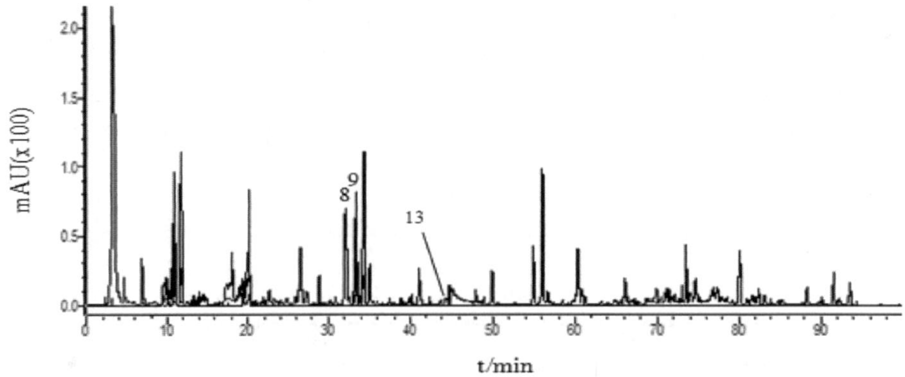

Figure 2. LCMS-IT-TOF anion chromatogram of *Ficus hirta* Vahl.

According to the results, in the positive ion mode, the m/z is $[M+H]^+$ or $[M+Na]^+$. In the anion ion mode, the m/z is $[M-H]^-$. Coumarin contains keto carbonyl, and the m/z of its fragments are $[M+H-CH_3]^+$, $[M+H-CO]^+$ and $[M+H-2CO]^+$, etc. The mass spectrum of peak 15 is shown in Figure 3.

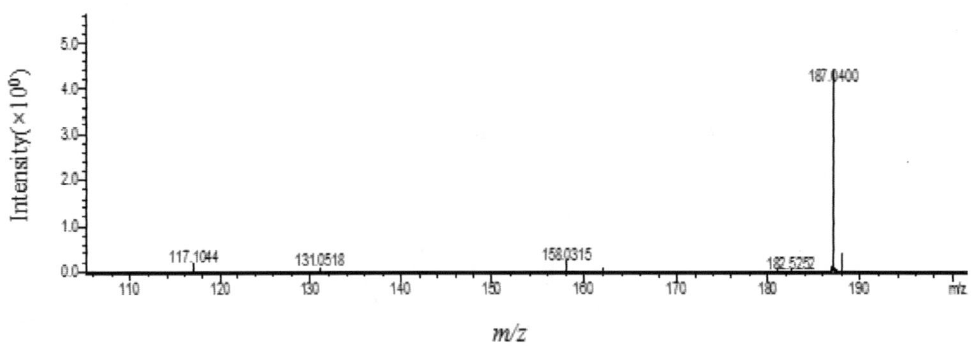

Figure 3. MS spectrum of peak of psoralen (m/z187).

The fragmentation process of psoralen by mass spectrometry is shown in Figure 4. The $[M+H]^+$ of peak 15 is 187. Its relative molecular mass is 186. The m/z of fragment $[M+H-CO]^+$ is 159. The m/z of fragment $[M+H-2CO]^+$ is 131. The substance can be identified as psoralen.

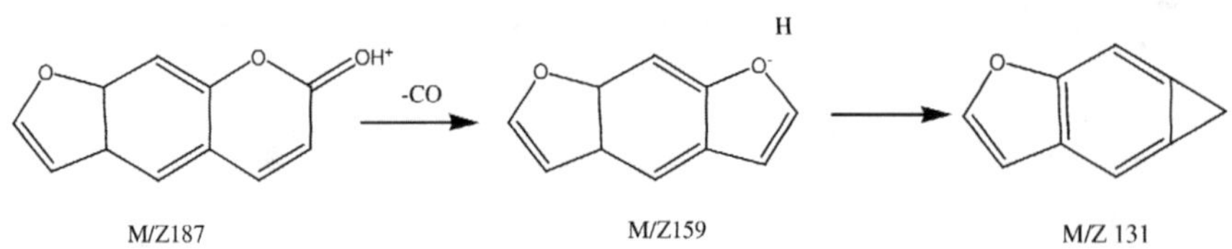

Figure 4. Proposed MS-MS fragmentation of psoralen.

3.4.2. GC-MS analysis of ethanol-extracts from Ficus hirta Vahl. GC-MS was used to detect the chemical components of the extract of *Ficus hirta* Vahl., The total ion chromatogram is shown in Figure 5. The content was calculate by the area normalization method of gasification products. The analysis and identification results is shown in Table 2.

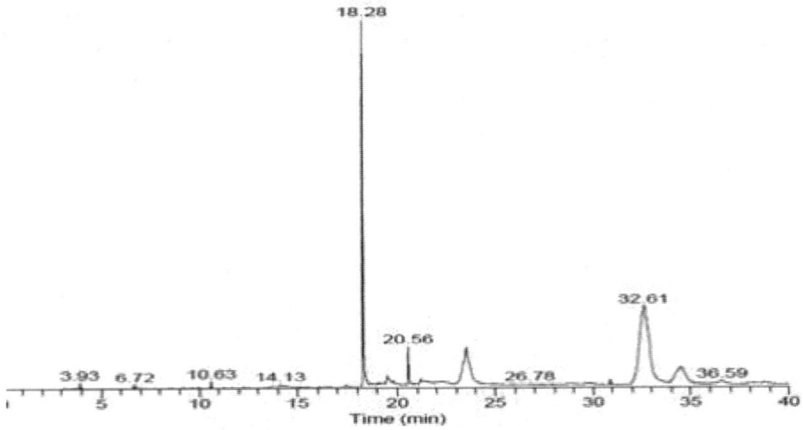

Figure 5. GC-MS total ion chromatogram of *Ficus hirta* Vahl.

Table 2. Identification results of *Ficus hirta* Vahl.

peak	T_R	Formula	Compound	relative content/%
1	10.630	$C_{11}H_{20}O_2$	2-ethylhexyl ester-2-Propenoic acid	0.46
2	18.280	$C_{11}H_6O_3$	Psoralen	25.56
3	19.480	$C_{16}H_{32}O_2$	n-Hexadecanoic acid	0.74
4	19.560	$C_{26}H_{44}O_5$	Ethyl iso-allocholate	0.92
5	20.550	$C_{12}H_8O_4$	Bergapten	2.88
6	21.200	$C_{19}H_{36}O$	2-Methyl-Z, Z-3,13-octadecadienol	1.20
7	23.530	$C_{29}H_{48}O$	Stigmasterol	3.53
8	32.600	$C_{29}H_{50}O$	22,23-dihydro- Stigmasterol	32.57
9	41.690	$C_{30}H_{48}O$	Amyrin	7.90

According to the results in Table 2, The ethanol extracts of *Ficus hirta* Vahl. include coumarins, flavonoids, sterols and other compounds. The compounds with higher content were 22, 23-dihydro-stigmasterol (32.57%), Psoralen (25.46%), Amyrin (7.90%), Stigmasterol (3.53%), Bergapten (2.88%), and so on.

4. Discussion

The analysis of chemical components of traditional Chinese medicine includes separation, analysis and identification, such as TLC, HSCCC, HPLC and HPCE. However, the chemical components of traditional Chinese medicine are complex, so the above methods are difficult to identify and analyze. The GC-MS technology combines the advantages of mass spectrometry and chromatography, and mass spectrometry has high sensitivity and specificity, which is very suitable for the analysis of complex samples such as traditional Chinese medicine[7]. In this study, the components of the extracts of *Ficus hirta* Vahl. were separated and analyzed, and 27 compounds were identified, including flavonoids, coumarins, organic acids, terpenes, polyphenols, alkaloids, anthraquinones and sterols. The active substance kaempferol could regulate the activity of pro-inflammatory enzymes and the expression of inflammation related genes, which has obvious anti-inflammatory effect[8]. Epicatechin significantly reduced the level of inflammatory factors released by RAW264.7 cells induced by LPS, and reduced the inflammatory response after cerebral hemorrhage[9]. Geranium could inhibit the migration, invasion and proliferation of gastric cancer cells[10].Ibuprofen could regulate NF-κB signal pathway, control the expression of cytokines, and partially inhibit the release of NO and PGE2[11], Psoralen inhibited the expression of inflammatory factors in human periodontal ligament cells induced by Pg-LPS, and could be used for the treatment and prevention of periodontitis[12]. Bergamot lactone significantly inhibited the recruitment of neutrophils and macrophages from zebrafish to the injured site, and promoted the clearance of neutrophils and macrophages from the wound site[13]. Apigenin inhibited the secretion of IL-10 and TNF-α and play an anti-inflammatory role[14].

It can be seen from the total ion chromatogram of *Ficus hirta* Vahl. that there are some chromatographic peaks with good response, and their mass can not be found in the database of known chemical components, which indicates that there are still a lot of unknown components of *Ficus hirta* Vahl. to be further studied.

5. References

[1] Zhang Z C, Liu J H, Pan Y A. Effects of *Ficus hirta* Vahl. and astragalus extract on cellularimmunity of immunosuppressed mice [J]. *Chinese folk medicine*, 2017, **26 (05)**: 58-59.

[2] Chen Q, Ye S X, Yu J. Antibacterial Activity of RADIX FICI HIRTAE by Chromotest Microassay[J]. *Medicinal Plant*, 2012, **3 (9)**: 13-16.

[3] Wang M, He R R, Li Y F, et al. Protective effect of water extract from *Ficus hirta* Vahl. on restrained stress liver injury [J]. *Chin Hosp Pharm J*, 2015, **35(6)**: 522-525.

[4] Zeng Y W, Liu X Z, Lv Z C, et al. Effects of *Ficus hirta* Vahl. (Wuzhimaotao) extracts on growth inhibition of HeLa cells[J]. *Exp. Toxicol. Path.* 2012, **64(7-8)**: 743-749.

[5] Wang X P, Duan L J, Huang X, et al. Protective effect of aqueous extract of *Ficus hirta* Vahl.on DNA damage of mouse bone marrow cells induced by - (60)Coγ ray [J]. *Chinese Modern Applied Pharmacy*, 2011, **28 (04)**: 284-287.

[6] Cheng J, Yi X, Wang Y, et al. Phenolics from the roots of hairy fig (*Ficus hirta* Vahl.) exert prominent anti-inflammatory activity[J]. *J. Funct. Foods*, 2017, **31**:79-88.

[7] Wang Y M, Zhang S J, Luo G A, et al. Study on phenylethanolic glycosides in Cistanche deserticola and its substitutes by LC/ESI-MS/MS [J]. *Acta Pharma Sinica*, 2000, **35 (11)**: 839-842.

[8] Devi K P, Malar D S, Nabavi S F, et al. Kaempferol and inflammation: From chemistry to medicine[J]. *Pharmacol. Res.* 2015, **99**: 1-10.

[9] Ruan H S, Mou J Z. Effect of epicatechin on lipopolysaccharide induced secretion of inflammatory factors in RAW264.7 cells [J].*Chinese Journal of ETMF*, 2017, **23(4)**: 159-163.

[10] Liang S, Jiang Z C, Feng J, et al. Inhibition of human gastric cancer cell migration and invasion by carmine radish Geranium [J]. *Food industry technology*, 2017, **38 (22)**: 1-4+10

[11] Wang Y. Study on the anti-inflammatory effect of isoeugenol in vivo and in vitro [D]. *Shandong university*, 2016.

[12] Li X T, Zhou W, Song Z C. Anti-inflammatory effects of psoralen and isopsoralen on humanperiodontal cells [J]. *Shanghai Jiaotong Univ. (Sci)*, 2018, **38 (2)**: 128-132.

[13] Yang Y, Zheng K, Mei W, et al. Anti-inflammatory and proresolution activities of bergapten isolated from the roots of *Ficus hirta* Vahl. in an in vivo zebrafish model[J]. *Biochem. Biophys. Res. Commun.* 2018, **496 (2)**: 763-769.

[14] Palacz-Wrobel M, Borkowska P, Paul-Samojedny M, et al. Effect of apigenin, kaempferol and resveratrol on the gene expression and protein secretion of tumor necrosis factor alpha (TNF-α) and interleukin-10 (IL-10) in RAW-264.7 macrophages[J]. *Retour Au Numéro*, 2017, **93**: 1205-1212.

IOP Conf. Series: Materials Science and Engineering **730** (2020) 012028 doi:10.1088/1757-899X/730/1/012028

Friction and Wear Mechanism of High Temperature Brake Friction Materials

Junjiao Han[1], Qian Qiu[1], Keyan Ning[2], Ming Han[2], Nan Zhang[1], Chaoyang Lang[1], Zhe Liu[1], Zhen Ji[1] and Chengchang Jia[1]

[1)] School of Materials Science and Engineering, University of Science and Technology Beijing, Beijing 100083, China
[2)] China North Vehicle Research Institute, Beijing 100072, China
Corresponding author: 1)Zhen Ji E-mail: jizhen@mater.ustb.edu.cn
2)Keyan Ning E-mail: kyning@noveri.com.cn

Abstract. In this paper, the microstructure and frictional wear mechanism of iron based high temperature brake friction materials prepared by powder metallurgy are studied. The results show that iron based friction materials are mainly composed of iron matrix, graphite and hard phase. Under the working condition of rotating speed of 6000r/min, the frictional wear mechanism is mainly abrasive wear and slight oxidative wear. Under the working condition of rotating speed of 8800r/min, the frictional wear mechanism is mainly abrasive wear, adhesive wear and oxidative wear.

1. Introduction

Brake is the most important safety device for traffic vehicles and transportation equipment. It uses the frictional properties of friction materials to convert kinetic energy into thermal energy for braking and stopping [1]. At present, with the increase of high mobility and power density of vehicles, resulting in more severe working conditions and frequent failure of brake friction devices, friction materials for high temperature braking have become one of the key technologies to improve the braking performance and reliability of vehicles.

At present, metal based friction materials are mainly used in high-speed and heavy load conditions. Metal-based friction materials include powder metallurgy and fused metal [2]. Powder metallurgy friction materials have been widely used in various large civil aircraft and high-performance military aircraft and other mechanical braking devices because of their advantages of high strength, long service life, good high temperature and good friction stability[3-7]. Iron-based friction materials have the largest amount of powder metallurgical friction materials, lower cost, good thermal stability, and strong adhesion resistance.

The components and friction velocity of friction materials will affect the friction and wear properties and mechanism of materials. Yu Xiao [8] et al. studied the influence of the friction components of Fe, Al_2O_3 and SiO_2 on the friction and wear properties of sintered alloys, the results show that the friction coefficient of the material with SiO_2 is the smallest and the wear amount is the largest. Chen Jie et al. [9] studied the mechanism of the influence of the friction surface temperature on the friction and wear properties of iron-based friction materials. The results showed that with the increase of friction speed, the temperature of the friction surface kept rising, and the friction coefficient and wear amount of the materials showed a trend of first decreasing and then increasing.

In this paper, the microstructure and wear mechanism of iron-based high-temperature brake friction materials prepared by powder metallurgy method under different working conditions are studied,

IOP Conf. Series: Materials Science and Engineering **730** (2020) 012028 doi:10.1088/1757-899X/730/1/012028

which provides the basis for the design and application of friction materials.

2. Experimental

The samples of iron-based powder metallurgy friction plate and dual plate are shown in Figure 1. In the experiment, the microstructure of iron-based high-temperature brake friction materials prepared by powder metallurgy was characterized by using quanta 650 scanning electron microscope (SEM) and energy dispersive spectrometer (EDS). Then, the friction and wear experiments of the samples were carried out by using mm3000 friction and wear testing machine at the speed of 6000r/min and 8800r/min respectively. Finally, the friction and wear mechanism of the material was analyzed by the macro and micro morphology of the wear surface and the wear debris of the sample.

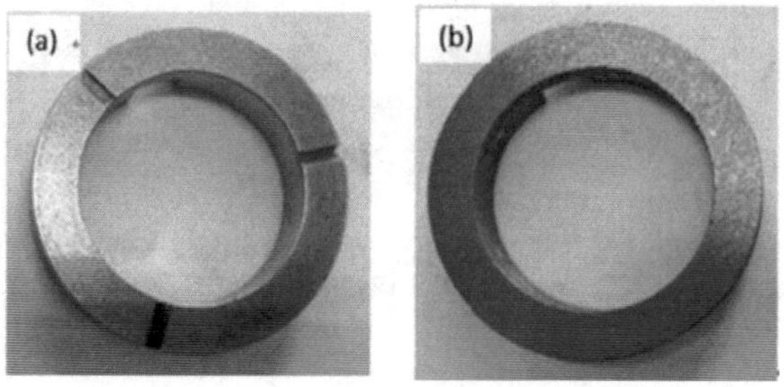

Figure 1. Fe-based powder metallurgy friction part(a) and dual part(b)

3. Result and Discussion

3.1. Microstructure

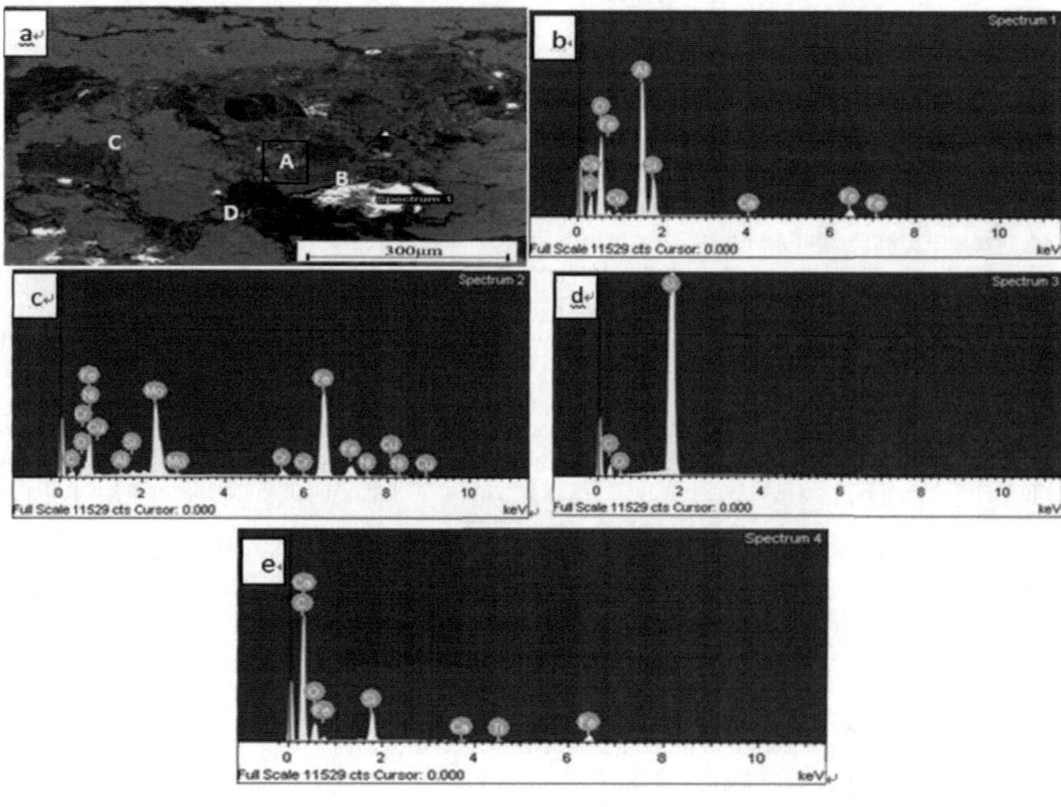

Figure 2. SEM of Fe-based friction material(a) and EDS of pint A, B, C, D(b),(c),(d),(e)

The iron-based friction material is mainly composed of iron matrix, graphite and hard phase, and its microstructure is shown in Figure 2. EDS spectrum analysis was performed on points A, B, C, and D in Figure 2 (a). See Fig. 2 (b), (c), (d) and (E) in turn. The light gray at point C is the iron matrix, which plays a connecting role and determines the strength of the friction material; the black area at point D is graphite, which plays a role in lubrication and wear reduction. The bright block area at point B is Al_2O_3, while the light black area at point A is SiC, which plays a role in increasing friction and improving wear resistance. The iron matrix is surrounded by graphite and SiC, and Al_2O_3 is embedded on the matrix.

3.2. Friction and Wear Mechanism

The macro morphology of the friction part under different working conditions is shown in Figure 3. It can be seen from Fig. 3 (a) that under the condition of 6000r/min, the friction surface of the sample has obvious furrows and slight spalling along the sliding direction. From Fig. 3 (b), it can be seen that there are obvious furrows and spalling along the sliding direction of the material friction surface under the condition of the rotation speed of 8800r/min.

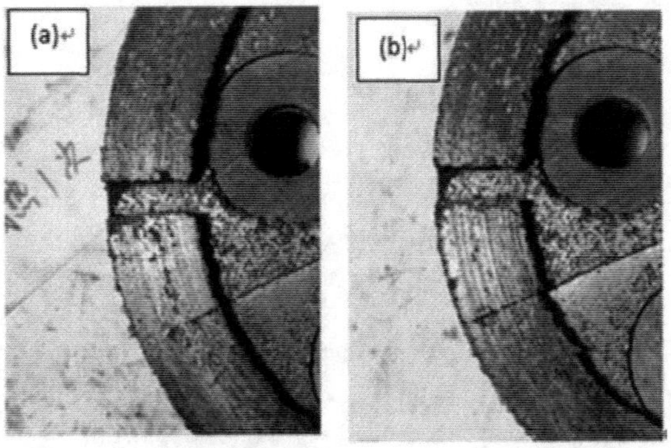

Figure 3. Macromorphology of friction parts under different working conditions (a)6000r/min,(b)8800r/min

The surface morphology and energy spectrum analysis of friction parts and dual parts under different working conditions are shown in Fig. 4 and Fig. 5 respectively. It can be seen from figure 4 (a) and figure 5 (a) that under the working condition of 6000r/min, scratches and slight peeling appear on the surface of the friction member, and the surface of the friction piece contains a small amount of oxygen element. Therefore, under the condition of 6000r/min, abrasive wear and slight oxidation occurred on the surface of the material. From Fig. 4 (b), (c) and Fig. 5 (b), (c), it can be seen that under the working condition of 8800r/min, there are sliding scratches and peeling off along the sliding direction on the surface of the friction piece, there are sliding scratches and peeling off along the sliding direction on the surface of the friction piece, and there are obvious scratches along the sliding direction on the surface of the dual piece, and the oxygen content of the surface of the friction piece and the dual piece is higher than that of the material under the working condition of 6000r / min, and the surface of the counterpart also contains the metal elements contained in the friction member, which proves under the condition of 8800r/min, ploughing, adhesion and oxidation occurred in the friction pair.

Figure 4. Surface Micromorphology of friction part and dual part under different working conditions (a)6000r/min, friction part,(b)8800r/min, friction part,(c)8800r/min, dual part

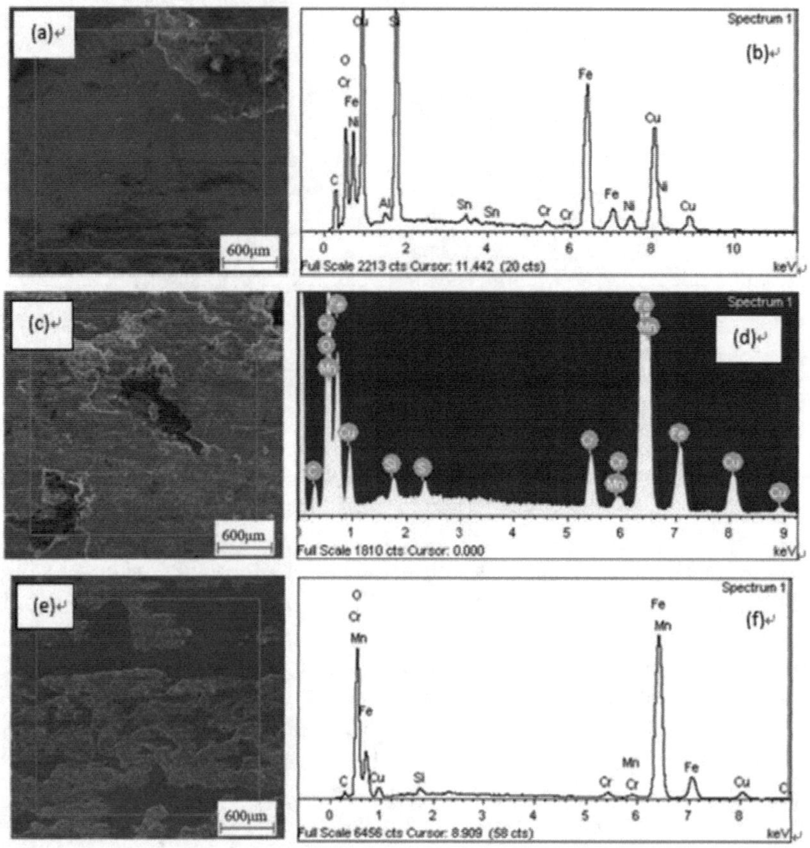

Figure 5. Surface Micromorphology-EDS analysis of friction part and dual part under different working conditions
(a)6000r/min, friction part,(b)8800r/min, friction part,(c)8800r/min, dual part

The morphology and energy spectrum analysis of wear debris under different working conditions are shown in Figure 6 and Figure 7, respectively. It can be known from Figure 6 (a) and Figure 7 (a) that under the working condition of 6000r/min, the wear debris generated by friction material braking has flakes and granules, and the size of the wear debris is less than 200 μ M. During the braking process, the hard particles fall off and produce granular wear debris. The flake wear debris is mainly caused by the failure to effectively generate continuous oxide layer on the wear surface, and the formed oxide layer does not combine with the substrate firmly, easy to fall off from the substrate in the subsequent braking. It can be known from Fig. 6 (b) and Fig. 7 (b) that under the working condition of

8800r/min, the wear debris produced by material braking mainly consists of particles and flakes, and the size of the wear debris is large and less than 500 μ M. The oxide layer is not firmly combined with the substrate, which is easy to fall off and form flake debris. The granular debris is formed by hard falling off.

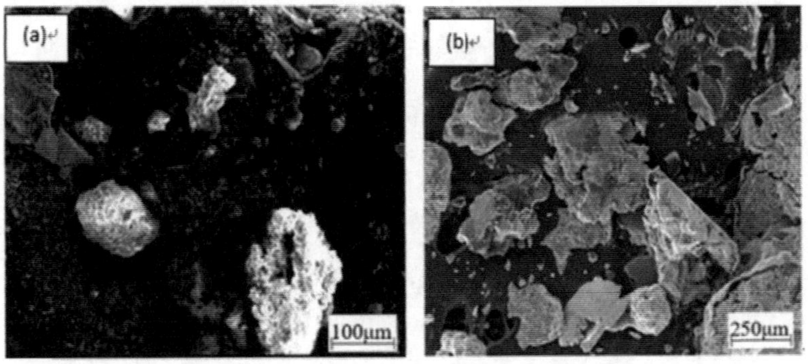

Figure 6. Debris morphology under different working conditions (a)6000r/min,(b)8000r/min

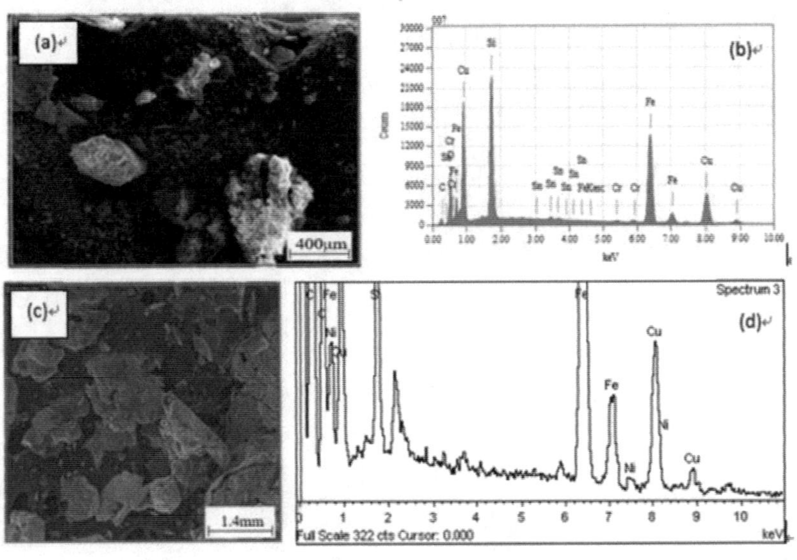

Figure 7. Debris morphology-EDS analysis under different working conditions (a)6000r/min,(b)8800r/min

To sum up: under the condition of rotating speed of 6000r / min, the friction and wear mechanism of the material is mainly abrasive wear and slight oxidation wear; under the condition of rotating speed of 8800r /min, the friction and wear mechanism of the material is mainly abrasive wear, adhesion wear and oxidation wear.

4. Conclusions

The iron-based friction material is mainly composed of iron matrix, graphite and hard phase. Under the working condition of the speed of 6000r/min, the friction and wear mechanism is mainly abrasive wear and slight oxidation wear. At the speed of 8800r/min, the friction and wear mechanism is mainly abrasive wear, adhesion wear and oxidation wear.

5. References

[1] Qingjin Zhang, Jiusheng Bao, Yan Yin, et al. Research and development status of brake friction materials [J]. Surface Friction and Wear and Lubrication, 2016, 45 (11): 32-40.

[2] Qingjin Zhang. Orientation preparation process and friction and wear performance of automotive hard magnetic brake friction material [D]. Beijing: China University of Mining and Technology, 2018.

[3] Xiufei Wang, Cailiu Yin. Application status of powder metallurgy friction materials and requirements for raw materials [J]. Powder metallurgy industry, 2017, 27 (3): 1-6.

[4] Mingfu Tan, Shangwen Huang, Xianjiao Liu, et al. Research on rotor brake materials of Zhiba multipurpose helicopter. Powder metallurgy technology, 1989,7 (3): 149-155.

[5] Translated by Xu Runze. Modern Friction Materials. Metallurgical Industry Press, 1983, 97-99.

[6] Locker K D. Friction materials-on overview. Powder Metallurgy,1992,35(4):253~255.

[7] Lianrong Zhang, Dongsheng Li, Julan Ye. Research on Material Selection of Civil Aviation Aircraft Brake Device. Materials Engineering, 1996, 5: 27-30.

[8] Xiao Yu, Zhimeng Guo, Jian Yang, et al. Effect of Fe content and friction components on properties of copper based PM friction materials [J]. PM technology, 2014, 1 (32): 43-47.

[9] Jie Chen, Xiang Xiong, Pingping Yao, et al. Study on the influence mechanism of friction surface temperature on the friction and wear properties of iron-based friction materials. PM technology, 2004, 22 (4): 223-227.

Study on Preparation, Microstructure and Properties of Micro-Arc Oxidation Ceramic Coating on AZ91 Magnesium Alloy in Phosphate Electrolyte

Xianming Chen and Yingying Fan

School of Electronic and Electrical Engineering, Zhaoqing University, Zhaoqing Avenue, Zhaoqing City, Guangdong Province, China
Email: cxm2050@163.com

Abstract. The micro-arc oxidation behavior of AZ91 magnesium alloy in alkaline phosphate electrolyte was studied. The effect of phosphate on the microstructure and properties of micro-arc oxide film was investigated and characterized by X-ray diffraction (XRD), transmission electron microscope (TEM), scanning electron microscope (SEM), energy dispersive spectrum (EDS) and salt spray test. The results show that phosphate concentration has an important effect on the surface morphology, phase structure， thickness and corrosion resistance of micro-arc oxidation film. With the increase of phosphate concentration from 5g/L to 30g/L, the film quality decreases, the surface roughness increases, the thickness decreases and the corrosion resistance decreases. So in phosphate electrolysis the lower phosphate concentration in the liquid system is helpful to the formation of better film quality.

1. Introduction

AZ91 is the most widely used magnesium alloy, because of its high specific strength and specific stiffness, good damping and shock absorption, good casting performance and easy to be recycled[1-3]. It is widely used in electronic, automotive, aviation and various consumer goods. However, the standard electrode potential of Mg is extremely low(E_0=-2.37V), so the magnesium alloys are vulnerable to corrosion and oxidation in the process of use[4-5]. In order to improve the corrosion resistance and oxidation resistance, now there are three ways to be used. One is the microalloying of magnesium alloys, the second is to develop various coating treatment techniques, the third is to design more reasonable component connection mode and component matching mode to reduce or avoid chemical corrosion. Micro-arc oxidation is a coating treatment technology, which can produce ceramic coating in situ on the surface of magnesium alloy, greatly improving the corrosion resistance and friction resistance of magnesium alloy. The study of micro arc oxidation technology in America, Japan and Germany began in the 70s of last century, but in China only in 90s of last century began to do this research work. At present, the research on micro-arc oxidation technology of Al, Mg, Ti alloys at home and abroad has been extensive and thorough. The research contents include micro-arc oxidation mechanism, electrolyte formula, electric parameters, oxidation process and so on[6-11]. Although a great deal of work has been done on the micro-arc oxidation technology, there is still a long way to go from the maturity of the technology and the large-scale industrial application. At present, only a few enterprises at home and abroad are popularizing and applying, such as ALGT, Microplasmic Corporation and Keronite, etc[12]. Therefore, it is the only way to promote the development of the technology by continuing to study the process parameters and mechanism. In this paper, AZ91 magnesium alloy was used as substrate to prepare ceramic coating in phosphate electrolyte, and the mechanism of film formation was discussed, and the relationship between microstructure and properties was established. The aim is to provide theoretical and experimental support for the development of this technology.

2. Materials and Experimental Procedures

The experimental material was AZ91 magnesium alloy, and its chemical composition(wt.%) shows down:Al:8.56,Mn: 0.82, Zn: 0.41, Mg: Bal.. The specimen was rectangular solid(50mm×30mm×4mm). Their treat procedure followed: magnesium ingot→cutting→drilling→rough grinding→fine grinding →polishing→cleaning→degreasing→micro-arc oxidizing→testing. The power supply style of micro-arc oxidizing was MOP-60 which researched by ourselves. Its power was 42KVA. Electrolytic solution was alkaline solution that its major composition was phosphate(Na_3PO_4:5-30g/L, NaOH:1-5g/L, KF:5-8g/L, $Na_3C_6H_5O_7$:0.5-2g/L, EDTA:0.5-2g/L). The whole reaction process was controlled by constant current, current density was 20A/dm^2, the reaction temperature was 20-40°C, the oxidizing time was 30min.

TEM analysis of micro-arc oxide film was carried out at JEM-2010 HRA produced by Hitachi, Japan. To get TEM thin foils, the samples were pre-thinned by electrolysis with single face firstly. The electrolytic double spray apparatus was Tenupol-2, the electrolyte composition was as follow:5.3gLiCl+11.2gMg$(ClO_4)_2$+100mL$C_6H_{14}O_2$ + 500mLCH_3OH. When spraying the temperature was controlled at about -30°C, the current was 0.12A, the voltage was 40-60V. After double jet thinning, the sample continued to be reduced by ion reducer machine whose style was GL6900, made in China. Thinning working voltage was 5-7KV, the current was 0.15-0.3mA, the incident angle was 15-20°.

The surface morphology of the micro-arc oxide film was observed on Leo 1530V with EDS analysis. The phase structures of the film determined with a X-ray diffractometer(XRD, X-pert MPD.Pro, Philips company, made in Holland). The film thickness was measured with an eddy current coating thickness meter(phynix, made in Germany). The corrosion resistance of the film was evaluated by neutral salt spray test. The test was carried out according to the national standard (GB/T10125-1997, artificial atmosphere corrosion Test—Salt spray Test), the machine style was FQYO/0A(produced by Shanghai Experimental instrument General Factory in China), the corrosion time was 36 hours. The surface roughness of the micro-arc oxidizing film was measured using surface profilometer(BMT expert3D, BMT Company, made in Germany).

3. Results and Discussion

3.1. Effect of Phosphate Concentration on the Process of Micro-Arc Oxidation

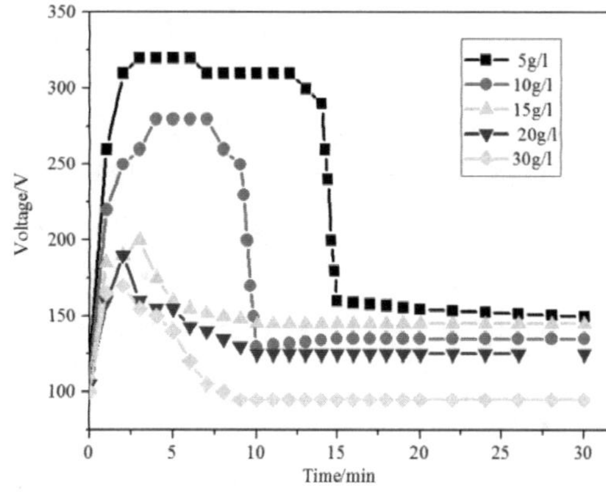

Figure 1. Effect of phosphate concentration on tank voltage in constant current mode

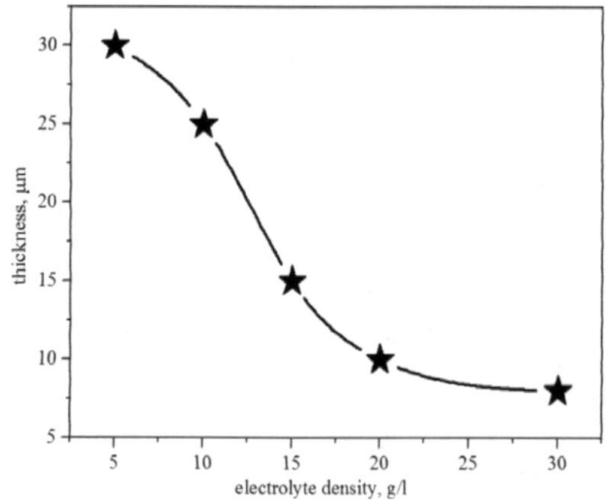

Figure 2. Effect of phosphate concentration on film thickness

Figure 1 studies the effect of phosphate concentration on tank voltage during micro-arc oxidation. According to the curve of the relationship between voltage and treatment time, the micro-arc oxidation process of AZ91 magnesium alloy can be divided into four stages.

The first stage corresponds to the rapid rising phase of the curve, which ranges from a few seconds to a few hundred seconds, which is related to the composition of the electrolyte. There were no arcs on the surface of the samples at this stage, but a large number of bubbles appeared, corresponding to the anodizing stage of the sample, the passivation layer was formed by anodic oxidation, which was a prerequisite for the appearance of micro-arc oxidation. The second stage of the curve is the peak platform stage, when the sample enters the micro-arc oxidation, a large number of small flicker and fast moving arc spots appear on the surface of the sample, accompanied by the explosion sound and a large number of bubbles. The length of the peak platform is determined by the process parameters. The third stage of the curve is the rapid drop in voltage. The duration of this phase is also determined by process parameters, ranging from tens of seconds to hundreds of seconds. At this time, the small arcs on the surface of the samples are less and less, accompanied by the appearance of some larger arcs, which move slowly or even gather in one place. The voltage of the fourth stage of the curve decreases to a certain value basically no change. At this time the surface of the sample basically can not see the arc spot, but there are bubbles coming out, this stage is the phase of extinguishing the arc.

It can be seen from the relationship curve that the concentration of phosphate has a significant effect on the oxidation tank voltage. First, it affects the striking voltage of micro-arc oxidation and the rising speed and duration of the first stage. With the increase of phosphate concentration, the starting arc voltage decreases from 310-320V with the concentration of phosphate is 5g/L to about 180-190V with the concentration of phosphate is 30g/L. One of the explanations is that the increase of ion concentration increases the conductivity of electrolyte, and the increase of conductivity is favorable to the decrease of starting voltage[13-14]. Second, the concentration of phosphate affects the peak platform survival time of micro-arc oxidation. With the increase of phosphate concentration, the peak platform survival time shortens. At 5g/L and 10g/L, the peak platform lasts longer, and the peak platform's duration is shorter than that of 15g/L. The peak platform stage is the normal growth stage of the micro-arc oxide film, and the longer the time is, the more favorable the growth of the film is. Third, the concentration of phosphate affects the voltage drop rate in the third stage of micro-arc oxidation, and decreases with the increase of the concentration. The high falling rate is beneficial to the quality of the oxide film and can reduce the damage caused by the large arc spot on the surface of the film. Fourth, the concentration of phosphate also affects the termination voltage of micro-arc oxidation, the general trend is that the termination voltage decreases as the concentration increases.

Fig.2 shows the relationship between phosphate concentration and film thickness. It can be seen that with the increase of phosphate concentration, the thickness of the film decreases. The thickest film is formed at 5g/L. It is in agreement with the results of the above voltage analysis that at 5g/L the peak platform survival time is longest, which is most favorable to the growth of the oxide film. Therefore, when phosphate is used as main film forming salt, low salt concentration should be selected for micro-arc oxidation.

3.2. Morphologies of MAO Coating

Figure 3 studies the effect of phosphate concentration on the morphology of the surface during micro-arc oxidation with current density 20A/dm^2 in constant current mode. It can be seen from the results that when phosphate concentration is low, the surface of the film is smooth, even and compact when the phosphate concentration is 5g/L and 10g/L(Fig.3a, Fig.3b), and when the concentration is relatively high(15g/L, 20g/L and 30g/L), the surface of the film is relatively rough, uneven and loose(Fig.3c, Fig.3d and Fig.3e).

Fig.4 shows the relationship between the roughness and the phosphate concentration. The surface roughness of the film increases with the increase of phosphate concentration, which is consistent with the observation by SEM. The results of the analysis in figure 1 show that the phosphate concentrations of 5g/L and 10g/L allow the sample to have a longer time of micro-arc oxidation at the peak platform stage, which corresponds to small and dense arcs. As a result, the surface of the film is smooth and compact, and the time of rapid voltage drop is very short, the corresponding larger arc spot appears in a shorter time, and the damage to the surface of the film is also relatively slight. Therefore, The higher the mass of the film formed in the micro-arc oxidation stage, the more favorable to reduce the destruction of the large arc spot. However, the peak platform duration of phosphate concentration of 15g/L,20g/L and 30g/L is shorter, which is not conducive to the growth of the film, and the time of falling stage is longer, the duration of large arc-spot is longer, and the damage effect of the film is greater[15-17]. The region marked by A , B, and C(Fig. 3c, Fig. 3d, Fig. 3e) should be the result of a large arc spot effect. Therefore, in order to obtain smooth, compact and high quality coatings in phosphate electrolyte, the micro-arc oxidation should be carried out at lower concentration.

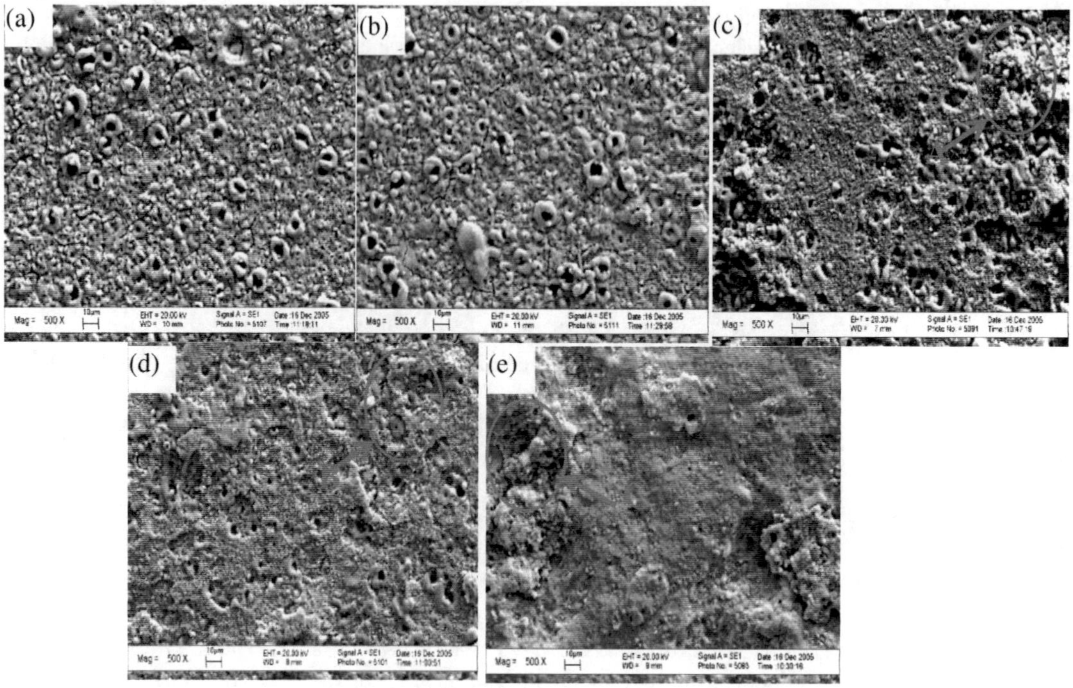

Figure 3. Effect of phosphate concentration on surface morphologies of MAO film (a)5g/L; (b)10g/L;(c)15g/L;(d)20g/L;(e)30g/L

3.3. Phase Structure Analysis of MAO Film

Figure 5 is the result of EDS analysis on the surface of the film. It is shown that the surface of the film is mainly composed of Mg, O, P, Al and a small amount of K, Na elements (the specific content is shown in Table 1). Among them, Mg and Al come from the matrix of the AZ91 magnesium alloy, P, K and Na come from the electrolyte.

XRD patterns of MAO films with different phosphate concentrations are shown in Fig.6. The micro-arc oxide films consist mainly of MgO, $Mg_3(PO_4)_2$, $MgAl_2O_4$ and Mg. The diffraction peak of MgO and $Mg_3(PO_4)_2$ intensity is strongest, combining the results of EDS, the two phases are the most abundant in the MAO coatings. However, the K and Na appeared in EDS analysis did not show the corresponding phase in XRD analysis. These two elements should be distributed in the surface layer of the film as the adsorbed form, but the amount of K and Na should be relatively small compared with the whole film layer, so there is no corresponding diffraction peaks.

3.4. TEM Analysis of MAO Coating

Figure 7 shows the TEM observation of the micro-arc oxidation film. Fig.7a is a blight field image, Fig.7.b is a dark field image(*g=(200)*), and Fig.7c is a selective diffraction ring and an amorphous diffraction disk. Through the standardization to the diffraction pattern, it is found that there are MgO and $MgAl_2O_4$ of crystalline and amorphous states in the region. The crystalline substance mainly exists in nanocrystalline and microcrystalline form in the film (Fig.7a and Fig.7b).

Micro-arc oxidation of magnesium alloys is a plasma discharge process, and the instantaneous temperature in the plasma region can reach 8000K[18]. The matter in this region will be melted and sintered by plasma discharge, which will be condensed to form sintered oxide in the electrolyte system. The temperature difference between the plasma region and the electrolyte is very large, so the cooling rate of the material formed in the plasma discharge region is very high, up to 10^8K/s [19]. At such a high cooling rate, the grain nucleation and growth time of the micro-arc oxide film is very short, according to nucleation and growth kinetics, the grains formed under such conditions are very small. Therefore, the grains formed in the process of micro-arc oxidation mainly exist in nanocrystalline and microcrystalline morphology.

According to TEM observation, there are some amorphous substances in the micro-arc oxide film. The formation condition of amorphous phase is also due to the high cooling rate. When the cooling rate reaches 10^6K/s, amorphous phase can be formed[20-22]. However, the cooling rate in the process of micro-arc oxidation can reach 10^8K/s, so there are some amorphous substances in the ceramic coating.

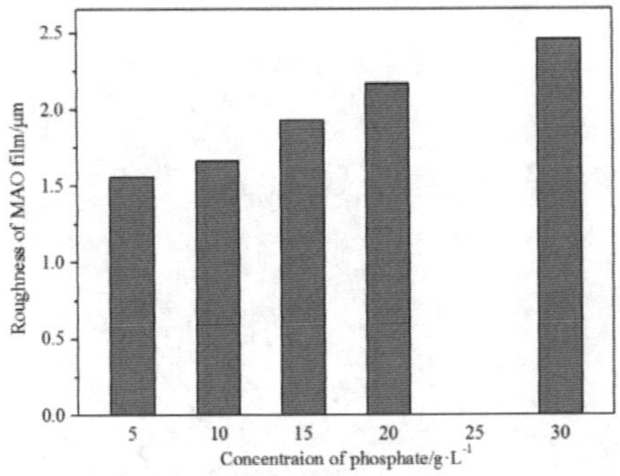

Figure 4. Effect of phosphate concentration on surface roughness of MAO film

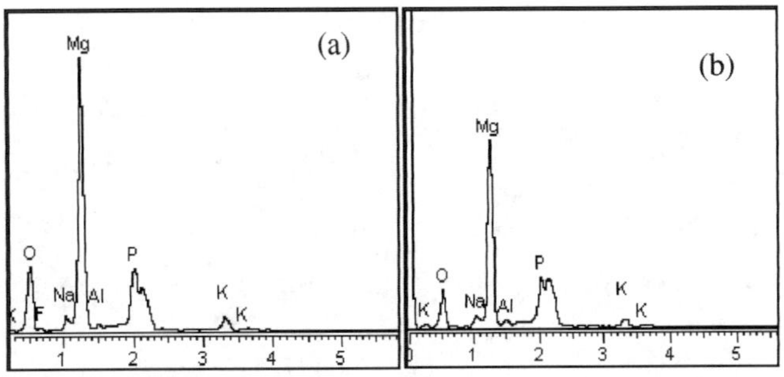

Figure 5. EDS spectra of MAO film (a) Na_3PO_4: 5g/L; (b) Na_3PO_4: 30g/L

Table 1. Content of related elements in EDS spectra

P(g/L)	Composition (wt.%)					
	Mg	O	P	Al	Na	K
5	46.12	34.21	11.07	2.43	2.72	3.45
30	47.23	34.26	9.88	2.65	2.75	3.23

3.5. Effect of Phosphate Concentration on Corrosion Performance

Figure 8 studies the effect of phosphate concentration on corrosion rate. The film was prepared by micro-arc oxidation of 30min with current density 20A/dm^2 in constant current mode. It can be seen from the diagram that the corrosion rate increases with the increase of phosphate concentration, lowest at 5g/L, highest at 30g/L. From the previous discussion, we know that the quality of the film is the best when phosphate concentration is 5g/L, so the corrosion resistance is also the best. The quality of the film formed at 30g/L is the worst, so the corrosion resistance is also the worst. Therefore, in order to obtain a high corrosion resistance film, a lower phosphate concentration should be selected.

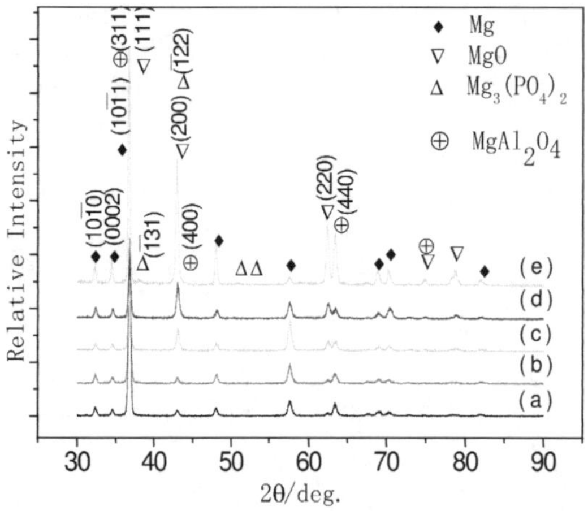

Figure 6. XRD patterns of MAO film (a)30g/L; (b)20g/L;(c)15g/L;(d)10g/L;(e)5g/L

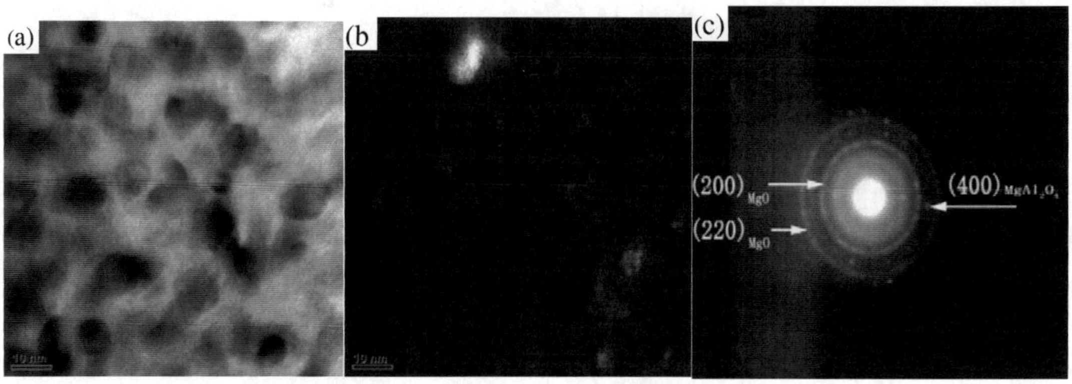

Figure 7. TEM image of MAO coatings(5g/L)
(a)TEM image of the mixed micro-crystalline and amorphous MgO(BF); (b)DF, g=(200); (c) a selective diffraction ring and an amorphous diffraction disk

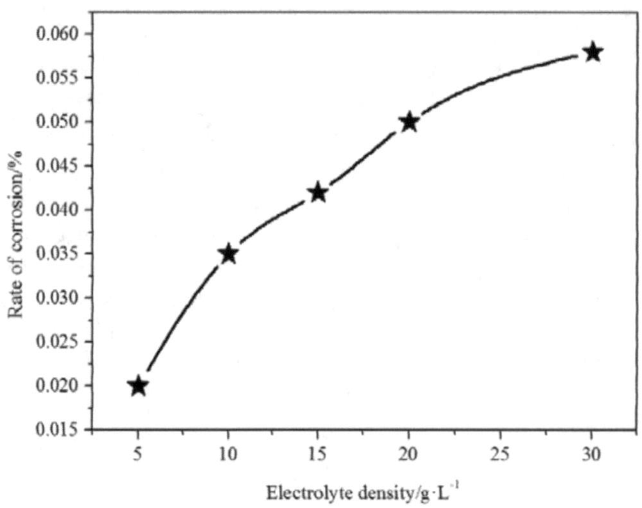

Figure 8. Effect of phosphate concentration on corrosion performance

4. Conclusions

(1) The process of micro-arc oxidation in constant current mode can be divided into four stages, which are fast rising stage, peak platform stage, fast descending stage and arc extinguishing stage. The survival time of the peak platform has an important effect on the quality and performance of the film.

(2)When micro-arc oxidation is carried out in phosphate system, the film formed with lower phosphate concentration had better quality, lower surface roughness and higher corrosion resistance.

(3)The MAO films prepared in phosphate electrolyte contains most crystalline substances and a small amount of amorphous substances, the crystalline substances exists in nanocrystalline and microcrystalline.

5. Acknowledgments

This project was supported by the post doctoral foundation of Central South University(126226).

6. References

[1] Wang L, Zhao Y Q, Chen H M 2018 Acta Metall. Sinica 31(1) 63
[2] Lin H, Yang M B, Tang H Prog. 2018 Nat. Sci.:Mater. Int. 1 66
[3] Zhang D D, Zhang D P, Bu F Q 2017 Mate. Sci. Eng. A 693(2) 51
[4] Sharifi P, Jamali J, Sadayappan K 2018 J. Mater. Sci. Techno. 34(2) 324
[5] Cui L Y, Liu H P, Zhang W L 2017 J. Mater. Sci. Technol. 33(1) 1263
[6] Zhao Y B, Shi L Q, Cui L Y 2018 Acta Metall. Sinica 31(2) 180
[7] Laleh M, Kargar F, Rouhaghdam A S 2012 J. Rare Earths 30 (12) 1293
[8] Wang P, Li J P, Guo Y C 2010 J. Rare Earths 28 (5) 798
[9] Wang Z X, Lv W G, Chen J 2013 Rare Metals 32 (5) 459
[10] Veys-renaux D, Barchiche C E, Rocca E. 2014 Surf. Coat. Technol. 251 (8) 232
[11] Rehman Z U, Jeong Y S, Koo B H 2015 Korean J. Mater. Research 25 (10) 509
[12] Guo X W, Guo J C, Zhang Z C 2017 Surf. Technol., 46(3) 53
[13] Jiang Z H, Li S, Yao Z P 2006 Mater. Sci. Techonol. 14(5) 460
[14] Jiang B L, Zhang X F 2005 Rare Met. Mater. Eng. 34(3) 393
[15] Dunleavy C S, Golosnoy I O, Curran J A2009 Surf. Coat. Technol. 203(22) 3410
[16] Nominė A, Troughton S C, Nomminė A V 2015 Surf. Coat. Technol. 269(1) 125
[17] Dunleavy C S, Curran J A, Clyne TW 2013 Applied Surf. Sci. 268(268) 397
[18] Krysmann W, Kurze P, Dittrich K H 1984 Cryst. Res. Techol. 973
[19] Yerokhin A L, Nie X, Leyland A 1999 Surf. Coat. Technol. 122(2-3) 73
[20] Wang Q, Li Q C , Chang G W 2003 J. Liaoning Inst. Technol. 23(5) 40
[21] Katgerman L, Dom F 2004 Mater. Sci. Eng. A 375–377(21) 1212
[22] Sugamata M, Hanawa S, Kaneko J 1997 Mater. Sci. Eng. S226-228(1) 861

IOP Conf. Series: Materials Science and Engineering **730** (2020) 012030 doi:10.1088/1757-899X/730/1/012030

Research on Cumulative Damage Evaluation Method for Structural Impact on Airborne Armored Vehicle in Landing Process

Xinhua Fu[1,*], Yong Lu[2], Jun Wang[1] and Xueyang Chen[1]

[1]Airborne Training Center, Guilin, Guangxi, 541003, China.
[2]Scholl of Computer Science, University of Electronic Science and Technology of China, Zhongshan Institute, Zhongshan, 528402, China.
*Email[1]: fuxinhua_003@163.com

Abstract. Since it was very costly and complicated to analyze cumulative damage caused by structural impact on airborne armored vehicle in landing process using real equipment airdrop test, finite element method(FEM) was taken to build a finite element(FE) model of airborne armored vehicle and airbag system to simulate the landing impact process of airborne armored vehicle. In combination with Lemaitre Damage Model and the damage evolution law of materials, the cumulative damage caused by structural impact of vehicle was calculated. The cumulative damage of structure of airborne armored vehicle in multiple landing processes was assessed, results of which were expected to provide theoretical guidance for the operational use and maintenance of airborne armored vehicle.

1. Introduction

Impact load was one of the main factors causing structural damage to airborne armored vehicle in the landing process. For airborne armored vehicle with thin-shelled and complicated structure, the impact produced in the landing and buffering period would cause plastic deformation and structural damage to its structure partially. Such structural damage, however, can be transmitted and cumulated due to the repeatable airdrop characteristic of airborne armored vehicle[1]. Undoubtedly, cumulative damage to the body of airborne armored vehicle would greatly affect its technical conditions during the subsequent service. At present, two methods are mainly adopted in researches on the cumulative damage caused by structural impact on airborne armored vehicle under the condition of landing impact, namely, real equipment airdrop test and numerical simulation. Influenced by factors such as climate and ground environment, the landing conditions for airborne armored vehicle varied widely along with the challenge of prediction. As a result, a huge investment in manpower, material and financial resources would be required to conduct real equipment airdrop tests under multiple landing conditions. At the beginning of developing and finalizing airborne armored vehicle, real equipment airdrop tests had been conducted, but the test data was too limited to meet research requirements for cumulative damage from structural impact[1-2]. In recent years, the development of computer technology and the theory of finite element have provided a new approach for researching the cumulative damage caused by structural impact on airborne armored vehicle in the landing process. With the method of finite element, a finite element model of airborne armored vehicle and airbag system was built, and a non-linear finite-element model was used to simulate the process of landing impact on airborne armored vehicle. Meanwhile, in combination with Lemaitre Damage Model and the damage evolution law of materials, the cumulative damage from structural impact was calculated, and an analysis was conducted toward the cumulative damage evaluation for the structure of airborne armored vehicle during repeated landing, so as to provide theoretical guidance for the operational use and maintenance of airborne armored vehicle.

2. Finite Element Analysis Model

2.1. Lemaitre Damage Model

Based on the theory of energy damage, Lemaitre presented a model for evaluating the structural damage under significant plastic deformation, which is presented as follows [3]:

$$D = D_R \frac{\varepsilon_p \varepsilon_t - \varepsilon_D}{\varepsilon_R - \varepsilon_D}$$

Where, D, D_R and ε_p represent damage variable, damage limit value and cumulative plastic strain respectively, ε_D and ε_R correspond to plastic strain of damage threshold value and damage limit value, and ε_t reflects the influence of triaxial stress ratio on material damage, which is called triaxial stress factor.

$$\varepsilon_t = \frac{2}{3}(1+v) + 3(1-2v)\left(\frac{\sigma_H}{\sigma_{eq}}\right)^2$$

Under uniaxial stress, $\sigma_H = \sigma / 3$, $\sigma_{eq} = \sigma$, $\varepsilon_t = 1$.

In Lemaitre Damage Model, the damage model parameters ε_D, ε_R and D_R can be obtained by measuring the variation of elasticity modulus of materials.

2.2. Analysis Process

The process of analyzing the cumulative damage caused by structural impact on airborne armored vehicle in landing process is shown in Figure 1.

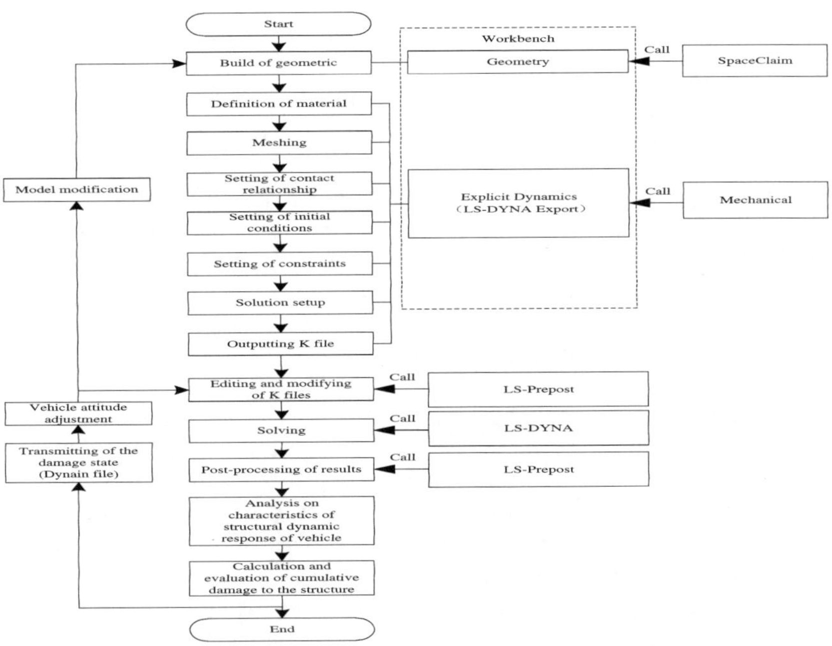

Figure 1. Process of Analyzing the Cumulative Damage caused by Structural impact on Airborne Armored Vehicle in Landing Process

(1) The establishment of File K for model analysis. Based on the software of ANSYS, the simulation and analysis model for the landing and buffering process of airborne armored vehicle was built in accordance with steps such as geometric modeling, material definition, meshing, contact relation setting, initial condition setting, constraints setting, solution setup, File K output, and File K edition and modification. Mesh refinement was conducted towards the parts and components such as the main support structure of vehicle body and the engine support. Johnson-cook Model was selected as the material model of vehicle body[4]. The established finite element model of airborne armored vehicle and airbag system is shown in Figure 2.

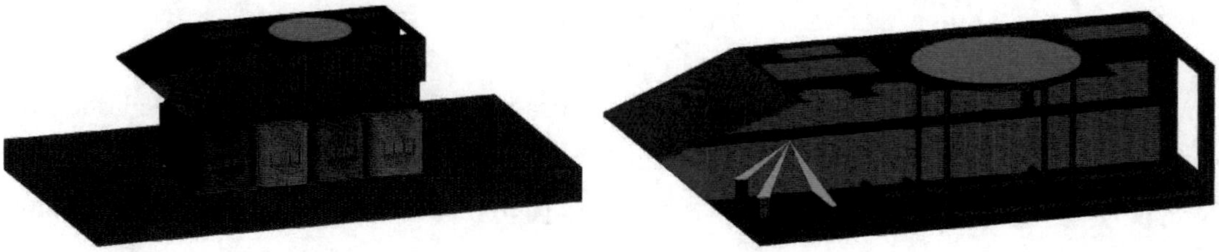

Figure 2. Finite Element Model of Airborne Armored Vehicle and Airbag System

(2) Solving of File K. LS-DYNA solver was called to solve the analysis model, that is, the modified File K.

(3) Viewing of simulation results. The output data obtained by simulation calculation was imported into LS-Prepost to view the results data and related curves, such as stress contours, stress change curve, and plastic strain contours.

(4) Analysis on characteristics of structural dynamic response of vehicle. Based on the simulation data, characteristics of structural dynamic response of vehicle in the landing process this time were analyzed.

(5) Transmitting of the damage state. The damage distribution of vehicle under the i-th landing impact was used as the initial condition of the $(i+1)$-th calculation, that is, file Dynain with the stress-strain data output from the i-th simulating calculation was imported into file K, the analysis model established in step (1), and the vehicle attitude was adjusted simultaneously.

(6) Calculation and evaluation of cumulative damage to the structure. Step (2) to (4) were repeated to obtain characteristics of structural dynamic response of vehicle in the $(i+1)$-th landing process. In combination with Lemaitre Damage Model and the damage evolution law of materials, cumulative damage to the structure was calculated.

(7) Similarly, by repeating step (5) to (6), cumulative damage to the structure of vehicle in the $(i+2)$-th landing process could be figured out via simulating calculation, thus obtaining the cumulative damage to the structure of vehicle under repeated airdrop, the process of which continued until the destroying threshold of components was met.

3. Simulating Calculation of Cumulative Damage

3.1. Simulation Results

Based on the established finite element model of airborne armored vehicle and airbag system, simulating calculation was carried out towards the landing and buffering process of airborne armored vehicle under typical airdrop conditions, so as to obtain the simulation results such as stress contours, stress change curve, and plastic strain contours. In combination with Lemaitre Damage Model and the damage evolution law of materials, the cumulative damage from structural impact was calculated, and an analysis was conducted towards the cumulative damage evaluation for the structure of airborne armored vehicle under repeated airdrop. When the vertical landing speed is 8m/s over a plain, the value of cumulative damage to typical structural parts of airborne armored vehicle in the landing process is shown in Table 1, and when the speed is 9m/s, the value is shown in Table 2, where the italic number indicates that the element number corresponding to the maximum stress value is

IOP Conf. Series: Materials Science and Engineering **730** (2020) 012030 doi:10.1088/1757-899X/730/1/012030

different from that corresponding to the maximum plastic strain on a certain part or component. If a certain component contains two element numbers, it suggests that the position of maximum stress is changed between the two consecutive times of airdrop.

Table 1. Cumulative Damage to Structure of Typical Vehicle Parts at the Vertical Landing Speed of 8m/s

Time of Airdrop	Typical Damage Part	Specific element number	Max stress value (MPa)	Max plastic stress (%)	Damage Value D	Variation D_{i+1}-D_i
The 1st time	Baseboard of armor plate	361583	216.88	0	0	0
	Bottom of the left upright in the front of vehicle	202515	304.95	0	0	0
	Bottom of the right upright in the front of vehicle	202183	348.21	0	0	0
	Bottom of the left upright at the back of vehicle	201906	320.58	0	0	0
	Bottom of the right upright at the back of vehicle	202330	361.42	0	0	0
	Bottom of the support at the back of engine	265729	137.14	0	0	0
	Bottom of the left support in the front of engine	265730	185.02	0	0	0
	Bottom of the right support in the front of engine	268979	177.47	0	0	0
The 2nd time	Baseboard of armor plate	361583	233.71	0	0	0
	Bottom of the left upright in the front of vehicle	202515	351.61	0	0	0
	Bottom of the right upright in the front of vehicle	202183	394.39	0	0	0
	Bottom of the left upright at the back of vehicle	201906	361.73	0	0	0
	Bottom of the right upright at the back of vehicle	202330	395.20	0	0	0
	Bottom of the support at the back of engine	265729	233.54	0	0	0
	Bottom of the left support in the front of engine	265730	211.85	0	0	0
	Bottom of the right support in the front of engine	268979	198.50	0	0	0
The 3rd time	Baseboard of armor plate	361583	223.85	0	0	0
	Bottom of the left upright in the front of vehicle	202515	331.76	0	0	0
	Bottom of the right upright in the front of vehicle	202183	398.65	0	0	0
	Bottom of the left upright at the back of vehicle	201906	338.29	0	0	0
	Bottom of the right upright at the back of vehicle	202330	374.89	0	0	0
	Bottom of the support at the back of engine	265729	200.02	0	0	0
	Bottom of the left support in the front of engine	265730	172.75	0	0	0
	Bottom of the right support in the front of engine	268979	268.01	0	0	0
The 4th time	Baseboard of armor plate	361583	180.25	0	0	0
	Bottom of the left upright in the front of vehicle	202515	304.23	0	0	0
	Bottom of the right upright in the front of vehicle	202183	325.80	0	0	0
	Bottom of the left upright at the back of vehicle	201906	296.96	0	0	0
	Bottom of the right upright at the back of vehicle	202330	306.15	0	0	0
	Bottom of the support at the back of engine	265729	224.54	0	0	0
	Bottom of the left support in the front of engine	265730	172.08	0	0	0
	Bottom of the right support in the front of engine	268979	192.31	0	0	0

IOP Conf. Series: Materials Science and Engineering **730** (2020) 012030 doi:10.1088/1757-899X/730/1/012030

Table 2. Cumulative Damage to Structure of Typical Vehicle Parts at the Vertical Landing Speed of 9m/s

Time of Airdrop	Typical Damage Part	Specific element number	Max stress value (MPa)	Max plastic stress (%)	Damage Value D	Variation D_{i+1}-D_i
The 1st time	Baseboard of armor plate	320806	487.50	0.46	0.017	-
	Bottom of the left upright in the front of vehicle	143924	513.52	1.50	0.055	-
	Bottom of the right upright in the front of vehicle	202183	518.64	0.92	0.034	-
	Bottom of the left upright at the back of vehicle	201906	522.67	0.97	0.035	-
	Bottom of the right upright at the back of vehicle	202330	525.42	1.01	0.037	-
		150147	458.97	0.03	0.001	
	Bottom of the support at the back of engine	265729	459.91	0.06	0.002	-
	Bottom of the left support in the front of engine	265730	497.84	0.61	0.022	-
	Bottom of the right support in the front of engine	268979	513.66	1.43	0.052	
The 2nd time	Baseboard of armor plate	320806	490.19	0.77	0.028	0.011
		289279	435.66	0.89	0.032	-
	Bottom of the left upright in the front of vehicle	143924	516.49	1.75	0.064	0.009
	Bottom of the right upright in the front of vehicle	202183	496.88	1.07	0.039	0.005
		137123	587.77	3.90	0.142	-
	Bottom of the left upright at the back of vehicle	201906	501.46	1.25	0.046	0.011
		137693	594.41	4.04	0.147	-
	Bottom of the right upright at the back of vehicle	202330	493.77	1.06	0.039	0.002
		150147	582.65	3.29	0.120	0.119
	Bottom of the support at the back of engine	265729	459.31	0.06	0.002	0
	Bottom of the left support in the front of engine	265730	510.94	1.10	0.040	0.018
	Bottom of the right support in the front of engine	268979	556.57	2.79	0.102	0.050
The 3rd time	Baseboard of armor plate	289279	538.61	2.55	0.093	0.061
	Bottom of the left upright in the front of vehicle	143924	614.84	4.59	0.167	0.103
	Bottom of the right upright in the front of vehicle	137123	695.77	6.20	0.226	0.084
	Bottom of the left upright at the back of vehicle	137693	683.89	5.46	0.199	0.052
	Bottom of the right upright at the back of vehicle	150147	756.34	7.25	0.264	0.144
	Bottom of the support at the back of engine	265729	478.95	0.36	0.013	0.011
	Bottom of the left support in the front of engine	265730	495.98	1.12	0.041	0.001
		265734	509.67	1.41	0.051	-
	Bottom of the support at the back of engine	268979	590.92	3.68	0.134	0.032
The 4th time	Baseboard of armor plate	289279	647.50	4.79	0.175	0.082
	Bottom of the left upright in the front of vehicle	143924	402.77	4.59	0.167	0
		143950	491.45	3.36	0.123	-
	Bottom of the right upright in the front of vehicle	137123	765.06	7.90	0.288	0.062
		137085	791.46	8.17	0.298	-
	Bottom of the left upright at the back of vehicle	137693	857.54	10.77	0.393	0.194
		137732	899.81	11.16	0.407	-
	Bottom of the right upright at the back of vehicle	150147	889.26	11.35	0.414	0.150
		150185	939.92	11.90	0.434	-
	Bottom of the support at the back of engine	265729	539.94	2.22	0.081	0.068
	Bottom of the left support in the front of engine	265734	556.19	2.84	0.103	0.052
	Bottom of the right support in the front of engine	268979	611.90	4.60	0.168	0.034

3.2. Result Analysis

(1) In case of repeated landing under normal conditions, that is, when the vertical landing speed was lower than 8m/s over a plain, the structural stress of vehicle would not exceed the yield limit of materials and no damage would be caused to vehicle structure, which implies that it was safe for airborne armored vehicle to implement repeated airdrop under normal conditions.

(2) When the vertical landing speed of vehicle was 9m/s, plastic strain would arise in vehicle structure, with the plastic strain contours of the whole vehicle structure shown in Figure 3, from which it could be seen that the majority of structural damages of the whole vehicle were subtle. Major

damages mainly concentrated around the root of the bottom of the four turret uprights (four uprights of the main support of vehicle body) and around the front-left vertex of the front-right engine support. The value of plastic strain on the left root of the bottom of front-left upright of the main vehicle support was 5.5%, and the value of plastic strain on the front-right vertex of the bottom of front-right engine support was 5.2%. Therefore, it could be said that the whole vehicle was relatively reasonable in structural design with full consideration of the huge load on parts such as the upright and the engine support where counter measures had been reinforced correspondingly. In consideration of increasing the strength of vehicle body and its impact resistance, it is recommended that the bottom of front-left upright and that of front-right engine support should be furthered reinforced.

(3) In repeated landing of airborne armored vehicle at a speed of 9m/s, by comparing the simulation results of repeated airdrop thereof, it could be seen that the distribution law of stress and plastic strain of the whole vehicle structure was similar, and with the increase of airdrop times, the maximum stress value, maximum plastic strain and damage value of the vehicle would all gradually rise, presenting a distinct accumulative effect of plastic strain. When airdrop was conducted for the fourth time, large plastic strain would occur to vehicle structure, with partial positions exceeding the plastic strain corresponding to the limit value of material damage and the damage value of the bottom of right upright in the front of vehicle and that of the left upright at the back of vehicle exceeding the secure threshold. It indicates that three times of airdrop could be executed at most for airborne armored vehicle at the vertical landing speed of 9m/s. Seen from the contours of effective plastic strain of whole vehicle structure in the 4th airdrop (Figure 4), the structure of vehicle had been severely deformed, which had affected the normal use of airborne armored vehicle. Therefore, the mission of airdrop failed.

Figure 3. Plastic Strain Contours of the Whole Vehicle Structure at the Vertical Landing Speed of 9m/s for the First Time

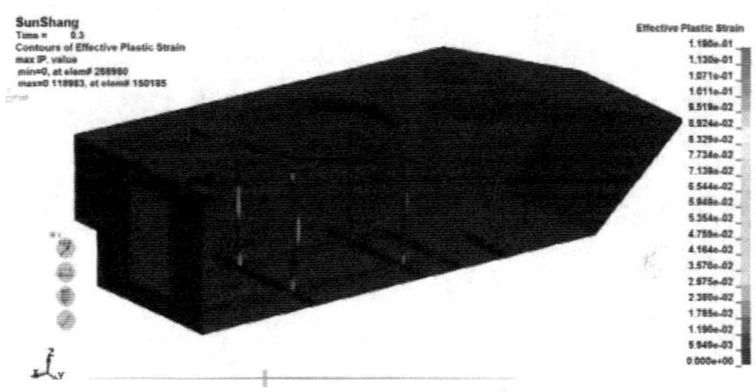

Figure 4. Plastic Strain Contours of the Whole Vehicle Structure at the Vertical Landing Speed of 9m/s for the Fourth Time

(4) In repeated landing of airborne armored vehicle at a speed of 9m/s, by comparing the stress change curve of the 143924th element which is at the bottom of the left upright in the front of vehicle (Figure 5), the stress change curve of the 150147th element which is at the bottom of the right upright at the back of vehicle (Figure 6), and the stress change curve of the 268979th element which is at the bottom of the right support in the front of engine (Figure7) in previous two times of airdrop, the stress state of these positions had changed already due to plastic strain in the first time of airdrop, while the existence of residual stress had led to initial stress and strain already at the beginning of the 2nd simulation of landing. In the front section of the curves, that is, before the landing of airborne armored vehicle, the stress change curve in the 1st airdrop was on monotonically increase, and the stress rapidly increased when the vehicle was about to touch the ground, making the curve relatively steep. While the stress change curve in the 2nd airdrop presents an oscillating phenomenon, but in the same way, the stress increased rapidly when the vehicle was about to touch the ground, making the curve relatively steep. The reason for oscillation of the stress change curve in the 2nd airdrop was that residual stress existed in the 1st airdrop. In the rail section of the curve, that is, after the landing of the vehicle, the stress change curves in the two times of airdrop both presented oscillation, with the impact

energy converted into structural deformation, where elastic deformation existed partially and the energy it stored could be dispersed through vibration.

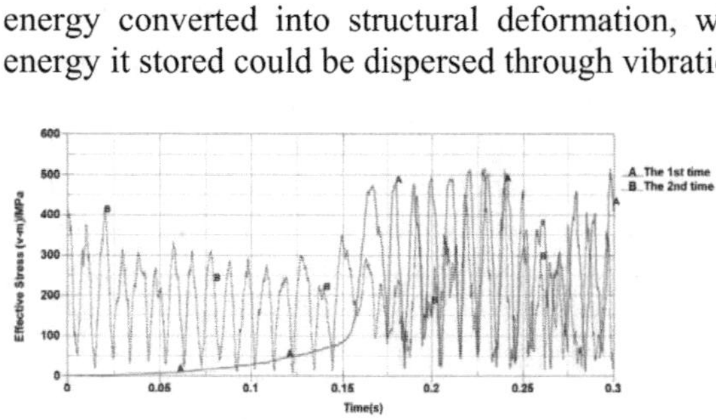

Figure 5. Comparison of the Stress Change Curve of the 143924th Element in Previous Two Times of Airdrop

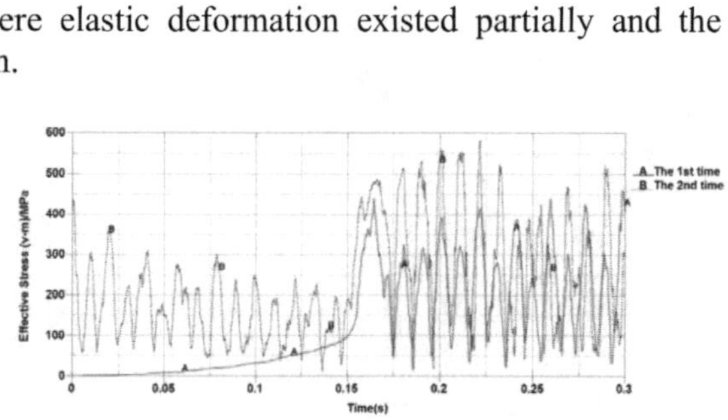

Figure 6. Comparison of the Stress Change Curve of the 150147th Element in Previous Two Times of Airdrop

(5) In repeated landing of airborne armored vehicle at a speed of 9m/s, by comparing the plastic strain change curve of the 143924th element at the bottom of the left upright in the front of vehicle (Figure 8), the plastic strain change curve of the 150147th element at the bottom of the right upright at the back of vehicle (Figure 9), and the plastic strain change curve of the 268979th element at the bottom of the right support in the front of engine (Figure10) in four times of airdrop, it could be found that the plastic strain value in the same position was gradually cumulated by a certain trend rather than remain unchanged, that is to say, the damage value was cumulated gradually. In each time of airdrop, the stress of the whole vehicle structure was usually the maximum when the vehicle was about to touch the ground. At this moment, a sudden increase would happen to the plastic strain. However, the increase of plastic strain value did not follow the linear law each time, which indicates that the damage value did not always increase linearly.

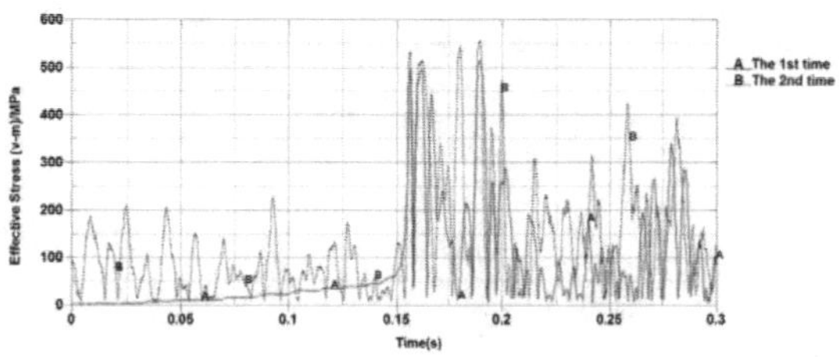

Figure 7. Comparison of the Stress Change Curve of the 268979th Element in Previous Two Times of Airdrop

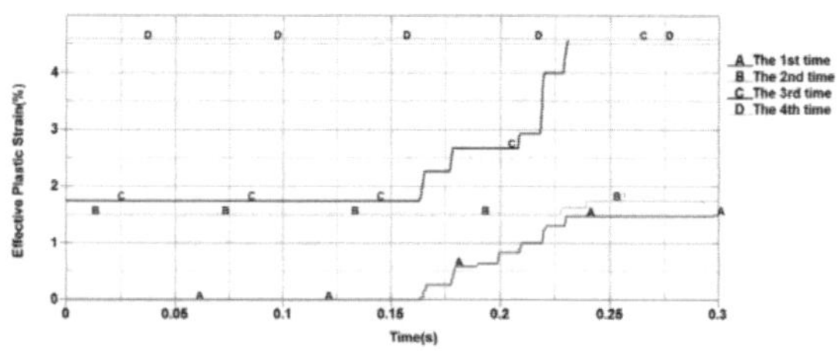

Figure 8. Comparison of the Plastic Strain Change Curve of the 143924th Element in Four Times of Airdrop

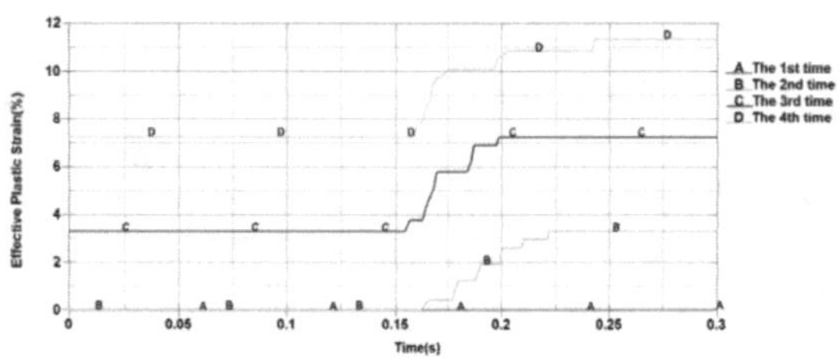

Figure 9. Comparison of the Plastic Strain Change Curve of the 150147th Element in Four Times of Airdrop

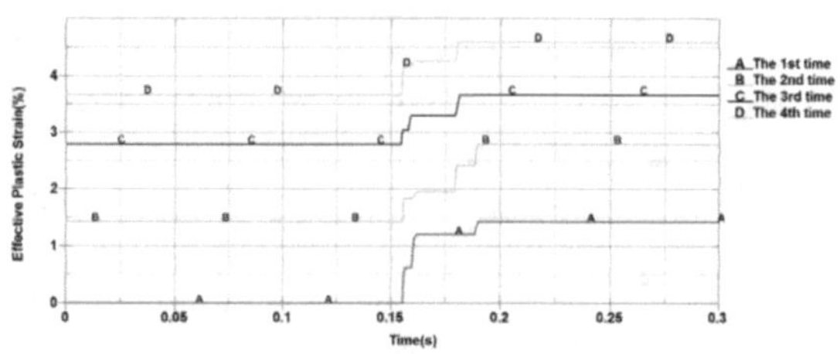

Figure 10. Comparison of the Plastic Strain Change Curve of the 268979th Element in Four Times of Airdrop

4. Conclusions

With finite element method, a finite element model of airborne armored vehicle and airbag system was established to simulate the landing process of airborne armored vehicle. Through simulating calculation of the landing and buffering process of airborne armored vehicle under typical airdrop conditions, an analysis was conducted towards the cumulative damage evaluation for the structure of airborne armored vehicle in repeated airdrop. Conclusions are obtained as follows: 1) when the vertical landing speed is lower than 8m/s over a plain, the structures stress in the vehicle would not exceed the yield limit of materials, thus no damage would be caused to the vehicle structure. Therefore, the airborne armored vehicle is secure in repeated airdrop under normal conditions; 2) Three times of airdrop can be executed at most for airborne armored vehicle when the vertical landing speed is 9m/s; 3) In repeated landing under the same airdrop conditions, increase of the value of structural damage is non-linear.

5. References

[1] Li Jian-yang, Wang Hong-yan, Rui Qiang, Hong Huang-jie, Zhang Fang. Research on Cumulative Damage Assessment Method for Airborne Vehicle at Landing Based on Finite Element Analysis [J]. Journal of System Simulation, 2014.1, 26(1)pp 208-214

[2] Li Jian-yang, Wang Hong-yan, Hao Gui-xiang. Simulation of Landing Process Based on Explicit Finite Element Method [J]. Journal of Academy of Armored Force Engineering, 2010.6, 24(3)pp 25-28

[3] J Lemaitre. A continuous damage mechanics model for ductile fracture[J]. Journal of Engineering Materials and Technology, Transactions of the ASME (s0094-4289), 1985, 107(1) pp 83-89

[4] LI Yun-fei, Zeng Xiang-guo, Liao Yi. Thermal-viscoplastic constitutive relation of Ti-6A1-4V alloy and numerical simulation by modified Johnson-Cook modal [J]. The Chinese Journal of Nonferrous Metals, 2017.7, 27(7)pp 1419-1425

IOP Conf. Series: Materials Science and Engineering **730** (2020) 012031 doi:10.1088/1757-899X/730/1/012031

Numerical Simulation of a Bionic Blade on Small-Scale Wind Turbine

Kuang-Yuan Kung[1], Jun-Liang Chen[2*], Cheng-Hsing Hsu[3] and Wei-Feng Tsai[3]**
[1] Institute of Mechanical and Electrical Engineering, Lingnan Normal University, Guangdong Zhanjiang 524048, China kykung@lingnan.edu.cn
[2*]Department of Mechanical Engineering, Wu Feng University, Chiayi 62153, Taiwan, ROC jlchen@wfu.edu.tw
[3**]Department of Mechanical Engineering, Chung Yuan Christian University, Chungli 32023 Taiwan, ROC chhsu88@gmail.com;chhsu@cycu.edu.tw

Abstract. This study presents the aerodynamic characteristics of a small-scale horizontal axis wind turbine (HAWT) with bionic blade contours. Taiwan incense cedar was chosen as imitations. On account of the contour of the winged disseminules and the ANSYS FLUENT 14, we designed the bionic blades to process the computations of aerodynamic characteristics of the bionic wind turbine blades. The aerodynamic characteristics and the wind/electricity power transform efficiency of the turbines were examined in a wind tunnel. The wind tunnel experimental outcomes showed that the wind/electricity transformation efficiency of the wind turbine setup could reach to 37%, and it is benefit environment and global energy management.

1. Introduction

The world seek to provide stable and sufficient quantity energy which is a important energy industry. Fossil oils was the primary energy source, but air pollution did harmful influences on people and environment through exhaust emissions. Green technologies developed progressively to energy management. The sources of wind energy is clean and inexhaustible which makes wind power a promising green power resources. The wind turbine is a key component in wind power facilities, either in vertical or horizontal configurations. A large-scale wind turbines are more available in areas with abundant wind resource[1]. But inferences of a large-scale wind turbine farm on the local climate in rising the local surface temperature and causing a larger rate of precipitation was dominated.[2-4] Hence, Tummala *et al.*[1]recommended a small-scale wind turbines installed of the decentralized grid system. The small-scale wind turbines can guarantee a better aerodynamic execution at poor wind resource areas with a lower wind velocity. Thus, the subject of the blade designs for a small HAWT is important. Blade designs will strongly influence the turbine performance and the wind/electricity transformation efficiency for electrical power generator.

The active blade design constantly referred to the NACA [7-9] or animals airfoils [10-11] for increasing the wind turbine efficiency in the wind/electricity power transformation. Based on the theoretical aerodynamic analysis[5-6], the same aerodynamic characteristics are presented with either an object moving with steady speed in stationary air or a fixed object in a uniform air with steady velocity under the steady air speed or a moving object. The equivalent law of aerodynamics such as birds, bats, and insects are furnished with wings, which provides energy to drive the air and produces the aerodynamic force needed for flight. Active wing configurations inspired engineers design and manufacture turbine blades of propeller or aircraft. Some researchers draw their attention on the maple samaras kinematic characteristics. Varshney et al.[11] experimentally revealed the dynamic characteristics of a dropping down maple samaras.

In this research, the aerodynamic characteristics of the bionic passive wing lift-driven (PWLD) blades are surveyed. The winged disseminules of the Taiwan incense cedar is chosen as the shape of bionic PWLD blades from the Calocedrus formosana[12]. The blade contour is mimicked and designed by using the seed wings contours, and examined methodically by ANSYS FLUENT 14 to obtain the essential aerodynamic characteristics. The wind tunnel experimentally tested the HAWT to show the relation between the efficiency of wind/electricity power transformation and aerodynamic characteristics.

2. Numerical Model Constructions

The wind will spread away the winged seeds when the Taiwan incense cedar cones are burst and rotational split. Figure 1 present the contours of the winged-like appendages sketched by an optical microscope. Then, the Creo software process the data to build the contour bionic blades models. IGES files reserved those models and fax the information into ANSYS FLUENT 14 to analog those aerodynamic properties. And, the bionic blades sample were put in the rectangular experimental tunnel with the size of 30 mm×40 mm×70 mm to process the numerical simulations. The tetrahedral meshes were generated by ANSYS FLUENT 14 automatically. Aluminum material are used in the simulation bionic blades.Turbulent air is trusted as the working fluid. The k-ε model is used to describe the turbulent flow. The inlet boundary condition and the atmospheric pressure is the outlet and lateral surface boundary conditions. The wall function of the ANSYS FLUENT 14 is used to process to kinematic properties of the bionic blade near-wall region. The convergence criterion will reach 10^{-6} after approximately 400 iterations.

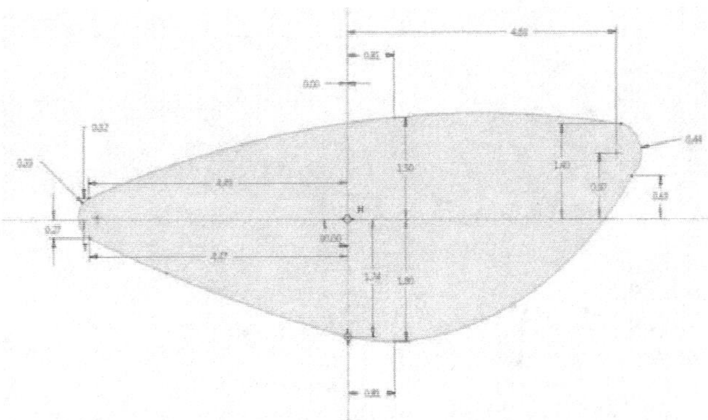

Figure 1. The winged seed shape of Taiwan incense cedar as original bionic blade contour

3. Wind Tunnel Experiment: Set Up and Results

For experimental testing, Taiwan incense cedar blades fabricate a small-scale HAWT based on the enlarged winged disseminule sketch. Each enlarged bionic blade was formed by a 0.3 mm thick aluminum plate and the enlarged blade span is 235 mm. A wood stick with 5 mm × 10 mm cross-sectional area was adhered to the rear blade surface to strengthen the bionic blade stiffness.

The three bionic blade and five bionic blade of the HAWT model were formed and investigated in a wind tunnel to explore the correspondence relationship between the driving wind speed and the wind/electricity power transformation efficiency. The schematic diagram of the wind tunnel is shown in Figure 2. The cross-sectional area of the test section is 800 mm × 800 mm. The examining wind velocity is set to be 3, 4, 5, and 6 m/sec respectively because of the average wind speed per year is 5 m/sec in Taiwan. The local pitch angle of the blades is 15°, 20°, 25°, and 30° respectively. The repeat-ability of the test results for HAWT model was confirmed by three times under the testing wind speed and pitch angle conditions.

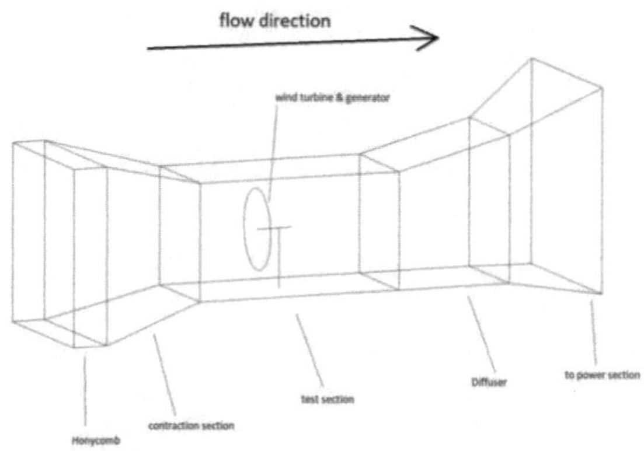

Figure 2. The wind tunnel diagram for testing HAWT model

Figure 3 depicts the relationship between the wind speed and the rotor power coefficient of the five-blade Taiwan incense cedar HAWT model. One can reveal that a maximum rotor power transformation efficiency is reached 37% while the pitch angle is 15° and the wind speed 4 m/sec. But, rotor power transformation efficiency decreases under 5 m/sec wind speed except the 20°pitch angle case. All the wind/electricity efficiency increase monotonically with wind velocity when the wind speed is below 4 m/sec. The results of a small-scale wind turbines presents a better aerodynamic execution at poor wind resource areas with a lower wind velocity.

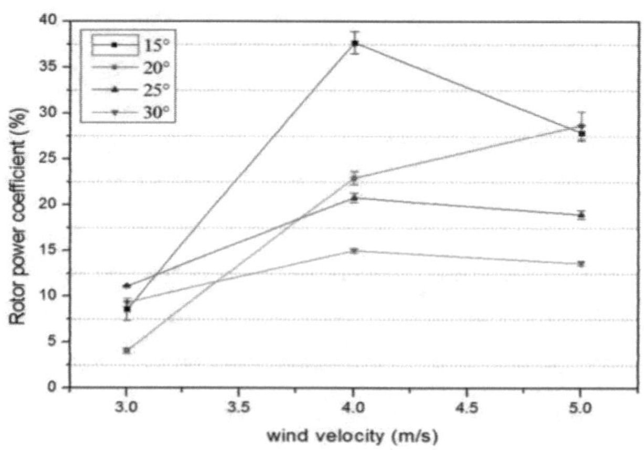

Figure 3. Relationship between the rotor power coefficient and wind speed

The tip speed ratio connect the relations of the pitch angles and wind/electricity efficiency of HAWT models. The corresponding relation between the tip speed ratios and the rotor power transform efficiency of the three bionic blade and five bionic blade of HAWT model is shown in Figure 4. For both three bionic blade or five bionic blade HAWT model, the values of the rotor power transformation efficiency is located from 25% to 37% as the tip speed ratio is within 2.5 to 3.5 and the pitch angle 15°or 20°. These results demonstrate excellent bionic blade aerodynamic properties. However, the wind/electricity power transformation efficiency of the five-blade HAWT model is higher than that of three-blade one.

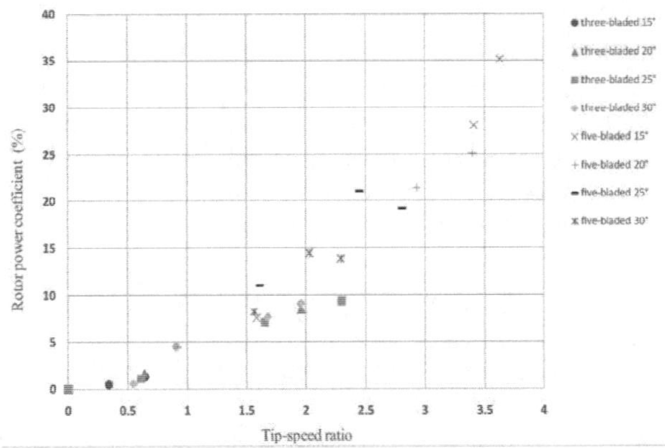

Figure 4. Relationship between the rotor power coefficient and tip-speed ratio of Taiwan incense cedar blades HAWT model

The bionic blades aerodynamic performances are depicted as following. Firstly, a constant wind speed of 3 m/sec is set. Although the wind speed per year is 5 m/sec in Taiwan, while a good weather often accompanies with a 3 m/sec wind speed. It is important that a small-scale HAWT can operate steadily in such a low wind speed. Based on the contour characteristics of the Taiwan incense cedar, figure 5 showed the variation of the lift and drag coefficients relating to the bionic blade angles of attack (AOA). As bionic blade AOA varied from 10° to 60°, the range of drag coefficient of the bionic blade is within 0.8 to 1.16. In general, active airfoils or blades with NACA have a narrow range of AOA. Figure 5 also indicated that a maximal lift coefficient value 0.2 was reached when its AOA is 10° in the case of Taiwan incense cedar bionic blades.

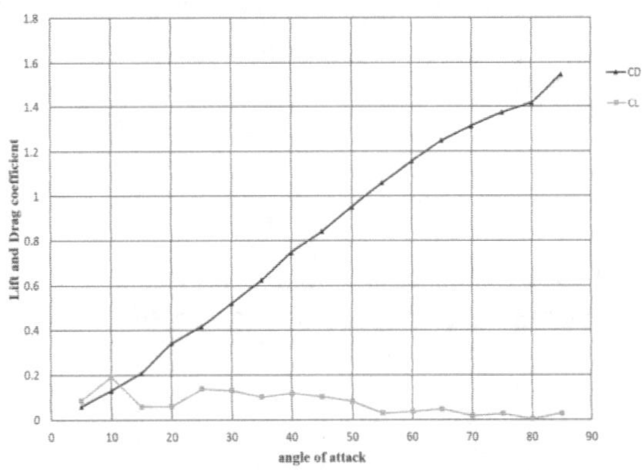

Figure 5. Changes between the angle of attack and the lift and drag coefficients

Under a driving wind speed is 3 m/sec, the corresponding relation of the lift-drag coefficient with the AOA of the bionic blades are showed. The maximal lift-drag ratio value of the bionic blade achieved at a smaller angle of attack. The maximal lift-drag ratio value is 1.5 of the Taiwan incense cedar bionic blades at AOA 5°. The spin mechanism of the winged disseminule might be related to the phenomenon. Based on the above numerical and experimental outcomes, one could reason out that the bionic blade aerodynamic performance is confident enough to guarantee a higher wind/electricity power transformation efficiency. The numerical results obtained from ANSYS FLUENT 14 also confirm the wind tunnel experimental results of the bionic blade lift/drag ratio under the same testing conditions.

4. Conclusions

Adapted the passive wing design turbine blade, one could match the requirement of the driven mechanism for wind turbine under unsteady flows. The imitated objects is the shapes of the winged disseminules of the Taiwan incense cedar, the designed bionic blades were based on their contours and analyzed the aerodynamic characteristics. The research and design of three bionic blade or five bionic blade configuration of wind turbine are carried out. In the wind tunnel experiment, a five bionic blade HWAT model with Taiwan incense cedar can generates the highest wind/electricity power transformation efficiency.

5. Acknowledgement

Present investigation was supported by Lingnan normal university under granted ZL2001.

6. References

[1] Tummala A, Velamati RK, Sinka DK, Indraja V, Krishna VH. A review on small scale wind turbines. Renew. Sust. Energ. Rev. 2016; 56: 1351-1371.

[2] Wang C, Prinn RG. Potential climate impacts and reliability of very large-scale wind farms. *Atoms. Chem. Phys.* 2010; 10: 2053-2061.

[3] Fiedler BH, Bukovsky MS. The effect of a giant wind farm on precipitation in a regional climate model. *Environ. Res. Lett.* 2011; 6: 045101.

[4] Kirk-Davidoff DB, Keith DW. On the climate impact of surface roughness anomalies. *J. Atoms Sci.* 2008; **65**: 2215-2234

[5] Karamcheti K. *Principles of ideal-fluid aerodynamics*. Reprint edition with corrections. Krieger: Florida, USA, 1980, pp. 278-311

[6] Abbott IH, von Doenhoff AE. *Theory of wing sections - including a summary of airfoil data.* Dover Publications Inc.: New York, USA, 1959.

[7] Karthikeyan N, Murugavel KK, Kumar SA, Rajakumar S. Review of aerodynamic development on small horizontal axis wind turbine blade. *Renew. Sust. Energ. Rev.* 2015; 42: 801-822.

[8] Lissaman PBS. Low-Reynolds-number airfoils. *Annu. Rev. Fluid. Mech.* 1983; 15: 223-239.

[9] Giguere P, Selig MS. Low Reynolds number airfoils for small horizontal axis wind turbines. *Wind Eng.* 1977; 21: 367-380.

[10] Singh RK, Ahmed MR, Zullah MA, Lee Y-H. Design of a low Reynolds number airfoil for small horizontal axis wind turbines. *Renew. Energ.* 2012; 42: 66-76.

[11] Song Q, Lubitz WD. Design and testing of a new small wind turbine blade. *J. Sol. Energ. Eng.* 2014; 136: 034502.

[12] Biodiversity Research Center, Academia Sinica ROC, Database of Native Plants in Taiwan, http://www.hast.biodiv.tw/Announce/newsC.aspx (accessed 15 February 2018).

Research on Static Characteristics of Steel-Concrete Composite Beam Cable-Stayed Bridge during Construction

Zeying Yang[1], Yongye Qu[1], Jianbo Qu[2], Yalei Zhang[1], Wenchao Qi[1] and Xinnan Xie[1]

[1]School of Civil Engineering, Shandong University. Jinan, Shandong, China
[2]Shandong Transportation Department. Jinan, Shandong, China
Corresponding author e-mail:253277272@qq.com.

Abstract. As the construction continues, the structural system and load of the cable-stayed bridge will change. At different construction stages, the forces and deformations of the structure will have their own characteristics. In this paper, in order to study the static characteristics of steel-concrete composite beam cable-stayed bridge during construction, the Jiqi Yellow River Highway Bridge is used as the engineering background, and the finite element modeling is carried out by midas. The main beam, cable tower and stay cable of the cable-stayed bridge are analyzed respectively. Among them, the static characteristics of steel and concrete bridge deck during construction are specially analyzed. In this way, it is ensured that all structures are in a safe state during the construction process, and then the construction

1. Introduction

The various states of the cable-stayed bridge are unstable during the actual construction process, so the simulation of the cable-stayed bridge is required. The bridge belongs to the steel-concrete composite beam structure. In particular, it is necessary to pay attention to the stress state of the bridge deck in each construction stage to prevent its damage.

2. Analysis of the Construction Phase

The deformation and stress of different parts of the cable-stayed bridge are different at each construction stage[1-5]. In this paper, the main beam, cable tower and stay cable of cable-stayed bridge are studied. Static characteristics of steel-concrete composite beam during construction were discussed.

2.1. Static Characteristics Analysis of the Main Beam

(1) Steel beam stress

The maximum tensile stress and maximum compressive stress of the steel beam, which is part of the composite beam, at each construction stage were summarized, as shown in Table 2-1.

IOP Conf. Series: Materials Science and Engineering **730** (2020) 012032 doi:10.1088/1757-899X/730/1/012032

Table 2-1. Maximum tensile stress and maximum compressive stress of steel beam

Construction stage	maximum tensile stress(Mpa)	maximum compressive stress(Mpa)
Tower, cable, beam 1	0	1.62
Tower, cable, beam 2	1.21	8.83
Tower, cable, beam 3	15.77	12.94
Tower, cable, beam 4	36.94	15.58
Tower, cable, beam 5	54.50	19.17
Tower, cable, beam 6	71.43	26.24
Tower, cable, beam 7	78.25	36.37
Tower, cable, beam 8	81.30	48.26
Tower, cable, beam 9	79.40	43.30
Tower, cable, beam 10	92.22	57.85
Tower, cable, beam 11	102.69	75.38
Tower, cable, beam 12	113.40	84.67
Tower, cable, beam 13	123.74	110.83
Tower, cable, beam 14	128.73	130.84
Tower, cable, beam 15	128.80	147.08
Tower, cable, beam 16	125.67	165.04
Tower, cable, beam 17	125.67	165.06
Tower, cable, beam 18	125.67	165.04
Tension pressured tendons on the side-span bridge deck	125.77	175.38
Shut mid-span	124.64	176.18
Tension pressured tendons on the mid-span bridge deck	120.40	176.22
Secondary weight	120.36	172.40
Bridge deck pavement	113.58	179.79

According to Table 2-1, the maximum tensile stress of the steel beam is 128.8 MPa, which occurs in the No. 15 beam section and meets the allowable stress of the steel. Compared with the stress in the construction process and the stress in the finished bridge stage, the stress distribution is very different. The steel beam may be pulled during the construction process, but may be compressed under the finished bridge stage. This is mainly due to the fact that during the construction of the cantilever, the loading and unloading of temporary loads such as cranes will have a certain impact on the steel beam.

Figure 2-1. Steel beam stress on finished stage

(2) Bridge deck stress

The stress of the bridge deck during the construction process and the finished bridge stage were summarized, as shown in Table 2-2.

Table 2-2. Maximum tensile stress and maximum compressive stress of brdige deck

Construction stage	maximum tensile stress(Mpa)	maximum compressive stress(Mpa)
Tower, cable, beam 1	0.65	1.52
Tower, cable, beam 2	1.15	10.54
Tower, cable, beam 3	1.07	10.68
Tower, cable, beam 4	1.10	11.22
Tower, cable, beam 5	1.14	11.91
Tower, cable, beam 6	1.46	12.82
Tower, cable, beam 7	1.44	13.65
Tower, cable, beam 8	1.27	14.33
Tower, cable, beam 9	1.31	15.08
Tower, cable, beam 10	1.43	15.69
Tower, cable, beam 11	1.5	16.28
Tower, cable, beam 12	1.72	16.97
Tower, cable, beam 13	1.89	17.61
Tower, cable, beam 14	2.21	18.22
Tower, cable, beam 15	2.39	18.80
Tower, cable, beam 16	2.77	19.37
Tower, cable, beam 17	2.77	19.37
Tower, cable, beam 18	2.77	19.37
Tension pressured tendons on the side-span bridge deck	1.7	19.37
Shut mid-span	1.67	19.39
Tension pressured tendons on the mid-span bridge deck	1.65	19.52
Secondary weight	1.72	19.53
Bridge deck pavement	1.99	20.92

During construction, almost all bridge decks have tensile stresses. If the bridge deck is more likely to be cracked during the construction process, it will have a certain impact on the overall structure of the bridge. Therefore, it is necessary to pay attention to the control of the stress of the bridge deck during the construction process.

The stress distribution of the bridge deck in each construction stage and the finished bridge stage is very different. Under the finished bridge stage, the bridge deck is all pressed, which has a great relationship with its force characteristics.

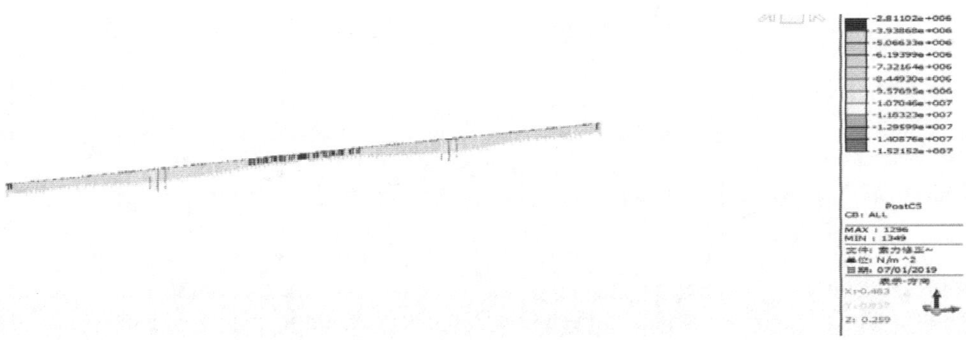

Figure 2-2. Bridge deck stress on finished stage

(3) Bending moment and deformation of the main beam

Since the steel beams and bridge decks are simulated separately, the bending moments in the finished bridged stage are separately analyzed and compared, as shown in Figure 2-3. In the common pier, the auxiliary pier and the mid-span, the steel beam exhibits negative bending moments, and the rests are positive bending moments. This is consistent with the deformation characteristics of the main beam in the finished bridge stage.

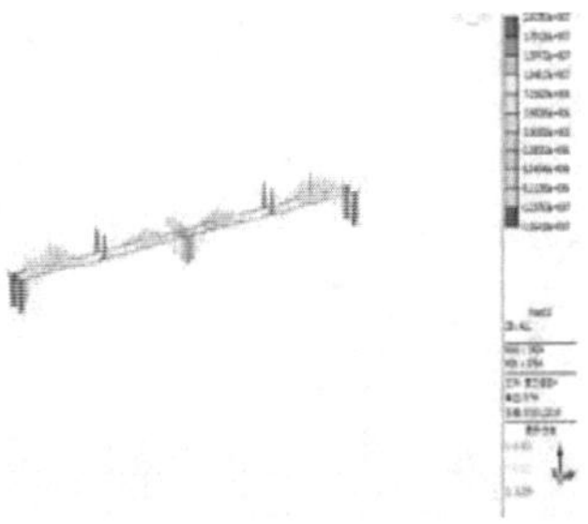

Figure 2-3. Steel beam bending moment on finished stage

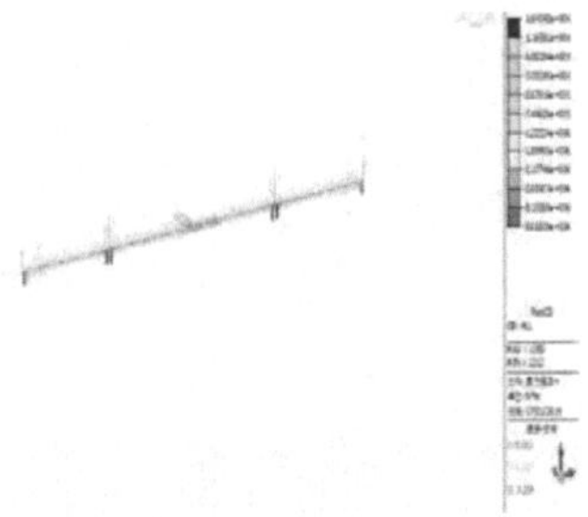

Figure 2-4. Bridge deck bending moment on finished stage

Figure 2-5. Main beam line type on finished stage

Figure 2-4 shows the bending moment of the bridge deck in the finished bridge stage. In the common pier, the auxiliary pier and the mid-span, the bridge deck exhibits a large negative bending moment and a positive bending moment, and most of the remaining bending moments are evenly distributed. The maximum positive bending moment of the bridge deck is 1643.4kN·m, and the maximum negative bending moment is 3610.3 kN·m. They all meet the requirements of the bending moment of the cable-stayed bridge design.

Figure 2-5 shows the deformation of the main beam in the finished bridged stage. Combined with the bending moment of the steel beam and the bridge deck, it can be seen that the deformation of the main beam is basically consistent with the bending moment diagram of the them. The main beam has a large deformation in the mid-span and the auxiliary pier, and the vertical maximum displacement is 41.0 cm.

2.2. Static Characteristics Analysis of the Cable Tower

By comparing the completion of the construction stage and the finished bridge stage, the axial force, bending moment and stress of the cable tower were analyzed.

(1)Axial force

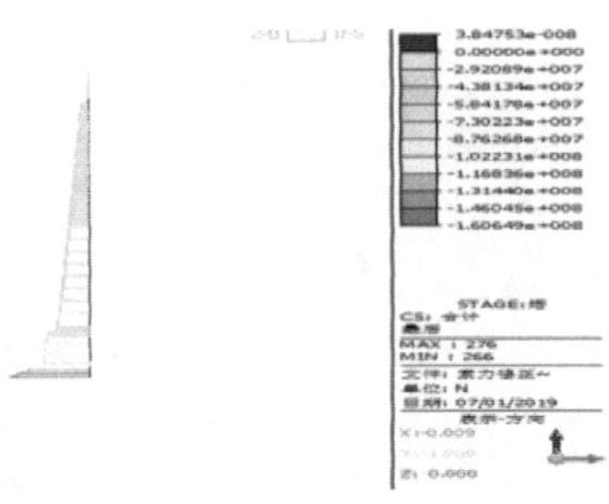

Figure 2-6. Tower axial force on completion stage

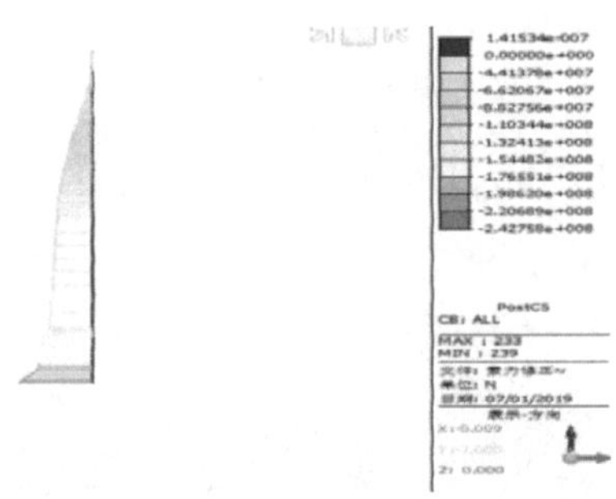

Figure 2-7. Tower axial force on finished stage

Figures 2-6 and 2-7 represent the axial forces of the tower under completion of the construction stage and finished bridge stage, respectively. At the completion of the construction, the maximum axial force of the tower occurs at the bottom of the tower, which is 160649kN. This is mainly because at this stage, the tower is only affected by its own weight, and the closer to the top of the tower, the smaller the axial force. Comparing the two axial force diagrams, the axial force of the cable tower in the finished bridge stage is larger than that of the completion stage, especially in the upper tower of the tower. This is mainly because of the influence of the cable force, the cable is mainly anchored in the upper tower column, and the component force of the cable force in the finished stage of the bridge will have a certain influence on the axial force of the cable tower.

(2)Bending moment and deformation

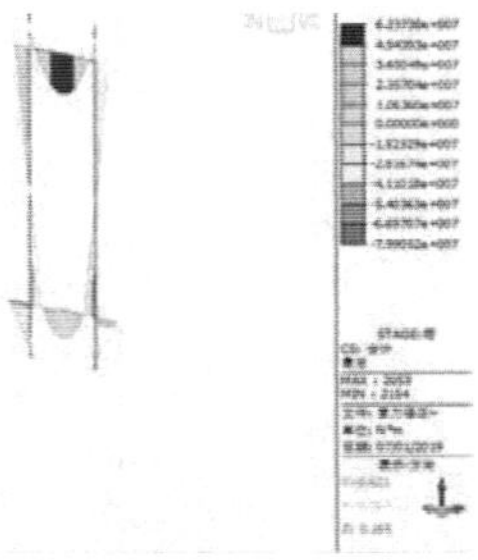

Figure 2-8. Bending moment on completion stage

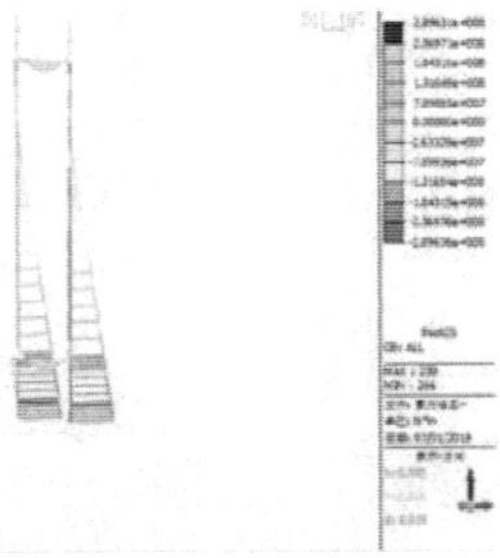

Figure 2-9. Bending moment on finished stage

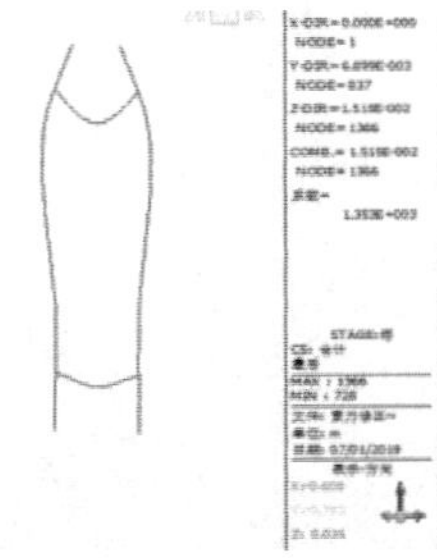

Figure 2-10. Deformation on completion stage

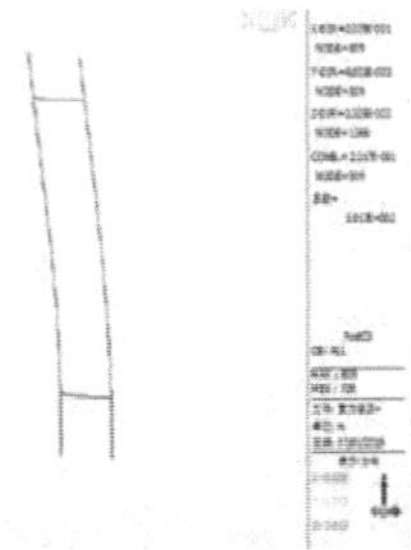

Figure 2-11. Deformation on finished stage

Figure 2-8 to Figure 2-11 show the bending moment and deformation of the tower during the completion stage and the finished bridge stage. The deformation of the tower at the completion stage is basically consistent with the bending moment, which is much smaller in value than the bending moment in the finished bridge stage. The maximum bending moment of the tower under the bridge stage occurs in the pile cap, the value is 289636 kN·m, and the distribution characteristics of the bending moment correspond to the deformation of the tower. The bending moment is mainly concentrated in the middle and lower tower columns, which is combined with the shape of the actual tower.

(3)Stress

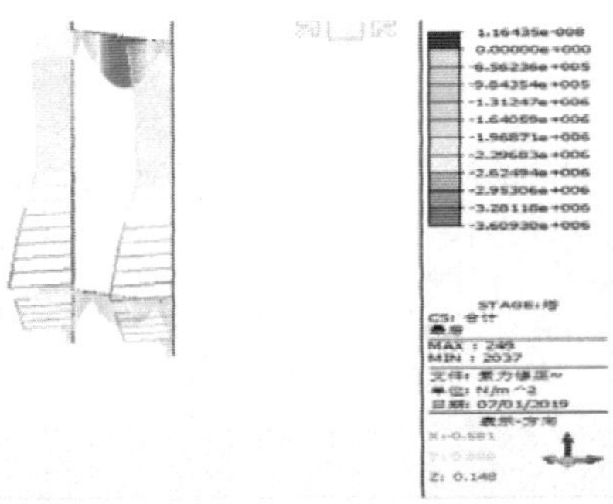

Figure 2-12. Stress on the completion stage

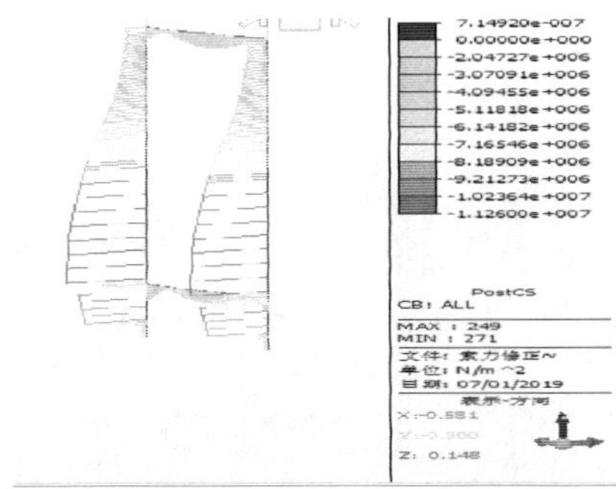

Figure 2-13. Stress on finished stage

According to the above figure, the maximum stress value of the tower is 3.6 Mpa in completion stage, which is much smaller than the maximum stress value of 11.3 MPa in the finished bridge stage. However, in two stages, the stress distribution characteristics of the tower are basically the same. In the completion stage, the stress distribution of the middle tower and the lower tower is relatively uniform, mainly due to the use of a variable cross-section design, the cross section from top to bottom is getting larger and larger. For the upper tower column, although the upper tower column received by the force component of the cable force is getting bigger and bigger from top to bottom, however, the upper column is also designed with a variable cross-section, so the stress distribution is also uniform.

2.3. Static Characteristics Analysis of Stay Cables

In addition to focusing on the static characteristics of the main beam and the tower, the static characteristics of the stay cable should also be analyzed. The stay cables are spatially fan-shaped and are arranged on both sides. Each cable surface is arranged with 16 cables in the side span and the middle span, and 7 different specifications of cables are used in the design.

Table 2-3. Stay cable force on finished stage

Stay cable	Cable tension(kN)	Stay cable	Cable tension(kN)	Stay cable	Cable tension(kN)	Stay cable	Cable tension(kN)
S16	5422.3	S8	3635.6	M16	5564.3	M8	4120.4
S15	5317.0	S7	3618.5	M15	5143.5	M7	3598.1
S14	5325.1	S6	3168.2	M14	5538.2	M6	3580.5
S13	5353.5	S5	3160.5	M13	5469.1	M5	3138.1
S12	4924.2	S4	2775.5	M12	5427.9	M4	3064.0
S11	4874.7	S3	2680.7	M11	4930.0	M3	2919.1
S10	4361.5	S2	2008.1	M10	4902.1	M2	2173.4
S9	4297.6	S1	1765.2	M9	4855.1	M1	1912.7

Table 2-3 shows the cable tension in the finished stage. It can be seen that the maximum cable tension occurs in M16 and the cable force value is 5564.3 kN. The cable tension of the cable stayed in the finished stage is gradually increased from the bridge tower to both sides, so it is necessary to use a variety of cable stays. For the same pair of stay cables, the cable tension in the mid span is greater than the cable tension in the side span, and the horizontal component of the cable force is greater than the side span in the middle span. Therefore, the tower will shift toward the mid-span, which is consistent with the deformation of the tower.

In addition to the analysis of the cable tension of the finished stage, we also analyze the cable tension during the construction stage of the No.5 and No.15 cable. The No. 5 cable is a double cantilever construction stage, and the No. 15 cable is a single cantilever constructionstage. Figure 2-14 and Figure 2-15 show the cable tension of No. 5 and No. 15 in the construction phase. It can be seen that at the beginning of the construction, the cable tension has a large change, and as the construction progresses, the cable tension of the side span and the middle span tends to be stable. n the bridge decking stage, the cable tension will change considerably due to the application of the load.

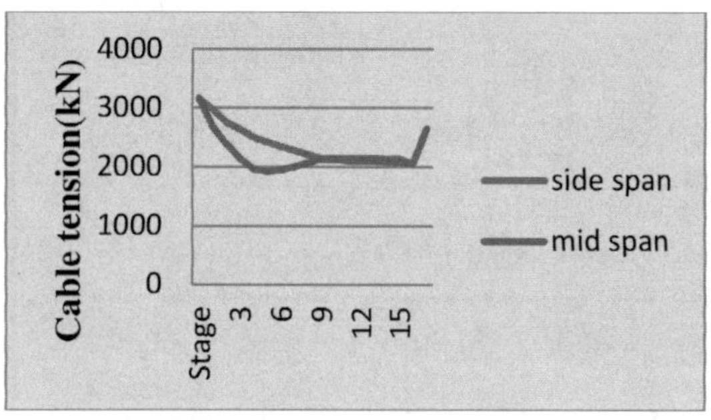

Figure 2-14. Cable tension of No. 5 stay cable on construction stage

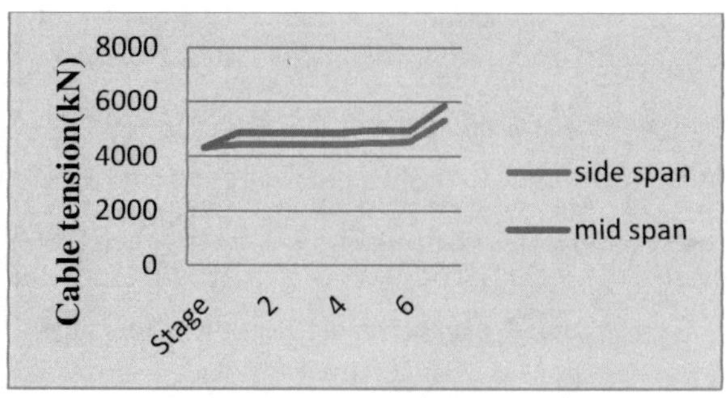

Figure 2-15. Cable tension of No. 15 stay cable on construction stage

3. Conclusion

(1)The stress distribution of the bridge deck in each construction stage and in the finished bridge stage is very different, we should pay particular attention to the stress control of the bridge deck during construction.

(2)The design of the tower with variable cross-section is conducive to the uniform distribution of the stress of the tower. The deformation of the tower is mainly concentrated on the top of the tower. In the construction, special attention should be paid to the monitoring of the displacement of the tower top.

(3)During the construction process, it is necessary to focus on monitoring the change of cable force. If it is necessary to adjust the cable force, it is necessary to monitor the stress variation of the main beam.

4. Acknowledgments

The research presented in this paper has been co-financed by Shandong Science and Technology Development Plan (2014GSF120015) and Shandong province Transportation Science and Technology Project (2013A12-03). Their financial support is gratefully acknowledged.

5. References

[1] Sun Caizhi, Zhao Lei, Chen Wenyuan. Static analysis on construction process of long-span composite cable-stayed bridge. Building Science Research of Sichuan, 2012.

[2] Yan Xingwei. Key Techniques for Construction of Steel-Concrete Composite Beam Bridges. Technological innovation and application, 2018.

[3] Wan Liang, Sun Jianyuan. Geometric Nonlinear Analysis of Large-span Extradosed Cable-stayed Bridge of Largest Single Cantilever Stage. Journal of Jiamusi University, 2014.

[4] Yang Yang. Study on the mechanical behavior of cable-stayed bridge structures based on finite element. Zhengzhou University, 2016.

[5] Shao Changyu. Study on Design and Mechanical Properties of Long Span Composite Beam Cable-Stayed Bridge. Bridge Construction, 2017.

IOP Conf. Series: Materials Science and Engineering **730** (2020) 012033 doi:10.1088/1757-899X/730/1/012033

Landscape Ecological Risk Assessment of Dianchi Lake Basin Based on Land Use Change

Zhiqiang Lin[1,2],Shuangyun Peng[1,2*],Shaojian Pu[1,2],Xiaoliang Ma[1,2],Yanlian Zhou[1]

[1]College of Tourism and Geography, Yunnan Normal University, Kunming 650500, China.

[2]Yunnan Geospatial Information Technology Engineering Technology Research Center, Kunming 650500. Fund Project: Project supported by the National Natural Science Foundation of China (41971369, 41561086, 41661082, 41461038), China.

*Corresponding Author: Shuangyun Peng. First author Email: 248972265@qq.com.

Abstract. In order to explore the impact of land use type change on landscape ecological risk, this paper uses Landsat remote sensing image of Dianchi Lake Basin in 2003-2018 as the basis data to interpret its land use type by ENVI, and combines ArcGIS and Fragstats4.2 according to landscape ecology theory. The various landscape ecological indices were calculated, and the spatial pattern and process characteristics of landscape ecological risk in Dianchi Lake Basin from 2003 to 2018 were analyzed. The results show that: (1) The landscape pattern of Dianchi Lake Basin has changed significantly in 2003-2018. (2) The landscape ecological risk level has a gradually increasing trend, that is, the proportion of the medium risk area to the high-risk area is relatively large. (3) From the perspective of spatial distribution, the overall risk distribution of Dianchi Lake Basin has an increasing trend from Dianchi Lake to Dianchi Lake.

1. Introduction

Ecological risk assessment (ERA) refers to the assessment method, evaluation system and assessment of the possibility of damage to the ecosystem function due to changes in the natural environment or human activities, resulting in the loss of the composition or structure of the ecosystem.[1] In recent years, due to frequent human development activities and severe ecological and environmental problems, ecological risk assessment has gradually become an important means to study and solve environmental problems.[2] How to scientifically analyze the changing process of ecological risks, explore its impact mechanism on ecosystems, evaluate its impact on human society and natural ecosystems, and achieve sustainable socioeconomic development have become hotspots in social science, geography, and ecology research one of the problems.[3]

With the development of society, the ecological environment has faced tremendous pressure in recent years. However, since the US Environmental Protection Agency proposed ecological risk assessment in 1990, ecological risk assessment has made great progress.[4] At home and abroad, the ecological risk of regional space is calculated mainly through the change of land use type and the landscape ecological correlation index. For example: Potter et al.[5] extracted landscape pattern indices from 73 different watersheds in North Carolina, and explored the relationship between landscape ecological risk and non-point source pollution ecological risk. Obery et al.[6] used the Codorus Creek basin in the United States as a research area to explore the risk effects of the landscape ecological pattern in the basin on the fragile ecological environment of the region. Liu et al.[7] analyzed the road-to-region ecological risk image and indicated the potential ecological risk of the Lancang River with the intensity and ecological vulnerability of different soil erosion. Lu et al.[8]

used the Xi River Basin as the research basis to explore the impact of land use change on the landscape ecology of the Xi River Basin from the aspects of land use change, landscape ecological risk and spatial correlation analysis. Wang et al [9] used the spatial and spatial correlation of landscape to evaluate the landscape ecological risk of the Bohai Sea Basin. Liang et al.[10] based on the transformation of land use types in the Manas River Basin in Xinjiang, used the landscape ecological index and spatial analysis method to construct an ecological safety index for the Manas River Basin, and made a safety evaluation.

As a pearl on the Yunnan-Guizhou Plateau, Dianchi Lake is the only lake in the nine high-altitude lakes in Yunnan Province that is adjacent to the big city. The landscape pattern of the Dianchi Basin has also undergone tremendous changes in recent decades. With the advancement of urbanization, the Dianchi Basin has become a key area for development, and the ecosystem has been affected and destroyed to varying degrees. Therefore, this paper uses the Dianchi Basin as a research area to construct a landscape ecological risk index for the study area, for 2003-2018. The landscape pattern is analyzed by ecological risk assessment, revealing the temporal and spatial variation characteristics of ecological risk in the study area, in order to provide theoretical, technical support and decision-making basis for ecological risk management of Dianchi Basin in Yunnan Province.

2. Research Area Overview

The Dianchi Lake Basin is located in Kunming City in the central part of Yunnan Province. The drainage area is 2,970km^2. The geographical coordinates are 102°30'-103°02' east longitude, 24°28'-25°23' north latitude, and the average elevation is about 1900m. (Figure 1). The landscape structure is rich in the basin. It has the largest plateau lake and pond in Yunnan Province, and has a relatively wide flat land. It is the economic, political and cultural center of Yunnan Province, with developed economy, large population and rapid urbanization.

3. Data Source and Processing

The remote sensing image data of this paper is from geospatial data cloud (http://www.gscloud.cn/) with a spatial resolution of 30m. Using ENVI5.3 software to preprocess remote sensing images, refer to the 2017 Land Use Classification Standards of the People's Republic of China, refer to existing domestic The land use cover classification system, combined with ArcGIS10.2 software, divides the study into six major landscape types. Combined with Fragstats4.2 software, the ecological risk index spatial distribution map of Dianchi Lake Basin was obtained, and its temporal and spatial variation characteristics of ecological risk were analyzed.

4. Research Methods

4.1. Analysis of Land Use Change Characteristics

In the transformation of land use types, the number of land use types is constantly changing, and the transformation between land types is also two-way. The land use change classification map and transfer matrix are used to analyze the land use change process. Use land use dynamics to measure the degree of change in land resource use types. The formula is as follows:

$$K = \frac{U_{ib} - U_{ia}}{U_{ia}} \times \frac{1}{T} \times 100\% \tag{1}$$

Where K is land use dynamics; U_{ia}, U_{ib} is the beginning and end area of a certain land type in the study area; T is the start and end time period of land use change.

4.2. Analysis of the Characteristics of Landscape Pattern Change

The landscape pattern is the result of the combination of natural and human factors in a certain period of time. The landscape pattern index can accurately reflect the landscape pattern change. According to the research content of this paper, the following indices were calculated using Fragstats4.2 software: plaque number (NP), plaque density (PD), landscape shape index (LSI), maximum plaque index (LPI),

Shannon diversity index (SHDI), Contagion Index (CONTAG), Perimeter Area Dimensionality (PAFRAC), Aggregation Index (AL), and Cohesion Index (COHESION).

4.3. Landscape Ecological Risk Analysis

4.3.1. Landscape ecological risk assessment unit division

Using ArcGIS10.2 software, a fishing net was created for the Dianchi Lake Basin to generate an ecological risk assessment unit. According to the average area of landscape ecological patches in Dianchi Lake Basin, combined with the scope and workload of the study area, a 2km×2km risk area sampling grid was used to generate 843 ecological risk assessment units, and then the landscape ecological risk index of each unit was calculated in turn, as the ecological risk value of the sample.

4.3.2. Landscape ecological risk index and model

In this paper, the spatial pattern of landscape is transformed into spatialized ecological risk variable by sampling method. An ecological risk assessment model based on multiple integrated landscape pattern indices is established by using landscape pattern index to quantitatively reflect the correlation between landscape pattern and ecological risk. The landscape ecological risk index ERI is constructed according to the area specific gravity of the land use type and the landscape loss index. The calculation formula is as follows:

$$ERI_n = \sum_{i=1}^{N} \frac{A_{ni}}{A_n} \times R_i \tag{2}$$

Where ERI_n is the landscape ecological risk index of the landscape ecological risk assessment unit n, Ani is the area of the i-type landscape in the landscape ecological risk assessment unit n, A_n is the total area of the landscape ecological risk assessment unit n, and R_i is the loss index of the i-type landscape, its calculation formula is as follows:

$$R_i = S_i \times F_i \times 10 \tag{3}$$

Among them, the calculation formula of the landscape structure index is as follows:

$$S_i = aC_i + bN_i + cD_i \tag{4}$$

In the formula, C_i is the landscape fragmentation index, N_i is the landscape separation index, D_i is the landscape dominance index. a, b, c is the weight of each landscape index, and a+b+c=1. According to the relevant research, the fragmentation index, the separation index, and the dominance index are assigned 0.5, 0.3, and 0.2, respectively.[11]

5. Results and Analysis

5.1. Analysis on the Characteristics of Land Use Change in Dianchi Lake Basin

5.1.1. Analysis on the change of land use structure in Dianchi Lake Basin

During the 15 years from 2003 to 2018, with the continuous development of the economic and social development of the Dianchi Lake Basin, the types of land use are constantly changing. It can be seen from Table 1 that there is a small increase in the water area within the basin, forest land and buildings. The increase in land area is the most obvious. The area of cultivated land, grassland and unused land has declined to varying degrees, with the most decline being cultivated land. In 2003, the cultivated land area was 686.72km^2, accounting for 23.12%, while the 2018 was down by 38.38km^2, accounting for 10.20%; the construction land increased by 49.04km^2, from the original ratio of 12.01% to 28.68%; grassland and unused land decreased from 750.43km^2 and 95.5km^2 to 268.78km^2 and 49.46km^2 respectively; the water area and forest area increased from 11.30% and 25.09% respectively to 11.45% and 38.95%.(Table 1) It can be seen from Figure 1 that the construction land is growing rapidly along the coast of Dianchi Lake and the eastern part of the basin, and the forest land has increased in the northern mountainous areas and the eastern and southern mountainous areas of the basin.

Table 1. Land use change in Dianchi Lake Basin

Years	Project	Water	Arable land	Forest	Grass	Building land	Unutilized land
2003	area/km^2	335.63	686.72	745.04	750.43	356.68	95.50
	rate/%	11.30	23.12	25.09	25.26	12.01	3.22
2018	area/km^2	340.09	303.00	1156.89	268.78	851.72	49.46
	rate/%	11.45	10.20	38.95	9.05	28.68	1.67

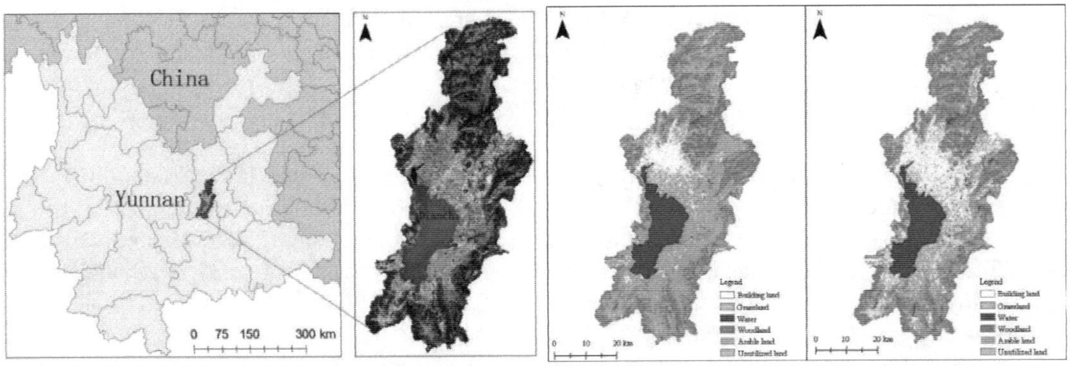

Figure 1. Research area and Classification of land use in Dianchi Lake Basin in 2003 and 2018

5.1.2. Analysis of land use transfer in Dianchi Lake Basin

Using ENVI5.3 to extract the land use transfer matrix and combine the matrix model to transfer the land type: from 2003 to 2018, there will be different degrees of change between various types of land, of which the area of cultivated land and grassland is greatly reduced, and the grassland has 153.31km^2 was transformed into construction land, 263.13km^2 was converted into forest land, 74.03km^2 was converted into cultivated land; cultivated land was converted into construction land with 267.64km^2, and 201.33km^2 was transformed into forest land. (Table 2)

Table 2. Dianchi Basin Land Use Conversion Matrix(km^2)

		Various land areas in 2018					
	Land use type	Building	Grass	Water	Forest	Arable	Unutilized
Various land areas in 2018	Building	---	20.01	0.31	4.15	3.78	2.92
	Grass	153.31	---	2.00	263.13	74.03	1.32
	Water	0.03	2.12	---	0.04	0.02	0.01
	Forest	67.25	6.33	2.21	---	6.79	7.21
	Arable	267.64	1.04	1.32	201.33	---	5.00
	Unutilized	21.02	0.65	0.83	22.99	18.99	---

5.1.3. Dynamic analysis of land use in Dianchi Lake Basin

Through the use of land use dynamic formulas in the continuous change of land use types in Dianchi Lake Basin, it is analyzed that the average annual reductions of cultivated land, grassland and unused land in Dianchi Lake Basin during the 15-years period from 2003 to 2018 are -3.74% and -4.30%, respectively. And -3.23%, the proportion of reduction is relatively large. Among them, the grassland decreased by 481.65km^2, the largest reduction; the construction land use area increased by 9.30% annually, which is the fastest growing among all land use types.

5.2. Analysis on the Change Characteristics of Landscape Pattern of Dianchi Lake Basin

5.2.1. Analysis of landscape pattern Index change on horizontal scale of landscape

The total number of patches in the Dianchi Lake Basin is increasing, from 23,876 in 2003 to 35,526 in 2018. The plaque density increased from 2.68 to 3.13, and the number of plaques per unit area continued to increase, showing fragmentation. The patch shape index increased from 44.98 in 2003 to 50.79 in 2018, and the degree of landscape irregularity increased. The maximum plaque index decreased from 30.74% to 22.10%, the dominance was decreasing, and the fragmentation was increasing. The Shannon diversity index increased from 1.32 to 1.37, the dominance of the landscape species decreased, and each patch type tended to be evenly distributed in the landscape. The spread has dropped from 62.12 to 60.36, the connectivity of the dominant patch types is poor, and the landscape landscape in the watershed is severely fragmented.

5.2.2. Analysis of plaque type landscape pattern index change

The changes in the plaque type index from 2003 to 2018 are shown in Table 3. (1) From 2003 to 2018, the number of patches in waters, cultivated land, woodland, and grassland increased, and the increase in the number of patches in waters and grassland was the most significant. (2) The water area index in the patch shape index of the Dianchi Basin from 2003 to 2018 increased from 5.31 in 2003 to 16.01 in 2018. Cultivated land and grassland increased from 76.29 and 52.69 in 2003 to 90.93 and 69.96 in 2018, respectively. (3) From 2003 to 2018, the largest patch reduction amount on the landscape scale of the Dianchi Basin was cultivated land, which dropped from 20.74% to 12.10%. The maximum patch index of construction land is rising, from 4.77% to 11.06%. (4) Among the types of land use in the Dianchi Basin from 2003 to 2018: the perimeter area fractal dimensions of water areas, cultivated land, forest land, grassland, building land, and unused land have all shown an increasing trend. (5) From 2003 to 2018, the aggregation index of water area, cultivated land, forest land, grassland, and unused land have all declined, and construction land has increased. (6) Except for construction land, the combination index of other land types has decreased, among which the grassland decreased by 3.67, the largest decline. Building land increased from 98.77 in 2003 to 99.63 in 2018.

Table 3. Changes in plaque type index in Dianchi Lake Basin from 2003 to 2018

Years	Landscape	NP	LSI	LPI(%)	PAFRAC	AI(%)	COHESION
2003	Water	1157	5.31	9.64	1.19	99.26	99.63
	Arable land	4731	76.29	20.74	1.32	94.12	99.86
	Wood land	4515	65.90	10.28	1.32	92.95	99.18
	Grass land	4282	52.69	10.48	1.26	75.49	86.37
	Building land	6653	66.52	4.77	1.27	88.25	98.77
	Unutilized land	2538	43.32	1.04	1.25	78.57	88.10
2018	Water	4517	16.01	10.22	1.23	97.55	99.07
	Arable land	7604	90.93	12.10	1.36	92.31	99.84
	Wood land	8577	66.15	13.68	1.36	92.87	98.76
	Grass land	6881	69.96	6.09	1.27	69.62	82.70
	Building land	5896	65.48	11.06	1.36	91.43	99.63
	Unutilized land	2051	34.34	0.80	1.28	77.76	87.06

5.3. Landscape Ecological Risk Assessment of Dianchi Lake Basin

5.3.1. Analysis on the change of landscape ecological loss index in Dianchi Lake Basin

Landscape ecological loss is an important indicator to measure landscape ecological risk. The loss is mainly composed of landscape fragmentation index, separation index and dominance index. In addition to construction land and unused land, landscape ecological fragmentation of waters, forest, arable land, and grassland are all increasing. The fragmentation index of waters has increased the most, from 0.05 in 2003 to 0.13 in 2018; The overall increase in land fragmentation reflects the increase in the number of patches in the landscape type of the Dianchi Lake Basin, and the patches show a trend of fragmentation, which promotes an increase in the ecological risk value of the river basin landscape.

5.3.2. Analysis on spatial variability of landscape ecological risk in Dianchi Lake Basin
Combined with previous research, the ecological risk index is divided into five levels, namely high ecological risk area, higher ecological risk area, middle ecological risk area, lower ecological risk area, and low ecological risk area. It can be seen from Figure 2 that from 2003 to 2018, most of the lower ecological risk areas and middle ecological risk areas were transformed into higher ecological risk areas and high ecological risk areas, and the overall trend of the risk index shifted to higher levels of risk. Ecological risk zone in the past 15 years, the landscape ecological risk index has continued to increase, and the overall regional environment has deteriorated. In the future land use process, the land use structure should be optimized, the construction land should be reasonably planned, a certain amount of grassland, cultivated land, and water area should be reserved, and the complete patches should be avoided as much as possible, and the land use type should be intensive.

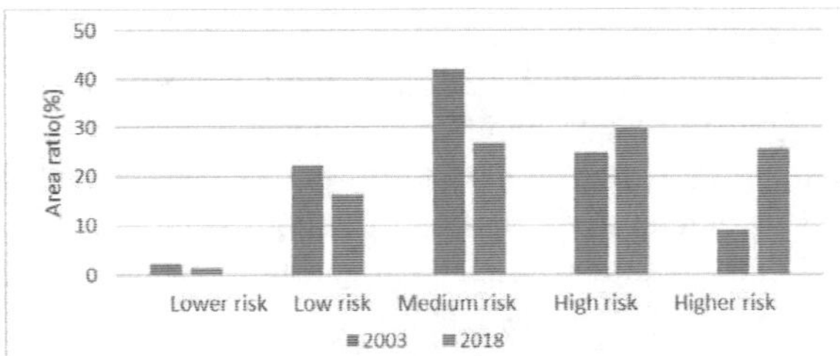

Figure 2. Proportion of landscape ecological risk ratio in Dianchi Lake Basin from 2003 to 2018

In 2003, the higher ecological risk and high ecological risk areas were mainly located in the mountainous areas around the Dianchi Basin. The middle ecological area is mainly located in the plain area in the east of Dianchi Lake, and the lower ecological risk areas and low ecological risk areas are mainly located in Kunming's main urban area and Dianchi Lake. In 2018, high ecological risk areas were mainly distributed around Dianchi Lake and newly developed small and medium-sized urban areas. Medium ecological risk areas were mainly distributed around forest areas and Kunming urban areas, and low ecological risk areas were distributed in central Kunming and Dianchi waters. (Figure 3)

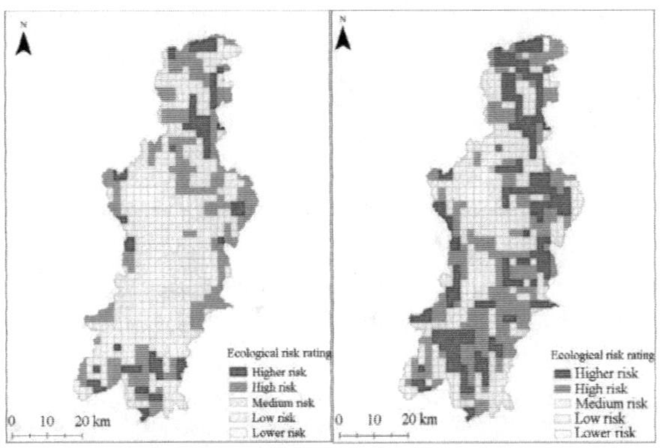

Figure 3. Distribution of ecological risk levels in Dianchi Lake Basin in 2003-2018

Combined with the land use classification maps, it can be seen that the spatial distribution of landscape ecological risk in Dianchi Lake Basin is affected by land use types and human activities. The ecological risk change of the landscape around Dianchi Lake was the most obvious in 2003-2018, and it changed from medium ecological risk and low ecological risk area to high and high ecological risk area. This was mainly due to the continuous development of the Dianchi Lake in the past 15 years: the construction of ecological parks. The development of tourism, the development of real estate, and the construction of university towns have led to a significant reduction in the area of cultivated land around Dianchi Lake. The landscape types have changed from cultivated land and grassland to construction land, landscapes have been fragmented, and the risk index has increased. The above results are consistent with the development of the Dianchi Lake Basin in the past 15 years. In the future, the development process around Dianchi Lake should increase planning efforts to promote the growth of construction land within a reasonable range, and protect basic cultivated land, original forest land and existing waters.

6. Conclusion

Based on the analysis of land use dynamic change and landscape pattern change in Dianchi Lake Basin, through the above research, mainly got the following conclusions: (1) During the period of 2003-2018, the landscape pattern of the Dianchi Lake Basin has undergone significant changes. (2) The landscape ecological risk level has a gradually increasing trend, that is, the proportion of the medium risk area to the high-risk area is relatively large. (3) From the perspective of spatial distribution, the overall risk distribution of Dianchi Lake Basin has an increasing trend. Aiming at the above research results, this paper explores countermeasures for ecological risk management in the study area. The Dianchi Lake Basin should optimize the land use pattern, increase vegetation coverage, continue to implement the policy of returning farmland to forests and grasslands, and increase landscape connectivity; it is forbidden to expand construction land to ecological function areas, water conservation areas, conservation forest areas, and key tourist attractions. Reduce regional ecological risks, strengthen soil environmental governance, and reduce the use of highly polluting pesticides and fertilizers.

7. References

[1] Fang G L, Xiang B, Wang B L, et al.Ecological risk assessment of human activities in the rapidly developing region of Southern Jiangsu Province-taking dantu district of Zhenjiang city as an example[J]. *Chinese Journal of Applied Ecology*, 2014, 25(04): 1076-1084.

[2] Zhao Y, Luo Z J, Cao L P, et al. Ecological risk assessment of Poyang Lake Basin Based on land use change[J]. *Journal of Jiangxi Agricultural University*, 2018, 40(03): 635-644.

[3] Li J L, Xu Y H, Yang L, et al. Study on the evolution of landscape ecological risk pattern in coastal zone of Zhejiang Province[J]. *Journal of Soil and Water Conservation*, 2016, 30(01): 293-299+314.

[4] Zeng J J, Zou M J, Guo J J, et al. Summary of research progress in ecological risk assessment[J]. *Environmental Monitoring Management and Technology*, 2017, 29(01): 1-5+10.

[5] Potter K M, Cubbage F W, Blank G B, Schaberg R H. A Watershed-scale model for predicting nonpoint pollution risk in North Carolina[J]. *Environmental Management*, 2004,34(1): 62-74.

[6] Obery A M, Landis W G. A regional multiple stressor risk assessment of the codorus creek watershed applying the relative risk model. Human and Ecological Risk Assessment[J]. *An International Journal*, 2002, 8(2): 405-428.

[7] Liu S L,Cui B S,Yang Z F. Evaluating the influence of road networks on landscape and regional ecological risk-A case study in Lancang River Valley of Southwest China[J]. *Ecological Engineering*, 2008, 34(2): 215-409.

[8] Lu L T, Zhang J, Sun C Z, et al. Landscape ecological risk assessment of Xihe River Basin based on land use change[J]. *Chinese Journal of Ecology*, 2018, 38(16): 5952-5960.

[9] Wang T, Zhang C, Yu X T, et al. Land use change and its impact on landscape ecological risk in the Bohai Basin[J]. *Chinese Journal of Ecology*, 2017, 36(07): 2003-2009.

[10] Liang E M, Zhang J M. Ecological security analysis of landscape pattern change in Manas River Basin in Xinjiang[J]. *Soil and Water Conservation Research*, 2016, 23(3): 170-175.

[11] Wang J, Cui B S, Yao H R, et al. Temporal and spatial differentiation characteristics of landscape ecological security in Lancang River Basin in Longitudinal Range Valley[J]. *Chinese Journal of Ecology*, 2008, 28(4): 1681-1690.

IOP Conf. Series: Materials Science and Engineering **730** (2020) 012034 doi:10.1088/1757-899X/730/1/012034

A Comparative Study of Toll Collection Systems between Philippines and Taiwan

Chiao Ven Huang[1], Mary Chanille Teaño[2] and Hui Ming Wee[1,*]

[1]Department of Industrial & Systems Engineering, Chung Yuan Christian University, Taiwan
[2]Department of Civil Engineering, Chung Yuan Christian University, Taiwan
*Email: weehm@cycu.edu.tw

Abstract. The paper focuses on transportation, an important economic activity among the components of a business logistics system. The objective of the paper is to compare the existing toll collection systems between a developing and a developed country to get some insights on transportation improvement. The study was undertaken to assist transportation planners, operators, investors, managers, and other related professionals in the field. The latest technologies in transportation, management and infrastructure can provide better logistics efficiency. The paper concludes that Taiwan's electronic toll collection system can solve the existing vehicular traffic flow problem in the Philippines.

1. Introduction

Various researches focused logistics in different applications since the beginning of its advancement in the 1950s. In the recent decades, logistics management evolved rapidly and became significant due to the trend of nationalization and globalization. It helped industries in managing their production, distribution processes, and in improving their efficiency and competitiveness in the market. Part of logistics management is ensuring the smooth flow of delivering the products. Transporting is necessary in the whole process of production, from manufacturing to delivery to the final consumers and returns. Hence, transportation system is a key element in a logistics chain.

In the field of transportation, there exists numerous types of infrastructure. One of them is an expressway, which is a high-speed road that connects a country's important cities. There is an expressway where you need to pay a fee to be able to travel on that road, and they are called Toll Roads. Toll fees are collected in toll roads to help recoup the cost of road construction and maintenance. They can be collected in three ways: (1) open, (2) close and (3) open road. Open system consists of mainline barrier toll plazas, close system has entry and exit tolls and open road consists no toll booths, only electronic toll collection gantries at entrances and exits. The type of system used, however, has an impact to flow of traffic and congestion.

The purpose of the study is to compare the toll collection systems between a developing country, Philippines, and a developed country, Taiwan, through collecting and analyzing their various practices regarding the implementation of such system from related websites and literatures. It is to provide a significant information, which could assist emerging countries with making progress in their transportation industry. The paper starts by establishing questions it aims to answer, specifying its significance, and setting its scope and delimitations. Afterwards, it presents the methodology used in the study followed by the review of literatures. It expresses how toll collection system affects the flow of transportation and the importance of transportation in logistics activities. Furthermore, the study independently presents the comparative analysis of the two countries in order to emphasize their current situations in terms of logistics and transportation. Finally, the paper discusses and concludes the differences in toll collection systems of the two countries and the potential of a developing country for further development in the industry of transportation.

1.1. Statement of the Problem

The researcher intends to address the following problems:

a. How do Philippines and Taiwan implement their toll collection systems for their respective tollways or expressways?
b. What are the challenges that affect the performance of toll collection system in the Philippines and how did Taiwan overcome these challenges?
c. What are the best practices that the Philippines can adapt from Taiwan to alleviate traffic problems?

1.2. Significance of the Study

The comparative study will be beneficial to the entire logistics and transportation industry especially the emerging countries. It can be a learning paradigm to both public and private sectors engaged in managing businesses and transportation systems. Planners, managers, and other professionals in the industry will be provided with insights related to the emerging technologies in transportation that could enable greater efficiency and more systematized operations in the supply chain. Considering Taiwan as a country that is more advanced in the industry of transportation, she can provide Philippines with techniques to minimize congestion and traffic problems. It is hoped that the comparative study will take lead to further enhancements of transport systems in the Philippines.

1.3. Scope and Delimitations of the Study

The focus of the researcher covers transportation. The geographical scope of the study is limited only within the island of Luzon in the Philippines, and Taiwan. Three of the current thirteen operational expressways in the Philippines are the focus of this research. Toll collection systems are evaluated based only on their efficiency and their impact in terms of vehicle flow.

2. Literature Review

Logistics plays a significant role in a country's economic development. An efficient logistics sector contributes to trade performance and economic development by lowering transaction costs and creating more customer value, thus providing firms with opportunities to increase their earnings and enhance their competitiveness (Tongzon, 2018). Therefore, a country's road infrastructure is an important element in the activities of logistics. Without well-developed transportation systems, logistics could not bring its advantages into full play (Taylor et al., 2005). Transportation is around one-third to two-thirds of the expenses in logistics costs. According to the investigation of National Council of Physical Distribution Management (NCPDM) in 1982, transportation averagely costs 6.5% of market revenue and 44% of logistics cost (Taylor et al., 2005). Vehicles used for the delivery of goods travel on the major road arterials, which are usually toll roads. Toll roads vary in systems used for collecting the fees. They can be collected manually wherein the car stops at toll plazas to pay cash or electronically wherein sensors are used to recognize vehicle details and fees are paid using cards. The electronic toll collection (ETC) system has the application of modern technology to increase safety and efficiency of transport systems. They are expected to greatly reduce service times at toll plazas by reducing transaction times compared to manual toll collection (Gugol et al., 2013). The efficiency of ETC systems have already been a subject of several studies. Take for example, the findings of Galanda et al. (2015), Masurekar et al. (2016), and Chauhan et al. (2018) in their respective studies. They have unanimously concluded that an ETC system, if implemented properly, has the potential, among other advantages, to generate a seamless movement of traffic. Karsaman et al. supported this claim thru his study in 2015. He stated that an ETC system helps in smooth and efficient payment transaction at toll plaza, without waiting in long queues on expressways. But perhaps, the most enlightening was the 2003 study of Padayhag et al. of Metro Manila's ETC system. In their study, he found out that the services of the toll plazas vary significantly by the modes of payment employed at the tollbooth. They have also concluded that ETC system is ten times faster than manual lanes. Indeed, the toll collection system significantly affects the flow of traffic in urban areas. They have a huge impact on the movement of goods in the supply chain. If a good transport system exists in logistics activities, then there would be a better logistics efficiency, reduced operation cost, and higher service quality.

3. Methodology

Data used in the study were gathered from related researches conducted by various individuals and agencies, such as Taiwan's Freeway Bureau and the Philippines' Department of Public Works and Highways. The data gathered were then, analyzed to evaluate the effectiveness of toll systems that currently exist in both countries.

3.1. Toll Collection System in Philippines

3.1.1. Selected Expressways

The expressway network that currently connects the Northern and Southern regions of Luzon thru Metro Manila, are composed of 10 controlled-access highways, owned and operated by private institutions. Eventually, thru the initiative of the Philippine government, these highways will be connected, along with other expressways that are currently in the drawing board, to form the Luzon Spine Expressway Network (or Philippine Expressway Network). In total, the Philippines has 13 operational expressways and others are under planning and construction. In order to facilitate the goal of this study, three expressways were selected for the analysis such as North Luzon Expressway (NLEx), Subic-Clark-Tarlac Expressway (SCTEx), and the Tarlac-Pangasinan-La Union Expressway (TPLEx). They are briefly described as follows.

North Luzon Expressway (NLEx)

The NLEx is a 2 to 8-lane limited-access toll expressway that connects Metro Manila to the provinces of the Central Luzon region in the Philippines. It was built in the 1960s and has a total length of 84 km. It is considered as one of the major logistics corridors according to the JICA High Standard Highway Study in 2010.

Subic-Clark-Tarlac Expressway (SCTEx)

The SCTEx is a four-lane expressway north of Manila in the Philippines. It was built in 2005 and currently the country's longest expressway at 93.77 km. The expressway also crosses four rivers in Central Luzon region.

Tarlac-Pangasinan-La Union Expressway (TPLEx)

The TPLEx is a four-lane expressway north of Manila in the Philippines. It connects central to northern Luzon, with its southernmost terminal located in Tarlac City. It has a total length of 88.85 km and crosses the three rivers within the province of Pangasinan.

3.1.2. Toll Collection Systems

All of the expressways in the Philippines collect tolls to finance the operation and maintenance, and recoup the investment during the construction of toll facilities. The expressways implement barrier tolling, a fixed-rate toll to collect toll at the plaza. Other expressways have already adopted the use of electronic toll collection (ETC) system.

NLEx has both an open and closed sections. The open section charges a flat toll rate based on vehicle class while the closed section is distance-based or charging based on the vehicle class and the distance traveled. To speed up the transactions, NLEx adopted the use of ETC system that was set up for Class 1 vehicles; prepaid magnetic cards were assigned for Class 2 and 3 vehicles. A unified ETC system, operated by Easytrip was then added to supplement the system.

3.1.3. Current Traffic Situation

The Philippines has a unique geographic position verging the key global and regional trade routes. For this reason, coupled with its historical ties to the United States, the country is being provided with a significant advantage in the competitive global transport and logistics.

To this date, the transportation in the Philippines is relatively underdeveloped. The land transit corridors still need development as congestion in and around the capital, Metro Manila, is extremely severe and of a great concern for the Philippine government. Companies and the people suffer from heavy traffic especially during working days and holidays. As a result, pick-up and delivery time of goods are sometimes experiencing delays, which could affect the companies' integrity. In addition, the heavy traffic is also risky for the perishable goods. In order to prevent this kind of situations, the logistics companies should stretch their pick-up and delivery time to give some allowance for heavy traffic and other unpredicted circumstances.

In 2015, traffic in Metro Manila has the "worst traffic on Earth" based on the Global Satisfaction conducted by Waze. Manila scored 0.4 out of 10 in the study, which was based on a 1-10 ranking system within six categories. Moreover, in 2017, as part of a study called "Unlocking Cities", a global firm, Boston Consulting Group (BCG), conducted a survey. It was done for a month and participated by around 300 respondents from Southeast Asian cities. The result placed Manila as 3rd worst in the region, with an average of 66 minutes stuck in traffic daily.

3.1.4. Urban Logistics Issues

In terms of logistics, Philippines ranked 60th among 160 countries based on the 2018 Logistics Performance Index conducted by World Bank. The overall score in this kind of survey reflects perception of a country's logistics based on some activity measures including its transport-related infrastructure. Currently, the Philippines is facing some urban logistics issues. One of these is its poor logistics infrastructure. Its road network has not been developed fast enough to meet the ever-increasing transport demand and has been focused on road widening and construction of grade-separations at intersections. With the rapid increase in the volume of vehicles and the interconnectivity issues on the existing expressways, traffic congestion will remain a serious problem (Castro, 2015). Another issue that the country is facing relates to the effect of this traffic problem. In order to ease traffic congestion, the truck ban policy is being implemented. The policy prohibits trucks with gross weights of more than 4.5 tons to travel along major arterial roads. In addition, there are only specific time schedules that these large trucks can travel. Consequently, several businesses are getting affected particularly when it comes to delivering their products. Since large trucks are prohibited in some areas, the truck drivers will use secondary road network, where routes are usually longer and this will directly result to increased travel distances. Also, the limited schedule due to truck restriction reduces the quantity of products to be delivered. Thus, trucking companies who want to maintain their service levels are forced to use smaller trucks that are exempted from the ban (Castro, 2005). For these reasons, the businesses suffer from the additional costs incurred in their operations.

3.1.5. Economic Impact

The Metro Manila traffic, which is the most critical problem of Philippines for so many years, has a huge impact in its economy. According to a study by Japan International Cooperation Agency (JICA) in 2012 entitled "Roadmap for Transport Infrastructure Development for Metro Manila and Surrounding Areas," the traffic snarls cost the country P2.4 billion ($53.3 million) per day and could increase to P6 billion ($133.2 million) per day by 2030 if the problem will remain unresolved. This cost amounts to over P20 billion per year in the greater Metro Manila area, which is equivalent to more than 10% of the country's GDP. On the other hand, the economic cost of truck ban based from the study conducted by Citigroup in 2014 amounts at as much as $7.3 billion. In response to this enduring problem, the Philippine government has responded with several road projects including the target of seamless electronic toll collection systems in all of its expressways.

3.2. Toll Collection System in Taiwan

3.2.1. Expressways in Taiwan
Taiwan's current tollway system is composed of ten National Freeways.

3.2.2. Toll Collection Systems
Over the years, various toll collection systems were launched in several toll stations like the Exact Change Toll Gate in February 1983 and the Toll Tickets Only Gate in December 1996. Taiwan then began launching the use of ETC systems in February 2005. For a long time, these ETC tollgates also had a flat-rate collection of toll fees. According to statistics, at the time of this flat-rate toll station, about 65% of cars did not go through toll stations by instead using alternative local roads. As a result, they did not have to pay tolls resulting a "Toll Equity" issue. To address the issue, the government decided to implement the distance-based toll collection system. However, if manual toll collection system would be adopted, there were land requisition and labor recruitment issues yet to be encountered. Therefore, a distance-based system, which combines electronic toll collection and technologies of computer and communication, became the solution. The old flat-rate manual and electronic toll stations were replaced by a distance-based pay-as-you-go all electronic collection on all of its major freeways in December 30, 2013. Taiwan government created a two-step transition to help them get accustomed to the new system. First, parts of the flat-rate tollgates at toll stations were changed to ETC system to acquaint the freeway users. After some time, the manual toll collection system was completely overhauled, allowing the distance-based ETC system a complete take over. When the freeway users completely adapted the new ETC system, the ETC utilization rate significantly increased. The ETC system can process about 16 million transactions a day in average as of February 2015, the tolling rate has reached 99.97%, and the vehicle detection accuracy rate has reached 99.9%. In summary, the distance-based ETC system is a milestone in Taiwan's transportation industry. It helps eliminate traffic bottlenecks and reduce travel time, as vehicles do not need to stop to pay for tolls when travelling on freeways. Most importantly, this advanced technology and the outstanding construction efficiency truly set models for other countries to learn from.

3.2.3. Taiwan's Recognitions
With the development of ETC distance-based toll collection system, Taiwan gained the following world records:

a. The first country in the world to transfer from flat-rate toll stations to distance-based pay-as-you-go tolling system on all of its freeways.

b. The first country to transfer from manual tolling to all electronic, multi-lane free flow tolling on all national freeways in the country.

c. The country that has the longest ETC freeway mileage in the world.

3.2.4. Urban Logistics
Taiwan, being an exported-oriented nation, has invested into intelligent and topnotch logistics management technologies. Considering these advantages, Taiwan is a natural partner in the logistics industry. In comparison to the Philippines' LPI ranking, Taiwan ranked 27th in World Bank's 2018 Logistics Performance Index. The major contributor in this achievement was the quality of its infrastructure. Because of this, Taiwan is able to provide convenience and efficiency in freight handling.

3.2.5. Economic Impact
Taiwan's freeways have contributed to its economic growth for the past decades. By now, they continue to contribute hugely especially with the development of the distance-based ETC system. It covers 1,050 km of freeway, making it the world's largest system of its kind. Since the introduction of this ETC system, motorists can travel at speed of 50 to 70 km/h on ETC lanes, leading to considerable fuel savings, shortened travel time, reduced CO2 emission and an eliminated cost of printing toll

IOP Conf. Series: Materials Science and Engineering **730** (2020) 012034 doi:10.1088/1757-899X/730/1/012034

tickets. Combining all these benefits, the total estimated cost of savings amounts to NT$2.4 billion (US$80 million) annually.

4. Conclusion and Future Research

Based on the comparative study, the following conclusions can be made:

1. The Philippines should learn from Taiwan in the use of a distance-based system which combines electronic toll collection and technologies to solve the toll equity issue and to shorten toll collecting time. Seamless toll collection technology should be imported from Taiwan.
2. The ETC system in Taiwan could alleviate queuing brought about by build-up of traffic at toll plazas due to the tedious nature of conventional transactions for toll plazas. Since the ETC system in Taiwan provides uninterrupted flow for vehicles; it could possibly solve traffic problems in the Philippines if implemented. This is significant especially when she has a plan to connect all of its expressways. If ETC system of Taiwan is used; vehicles no longer need to stop at toll plazas to pay tolls. By this way, vehicle queueing can be eliminated.

The study only focused on the review and comparison of existing toll collection systems in the Philippines and Taiwan. A more comprehensive study of toll collection systems could be the subject for future works. The future researcher may include the feasibility of Taiwan's ETC systems in the Philippines. In particular, the cost of existing ETC systems and the construction cost of toll plazas in the Philippines.

5. References

[1] Castro, J. T. (2015). "Urban Logistics in the Greater Manila Area (GMA)". Conference Paper, December 2015.

[2] Chauhan, S. et al. (2018). A Comparative Study of Toll Collection Systems in India. *International Journal of Research in Engineering, Science and Management, 1*(4), pp. 72-74.

[3] *Far Eastern Electronic Collection Co., Ltd.* (n.d.) Retrieved May 3, 2019, from https://www.fetc.net.tw

[4] Galande, S.D. et al. (2015). Automated Toll Cash Collection System for Road Transportation. *International Journal of Computer Science and Mobile Computing, 4*(2), pp. 216-224.

[5] Karsaman, R. H. et al. (2015). "A Comparative Study on Three Electronics Toll Collection Systems in Surabaya", *International Conference on Information Technology Systems and Innovation*, November 2015, Bandung, Indonesia.

[6] Liu, P. (2015, April 15). *Taiwan's eTag ETC System Gains International Fame*. Retrieved from https://topics.amcham.com.tw

[7] Masurekar, D. & Mehendale, R. (2016). A Comparative Study of Different Technologies for Electronic Toll Collection System. *International Journal of Innovative Research in Computer and Communication Engineering, 4*(2), pp.1532-1538.

[8] *Wikipedia.* (n.d.). Retrieved May 25, 2019, from wikipedia.com

[9] Yue, W. L. (2005). The role of transportation in logistics chain. Proceedings of the Eastern Asia Society for Transportation Studies, Vol. 5, pp. 1657 – 1672.

Two-rotation Orthogonal Combination Design Optimizes the Process Parameters of Fresh-cut Rosa Green

Peixin Han, Xiaoguang Feng, Xin Zhao, Huijun Liu and Xiangning Chen*

Department of Food Science and Engineering, Beijing University of Agriculture, Beijing 102206, China
Email: 740828870@qq.com

Abstract. In order to improve product quality, extend the shelf life of fresh-cut Rosa green, and reduce the loss rate of processing and circulation, the secondary rotation-orthogonal composite design is used to optimize the process parameters of ready-to-eat fresh-cut Rosa green. The results showed that weight loss rate, conductivity, POD, Vc, total number of colonies, nitrite, and chromaticity value a were significantly correlated with storage time ($p<0.01$).Principal component analysis showed that the two principal components accounted for 96.215% of the contribution rate. Fresh cut Rosa green quality evaluation value comprehensive value Y = -0.371 Chromaticity value a - 0.373 Weight loss rate -0.357 Total number of colonies -0.005 POD + 0.374 Vc - 0.333 Relative conductivity - 0.269 Nitrite. The secondary disinfection sodium hypochlorite concentration, the ratio of O_2 (v / v) and CO_2 (v / v) and the storage temperature were studied to influence the comprehensive quality of fresh cut Rosa green during storage. The results showed that the optimal process parameters were: secondary disinfection sodium hypochlorite concentration was 31.43ppm, O_2 concentration was 3.04%, CO_2 concentration was 5.55%, and storage temperature was 3.68°C. Fresh cut Rosa green processed under the optimum process parameters can be stored for 15 days. This study has positive significance for the post-harvest processing of fresh cut Rosa green.

1. Introduction

Rosa green is a leafy lettuce. Widely planted all over the world and loved by consumers. It contains a variety of camp substances and antioxidants, which can promote blood circulation and diuretic effect in the body.

In order to avoid the problem of rottenness and browning caused by improper technology during the processing of fresh Rosa Green, reduce the loss rate, ensure that fresh-cut Rosa Green has good quality in circulation, and maintain an appropriate degree of freshness. It is necessary to optimize the key processes of fresh-cut Rosa Green during processing.

2. Materials and Methods

2.1. Materials and Sample Preparation

Fresh-cut Rosa Green after 110ppm sodium hypochlorite disinfection solution and air-packing treatment is used as a key indicator to evaluate the shelf life. Untreated Rosa Green is used to optimize the process parameters of fresh-cut Rosa Green. All Rosa greens are supplied by the Yang zhen base of Beijing Yu nong Quality Agricultural Products Planting Company. Harvested in the morning and shipped to the laboratory throughout the cold chain within 3 hours. Pack 50g of Rosa green high density polyethylene film (HDPE, length 290mm, width 280mm, film thickness 37.1μm, air permeability coefficient: CO_2 4.5×10^4, O_2 2.6×10^4, N_2 2.2×10^{-2}), seal and place Store at 4 °C. Rosa green filled with air was placed in the refrigerator as a control group, and samples were taken and tested every 2 days.

2.2. Experiment Method
With reference to the method of AGÜERO et al.10 professionals were evaluated from various aspects such as color, odor, and organization status[1].The weight loss rate was measured according to the method of Rebogile et al[2].The relative conductivity was measured according to the method of Agüero et al.the PPO was measured according to the method of Min Li et al[3]. Dichloroindophenol titration method for determination of vitamin C, Shi-peng Liu et al[4]. For determination of SOD, portable handheld refractometer for total soluble solids, GB 4789.2-2016 for determination of total colony count, Determination of nitrite, determination of soluble protein by the method of He-sheng Li, et al. Measurement of the color of Rosa green (L, a *, b *) using the DC-P3 full-automatic colorimeter, and determination of CO_2 using a fruit and vegetable respirometer Concentration, calculate the breathing rate[5].

2.3. Secondary Rotation-Orthogonal Compound Experimental Design
In order to optimize the fresh-cut Rosa green processing technology, the key steps and factors that affect its quality are selected: secondary disinfection sodium hypochlorite concentration, spin-drying time, ratio of O_2 and CO_2, and storage temperature as factors. Cross compound experimental design (Table 1) set 5 levels for each factor.

Table 1. Five-factor quadratic regression orthogonal rotation combination test design factor horizontal coding table

coding	X1 Degree of sodium hypochlorite (ppm)	X2 Drying time (S)	X3 O_2 ratio (%)	X4 CO_2 ratio (%)	X5 Storage temperature
2	60	90	6	10	32
1	45	75	4.5	7.5	24
0	30	60	3	5	16
-1	15	45	1.5	2.5	8
-2	0	30	0	0	0

2.4. Data Analysis
Correlation analysis and principal component analysis were performed using SPSS 19.0. Model significance was evaluated using DPS.

3. Results and Discussion

3.1. Determination of Fresh-Cut Rosa Green Indicators during Storage
It can be seen from Table 2. The sensory quality of fresh-cut rosa green showed a downward trend, and the decline trend was larger after 4 days, indicating that the quality of fresh-cut rosa green began to deteriorate after 4 days. As the storage time increases, the color of fresh-cut vegetables begins to darken and brown. Relative conductance increased at 2-6d, and cell membrane integrity was impaired. During storage, the respiratory rate showed an increasing trend, and the total number of colonies gradually increased, resulting in a continuous increase in the nitrite content in fresh-cut Rosa green. PPO is one of the main enzymes that cause enzymatic browning of lettuce. During storage, PPO and SOD showed a downward trend, and PPO activity in leaf veins was 5.7 times that in leaves. POD is on the rise, and the browning potential is highly correlated with POD activity. Protein content is closely related to plant senescence, and is an important indicator for measuring the total metabolism of plants. The soluble protein content at 10d decreased by 67% compared with the starting point of storage, indicating that fresh-cut Rosa green began to age during storage, the protein decomposition rate was high, and the protein content showed a downward trend.

Table 2. Changes in Fresh-Cut Rosa Green Index Values During Storage

Storage time (d)	Sensory analysis	Weight loss (%)	Relative conductivity (%)	POD (U/g)	Vc (mg/g)	SOD (U/g)	Respiratory rate(mg.co_2^{-1}.h^{-1}.)	Total colony (logCFU/g)	L	b *	a*	PPO (U/g)	Nitrite (μmol.g^{-1})	Soluble solids (%)	Soluble protein (mg/g)
0	9	0	7.2	90.3	7.6	243.3	16.7	3.8	56.9	34.2	19.1	247.3	0.36	0.87	16.95
2	8.6	1.5	11.3	89.7	6.87	148.7	13.4	3.9	56.7	34.9	16.3	176.4	0.61	0.92	17.6
4	7.9	2	12.4	92.1	6.52	271.2	14.1	4.2	56.4	34.7	17.2	261.6	1.52	1	17.37
6	6.2	3.2	12.9	109.8	5.63	121.5	17.3	4.8	54.3	34.8	15.7	149.8	1.83	0.93	18
8	4.9	4.6	17.1	129.8	4.71	87.2	17.1	6.6	53.1	35.4	14.4	87.5	2.47	0.83	22.08
10	3.3	5.1	25.2	168.3	4.32	44.9	24.3	8.1	51.2	36.3	13.8	52.3	3.43	0.74	27.33

3.2. Correlation Analysis between Indicators of Fresh-Cut Rosa Green and Shelf Life

Correlation analysis between fresh-cut Rosa green indicators and storage time was performed by SPSS. The Pearson correlation coefficient ranged from -1 to 1. The larger the absolute value, the stronger the correlation. From Table 3, it can be seen that the storage time is extremely significantly related to weight loss rate, electrical conductivity, POD, Vc, total number of colonies, nitrite, a value ($p < 0.01$), and sensory PPO, SOD, soluble solids, soluble proteins, The values were significantly correlated ($p < 0.05$) and were not related to L value and respiratory rate. Sort the extremely significant shelf-life related indicators: weight loss rate> total number of colonies> POD> Vc> conductivity> nitrite> a value.

Table 3. Correlation between the indicators of fresh cut Rosa green and shelf life

		Storage time	Sensory analysis	Weight loss	Relative conductivity	PPO	POD	VC	SOD	Soluble solids	Soluble protein	Total colony	Nitrite	L	a	b	Respiratory rate
Storage time	Pearson Relevance	1	-.953	-.997	-.968	-.916	-.977	-.974	-.956	.927	-.923	-.979	.962	-.481	.954	.962	.657
	Saliency		*	**	**	*	**	**	*	*	*	**	**		**	*	

3.3. Establishment of Fresh-cut Rosa Green Evaluation System

It can be known from Table 4 that two principal components can be obtained through principal component analysis, and the cumulative contribution rate is 96.215%, indicating that these two principal components represent almost all the information.

Table 4. Total variance of key indicators of fresh cut Rosa green

Ingredients	Initial eigenvalue			Extract square sum load		
	total	variance%	accumulation %	total	variance%	accumulation %
1	5.645	80.644	80.644	5.645	80.644	80.644
2	1.090	15.571	96.215	1.090	15.571	96.215
3	0.210	3.003	99.218			
4	0.055	0.782	100.000			
5	1.94×10^{-16}	2.774×10^{-15}	100.000			
6	-7.18×10^{-17}	-1.026×10^{-15}	100.000			
7	-2.84×10^{-16}	-4.061×10^{-15}	100.000			

The principal component analysis showed that: principal component 1 = 0.414 b +0.408 c -0.162 d-0.398 e +0.415 f +0.401 a +0.378 g; principal component 2 = 0.1601 b +0.0957 c +0.872 d -0.249 e -0.089 f + 0.213 a-0.297 g. Comprehensive score at different storage times = -0.838 principal component 1-0.162 principal component 2 to obtain the comprehensive value of the fresh-cut Rosa Green evaluation system at different storage times Q = -0.371 a-0.373 b -0.357 c -0.005 d +0.374 e -0.33 3f -0.269 g (a—color value a; b—weight loss rate; c—total colony; d—POD; e—Vc;

f—conductivity; g—nitrite).

It can be seen from Table 5 that the overall score of the fresh-cut Rosa Green evaluation system is positive and decreases with time. The comprehensive score of the Fresh-cut Rosa Green Evaluation System is 0.625 at 0d, and drops to 0.457 at 8d, and the sensory evaluation shows 8d Losing the value of the product, you can get a good value for the fresh-cut Rosa Green with a comprehensive value between 0.457 and 0.625. The lower the comprehensive value, the better the quality.

Table 5. Comprehensive evaluation value of fresh cut Rosa green during the shelf life

Storage time	Main component 1	Main component 2	Comprehensive value
0	-1.131	1.99	0.625
2	-1.10	2.04	0.588
4	-1.08	2.10	0.560
6	-1.068	2.12	0.518
8	-0.878	1.70	0.457

3.4. Significance Test of Two Regression Models

The DPS data processing system was used to perform quadratic regression rotation fitting on the experimental data. As can be seen from the analysis of variance in Table 6, the total regression reached a significant level, which proved that the experiment was effective. From the test results, x1, x2, x3, and x4 'are all less than 1, indicating hat the effect is not significant. The FLF obtained through the fitness test is less than 1, indicating that the equation has a good fit. The F value shown in Table 7 shows that the model is extremely significant ($p < 0.01$), and the comprehensive value of fresh-cut Rosa green was accurately predicted on the 4th day.

The regression model is:

Y=-0.5843-0.0017X4+0.003X5+0.0049X1X2+0.00199X1X3+0.0031X1X4+0.0165X1X5+0.0057X2 X3-0.0031X2X4+0.0011X2X5-0.0011X3X4+0.0027X3X5+0.005X4X5-0.0039X12-0.0094X22+0.00 34X32+0.0073X42 The results showed that the storage temperature had a significant effect on the comprehensive value of fresh-cut Rosa green during storage ($p < 0.01$). The order of the effects of the five processes on the quality of fresh-cut Rosa green was storage temperature> CO_2 ratio> drying time> O_2 ratio> secondary disinfection sodium hypochlorite concentration.

Table 6. Variance analysis of five-element quadratic orthogonal rotation combination design for fresh-cut Rosa green processing optimization experiment

project	sum of squares(SS)×10^{-4}	Degrees of freedom(df)	Mean square(MS)×10^{-4}	F value	Saliency
x_1	0.00865	1	0.00865	<1	ns
x_2	0.0792	1	0.0792	<1	ns
x_3	0.0402	1	0.0402	<1	ns
x_4	0.718	1	0.718	8.21*	p<0.5
x_5	2.24	1	2.24	25.6**	p<0.01
x_1x_2	3.87	1	3.87	44.3**	p<0.01
x_1x_3	0.639	1	0.639	7.30*	p<0.5
x_1x_4	1.54	1	1.54	17.6**	p<0.01
x_1x_5	43.6	1	43.6	499**	p<0.01
x_2x_3	5.2	1	5.2	59.4**	p<0.01
x_2x_4	1.59	1	1.59	18.2**	p<0.01
x_2x_5	0.217	1	0.217	2.49	
x_3x_4	0.197	1	0.197	2.25	
x_3x_5	1.25	1	1.25	14.3**	p<0.01
x_4x_5	4.27	1	4.27	48.8**	p<0.01
x_1'	4.67	1	4.67	53.3**	p<0.01
x_2'	26.7	1	26.7	305**	p<0.01
x_3'	3.66	1	3.66	41.7**	p<0.01
x_4'	0.0482	1	0.0482	<1	ns
x_5'	16.2	1	16.2	185**	p<0.01
return(SS_R)	5.79	20	0.289	3.31*	p<0.5
Remaining(SS_r)	0.963	11	0.0875		
error(SS_e)	0.0153	5	0.00305	<1	ns
Lack of fit(SS_{LF})	0.947	6	0.158	<1	ns
Total return(SS_y)	6.75	31			

Note: ns means the difference is not significant, * means significant, ** means extremely significant

3.5. Optimal Parameter Combinations

The optimal solution is calculated by using the comprehensive value of 4d: x1 = 0.0477, x2 = -0.0036, x3 = 0.0129, x4 = 0.111, x5 = -1.5397 The best process for processing and storing fresh-cut Rosa green is obtained: secondary disinfection sodium hypochlorite The concentration was 31.43ppm, the drying time was 59.89s, the O_2 concentration was 3.04%, the CO_2 concentration was 5.55%, and the storage temperature was 3.68°C.

4. Conclusion

The weight loss rate, relative conductivity, POD enzyme activity, Vc content, the total number of colonies, nitrite, and a value were extremely significantly correlated with the storage time in the previous indicators. Comprehensive value = -0.371 a value -0.373 weight loss rate -0.357 total colony -0.005 POD +0.374 Vc -0.333 relative conductivity -0.269 nitrite. Different factors will interact and affect the shelf life of fresh-cut Rosa Green. The best fresh-cut technology and storage conditions for Rosa Green are established by calculation and calculation: the concentration of sodium hypochlorite in secondary disinfection is 31.43ppm, and the drying time is 59.89s. The O_2 concentration was 3.04%, the CO_2 concentration was 5.55%, and the storage temperature was 3.68°C. It can be stored up to 15 days through processing under optimal conditions. The comprehensive evaluation value is 0.449, which still has commercial value, which is 7 days higher than the original process.

5. References

[1] AGÜERO M V, PONCE A G, et al. Lettuce quality loss under conditions that favor the wilting phenomenon[J]. Postharvest Biology and Technology, 2011,59(2):124-131.

[2] Rebogile, R. M., Olaniyi, A. F., & Umezuruike, L. (2016). Influence of packaging system and long term storage on physiological attributes, biochemical quality, volatile composition and antioxidant properties of pomegranate fruit. Opara Scientia Horticulturae, 211(1): 140-151.

[3] Li Min, Wang Weihua. Effects of High Temperature Stress on Protective Enzyme Activity and Membrane Permeability of Spinach Leaves [J]. Acta Horticulture, 2004, 31 (1): 99-100.

[4] Liu Shipeng, Pang Xiaoliang, Cao Juanyun, Xue Hao, Chen Zongli. Determination and Content Analysis of Superoxide Dismutase in Fresh Jujube Fruits [J]. Hunan Agricultural Science, 2012 (15): 36-38.

[5] Dalian Institute of Light Industry. Food Analysis [M]. Beijing: China Light Industry Press, 1994.

Positive Effect of Chaiqin Wendan Decoction on Anxiety and Depression Model Rats

Xiaobing Li[1*], Xinmin Li[2], Ruijuan Cai[2], Binbin Qi[2], Yuzhu Li[2],Tian Wang[2], Wen Liu[2], Wenyu Chen[3], Yan Chen[3], Mengge Ma[3], Shicheng Ren[3], Xiaomei Liu[3], Xuejiao Liang[3], Dexing Wang[3], Guangbin Shen[3], Huxiao Zhang[3] and Bing Zhang[3]

1 Henan University of Chinese Medicine, Zhengzhou, China
2 The first Clinical Medical College of Henan University of Chinese Medicine, Zhengzhou, China
3 The Second school of clinical medicine of Henan University of Traditional Chinese Medicine, Zhengzhou, China
*Corresponding author e-mail:baishaoyao@163.com

Abstract. Objectives: To investigate the effect of Chaiqin Wendan Decoction (CWD) and mechanisms for anxiety and depression-like behavior in rats. **Methods:** We divide SD male rats into four groups randomly (blank group, model group, diazepam group and Chinese medicine group).The anesity and depression model rats were induced by chronic unpredictable emotional stress. Chinese medicine group and diazepam group were given CWD and diazepam respectively.The model group and the blank group were given the same amount of normal saline per day.The general state of the rats was recorded, and an open field trial and operation of the Morris water maze test were performed.the concentration of 5-HT was measured. **Results:** Compared with the model group, the body weight, horizontal movement distance score, number of upright, and target quadrant percentage of the rats in the Chinese medicine group increased significantly ($P < 0.05$), and the latency decreased significantly ($p < 0.05$). The concentration of 5-HT in the Chinese medicine group and diazepam group was significantly decreased by ELISA kit ($P<0.05$).**Conclusion:**WD can significantly promote the learning and cognitive ability of rats with anxiety and depression. Its mechanism of action may be closely related to the ability of CWD to reduce 5-HT concentration.

1. Introduction

Anxiety and depression are the two most common neuropsychiatric diseases in today's society. They are highly related and often occur at the same time, causing a great financial burden for patients. [1] WHO epidemiological survey shows that the proportion of co-diseases can be as high as 70 %, and nearly 80 % of depression patients are accompanied by anxiety symptoms. Anxiety symptoms can not only cause resistance to depression treatment, but also increase their disability rate [2].

Currently, more and more modern medicines, tricyclic antidepressants, for example, have been used to treat anxiety and depression. However, these drugs have kinds of side effects and high price, which greatly delay patients' treatment [3].

CWD, a famous Chinese prescription, origins from *Three Causes——Diseases and Syndrome* (a traditional Chinese medicine masterpiece). Currently, it has been demonstrated with obvious effects on the treatment of anxiety and depression [4-5].

In this study, we used anxiety model rats introduced by unpredictable mild stress to research the effect of CWD on their behaviors and inhibitory neurotransmitter.

2. Materials and Methods

2.1. Animals
The SD rats, body weight (180 ~ 200) g, are male and SPF grade. And all rats received standard feed and drink water.

2.2. Experimental Materials

2.2.1. Drugs and Reagents
Experimental reagent:*5-HT kit(Shanghai Fanke Biotechnology Co., Ltd., 96T);* compound diazepam tablets (0.25mg/tablet,10 tablets, Jining Ankang PharmaceuticalCo., Ltd., Chinese medicine quasi-word H10970219,); Bupleurum 12g; Scutellaria 10g; Qing Pinxia 12g; Fructus Aurantii Immaturus 12g; Bamboo shaving10g; dried orange peel 6g; Poria cocos 20g; raw licorice 3g (Guangdong Tianjiang Fenjian granules).

2.2.2. Experimental Equipment
Open field box, constant temperature bath pot, and small centrifuge desktop.

2.3. Methods

2.3.1. Grouping and Administration
Twenty male rats were divided into 4 groups randomly: blank group, model group, Chinese medicine group (11.5 $g{\cdot}kg^{-1}$) and diazepam group (0.26$mg{\cdot}kg^{-1}$) .The blank group and the model group were respectively given the same volume of distilled water at 0.4 ml/kg/d. The other groups were given corresponding drugs at 2 ml/kg/d by irrigation stomach.

2.3.2. *Model Preparation*
The model rats were established by chronic unpredictable emotional stress. Except the blank group, other three groups were received nine kinds of unpredictable stress in 7 days .The rats received 2 or 3 kinds of random stress stimulation a day.

2.4. Observations

2.4.1. Body Weight
Record daily body weights of rats.

2.4.2. Open-field Test
The rats were placed in the bottom of the open box. Record the horizontal movement of the rats and the vertical movement of the rats. And the total score is used to evaluate the change of their exploratory behavior.

2.4.3. Morris Water Maze
The water maze device is divided into four quadrants of equal size. The central position of each quadrant wall above the water surface is marked with chalk drawing of 4 different eye-catching marks. A black circular platform below the water surface, and was placed in the center of the fourth quadrant with the same position.

2.4.4. Place Navigation Test
In each experiment, the time of discovering the hidden platform (escape latency) and distance was followed by the intelligent video tracking system. The training lasted for 5 days, once a day [6].

2.4.5. 5-HT
The concentration of 5-HT in rat hippocampus was strictly determined according to the ELISA kit

instructions.

2.5. Statistical Analyses

Numerical data were expressed as the mean ± standard deviation (SD). Statistical analyses were performed using SPSS 17.0 for Windows (SPSS, China, IL, and U.S.A.). *P*<0.05 was considered statistically significant.

3. Results

3.1. Effect of CWD on the General State of Anxiety Model Rats

Compared with the model group, the weight of the diazepam group and the Chinese medicine group increased significantly (P<0.05) (Table1).

Table 1. Effect of Chaiqin Wendan Decoction on the Weights of Anxiety Model Rats (unit:g, n=3, $\bar{x}$ ±s)

Group	weights
Blank group	256.00±10.82
Model group	183.00±4.00*
Diazepam group	208.33±11.24#
Chinese medicine group	212.33±2.52#

Note: Compared with the blank groups, *P<0.05; Compared with the model groups, #P <0.05.

3.2. Effect of CWD on Open-field Test Level Score of Anxiety Model Rats

Compared with the model group, the horizontal movement distance score of the diazepam group and Chinese medicine group was significantly increased (*P*<0.05) (Table2).

Table 2. Effect of CWD on Open-field Test Level Score of Anxiety Model Rats (unit: score, n=3, $\bar{x}$ ±s)

Group	Level the score
Blank group	44.33±10.02
Model group	13.00±4.58*
Diazepam group	29.67±8.50#
Chinese medicine group	33.67±7.77#

Note: Compared with the blank groups, **P*<0.05; Compared with the model groups, #*P* <0.05.

3.3. Effect of CWD on Open-field Test Upright Times of Anxiety Model Rats

Compared with the model group, the upright times of the diazepam group and the Chinese medicine group were obviously increased (*P*<0.05) (Table 3).

Table 3. Effect of CWD on Open-field Test Upright Times of Anxiety Model Rats (unit: score, n=3, $\bar{x}\pm s$)

Group	Upright times
Blank group	11.67±1.15
Model group	4.33±1.53*
Diazepam group	8.67±2.52#
Chinese medicine group	11.00±1.00#

Note: Compared with the blank groups, **P*<0.05; Compared with the model groups, #*P* <0.05.

3.4. Effect of CWD on Water Maze Incubation Period in Anxiety Model Rats
Compared with the model group, the latency of the diazepam group and the traditional Chinese medicine group was significantly decreased ($p < 0.05$) (Table 4).

Table 4. Effect of CWD on Water Maze Incubation Period in Anxiety Model Rats (unit: s, n = 3,$\bar{x}\pm s$)

Group	Latency
Blank group	11.00±1.00
Model group	36.33±4.04*
Diazepam group	8.33 ±2.08#
Chinese medicine group	8.00 ±1.00#

Note: Compared with the blank groups, **P*<0.05; Compared with the model groups, #*P* <0.05.

3.5. The Effect of CWD on the Percentage of the Water Maze Target Quadrant in the Model of Anxiety Model
Compared with the model group, the percentage of target quadrant in the diazepam group and the traditional Chinese medicine group was significantly increased ($p < 0.05$) (Table5).

Table 5. Effect of CWD on the Percentage of Water Maze Target Quadrant in Anxiety Model Rats (unit: %, n= 3, $\bar{x}\pm s$)

Group	percentage of target quadrant
Blank group	28.20±2.27
Model group	6.94 ±1.45*
Diazepam group	23.18±0.94#
Chinese medicine group	24.76±1.05#

Note: Compared with the blank groups, **P*<0.05; Compared with the model groups, #*P* <0.05.

3.6. *Effect of CWD on the Concentration of 5-HT in Hippocampus of Anxiety Model Rats*
Compared with the model group, the concentration of 5-HT in hippocampus of the diazepam group and the Chinese medicine group decreased significantly ($P<0.05$) (Table 6).

Table 6. Effect of CWD on the Concentration of 5-HT in Hippocampus of Anxiety Model Rats (unit: ng/ml, n=3,$\bar{x}$±s)

Group	5-HT
Blank group	102.34±4.30
Model group	128.78±0.61*
Diazepam group	106.50±4.23#
Chinese medicine group	106.52±0.94#

Note: Compared with the blank groups, *P<0.05; Compared with the model groups, #P <0.05.

4. Discussion

In our experiments, body weight in the model group rats was significantly reduced, the reason is anxiety and depression can suppress appetite, which is consistent with clinical observations.[7] Compared with the model group, body weight in the Chinese medicine group rats was obviously increased .We conclude that CWD does opposite anxiety and depression-like behavior in rats.

In the open field experiment, the horizontal distance score and the number of upright times in the anxiety and depression model rats should be decreased. [8] In our experiments, the scores of the Chinese medicine group and the diazepam group were significantly higher than those of the model group. Therefore, we can conclude that CWD has a positive effect on anxiety and depression.

In the Morris water maze test, compared with the model groupthe, latency of the Chinese medicine group and the diazepam group was significantly reduced, and the target quadrant percentage was significantly increased. We know that the longer the incubation period, the lower the target quadrant percentage, the higher the anxiety level, and the worse the corresponding learning and memory skills. [9-10]. This indicates the improvement of CWD on the learning and memory ability of anxious rats.

A large number of facts have proven to reduce the anti-anxiety effect of 5-HT drugs. In our experiments, it was clear that the 5-HT concentration in the diazepam and Chinese medicine groups was significantly lower than in the model group, indicating that diazepam and CWD can achieve anxiolytic and depression by reducing the concentration of 5-HT. effect.

5. Acknowledgments

Supported by a project grant from The National Natural Science Foundation of China (Grand No.81603527), Science and technology project of Henan Province (Grand No.162102310466), Key scientific research projects of Henan Province College sand Universities (Grand No.16A360010), Henan University of Traditional Chinese Medicine Scientific and technological innovation talent support program (Grand No.2015XCXRC05). Science and technology project of Zhengzhou City (Grand No.20150310), The Young Core Teacher of Henan Province (Grant No. 2016GGJS-080).

6. References

[1] Health Quality Ontario. Psychotherapy for major depressive disorder and generalized anxiety disorder: a health technology assessment [J]. Ontario Health Technology Assess Ser, **2017**, 17 (15):1-167.
[2] Meier SM, Petersen L, Mattheisen M, et al. Secondary depression in severe anxiety disorders: a population -based cohort study in denmark [J]. Lancet Psychiatry, 2015, 2(6):515-523.
[3] Zhao Hongqing, Lei Chang, Yang Wei, He Wenlong, Zhang Sijing, Han Yuanshan, Wang Yuhong. Effects of venlafaxine on anxiety and depression-like behavior in rats with chronic restraint stress [J/OL]. Chinese Journal of Comparative Medicine:1-6 [2019-05-24].http://kns.cnki.net/kcms/detail/11.4822.R.20190514.1722.015.html.
[4] Liu, Peng, Xiao crystal segment Chaiqin Wendan Treating anxiety with insomnia 30 cases [J]

Chinese Medicine Science, 2011,1 (05): 69 + 89.
[5] DOU Chunxia. Clinical observation of Jiawei Chaiqi Wendan Decoction in the treatment of post-stroke depression [J]. Chinese Medicine Science, 2015, 22(04):413-414.
[6] Shan Xin, Kong Zhouyang, Wang Sujuan, etc. study on the anti-anxiety activity of spider incense extract and Total Valeriana. J. Chinese Traditional and Herbal Drugs, 2016, 4, 1361-1365.
[7] Hong Deng, Qi Yaling, Zhang Yanhui, et al. Simple and easy to operate rat CUMS depression model construction method [J]. China Health Industry, 2011, 8 (24): 3-4.
[8] Huang Kuan, Ye Lizhi, Li Wei, et al. Effect of high salt diet on behavior of mice with menopausal anxiety [J]. Hunan University of Traditional Chinese Medicine, 2013, 29 (12): 130-132.
[9] Mineur YS, Obayemi A,Wigestrand MB, et a. Cholinergic signaling in the hippocampus regulates social stress resilienceanxiety-and depression-like behavior[J].Proc Natl Acad SciUSA,2013,110 (9):3573-3578.
[10] Garner B, Wood SJ, Pantelis C, et al. Early maternal deprivation reduces prepulse inhibition and impairs spatial learning ability inadulthood: no further effect of post-pubertal chronic corticosterone treatment [J]. Behav Brain Res, 2007, 176(2): 323-332.

IOP Conf. Series: Materials Science and Engineering **730** (2020) 012037 doi:10.1088/1757-899X/730/1/012037

Conventional Bridge Damage Identification Based on BP Neural Network

Zeying Yang[1], Yalei Zhang[1], Yaping Wang[2], Ning Wang[2], Haina Cui[2] and Yongye Qu[1]

1. School of Civil Engineering, Shandong University, Jinan, Shandong 250061, China
2. Shandong Huawei Project Supervision Consulting Center 261061, China
Corresponding author. E-mail address: 1197598624@qq.com

Abstract. Aiming at the problem of insensitivity of damage identification in conventional bridges, the application of BP neural network in long-span bridges is studied, and it is extended to the field of damage identification of conventional bridges. The finite element models of three-span continuous variable cross-section box girders under intact and damaged conditions are established by Midas civil, and the eigenvalues of bridges under different conditions are analyzed. It is found that in conventional bridges, the sensitivity to structural damage is: mode > vertical displacement > natural frequency. The parameterized natural frequencies and modes of structures are used as input of BP neural network, and the damage location and degree are used as output to train the neural network. Then, the damage location and degree are identified under different working conditions, and the results show that the recognition effect is not satisfactory. Analyzing the reason, the natural frequency and modal shape of the structure will change when the conventional bridge is damaged, but the deformation value is very small. When training the BP neural network, it is easy to appear over-fitting, which results in poor recognition effect. Therefore, it is difficult to identify structural damage by using BP neural network through acquiring the characteristics of conventional bridges. It is still necessary to study appropriate damage identification parameters and methods that can be applied to conventional bridges.

1. Introduction

According to the proportion of dangerous bridges, conventional bridges are much higher than super-long-span bridges, and most of the bridges collapsed during operation are concentrated in conventional girder bridges[1]. The damage identification method based on neural network has been widely concerned and studied in engineering structures [2-5]. Taking three-span continuous variable cross-section concrete box girder as an example, this paper studies the application of neural network method in damage identification of conventional girder bridges, in order to promote the establishment of conventional bridge health monitoring system.

2. Paper Finite Element Model of Three-Span Continuous Variable Section Concrete Box Girder

Based on a three-span continuous variable cross-section box girder, this paper establishes a finite element model. The width of bridge deck is 6.5m, and the layout of bridge type is 37.5m+65m+37.5m=140m. Vehicle load grade: highway grade I; bridge seismic fortification intensity: VI; peak acceleration coefficient of ground motion is 0.05g. The top section of the pier is 3.5 m high, as shown in Figure 1. The beam end and mid-span section are 1.8m high, as shown in Figure 2. The finite element models of three-span continuous variable cross-section box girders based on Midas civil

IOP Conf. Series: Materials Science and Engineering **730** (2020) 012037 doi:10.1088/1757-899X/730/1/012037

are presented in Figure 3, which are subdivided into left and right spans, middle piers and middle spans.

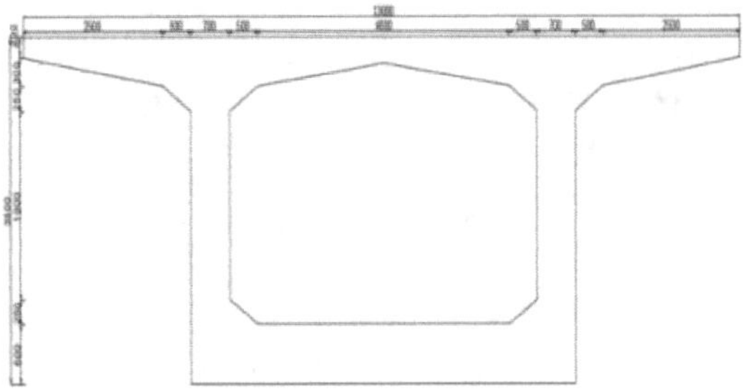

Figure 1. Pier Top section

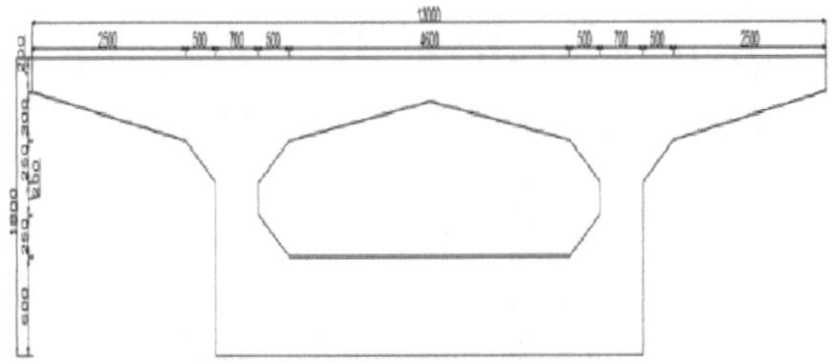

Figure 2. Beam End and Mid-span sections

Figure 3. Finite element model of continuous variable cross-section box girder with three spans

3. Damage Simulation

Next, damage simulation of the three-span continuous variable cross-section box girder is carried out, which is divided into the selection of damage location and the simulation of damage degree. Damage mainly includes breathing crack, concrete carbonization, steel bar and concrete peeling, steel bar corrosion and so on. Under the action of reciprocating load of high-speed transport vehicle, when the damage of breathing crack reaches a certain degree, concrete will peel off steel bar, steel bar will be corroded, accompanied by fatigue damage, and ultimately fracture. In this paper, the damage locations are arranged in the middle of the left and right span, the top of the pier and the middle span, respectively. As shown in Figure 4, the shadow part is the location of the crack damage on the concrete beam bridge. The damage condition is divided into single position damage and double position damage. The single damage location is taken as the research object, and the damage condition is set as shown in Figure 5.

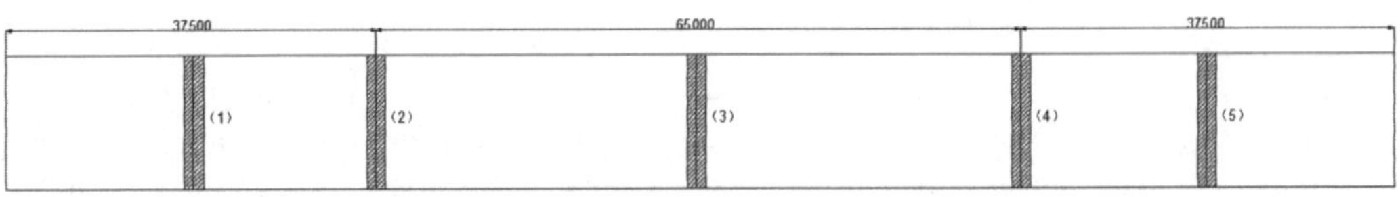

Figure 4. Crack damage of continuous variable section box girder (overlooking)

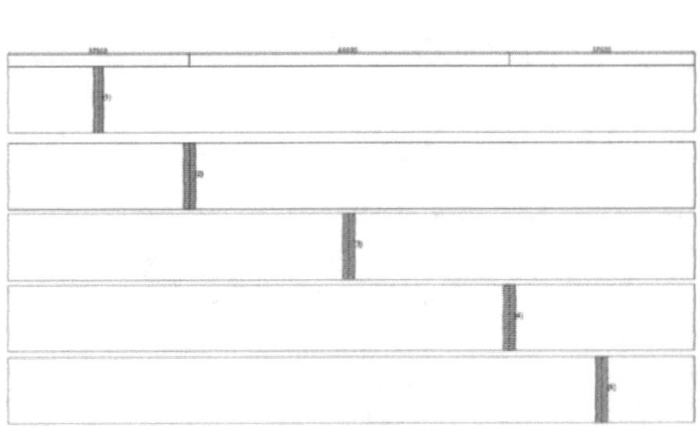

Figure 5. Single damage location

Taking the top view of three-span continuous variable cross-section box girder, in order to facilitate the output of damage location identification, from the left span to the right span, the above five damage locations are sequentially arranged into 1-5, expressed by G1-G5. Generally, the crack damage of bar structure and frame structure can be simulated by changing the section area and elastic modulus of material. The crack damage of the bridge is simulated by changing the elastic modulus of the concrete of the three-span continuous girder bridge. In order to simulate different degrees of crack damage, the elastic modulus of C40 concrete is reduced to 90%, 85%, ... , 25% , 20% respectively. Indicating that the degree of crack damage is 10%, 15%, 15%, ... , 75% and 80%. D1 ~ D15 are used to express them respectively. D0 indicates that the structure is in good condition. The location and degree of damage can be expressed by D#-G#, for example, D1-G1 for 10% of the damage under condition 1.

4. Modal Analysis

After the damage of engineering structure, its dynamic characteristics will change accordingly. The change of structural dynamic characteristics can be represented by observing the change of natural frequencies and mode shapes of the structure. In order to observe this change more clearly and to study the sensitivity of natural frequencies to damage, the first 7 natural frequencies under different working conditions are selected. The natural frequencies and modes described in this paper refer to the vertical vibration of bridge structures. Taking D1-D15 working conditions under D0 and G1 as examples, the first eight natural frequencies of three-span continuous variable cross-section concrete box girders are shown in Table 1.

Table 1. First 7 Natural Frequencies of Bridge Structures under 15 Working Conditions of D0 and G1

Working Conditions	Frequency /Hz						
	1 order	2 order	3 order	4 order	5 order	6 order	7 order
D0	1.625732	3.325308	4.280689	5.567203	6.021167	7.704737	9.8668
G1-D1	1.625709	3.325237	4.280618	5.567173	6.021126	7.704722	9.866782
G1-D2	1.625696	3.325195	4.280576	5.567156	6.021103	7.704714	9.866772
G1-D3	1.625681	3.325149	4.28053	5.567136	6.021076	7.704704	9.86676
G1-D4	1.625664	3.325096	4.280477	5.567114	6.021045	7.704693	9.866746
G1-D5	1.625645	3.325036	4.280417	5.567089	6.021011	7.704681	9.866731
G1-D6	1.625623	3.324968	4.280348	5.567059	6.020971	7.704667	9.866713
G1-D7	1.625597	3.324888	4.280269	5.567025	6.020925	7.704651	9.866692
G1-D8	1.625567	3.324794	4.280175	5.566985	6.020871	7.704631	9.866668
G1-D9	1.625531	3.324683	4.280064	5.566936	6.020807	7.704609	9.866639
G1-D10	1.625488	3.324548	4.279929	5.566877	6.020729	7.704581	9.866604
G1-D11	1.625434	3.324381	4.279763	5.566803	6.020633	7.704546	9.86656
G1-D12	1.625366	3.32417	4.279552	5.566708	6.020511	7.704503	9.866504
G1-D13	1.625278	3.323894	4.279277	5.566581	6.020351	7.704445	9.86643
G1-D14	1.625156	3.323516	4.278901	5.566404	6.020133	7.704367	9.866328
G1-D15	1.62498	3.322969	4.278359	5.566138	6.019818	7.704252	9.866177

From Table 1, we can see that under the same damage condition, the natural frequency of concrete bridge structure increases with the increase of modal order; the natural frequency of concrete bridge structure decreases after damage occurs; after damage occurs, the higher-order frequency decreases more than the lower-order frequency, which indicates that the higher-order frequency change is more sensitive to structural damage. With the increase of damage degree, the natural frequency decreases, but the change is only about 3%.

When the structure is damaged, the mode shapes will change accordingly. The following is an illustration of the mode change under D0 and G1 conditions. For the convenience of viewing the changes of the modes of the damaged parts, the specific values will be shown in Table 2.

Table 2. The first five modes of bridge structure under D0 and G1-D15

Working Conditions	Modal shape/m				
	1 order	2 order	3 order	4 order	5 order
D0	1.23E-05	2.73E-05	3.08E-05	9.34E-06	2.12E-05
G1-D15	1.3E-05	2.83E-05	3.17E-05	9.39E-06	2.40E-05
Variation	5.7%	3.7%	3%	0	13.2%

From the above table, it can be seen that the first five modes of bridge structure simulated by Midas are relatively small; the change of modes is more obvious than the change of natural frequencies after the damage of the structure, which shows that the modes are more sensitive to the structure, but the maximum change is about 10%. In addition, the displacement changes of damage location under simulated actual loads are obtained. The following examples are D0 and G1-D15, as shown in Table 3.

Table 3. Displacement of Damage Location of Bridge Structures under D0 and G1-D15

Working Conditions	Displacement /mm				
	Element 51	Element 52	Element 53	Element 54	Element 55
D0	-17.48258	-16.0755	-14.6374	-13.1805	-11.7218
G1-D15	-17.56168	-16.1459	-14.7006	-13.2368	-11.7713
Variation	0.452%	0.438%	0.432%	0. 427%	0. 422%
Working Conditions	Displacement /mm				
	Element 56	Element 57	Element 58	Element 59	Element 60
D0	-10.27559	-8.85377	-7.46141	-6.10481	-4.78921
G1-D15	-10.31793	-8.88987	-7.49163	-6.1294	-4.8084
Variation	0.412%	0.408%	0.405%	0.403%	0.401%

From Table 3, it can be seen that the absolute value of the vertical displacement of the damaged position increases under the action of environmental loads after the bridge is damaged. Compared with the natural frequencies and modes of the structure, the displacement change is not particularly sensitive to the damage of the structure at the same damage degree. The location and degree of damage can be predicted by placing displacement sensors on the bridge.

5. Damage Identification of Concrete Beam Bridge Based on BP Neural Network

5.1. Input and Output of Neural Networks

According to the results of finite element modal analysis, the change of natural frequencies of damaged structures is small, and only the change of higher natural frequencies is large. This paper expects to identify the damage location and the damage degree of bridge structure through BP neural network. When the damage degree is large, the modal shape of bridge structure changes about 10%, so the modal shape can be normalized as input parameters. Taking the first five modes of working condition G1-D15 as an example, the first step is to increase the mode value by five orders of magnitude for the convenience of the whole. Then the mapminmax function in MATLAB is used to normalize the mode and map it to the interval [-1, 1] as shown in Table 4.

Table 4. Normalization of modal mode shapes of bridge structures under G1-D15

Normalization	1 order	2 order	3 order	4 order	5 order
Pre-treatment	1.30	2.83	3.17	0.94	2.40
Post-treatment	-0.68	0.70	1.00	-1.00	0.31

After normalizing the first five modes of damage location of three-span continuous variable cross-section box girders under different working conditions, the damage response is taken as the input of the network and the output of the network. In this paper, it is expected that BP neural network can be used to identify the damage location and degree of bridge structure at the same time. The output of the network is parameterized to Ti, j (i = 1~5, j = 2~16). Five values of I correspond to the damage location of the bridge structure (damage location is 1, non-damage location is 0), and the corresponding damage degree of j to the bridge structure (near the preset damage level is 1, the others are 0). For example, the output vector of G1-D15 is shown in equation (1), which indicates that the damage location is in the middle of the left span and the damage degree is 80%.

$$T_{1,16} = \begin{bmatrix} 1 & 0 & 0 & \cdots & 1 \\ 0 & 0 & 0 & \cdots & 0 \\ 0 & 0 & 0 & \cdots & 0 \\ 0 & 0 & 0 & \cdots & 0 \\ 0 & 0 & 0 & \cdots & 0 \end{bmatrix} \quad (1)$$

5.2. Construction of Network Structure

Any continuous function in the closed interval can be approximated by BP network with a single hidden layer, that is, a three-layer BP network can complete any mapping from n-dimension to m-dimension. Therefore, a three-layer BP neural network is constructed to deal with the non-linear mapping relationship between structural damage and structural modal. The difficulty of building BP neural network is to determine the number of neurons in the hidden layer, which generally needs the experience of the designer and many experiments to determine. It is difficult to find an ideal analytic formula to obtain. There is a direct relationship between the number of hidden layer units and specific problems, and between the number of input and output units. Too few hidden units will lead to poor fitting effect and poor recognition results; too many hidden units will lead to long learning time, but the error will not necessarily reduce, and may lead to poor fault tolerance, unable to identify samples that were not seen in the past. Therefore, it is necessary to determine the optimal number of hidden elements. In this paper, the initial number of hidden elements is determined according to the following formula (2), and then the optimal number of hidden elements is adjusted repeatedly.

$$n_1 = \sqrt{n+m} + a \quad (2)$$

Among them, n_1 is the number of hidden units and n is the number of input units. In this paper, let $n = 15$. m is the number of output neurons, $m = 2$ and a is the constant between [1,10]. After debugging, $a = 5$ is chosen in this paper. The number of hidden units n_1 is taken as an integer and eventually 10.

5.3. Network Training and Its Effect

Data from D1 to D14 under G15 are selected as training samples, and data from D15 under G1 to G5 are selected as test samples. Generally, the more damage cases are considered, the more training samples are obtained and the more accurate the recognition is, but the training samples will increase infinitely. In fact, the neural network has a strong generalization ability. As long as the training samples have a certain scale and high quality, it is not necessary to train all the samples.

In practical engineering, different network models can get different results of structural damage identification. The BP neural network with single hidden layer is simple to construct and easy to implement. It can approximate any non-linear continuous function. When training the BP neural network, the damage degree of each working condition is close (the damage degree increases from 10% to 5% as a gradient to 80%). The modal parameters simulated by the cable element have little difference under each working condition, and the over-fitting situation appears (as shown in Figure 6). The recognition result is quite different from the expected value.

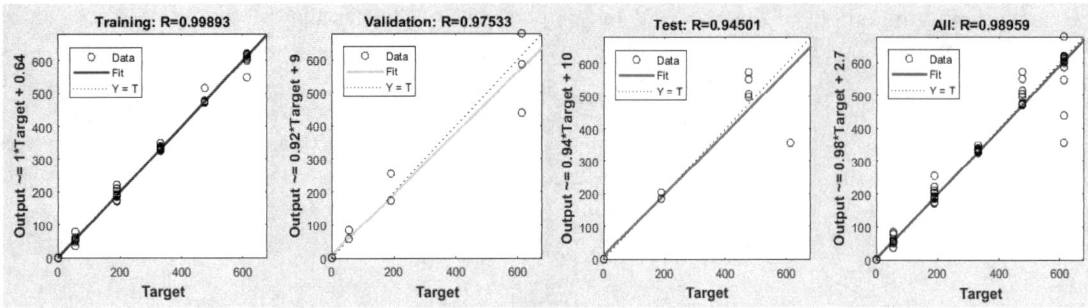

Figure 6. The fitting effect of BP neural network

6. Conclusion

In this paper, the finite element model is established to obtain the modal parameters under various working conditions, and the damage identification using neural network is still limited. When conventional bridges suffer from micro-damage, the change of modal parameters is very small, and the displacement and deformation of structures are very small. It is difficult to identify the damage. Only when there is a large damage in the bridge structure, can the damage be measured accurately. In the conventional bridge damage identification, how to accurately identify the damage degree by BP neural network is still a difficult problem.

7. Acknowledgments

The research presented in this paper has been co-financed by Shandong province Transportation Science and Technology Project (2012A18), (2013A08-03), (2013A12-03). Their financial support is gratefully acknowledged.

8. References

[1] C. Farrar, S. Doebling, D. Nix, Vibration-based structural damage identification, Philos. Trans. Roy. Soc. Lond. A: Math., Phys. Eng. Sci. 359(1778)(2001)131-149.

[2] Luca Facchini, Michele Betti, Paolo Biagini. Neural network based modal identification of structural systemsthrough output-only measurement. Computers and Structures 138 (2014) 183–194.

[3] Jordan C.Weinstein; Masoud Sanayei, M.ASCE; and Brian R.Brenner, F.ASCE. Bridge Damage Identification Using Artificial Neural Networks. J. Bridge Eng., 2018, 23(11): 04018084.

[4] HU Qin, XU Wei, GAO Fei, et al. Damage Identification of CA Mortar Layer of CRTS I Slab TrackBased on BP Neural Network[J]. Journal of Civil Engineering and Management, 2018, 35 (05): 87-93.

[5] LI Xue-song, MA Hong-wei, LIN Yi-zhou, et al. Structural damage identification based on convolution neural network[J]. Journal of Vibration and Shock, 2019, 38 (01): 159-167.

Numerical Model Progress of Liquid Fuel Micro-droplets Evaporation

Yang Yu[1] and Xiao Zhang*, [2]

[1]School of Mechanical Engineering and Automation, Northeastern University, 110819, Shenyang
[2]School of Ocean, Yantai University, 264005, Yantai
zxzn2004@163.com

Abstract. The evaporation of spray injected in material and mechanical engineering has significant effect on mixture formation, emissions and power output. In fact, for only micro-droplets in regular arrangement, fluid motion, supercritical and various shapes changed with time and so on, these common phenomena are quite complex and difficult to be understood and simulated. Especially, the studies of latest decade can show the development level of the relevant models, which will be reviewed in this paper. Some typical traditional quasi- and multi-dimensional models predicting various evaporation characteristics are reviewed for the newest developments to understand much more complex species, momentum and heat transport phenomena through the gas-liquid interface in industrial sprays with real operating conditions. Some experiments used to compare with these models developed are analyzed to reveal some aspect to be improved both for experiment and model. As can be seen from the comparison of the multidimensional results with the quasi-dimensional ones, the more comprehensive models are apparently in favor of understanding the micro-droplet evaporation better. Furthermore, the micro-droplet group shows more complex evaporation process than that of isolate micro-droplet.

1. Introduction
The conventional spray combustion is still the most effective way to provide energy in internal combustion engine. The conceptual model with relative experiments has been revealed the mechanism of soot emission due to dense premixed mixture produced by the conventional spray. So the production of the dense mixture has significant effect on soot, which is influenced by evaporation of spray injected. Especially, the exhaust gas recirculation (EGR) can lowing soot emission due to high proportion of premixed mixture which is confirmed by the conceptual model with relative experiments. The species of EGR inevitably affect the evaporation of the fuel. And if the evaporation can be controlled by some technique, the mixing of fuel and oxygen can be improved so that the dense mixture and soot are lowered. So the evaporation of spray needs to be studied further [1-11].

2. Model of Micro-droplet Evaporation
Ragab R and Wang T [1] develop a three-zone model to simulate the evaporation: (a) the droplet zone with water liquid mass fraction of 1, (b) the shell zone around the droplet with mass source for water vapor and energy sink to simulate liquid evaporation, and (c) the outer domain represents the rest of the flow field. So interface boundary conditions are used between the droplet, the shell, and the domain. And these three distinct cell zones are connected via non-conformal grid interfaces to allow the transport equations to be solved in the whole domain. The fluid and temperature fields are given more detail, though in fact the internal flow of droplet must be influenced by outer domain. Especially,

they thought Biot number should be studied further which can help provide an idea of how thermal resistance ratio would affect heat flow distribution.

To be capable of considering the integral computation region, Strotos et al [2] modeled the evaporation of droplet in using volume of fluid method to track the interface of fuel and air. They found that the velocity streamline is regular under convention flow, but the velocity streamline are not very regular relatively which has two different parts. It's known that the internal circulation has important effect on the evaporation. And the three-dimensional simulation can consider this effect, which should further be studied such as velocity, direction, location and area. For both conditions, the high concentration regions of vapor occur at the downstream of the flow direction. It means that the vapor also flow radically and downwardly. Also, it's interesting that below the droplet there is backflow. That's quite different from the other simulation and the phenomenon of the spray. The backflow may be due to the suspender existing redundantly. However, the results were still close to the experiments in fig 1.

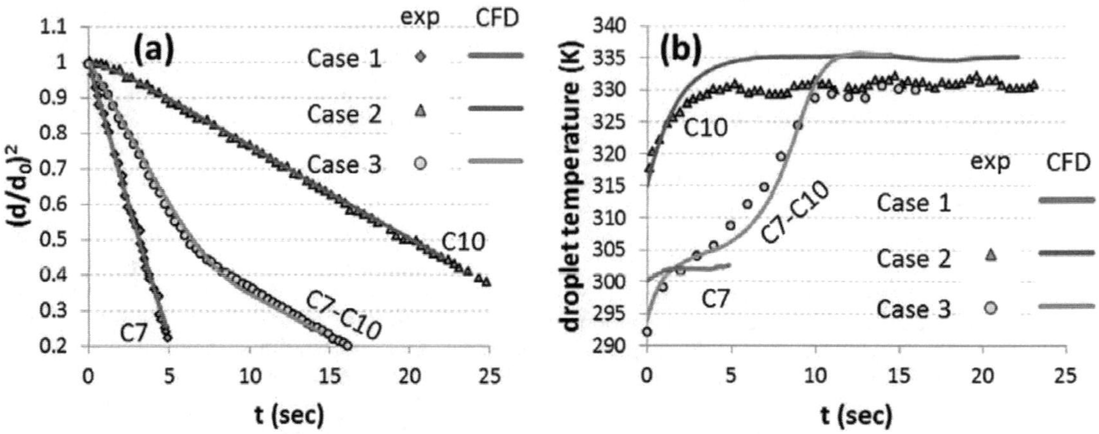

Figure 1. Model predictions for the experimental data[2]

Saufi A E et al [3] designed a camber with different thermocouple to realize the experiments in a wide range of operating conditions, controlling the heating rate (up to 1000 K/s), the pressure (from 0.1 kPa to 10 MPa) and the atmosphere composition (usually nitrogen or air). The suspender using one thermocouple for droplet was hanged on the top face. The temperature field showed the temperature was very hard to be controlled though using three thermocouples. From the other new study, Naito Y et al [4] established a vessel only with one thermocouple and adiabatic material wall. So the current equipments are hard to keep constant temperature for several seconds. Rehman H L et al [5] studied the outcome of varying support size, support material, ambient temperature, ambient velocity and droplet size on the droplet evaporation process. The average over estimation of mean droplet evaporation rates in the absence of heat conduction for linear extrapolation are found to be 30% and 8% for thermocouples and glass fibers respectively. This non-linearity behavior in droplet evaporation rate curves was more profound for thermocouple than glass fiber due to higher thermal conductivity as shown in fig 2. So the equipment should be organized carefully to compare with the mathematical model.

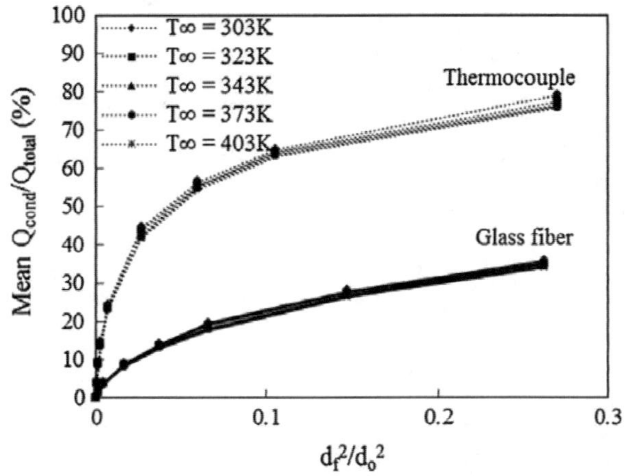

Figure 2. Variation of heat transfer through conduction[5]

Saufi A E et al [3] also adopted VOF to simulate the droplet evaporation, considering non-ideal mixtures evaluated with the UNIFAC approach. The results of the acetic acid mass fraction in the mixture shows the vapor distributed uniformly around the droplet that is quite different from the pure matter. In Naito Y et al' study [4], they only use quasi-dimensional vaporization model for the multi-component. And for temperature distributions inside the droplet, a quadratic polynomial distribution was introduced which is common in three-dimensional simulation. The different types of enthalpy diffusion term in heat flux were discussed. Muthuramalingam V P and Karlsson A [6] considered the non-ideal gas-liquid equilibrium utilizing the NRTL method in quasi-dimensional vaporization model which was incorporated into CFD code. The results showed that the vapor mass fractions were higher in non-ideal cases than in ideal cases.

Recently, differing from the traditional way with several common methods above, the sphere of droplet is queried and become another way to improve the vaporization model [7]. Though the shapes of droplets in spray are complex and change frequently, special droplet shapes approximating to prolate and oblate spheroids were still studied to find some way for first step. They use ellipsoidal coordinates to solve the equations system of original vaporization model. So the shape of droplet was controlled by eccentricity. were integrated using the finite-element-based PDE modules of COMSOL Multiphysics including Moving Mesh The temperature fields also presented ellipsoidal distribution. Note that the temperature is lower around the short axis, which should be given the explain.

The microcosmic methods for kinetic and molecular dynamics appear for a long time. Sazhin S S et al [8] brought forward the method combined kinetic model with traditional quasi-dimensional vaporization model in fig 3. It is shown that in all cases the kinetic effects lead to a decrease in droplet surface temperature and an increase in the evaporation time. The kinetic effects on the droplet evaporation time are shown to increase with increasing gas temperatures. There usually existed a series of assumptions at molecule level and the simulating results have not shown more advantages than the traditional ones. Especially, the quantum-chemical effects on the evaporation coefficient for realistic multi-component fuel droplets have not been investigated.

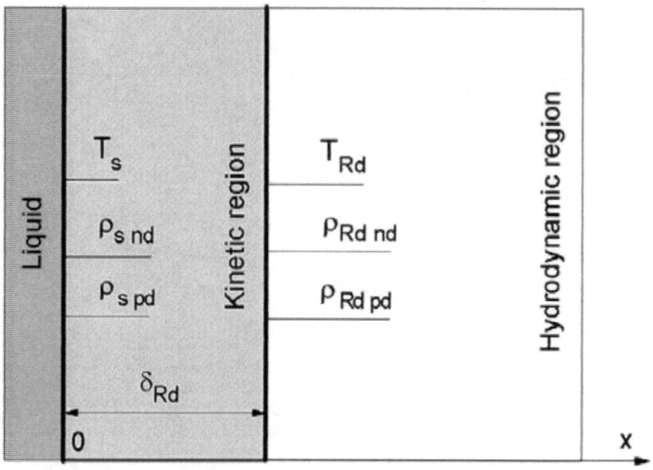

Figure 3. Variation of heat transfer through conduction[8]

The lattice Boltzmann method (LBM) is the no tracking methods without the need for generating an interface which is tried to compute the thermal multiphase fluid with evaporation [9]. A multiple relaxation time and exact difference method scheme was considered to simulate the heating and vaporization of a droplet under forced convection. The density near the gas-liquid interface is treated with one function of some given thick value, and the computations adopted two dimensional planar coordinate with 300×300 lattice to simulate the circle in the central section of droplet. However it is found that the method shows good agreement with the well-known D^2 law that prescribes the evaporation rate.

Zoby M R G [10] suggested examining the effects of the complex interactions of droplet arrays and turbulence on evaporation rates in inert and reactive turbulent environments using Direct Numerical Simulations (DNS). In particular the local temperature affects the evaporation rates of droplets are hard to be studied in previous works. Other than the above studies, the Level Set method has been combined with the Ghost Fluid method, which is not inherently mass conserving but can obtain smoother gas-liquid surface. They found that when combustion occurs as a group phenomenon, a reduction in the evaporation rates is observed. Evaporation processes of some droplets are quite different from that of isolate droplet. And turbulence does not seem to have significant effects on evaporation rates under any of the conditions.

Except for the above methods, the phase-field also can be used to handle multiphase fluid problem. Safari H et al [11] combined the phase-field with the level set to compute the evaporation. The modified convective Cahn-Hilliard equation is employed to reconstruct the dynamics of the interface topology, which was added a source term considering the phase change within the interface. They set the initial droplet temperature and the inflow air temperature to the range near the room temperature. For the laminar forced convective case, especially, the minimum temperature point pushes to the aft section of the droplet and eventually the cooled liquid is drawn into the droplet interior due to the internal circulation.

3. Conclusions

In fact, though providing enough description of evaporation process, the above results of these present models should be still compared with more detailed experimental data to validate various hypotheses simplifying reality. Shadowgraphs of the evaporating micro-droplets have been acquired, though temporally and spatially resolved composition profiles were measured along a line through the micro-droplets using one-dimensional Raman spectroscopy. It was quite clear that the surface regression rate is not a constant but a non-linear decrease. And if further considering the solubility of gas into liquid under not only supercritical condition but also the wider one and multicomponent' evaporation, the study of evaporation still need noticeable progress.

4. Reference

[1] Ragab, R, Wang T. An investigation of liquid droplet evaporation model used in multiphase flow simulation[C]. ASME 2012 International Mechanical Engineering Congress and Exposition, 2012, 7: 485-497.

[2] Strotos G, Malgarinos I, Nikolopoulos N, Gavaises M. Predicting the evaporation rate of stationary droplets with the VOF methodology for a wide range of ambient temperature conditions[J]. International Journal of Thermal Sciences, 2016, 109:253–62.

[3] Saufi, A.E, Calabria, R., Chiariello, F., Frassoldati, A., Cuoci, A., Faravelli, T., Massoli, P. An experimental and CFD modeling study of suspended droplets evaporation in buoyancy driven convection [J]. Chemical Engineering Journal, 2019, 375:122006.

[4] Naito Y, Hashimoto N, Fujita O. Development and validations of droplet evaporation model for low ignitibility fuel considering inner distributions under atmospheric pressure[C]. 12th Asia-Pacific Conference on Combustion, 2019.

[5] Rehman H L, Weiss J, Seers P. Effect of heat conduction on droplet life time and evaporation rate under forced convection at low temperatures [J]. Experimental Thermal and Fluid Science, 2016, 72:59–66.

[6] Muthuramalingam V, Karlsson A. Influence of Considering Non-Ideal Thermodynamics on Droplet Evaporation and Spray Formation for Gasoline Direct Injection Engine Conditions Using VSB2 Spray Model[C]. SAE Technical Paper, 2018: 2018-01-0181.

[7] Zubkov V S, Cossali G E, Tonini S, Rybdylova O, Crua C, Heikal M, Sazhin S S. Mathematical modelling of heating and evaporation of a spheroidal droplet[J]. International Journal of Heat and Mass Transfer, 2017, 108: 2181-2190.

[8] Sazhin S S, Shishkova I N, Qubeissi M A. A self-consistent kinetic model for droplet heating and evaporation [J]. International Journal of Heat and Mass Transfer, 2016, 93: 1206-1217.

[9] Albernaz D, Do-Quang M, Amberg G. Multirelaxation-time lattice Boltzmann model for droplet heating and evaporation under forced convection [J]. Physical Review, 2015, 91: 043012.

[10] Zoby M R G, Kronenburg A, Navarro-Martinez S, Marquis A J. Assessment of Conventional Droplet Evaporation Models for Spray Flames[C]. 14th Annual Results and Review Workshop on High Performance Computing in Science and Engineering, 2011: 209-227.

[11] Safari H, Rahimian M H, Krafczyk M. Consistent simulation of droplet evaporation based on the phase-field multiphase lattice Boltzmann method [J]. Physical Review, 2014, 90: 033305.

IOP Conf. Series: Materials Science and Engineering **730** (2020) 012039 doi:10.1088/1757-899X/730/1/012039

On Broadband Nonlinear Piezoelectric Energy Harvesting Techniques

Chuahua Sun[1,2] and Guangqing Shang[1]

1 Department of Mechanical and Electrical Engineering, Suzhou Vocational University, Zhineng Rd.106, Suzhou, China
2 3C-Product Intelligent Manufacturing Engineering Technology Research and Development Center of Jiangsu Province, Zhineng Rd.106, Suzhou, China.
Email: chh_sunny@163.com

Abstract. With the development of portable electronic devices, internet of things, micro-electromechanical system (MEMS) and wireless sensor networks, mirco sources and self-powered devices have been attracted great attention. It is an effective way to supply mirco power or realize the miniaturization of the device via the piezoelectric energy harvesting effect. Therefore, research on piezoelectric energy harvesting techniques has been highly valued. To enhance adaptability and improve harvesting efficiency, the broadband nonlinear piezoelectric energy harvesting technique turns to be a research hotspot. The current status of broadband nonlinear piezoelectric energy harvesting technique with nonlinear stiffness was reviewed according to methods of realizing nonlinearity. The characteristics and existed problems of each kind of nonlinear harvester were then concluded. The nonlinear piezoelectric energy harvester with nonlinear stiffness has a good prospect if its advantages can be given full use.

1. Introduction
As reported in the published literature, nonlinearities in energy harvesters are mainly manifested in the following two aspects: nonlinear stiffness and nonlinear piezoelectric coupling. Compared to the nonlinear piezoelectric coupling which results from piezoelectric materials, the nonlinear stiffness of a harvester is relatively easier to be achieved and controlled. Until now, three methods have been reported to realize nonlinear stiffness [1-3]: nonlinear magnetic force, piecewise-linear stiffness and nonlinear large strain. Therefore, this paper reviews recent advances in designing broadband piezoelectric energy harvester (PEH) with nonlinear stiffness based on the above methods. The paper will include the following sections: three realization ways of nonlinear stiffness, comments on nonlinear PEH and conclusion.

2. PEHs with Nonlinear Stiffness
A PEH with nonlinear stiffness has a wider frequency band and triggers a larger voltage, so this kind of PEH can more easily adapt to the environment and harvest more energy from vibration Comprehensive analysis of the existing literals, there exist three methods to realize nonlinear stiffness: nonlinear magnetic force, piecewise-linear stiffness and nonlinear large strain. The following section will develop the contents according to the three methods.

2.1. PEHs Excited by Nonlinear Magnetic Force
The magnets are initially introduced to tune resonance in PEH. With the furtherly research, these magnets also induce changes in both linear and nonlinear stiffness. The nonlinear behavior becomes

IOP Conf. Series: Materials Science and Engineering **730** (2020) 012039 doi:10.1088/1757-899X/730/1/012039

apparent when the harvester experiences oscillation with significant amplitude. Consequently, it takes significantly effect on widening frequency broadband and improving the harvesting efficiency. PEHs with monostable or bistable nonlinear motion are described below.

PEHs with monostable nonlinear motion. The hardening mechanism of nonlinear monostable harvester was investigated by Ramlan et al.[4]. The study revealed that the maximum amount of power harvested by PEH with a hardening stiffness was the same as one by a linear PEH, regardless of the degree of nonlinearity. However, this kind of device had such a larger bandwidth that the significant power can be harvested over a wider resonant frequency. Typical frequency-response curve for a hardening system was shown in **Figure 1**.

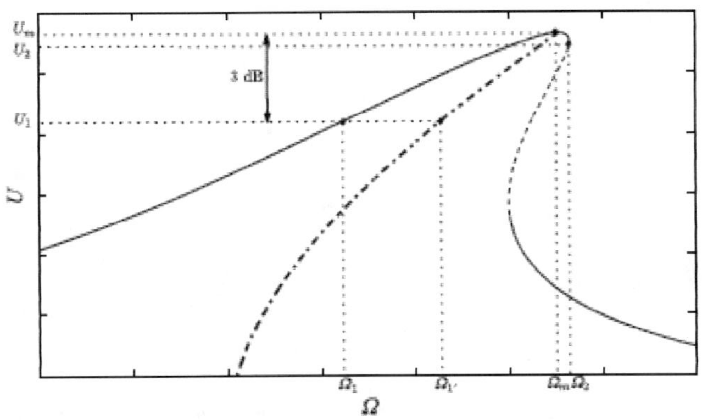

Figure 1. Typical frequency-response curve for a hardening system[4]

Another monostable device in **Figure 2** was proposed by Stanton et al.[5]. Both hardening and softening responses may occur by tuning the nonlinear magnetic interactions around the end mass. The result showed that as compared to the linear PEH, both a wider bandwidth and a better performance could be obtained by a PEH with monostable motion. The reason lied in that the high-energy attractor was captured through a linearly decreasing or increasing frequency sweep. Hence the bandwidth for the softening and hardening cases could be improved. However, it is difficult to guarantee such conditions in practical application, particularly in the low-energy branch and the requisite actuation energy. The problem of offering the required momentary perturbation did not address in many previous reported studies. Fortunately, a fast pulse perturbation system was used to realize the transition from low-energy to high-energy solutions in Ref.[6-7]. The output power of the nonlinear PEH was 16.5 times of that of the linear one and two steady-state harmonic solutions at the maximum output power in the PEH was obtained.

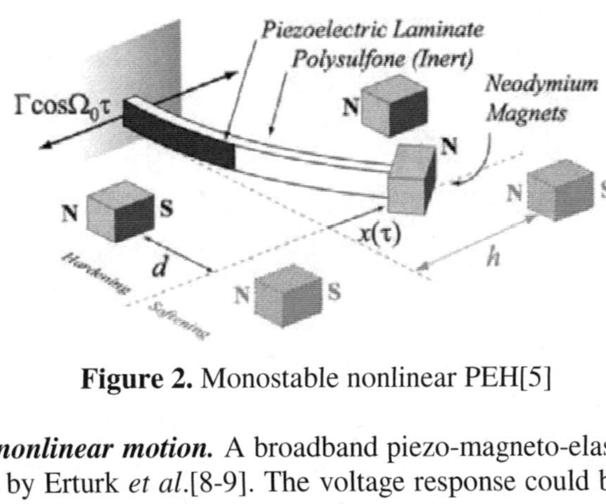

Figure 2. Monostable nonlinear PEH[5]

PEHs with bistable nonlinear motion. A broadband piezo-magneto-elastic generator, as shown in **Figure 3**, was designed by Erturk *et al.*[8-9]. The voltage response could be chaotic strange attractor motion or large-amplitude periodic motion when an initial deflection at one of the stable equilibriums was offered under small or large harmonic excitations. While applying a disturbance or equivalently an initial velocity condition, the large-amplitude periodic motion could also be obtained under small excitation level. On the other hand, with a clear advantage over the linear piezo-elastic configuration (with two magnets removed), a large-amplitude response as well as broadband performance may be acquired at off-resonance frequencies.

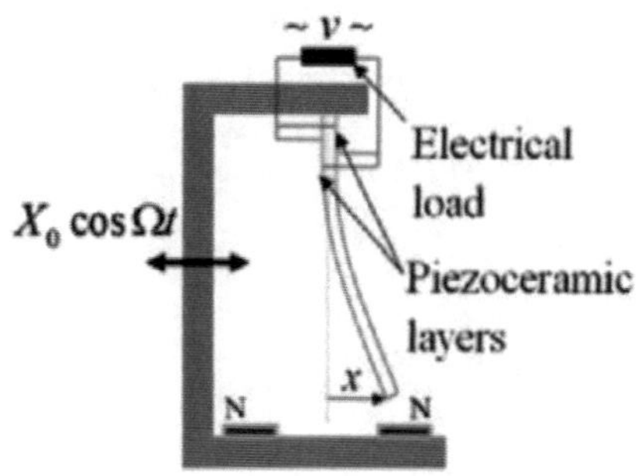

Figure 3. The piezo-magneto-elastic generator [8-9]

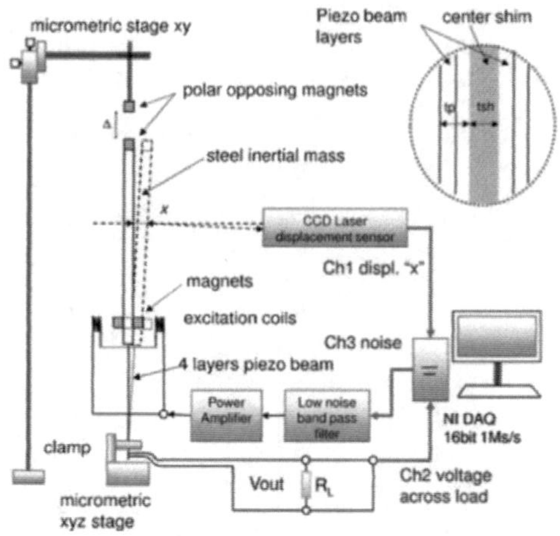

Figure 4. A piezoelectric inverted pendulum [10]

By using the bistable mechanism, a PEH similar to pendulum was realized by Cottone *et al.* [10], as shown in **Figure 4**. The pendulum swung with small oscillations around each equilibrium or with large excursions from one equilibrium position to another excited by the random vibration. In a specific Δ and noise level, the maximum deflection of the pendulum displacement of x_{RMS} could be reached to generate the maximum power which was up to 4-6 times of the linear PEH.

The idea was extended to study energy harvesting performance of bistable cantilevers with repulsive magnets under wide-spectrum vibrations. In these studies[11-13], the critical issue on how to enable the harvester to readily transit between the two stable states was involved. For the broadband energy harvesting, stable transition between two states was dependent on the excitation amplitude, frequency, and the extent of nonlinearity.

2.2. *PEHs with Piecewise-Linear Stiffness*

Other than the Duffing-type oscillator, a piecewise-linear stiffness is also introduced to realize the nonlinearity of PEH. The common configuration is to use mechanical stoppers, as shown in **Figure 5**.

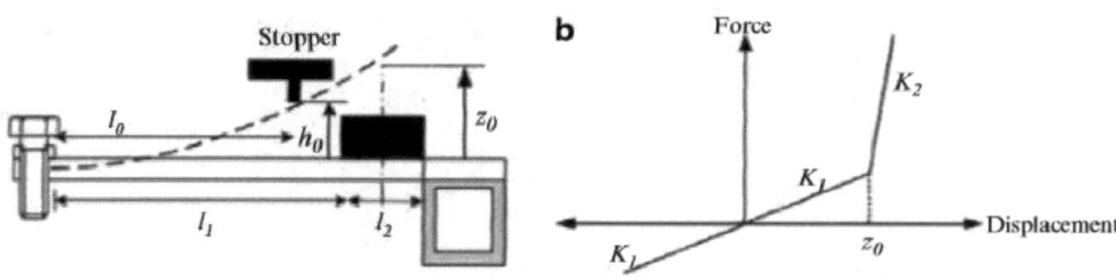

(a) Typical PEH with mechanical stopper (b)Piecewise-linear stiffness with Fig.5(a)

Figure 5. A typical PEH and stiffness with a mechanical stopper

The benefit of a PEH with a stopper was investigated by Soliman *et al.*[14]. The analytical, numerical and experimental results were showed in **Figure 6**. It could be seen that vibration frequency randomly changed in a range of 13.8Hz around the natural frequency. The overall collected energy was up to 30% due to a larger bandwidth. However, the power output of this kind of PEH was smaller than a PEH without stopper owing to the stop absorbing most of the vibration energy.

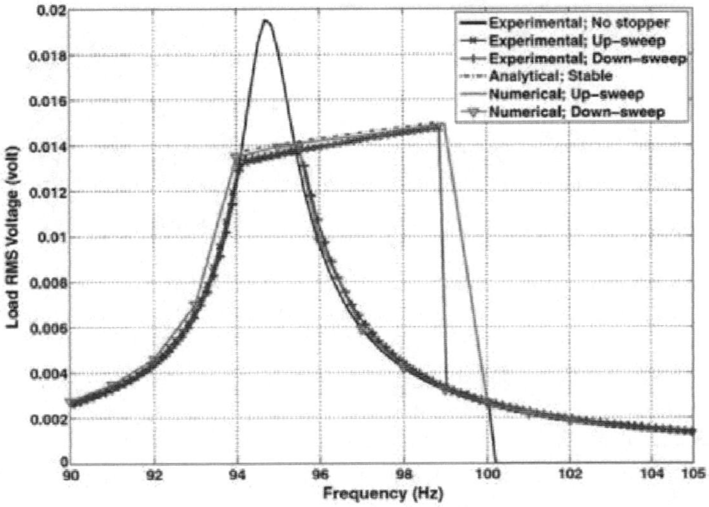

Figure 6. Analysis, numerical and experimental frequency response [14]

To solve the above question, an optimization procedure for a harvester with stopper was presented by researchers [15]. The studied result revealed that two factors dominated the performance of such a device were listed in the following: the stiffness ratio and the velocity of the beam at the impact point. The two factors are controlled by the stopper height and the offset distance of the stopper from the cantilever support. As long as the two factors are reasonably selected, higher capture energy can be obtained in a wider frequency range with this kind of PEH.

Based on the two-degree-of-freedom PEH with elastic base, Halim *et al.* [16] proposed a new configuration of enlarging the base structure and making the end mass to collide with the base during vibration, as shown in **Figure 7**. The bandwidth of the improved PEH is 2.5 times that of one without collision under the same conditions. Meanwhile, the base is similar to the stopper, and the collision behavior effectively broadens the bandwidth.

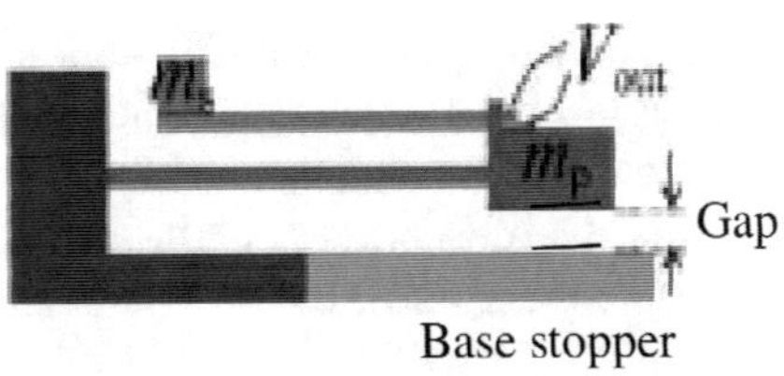

Figure 7. 2DOF piecewise-linear PEH[16]

2.3. PEHs with Nonlinear Large Strain

Hajati *et al.*[17] introduced the non-linear stiffness by using tensile strain of the fixed beam of the MEMS, and proposed an ultra-wideband PEH, as shown in **Figure 8**. The study revealed that frequency bandwidth and energy density of the PEH were one order of magnitude higher than those ones reported previously.

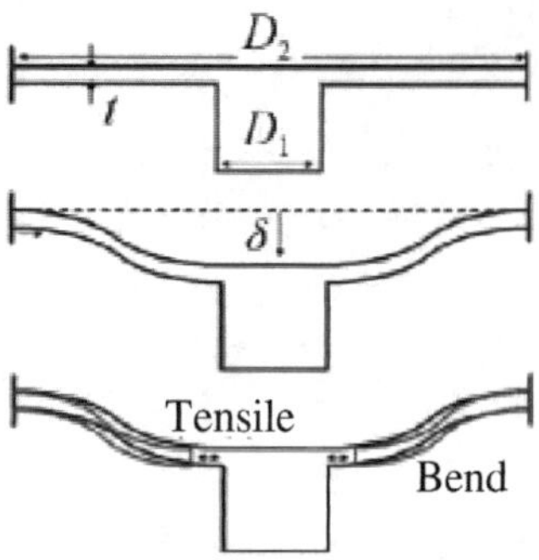

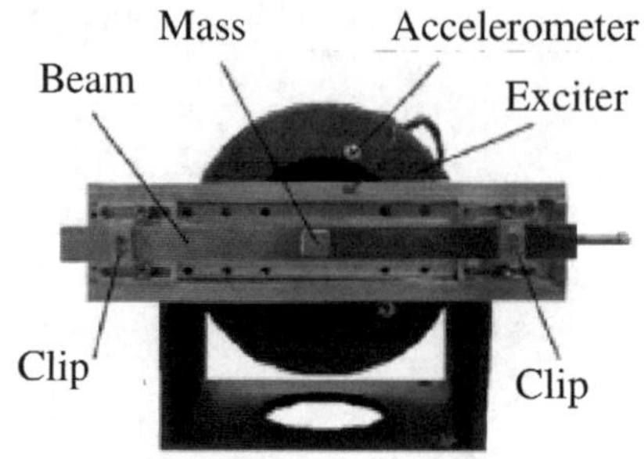

Figure 8. An ultra-wideband PEH[17] **Figure 9.** Post-buckled PEH[18]

Sneller *et al.*[18] proposed one nonlinear post-buckled PEH with buckling beam in **Figure 9**. A mass was put on the middle of the beam. Jumping phenomenon occurred under harmonic excitation. Two equilibrium positions existed in the system. Experimental result showed that the additional mass was helpful to broaden the working frequency of PEH and the jumping motion makes PEH to produce larger amplitude oscillation and power generation. Besides, increasing the mass helps to reduce the threshold of exciting force for induced jump.

Cottone *et al.*[19] studied the generation characteristics of PEH with buckling beams under broadband random excitation. The result shows that output power of PEH is more than 10 times that of the non-buckling state.

3. Conclusions

In practical scenarios, the ambient vibration source is frequency-variant or random. Therefore, the drawback of narrow bandwidth of linear PEHs limits their application. To overcome this drawback, PEHs with nonlinear stiffness were proposed. The typical nonlinear PEHs can be classified into three kinds: a Duffing-type oscillator with cubic nonlinear stiffness typically introduced by using magnets, a piecewise-linear oscillator with nonlinearity caused by a mechanical stopper, a piezoelectric cantilever beam with nonlinear large strain. Each kind of PEH may be preferable in some specific application conditions. For nonlinear PEH with wider broadband and higher efficiency, characteristics of ambient vibration source should be taken into account, which include the type of excitation (periodic or stochastic), the variation of frequency (infrequent or frequent), the excitation level and the targeted frequency range, *etc*[20]. In the small exciting environment, the output frequency response of the nonlinear PEH is similar to that of the linear one, and nonlinear advantages are not obvious.

Although the nonlinear PEHs exhibit broadband characteristics in the sweeping process, most of them require large exciting acceleration. Only when the nonlinearity of PEHs is excited, can a wider frequency band be obtained. There are multiple solutions to causing the nonlinearity of PEHs in a certain frequency range or under excitation conditions, it is worth studying how to keep the harvester's stability in the high energy orbit (especially for PEHs with bistable characteristic).

4. Acknowledgments

This work was supported by "3C-Product Intelligent Manufacturing Engineering Technology Research and Development Center of Jiangsu Province" from Jiangsu Education Department and "New Driving and Ambient Energy Harvesting Innovation Team Based on Functional Materials" from Suzhou Vocational University.

5. References

[1] Xu Zhenlong, Shan Xiaobiao, Xie Tao. A review of broadband piezoelectric vibration energy harvester. Journal of vibration and shock., 2018,37(8): 190-199,205
[2] Erturk A, Inman DJ (2008) Issues in mathematical modeling of piezoelectric energy harvesters[J]. Smart Mater Struct, 17:065016

[3] Gammaitoni L, Neri I, Vocca H (2009) Nonlinear oscillators for vibration energy harvesting. Appl Phys Lett 94:164102

[4] Ramlan R, Brennan MJ, Mace BR, Kovacic I (2010) Potential benefits of a non-linear stiffness in an energy harvesting device. Nonlinear Dyn 59:545–558

[5] Stanton SC, McGehee CC, Mann BP (2009) Reversible hysteresis for broadband magnetopiezoelastic energy harvesting. Appl Phys Lett 95:174103

[6] Sebald G, Kuwano H, Guyoma R D, et al. Experimental duffing oscillator for broadband piezoelectric energy harvesting[J]. Smart Materials and Structures, 2011, 20(10) :102001.

[7] Sebald G, Kuwano H, Guyoma R D, et al. Simulation of a duffing oscillator for broadband piezoelectric energy harvesting [J]. Smart Materials and Structures, 2011, 20(7) :075022.

[8] Erturk A, Hoffmann J, Inman DJ (2009) A piezomagnetoelastic structure for broadband vibration energy harvesting. Appl Phys Lett 94:254102

[9] Erturk A, Inman DJ(2011) Broadband piezoelectric power generation on high-energy orbits of the bistable Duffing oscillator with electromechanical coupling. J Sound Vib 330:2339-2353

[10] Cottone F, Vocca H, Gammaitoni L (2009) Nonlinear energy harvesting. Phys Rev Lett 102:080601

[11] Ferrari M, Ferrari V, Guizzetti M, Andò B, Baglio S, Trigona C (2010) Improved energy harvesting from wideband vibrations by nonlinear piezoelectric converters. Sens Actuators A 162:425–431

[12] Lin J, Alphenaar B (2010) Enhancement of energy harvested from a random vibration source by magnetic coupling of a piezoelectric cantilever. J Intell Mater Syst Struct 21:1337–1341

[13] Andò B, Baglio S, Trigona C, Dumas N, Latorre L, Nouet P (2010) Nonlinear mechanism in MEMS devices for energy harvesting applications. J Micromech Microeng 20:125020

[14] Soliman MSM, Abdel-Rahman EM, El-Saadany EF, Mansour RR (2008) A wideband vibration-based energy harvester. J Micromech Microeng 18:115021

[15] Soliman MSM, Abdel-Rahman EM, El-Saadany EF, Mansour RR (2009) A design procedure for wideband micropower generators. J Microelectromech Syst 18:1288-1299

[16] Halim M A, Park J Y. Low frequency vibration energy harvester using stopper-engaged dynamic magnifier for increased power and wide bandwidth. Journal of Electrical Engineering and Technology, 2016,11(3):709-716.

[17] Hajati A, Kim S G. Ultra-Wide Bandwidth piezoelectric energy harvesting. Appl Phys Lett 2011, 99(8) : 083105.

[18] Sneller A J, Cette P, Mann B P. Experimental investigation of a post-buckled piezoelectric beam with an attached central mass used to harvest energy[C]//Proceedings of the Institution of Mechanical Engineers, Part I: Journal of Systems and Control Engineering,Durham,North Carolina, 2011: 497-509.

[19] Cottone F, Gammaitoni L, Vocca H, et al. Piezoelectric buckled beams for random vibration energy harvesting[J]. Smart Materials and Structures ,2012,21(3) : 035021.

[20] Niell Elvin, Alper Erturk. Advances in Energy Harvesting Methods: pp 17-61. Springer Science + Business Media New York, 2013

IOP Conf. Series: Materials Science and Engineering **730** (2020) 012040 doi:10.1088/1757-899X/730/1/012040

Photochemical Preparation of Selenium-rGO Composite for Application in Photocatalytic Degradation

Wenyu Zhong[1], and Fengqiang Sun[2,3,*] and Guping He[1]

[1]School of Chemistry and Environment, South China Normal University, Guangzhou, China

[2]Key Laboratory of Theoretical Chemistry of Environment, Ministry of Education, South China Normal University, Guangzhou, China

[3]Guangzhou Key Laboratory of Materials for Energy Conversion and Storage, Guangzhou, China

*Corresponding author e-mail: Fengqiang Sun: fqsun@scnu.edu.cn

Abstract. Se-rGO composite photocatalyst was prepared by photochemical method and improved Hummers method. The morphology and structure of the photocatalyst were characterized by SEM and XRD. The photocatalytic properties of the composite were studied by degrading RhB under UV light. The results showed that the composite retained the original mophorlogy of selenium and graphene oxide after compounding, where a large number of spherical Se particles are clustered and loaded on the surface of GO layers; hexagonal phase is observed in the pattern of pure Se, and the crystalline structure remains unchanged after compounding GO. The photocatalytic activity of the composite is more than twice as high as that of the pure Se. Moreover, the photocatalytic activity of the photochemical prepared Se-rGO composite is much higher than that of the mechanically mixed Se/GO. Se-rGO composite has good photocatalytic performance and the degradation process conforms well to the first-order reaction kinetics. The degradation rate of RhB solution can reach 76.8% after 120 minutes of UV irradiation.

1. Introduction

With the continuous development of science, technology and economy, environmental pollution caused by industrial production is becoming more and more serious, especially water pollution. Industrial dyes such as rhodamine B (RhB) and methylene blue (MB), having high content of organic matter and strong toxicity that will cause irreversible damage to the water system if discharged into the water without proper treatment. Therefore, water pollution treatment has become the focus of global pollution control, and the methods that commonly used to treat that mainly including physical adsorption, chemical flocculation and biodegradation, but there are many shortcomings such as long degradation cycle or low efficiency. Since reported by Bard in 1977 [1], photocatalytic degradation of wastewater has attracted wide attention in academic field. Semiconductor materials have good light response and suitable band gap, which enable them to make full use of ultraviolet or visible light. Semiconductor photocatalytic technology have unique advantages such as easy operation, low energy consumption, environmentally friendly and wide application.

Selenium (Se), belongs to chalcogenide elements, is a non-toxic non-metallic semiconductor with narrow band gap (about 1.7eV) that widely used in medicine, chemical industry, electrical and electronic industries, ceramics, metallurgy and glass manufacturing. In 1873, electrical resistance changes under light irradiation were first found in selenium [2]. After that, researchers have carried out a lot of research on the photoelectric properties of selenium. Selenium is also widely used in laser printing, photovoltaic cells, fuel cells and X-ray instruments. Chen Y [3], Che L [4] and Xie W [5] et. al reported the photocatalytic research of Se. At present, the commonly used methods to synthesize selenium are chemical reduction and laser ablation, which involve strong reductants and complex reaction conditions. Therefore, an easy-to-operate and environmentally friendly approach to synthesize selenium is necessary.

In recent years, graphene (GR) has become one of the most attractive materials in the fields of sensors [6] and photocatalyts [7]. There are abundant π bondsin GR and the π electrons can move freely in two-dimensional plane, so the surface of GR has efficient electron transfer which gives GR good conductivity. As an important derivative, reduced graphene oxide (rGO) has very high specific surface area and a large number of π-π binding sites. It can enrich organic macromolecules on the surface of photocatalysts, which is conducive to the photocatalytic reaction, therefore improved the photocatalytic efficiency. In the past few years, Pd/GO, Au/GO and ZnO/GO [8-10] have been synthesized constantly, and they all showed good photocatalytic activity.

In this work, $SeCl_4$ solution was used as precursor and GO was used as raw material to prepare the in-situ Se-rGO composite by one-pot photochemical method under mild reaction conditions. Se and rGO were successfully compounded under UV irradiation at room temperature. RhB solution was used to simulate organic wastewater for the measurement of photocatalytic performance. The photocatalytic properties of the Se-rGO composite, pure Se and mechanically mixed Se/GO were compared and the reaction kinetics were also discussed. In the process of photodegradation, most of the photogenerated electrons will be transferred to rGO, which promotes the separation of the photogenerated electrons and holes and greatly inhibits the recombination of them, thus improves the photocatalytic performance of elemental selenium.

2. Experiment Section

2.1. *Materials*

Selenium dioxide (SeO_2, AR, 99%), isopropanol (AR, 99%), hydrochloric acid (AR, 7%) and rhodamine B (RhB, AR, C=$1.9*10^{-5}$mol/L). Graphene oxide was synthesized by modified Hummers method with the solid content of 0.086 mg/ml.

2.2. *Synthesis of Se-rGO Composite*

Se-rGO composite is synthesized by one-pot photochemical method. The precursor solution is formed by dissolving selenium dioxide in a mixture of hydrochloric acid and isopropanol, followed by dropping graphene oxide aqueous solution into. The mixture was then irradiated by two ultraviolet lamps (8w, 254nm) for 8 hours under constantly magnetic stirring. The red-grey solid was then collected and washed with deionized water and ethanol several times before drying at room temperature.

2.3. *Characterization*

The morphologies and size of selenium, graphene oxide and the Se-rGO composite were observed by scanning electron microscopy (SEM, Carl Zeiss Gemini 500). The composition information was characterized via X-ray powder diffraction (XRD, BRUKER D8 ADVANCE D/max2200, with Cu-Kα radiation) and the diffraction peaks and the indices of crystallographic plane were then analyzed by MDI Jade 6.5.

2.4. *Measurement of Photocatalytic Activity*

The degradation rate of RhB under ultraviolet light within limited time was tested to evaluate the photocatalytic performance of the composite. 12.5 mg of the Se-rGO composite was added to 100 ml RhB aqueous solution and treated ultrasoundly for 15 minutes to make the composite disperse evenly. The solution was placed in a dark box under continuously stirring for 30 minutes to achieve the adsorption-desorption equilibrium between the dye and the catalyst. After the photodegradation began, 5 ml solution was taken out every 20 minutes and centrifuged at 4000 rpm for 2 minutes, then the absorbance of supernatant was measured at 553 wavelength. The photodegradation rate is calculated

IOP Conf. Series: Materials Science and Engineering **730** (2020) 012040 doi:10.1088/1757-899X/730/1/012040

by the following formula:

$$\eta = \frac{C}{C_0} * 100\%$$

In the formula, η is the photodegradation rate, C_0 is the concentration of RhB after adsorption, and C is the concentration of RhB at photocatalytic reaction reaches time t. The photocatalytic properties of selenium, graphene oxide and mechanically mixed Se/rGO were also tested for comparison.

3. Results and Discussion

3.1. Composition and Morphology

Figure 1 shows the SEM and XRD information of Se, GO and the Se-rGO composite. Figure 1a shows that selenium is presents as spherical particles with diameter ranging from 0.5-1.5 μm, and some of the Se spheres are fused and aggregated. Graphene oxide showed the usual morphology as thin-layer folded structure (Figure 1b). After compounding, the original morphology of selenium spheres and GO are basically retained and GO is coated on the surface of selenium spheres (Figure 1c).

Figure 1d shows the XRD patterns. The prepared GO has a characteristic peak of 2**θ**=12.5°. The broad diffraction peaks of Se observed at 2**θ**=23.52°, 29.70°, and 51.72° correspond to the (100), (101) and (201) planes of hexagonal selenium (JCPDS card NO. 06-0362). No other diffraction peaks are observed in the XRD spectra, which indicates that the as-prepared samples have high purity. The characteristic peaks of composite are almost the same as those of elemental selenium, indicating that combination with graphene oxide does not change the crystalline structure of selenium.

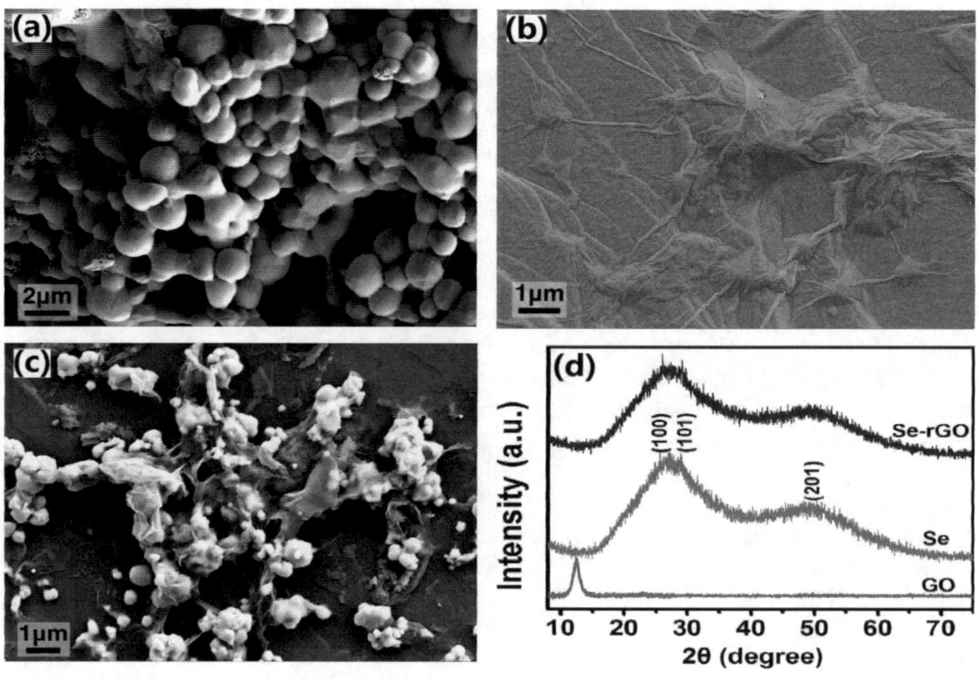

Figure 1. The SEM images of (a) pure Se, (b) pure GO, (c) Se-rGO composite; and (d) XRD patterns of Se, GO and Se-rGO composite.

3.2. Growth Mechanism

Simply put, SeO_2 is dissolved in hydrochloric acid solution to form $SeCl_4$, then reacts as follows under the driving of UV light:

$$SeO_2 + HCl \rightarrow SeCl_4 + H^+ \quad (1)$$

$$Se^{+\mathrm{IV}}Cl(aq) + hv \rightarrow Se^{+\mathrm{II}}Cl(aq) + Cl \bullet \mathrm{(aq)} \quad (2)$$

$$Cl \bullet \mathrm{(aq)} + \mathrm{H}Scav(aq) \rightarrow \mathrm{Scav} \bullet \mathrm{(aq)} + \mathrm{HCl(aq)} \quad (3)$$

$$2Se^{+\mathrm{II}}Cl_2 \rightarrow Se^0(solid) + Se^{+\mathrm{IV}}Cl_4 \quad (4)$$

Graphene oxide absorbs electrons under ultraviolet radiation and partially reduces to form reduced graphene oxide. Reduced graphene oxide contains abundant hydrophilic oxygen-containing functional groups such as hydroxyl group and epoxy group [11] that can provide sites for the growth of selenium, a tightly bounded in-situ Se-rGO composite was thus prepared. The growth process is shown in Figure 2.

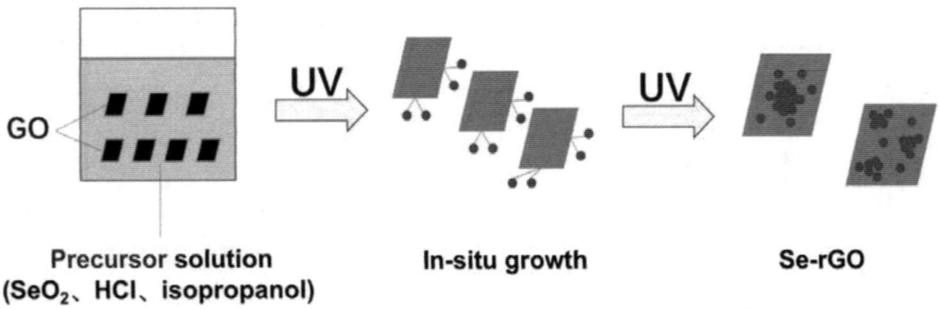

Figure 2. Schematic view of the formation mechanism of the composite Se-rGO.

3.3. Photocatalytic Activity

In order to investigate the photocatalytic properties of the composite, the degradation of RhB by elemental Se, pure GO, Se-rGO composite and mechanically mixed Se/GO was examined, and the reaction rates were calculated and compared. Before the photodegradation started, the catalyst was added to RhB aqueous solution, and the mixture was treated by ultrasound to make the catalyst disperse evenly. The mixture was then stirred in darkness for 30 minutes to achieve an adsorption-desorption equilibrium. The results show that all samples have a certain degree of adsorption on RhB aqueous solution. The degradation rate curves of RhB in 120 minutes are shown in Figure 3a and the corresponding degradation efficiency is summarized in Figure 3b. The photocatalytic degradation rate of RhB with Se-rGO was the highest with 76.8%, and that of pure Se and mechanically mixed Se/GO was 31.1% and 19.5%, respectively. Pure GO did not show degradation ability. Compared with the degradation of RhB by in-situ composite Se-rGO and mechanically mixed Se/GO, it can be concluded that the combination of the two components is tighter and the built-in electric field is more conducive to charge transfer and electron hole separation [12], which makes the higher photocatalytic activity of Se-rGO. In addition, the absorbance of RhB aqueous solution increased when pure GO or mechanically mixed Se/GO was added, indicating that the desorption between dyes and catalysts occurred after illumination, which also contributed to the poor photocatalytic performance. Kinetic analysis of all degradation reactions was carried out (Figure 3c). The corresponding equation was suggested as follows:

$$\ln\left(\frac{C_0}{C}\right) = kt \ [13]$$

k represents the apparent reaction rate (ARR) constant. It can be seen that the catalytic behavior of all catalysts conforms well to the first order kinetics. The ARR constants of all samples are summarized in Figure 3d, and Se-rGO has the highest value of 0.0218 min^{-1}. The ARR of pure Se, pure GO and mechanically mixed Se/GO are 0.0119 min^{-1}, 0.0100 min^{-1} and 0.00010 min^{-1}, respectively.

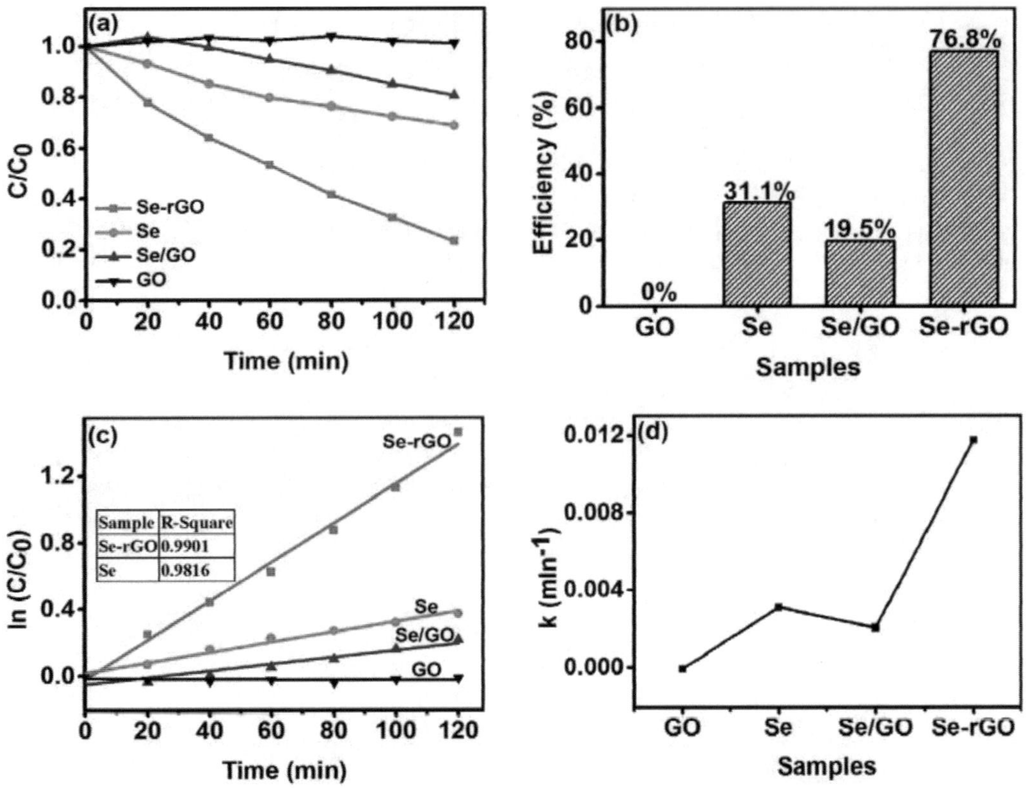

Figure 3. The photocatalytic activities of Se, rGO, Se-rGO and Se/rGO. (a) The degradation rate curves of RhB in 120 minutes; (b) The degradation efficiency of RhB in 120 minutes; (c) First-order kinetic simulations of the degradation processes of different materials; (d) Summary of the ARR constants.

3.4. Photocatalytic Mechanism

The typical photocatalytic mechanism is that the free radicals with photoactivity functioning as the active catalytic species, generally considered as hydroxyl radicals (•OH) and superoxide radicals ($•O_2^-$) [14]. In order to verify the existence of these free radicals, free radical scavenging experiments were carried out. Methanol (as h^+ scavenger) and tert-butanol (as •OH scavenger) were added to RhB aqueous solution respectively in the presence of Se-rGO, and pure nitrogen (as $•O_2^-$ scavenger) was also bubbled into. The results of controlled photocatalytic experiments are shown in Figure 4a. Compared with the solution without any scavenger, the addition of methanol or nitrogen almost completely inhibited the degradation of RhB, while the degradation rate of RhB decreases significantly when the solution was added with tert-butanol. The results showed that the main active species in the photocatalytic degradation process were •OH and $•O_2^-$. Therefore, the catalytic mechanism can be expressed as shown in Figure 4b. When irradiated by ultraviolet or visible light, holes and electrons will be generated in Se-rGO. In the following reaction process, photogenerated electrons react with dissolved oxygen to produce superoxide radicals, which will be further converted into hydroxyl radicals, and H_2O reacts with photogenerated holes h^+ to produce hydroxyl radicals. Hydroxyl radicals possess strong oxidation that can degrade dye molecules into carbon dioxide and water [15]. Due to the good conductivity of graphene, a large number of photogenerated electrons are transferred from Se to rGO, which greatly inhibits the electron-hole recombination rate and enhances the photocatalytic ability.

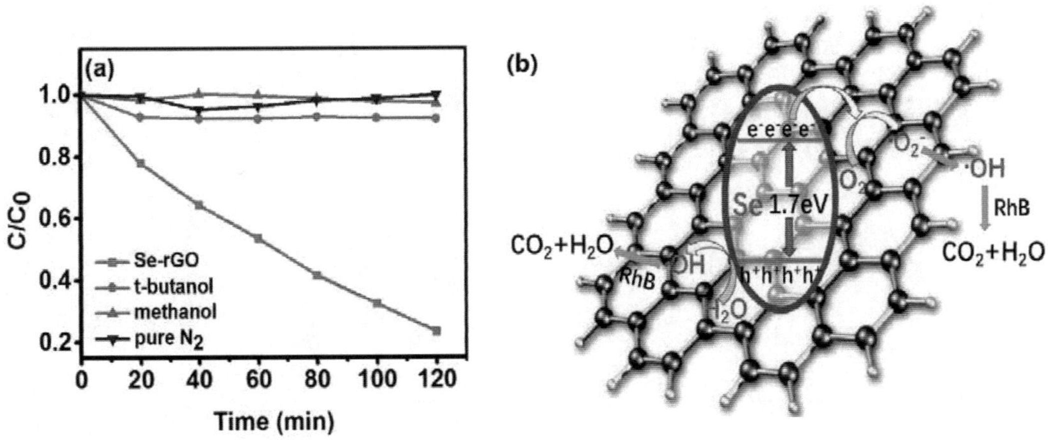

Figure 4. (a) The photodegradation behavior of the RhB aqueous solution in the presence of Se-rGO under varied controlled conditions; (b) The schematic view of proposed mechanism of the RhB degradation in the presence of Se-rGO.

4. Conclusion

In summary, the Se-rGO composite photocatalyst was prepared by a simple photochemical method. After in-situ compounding with graphene, the photogenerated electron-hole recombination rate decreased, which resulting a significant improvement in the photocatalytic performance of the Se-rGO compared with that of pure Se and mechanically mixed Se/GO. The degradation rate of RhB reaches 76.8% within 120 minutes of low-power UV irradiation. The mechanism study reveals that the hydroxyl radicals (•OH) and superoxide radicals (•O_2^-) are the main active species in degradation. Photochemical method is a promising method in the synthesis of catalysts, and the Se-rGO composite have broad applications in photocatalysis, sensors and batteries.

5. Acknowledgements

This work co-supported by the National Natural Science Foundation of China (No. 21571068), the Natural Science Foundation of Guangdong Province (No. 2015A030313387) and the Science and Technology Program of Guangzhou (No. 201607010301).

6. References

[1] Mahoney L, Wu C, KibomboH, Thiruppathi E, Baltrusaitis J, Rasalingam S and Koodali R2013 *J. Microporous and Mesoporous Materials.* **170** 211-225
[2] Remitz KE, Neuroth N and Spert B 1991 *J. Mater. Sci. Eng.* **139** 413–6
[3] Chen Y, Wang L, Wang W and Cao M 2017 *J. Materials Chemistry and Physics.* **199** 416-423
[4] Che L, Dong Y, Wu M, Zhao Y, Liu L and Zhou H 2017 *J. ACS Sustainable Chem. Eng.* **5** 2535–43.
[5] Xie W, Li R and Xu Q 2018 *J. Scientific Reports* **8** 8752
[6] Chen W, Deng F, Xu M, Wang J, Wei Z and Wang Y 2018 *J. Sensors & Actuators: B. Chemical* **273** 498–504
[7] Yu Y, Yang Q, Yu X, LuQ and Hong X 2017 *J. Chemistry Select* **2** 5578-86
[8] a) Fakhri P, Nasrollahzadeh M, Jaleh B 2014 *J. RSC Adv.* **4** 48691–697; b) Omidvar A, Jaleh B and Nasrollahzadeh M 2017 *J. Colloid Interf. Sci.* **496** 44–50
[9] a) JalehB, Khalilipour A, Habibi S, Niyaifar M and Nasrollahzadeh M 2017*J. Mater. Sci. Mater. Electron.* **28** 4974–83; b) Maryami M, Nasrollahzadeh M, Mehdipour E, Sajadi S 2016 *J. Int. J. Hydrogen Energy.* **41** 21236–245
[10] M. Nasrollahzadeh, B. Jaleh and A. Jabbari 2014 *J. RSC Adv.* **4** 36713–720
[11] Wang Y, Tan G, Li B, Dang M, Lv L, Wang M, Zhang D, Ren H and Xia A 2019 *J. Applied Surface Science* **489** 1–12

[12] Shinde S, Sami A and Lee J *2016 J. Carbon* **96** 929-936
[13] Sudrajat H and Hartuti S 2018. *J. Colloid and Interface Science.* **524** 227-235
[14] Chiou Y and Hsu Y 2011 *J. Applied Catalysis B: Environmental* **105** 211–219
[15] Chen Y, Sun F, Huang Z, Chen H, Zhuang Z, Pan Z, Long J and Gu F 2017 *J. Applied Catalysis B: Environmental* **215** 8–17

Oxidative Leaching of Vanadium from Vanadium-chromium Reducing Residue with MnO_2

Hao Peng, Liu Yang, Ya Chen and Jing Guo

College of Chemistry and Chemical Engineering, Yangtze Normal University, Chongqing, PR China
cqupenghao@126.com

Abstract. This paper focused on the leaching process of vanadium from vanadium-chromium reducing residue in alkaline medium with MnO_2. The effects of serval parameters like reaction time, reaction temperature, dosage of MnO_2, dosage of NaOH, and liquid-to-solid ratioon the leaching process had been studied. The results indicated that MnO_2 was an efficient oxidant for vanadium leaching out. The leaching efficiency of vanadium was up to 97.25% at reaction temperature of 90 °C, reaction time of 60 min, dosage of MnO_2 at 50 wt.%, concentration of NaOH at 30 wt.% and liquid-to-solid at 5:1 mL/g.

1. Introduction

The vanadium and its compounds were called "vitamins of modern industry" and widely used in many fields, like catalyst, steel and petrochemical industry, catalyst, iron steel [1-6]. The vanadium valence could be +2, +3, +4 and +5 due to its electronic configuration of $[Ar]3d^34s^2$, and +5 was the most common oxidation states [7]. In China, the main sources of vanadium were vanadium-titan magnetite ore [8] and stone coal [9] and the vanadium was almost existed in low valence. In the vanadium producing process, a large amount wastewater containing V(V) and Cr (VI) was produced. The wastewater was reduced and formed the precipitate containing low valence chromium and vanadium, which was usually called vanadium-chromium reducing residue (VCRR). More than 500 thousand tons of the VCRR were produced. The handling of these residues had become a thorny problem in many plants in China.

Many hydrometallurgy technologies had been investigated to recover vanadium from the vanadium-chromium reducing residue, directing leaching with NaOH [10], alkaline leaching with electric field [11, 12]. The sub-molten salt (SMS) technology had been successfully applied to leach out vanadium in concentrated alkaline solution[13-16]. Above 95% vanadium could be leached out with hazardous gas or toxic tailings. But some problems like high alkaline consumption, high energy cost and equipment corrosion were still unsolved. The treatment of VCRR was still a waste resource and also resulting in environmental pollution.

In this paper, MnO_2 as an oxidant was used to enhance the alkaline leaching process of vanadium. The effect of reaction time, reaction temperature, dosage of MnO_2 dosage of NaOH and liquid-to-solid ratio on the leaching efficiency of vanadium were investigated.

2. Experimental

2.1. Imental Materials

The VCRR was collected from an iron and steel mill from Pan gang Group Co., Ltd., Panzhihua, Sichuan Province, China. Before the experiment, the residue dried and grounded to suitable particles.

IOP Conf. Series: Materials Science and Engineering **730** (2020) 012041 doi:10.1088/1757-899X/730/1/012041

The chemical composition of the residue was listed in Table 1 and the phase composition was detailed in Figure 1. All the reagents used in the experiments were analytical grade.

The experimental process was the same as the recent studies [11]: a predetermined amount of NaOH and deionized water was added to the beaker to produce homogeneous slurry under constant stirring. The slurry was heated to a predetermined temperature. Next, the residue was added to the reactor, and then the amount of MnO_2 was added. After the required contact time, the filtrate was separated from the residue by vacuum filtration. The concentration of vanadium in the filtrate was determined by inductive couple plasma-optical mission spectrometry (ICP-3000).

The leaching efficiency of vanadium (η_V) was calculated in Equation (1), as followed:

$$\eta = \frac{V \cdot C}{m\omega} \times 100\% \quad (1)$$

Where, *C* is the concentration of vanadium in the filtrate, g/L; *V* is the volume of the filtrate, mL; ω is the mass fraction of vanadium in the vanadium-chromium reducing residue; *m* is mass of the vanadium-chromium reducing residue used in the leaching experiments, g.

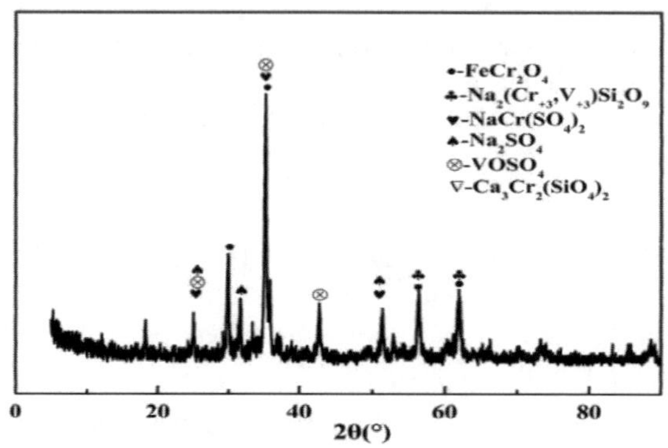

Figure 1. The XRD pattern of the original residue

Table 1. Composition of the original vanadium and chromium residue (%, wt)

Component	O	Cr	Si	Na	S	V
Amount (wt.%)	41.44	18.80	11.30	10.93	10.64	3.11
Component	Ca	Cl	Fe	K		
Amount (wt.%)	1.94	1.06	0.37	0.17		

3. Results and Discussions

3.1. Reaction Mechanism

The vanadium existed as low valent in vanadium-chromium reducing residue according to the results showed in Figure 1, and it was hard to leach out in alkaline medium [10]. It was known that $E^0_{MnO2/Mn2+}$ =1.21 V and $E^0(VO_2^+/VO^{2+})$ =1.00V, and therefore MnO_2 might be used to oxidize low valence vanadium in the leaching process [19, 20]. Thus, the main reaction during the oxidative leaching process was between the low valent vanadium and MnO_2 like Equation (2) and (3). The ΔG^θ_T of Equation (2) and (3) at different temperatures could be calculated with $\Delta_f H^\theta_{298}$, S^θ_{298} and C_p at 298 K [21, 22]. The results showed in Figure 2 showed that the ΔG was negative which indicated that the reaction of Equations (2) and (3) were easy to happen and the oxidative leaching of vanadium with MnO_2 was feasible.

$$\mathbf{MnO_2 + V^{3+} + 6OH^- = Mn(OH)_2 + VO_4^{3-} + 2H_2O} \quad (2)$$

$$MnO_2 + 2VO^{2+} + 10OH^- = Mn(OH)_2 + 2VO_4^{3-} + 4H_2O \quad (3)$$

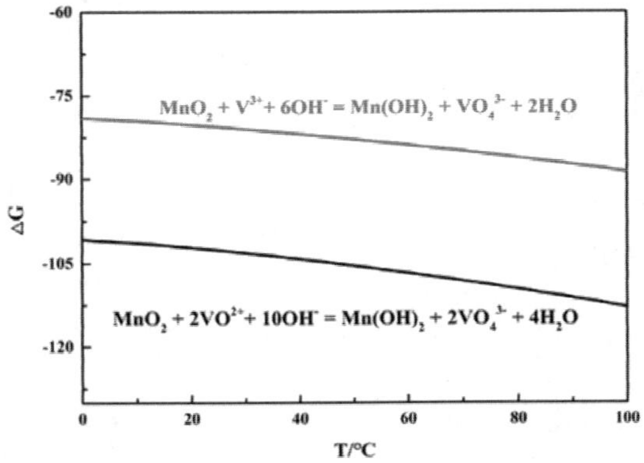

Figure 2. Relationship between ΔG and temperature of oxidation of vanadium with MnO_2

3.2. *Effect of Dosage of* MnO_2

Based on the above, the oxidation reaction of low valence vanadium and MnO_2 was easy to happen in thermodynamics. The effect of MnO_2 dosage on the leaching efficiency of vanadium was investigated as other reaction conditions kept as constant: reaction time of 60 min, reaction temperature of 90 °C, liquid-to-solid ratio of 5:1 mL/g, and dosage of NaOH at 30 wt.%. The results were shown in Figure 3.

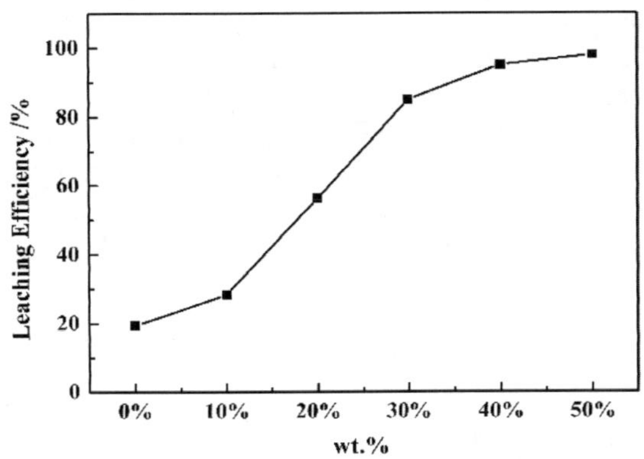

Figure 3. Effect of dosage of MnO_2 on the leaching efficiency of vanadium

The leaching efficiency of vanadium was increased linearly with the increase of MnO_2. The leaching efficiency of vanadium was increased from 19.38% without MnO_2 to 97.25% at dosage of MnO_2 at 50 wt.%. The vanadium in low valences was oxidized to high valence by MnO_2, and then dissolved in alkaline medium and achieved high leaching efficiency. The results confirmed that MnO_2 could oxidize low valence vanadium to high valence and contributing to high leaching efficiency of vanadium.

3.3. *Effect of Dosage of NaOH*

Results shown in Figure 4 summarized the effect of dosage of NaOH on the leaching process as the other reaction conditions kept as: reaction time at 60 min, reaction temperature of 90 °C, liquid-to-solid ratio of 5:1 mL/g and dosage of MnO_2 at 50 wt.%. The concentration of NaOH was set as 0 wt.%, 10 wt.%, 15 wt.%, 20 wt.%, 25 wt.% and 30 wt.%.

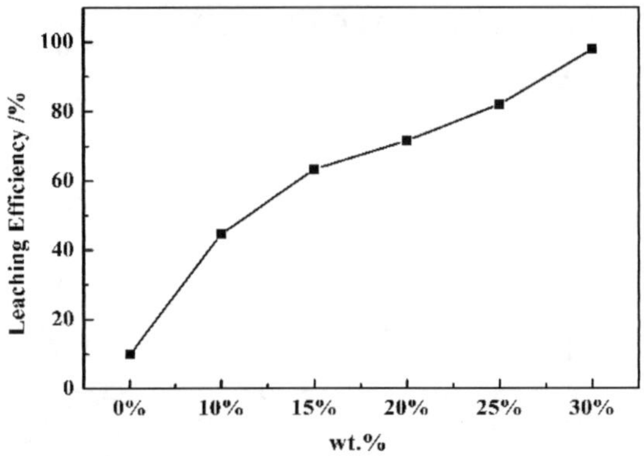

Figure 4. Effect of dosage of NaOH on the leaching efficiency of vanadium

The leaching efficiency of vanadium was increased with the increasing of dosage of NaOH. About 9.94% vanadium leached out in neutral medium with oxidation of MnO_2. And the leaching efficiency increased up to 97.25% at NaOH of 30 wt.%. The low valence vanadium was hard to leach out in alkaline medium, while the high valent vanadium was easy to dissolve in alkaline medium and achieve high leaching efficiency. The high concerned NaOH solution was beneficial for high valent vanadium leaching out. As the leaching efficiency had up to 97.25%, thus the dosage of NaOH at 30 wt.% was chosen as optimal condition for further experiments.

3.4. Effect of Reaction Temperature

Reaction temperature had significant meaning in the diffusion process. The higher temperature increased the activity of atoms and molecules, and also the reaction rate [23]. Figure 5 summarized the effect of reaction temperature on the leaching efficiency of vanadium on standard conditions: reaction time of 60 min, liquid-to-solid ratio of 5:1 mL/g, dosage of MnO_2 at 50 wt.% and concentration of H_2SO_4 at 30 wt.%. The leaching efficiency of vanadium was improved by increasing the reaction temperature. 97.25% of vanadium could leach out at a reaction temperature of 90 °C and therefore 90 °C was chosen for optimal reaction temperature in further experiments.

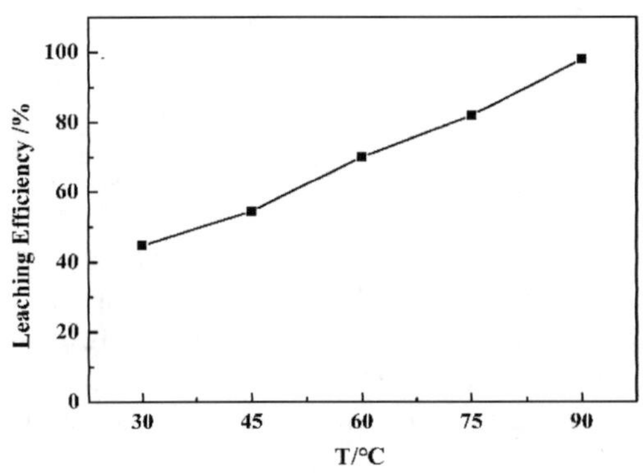

Figure 5. Effect of dosage of reaction temperature on the leaching efficiency of vanadium

3.5. Effect of Reaction Time

Figure 6 showed the leaching efficiency of vanadium as reaction time ranged from 10 min to 60 min under selected conditions: reaction temperature of 90 °C, liquid-to-solid ratio of 5:1 mL/g, dosage of MnO_2 at 50 wt.% and concentration of NaOH at 30 wt.%. Increasing reaction time could promote the reaction between MnO_2 and low valent vanadium in alkaline medium and achieve high leaching efficiency. The leaching efficiency of vanadium was increased from 49.69% at 10 min to 97.25% at 60 min. which indicated that 60 min was enough to leach out vanadium.

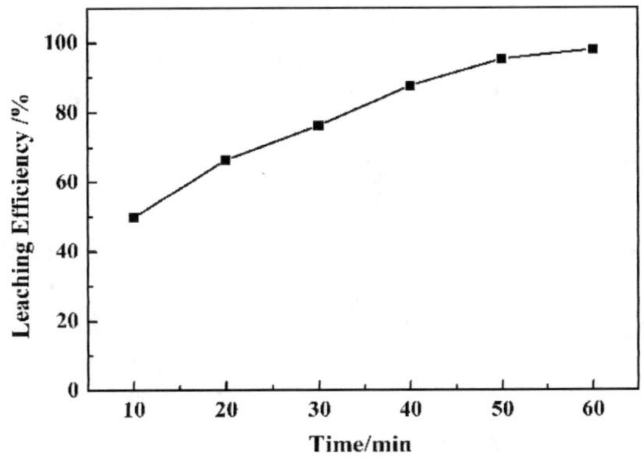

Figure 6. Effect of dosage of reaction time on the leaching efficiency of vanadium

3.6. *Effect of Liquid-To-Solid Ratio*

The liquid-to-solid ratio showed a significant effect on the leaching process due to the filtrate volume and purpose product concentration in the leaching process [23]. A series of experiments were conducted to investigate the effect of liquid-to-solid ratio on the leaching efficiency of vanadium. The liquid-to-solid ratio was set as 3:1, 4:1, 5:1, 6:1 and 7:1 mL/g, and other reaction conditions were kept as constant: reaction temperature of 90 °C, reaction time of 60 min, dosage of MnO_2 at 50 wt.% and concentration of NaOH at 30 wt.%. The results were shown in Figure 7.

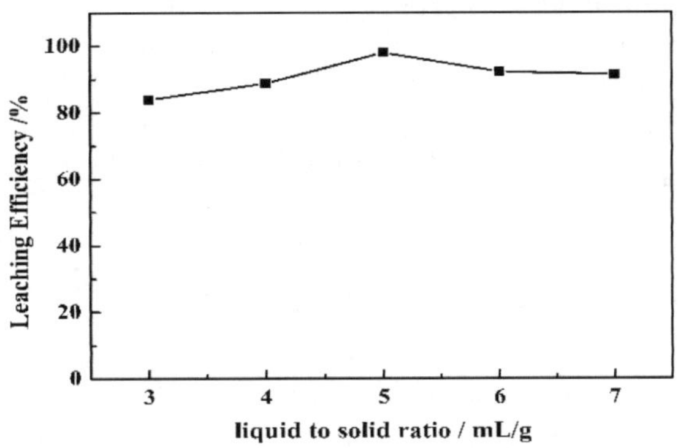

Figure 7. Effect of dosage of liquid-to-solid ratio on the leaching efficiency of vanadium

The leaching process was inefficient for low liquid-to-solid ratio as the viscosity of reaction medium was large, just 83.85% of vanadium were leached out in liquid-to-solid at 3:1 mL/g. The leaching process favoured higher liquid-to-solid ratio, where 97.25% of the vanadium was obtained at a liquid-to-solid of 5:1 mL/g. While further increase of the liquid-to-solid ratio had negative effect on the leaching process. Therefore, a liquid-to-solid of 5:1 mL/g was recommended for future use.

Based upon the above experimental results, the optimal reaction conditions for vanadium leaching out was: reaction temperature of 90 °C, reaction time of 60 min, dosage of MnO_2 at 50 wt.%, concentration of NaOH at 30 wt.% and liquid-to-solid at 5:1 mL/g. The leaching efficiency of vanadium was up to 97.25% under optimal reaction conditions.

4. Conclusions

This paper focused on the oxidative leaching process of vanadium with MnO_2 from vanadium-chromium reducing residue. Low valence vanadium was oxidized to high valent by MnO_2 in the alkaline leaching process, which contributed to high leaching efficiency of vanadium and it was up to 97.25% under optimal reaction conditions: reaction temperature of 90 °C, reaction time of 60 min, dosage of MnO_2 at 50 wt.%, concentration of NaOH at 30 wt.% and liquid-to-solid at 5:1 mL/g.

5. Acknowledgements
This work was supported by the Science and Technology Research Program of Chongqing Municipal Education Commission (No. KJQN201901403) and the Chongqing Science and Technology Commission (No. cstc2018jcyjAX0018).

6. References
[1] M. H. Anjass, K. Kastner, F. Nagele, M. Ringenberg, J. F. Boas, J. Zhang, A. M. Bond, T. Jacob, C. Streb 2017, *Angew Chem Int Ed Engl*, **56** 14749-52.
[2] Z. Wei, D. Liu, C. Hsu, F. Liu 2014, *Electrochemistry Communications*, **45** 79-82.
[3] M. B. Smirnov, V. Y. Kazimirov, R. Baddour-Hadjean, K. S. Smirnov, J.-P. Pereira-Ramos 2014, *Journal of Physics and Chemistry of Solids*, **75** 115-22.
[4] V. Y. Zadorozhnyy, S. N. Klyamkin, M. Y. Zadorozhnyy, O. V. Bermesheva, S. D. Kaloshkin 2014, *Journal of Alloys and Compounds*, **586** S56-S60.
[5] V. Efremenko, K. Shimizu, Y. Chabak 2013, *Metall and Mat Trans A*, **44** 5434-46.
[6] M. Lübke, N. Ding, M. J. Powell, D. J. L. Brett, P. R. Shearing, Z. Liu, J. A. Darr 2016, *Electrochemistry Communications*, **64** 56-60.
[7] E. Del Carpio, L. Hernández, C. Ciangherotti, V. Villalobos Coa, L. Jiménez, V. Lubes, G. Lubes 2018, *Coordination Chemistry Reviews*, **372** 117-40.
[8] E. Janssens, S. M. Lang, M. Brummer, A. Niedziela, G. Santambrogio, K. R. Asmis, J. Sauer 2012, *Physical chemistry chemical physics : PCCP*, **14** 14344-53.
[9] D. He, Q. Feng, G. Zhang, L. Ou, Y. Lu 2007, *Minerals Engineering*, **20** 1184-86.
[10] Y. Kang, Z. Xiaoyun, T. Xueda, Y. Yonglong, C. Yanbo 2010, *Hydrometallurgy*, **103** 7-11.
[11] H. Peng, Z. Liu, C. Tao 2015, *Journal of Environmental Chemical Engineering*, **3** 1252-57.
[12] Z. Liu, A. Nueraihemaiti, M. Chen, J. Du, X. Fan, C. Tao 2015, *Hydrometallurgy*, **155** 56-60.
[13] L. Huibin, D. Hao, W. Dawei, W. Shaona, Z. Shili, Z. Yi 2013, *Transactions of Nonferrous Metals Society of China*, **23** 1489-500.
[14] L. Biao, D. Hao, W. ShaoNa, Z. Yi, Z. ShiLi, L. LanJie, C. DongHui 2013, *AIChE Journal*, **59** 541-52.
[15] X. Hongbin, Z. Shili, Z. Yi, L. Zi, W. Z 2005, *Minerals Engineering*, **18** 527-35.
[16] Z. Hai, X. Hongbin, Z. Xiaofei, Z. Yang, Z. Yi 2014, *Hydrometallurgy*, **142** 47-55.
[17] H. Peng, Z. Liu, C. Tao 2016, *Energy & Fuels*, **30** 7802-07.
[18] H. Peng, Z. Liu, C. Tao 2018, *Scientific Reports*, **8** 426.
[19] M. Wang, B. Chen, S. Huang, X. Wang, B. Liu, Q. Ge, S. Xie 2017, *Hydrometallurgy*, **171** 116-22.
[20] B. Chen, S. Huang, B. Liu, Q. Ge, M. Wang, X. Wang 2017, *Journal of Environmental Chemical Engineering*, **5** 4702-06.
[21] H. Peng, J. Guo, Z. Liu, C. Tao 2018, *SN Applied Sciences*, **1**.
[22] H. Peng, J. Guo, G. Li, Q. Cheng, Y. Zhou, Z. Liu, C. Tao 2019, *Water science and technology : a journal of the International Association on Water Pollution Research*, **79** 366-74.
[23] H. Peng, J. Guo, X. Zheng, Z. Liu, C. Tao 2018, *Journal of Environmental Chemical Engineering*, **6** 5119-24.

IOP Conf. Series: Materials Science and Engineering **730** (2020) 012042 doi:10.1088/1757-899X/730/1/012042

Critical Technologies for Development of Turbine Blade Design Inspection & Integration Software

Wei Wang and Xiaolong Xiao

School of Mechanical Engineering and Automation, Beihang University, room New Main Building D315, 37 Xueyuan road, Haidian district, Beijing, China.
Email: jrrt@buaa.eud.cn

Abstract. Unstable design quality, absence of design and processing assessment tools, and insufficient development software and engineering data management method have already emerged as projecting problems in practical turbine blade design work. The paper proposes blade model inspection scheme based on profile waviness adjustment algorithm, and developed turbine blade design and processing quality assessment add-ins based on secondary development of UG software. All these add-in modules as well as some commercial software are integrated in ANSYS Workbench for unified management of design flow and sharing of engineering data in ANSYS Workbench. The obtained tools are applied into practical turbine blade design workflow and helped the reduction of manual labour.

1. Introduction

The surface of the aero-engine turbine blade is a complex free-form surface. The surface with good smoothness can reduce the difficulty for manufacturing process, simplify processing scenarios cut down manufacturing cost [1]. And the surface quality, which is a direct reflection for the quality of blade's final geometry, has great impact on its performance [2]. So in the blade design work, it's important to guarantee the design data fully exploited by manufacturing process, and export smooth surface for the downstream analysis as well as CAM process, thus reduce the possible labour caused by extra communication between design and manufacturing department, which tends to arise when the blade geometry data cannot be used in machining.

Based on the survey work by authors, the existing turbine blade design process has the following problems at present: the design flow is complicated and not unified, involving numbers of designers and tools. The primary design results often need corrections, but these corrections and further inspections heavily rely on manual operation, resulting in turbine blade design's unstable quality. What's more, turbine blade design is a multi-disciplinary process involving industrial software aimed at different problems, the integral application of these software requires standard design specifications and unified design data management [3], which is also in shortage for real work.

Therefore the development of specialized software tools for turbine blade automatic adjustment and quality assessment, as well as software integration can make up for the shortcomings in current work. This scheme can integrate the actual turbine blades' design flow to achieve the solidification of turbine blade design processing and the unified management of engineering data, on the basis of standard quality assessing methods and uniform model correction tools, the design result can reach consistent quality. This paper is addressing this problem and offer an overall technology approach for improving the current turbine blade design process. The essentials of our work can be divided into the following three parts: automatic adjustment of the blade profile, design and processing quality assessment, and turbine blade design as well as quality assessment environment integration.

2. Turbine Blade Profile Adjustment

The turbine blade profile curve can be generated by interpolating the blade profile coordinate data in the UG software. The blade profile coordinate data only contains the position information of the discrete blade profile, and lacks the definition of the parameterization method. The unqualified interpolation curve parameterization method will cause the newly generated blade profile spline to have waviness, which can be judged by the curvature vector distribution of the resulting profile spline, as shown in figure 1. The blade profile spline is used to construct key aerodynamic surfaces whose waviness will increase the energy consumption and degrade the performance of the engine. In the traditional design process, the designer's manual work is needed to adjust the position of the profile spline's interpolation point to eliminate the waviness, making turbine blade design model with uncertainty of the result and often need extra rework.

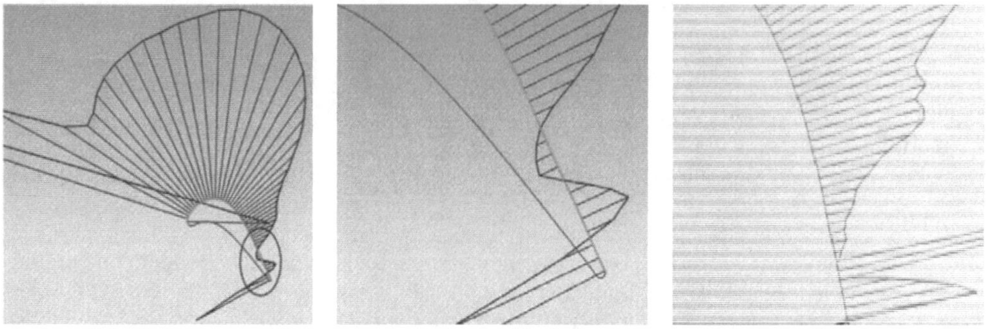

(a) (b) (c)

Figure 1. Contrast figure before and after manual adjustment of blade profile spline waviness

To handle this situation, we have developed automatic adjustment method of blade profile spline, it is based on the convexity preserving property of NURBS curves [4], which analyses the distribution of the curvature of blade profile spline, and adjusts the control polygon of blade profile spline to make it convex, thus achieving automatic elimination of blade profile spline waviness. The specific workflow is as follows:

- According to the blade profile coordinate data, the interpolation spline is calculated by using the chord-length parameterization method.
- Determine the sampling point density according to the length of the spline, calculate the curvature vector at the sampling point, as in equation (3), and judge whether the direction of the curvature vector of each sampling point satisfies the non- waviness condition. As in equation (1), P_i is a control vertex whose index value is i, k represents a k -th derivative, p is spline's degree, and u_i represents a knot node vector node whose index value is i. As in equation (3), $N_{i,p-k}(u)$ is a $p-k$ -degree B-spline basis function with an index value of i at the parameter u, and $C^{(k)}(u)$ is the k -th order of the spline at the parameter value u [5].

$$P_i^{(0)} = P_i \,, k = 0 \,;\; P_i^{(k)} = \frac{p-k+1}{u_{i+p+1}-u_{i+k}}\left(P_{i+1}^{(k-1)} - P_i^{(k-1)}\right) , \mathrm{k} > 0 \tag{1}$$

$$C(u) = C^{(0)}(u) = \sum_{i=0}^{n} N_{i,p}(u) P_i^{(0)} \tag{2}$$

$$C^{(k)}(u) = N_{i,p-k}(u) P_i^{(k)} \tag{3}$$

- The control vertices near the abnormal curvature vector are adjusted along the curvature vector direction to satisfy the control polygon convexity preservation condition. The new control vertices are used to construct the profile spline, as shown in figure 2. The blade profile spline is interpolated at 99 points, the maximum adjustment distance at the trailing edge of the profile spline is 0.02083 mm, and the number of adjusted control vertices is 26.

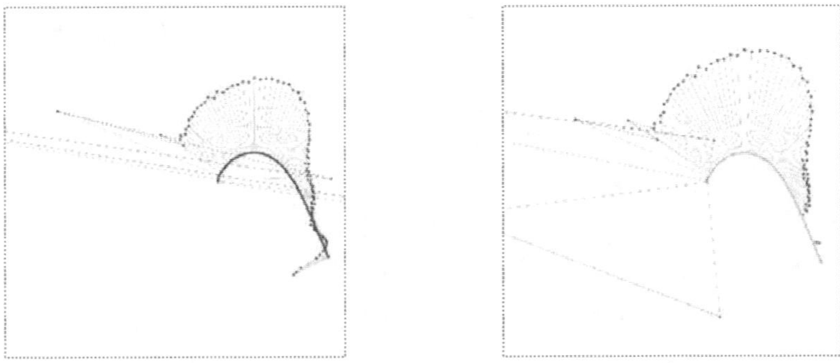

(a) (b)

Figure 2. Contrast figure before and after the automatic adjustment of blade profile spline waviness.

3. Turbine Blade Design and Processing Quality Assessment

The quality assessment of turbine blades consists of turbine blade design model surface quality assessment and surface processing quality assessment. Traditional design model surface quality assessment methods include curvature analysis method, illumination method and contour analysis method, etc. [6], which are used to obtain the inflection point of turbine blade body section, the concave-convex distribution and the shape change of the surface. These methods are intuitive but lack quantitative results, and leaves further work unclear to deploy. Turbine blade processing quality assessment generally uses three coordinate measuring instrument or optical scanning system to obtain measurement data, and the measurement data can be imported into UG software to compare with the design model to calculate the shape and position error, while there is a lack of unified method for evaluating the surface waviness of turbine blade. In workshop the blade design and processing quality assessment process relies on the experiences of the workers, to provide standard tools and methods for the assessment is the pursuit of practical work for increasing the competence. So in our work we developed the turbine blade quality assessment tools based on secondary development of UG software.

3.1. Design Quality Assessment Method

The assessment of turbine blade design quality work is mainly by the calculation of blade surface waviness. Based on the UG/Open API library, a specialized add-in for surface waviness assessment of turbine blade model was developed, which is used to extract and quantify the fluctuation information of turbine blade surface to transform the assessment rules into an objective manner.

The specialized software extracts the section information of the blade body in two orthogonal directions, and uses the curvature analysis method to calculate the inflection point information of turbine blade profile (see figure 3)and the stacking curves(see figure 4) based on the curvature vector distribution. The position and number of spline inflection points of the turbine blade profile and turbine blade stacking curves are output as the assessment result of the model design quality.

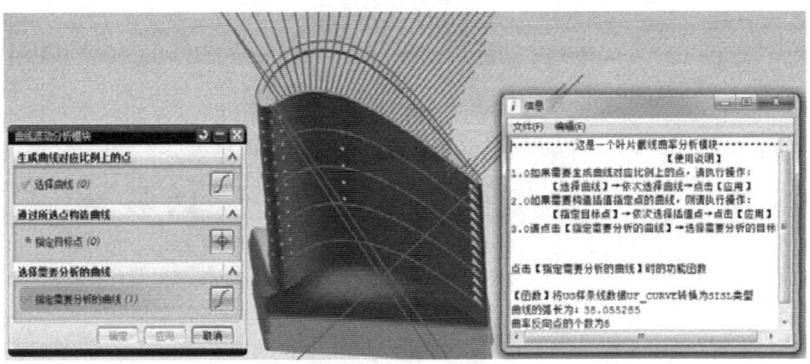

Figure 3. Assessment of turbine blade profiles.

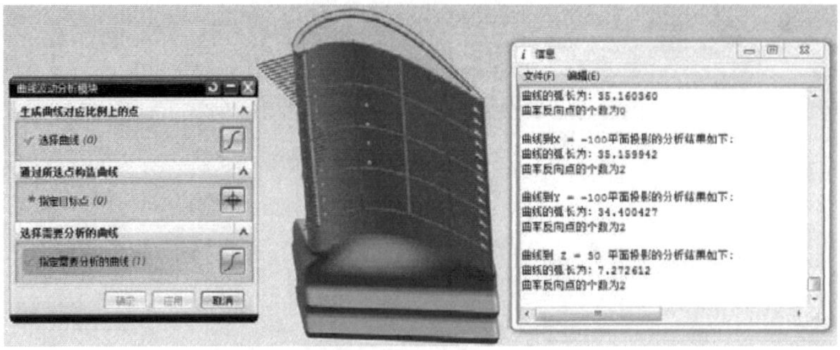

Figure 4. Assessment of turbine blade stacking curves.

In addition, based on the UG/Open API library, the feature parameter extraction specialized for the blade model is developed, and the turbine blade gravity centre parameter and the centre parameters of each section of turbine blade can be extracted, as shown in figure 5. These functions can further help to find the problems in the design model.

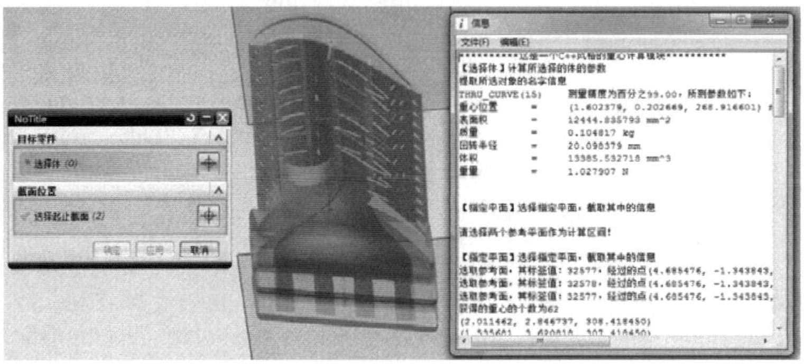

Figure 5. Turbine blade centre of gravity and central parameters of each section.

3.2. Processing Quality Assessment Method

The assessment of turbine blade processing quality work is mainly by the calculation of the surface waviness parameters. The raw processing data is measured by a coordinate measuring machine and the coordinate registration work is not considered in this paper. Based on the UG/Open API library, blade processing quality assessment tool is developed. In the UG environment, the deviation information between the coordinate measured and the turbine blade CAD model can be compared, and the surface's waviness can thus be revealed. This function is realized by calculating the relative distances between the measurement data picked up by the user and the theoretical blade profile spline according to the blade section profile. In the calculation process, we get X_i which is the nearest point to P_i on theoretical blade profile spline (see figure 6), and d_i which is the signed distance between X_i and P_i. The sign of d_i represents the relative position between P_i and theoretical blade profile spline. The arc length corresponding to X_i is obtained as l_i.

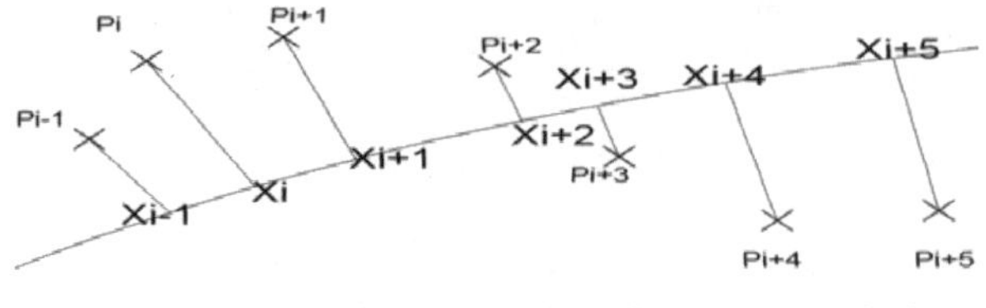

Figure 6. Arc length and relative distance calculation.

We take l_i as abscissa and d_i as the ordinate, and fit those coordinates (l_i, d_i) with B-spline which is used to calculate the inflection points I0, I1, I2, etc. (see figure 7). The absolute value of the d-

coordinate and l-coordinate difference between the adjacent inflection points are taken as the amplitude and wavelength of the waviness. The ratio of the amplitude to the wavelength is calculated as the change rate of the waviness, and the fluctuation of the output wavelength between 1 millimetre and 10 millimetre is considered as the waviness information.

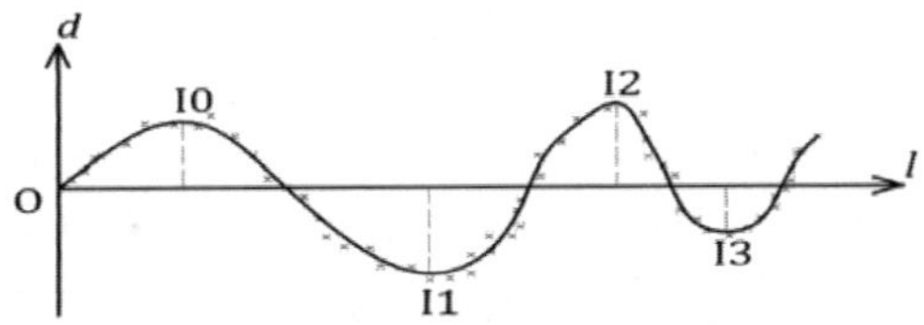

Figure 7. Waviness amplitude and wavelength calculation

As is shown in figure 8, the processing quality assessment data is 392 points.

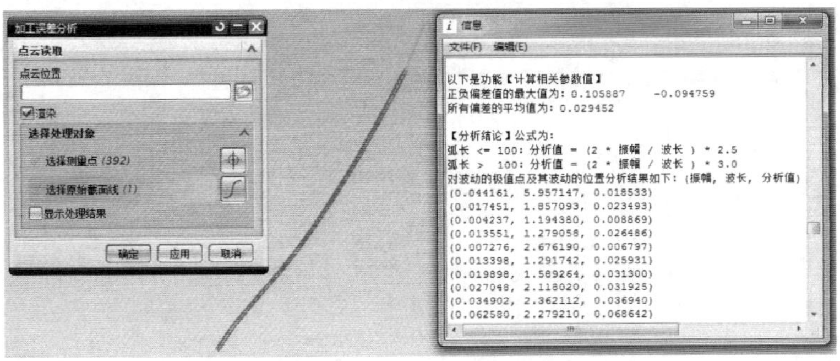

Figure 8. Blade processing quality assessment

4. Turbine Blade Design and Quality Assessment Environment Integration

Turbine blade design results can be obtained from a sequential engineering or reverse engineering workflow. A variety of software maybe used, such as UG(for design), Geomagic(for point cloud processing), ANSYS(for performance analysis) and so on. The real work process can be very complex and much work is devoted to the data conversion in different software or the function shift between different platforms. In order to regulate the work process and coordinate the software tools, we use ANSYS workbench as the basis for integrating cooperation of different software and manage the overall dataflow. The workflow of each software module and the data file management logic are shown in figure 9.

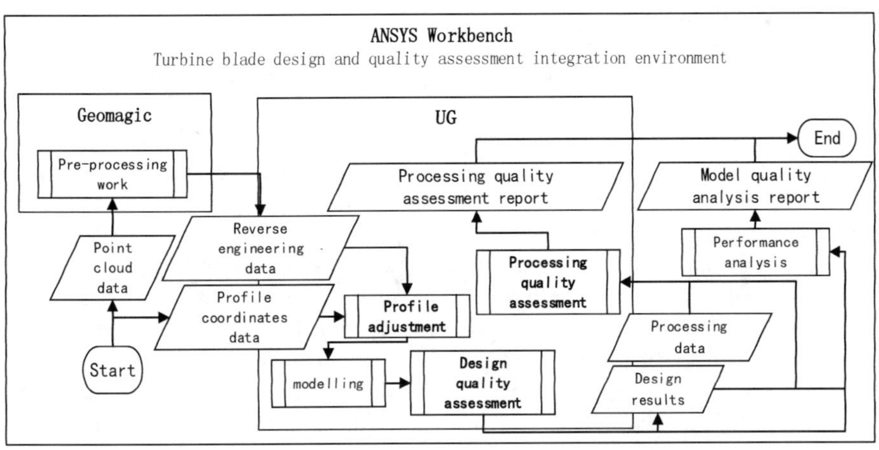

Figure 9. Turbine blade modelling and testing standardization process

During the reverse engineering step, the designer first enable the Geomagic by Workbench to complete the pre-processing work from the turbine blade point cloud data, and the output data from Geomagic will be automatically transmitted to UG. After the data has entering the UG, the user adjusts the blade profile spline waviness by adopting the blade profile spline waviness correction add-in, and completes the blade modelling work, generating the model quality analysis report and turbine blade CAD model that meets design requirements. The turbine blade model data output by UG will be automatically transferred to the ANSYS simulation analysis software, and the user can complete the simulation analysis there. After completing the machining of the turbine blade parts, the user can compare the deviation of the part processing measurement data with the CAD model in UG, and output the blade processing quality assessment report through the blade processing quality assessment tool. Based on the secondary development of ANSYS Workbench, the development of turbine blade design environment integration tools is realized. The workflow and file management of the relevant software is customized using XML and Iron Python. The module operation is shown in figure 10.

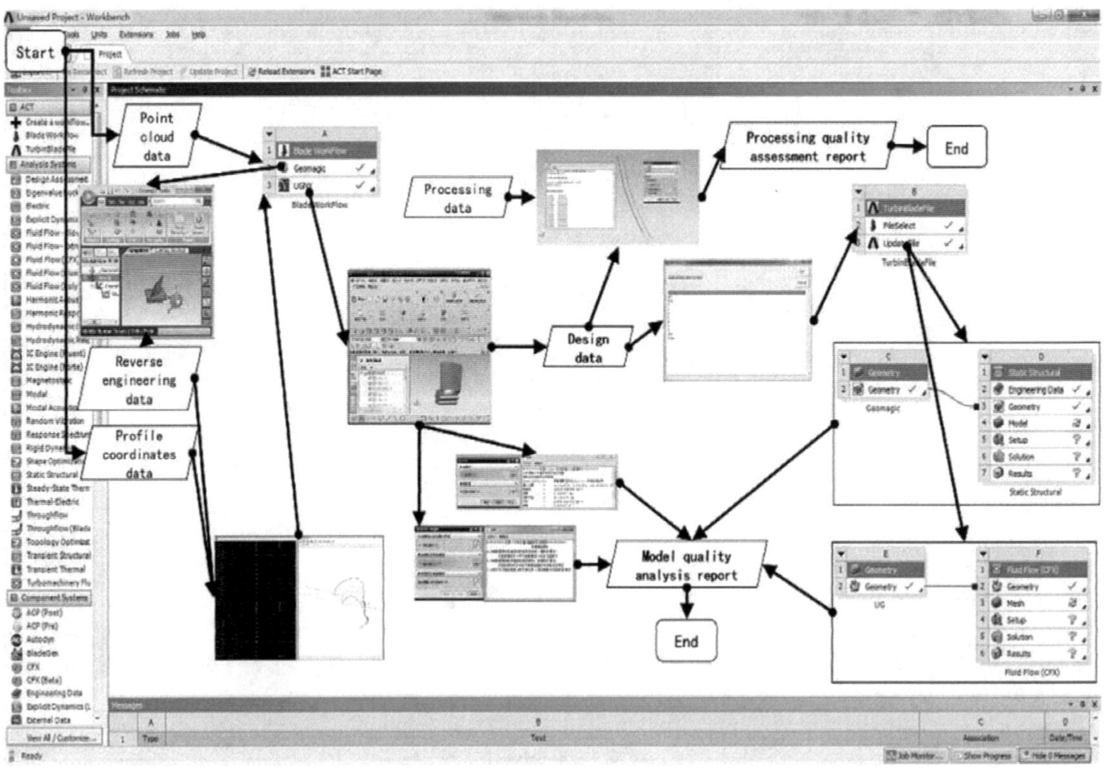

Figure 10. Turbine blade design and inspection software integration platform

5. Conclusion

The existing CAD software are usually for general design work, while in turbine blade design, special concerns will arise based on the relevant plant's existing habits and work requirements. So add-ins of some technical functions are heavily needed in real work. In this paper, functional modules aimed at guaranteeing the design models' uniforming quality as well as machining quality inspection are developed, these are based on general-purpose CAD software UGS, and can strengthen the primary functions of existing system. Also an integrated platform for the data sharing and workflow management is developed based on the newly developed add-ins. The work is highly focused on practical applications and oriented for real problem solution. Future work will explore the more close combination between different software used for blade design from more essential level, such as geometry representation, modelling history analysis and so on.

6. Acknowledgements
This work is supported by the Natural Science Foundation of China (Project Nos. 61572056 and 61972011)

7. References
[1] Zhang W L, Yuan Q H, Huang Z G and He Y L 2007 Research and Application on Smoothing Evaluation Methodology for Car- Body Surface Modeling *J Automobile Technology.* **05** pp 10-14
[2] Luo C 2017 Research on Measurement Method and Realization Technology of Aeroengine Blades *D Chongqing University of Technology* chapter 1 pp 1-10
[3] Xia W 2004 Research on CAD system of turbine blade modeling D Northwestern Polytechnical *University* chapter 1 pp 1-5
[4] Liu C Y 2006 A Geometric Proof of the Convexity Preserving Property of NURBS Curves *J Chinese Quarterly Journal of Mathematics* **01** pp 44-48
[5] Shi F Z 2001 *CAGD & NURBS* (Beijing: Higher Education Press) chapter 7 pp 237-249
[6] Yue L L 2014 Research on turbine blade Reconstruction and DataMatrix Fairing *D Central South University* 2 pp 16-17

Biomass Derived Fuels for Gas Turbine Application

Taiyan Zhang[1]*, Yongling Yao[1], Chengbin Lu[1], Bin Xu[1] and Dongyang He[2]

[1] Jiangsu Frontier Electric Power Technology Co., Ltd., Nanjing 210036, China
[2] School of Energy and Power Engineering, Nanjing Institute of Technology, Nanjing 211167, China
*Corresponding author e-mail: 15905166939@163.com

Abstract. The use of biomass-derived fuels as an alternative energy has become a key solution to solve fossil fuels shortage and reduce greenhouse gas emission. With the development on biomass energy conversion, biofuels show great improvement in product yield, fuel quality and combustion characteristics, making it possible for gas turbine applications. The stable combustion flame of the gas turbine makes it adaptable to various biofuels. The main research of utilizing biomass-derived fuels in gas turbine in recent years was summarized. Characteristics and combustion feasibility of typical biofuels were discussed including both liquid phase biofuels and gas phase biofuels. The combustion performance, pollutant emission and combustion optimization of biofuels combustion in gas turbines was analysed. The main problems and effective solutions for biofuel applications were summarized, providing guidance for clean and efficient use of biofuels in gas turbines.

1. Introduction

Currently, besides coal, oil and natural gas, biomass is the fourth largest energy source in the world, showing abundant source, great sustainability and good greenhouse gas reduction effect. Therefore, biomass can be used as a substitute for fossil fuels. As a new type of internal combustion engine, gas turbine shows stable combustion flames, fast start-up and high thermal efficiency. The strong mechanical structure, wide power range and flexible combustion system of gas turbine all contribute to the application of biofuels. The application of alternative biofuel on gas turbine can not only control CO_2 emissions but also limit NO_x and soot emissions, achieving dual benefits of environmental protection and economy. However, due to the difference of biofuels with conventional fossil fuels in viscosity, composition and heating value, the use of biofuels on gas turbines still faces problems such as ignition difficulty, poor atomization effect and coking and corrosion of equipment. In this paper, the typical biomass-derived fuels used in gas turbines in recent years are summarized for analyzing the main characteristics, application status and key technical problems in biofuel application.

2. Liquid Phase Biofuel

Liquid biofuels could be mainly classified as vegetable oil and bio-oil. A typical liquid biofuel production path is shown in Figure 1.

2.1. Vegetable Oil

Pure vegetable oil is low in sulfur content, non-toxic and biodegraded. The combustion of vegetable oil is benefited to reduce sulfur oxides, CO and polycyclic aromatic hydrocarbons (PAH). Cavarzere et al. [1] analyzed the characteristics of rapeseed oil, sunflower oil and soybean oil and verified the application of vegetable oil on gas turbine. However, pure vegetable oil combustion would increase fuel injection pressure exponentially, resulting in poor atomization and burner ash accumulation. Fuel

IOP Conf. Series: Materials Science and Engineering **730** (2020) 012043 doi:10.1088/1757-899X/730/1/012043

pre-heating, dilution/pre-mixing, micro-emulsification and transesterification can reduce vegetable oil viscosity. F. Chiariello et al. [2] mixed kerosene with sunflower oil and rapeseed oil respectively, and tested the combustion and emission of them in a micro-turbine. It was found that low addition of vegetable oil could increase particulate matters emission with little effect in CO and NO emission. A. Hoxie et al. [3] mixed soybean oil with low-sulfur diesel at a maximum volume ratio of 75%. They confirmed composite fuel oil can be stably and efficiently burned in gas turbines.

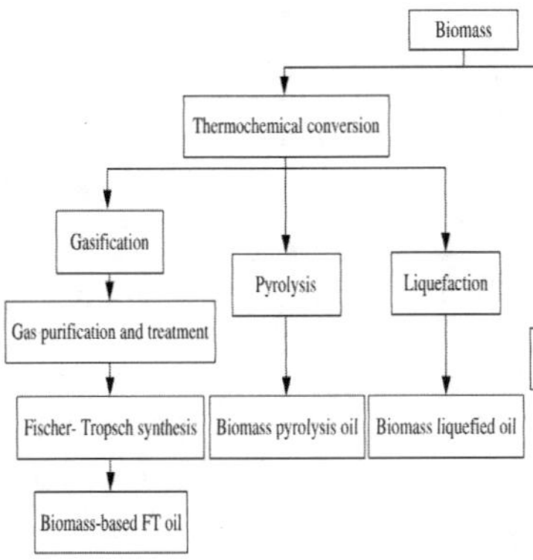

Figure 1. Liquid biofuels production [4].

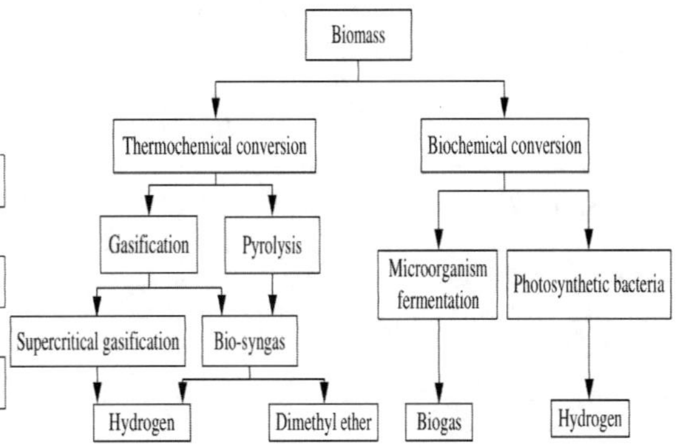

Figure 2. Gas biofuels production.

2.2. Biodiesel
Biodiesel is a fatty acid methyl ester or ethyl ester formed from animal or vegetable oil through ester conversion with methanol or ethanol. KK.Gupta et al. [5] found by blending 20% biodiesel in diesel, CO and particulate matters could be decreased by 12% and unburned hydrocarbon could be declined by 20%. Although the addition of biodiesel decreases combustor temperature, the high oxygen content of biodiesel could lead to lean combustion and improve thermal efficiency. Zhang et al. [6] found the combustion efficiency could still reach more than 99% with biodiesel and the influence of biodiesel could be limited with increasing oil/gas ratio. Habib et al. [7] found biodiesel reduced static thrust of gas turbine and CO and NO emission. Chong et al. [8] detected a wider flame reaction zone, higher spray droplet density and higher spray volume flow rate in the combustion of rapeseed biodiesel than ordinary diesel. Thus, soot and NO_x could be reduced by combustion. Besides, the blending ratio could be enhanced to 50% and the thermal efficiency could reach about 26% by biodiesel preheating [9, 10].

2.3. Bioethanol
Bioethanol refers to alcohol produced by microbial fermentation or thermochemical conversion. It has a low flash point and cetane number, making it easy to ignite for clean combustion. There are no atomization problems for bioethanol [11].

Lupandin et al. [12] carried out multiple biofuels combustion on an OGT2500 gas turbine. The modified gas turbine can adapt to bio-oil, bioethanol and biodiesel. Among these biofuels, bioethanol resulted low NO_x, CO, and SO_x emissions below the government limits. J.A. Alfaro-Ayala et al. [13] compared the combustion of natural gas and bioethanol in gas turbines. The inlet temperature, power generation and efficiency of gas turbines for bioethanol were slightly lower with same amount of fuel. Therefore, under the same load, bioethanol fuel consumption is higher than natural gas. But NO_x emission of bioethanol was only 65% of natural gas. Sallevelt et al. [14] carried out combustion of bioethanol and diesel in a 2MW OP16 gas turbine, and simulated the atomization of bioethanol combustion. The results show that CO emissions of bioethanol are comparable to diesel and the NO_x emissions are low.

2.4. Bio-oil
Bio-oil is liquid fuel made from biomass thermochemical conversion. The chemical composition of bio-oil is relatively complicated with high oxygen and impurity content, high viscosity and corrosivity, making it difficult for gas turbine application [15].

Pedro L. Cruz et al. [16] evaluated biomass gasification-Fischer-Tropsch synthesis combined cycle by exergy analysis. They studied the behavior of bio-oils derived from Fischer-Tropsch synthetic in gas turbines, and proposed optimum design for improving exergy efficiency. M.E Boucher et al. [17] evaluated the viscosity, stability, combustion reactivity and aging characteristics of biomass pyrolysis oil. They added methanol to the pyrolysis oil to promote its application in gas turbines. Marco Buffi et al. [18] prepared a low-viscosity, high-organic bio-oil based on the liquefaction with an acidified glycol solvent. They used it in a 25kW micro gas turbine by mixing bio-oil with ethanol, increasing CO emission and decreasing NO_x emission. The components and fuel characteristics of diesel, aviation kerosene, and typical liquid phase biomass-derived fuels are shown in Table 1.

Table 1. Liquid biofuel properties [19-20].

Properties	Diesel	JET-A	Straight vegetable oils	Bio-diesel	Bio-ethanol	Pyrolysis oil
Density (kg/m³)	827.4	807	900~940	860~900	794~810	984~1250
Kinetic viscosity (Cst, 40 °C)	1.728 3	0.88	30~40	3.5~5	1.4~1.7	32~45
Flash point (°C)	44~50	38	230~280	120~180	13	56~130
Cloud point (°C)	-6	——	-4~14	-3~-12	——	——
Pour point (°C)	-16	-47	12~10	-15~5	-117	-35~-10
Lower calorific value (MJ/kg)	43	43.23	38~39	39~41	25~26	13~18
Ignition temperature (°C)	250	220	325~370	177	423	580
Cetane no.	45~55	55	37~42	48~60	8	10
Stoichiometric air/fuel ratio	14.6	14	13.8	13.8	9.79	34:1

3. Gas Phase Biofuel

As shown in Figure 2, gas phase biofuels which are produced by biochemical or thermochemical conversion, mainly contain biogas, biomass syngas, dimethyl ether and hydrogen.

3.1. Biogas

Biogas is a kind of gas mainly composed of CH_4 and CO_2 produced by microbial fermentation under anaerobic conditions. Its calorific value is lower than natural gas. Bruno et al. [21] used a biogas-burning micro-turbine to build a sludge treatment tri-generation system. Many researchers have further confirmed the economic and environmental benefits of gas turbine biogas combustion [22, 23]. Nikpey et al. [24] conducted a mixed combustion test of biogas and natural gas based on a regenerative micro gas turbine, verifying that biogas can be stably burned with natural gas in an unmodified micro gas turbine. Compared with natural gas, the electrical efficiency and operating parameters are almost unchanged. Besides, CO_2 emissions can be reduced by 19% at full load. Somehsaraei et al. [25] built a thermodynamic model to analyse the effects of biogas combustion on gas turbine performance based on a 100kW micro gas turbine. Biogas combustion decreases gas turbine mass flow, pressure ratio and surge margin compared with nature gas. Electrical efficiency and heat recovery efficiency of heat exchangers decreased as the CH_4 content of biogas decreased.

3.2. Biomass Syngas

Syngas containing H_2, CO, CO_2, CH_4, etc. can be obtained by biomass pyrolysis or gasification at high temperature, requiring independent compressors for air and gas compression during combustion. Design and modification of gas mass flow measurement system, gas delivery and injection system, fuel nozzle and combustion chamber are needed. The power consumption of gas turbine shafts also increases accordingly [26, 27]. Adouane et al. [28] used a 1.5MW pressurized fluidized bed to prepare biomass syngas and achieved stable turbulent combustion in a gas turbine. The conversion efficiency of fuel NH_3 to NO_x was 10-60%. Sukumaran et al. [29] simulated the combustion of different biomass syngas and analyzed the key factors affecting the NO_x emissions and the mechanism, guiding the design of low NO_x syngas burners. Cameretti et al. [30] calculated the combustion efficiency and pollutant emissions of biomass syngas in micro gas turbines, proposing a coordinated improvement for combustion efficiency and NO reduction. Walter et al. [31] achieve stable co-combustion of biomass syngas and natural gas in gas turbines based on the reformation of the fuel supply system.

3.3. Dimethyl Ether

Dimethyl ether (CH_3OCH_3) has a simple structure, non-toxicity, high cetane number and strong environmental friendliness. Dimethyl can quickly evaporate and form a good fuel-air mixture to accelerate combustion. Lee et al. [32, 33] used GE7EA gas turbines to perform dimethyl ether combustion. They verified dimethyl ether can be used as clean gas turbine fuel. The NO_x emission is significantly lower than methane. Compared with methane, dimethyl ether has a lower spontaneous combustion temperature, a higher flame propagation speed and a greater risk of spontaneous combustion during premixing. Saitou et al. [34] developed a multi-cluster burner, which can significantly reduce the risk of spontaneous ignition and tempering of dimethyl ether. Lee et al. [35] designed and manufactured a new type of dimethyl ether combustion nozzle, which can not only prevent spontaneous combustion and tempering, but also reduce NO_x and CO emissions. Namasivayam et al. [36] used dimethyl ether as the ignition fuel for gas turbines. Results showed that dimethyl ether can reduce NO_x by about 20% at low load. Kang et al. [37] studied the effect of dimethyl ether addition on the combustion of turbulent methane/air jet diffusion flames and found the addition of dimethyl ether changed the flame structure and reduce soot and NO_x emissions.

Table 2. Gas biofuel properties.

Properties	Methane	Propane	Biogas	Biomass syngas	DME	Hydrogen
Chemical composition	CH_4	C_3H_8	CH_4, CO_2	CO, H_2, CH_4, CO_2, N_2	C_2H_6O	H_2
15°CDensity (kg/m^3)	717	490	0.65-0.91	765~785	670	70.8
Boiling point (°C)	-161	-42	-126~-162	-192	-25	-252.8
0°CVapor pressure at 0°C (MPa)	24.6	0.93	——	——	0.61	——
Flammable limits in air (%)	5~15	2.1~9.4	7~26	5~32	3.4~17	4~75
Ignition temperature (°C)	537	470	632~813	——	235	500
Max. burning velocity (cm/s)	37	43	——	——	50	289
Stoichiometric air/fuel ratio	16.9	15.7	10~17	9.1	9.0	34.1
Lower calorific value (MJ/kg)	49	46.3	20~28	10~18	28.8	120
Cetane no.	0	5	——	70	55~60	——

3.4. Hydrogen

Hydrogen can be produced by biomass pyrolysis, gasification and microbial fermentation. Only thermal NO_x is formed due to the presence of air. Hydrogen combustion has a high combustion speed and extremely high diffusivity. Hydrogen can be fully mixed with air, reaching a combustion temperature of 2300 K and a combustion efficiency of 100% [38]. Due to the rapid combustion reaction, premixed combustion is still the main use of hydrogen on gas turbines. Lee et al. [39, 40] tested H_2 and CO combustion in gas turbines. They found the gas turbine requires only a small amount of modification to stabilize combustion. The ratio of H_2 and CO directly affects syngas combustion. Premixing H_2 had an effect on the gas turbine operating amplitude, frequency and instability pattern. After premixing, the flame speed stage temperature changed and formed a variety of instability modes. York et al. [41] carried out experiments on premixed combustion of high-hydrogen fuel in gas turbines based on self-designed premixed combustion nozzles. Hydrogen fuel can achieve stable combustion without tempering and generate carbon-free flue gas. Taamallah et al. [42] and Kahraman et al. [43] evaluated hydrogen premixed combustion stability and NO_x emission based on simulation. The results were compared with conventional fuels to support optimizing flame structure and combustion dynamics.

The composition and fuel characteristics of methane, propane and typical gas phase biomass-derived fuels are shown in Table 2.

4. Optimization of Biofuel Application on Gas Turbine

Compared with conventional fossil fuels, the combustion of biofuels in gas turbines is still in its infancy. Further improvements need to be made in terms of combustion stability, fuel applicability and combustion solutions.

4.1. Fuel Modification and Upgrading

Compared with conventional fuels, biofuels still show differences in evaporative atomization effect, droplet size, flame stability, turbine inlet temperature and pollutant emission concentration. Biofuel upgrading can make biofuel closer to conventional fuels, improving combustion efficiency and avoiding major fuel system changes. Chan et al. [44] used hydrogenated vegetable oil (HVO) on turbofan engines. They found that compared with conventional Jet A-1 fuel, NO_x emissions of HVO combustion decreased significantly. Under 80% and 95% load idling conditions, the NO_x emissions of 50% HVO/Jet A-1 mixed fuel can be reduced by up to 0.3g/kg. Buffi et al. [45] observed through the optical swirl burner that the HVO fuel combustion exothermic temperature is very uniform. Thus the HVO combustion pollutant emissions are low.

4.2. Combustion Equipment Improvement

Due to the differences of biofuels and conventional fuels, the air-fuel ratio, fuel atomization effect and combustion stability of biofuels in gas turbine are different from conventional fuels. The gas turbine fuel supply system, injection system and combustion system need to be designed and modified. Prussi et al. [46] reformed commercial gas turbine fuel pipelines, realizing switch between diesel and pure vegetable oil. Under the same load, the fuel consumption of vegetable oil was only increased by 5%. The output pressure of fuel pump was increased by 30%. Al-attab et al. [47] constructed a dual fuel supply system which can simultaneously use liquefied petroleum gas and biomass syngas, achieving stable combustion for biomass syngas alone or the mixture with liquefied petroleum gas. Asai et al. [48] optimized the structure of burner and combustion chamber by single-nozzle fuel-air mixing experiments and burner combustion experiments for reducing NO_x, CO and unburned hydrocarbon. The stability of the combustion system and emission is optimized greatly.

4.3. Combustion Technology Optimization

Fuel preheating can reduce viscosity to conventional fuel levels and improve evaporative atomization. Sallevelt et al. [49] can significantly reduce vegetable oil viscosity to 1/3 by increasing fuel injection temperature of micro gas turbines, reducing CO emissions by 28% while improving combustion efficiency. Chiaramonti et al. [9] also found the CO emissions of biodiesel were higher than that of diesel without preheating. By preheating biodiesel to 120°C, CO emissions could be reduced by 63-64%.

The premixed and pre-evaporated lean combustion of biofuel could assist atomization and evaporation of fuel by forming a vortex air flow in the premixed tube. Jozsa et al. [50] performed a spectral analysis on the premixed combustion of rapeseed oil in a gas turbine. They found that controlling the number of premixed swirls can achieve stable combustion of rapeseed oil. The levels of NO_x and CO emissions are close to diesel. Chen et al. [51] used lean-burn premixed enhanced combustion nozzles to test the combustion of biogas in gas turbines. PIV observed good biogas flow in the nozzles and strengthened vortex flow.

5. Conclusion

The use of alternative biofuels for gas turbines can not only alleviate the shortage of fossil fuels, but also broaden the scope of biofuels application and improve the overall energy structure. Potential biofuels include vegetable oil, biodiesel, bioethanol, dimethyl ether and hydrogen. Biofuels can be used stably in gas turbines by modification, upgrading or mixing with fossil fuels. The fuel consumption rate is slightly higher than fossil fuels at the same load and the temperature of combustion chamber inlet is lower. The application of most biofuels can reduce SO_x, CO, NO_x and soot in gas turbines. The combustion efficiency is equivalent to that of fossil fuels. Fuel upgrading, combustion system modification, and optimization of combustion technology can all promote the stable application of biofuels in gas turbines. Premixed and pre-evaporated lean combustion are currently effective biofuel combustion application methods.

6. Acknowledgement

This work was funded by the Challenge Cup Competition Support Project of Nanjing Institute of Technology (TZ20190005).

7. References

[1] Cavarzere A, Morini M, Pinelli M, Spina P R, Vaccari A and Venturini M 2014 *Energy Procedia* **45** 91-100.
[2] Chiariello F, Allouis C, Reale F and Massoli P 2014 *Experimental Thermal and Fluid Science* **56** 16-22.
[3] Hoxie A and Anderson M 2017 *Renewable energy* **101** 886-893.
[4] Damartzis T and Zabaniotou A 2011 *Renewable and Sustainable Energy Reviews* **15**(1) 366-378.
[5] Gupta K, Rehman A and Sarviya R 2010 *Iranica Journal of Energy & Environment* **1**(3) 205-210.
[6] Zhang S J, Ma H Y, Jia C Y and Jin G 2016 *Journal of Aerospace Power* (7) 1562-1568.
[7] Habib Z, Parthasarathy R and Gollahalli S 2010 *Applied Energy* **87**(5) 1701-1709.
[8] Chong C T and Hochgreb S 2017 *Applied Energy* **185** 1383-1392.
[9] Chiaramonti D, Rizzo A M, Spadi A, Prussi M, Riccio G and Martelli F 2013 *Applied Energy* **101** 349-356.
[10] Nascimento M A, Lora E S, Correa P S, Andrade R V, Rendon M A, Venturini O J and Ramirez G A 2008 *Energy* **33**(2) 233-240.
[11] Balat M and Balat H 2009 *Applied Energy* **86**(11) 2273-2282.
[12] Lupandin V, Thamburaj R and Nikolayev A 2005 *American Society of Mechanical Engineers.*
[13] Alfaro-Ayala J, Gallegos-Muñoz A, Uribe-Ramirez A and Belman-Flores J 2013 *Applied Thermal Engineering* **61**(2) 481-490.
[14] Sallevelt J L, Pozarlik A K, Beran M, Axelsson L-U and Brem G. 2014 *Journal of engineering for gas turbines and power* **136**(7) 071501.
[15] Ail S S and Dasappa S. 2016 *Renewable and sustainable energy reviews* **58** 267-286.
[16] Cruz P L, Iribarren D and Dufour J 2017 *Fuel* **194** 375-394.
[17] Boucher M, Chaala A and Roy C 2000 *Biomass and Bioenergy* **19**(5) 337-350.
[18] Buffi M, Seljak T, Cappelletti A, Bettucci L, Valera-Medina A, Katrašnik T and Chiaramonti D 2018 *Applied Energy* **230** 1193-1204.
[19] Zhang Z, Chen L, Lu Y, Roskilly A P, Yu X, Smallbone A and Wang Y 2019 *Fuel* **239** 1351-1362.
[20] Gupta K, Rehman A and Sarviya R. 2010 *Renewable and Sustainable Energy Reviews* **14**(9) 2946-2955.
[21] Bruno J C, Ortega-López V and Coronas A 2009 *Applied Energy* **86**(6) 837-847.
[22] Basrawi F, Ibrahim T K, Habib K, Yamada T and Idris D M N D 2017 *Energy* **124** 238-248.

[23] Kang J Y, Kim T S and Hur K B 2014 *Energy* **67** 309-318.
[24] Nikpey H, Assadi M, Breuhaus P and Mørkved P 2014 *Applied Energy* **117** 30-41.
[25] Somehsaraei H N, Majoumerd M M, Breuhaus P and Assadi M 2014 *Applied Thermal Engineering* **66**(1-2) 181-190.
[26] Rodrigues M, Walter A and Faaij A 2007 *Energy Conversion and Management* **48**(4) 1289-1301.
[27] De Vries J and Petersen E 2007 *Proceedings of the combustion institute* **31**(2) 3163-3171.
[28] Adouane B, Hoppesteyn P, de Jong W, van der Wel M, Hein K R and Spliethoff H 2002 *Applied Thermal Engineering* **22**(8) 959-970.
[29] Sukumaran S and Kong S-C 2013 *Combustion and Flame* **160**(10) 2159-2168.
[30] Cameretti M C and Tuccillo R 2015 *Applied Thermal Engineering* **89** 280-290.
[31] Walter A and Llagostera J 2007 *Energy Conversion and Management* **48**(11) 2888-2896.
[32] Lee M C, Seo S B, Chung J H, Joo Y J and Ahn D H 2008 *Fuel* **87**(10-11) 2162-2167.
[33] Lee M C, Seo S B, Chung J H, Joo Y J and Ahn D H 2009 *Fuel* **88**(4) 657-662.
[34] Saitou T, Miura K, Inoue H, Kobayashi N, Suzuki S-I 2005 *American Society of Mechanical Engineers*.
[35] Lee M C and Yoon Y 2012 *Fuel* **102** 823-830.
[36] Namasivayam A, Korakianitis T, Crookes R, Bob-Manuel K and Olsen J 2010 *Applied Energy* **87**(3) 769-778.
[37] Kang Y, Shuang W, Jiang X, Song Y, Sun S, Zhang P, Sun Y, Lu X, Wang Q and Gou X 2017 *Fuel Processing Technology* **159** 421-435.
[38] Juste G 2006 *International Journal of Hydrogen Energy* **31**(14) 2112-2121.
[39] Lee M C, Seo S B, Chung J H, Kim S M, Joo Y J and Ahn D H 2010 *Fuel* **89**(7) 1485-1491.
[40] Park J and Lee M C 2016 *International Journal of Hydrogen Energy* **41**(18) 7484-7493.
[41] York W D, Ziminsky W S and Yilmaz E 2013 *Journal of Engineering for Gas Turbines and Power* **135**(2) 022001.
[42] Taamallah S, Vogiatzaki K, Alzahrani F, Mokheimer E, Habib M and Ghoniem A 2015 *Applied Energy* **154** 1020-1047.
[43] Kahraman N, Tangöz S and Akansu S O 2018 *Fuel* **217** 66-77.
[44] Chan T W, Chishty W A, Canteenwalla P, Buote D and Davison C R 2016 *Journal of Engineering for Gas Turbines and Power* **138**(1) 011506.
[45] Buffi M, Valera-Medina A, Marsh R, Pugh D, Giles A, Runyon J and Chiaramonti D 2017 *Applied Energy* **201** 84-93.
[46] Prussi M, Chiaramonti D, Riccio G, Martelli F and Pari L 2012 *Applied Energy* **89**(1) 287-295.
[47] Al-Attab K and Zainal Z 2014 *Applied Thermal Engineering* **70**(1) 61-70.
[48] Asai T, Dodo S, Koizumi H, Takahashi H, Yoshida S and Inoue H 2011 *American Society of Mechanical Engineers*.
[49] Sallevelt J, Gudde J, Pozarlik A K and Brem G 2014 *Applied Energy* **132** 575-585.
[50] Józsa V and Kun-Balog A 2017 *Fuel Processing Technology* Stability and emission analysis of crude rapeseed oil combustion **156** 204-210.
[51] Chen J, Wang Y, Liu H and Weng Y 2017 *Applied Energy* **121** 90-102.

Effect of Nozzle Structure Parameters on Steady Emission Characteristics in an Opposed-Piston Two-Stroke Diesel Engine

Shangxue Wang, Zhijun Zhang, Jiangfeng Dong and Wenyan Xu

North China Engine Research Institute, No.96 YongJin Road, BeiChen District, TianJin, China.
Email: wsx173@163.com

Abstract. According to 8-Mode cycle of measurement methods for non-road diesel engine of China III from GB20891-2014, the effects of the injection parameters on steady emission characteristics were researched on an opposed-piston two-stroke(OP2S) diesel engine. Test results show that nozzle-tip-protrusion(NTP) can't make a big influence on emission characteristics, but all of the CO, NOx and PM emissions decrease obviously with the increase of injector hole diameter. Moreover, there exists an optimal nozzle hole cone angle, which can lead to a least composite rate emission. Compared with the original engine, the gross emissions of NOx and HC decrease approximately by 17.2% for improved OP2S engine after nozzle parameters optimization matching, and the improved OP2S engine is able to meet non-road diesel engine of China III emission requirement.

1. Introduction

With the strengthening of people's environmental awareness, engine emission standards are becoming more and more strict. At present, in the development and production of non-road vehicles, the concept of developing "green" automobile products has been gradually concerned by the international community. Under the pressure of environmental protection, the emission regulations of non-road vehicles in many countries and regions limit the emission of different harmful emissions. In 2014, China III emission standard of non-road mobile machinery diesel engines made stricter provisions on the total emission limit of "NOx+HC"[1].

In the current power products, OP2S diesel engine has been considered as a third compromise in addition to hybrid and clean diesel engine. Advanced propulsion technologies of California has demonstrated how to improve the smoke emission of OP2S diesel engines[2]. The research of Chinese scholars on OP2S diesel engine mainly focuses on the design of key parts, in-cylinder scavenging, in-cylinder gas flow, in-cylinder combustion and other aspects[3-6], while the research on the emission characteristics of OP2S diesel engine has not been involved.

Fuel injection quality has a very important influence on engine combustion and engine performance[7]. Studies show that among the many factors that affect the combustion and emission characteristics of diesel engines, 60%~70% come from fuel injection, so the improvement of fuel injection system is the key to reduce the emission of diesel engines. Due to the particularity of the combustion chamber structure of OP2S diesel engine, its fuel injection and atomization process is different from the middle arrangement structure of single injector of traditional diesel engine. Therefore, when studying the combustion and emission characteristics of OP2S diesel engine, the influence of fuel injection system parameters should be fully considered.

In this paper, the impact of nozzle parameters on the emission characteristics of OP2S diesel engine will be tested, and through the comprehensive optimization of nozzle parameters, the emission index of engine in steady state condition will meet the emission standards of non-road diesel engines in China, which will lay a foundation for the emission upgrading and performance improvement of OP2S diesel engine in the future.

2. Test System

The test was carried out on a 1.2L OP2S diesel engine. The nominal power of the diesel engine is 32kW, the nominal speed is 1800r/min, and the geometric compression ratio is 22:1. The engine adopts single-pump and double-common rail oil supply system. According to the opening of the pressure regulating electromagnetic valve on the pump, the high-pressure oil pump provides quantitative and constant pressure high-pressure oil, which is fed into two common rail pipes through the three-way valve. The two injectors connected with each common rail pipe are arranged relatively around the cylinder and perpendicular to the center line of the cylinder. The three nozzle holes of the injector are arranged uniformly around the nozzle, and the elliptical combustion chamber composed of internal and external piston tops forms a combined jet combustion structure, as shown in figure 1.

Figure 1. Combustion chamber of OP2S engine.

The main technical parameters of the oil supply system are shown in table 1.

Table 1. Key technical parameters of the oil supply system.

Element	Index
Nozzle orifice number	3
Nozzle diameter	0.29mm
Nozzle angles	17°
NTP(nozzle-tip-protrusion)	2.7mm
Injection pressure	160MPa
Common rails volume(a single pipe)	75ml

Layout of the emissions test system is shown in figure 2: the eddy current dynamic dynamometer, control cabinet and KAMA-4000 fuel consumption meter produced by AVL company in Austria are adopted; 7613C-KiBOX explosion pressure sensor is used to measure the explosion pressure in the cylinder; The DIGAS-4000 emission analyzer produced by AVL in Austria is used to measure NOx, HC and CO in exhaust gas. The smoke is measured by AVL-439 extinction type smoke meter.

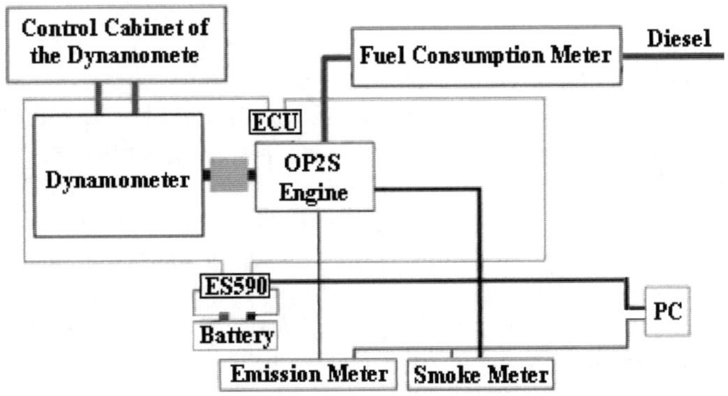

Figure 2. Schematic diagram of emission test system.

In order to ensure the reliability of test data, test boundary conditions and test process parameters should be strictly controlled, including the following parameters:

- Engine fuel temperature is about 35°C~40°C;
- Inlet temperature is about 50°C~80°C;
- Water temperature difference between inlet and return is not more than 8°C;
- Exhaust temperature is not more than 650°C;
- Maximum combustion pressure in cylinder is not more than 10.5MPa.

3. Emission Test Method

The emission test was carried out by using non-road diesel engine of China III emission requirement. The maximum permissible error of rotation speed is controlled within 50r/min, and the maximum error of torque is controlled within 2% by "8-Mode" test method. Run in each condition for 6 min, and complete the condition conversion within 1 min; Exhaust measurement including PM sample collection, 5 min in total. Test conditions and weighting coefficients are shown in table 2.

Table 2. Emission test condition cycle of OP2S diesel engine.

Test Condition Number	Speed	Load Percentage	Weighting Coefficient
1	Nominal Speed	100%	0.15
2	(1800r/min)	75%	0.15
3		50%	0.15
4		10%	0.1
5	Middle Speed	100%	0.1
6	(900r/min)	75%	0.1
7		50%	0.1
8	Idle Speed(700r/min)	0	0.15

"8-Mode method" uses the weighted ratio emission to evaluate the emission level of diesel engine. The weighted ratio emission of a certain harmful substance X is defined as:

$$X = \frac{\sum_{i=1}^{8} [X_{mass}]_i \cdot WF_i}{\sum_{i=1}^{8} Pe_i \cdot WF_i} \tag{1}$$

Where, X_{mass} is the weighted emission mass flow of the harmful substance; WF_i is the weighting coefficient; i is the operating condition serial number; Pe is the effective power.

4. Test Results and Analysis

4.1. Effect of Nozzle Diameter on Emission

The nozzle diameter determines the flow characteristics of nozzle when the number of nozzle orifice and back pressure of nozzle are constant. If the total flow area of the jet hole is too large, it will reduce the injection pressure and injection rate, shorten the injection duration, deteriorate the atomization quality of injection. A small flow area of the jet orifice will lead to high injection pressure and abnormal injection. The total flow area is directly determined by the nozzle diameter when the needle valve lift rule is fixed, which affects the fuel injection rule and ultimately affects the combustion characteristics and emissions of the engine. Figure 3 is the test result of the influence of nozzle diameter on the engine emission under eight working conditions.

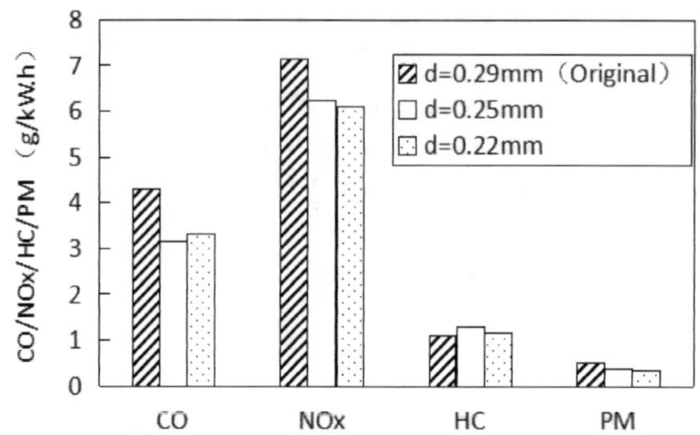

Figure 3. Influence of nozzle diameter on average specific discharge.

It can be seen from figure 3 that NOx, CO and PM emission are all decrease significantly with the decrease of nozzle diameter. This is because a smaller nozzle diameter will lead to a faster spray speed of fuel droplet, and the spray is evenly meticulous, and the same time the oil beam form around back-flow zone is strong, which can enhance the fuel vapor entrainment ability, and improve the utilization rate of air and fuel vapor heat utilization. In the end, the generation of CO and PM will reduce, but small diameter nozzle can reduce initial fuel injection quantity, and make the premixed combustion rate reduced. So the pressure rise rate and the maximum combustion temperature is reduced, thus effectively inhibits NOx generated; At the same time, the reduction of premix combustion ratio will increase the diffusion combustion ratio of fuel, and the inevitable collision between fuel injection and combustion chamber wall surface will intensify the cracking of small particle size fuel at higher wall temperature, thus causing HC emission to increase to a certain extent.

4.2. Effect of Angle of Nozzle on Emission

Nozzle angle is defined as the angle between the axis of nozzle orifice center and the center line of nozzle. It directly affects the position of the jet beam, and affects the subsequent combustion stability and emission characteristics. At the same time, the angle of nozzle orifice designed should be matched with the combustion chamber shapes.

To obtain a higher air utilization rate for OP2S diesel engine, the angle of nozzle orifice must be adjusted and optimized in order to matches the elliptic combustion chamber structure, the contrapuntal injector, the lateral combination injection cylinder liner. In addition, while the fuel beam penetrates through the external radial volume, it is necessary to prevent the fuel from condensing in the center of the combustion chamber and ensure that the fuel injection has less contact with the cylinder liner.

Figure 4 shows the test results of three different orifice angles(17°, 20°, 23°)on emission characteristics of OP2S diesel engine. As can be seen from the figure, the HC, CO and PM emission reached the lowest when the angle of the nozzle is 20°, while the NOx emission is the highest at this time.

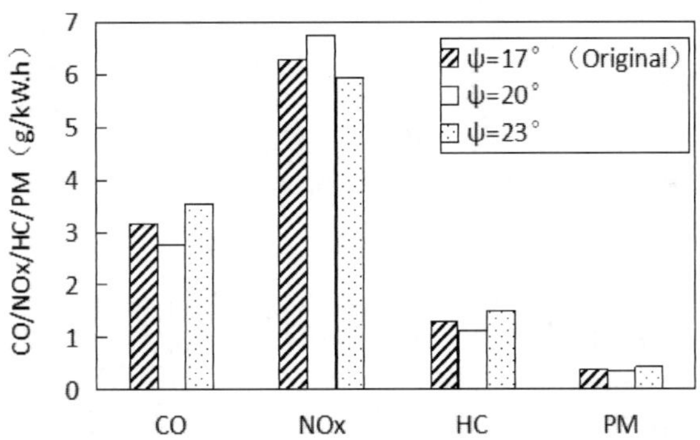

Figure 4. Influence of angle of nozzle on emission.

17°for nozzle angle, the jet oil mainly concentrates in the centre of the combustion chamber, oil beams barriers make it difficult to penetrate between external combustion chamber of transverse space. It is not conducive to the use of air to reach the best state of oil and gas mixing, which lead to an incomplete combustion, and HC, CO and PM emissions is increase. At the same time, the poor quality of oil and gas mixture made formed in the early years of the burning combustible mixture is less, so in the cross area of injection beam, the mixture of oxygen concentration is low, and reduce the burning zone NOx generation.

When the nozzle angle increases to 23°, the oil beam arrives at the top of the piston earlier in middle fuel injection period, where more and more fuel accumulates. Since the temperature of piston top surface increases, the rate of fuel evaporation increases. At this time, fuel vapor will decompose and form incomplete oxidation products of unburned hydrocarbon and carbon smoke, leading to HC emission increase and PM increase. In addition, the “air stagnation zone” is formed between the two umbrella shaped oil bundles, which result in reducing air utilization in the middle of the combustion chamber and a lower concentration of oil-gas mixture, and it will also lead to more HC, CO and PM generation and less NOx generation after combustion.

When the nozzle angle is 20°, the oil beam collide at the middle section of the combustion chamber, further crushing the oil droplets, promoting the air disturbance and enhancing the atomization of fuel, so that the combustion is more complete and the maximum combustion temperature in the cylinder increases, so HC, CO and PM emissions decrease and NOx emissions increase.

4.3. The Effect of NTP on Emission

Nozzle angle and NTP directly affect the spatial distribution of fuel in the cylinder, and thus the concentration distribution of the mixture, combustion process and emission generation. As the air flow movement in the cylinder of OP2S diesel engine is very complicated, coupled with the reciprocating motion of the piston, the position of the injector nozzle in the cylinder will affect the distribution of the spray body in the combustion chamber to a certain extent, resulting in different combustion and emission characteristics.

By adjusting the number of gaskets, the protruding height of NTP is changed. Figure 5 shows the overall emission test results corresponding to different NTP.

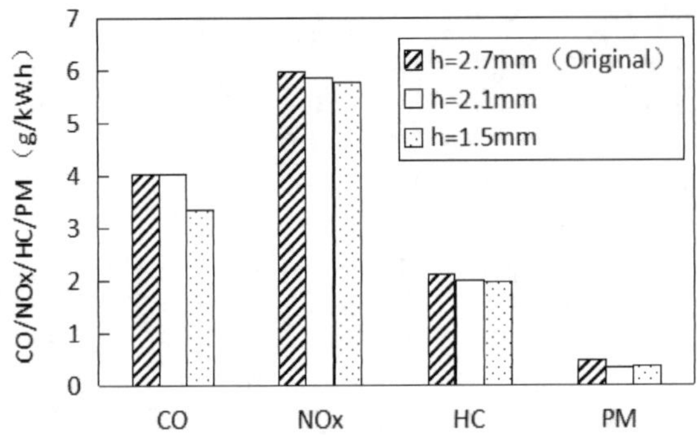

Figure 5. Influence of NTP on average specific emission.

As can be seen from figure 5, the smaller the NTP is, the less the CO emission, but the NOx, HC and PM emissions decrease unobviously in “8-Mode” test results. When NTP is 1.5mm, the emission characteristics of OP2S engine are the best.

5. Comprehensive Optimization of Nozzle Parameters

Because the emission standard of China III emission requirement for non-road mobile mechanical is the limit of NOx and HC emissions, so in order to make the OP2S diesel engine meet the regulatory standards, the above nozzle parameters need to be optimized. The final matched test scheme is: nozzle diameter is 0.22mm, nozzle angle is 20°, NTP is 1.5mm.

The test was carried out again according to the GB20891-2014 steady state test cycle test method, and the measured “8-Mode” average specific emission results are shown in table 3.

Table 3. Exhaust results before and after optimization of nozzle parameters.

Emissions	Phase III emission standards (g/kW.h)	Original machine emission value (g/kW.h)	Optimized engine emission value (g/kW.h)
CO	5.5	4.40	3.49
NOx	—	7.28	5.27
HC	—	1.11	1.68
NOx+HC	7.5	8.39	6.95
PM	0.60	—	0.37

From table 3 as you can see, compared with the original engine, the harmful emissions of OP2S diesel engine was effectively controlled after the nozzle parameters were optimized. In “8-Mode” test results, although the HC emissions increased, but the total emissions of “NOx + HC” decreased by 17.2%, and the improved emission levels of the machine can satisfy China III emission requirements of nor-road mobile machanical.

6. Conclusion

- The comprehensive specific emission of “8-Mode steady-state test” of OP2S diesel engine is more sensitive to the influence of nozzle diameter and nozzle angle, but less affected by NTP. By reducing the nozzle diameter, average specific emission of NOx, CO and PM can be reduced.
- The nozzle angle should match with the elliptical combustion chamber structure. When the nozzle angle is 20°, comprehensive emission of OP2S diesel engine is lowest in “8-Mode test” average specific emission results.
- After the optimization of nozzle parameters, the total emission of “NOx+HC” of OP2S diesel engine is reduced by 17.2% compared with that before optimization.

7. References

[1] Emission limits and measurement methods of exhaust pollutants from diesel engines for non-road mobile machinery (China's III/IV).GB20891-2014

[2] Regner G and Herold R. 2011The Achates Power Opposed-Piston Two-Stroke Engine: Performance and Emissions Results in a Medium-Duty Application [J]. SAE Int J Engines 4(3):p 2726-2735

[3] Chen W T, Zhuge W L and Zhang Y J. Simulation on the scavenging process of the opposite two-stroke diesel engine[J]. Journal of Aerospace Power, 2010, 25(6)p 1322-1326

[4] Xu H J, Song J O and Yao C D. Simulation on In-Cylinder Flow on Mixture Formation and Combustion in OPOC Engine[J]. Transactions of CSICE, 2009(27), 5 p 395-400

[5] Zhang Z Y, Zhao C L and Zhang F J. Study on Optimizing Method of Scavenging System of an Oppose-Piston Two-Stroke Diesel Engine[J]. Chinese Internal Combustion Engine Engineering. 2015(36),5 p 90-95

[6] Zhang Z Y, Zhao C L and Zhang F J. Simulation of the Effect of Fuel Injection Rate-shape on Combustion Process of an Opposed-Piston Two-Stroke Diesel Engine[J]. Chinese Internal Combustion Engine Engineering. 2015(36),6: p 32-37

[7] Hao L J. Research on diesel electronic pump nozzle system, fuel injection and spray characteristics[D].Chang chun: Jilin Industry University,1996

IOP Conf. Series: Materials Science and Engineering **730** (2020) 012045 doi:10.1088/1757-899X/730/1/012045

Analysis of the Influence of Coagulant on Phosphorus in Landscape Water Environment

Chunhong Su

Beijing Vocational College of Agriculture, Beijing 102442, China.
Email: suchunh@sina.com

Abstract. This study selected three kinds of inorganic polymer coagulant: polyferric sulfate, polyaluminium chloride, polyaluminium ferric chloride, two inorganic low molecular coagulants: aluminum chloride and aluminum sulfate. The coagulation experiment of lake water without "water bloom" was carried out. The results were analyzed by multivariate linear regression and principal component regression analysis. It showed that the effects of the five coagulants on the phosphorus were the same when the landscape water was treated with different dosages of coagulants. The sequence was: total nitrogen (TN) > dosage > ammonia nitrogen > CODcr > turbidity > PH. Among them: the highest effect of total nitrogen on phosphorus was aluminum chloride (71.73%). The most obvious effect of dosage on phosphorus was polyaluminium chloride (34.68%). The most significant effect of ammonia nitrogen on phosphorus was polyaluminum chloride (16.98%).

1. Introduction

The shortage of water resources is a major problem in China. Recycled water is widely used in landscape water environment, which solves the need of water shortage for urban water environment. It also was the best way to restore the natural restoration of water ecological cycle [1]. Phosphorus is a restrictive nutrient for algae growth in landscape water environment, playing an important role in the outbreak of "water bloom" and the growth process of phytoplankton, and is one of the main reasons for eutrophication in landscape water environment [2].

Controlling the concentration of phosphorus in landscape water environment, breaking the ratio of nitrogen and phosphorus between 10 and 20, preventing water bloom production, coagulant is the most direct, rapid and effective method. Taking the artificial lake of Beijing vocational college of agriculture and the lake of xiaoyue lake of yongding river -- the lake without water bloom as test water, the author analyzed the treatment effect of phosphorus in landscape water environment with different coagulants, studied the influence of various pollutants and different amounts of coagulant on phosphorus in water environment, and provided reference for landscape water environment management.

2. Materials and Methods

2.1. Test Materials

2.1.1. Test Water. Taking the artificial lake of Beijing Agricultural Vocational College and the lake of Yongding River without water blooms as test water.

2.1.2. Coagulant Selection. Based on the principle of wide application in the market, low cost and good effect, three industrial inorganic polymer coagulants are selected: polyferric sulfate, polyaluminum chloride, polyaluminum ferric chloride, and two inorganic low molecular coagulants: Aluminum chloride, aluminum sulfate. Coagulation experiments were carried out according to different dosages and different water quality.

2.1.3. Test Method. The experiment was carried out with a beaker of a programmable six-step mixing and coagulation experiment device. The water was stirred and separated into six of the 1000 ml beaker, open the mixer switch, lowered stirrer, and graduated pipette will make good coagulant added to upper part of the dosing tube instrument in four stages: the first phase of speed (G) 300 RPM for 1 minute, the second stage is 132.2 RPM, working for 1 minute, the third stage G value is 28.8 RPM, working for 10 minutes, and stage 4 G value is RPM for 20 minutes. We took supernatant through sampling mouth, suveied and evaluated the water quality indicators.

2.2. The Research Methods
The method of multivariate linear regression and principal component regression analysis in DPS data processing system is adopted [3]. Its basic principle is to deal with collinear relations of various influencing factors by the situation of high correlation (complex collinear) between independent variables. The effects of main test items and different coagulants on phosphorus in reclaimed water environment were studied.

3. Results and Discussions

3.1. Analysis of Phosphorus Removal Effect of Coagulant
Using polyferric sulfate, polyaluminium chloride and polyaluminium ferric chloride, aluminium sulphate, press 5, 10, 15, 20, 30, 40, 50, 60, 70, 80, 90 (mg/L) concentration of coagulant dosage, aluminium chloride respectively by 1, 2, 4, 8, 16, 32, 64 (mg/L) concentration of coagulant dosage, coagulation test results as shown in figure 1[4].

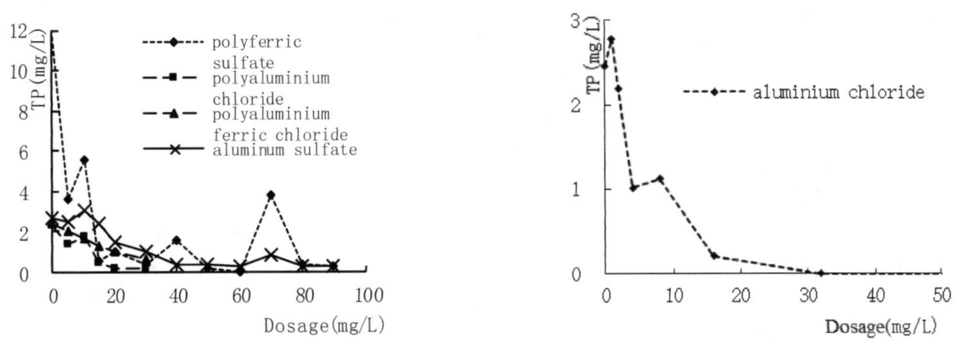

Figure 1. Phosphorus removal effect of different dosage

The above experimental results show that the amount of phosphorus in water gradually decreases with the increase of dosage of polyaluminium chloride and polyaluminium ferric chloride. Polymeric ferric sulfate, aluminum sulfate and aluminum chloride decrease the amount of phosphorus in water with the increase of dosage.

In order to further study the experimental results of the five reagents, the phosphorus removal percentage was analyzed, as shown in figure 2.

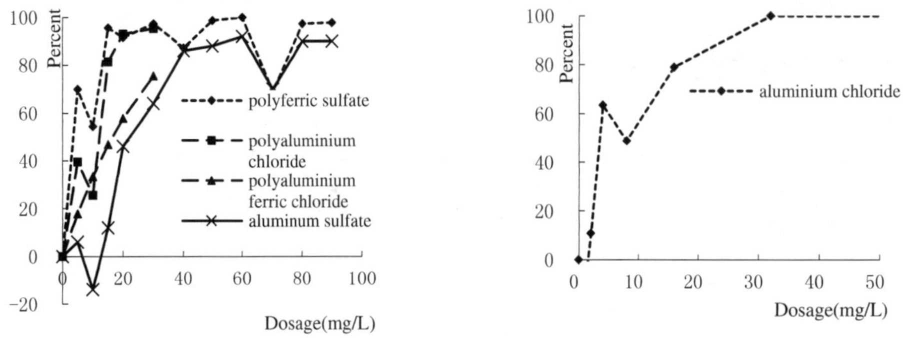

Figure 2. Percentage of total phosphorus affected by different dosage

The above figure shows that aluminum chloride has the best effect of removing phosphorus with the same dosage among the five reagents, and the removal rate of phosphorus in water reaches 100 % when the dosage is 32mg/L. The second best effect is polyferric sulfate. When the dosage is 15 mg/L, the removal rate of phosphorus is 95.58%, while when the dosage is 30 mg/L, the removal rate of phosphorus reaches 97.35%. Thirdly, polyaluminium chloride, when the dosage is 15 mg/L, the removal rate of phosphorus is 81.40%, while when the dosage is 30 mg/L, the removal rate of phosphorus is 95.35%. Moreover, the polyaluminum ferric chloride, when the dosage is 15 mg/L, the removal rate of phosphorus is 46.67%, while when the dosage is 30 mg/L, the removal rate of phosphorus is 75.56%. The worst effect was aluminum sulfate, the removal rate of phosphorus was 12.0 % when the dosage was 15 mg/L, and 64.00 % when the dosage was 30 mg/L. At the same time, with the addition of aluminum ferric chloride, the phosphorus removal effect of the other four coagulants fluctuated with the increase of dosage. When the dosage of aluminum sulfate was 10 mg/L, the phosphorus removal rate was -14.0 %.

3.2. Analysis of Factors Affecting Phosphorus Removal by Coagulant

Because of different coagulation reagents and their input amount, the effect of phosphorus treatment in water environment is different. So we used linear regression and principal component regression analysis with the results of three kinds of polymeric coagulants and two kinds of molecular coagulant test results, to seek different coagulants, the PH, ammonia nitrogen, CODCr, total nitrogen and the amount of pharmaceutical inputs with the change of phosphorus and analysis of its remarkable.

Five kind of coagulant with different coagulant dosage, PH, ammonia nitrogen, CODcr, total nitrogen and phosphorus by principal component regression linear regression analysis, the significance test is made to the regression model, the model factor regression coefficient significance test results: the regression model of F values were greater than $F_{0.05}$, correlation coefficient $R > R_{0.05}$, model fitting is good.

The six main influencing factors of drug dosage, PH, turbidity, ammonia nitrogen, CODcr and total nitrogen were calculated respectively. The characteristic values and percentages of the factors affecting total phosphorus in the five coagulants are shown in Table 1.

Table 1. Characteristic values and percentages of the effect of input factors on total phosphorus

Coagulant	Influencing factors	Characteristic values	Percent %	Cumulative percentage %
polyferric sulfate	TN(*mg/L*)	4.27	71.09	71.09
	Dosage(*mg/L*)	0.93	15.50	86.60
	Ammonia Nitrogen(*mg/L*)	0.56	9.35	95.95
	CODcr(*mg/L*)	0.16	2.74	98.70
	Turbidity(FNU)	0.06	0.95	99.65
	PH	0.02	0.35	100.00
polyaluminium chloride	TN(mg/L)	2.62	43.59	43.59
	Dosage(mg/L)	2.08	34.68	78.27
	Ammonia Nitrogen(mg/L)	1.02	16.98	95.25
	CODcr(mg/L)	0.21	3.55	98.79
	Turbidity(FNU)	0.06	1.08	99.87
	PH	0.01	0.13	100.00
polyaluminium ferric chloride	TN(mg/L)	3.96	66.03	66.03
	Dosage(mg/L)	1.75	29.22	95.25
	Ammonia Nitrogen(mg/L)	0.21	3.55	98.80
	CODcr(mg/L)	0.06	1.08	99.88
	Turbidity(FNU)	0.01	0.12	100.00
	PH	0.00	0.00	100.00
aluminum chloride	TN(mg/L)	4.30	71.73	71.73
	Dosage(mg/L)	0.92	15.37	87.10
	Ammonia Nitrogen(mg/L)	0.60	9.92	97.03
	CODcr(mg/L)	0.16	2.66	99.69
	Turbidity(FNU)	0.02	0.31	100.00
	PH	0.00	0.00	100.00
aluminum sulfate	TN(mg/L)	3.00	50.04	50.04
	Dosage(mg/L)	1.21	20.09	70.13
	Ammonia Nitrogen(mg/L)	0.90	14.96	85.09
	CODcr(mg/L)	0.52	8.62	93.72
	Turbidity(FNU)	0.31	5.17	98.89
	PH	0.07	1.11	100.00

According to the above table, the order of influence factors on total phosphorus in each coagulant is the same, and the order is: total nitrogen (TN) > dosage > ammonia nitrogen > CODcr > turbidity > PH.

The above table shows that the total nitrogen (TN) affects phosphorus in the range of 71.73 ~ 43.59 (%), the dosage of 34.68 ~ 14.96 (%), ammonia nitrogen accounted for 3.55 ~ 16.98 (%), CODcr accounted for 1.08 ~ 8.62 (%), turbid The degree is 0.12 to 5.17 (%), and the PH value is 0.00 to 1.11 (%). It can be seen that the influence value of each impact factor on phosphorus is different under the different percentage. Among them, aluminum chloride has the highest effect on phosphorus, accounting for 71.73%, followed by polyferric sulfate, accounting for 71.09%. The weakest effect is polyaluminium chloride, accounting for 43.59%.

The most obvious effect of the dosage on phosphorus is Polyaluminium chloride, accounting for 34.68%, followed by polyaluminum ferric chloride, accounting for 29.22%, the most important is aluminum sulfate, accounting for 14.96%; the most significant effect of ammonia nitrogen on phosphorus is polyaluminium chloride, accounting for 16.98%, after that is a polyferric sulfate, accounting for 14.96%. The weakest effect is polyaluminium ferric chloride, accounting for 3.55%.
The results showed that the turbidity and pH had the weakest effect on phosphorus. In order to serve the production better.The turbidity and pH of the coagulant were less than 10% were eliminated, and the principal component regression was combined. Linear regression simplified analysis, significant regression test of regression coefficient of each model factor: F in regression model was greater than

$F_{0.05}$, significant level P <0.0001; correlation coefficient R>$R_{0.05}$, model fit was still good. According to the calculation results, the order of influence on phosphorus in each coagulant is the same, and the order is still: total nitrogen (TN)> dosage> ammonia nitrogen> CODcr.

4. Conclusion

It can be seen that the removal effect of these five coagulants on phosphorus is: aluminum chloride > polyferric sulfate > polyferric chloride > polyferric aluminum chloride > aluminum sulfate.

The amount of phosphorus in water delinces with the increase of dosage of polyaluminum chloride and ferric chloride. The amount of phosphorus in water decreases with the increase of dosage of polyferric sulfate, aluminum sulfate and aluminum chloride, but when the dosage of aluminum sulfate is 10mg/L, the amount of phosphorus in water decreases. Compared with raw water, the amount of phosphorus increased by 14.0%.

After five kinds of coagulant use different dosage to treat the water quality of landscape environment, The main pollution indexes in the water environment have the same order of total phosphorus, and the sequence is: Total nitrogen (TN) and dosage of ammonia nitrogen >codcr> turbidity >PH.

The highest impact of total nitrogen on phosphorus was aluminum chloride, accounting for 71.73 %, followed by polyferric sulfate, accounting for 71.09 %, and the weakest impact was polyaluminium chloride, accounting for 43.59 %.

The most obvious effect of the dosage on phosphorus is Polyaluminium chloride, accounting for 34.68%, followed by polyaluminum ferric chloride, accounting for 29.22%, the most important is aluminum sulfate, accounting for 14.96%; the most significant effect of ammonia nitrogen on phosphorus is polyaluminium chloride, accounting for 16.98%, followed by it is a polyferric sulfate, accounting for 14.96%. The weakest effect is polyaluminium ferric chloride, accounting for 3.55%.

5. References

[1] LI Xin, HU Hongying, YANG Jia, BAI Yu. Method to establish the nitrogen and phosphorus standard of reclaimed water reused in landscape. J. *Ecology and Environmental Sciences*2009, **18(6)**:2404-2408.

[2] ZHANG Bo, LI Yong-feng, JIANG Xia, WANG Shu-hang, HU Jia-chen. Influences of environmental governance project on spatial distribution of phosphorus in Lihu Lake. J. *China Environmental Science*, 2013, **33(07)**:1271-1279.

[3] Tang, Q.-Y. and Zhang, C.-X. *DPS Data Processing System*. M. Science Press. 2007.

[4] GAO Xiu-qing, SU Chun-hong, JIAO You-quan, WU Xiaosu. Effects of Five Kinds of Coagulants in Controlling Water Environment Deterioration .J. *China Water & Wastewater*, 2016, **07**:89-91.

Progress of Airborne Lidar of Ocean Chlorophyll Observations Including Algorithm and Instruments

Yongzhi Li[1], Kun Yu[1],Wenkai Zhang[1], Dan Li[1], Zheng Zhao[1], Xiao Jiang[1], Dingjian Jin[1], Boxuan Shang[1] and Houmao Wang[2,3,4]

[1] Aviation Geophysical Remote Sensing Center of Land and Resources, China Geological Survey. NO. 31, Xueyuan road, Haidian district, Beijing, China.
[2] National Space Science Center, Chinese Academy of Science. NO.1, Nanertiao, Zhongguancun, Haidian district, Beijing, China
[3] Beijing Key Laboratory of Space-based Space Environment Exploration. Beijing, China.
[4] Key Laboratory of Science and Technology on Environmental Space Situation Awareness, CAS. Beijing, China.
Email: hmwang@nssc.ac.cn

Abstract. Airborne lidar is widely used to measure water depth, organic matter, chlorophyll, seabed topography and seabed types et al. The water depth observation has been studied for many years by many scientists, while the Chlorophyll retrieval by airborne lidar is studied by a few scientists. In this paper, the progress of measurements of Chlorophyll by use of airborne lidar is discussed, which includes the progress of retrieval algorithm, international instruments, and Chinese instruments.

1. Introduction

Chlorophyll is the main element for photosynthesis of marine phytoplankton. According to its content, the concentration of marine phytoplankton can be retrieved directly. Chlorophyll-a is the ultimate conversion of light energy of other chlorophyll, so it is the most important parameter which represents the spatial distribution of marine phytoplankton.

For the vast area and rapid change of the ocean, it will take a lot of manpower, material, and financial resources to rely on field measurements alone. Besides, it maybe need a long period. However, airborne or space-borne optical remote sensing can solve this problem, and it can obtain chlorophyll concentration of a regional/global area in a short time.

Optical remote sensing of Chlorophyll-a mainly includes passive optical observation and active optical observation, while passive observation mainly includes ocean color remote sensing [1-2] and fluorescence spectrophotometry [3]. They play an important role in marine chlorophyll remote sensing, but only carried out in the daytime.

As an active optical detection method, lidar can measure Chlorophyll-a in day and night time. Besides, it can penetrate thin cloud and aerosol, which can obtain much more data [4]. Lidar is an efficient and practical ocean optical observation technology [5-6], which can simultaneously measure optical parameters through several bands such as Raman scattering and chlorophyll fluorescence signals of ocean water. Based on the optical parameters, the chlorophyll concentration can be retrieved exactly.

2. Progress of Chlorophyll-A Retrieval Algorithm

Please follow these instructions as carefully as possible so all articles within a conference have the same style to the title page. This paragraph follows a section title so it should not be indented.

2.1. Progress in Abroad

The fluorescence radiation of chlorophyll in seawater will be generated by laser irradiation. There is a linear correlation between the intensity of the fluorescence radiation and the concentration of chlorophyll. So the lidar fluorescence equation of chlorophyll is as follows [7]:

$$\mathrm{d}P_F = \frac{K_F n(z)\sigma \exp(-\int_0^z(\gamma_L(z)+\gamma_F(z)))}{(z+mR)^2}dz \tag{1}$$

Where R is lidar altitude, σ is the absorption cross section, γ_L is the effective attenuation coefficient of laser seawater, m is the refractive index of seawater, n(z) is chlorophyll concentration, γ_F is the effective attenuation coefficient of fluorescent seawater. The constant parameter of K_F includes instrument parameters, laser intensity P0, surface albedo, and atmospheric attenuation coefficient etc.

$$K_F = \frac{P_0\xi A_R\Delta\lambda_D(1-\rho_L)(1-\rho_\mu)\exp(-(\beta_L+\beta_\mu)R)}{4\pi\Delta\lambda_F} \tag{2}$$

In equation (2), ξ if the signal receiving efficiency of the lidar instrument, A_R is the receiving area of the telescope, $\Delta\lambda_D$ represents the receiving wave width, ρ_L is the sea surface reflectivity of the laser, ρu is the sea surface reflectivity of the fluorescent, β_L is the extinction coefficient of the laser in the atmosphere, the extinction coefficient of the fluorescence in the atmosphere, and $\Delta\lambda_D$ represents the width of fluorcsccncc light. Then, equation (1) can be integrated:

$$P_F = \frac{K_F n(z)\sigma}{m^2R^2(\gamma_L+\gamma_F)} \tag{3}$$

According to equation (3), the intensity of chlorophyll fluorescence signal increases linearly with the increase of chlorophyll concentration and decreases linearly with the decrease of chlorophyll concentration. However, due to the disturbance of Raman scattering generated by water and instability of the instrument etc, the linear relationship will be changed, which will lead to the decrease of the inversion accuracy directly. Therefore, it is necessary to measure the Raman scattering signal of water by use of another band to eliminate its influence. In 1975, Fadeev et al [8] introduced firstly Raman scattering of seawater to chlorophyll fluorescence signal equation, and corrected the fluorescence signal. Then, Bristow [9-10], Poole [11] etc corrected the signal by Raman scattering respectively. The lidar Raman scattering equation is as follows:

$$P_R = \frac{K_R n_W \sigma_W}{m^2R^2(\gamma_L+\gamma_R)} \tag{4}$$

The parameters in equation (4) are similar to those in equation (3), but the independent variable should be changed from chlorophyll to water. Therefore, the ratio can be derived from equation (3) and (4):

$$\text{Ratio} = \frac{P_F}{P_R} = Cn_{Cl} \tag{5}$$

In equation 5, n_{Cl} is chlorophyll concentration, and C is correction coefficient. It shows that the chlorophyll concentration can be retrieved by use of Raman scattering assuming that the system and environmental parameters remain constant during the measurements.

On the basis of Raman correction, Exton[12] et al improved the inversion accuracy of chlorophyll concentration by avoiding from the influence of atmospheric Mie scattering. The Mie scattering equation is as follows:

$$P_S = \frac{K_S n_S \sigma_S}{m^2R^2 2\gamma_L} \tag{6}$$

Based on the equation (3), (4), and (6), we can obtain the following equations:

$$\frac{P_S}{P_R} = \frac{K_S}{K_R}\frac{n_S\sigma_S}{n_W\sigma_W}\frac{\gamma_L+\gamma_R}{2\gamma_L} \tag{7a}$$

$$\frac{P_F}{P_R} = \frac{K_S}{K_F}\frac{n_F\sigma_F}{n_W\sigma_W}\frac{\gamma_L+\gamma_R}{\gamma_L+\gamma_F} \tag{7b}$$

The ratio in equation (6) can minimize the disturbance of other factors, and it is not affected by laser intensity and chlorophyll height. Therefore, it can be better applied to chlorophyll inversion in shallow sea area (near land).

2.2. Progress in China

In 2007, Ocean University of China[7] independently developed the first Airborne Ocean Lidar System in China. Based on the system, the experiments were carried out nearby Yantai and Weihai. A new method was proposed to extract peak signal by use of least square fitting method. The concentration of chlorophyll-a was obtained based on Raman correction method. Besides, the attenuation coefficient of sea water was obtained. The retrieval results are consistent with the measurements of the same period in previous years.

In 2014, Ocean University of China [13] extracted and corrected underwater scattering signals based on CALIPSO satellite-borne lidar data. The correlation coefficients were obtained using the regression analysis between 532 nm backscattering signal and chlorophyll concentration. Then, the chlorophyll concentration was deduced based on the coefficients. The retrieval results were compared and consistent with MODIS chlorophyll concentration, and the averaged deviation is 0.052.

3. Progress of Instruments

3.1. Progress in Abroad

In 1970, Hickman et al. [14] used nitrogen laser to observe chlorophyll in laboratory. It was found that the fluorescence band of chlorophyll-a was near 680 nm. It was also found that chlorophyll concentration < 6 ug/L could be accurately measured. Then, Friedman et al. [15] found that algae species can be effectively distinguished by the fluorescence characteristics of laser echoes. Meanwhile, chlorophyll concentration below 1 ug/L can be obtained from 500 m distance by 100 KW laser. In 1973, Hongsuk [16] et al used the lidar to detect chlorophyll concentration. The laser system has a pulse energy of 0.25 J (within 300 nanoseconds) and a frequency of 2 pulses per minute. The system consists of two main parts: one part includes a control board, a converter, a signal processing and recording unit. Another part includes a laser transmitter, a photoelectric tube, and telescopes etc. In October 1972, the experiments were carried out using the laser system in Lake Ontario and other water areas, and the spatial distributions of chlorophyll concentration were obtained.

In the late 1970s, NASA and NOAA developed jointly an AOL (Airborne Oceanographic Lidar) system for chlorophyll concentration measurements. The AOL laser system is a double frequency Nd:YAG laser with an output wavelength of 532.1 nm. In June 1980, five experiments were carried out subsequently using AOL to observe and retrieve chlorophyll concentration at sea surface (5 m thick), and the distribution of chlorophyll concentration were obtained in these areas [17]. After that, the AOL system has been upgraded for several times, and up to the AOL-III system until now [18]. Compared with previous versions, AOL-III system mainly improves in weight, volume, and power consumption. Besides, much more new technologies were also applied to components of AOL-III system, such as optical fiber, spectrometer-detector optical chain, miniaturization of photomultiplier tube, dual-wavelength identical excitation source, and a new optical receiver. Finally, the main improvement of the system is the use of a dual-wavelength laser excitation source: one is 532 nm, the other is 355 nm (Table 1).

Table 1. The receiving wavelength of AOL-III system

Excited wavelength	Receiving wavelength	
	Wavelength center	Wavelength width
355	355	--
355	402	7
355	450	50
532	532	7
532	560	12
532	590	17
532	650	20
532	685	25

In the early 1990s, the University of Germany Odenbao [19] measured hydrological parameters of the European coast using lidar, whose system used EMG 100 with a maximum energy of 10 mW and a laser pulse width of 12 ns at 308 nm. In addition, there was a 450 nm laser wavelength with a pulse width of 6 ns. The system can be also used with infrared/ultraviolet scanners to detect phytoplankton.

In 1995, Barbini [20] in Italy observed the distribution of chlorophyll-a, CDOM in Lake Xiehu of Venice, and phytoplankton in Antarctic waters based on a triple-frequency Nd:YAG pulse laser. The 355 nm laser is mainly used to observe CDOM and chlorophyll, whose pulse energy is 30 mJ and pulse length is 10 ns. While the 266 nm laser is mainly used to observe organic matter and oil spill, whose pulse energy is 10 mJ and pulse length is 2 ns (Table 2).

Table 2. The lidar for chlorophyll-a in Italy

Parameters	Value
Excited wavelength	355 nm
Pulse width	10 ns
Receiving wavelength	402\450\650\690
Pulse energy	30 MJ

In 2002, Korea [21] detected the concentration of chlorophyll-a and CDOM in the ocean surface based on a marine fluorescence lidar system. The system mainly consisted of a flip mirror, a telescope, a polychromatic instrument, a CCD detector, a laser cooling system, a laser light source, and a double and triple frequency Nd:YAG etc. The laser had 8 ns pulse width, 25 Hz repetition rate. The energy of the 1064 nm laser pulse is 600 mJ, that of the 532.1 nm laser pulse is 280 mJ, while 354 nm laser pulse has 100 mJ energy. The system can be installed on aircraft or ships to monitor sea water, oil spill, and land vegetation. Besides, due to the similar laser spectral characteristics, the system can be used not only for pollution monitoring in coastal waters, but also for monitoring global forests or urban parks.

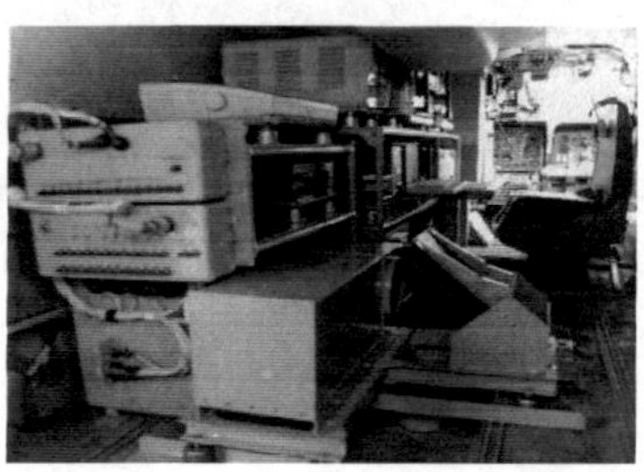

Figure 1. Radar System on Helicopter Developed in Korea

3.2. Progress in China

To date, the study of marine chlorophyll measurements based on Airborne Radar started in early 21st century, and just a few studies had been carried out. Ocean University of China [22-23] has developed the first airborne marine lidar in China, which was mainly used to measure the chlorophyll-a concentration of the ocean surface. It has been carried out for observation experiments in many areas. The excitation wavelength of the airborne marine fluorescent lidar is 355 nm, the single pulse energy is 80 mJ, the pulse width is 8-10 nm, and the telescope aperture is 200 mm. The receiving wavelengths are 404 nm and 685 nm. The 440 nm is mainly used to measure Raman scattering of sea water, while 685 nm is mainly used to obtain chlorophyll fluorescence signal.

Then, a new three-channel Ocean Lidar System TOLRSS (Three-channel Oceanographic Lidar Remote Sensing System) was built by the Ocean University of China and Ludong University [24]. The system used a double-frequency Nd:YAG pulse laser with a maximum single pulse energy of about 5 mJ and a pulse width of about 7 ns. It used 685 and 650 nm to measure the chlorophyll fluorescence signal and Raman scattering respectively, and then the chlorophyll concentration was retrieved using the signal of the two bands. Besides, the system also acquires elastic scattering signal using 532 nm, and retrieves suspended matter concentration combining with 650 nm Raman channel (Table 3).

Table 3. TOLRSS parameters of the Ocean University of China

Parameters	Value
Excited wavelength	532 nm
Pulse width	<10 ns
Receiving wavelength	532\650\685
Power	<400 W
One pulse energy	<5 MJ

4. Discussion

The research of marine chlorophyll detection by airborne lidar has been carried out for decades. Compared with the traditional field measurements, it can detect a wide area in a short time. Besides, it can observe continuously in day and night time. However, there are still some problems to be solved:

(1) It is greatly influenced by sea water floating matter, sea water pollution, and sea water transparency, especially for case-ii water area. The retrieval accuracy of chlorophyll will be reduced greatly due to more suspended substances and complex bottom topography and types of sea.

(2) It is greatly affected by waves. The size and shape of wave will greatly affect the echo signal of lidar, especially the peak information. To some extent, this will seriously interfere with the accurate extraction of chlorophyll echo signal, and then affect the retrieval accuracy of chlorophyll concentration.

Although there are still some problems to be solved urgently, the retrieval accuracy of chlorophyll concentration has been improved gradually with the continuous innovation of engineering technology and inversion algorithm. So the lidar has been used more frequently for remote sensing of marine chlorophyll. Besides, the active technology of lidar can also combined with other detection methods (such as multi-spectral passive remote sensing) to obtain much more remote sensing data and observation parameters, which can realize the complementarity of various data, and to improve the accuracy of chlorophyll observation.

5. Acknowledgement

Foundation: Investigation and application of aerophysical geophysics remote sensing in Bohai coastal zone (DD20160150)

6. References

[1] Hui Gao, Hui Zhao, Chunyan Shen 2017 Progress in ocean color remote sensing of chinese marginal seas *International journal of Ecology* **6** 82-92

[2] Behrenfeld M J, Westberry T K and Boss E S 2008 Satellite-detected fluorescence reveals global physiology of ocean phytoplankton *Biogeo sciences Discussions* **5** 779-94

[3] Song K, Li L and Wang Z 2012 Retrieval of total suspended matter (TSM) and chlorophyll-a (Chl-a) concentration from remote-sensing data for drinking water resources *Environmental Monitoring & Assessment* **184** 1449

[4] Li Xiaolong, Chen Yonghua and Yu Fei 2017 Comparison of analysis of inversion models for water optical property parameters by ocean lidar *ACTA OPTICA SINICA* **37** 35-41

[5] He yan and Wu Dong 2004 Performance Evaluation of airborne ocean lidar for measuring chlorophyll -a, suspended matter and coastal water depth in the east China sea *PERIODICAL O F O CEAN UNIV ERSITY O F CHINA* **34** 649-54

[6] Luan xiaoning, Li Jingwen and Guo Jinjia 2014 Ocean lidar for fishery resources survey and ecological environment monitoring *ACTA LASE R BIOLOGY SINICA* **23** 534-41

[7] Wei Zhiqiang, Zhang Kailin and Wu Dong 2007 Algorithmic research for measuring chlorophyll-a concentration in the sea surface layer by using airborne ocean fluorescence lidar *PE RIODICAL OF OCEAN UNIVERSIT Y OF CHINA* **37** 157-62

[8] Fadeev V V, Rubin Kust S V and Petrosyan V S 1975 In "Complex investgation of ocean nature" *The 1st Conf.of Moscow State university on problems of world ocean* 213-20

[9] M Bristow, D Nielsen and D Bun 1979 Airborne laser fluoro-sensor of surface water chlorophyll-a *Environment Protection Agency Las Vegas* **EPA-600** 79-048

[10] Bristow M, Nielsen D and Bundy D 1981 Use of water Raman emission to correct airborne laser fluorescence data effects of water optical attenuation *Appl. Opt.* **20** 2889-906

[11] Poole L R and Esaias W E 1982 Water Raman normalization of airborne laser fluorosensor measurement s :a computer model study *Applied Optics* **21** 3756-761

[12] Exton R J, Hought on W M and Esaias W 1983 Laboratory analysis of techniques for remote sensing of estuarine parameters using laser excitation *Applied Optics* **22** 54-64

[13] Yan Huaibin and Wu Dong 2014 Retrieval of chlorophyll concentration in open sea by using CALIPSO lidar signal below the sea surface *Master Dissertation*

[14] Hickman G. Daniel and Moore Richard B 1970 Laser Induced Fluorescence in Rhodamine B and Algae. Sparcom paper presented at 13th Conference on Great Lakes Research (Buffalo, New York)

[15] Friedman E. J. and Hickman G. D. 1972 Laser Induced Fluorescence in Algae: A New Technique for Remote Detection **NASA CR-62090**

[16] Kim H H 1973 New algae mapping technique by the use of an airborne laser fluorosensor *Applied Optics* **12** 1454-9

[17] F E Hoge and R N Swift 1981 Application of the NASA airborne oceanographic lidar to the mapping of chlorophyll and other organic pignents *Appl. Opt.* 349-74

[18] C Wayne Wright and Frank E Hoge 2001 Next-generation NASA airborne oceanographic lidar system *Appl. Opt.* **40** 336-42

[19] Hengstermann T and Reuter R 1990 Lidar fluoro-sensing of mineral oil spills on the sea surface *Appl. Opt.* **29** 3218-27

[20] Barbini R, Colao F and Fantoni R 1999. Shipborne laser remote sensing of the Cenice lagoon *International Journal of Remote Sensing* **20** 2405-21

[21] Lee K J, Park Y and Bunkin A 2002 Helicopter based lidar system for monitoring the upper ocean and terrain surface *Appl Opt* **41** 401-6

[22] Zhang Kailin, He Yan and Li Zhigang 2006 Simulation and Experiment for Measuring Chlorophyll-a with Airborne Fluorescence Lidar *PE RIODICAL OF OCEAN UNIVERSIT Y OF CHINA* **36** 790-4

[23] Liu Z, Ma S and Wang X 2008 Field detection of chlorophyll-a concentration in the sea surface layer by an airborne oceanographic lidar *Journal of Ocean University of China* **7** 108-12

[24] Feng Zhang, Jinjia Guo and Zhigang Li 2018 Preliminary measurement of chlorophyll a in offshore surface based on shipborne lidar *Laser & Optoelectronics Progress*

IOP Conf. Series: Materials Science and Engineering **730** (2020) 012047 doi:10.1088/1757-899X/730/1/012047

Analysis of Climate and Water Resources Characteristics in Chongqing

Rong Chen, Jinglan Wang and Weihua Zhang*

College of Resources and Environment, Southwest University, Chongqing, China.
*Email: swuwater@126.com

Abstract: Due to the impact of climate change, the spatial and temporal distribution of water resources has also changed accordingly. We want to know the response of Chongqing's water resources to climate change. By using the methods of linear regression and cumulative anomaly, this paper analyses the meteorological data of Chongqing from 1951 to 2018 and the climatic and hydrological conditions in recent years. It is found that Chongqing is in a period of high temperature and rainy weather under the influence of global warming, and is very prone to drought and flood disasters. This study can show the climate and water resources changes in Chongqing in the future to some extent and help the local authority for designing and operating the hydrological sensitive facilities.

1. Introduction

Climate change will inevitably cause temporal and spatial changes in water resources, and then affect the ecological environment and socio-economic development. To study the response of hydrology and water resources to climate change is of great significance in the long-term development and utilization of water resources, alleviating the contradiction between supply and demand of water resources and promoting regional economic development [1]. In the past hundred years, the earth's climate is experiencing a significant change characterized by global warming. The IPCC Fourth Climate Change Assessment Report pointed out that in the past 100 years (1906-2005), the global average surface temperature has increased by 0.74°C. This global warming is caused by natural climate fluctuations and the enhanced greenhouse effect of human activities[2].

Under the background of global warming, climate change also has obvious regional differences [3]. Chongqing is located in the sensitive area of climate change, with complex topographic conditions, high incidence of local abnormal climate and frequent disasters. Especially in September 2004, the catastrophic rainstorm in Kai County, Chongqing, the high temperature and drought in summer 2004, and the once-in-a-century floods in some parts of Chongqing in July 2007, has already affected the region. Regional economy and people's lives have a serious impact. Previous studies have shown that Chongqing's climate change trend is unique compared with global and national climate change, and its warming lags behind. After 1990s, Chongqing showed a more significant warming trend [3].

Chongqing is located in the southwest of China, between 28 10'N to 32 13'N, 105 17'E to 110 11'E, with a total area of 824,000 km^2. Chongqing is situated in the southeastern part of Sichuan Basin in the upper reaches of the Yangtze River, surrounded by mountains and spanning the transition zone between the plain of the middle and lower reaches of the Yangtze River and the Qinghai-Tibet Plateau. The landform is complex and changeable. Chongqing is in low latitude and affected by the subtropical monsoon and topography. It belongs to the subtropical monsoon humid climate. There are many rivers in Chongqing, which belong to the Yangtze River system. Precipitation plays a decisive role in the seasonal variation of river runoff[4]. The scope of the study area and the meteorological stations are shown in figure 1.

Figure 1. Scope of study area and distribution of meteorological stations

2. Data and Methods

2.1. Data
The precipitation and temperature data selected in this paper are from the climate data provided by the China Meteorological Data Network from the China Surface International Exchange Station, Shapingba and Youyang Meteorological Station in Chongqing (As shown in figure 1). The time span is 68 years (1951-2018).

2.2. Methods
(1)The linear trend of precipitation and temperature is estimated according to time series, and a linear regression equation is established: y=bx+c. In the formula, y is the precipitation, b is the trend of precipitation, x is the time and c is the constant. When b > 0, the precipitation shows an increasing trend; when b < 0, the precipitation shows a decreasing trend.

(2)At the same time, draw the cumulative anomaly curve of precipitation and temperature series. The cumulative anomaly at a given time is expressed as: $X=\sum_{i=1}^{t}(x_i-\bar{x})$, t=1,2,3......n. The cumulative anomaly curve [4] can be drawn by calculating all the anomaly values of the series.

(3)Because there are some missing or missing data of precipitation and temperature in the two stations, linear interpolation of adjacent data is used to complete them.

3. Results and Analysis

3.1. Characteristics of Precipitation Variation
From 1951 to 2019, the annual precipitation in Chongqing fluctuated greatly. The annual average precipitation was 1224.8 mm and the annual standard deviation was 165.7 mm. Figure 2 shows that in the 68 years from 1951 to 2018, the annual precipitation was stable between 850 mm and 1600 mm, but the linear trend was not obvious because R^2 is too small(R^2 = 0.0026). The maximum value was 1539.3 mm in 2016 and the minimum value was 896.1 mm in 2011.

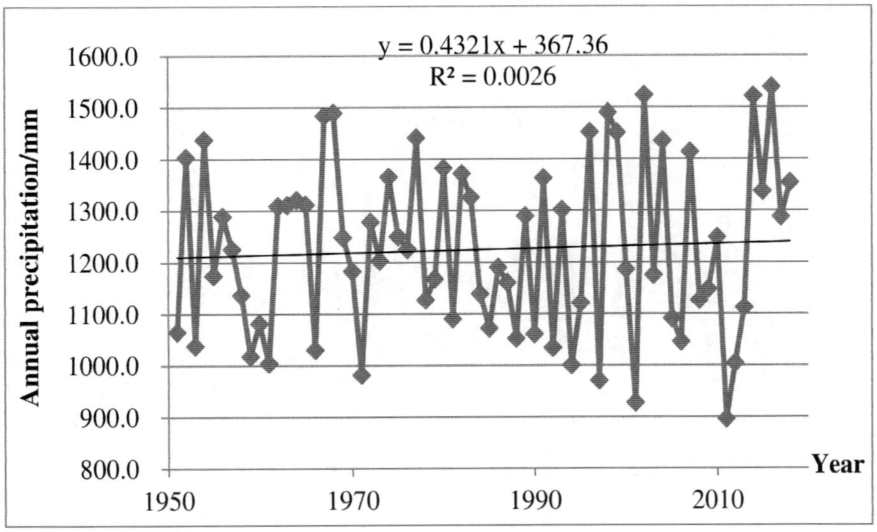

Figure 2. Annual precipitation change of Chongqing from 1951 to 2018

As shown in figure 3, from the annual precipitation cumulative anomaly curve, it can be found that there are several distinct high and low water years in these 68 years from 1951 to 2018. The low-water year group is from the mid-1950s to the early 1960s, the early 1980s to the late 1990s, and the beginning of the decade to the beginning of the decade; the high-water year group is from the early 1960s to the early 1980s, the middle and late 1990s to the beginning of the century and the beginning of the decade to the present.

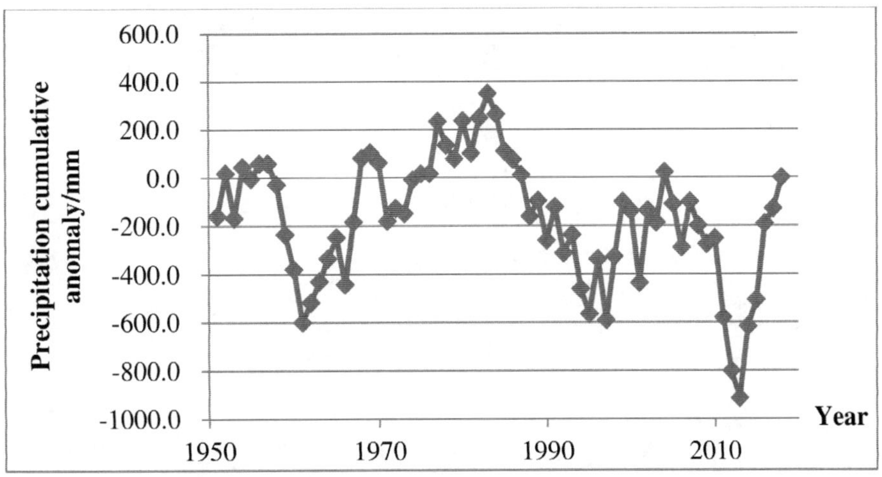

Figure 3. Precipitation cumulative anomaly curve

3.2. Characteristics of Temperature Variation

From figure 4, it can be seen that the annual average temperature in Chongqing has increased steadily and slightly in the past 68 years (the variation rate is 0.0093°C per a), the annual average temperature is 16.7°C, and the annual standard deviation is 0.4°C. The highest temperature in history appeared in 2006, which was 17.6°C. The lowest temperature of 16.1°C appeared in 1989, 1984, 1982, 1976 and 1954. The difference between the highest temperature and the lowest temperature was 1.5°C.

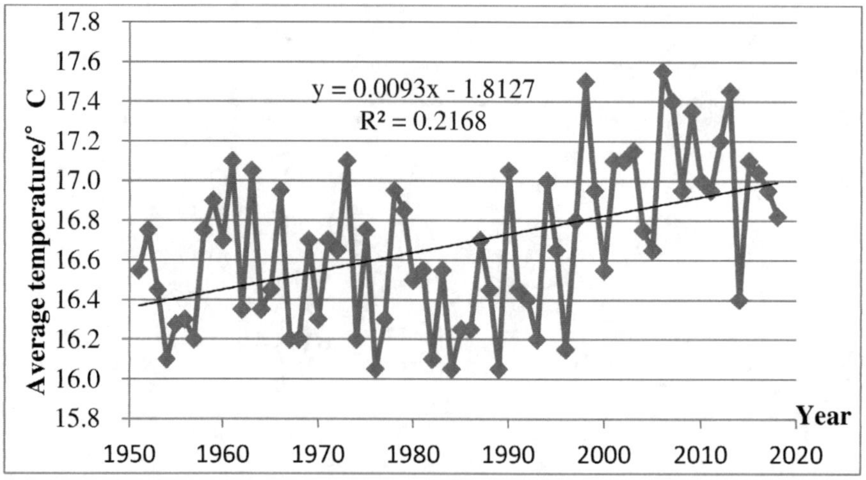

Figure 4. Annual average temperature change of Chongqing from 1951 to 2018

As can be seen from figure 5, the cumulative anomaly curve of average temperature presents a "V" shape on the whole, with only one inflection point in the mid-1950s. From the early 1950s to the mid-1950s, there was a very short period of low temperature, and from the mid-1950s to the early 1960s, it was also a very short period of high temperature. Then from the early 1960s to the mid-1990s, it was a long period of low temperature, and from the mid-1990s to now, it has been in a high temperature period, indicating that Chongqing is currently in a relatively warm period, the warming trend is very obvious, which also responds to the global warming situation in recent years.

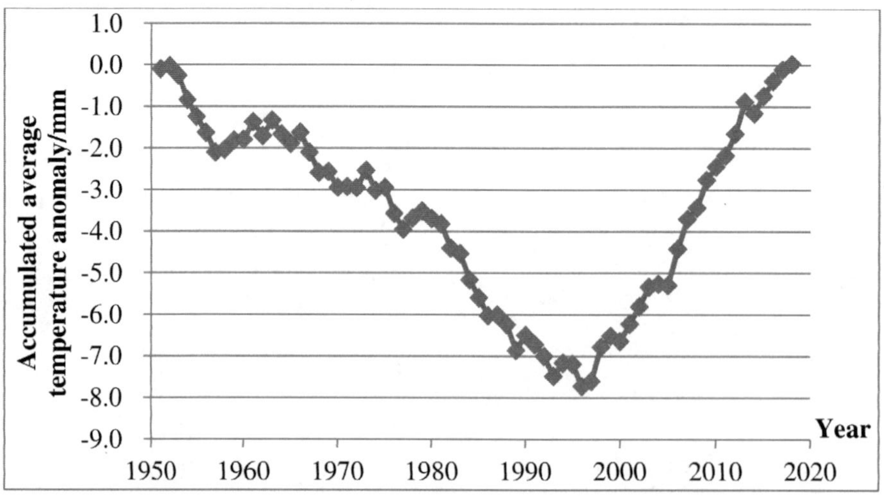

Figure 5. Accumulated anomaly curve of average temperature

3.3. Effects of Precipitation and Temperature on Drought and Flood Disasters

As can be seen from figure 6, although the annual precipitation in Chongqing is very large, the annual distribution is extremely uneven. The precipitation is mainly concentrated in May-September, and the high temperature weather also occurs in July-August. When the precipitation reaches its peak in June, the temperature reaches its peak in July, which well reflects the characteristics of the same period of rain and heat in Chongqing.

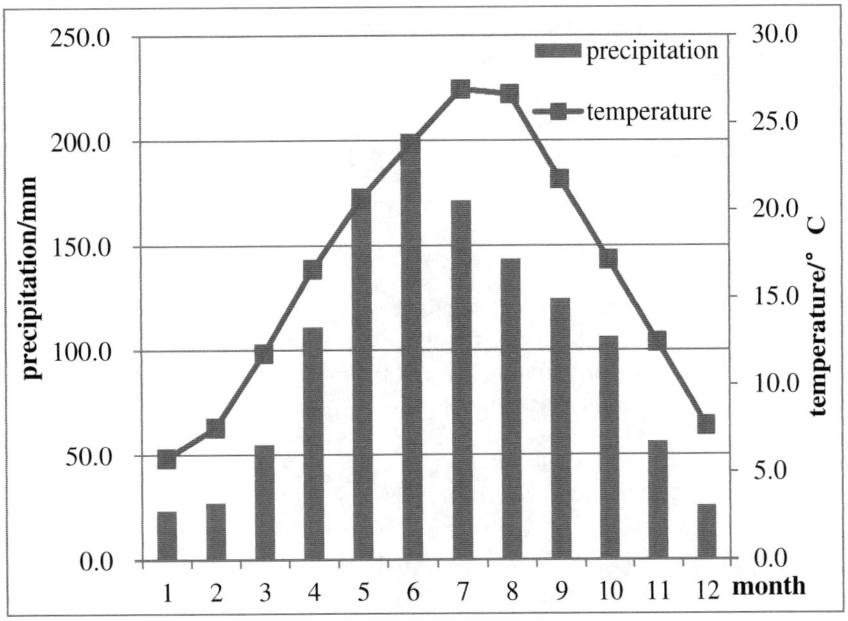

Figure 6. Monthly variation of precipitation and temperature in Chongqing from 1951 to 2018

Because of the concentration of precipitation, the instantaneous rainstorm is easy to cause disasters, so the rainstorm is the most important factor for the formation of urban floods in Chongqing. From the records of Shapingba Meteorological Station in 1951-2018, the extreme precipitation in the main urban area of Chongqing has changed remarkably in the last ten years from the end of the 20th century to the beginning of the 21st century. On July 21, 1996, the rainfall was 206.1 mm, which was the historical maximum since Shapingba Meteorological Station was built in the main urban area. Although the rainfall of 190.1 mm on June 14, 2002 failed to break the record of 1996, it was still destructive and caused great losses. The rainfall on July 17, 2007 was 266.6 mm, which was the highest record recorded by meteorological data in the main urban area of Chongqing in the past 115 years. It broke the record of one-day precipitation set in 1996, causing mountain torrents and urban waterlogging, resulting in serious economic losses and casualties[5, 6]. And these floods almost occurred in May to September when precipitation was concentrated.

Because the precipitation in the year is mainly concentrated in May-September, and the precipitation from November to March is very little, Chongqing will suffer a large area of drought in these months in a year. It also shows that winter drought and spring drought are more common in Chongqing. At the same time, Chongqing is in a high temperature period from June to August. Although the precipitation during this period is relatively rich, there will also be severe summer drought in some area. Due to the change of precipitation, there are floods in some areas every year, and the drought with the greatest impact is spring drought, summer drought and autumn drought, and the duration of drought and flood is long. In the past 40 years, drought is more serious than flood in Chongqing[7-10].

4. Conclusion

(1)The annual precipitation in Chongqing has fluctuated much in the past 68 years, but there are several distinct groups of high and low water years, which span about 10 years. In the past 10 years, Chongqing has been in a high water year group. I think in the next few years, Chongqing may enter a low water years period.

(2)The annual average temperature in Chongqing has a slight upward trend in the fluctuation of 68 years. It is a long low temperature period from 1960 to 1996 and a long high temperature period since 1996 to now, which also corresponds to global warming. And I think the temperature in Chongqing will keep going up in the future.

IOP Conf. Series: Materials Science and Engineering **730** (2020) 012047 doi:10.1088/1757-899X/730/1/012047

(3)Although the annual precipitation in Chongqing is very large, the annual distribution of precipitation is extremely uneven, and the precipitation is concentrated in May to September, which is very prone to flood disasters. While in November and March of the following year, winter drought and spring drought are prone to occur because of the low precipitation.

(4) In the future, the temperature in Chongqing is likely to continue to rise, and the precipitation is likely to continue to be low, which will lead to frequent droughts, so we should make preparations in advance.

5. Acknowledgement
This research is financially supported by “Southwest University Teaching Reform Program No. 2018JY092”, "Fundamental Research Funds for the Central Universities No. XDJK2019F002” and “Chongqing social livelihood key research and development projects, No.cstc2018jscx-mszdX0052”, and the supports were gratefully acknowledged.

6. References
[1] Wang L and Liu H L 2007 *Hydrology.* **(03)** 71-74+38.
[2] Bai Y, Gao Y H, Zhang Y, Li Y H and Wang Z 2010 *Meteorol. Month.* **36** 47-54.
[3] Duan X P, Wei X P and Liu X 2018 *J. Chongqing. Norm. Univ.* **35** 40-48+144.
[4] Liu B, Chen L Q, Zhou S and Zhang T Y 2018 *Resour. Environ. Yangtze. Basin.* **27** 1333-41.
[5] Tang Y X 2007 *Torr. Rain. Disas.* **(03)** 287-8.
[6] Zhang Z X, Xu G and Zhang Y H 2011 *J. Chongqing. Three-Gorges. Univ.* **27** 58-62.
[7] Huang Y H, Xu C, Yang H J, Wang J H, Jiang D and Zhao C P 2015 *Sustainability* **7** 13597-609.
[8] Zhao W, Zhang Y and Zhang Z H 2016 *Res. Soil. Water. Conserv.* **23** 192-8+203.
[9] Xu Y, Xu Y P, Wu Y F, Xu G and Wu L 2016 *Res. Soil. Water. Conserv.* **23** 363-8.
[10] Wang L and Ta Y E 2006 *J. Shihezi. Univ.* **(05)** 599-604.

IOP Conf. Series: Materials Science and Engineering **730** (2020) 012048 doi:10.1088/1757-899X/730/1/012048

Hydrochemical Types of Karst Groundwater in Tailai Basin

Jianguo Feng, Deshuai Ji, Zongjun Gao, Tongmin Lu and Minghao He

College of Earth Science and Engineering, Shandong University of Science and Technology, Qingdao, China.
*Email: fengjianguo20316@sohu.com

Abstract. Karst water resources is about 24% of the total amount of groundwater resources that can be exploited in China. Among the 15 groundwater sources in Tailai Basin, there are 12 karst water sources, accounting for 80%. With the survey and evaluation data of groundwater sources in Shandong Province, hydrochemical types analysis of karst water is carried out by Schukalev classification and The trilinear chart by Piper, and it is of great significance to development and utilization of karst water. The hydrochemical types of karst water is dominated by HCO_3•SO_4-Ca water. However, if the naming of hydrochemical types involve nitrate, excess 39% of the karst water hydrochemical types changes. In addition, historical data shows that the concentration of nitrate in karst water has gradually increased, so it is recommended to increase nitrate indicators in the classification of hydrochemical types.

The type of groundwater hydrochemical is the concentrated reflection of groundwater chemical composition and one of the important contents of the study of groundwater hydrogeochemical characteristics. In China, the distribution of karst water is extensive, which plays an important role in solving some urban, industrial and agricultural water problems. The chemical characteristics of karst water are mainly affected by aquifer lithology, groundwater burial conditions, human activities, etc. Many experts and scholars at home and abroad have carried out special research in related fields.

In the Tailai Basin described in this paper is bounded by the Taishan Mountains in the north, the Mengshan Mountain and the remaining veins in the south, the Lushan Mountain in the east, and the east bank of the Dongping Lake in the west. Its extreme geographical coordinates are: east longitude 116°20′52″—117°58′57″, north latitude 35°43′05″—36°34′15″, area 8173 km^2, administrative divisions belong to Tai'an City and Jinan City.

1. Hydrogeological Conditions of Karst Water

1.1. Carbonate Fractured Karst Water

The aquifer group consists of the Middle Ordovician and the Upper Cambrian Fengshan Formation limestone, dolomitic limestone and argillaceous limestone developed in the fractured karst, mostly distributed in the north of each fault block, and the exposed area Most of them are low hills. Until the mountain slope is buried or concealed under the Carboniferous, Permian and Mesozoic and Cenozoic, the limestone is thick and widely distributed, its fractured karst develops, connectivity is strong, groundwater hydraulics Close contact. However, due to the influence of structure, geomorphology and lithology conditions, the occurrence and water-richness of groundwater are very different. Generally, the water yield property in the bare area is poor, and the water yield of a single well is less than 500 m^3/d; the water yield property in the buried area is good, the water yield of a single well is 500-2000 m^3/d, the karst development is weakened with the increase of the burial depth, and the water yield property is relatively poor; in the Quaternary concealed area, the water yield property is strong, and the water yield of a single well is generally 1000-5000 m^3/d, which can often form a water-rich section or a strong water-rich zone with water supply significance. Under the mining conditions, the hydraulic connection between the fractured karst water in these areas and the Quaternary pore water in the upper part is very close, and the Quaternary pore water often becomes the main source of supply for the fractured karst water. The Quaternary sand gravel layer in the water source areas of Tai'an City, Jiuxian Township, Niuquan town, Fangxia town and other water sources directly contact with limestone to form a "skylight".

1.2. Carbonate Rock Clastic Rock Karst Fissure Water

The aquifer is composed of the limestone shale from the Lower Cambrian Shantou Formation to the Upper Cambrian Changqing Formation, and is mostly distributed in the north of the fault block and the northern edge of the basin. Due to the limestone and shale interbedded thin layer of limestone interbed, the groundwater recharge conditions are poor, the fractured karst is not developed, the water yield property is weak, and the water yield of a single well is generally less than 500 m^3/d. In the thick limestone distribution area of Shantou Formation and Zhangxia Formation, the fracture karst is relatively developed, and the water-rich water is relatively strong. The water yield of a single well is more than 1000 m^3/d.

2. Characteristics of Hydrochemical Types of Karst Water

2.1. Statistics of Water Quality Test Results

In September 2017, 90 samples of karst water were taken. The statistics of 10 indicators including pH, K^+, Na^+, Ca^{2+}, Mg^{2+}, Cl^-, SO_4^{2-}, HCO_3^-, NO_3^-, and total dissolved solids are shown in Table 1.

Table 1. Statistics of Water Quality Test Results

Number	Detection index	Min	Max	Mean	Std
1	pH	6.90	8.10	7.49	0.20
2	K^+	0.20	8.33	1.41	1.44
3	Na^+	3.67	253.85	27.59	31.61
4	Ca^{2+}	33.87	529.26	150.30	66.59
5	Mg^{2+}	1.46	78.99	25.44	11.18
6	Cl^-	8.51	274.76	63.96	49.06
7	SO_4^{2-}	18.25	1025.44	151.43	137.82
8	HCO_3^-	55.20	484.56	276.28	88.91
9	NO_3^-	<0.01	443.90	83.99	64.95
10	STD	185.00	2785.00	657.59	329.57

2.2. Hydrochemical Types

Groundwater has different classification methods according to its chemical composition, most of which are divided by the relative content and relationship between anions and anions, such as S Shukalev classification and The trilinear chart by Piper, etc. The hydrochemical type of each karst water sample was calculated by programming, and the statistics were shown in Figure 1.

IOP Conf. Series: Materials Science and Engineering **730** (2020) 012048 doi:10.1088/1757-899X/730/1/012048

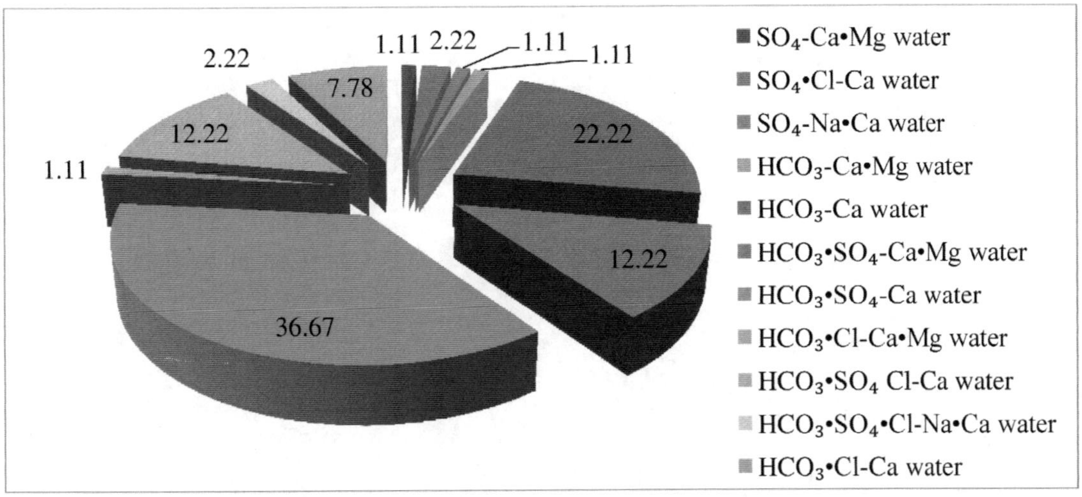

Figure 1. Pie chart of Karst Groundwater Hydrochemical Types

In the 90 water samples of karst waters, there are 11 types of groundwater hydrochemical, including 33 HCO_3•SO_4-Ca waters, accounting for 36.67% of the karst water samples. And there are 20 HCO_3-Ca water, accounting for 22.22% of the quantity of karst water samples. In terms of anions, there are 86 water samples with milligram equivalent percentage of bicarbonate over 25%, accounting for 95.56% of the total karst water samples. There are 62 water samples whose milligram equivalent percentage of sulfate is more than 25%, accounting for 68.89% of the total karst water samples. And there are 23 water samples with the milligram equivalent percentage of chloride over 25%, accounting for 25.56% of the total karst water samples. In terms of cations, there were 90 water samples with the milligram equivalent percentage of calcium over 25%, accounting for 100.00% of the total number of karst water samples. There are 14 water samples whose milligram equivalent percentage of magnesium is more than 25%, accounting for 15.56% of the total karst water samples. And there are 2 water samples whose sodium milligram equivalent percentage is more than 25%, accounting for 2.23% of the total karst water samples.

According to the statistical table of hydrochemical types of karst water, the proportion of the top three hydrochemical types of karst water in the Tailai Basin is more than 70%, and the proportion of HCO_3•SO_4-Ca water is the largest. Secondly, the percentage of HCO_3-Ca water in the quantity of karst water types is the second. The percentage of HCO_3•SO_4•Cl-Ca water in the quantity of karst water types ranks the third (Table 2).

According to the statistical analysis of the proportion of the hydrochemical types of different ions in karst water, bicarbonate and sulfate are the main anions, and the proportion of the two is relatively close; in terms of cations, calcium dominates.

Table 2. Statistics of Karst Groundwater Hydrochemical Types for different ions

Item	Anions			Cations		
	HCO_3^-	SO_4^{2-}	Cl^-	Ca^{2+}	Mg^{2+}	Na^+
Quantity	86	62	23	90	14	2
Percentage	95.56	68.89	25.56	100	15.56	2.23

The trilinear chart by Piper visually shows the hydrochemical characteristics of karst water in the Tailai Basin (Figure 2). If the water sample point falls in the diamond zone 1, the content of alkaline earth metal ions Ca^{2+} and Mg^{2+} exceeds the alkali metal ion K^+ and Na^+ content. Water sample points appeared in both the zone 3 and the zone 4, and most of them were in the zone 3, indicating that most of the water samples had weak acid roots HCO_3^- exceeding the strong acid roots SO_4^{2-} and Cl^-.

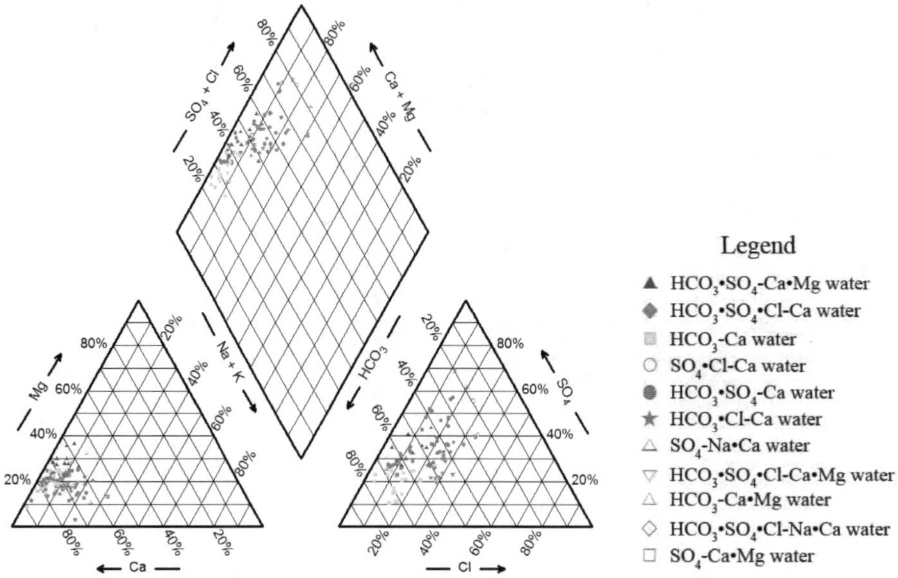

Figure 2. The trilinear chart by Piper of Karst Groundwater

3. Analysis of Hydrochemical Types of Karst Water

During data analysis, it was noted that the NO_3^-concentration of water sample no. J026 was 443.90 mg/L, and the average value of NO_3^-concentration of all karst water samples was 83.99 mg/L, and the high concentration of NO_3^-concentration may be related to the transitional application of nitrogen fertilizer. Therefore, NO_3^- is considered to be introduced into the hydrochemical type of karst water.

After adding NO_3^- in the anion to participate in the naming of hydrochemical types, the milligram equivalent percentage of bicarbonate, sulfate and chloride all decreased to varying degrees, and the naming of a total of 28 water samples changed, covering 31.1% of the total water samples.

As a result of the participation of NO_3^- in naming, groundwater hydrochemical types change as follows:

(1) Some of the anions participating in the naming cannot participate in the naming because the milligram equivalent percentage is less than 25%. For example, in B10 water sample, the milligram equivalent percentage of bicarbonate was changed from 43.96% to 33.22%; the milligram equivalent percentage of sulfate was changed from 29.50% to 22.28%; the milligram equivalent percentage of chloride was changed from 26.53% to 20.08%; and the milligram equivalent percentage of nitrate was 24.42. %, the hydrochemical type is changed from HCO_3•SO_4•Cl-Ca water to HCO_3•SO_4-Ca water. The B24 water sample was changed from HCO_3•SO_4-Ca water to HCO_3-Ca water.

(2) Some of the anions participating in the naming are replaced by NO_3^-. For example, in the case of B9 water sample, the hydrochemical type without considering NO_3^- is HCO_3•SO_4-Ca•Mg water, the milligram equivalent percentage of bicarbonate is 44.83%, the milligram equivalent percentage of sulfate is 31.97%; After adding NO_3^-, the milligram equivalent percentage of bicarbonate is 25.45%, the milligram equivalent percentage of sulfate is 18.14%, the milligram equivalent percentage of nitrate is 43.23%, and the water chemical type is changed to HCO_3•NO_3-Ca•Mg water. The B42 water sample was changed from HCO_3•SO_4•Cl-Ca water to HCO_3•SO_4•NO_3-Ca water.

4. Conclusion

(1) The hydrochemical type of karst groundwater in Tailai basin is mainly HCO_3•SO_4-Ca water. The sum of HCO_3•SO_4-Ca water, HCO_3-Ca water and HCO_3•SO_4•Cl-Ca water accounts for 71.11% of the total karst water, indicating that the water-bearing system of karst water determines the hydrochemical characteristics of karst water to some extent.

(2) In karst water of Tailai basin, the anions are mainly bicarbonate and sulfate, and the proportions

of the two are relatively close; Cations are calcium dominated absolutely.

(3) After 2000, the concentration of anion components in karst water, such as bicarbonate, sulfate, chloride and nitrate, increased significantly. The proportion of nitrate in anion in karst water cannot be ignored, which has an important influence on the type of hydrochemical, and it is suggested that nitrate index should be added into the classification of hydrochemical in the future.

(4) If nitrate is used as the classification index of hydrochemical types, how to express hydrochemical types in the form of symbols and graphs and theoretically reflect the different characteristics of different types still needs further work.

5. Acknowledgement

Groundwater source survey and evaluation of Shandong Province (SDZS-2016-GTT02).

6. References

[1] Dong W H, Su X S, Hou G C, Lin X Y and Liu F T 2007 Distribution Law of Groundwater Hydrochemical Type in the Ordos Cretaceous Artesian Basin *Journal of Jilin University(Earth Science Edition)* **37** 288-292.

[2] Feng J G 2003 Hydrochemical Type Classification-A Case Study of Cretaceous Huanhe Group Groundwater of Ordos Basin *Chang'an University*.

[3] Zhou X and Ye Y H 2014 Improvement and Application of Schukalev Groundwater Hydrochemical Classification Method: taking Groundwater in Jinjiang City, Fujian Province as an Example *Resources Survey and Environment* **35** 299-304.

[4] Liang Y P and Wang W T 2010 The Division and Characteristics of Karst Water Systems in Northern China *Acta Geoscientica Sinica* **31** 860-868.

[5] Liang X, Zhang R Q, Niu H, Jin M G and Sun R L 2012 Development of the Theory and Research Method of Groundwater Flow System *Geological Science and Technology Information* **31** 143-151.

[6] Zhang Y, Dong W H, Li M Z, Zheng Z X, Shi X F and Guo Z X 2011 Study on Distribution Characteristics of TDS and Hydrochemical Type of Shallow Groundwater in Henan Plain *JOURNAL OF CHINA HYDROLOGY* **31** 79-83,68.

[7] Ye Y H, Zhou X and Liu L 2015 Evolution and genesis of groundwater hydrochemical types in Yingtan of Jiangxi *Journal of Guilin University of Technology* **35** 269-273.

[8] Claudio E M, Matthias R, Mauricio T and Malcolm E C 2015 Hydrochemical evolution and groundwater flow processes in the Galilee and Eromanga basins, Great Artesian Basin, Australia: A multivariate statistical approach *Science of The Total Environment* **508** 411-426.

[9] Yue D D and Su X S 2016 Distribution Law of TDS and Groundwater Hydrochemical Type in the Downstream Plain of Yishu River *water saving irrigation* **41** 77-80.

[10] Yang Q C, Wang L C, Ma H Y, Yu K and Martín J D 2016 Hydrochemical characterization and pollution sources identification of groundwater in Salawusu aquifer system of Ordos Basin, China *Environmental Pollution* **216** 340-349.

[11] Song C, Wang P, Han G L and Shi Y C 2017 The hydro-chemical characteristics of shallow groundwater in loess tableland and its implication to carbon cycle *South-to North Water Transfers and Water Science & Technology* **15** 121-126

[12] Wang H F, Wang H Y, Li C Y and Mu Z 2017 Distribution of hydrochemical type and genesis of pollutant of groundwater in Handan city *Journal of Hebei University of Engineering (Natural Science Edition)* **34** 53-56,65.

[13] Cansu Y and Sakir S 2017 Hydrogeological and hydrochemical studies of the Kaman-Savcili-Büyükoba (Kirsehir) geothermal area, Turkey *Geothermics* **65** 99-112.

[14] Chafouq D, Mandour A E, Elgettafi M, Himi M, Chouikri I and Casas A 2018 Hydrochemical and isotopic characterization of groundwater in the Ghis-Nekor plain (northern Morocco) *Journal of African Earth Sciences* **139** 1-13.

[15] Liu R F 2006 The Feature of Water Environment Evolvement of the Dawen River Basin and the Appraisement of Its Influence to the East Route of the South to North Water Transfer Project *Qingdao: Shandong University of Science and Teehnology*,2006.

[16] Li B 2007 Study on the Groundwater Environment Simulaition of City's Wellfield and the Countermeasure of Feeding Water Supplement *Tai'An: Shandong Agricultural University*.

[17] Zhang Z X, Zhu Z H, Liu C E and Luo F 2014 Predication on Water Quality Evolution of Typical Karst Groundwater Source Area in Shandong Province *Land and Resources in Shandong Province* **30** 29-33,38.

[18] Wei X Y, Zhang B X, Li W L, Liu D M and Zhang J S 2015 Research on the Numerical Simulation of Karst Groundwater in Feicheng Basin *China Rural Water and Hydropower* 59-64.

[19] Zhang R Q, Liang X, Jin M G, Wan L and Yu Q C 2011 Foundation of Hydrogeology *Beijing: Geological Publishing House* 66-69.

[20] Gao Z J, Zhang H Y, Sun M X, Xu H B and Yao D X 2017 The chemical characteristics and causes of shallow groundwater in the Rizhao coastal area of Shandong province are discussed *Ground water* **39** 1-4.

[21] Wang Z L, Gao Z J, Xu Y and Li L 2013 Hydrochemical Characteristics of Karst Water in Jinan Spring Region *Land and Resources in Shangdong Province* **29** 27-29.

[22] Gao Z J, Sun W G Tang M S, Wu X L, Liu M, Han Y L and Cheng C L 2001 Relation between Karstic Water Exploitation and Geological Environment in Tai'an -Jiuxian Water Source Area *Geology of Shandong* **17** 86-91.

[23] Liu L C, Chen H H, Ma Z M and Gao Z J 2003 Simulation and analysis of karst water environmental system in Taian city *Journal of Hydraulic Engineering* 107-111,116.

[24] Wang M and Gao Z J 2009 Study on Groundwater Environment in Dongwu Water Sources Area, Taian City Based on GMS *Journal of Shandong University of Science and Technology (Natural Science Edition)* **28** 20-24.

IOP Conf. Series: Materials Science and Engineering **730** (2020) 012049 doi:10.1088/1757-899X/730/1/012049

Construction of Air Pollution Monitoring and Pre-Warning Platform—Taking Daming County, Handan City, Hebei Province as an Example

Huan Qian[1], Jianwei Yue[2,5], Jingze Li[3] and Kesong Ding[4]

[1] Faculty of Geographical Science, Beijing Normal University, No.19, XinJieKouWai St., HaiDian District, Beijing 100875, P. R. China.
[2] Faculty of Geographical Science, Beijing Normal University, No.19, XinJieKouWai St., HaiDian District, Beijing 100875, P. R. China.
[3] Faculty of Geographical Science, Beijing Normal University, No.19, XinJieKouWai St., HaiDian District, Beijing 100875, P. R. China.
[4] Faculty of Geographical Science, Beijing Normal University, No.19, XinJieKouWai St., HaiDian District, Beijing 100875, P. R. China.
[5,*] Email: 08108@bnu.edu.cn

Abstract. Based on GIS technology, computer hardware and software and network communication technology, an air pollution monitoring and pre-warning platform is constructed. It helps to monitoring air pollution situation in Daming county and analysis the data from monitoring sites about AQI, $PM_{2.5}$ and other five kinds of air quality index. By using this platform, people can intuitively find where is the main air pollution source and the air quality forecast around all county.

1. Introduction

With the continuous improvement of people's awareness of environmental protection, air pollution has become the focus of public attention. China's atmospheric environmental pollution problems are intensifying. Doing a good job in monitoring the atmospheric environment, timely grasping the changes in the quality of the atmospheric environment, and early detection of changes in the types of atmospheric pollutants and fluctuations in the amount of pollutants can help relevant department management personnel shall carry out prevention and control and governance work, prepare for responding to air pollution, and better protect our atmospheric environment. The pollution information publishing service based on the data network sharing mechanism and the research and development using Web information technology is gradually emerging [1-4].

Typical domestic systems include air quality release platforms, blue maps, and genuine gas networks.

However, the above system only provides real-time pollution information and simple reminder service. The pollution and environmental governance cannot be combined. The function of analyzing the pollution source traceability and atmospheric environmental treatment assessment for the management organization is not comprehensively considered.

From the complete air quality analysis chain of "Monitoring-Tracking-Governance", the development of the air pollution monitoring and pre-warning platform in Daming County, Handan City can provide dynamic simulation of governance assessment and target setting and simulation index changes to promote air pollution.

2. Overview of Platform

The platform is based on GIS technology, computer hardware and software and network communication technology. And it combines functions such as atmospheric environment monitoring, fine pollution mapping, air-time statistical analysis of air pollution, and air pollution source tracing and analysis. In platform, the air pollution monitoring data is the center.

3. Design of Platform Function

In order to fulfill the requirements of air environment management, combined with the characteristics of atmospheric environmental management, the platform has been designed including site monitoring, global mapping and query, spatial traceability and analysis, evaluation for governance effect and visual forecasting. These five subsystems of platform can achieve integrally logic for air pollution monitoring and governance.

3.1. Home Page

Based on the statistical analysis of the database-related site monitoring data, the overall situation of the atmospheric environment in Daming County is shown on home page.

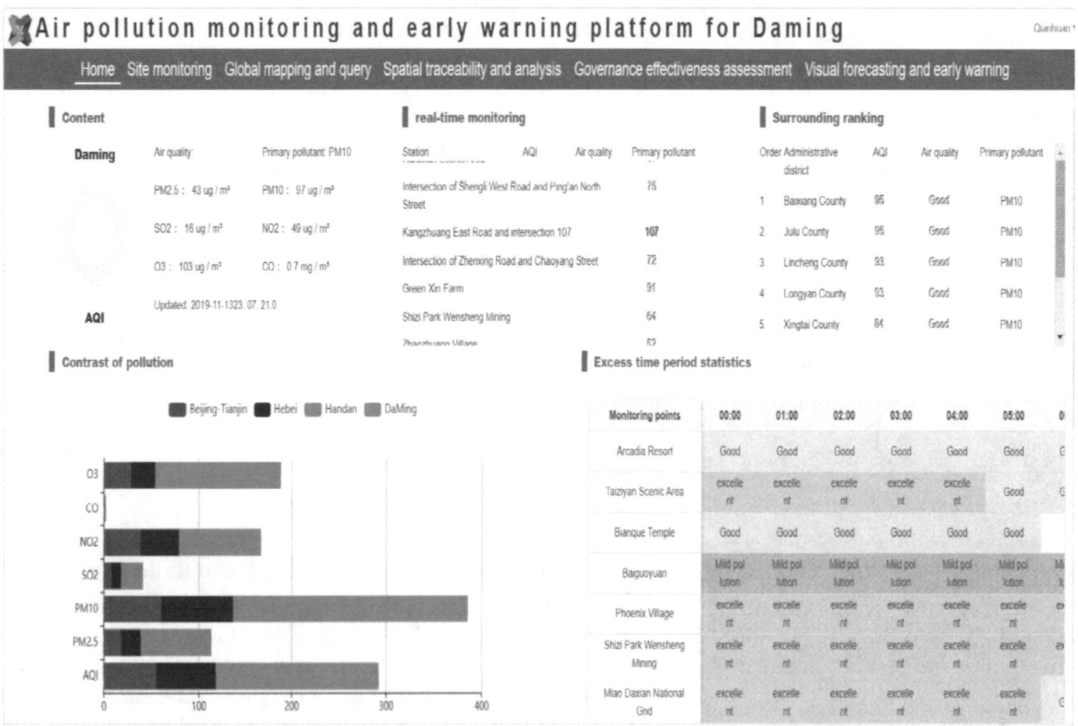

Figure 1. The home page of platform

3.2. Site Monitoring

Based on the sites monitoring data, the air environment of sites can be shown and analyzed on platform.

On platform, monitoring data of sites can be visualized and compared with around others by using AQI in time. And the air quality of each site is sorted according AQI and other six air pollutant concentration values. While the monitoring index is beyond warning line, platform can show it for people by using different color. In addition, platform links to all monitoring devices. People can open the monitoring video at any time.

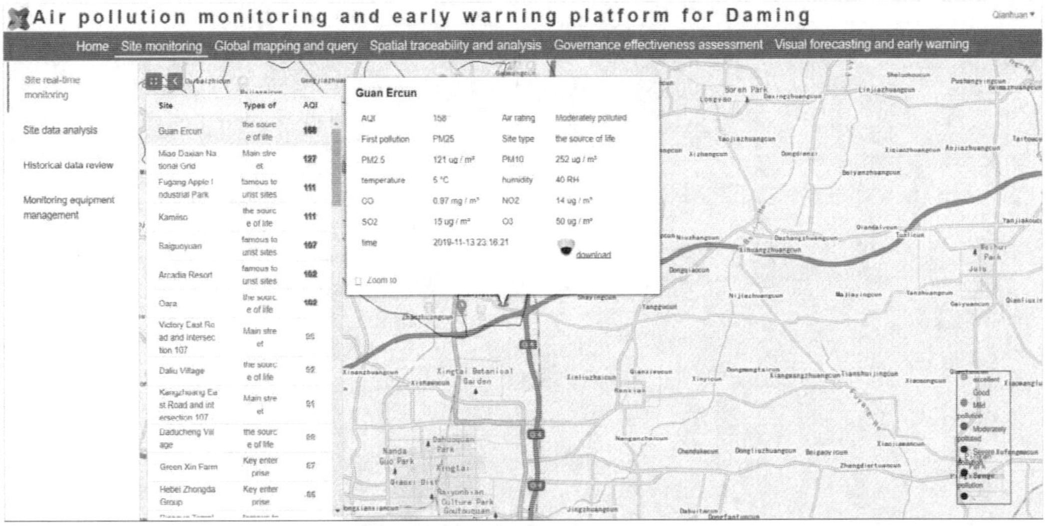

Figure 2. Query for monitoring sites

3.3. *Global Mapping and Query*

By applying regression model of land using and monitoring data, platform can simulate AQI and other index for whole county with thermal map displaying. Here we use the Kriging interpolation method. Kriging is a regression algorithm for spatial modeling and prediction (interpolation) of stochastic processes/destinations based on covariance functions. the Kriging method can give the Best Linear Unbiased Prediction (BLUP). And it can also calculate proportion of air quality and each main air pollutant in whole county. Platform can display these results by visual chart. The statistical results can be using for global mapping with different color for these air quality index.

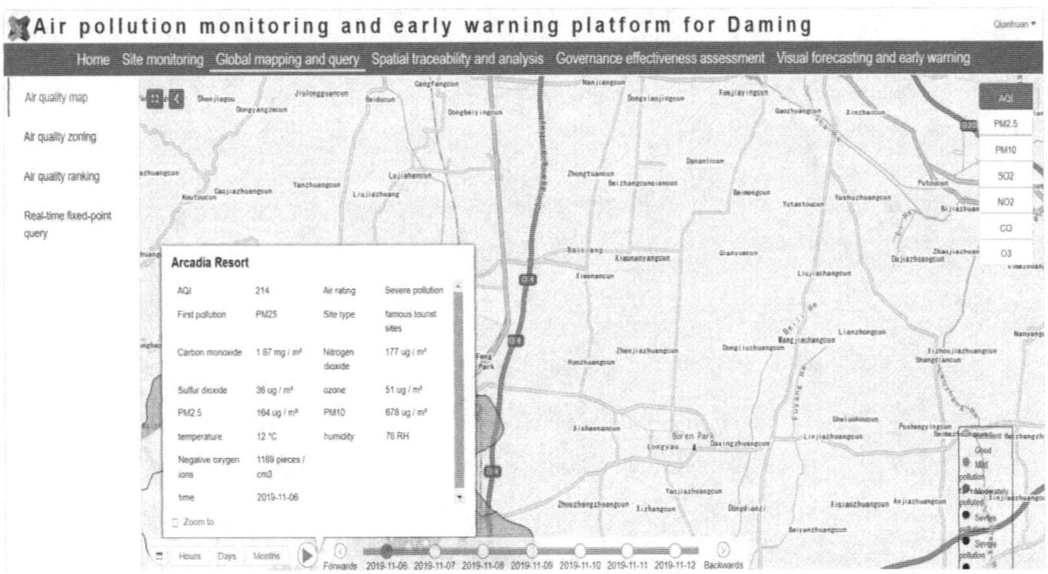

Figure 3. Air quality map of Daming County

3.4. *Spatial Traceability and Analysis*

Based on global mapping results, people can trace main air pollution resource such as enterprises, construction sites and others. Platform can statistic air pollution situation of these area and provide pre-warning. Platform displays results by charts and sorts. Besides, platform can also monitor live area and main streets for people. In this way, governance of air pollution can be more targeted.

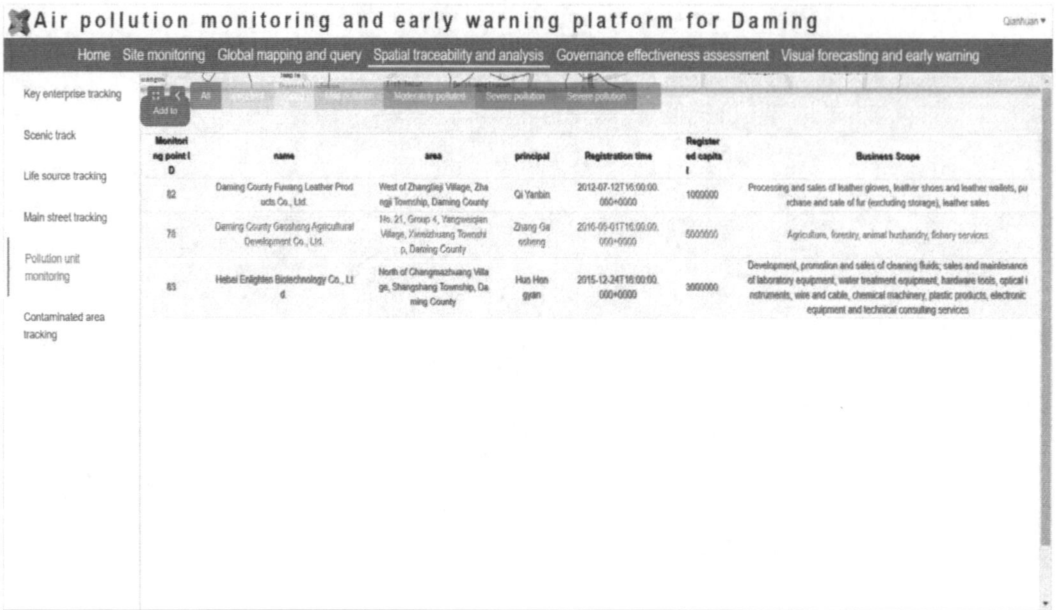

Figure 4. Statistical results for all index

3.5. *Evaluation for Governance Effect*

In this module, platform can display the governance effect by comparing change between air pollution sources before governance and after. The index change of air pollution sources is displayed by line chart and sorted with other sources.

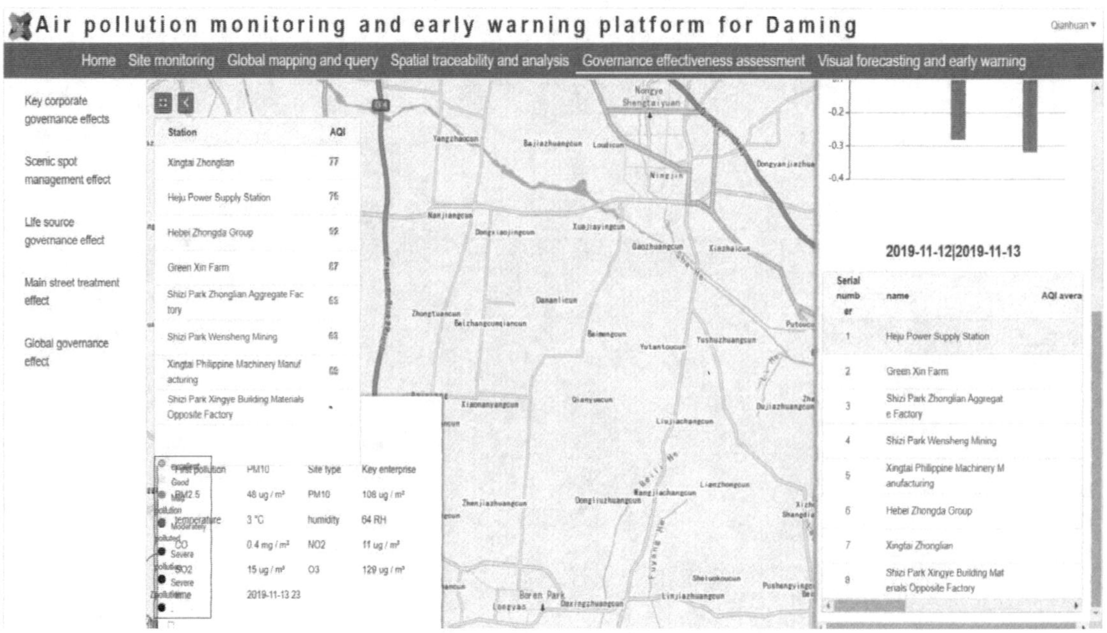

Figure 5. Evaluation for governance effect

3.6. *Visual Forecasting*

Using the predictive model, a 24-hour fine forecast and a 5-day forecast for Daming County are conducted. And platform can alert the heavily polluted areas.

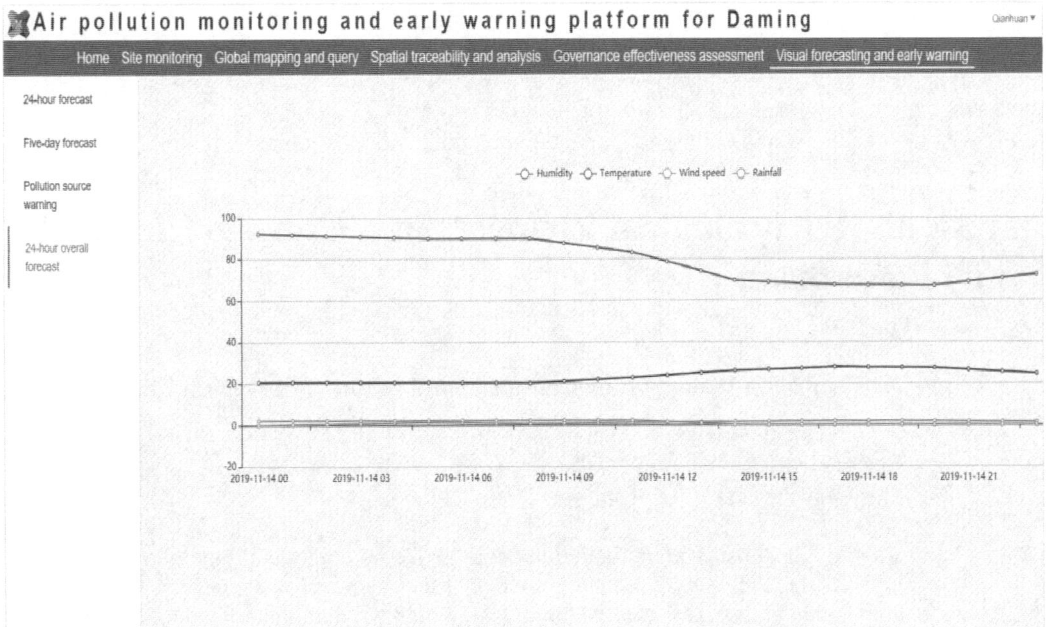

Figure 6. Air quality forecast for Daming country

4. Summary

By construct the platform, air pollution can be effectively monitoring and immediately pre-warning. The monitoring data collected by ground monitoring devices can be used to statistics and analysis. In visualization methods, people can intuitive find where the air pollution beyond standard or the pollution source from. The platform provides to government a new chain which is "monitoring-analysis-governance" by using GIS technical and computer combined.

5. Acknowledgement

Project Source: National Key R&D Program: Key Technologies for Emergency Assessment and Dynamic Decision Support for Major Disasters (Code: 2017YFB0504102)

6. References

[1]. Bermúdez, F.M., Laxe, F.G. & Aguayo-Lorenzo, Assessment of the tools to monitor air pollution in the Spanish ports system, Air Qual Atmos Health (2019) 12: 651. https://doi.org/10.1007/s11869-019-00684-x.

[2]. Girard S.F., Lates V., Lefrançois P., Sojic N., Clement F., Arbault S. Reactive Oxygen Species Generated by Cold Atmospheric Plasmas in Aqueous Solution: Successful Electrochemical Monitoring in Situ under a High Voltage System. [J]. Analytical chemistry,2019,91(13):8002-8007, https:// doi.org/10.1021/acs.analchem.9b01912.

[3]. Yang N, Zhang H, Zhang S, Yang Y, Pang Y, Liu X, Hou X. Monitoring and Control of Medical Air Disinfection Parameters of Nosocomial Infection System Based on Internet of Things. J Med Syst. 2019 Mar 27; 43:126.

[4]. Pérez E., Laura, García C, Maria A., Isabel S., María F., Cañizo A., Cerdá J. A., Fernandez B., Torre M., Ordoñez J., Agustin J.C., Digital thoracic drainage, a new system to monitor air leaks in pediatric population. [J]. Journal of pediatric surgery,2019. 54(4): 693-695, https://doi.org/10.1016/j.jpedsurg.2018.10.068

IOP Conf. Series: Materials Science and Engineering **730** (2020) 012050 doi:10.1088/1757-899X/730/1/012050

Analysis of Meteorological Elements and Climate Change in Poyang Lake Basin

Jie Wen[1],Yilin Wang[1], Guangru Sun[1], Shujing Liu[1] and Weihua Zhang[1,2*]

[1]College of Resources and Environment, Southwest University, No.2 Tiansheng Road, Beibei District, Chongqing, China.

*Email: swuwater@126.com

Abstract: In our study, five stations of Poyang Lake were selected: Nanchang, Boyang, Jingde, Lushan and Jing'an. The three observational values of temperature, precipitation and evaporation in these five stations from 1988 to 2017 were analyzed. Through average, extreme value, linear regression and cumulative anomaly analysis, the interannual temperature variation trend of the five stations is consistent and is expected to continue to rise over the next 20 years; precipitation shows different changes in different periods; evaporation is decreasing year by year. This study is of great significance to the maintenance of ecological function and security of Poyang Lake.

1. Introduction

Global climate change refers to a significant change in the statistical sense of the average state of climate or a long period of climate change (typically 30 years or more). In addition to having a bad impact on natural and social ecology, it has also brought adverse impacts on hydrological and water resources systems. It create the distribution of precipitation is more uneven, the intensity of rainstorm is increased, rainstorm,and flood occur frequently, which leads to the imbalance of seasonal flow[1].

In the past 50 years, the climate of the Poyang Lake Basin has undergone significant changes, which will affect the variation of water volume and evapotranspiration in the basin[2]. In order to maintain the ecological function and safety of Poyang Lake, it is of great significance to study the process of ecological climate change in typical wetlands under changing environment and the Countermeasures for its regulation and control [3]. Min Qian et al. [4] used folding algorithm to analyze the evaporation characteristics of Poyang Lake in the past 50 years. Guo Hua et al. [5] used M-K trend test to analyze the change trend of evaporating dish evaporation and reference crop evapotranspiration in Poyang Lake from 1961 to 2000. Liu Jian et al. [6] applied the complementary correlation evaporation model to estimate the actual evaporation and variation of stations in the Poyang Lake Basin from 1955 to 2001. In the previous studies, many analyses of climate change in Poyang Lake were done, but not many people studied the relationship between climate factor variables. So in this study, we want to know how climate elements interact with each other.

1.1. Profile of the Study Area

The Poyang Lake basin is located in a subtropical humid monsoon climate zone with an average annual temperature of 17.1 °C. The average annual precipitation is 1632 mm, with obvious seasonal and regional differences [7]. The precipitation in the main flood season from April to June accounts for 44% of the whole year, and the precipitation from the dry season from October to February is only 22.4% That has led to uneven distribution of runoff in the Poyang Lake Basin during the year.

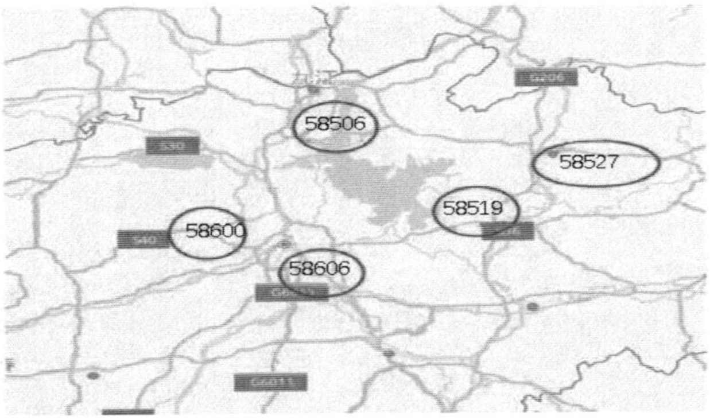

Figure 1. Geographical location of Poyang Lake Basin

2. Data Collection and Processing

2.1. Data Collection
Considering the evenness of the distribution of stations in the research area, 5 national reference stations in the watershed were selected, they are Nanchang station (58606), Boyang station (58519), Jingdezhen station (58527), Lushan station (58506), Jingan station (59600), respectively. Daily data of temperature, rainfall and evaporation in the five stations from 1988-2017 were collected(Figure 1).

2.2. Data Processing

2.2.1. Data pre-processing. After we collected the data, we pre-processed the data and modified some errors in the data. In the case of data loss, we use interpolation to simulate the missing data. However, when the value of the data is too small to process, we replace it with zero.
2.2.2. Calculation and drawing figures. Calculated and got monthly data and yearly data according to the daily data; *Made figures according to the monthly and yearly data;*
Perform regression analysis and dispersion analysis to make clear how the factors changed with time going by.

3. Analysis and Result

3.1. Temperature Analysis
From Figure 2 it can be seen that the temperature from January to July was rising, and then the temperatures decreased month by month. The highest temperatures at all five stations occurred in July and the lowest temperatures occur in January. Except for December, trend value of other months are positive, which represents temperature in those months have risen with years going by. According to the trend value, we can also find that the temperature in spring and fall increased more than that in 1988, and the temperature in spring was higher than that in autumn during the last 30 years.

From Figure 3, It can be seen the annual average temperature increased or decreases, but on the whole, the temperature of each station shows a growing trend.The annual average temperature of stations 58527, 58600, 58606 and 58519 ranged from 17°C to 19°C from 1988 to 2017, but the annual average temperature of stations 58506 ranges from 11°C to 13°C from 1988 to 2017. We can clearly see that the temperature of station 58506 was lower than that of the other four stations, the reason why the temperature of 58506 station is lower than other stations is that this station is located in Lushan Mountain[8], the elevation is higher, and the temperature is lower than other stations due to the influence of terrain.respectively. From Figure 4, we can see there is a good linear relationship between temperature and year. It tells us temperature is on the rise during the past 30 years.

IOP Conf. Series: Materials Science and Engineering **730** (2020) 012050 doi:10.1088/1757-899X/730/1/012050

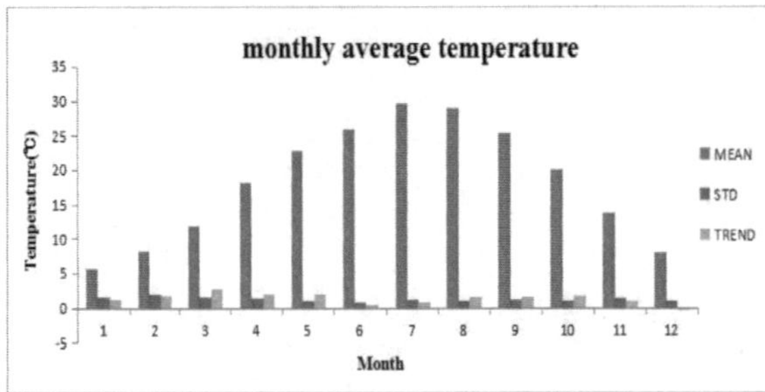

Figure 2. Monthly Average Temperature

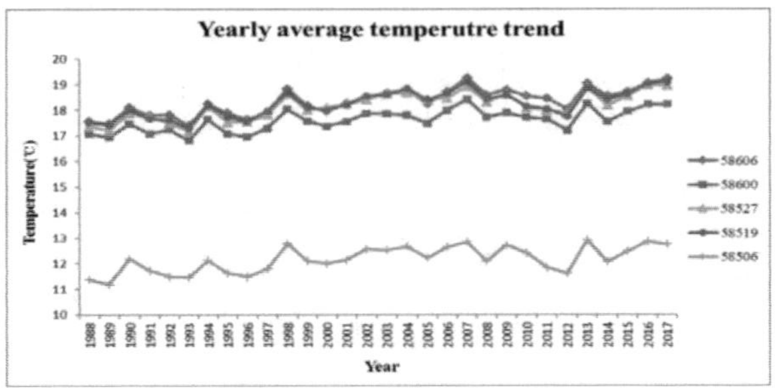

Figure 3. Inter-annual Variation of Thee Tmperature in Each Station

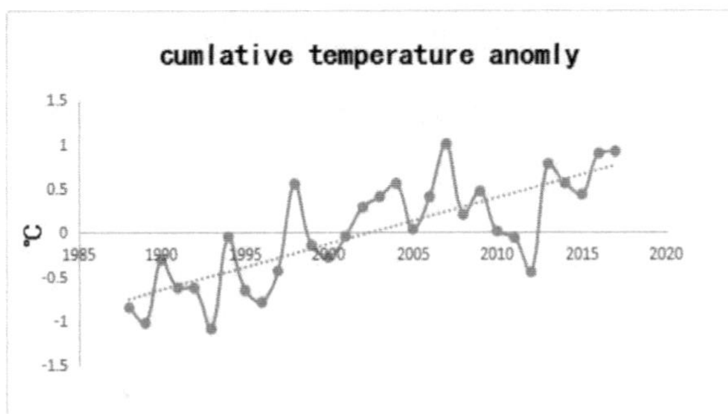

Figure 4. Cumlative Temperature Anomly

3.2. Evaporation Analysis

The evaporation capacity of the stations 58519, 58527 and 58606 has been largely missing since 2002, so the data from 1988 to 2001 are taken as the representative to analyze.Overall, the monthly evaporation at the five stations ranged from 489 mm to 2383 mm.The highest monthly evaporation occurred in June with a value of 2382 mm,while the lowest monthly evaporation occurred in January with a value of 354 mm.Trend showed negative numbers for five stations in October, November and December.It can be seen that under normal circumstances,the evaporation volume tends to decline in winter of future years(Figure 5).

According to the annual evaporation volume, the maximum annual evaporation volume is 1780.2mm, and the minimum annual evaporation volume is 789.4mm. The years in which the maximum evaporation capacity of the five stations appeared were different. The station 58506 was in 2004 , the station 58519 was in 1990, the station 58527 was in 2001. In terms of the minimum evaporation, station 58519 and station 58600 both appeared in 1999. station 58527 and station 58606 both appeared in 1993. station **598506** appeared in 2012(Figure 6) .

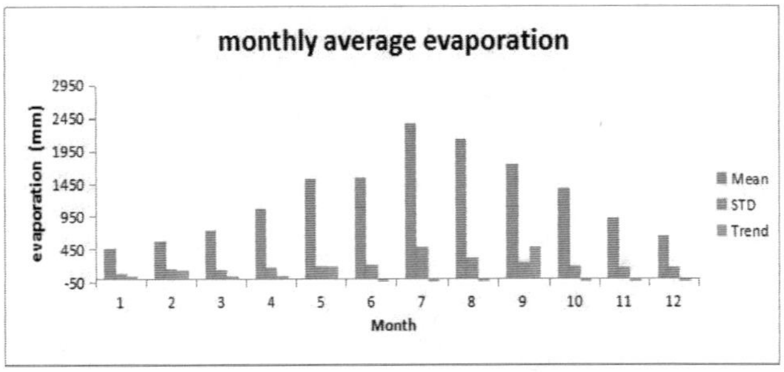

Figure 5. Monthly Average Evaporation

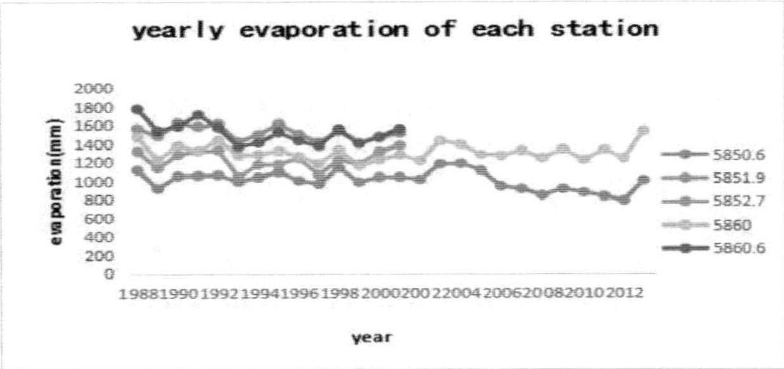

Figure 6. Inter-annual Variation of The Evaporation in Each Station

3.3. Precipitation Analysis

Ological stations in the Poyang Lake Basin from 1988 to 2017, we can know the characteristics of monthly average precipitation changes in this basin. The average precipitation in June was 662.72mm, accounting for 18.4% of the annual precipitation, which was the average maximum precipitation month; the average precipitation in December was 108.48mm, accounting for 3.0% of the annual precipitation, which was the smallest average precipitation. From January to June, the monthly average precipitation increased month by month, while precipitation from July to December decreased.In addition, due to the summer season from June to September, extreme precipitation weather was very prone to occur[9], resulting in a relatively large standard deviation of monthly average precipitation.By analyzing the trend value, we can find that the average monthly precipitation of 1-5 month is usually a negative value(Figure 7).

The average annual precipitation in the Poyang Lake Basin from 1988 to 12017 was 3587.1 mm, the highest precipitation in 1998 was 4762.8 mm, and the lowest precipitation in 2007 was only 2432.8 mm.Using the cumulative anomaly to analyze the annual precipitation characteristics of the Poyang Lake Basin from 1988 to 2017, we can see that the cumulative anomaly curve of precipitation in Poyang Lake area from 1988 to 2017 has increased in 1992-1999. And the trend of the curve in 2014-2015 has also increased, and the precipitation has also increased. In 1988-1992 and 1999-2009, the trend of the curve decreased, which was also the decrease in precipitation(Figure 8).

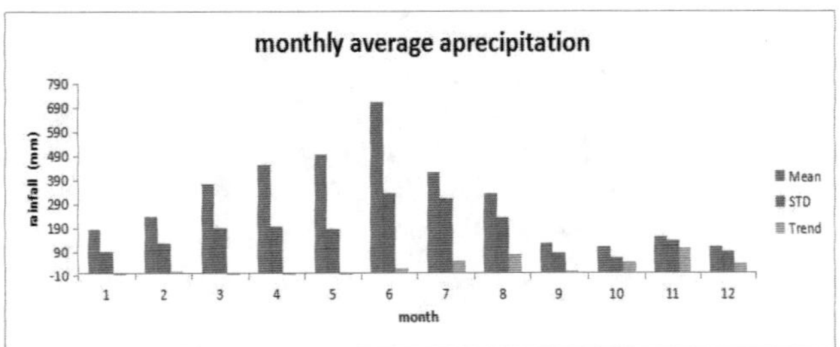

Figure 7. Monthly Average Precipitation

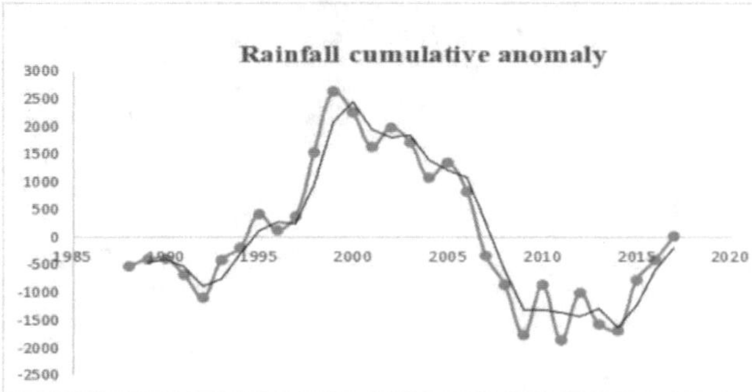

Figure 8. The Cumulative Anomaly Four Station

3.4. Comparison Between Climate Factors

As is shown in Figure 9, both precipitation and temperature went up and then went down. For temperature, it is higher in fall than that in spring, but for precipitation it is another way around.From Figure.11, we can see, annual distribution of temperature and evaporation are almost as same as Figure 10. We can understand why Figure 11 is similar to Figure 10. In our work, there are some differences with the previous research and questions about the results we got. We will try to explain them at this part.Compared to other 4 stations, station 58506 has lower temperature, more precipitation and higher evaporation. This is because the station is located in Lushan Mountain, its elevation is higher than other stations. Since the data of evaporation is taken from the evaporating dish experiment[10], it is much higher than the actual natural evaporation. The rainfall is concentrated in April-June because the area is heavily affected by plum rain.

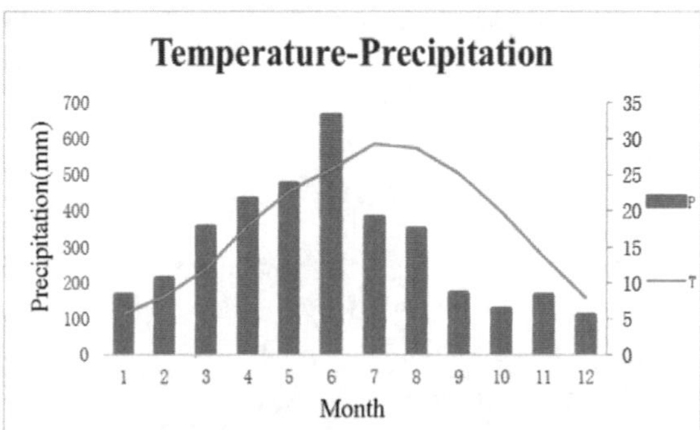

Figure 9. Comparison of Distribution of Temperature and Precipitation

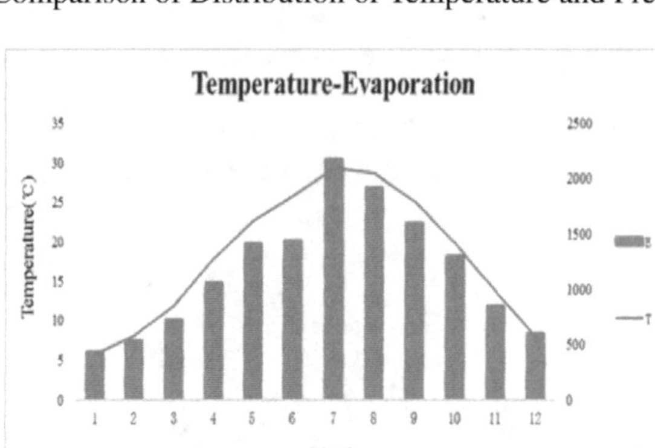

Figure 10. Comparison of l distribution of temperature and evaporation

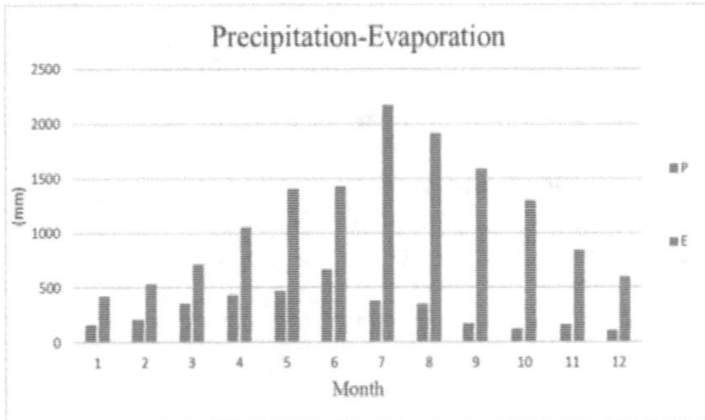

Figure 11. Comparison of distribution of evaporation and precipitation

4. Conclusion

From the above analysis we can get the following conclusions.Temperature went up from January to July and then went down. The highest temperature appears in July and the lowest occurred in January. Temperature in fall is higher than that in spring.Precipitation is mainly concentrated in March to June and the highest Precipitation occurred in June. The dry season is from September to December.The annual distribution of evaporation varies with temperature.The inter-annual trend of temperature of the five stations are the same. Temperature went up from 1988 to 1997, and increased in the following 20 years. It went down from 1988 to 1995, increased during 1995 to 2004 and decreased in the following 13 years.In the past 30 years, evaporation has been declining year by year. This should be caused by the combined effects of changes of temperature and precipitation.

5. Acknowledgement

This work was financially supported by Southwest University Core Curriculum Program "Principles of Hydrology", Chongqing key research and development projects of social and livelihood (No.cstc2018jscx-mszdX0052),and the projects "Design of optimal allocation and management scheme under limited water resources" supported by Southwest University's Training Program of Innovation and Entrepreneurship for Undergraduates(X201910635228). The supports are gratefully acknowledged.

6. Reference

[1] Liu J Y, Zhang Q, Deng X Y, Ci H and Zhang X H 2016 Quantitative analysis of the impact of climate change and human activities on the runoff process of poyang lake basin Lake Science 28(02):432-443
[2] Wei L 2014 Study on the ecological and hydrological process of the typical wetland of Poyang Lake under changing environment and its regulation and countermeasure Wuhan University,2014
[3] Min Q and Liu Y 2006 Calculation and trend analysis of evaporation of water surface in Poyang Lake from 1955 to 2004 *Lake Science*,2006(05):452-457
[4] Guo H, Jiang T, Wang G J, Su B D and Wang Y J 2006 Analysis of climate change trends and mutations in Poyang Lake Basin from 1961 to 2003 *Lake Science*,2006(05):443-451
[5] Liu J Y, Zhang Q, Deng X Y, Ci H and Zhang X H 2016 Quantitative analysis of the impact of climate change and human activities on the runoff process of poyang lake basin *Lake Science*,2016,28(02):432-443
[6] Zhan M J, Yin J M and Kong P 2013 Study on the facts of climate change in Poyang Lake Basin *Study on Meteorology and Disaster Reduction*,2013,36(03):18-24
[7] Sun S L, Zhou S H, Song J, Shi J H, Gu R Y and Ma F M 2010 The change characteristics and causes of evaporation of evaporative dishes in Jiangxi Province *Agricultural Engineering Journal*,2010,26(09):59-65
[8] Ye X, Zhang Q, Liu J et al 2013 Distinguishing the relative impacts of climate change and human activities on variation of streamflow in the Poyang Lake catchment, China *Journal of*

Hydrology, 2013, 494: 83-95

[9] Jin B S, Nie M, Li Q, Chen J K and Zhou W B 2012 Basic characteristics, challenges and key scientific problems in poyang lake basin *Resources and Environment in the Yangtze River Basin*,2012,21(03):268-275

[10] Xu X N, Gao J H and Jia J J 2015 Quantitative estimation of the effects of climate change and human activities on the amount of sand into the lake in the Poyang Lake basin *Geography*, 2015, 34(5): 838-850

IOP Conf. Series: Materials Science and Engineering **730** (2020) 012051 doi:10.1088/1757-899X/730/1/012051

The Numerical Simulation of Gas-Liquid Two Phase Flow in Annular Gap Venturi of Converter Wet Dust Removal

Mengmeng Ye, Mei Liang, Fuping Qian* and Jinli Lu

School of Civil Engineering and Architecture, Anhui University of Technology, Ma'anshan 243002, China.
*Email: fpingqian@163.com

Abstract. Based on the CFD-DPM method, the numerical simulation of gas-liquid two phase flow in annular gap venturi with different annular gap trip and liquid/gas ratio at the same gas flow rate was carried out in this paper. The results show that the pressure drop at the contraction section of venturi accounts for about 75% of the total pressure drop, and the pressure drop at the throat is about 20% of the total pressure drop. The average velocity of flue gas at the throat ranges from 106 to 192m/s. The total pressure drop of the venturi is proportional to the liquid/gas ratio at the constant annular gap trip, and the larger the annular gap trip, the greater the effect of the liquid/gas ratio on the pressure drop. As the annular gap trip is 800~900mm, the effect of liquid/gas ratio on the pressure drop is significant. The total pressure drop of the venturi and the droplets concentration at the venturi inlet are proportional to the annular gap trip at the constant liquid/gas ratio, and the growth rate of the pressure drop is greater with the annular gap trip as the annular gap trip is larger than 700mm. In addition, the growth rate of the droplets concentration at the venturi outlet also increases with the annular gap trip.

1. Introduction

A large amount of gas is generated during the smelting in the converter, and due to much dust therein, it must be purified to be effectively utilized. OG (Oxygen Converter Gas Recovery) device is an important facility for converter gas recovery. The operation of ascentional annular gap venture with adjustable throat directly affects the resistance and dust removal efficiency of the system. In recent years, many studies have been done on multiphase flow in venturi tube based on CFD method. Pak and Chang[1] used CFD software to conduct numerical simulation of gas-liquid-solid three phase flow in Pease - Anthony type venturi based on euler-lagrange method, and analyzed its pressure drop and dust removal efficiency. Lu Tao[2] et al. conducted three-dimensional numerical simulation of the flow and heat transfer process in the two-stage venturi of the third-generation OG system. Yu Xiaodong[3] carried out CFD simulation on the gas-liquid two phase complex flow problem of the descentional annular gap scrubber of the fourth generation OG system,revealingthe characteristics of multi-phase flow field and the influence of various influencing factors on the resistance in the annular gap scrubber of the converter. Guerra [4] et al. studied the factors affecting the effect of liquid spraying through multiple orifices in venturi scrubber by experiments and CFD simulation; Sharifi[5] et al. accurately predicted the pressure drop in the venturi by combining the population balance equation with CFX. Zhang Xu [6] et al. used CFD software to calculate the mutual coupling between continuous phase flue gas and discrete phase water droplets in a diameter-adjustable venturi under actual working conditions. Most of the venturis in the above documents are of the descentional type, that is, the flue gas flows from top to bottom. In this paper, the ascentional type annular gap venturi with adjustable throat was studied, and the flue gas flows from bottom to top, and the scrubber cone is ascentional. Compared with the venturi of previous generations of OG systems, it reduces system resistance and increases dust removal efficiency. Therefore, in-depth study and analysis of its internal multiphase flow are of great significance for optimizing the structure of the new OG system.

In this paper, the CFD-DPM method was used to numerically simulate the gas-liquid two phase flow in annular gap venturi. Under the same gas flow rate, the influence of different annular gap trip and different liquid/gas ratios on the pressure drop in the venturi and the droplet concentration at the venture outlet was analyzed. The obtained results have certain reference value for the optimal design of the annular gap venture in the new OG system.

2. Numerical Calculation Model of Annular Gap Venturi

2.1. Modeling and Meshing

Figure 1 shows a three-dimensional model of an annular gap venturi (the annular gap trip is 800 mm). The dust-containing gas enters from the bottom, and the middle part is a scrubber cone that can move up and down to adjust the opening rate of the venturi throat.

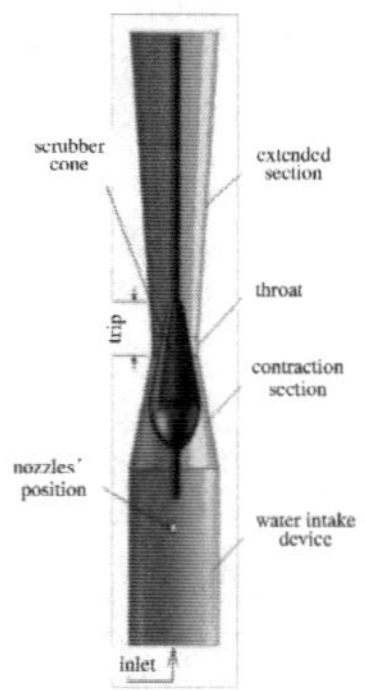

Figure 1. Annular gap venture

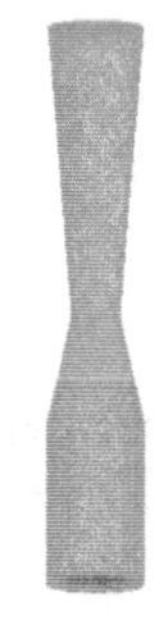

Figure 2. Mesh of the annular gap venture

Tetrahedral meshed by ICEM CFD, five mesh models with different annular trips were obtained. Figure 2 showsthe mesh of the annular gap venturiwith annular gap trip of 800mm. In order to improve the calculation accuracy, local encryption was used on the inlet and outlet of the annular gap venturi and the scrubber cone. Total number of grids: 314858, total number of nodes: 55786, grid quality: Min = 0.374, max = 0.999, average = 0.776.

2.2. Numerical Methodology

The furnace gas contains soot particles and flue gas. The soot particles mass concentration is about 4000 kg/m^3, and the flue gas mass concentration 1.365kg/m^3; generally, dust mass concentrationin the furnace gas is about 0.1 kg/m^3, which shows that the volume occupied by the soot particles is much smaller than the volume of the flue gas, so the influence of the soot particles on the flow and heat transfer could beignored[7]. Since the volume fraction of the spray droplets is much smaller than that of the flue gas, the flue gas is regarded as the continuous phase, and the spray droplets the discrete phase, and the two phases are coupled and calculated. The continuous phase adopts the Euler method to solve the NS equation. The continuous phase mathematical model includes the continuity equation, the momentum equation, the turbulent energy k equation, the turbulent dissipation ε equation and the energy equation. The discrete phase spray droplets adopt the Lagrangian particle orbit model.

The continuity equation, the momentum equation, and the energy equation can be expressed by a general formula [8]:

$$\frac{\partial(\rho\Phi)}{\partial t}+\nabla(\rho u\Phi)=\nabla(\Gamma grad\Phi)+S \qquad (1)$$

Where Φ is a general variable, which can represent u, v, w, T and other dependent variables, Γ is a generalized diffusion coefficient; S is a generalized source term. The four terms in the equation represent the transient term, the convection term, the diffusion term, and the source termrespectively.

The Realizable k-ε model was selected as a turbulence model, and the turbulent energy k equation and the turbulent dissipation ε equation are as follows [9]:

$$\frac{\partial}{\partial t}(\rho k)+\frac{\partial}{\partial x_i}(\rho k u_i)=\frac{\partial}{\partial x_j}\left[\left(\mu+\frac{\mu_t}{\sigma_k}\right)\frac{\partial k}{\partial x_j}\right]+G_k+G_b-\rho\varepsilon-Y_M+S_k \tag{2}$$

$$\frac{\partial}{\partial t}(\rho\varepsilon)+\frac{\partial}{\partial x_i}(\rho\varepsilon u_i)=\frac{\partial}{\partial x_j}\left[\left(\mu+\frac{\mu}{\sigma_\varepsilon}\right)\frac{\partial \varepsilon}{\partial x_j}\right]+C_{1\varepsilon}\frac{\varepsilon}{k}(G_k+C_{3\varepsilon}G_b)-C_{2\varepsilon}\rho\frac{\varepsilon^2}{k}+S_\varepsilon \tag{3}$$

Where G_k and G_b represent the turbulent flow energy produced by the laminar velocity gradient and the buoyancy force, respectively. YM is the wave effect generated by transitional diffusion in compressible flow. $C_{1\varepsilon}$, $C_{2\varepsilon}$, $C_{3\varepsilon}$ are constant, and σ_k and σ_ε are the turbulent Prandtl numbers in the k and ε equations, and S_k and S_ε are user-defined values.

The motion equation of the droplet:

$$\frac{du_p}{dt}=F_d\left(u-u_p\right)+\frac{g\left(\rho_p-\rho\right)}{\rho_p}+F \tag{4}$$

Where $F_d(\mathrm{u}-u_p)$ is the unit mass drag force of the flue gas on the droplets (in this paper, spherical law is used, i.e. the droplet is assumed to be spherical particle).

$$F_d=\frac{18\mu}{\rho_p d_p^2}\frac{C_d \mathrm{Re}_p}{24} \tag{5}$$

Where u_p, g, u, ρ_p, F, C_d and R_{ep} are the droplet velocity, gravitational acceleration, droplet density, additional mass force (ignored herein), drag coefficient, and droplet relative Reynolds number.

Within the same time step, the equation was resolved by the SIMPLE algorithm. The pressure, momentum, turbulent kinetic energy and turbulent diffusivity are all in the second-order upwind discrete scheme. On the basis of the first-order upwind discrete scheme, this format considers the influence of the curvature of the curve about the distribution of physical quantities among nodes, which helps to improve the accuracy and stability of the calculation. The residual convergence criterion for continuity, velocity, turbulent kinetic energy, and turbulent diffusivity are all set to 0.001.

For the discrete phase, in order to accurately describe the sprayingprocess, the gravity, drag force and droplets breakage are considered in the simulation calculation. Taking the droplet group as a discrete system, the orbit of the discrete phase particles is obtained by differential equation of particle force in integral Lagrangian coordinate system. The droplets are tracked by an unsteady discrete phase model [10], and random orbital model is used to consider the turbulent diffusion of the particles. The random walk model in the tracking model of the particles is used to consider the interaction between the particles and the discrete vortices of the fluid.

2.3. Boundary Conditions

The external atmospheric pressure is 101325 Pa. The flue gas rate at the Venturiinlet is 105 000 m^3/h, and the venturi nozzle with forward spray mode is located on the central axis of the water intake device. The specific simulation settings are shown in Table1 and Table 2.

Table 1. Continuous phase boundary conditions

Items	Continuous phase boundary conditions
Inlet boundary type	Velocity-inlet
Inlet Velocity	18.96m/s
OutletBoundary Type	Pressure-outlet.
Outlet Pressure	0
Turbulent Intensity	5%
Turbulence Viscosity Ratio	10
Wall Boundary Type	No-slip Wall; Adiabat; Particle Escape
Inlet and Outlet Temperature	300K

Table 2. Discrete phase model settings

Items	DPM settings
Nozzle Type	Pressure Swirling Nozzle
Droplet Type	Droplet
Water Temperature	300K
Pressure	1MPa
Nozzle Diameter	5mm

3. Numerical Simulation Results and Analysis

3.1. Flow Field Distribution

The velocity and pressure contour on the symmetrical plane of Venturi with different annular gap tripswere obtained by numerical simulation, as shown in Figure 3 (taking 800mmannular gap trip as an example). As can be seen from Figure 3, in general, the pressure dropsat the contraction section, where the pressure drop accounts for about 75% of the total pressure drop.The pressure drop at the throat is about 20% of the total pressure drop. The pressure value reaches a minimum at the throat and partially builds up in the extended section. This is becauseas the cross-section area of the shrinkage section decreases, the flue gas velocity increases.The flue gas velocity at the throat is accelerated to its maximum value. The static pressure decreases with the increase of flue gas velocity. Then the flow velocity decreases gradually with the increasing cross-section area of the expansion section, and the static pressure rises to a certain extent. The average flue gas velocity at the throat reaches 106-192 m/s in the range of annular gap trips selected in this paper.

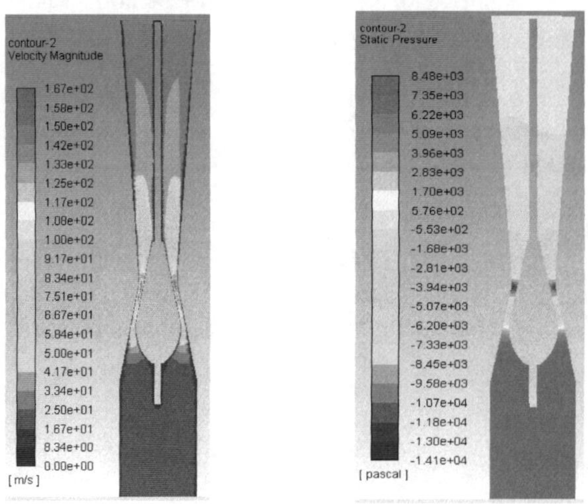

(a) (b)

Figure 3. Velocity and pressure contours of annular gap venturi

3.2. Effect of the Annulargap Tripon Average Throat Velocity and Droplet Concentration at Venturi Outlet

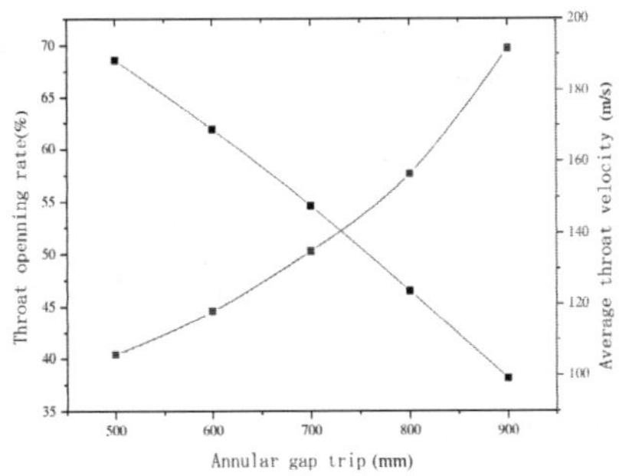

Figure 4. The variation of throat opening rate and average throat velocity versus the annular gap trip

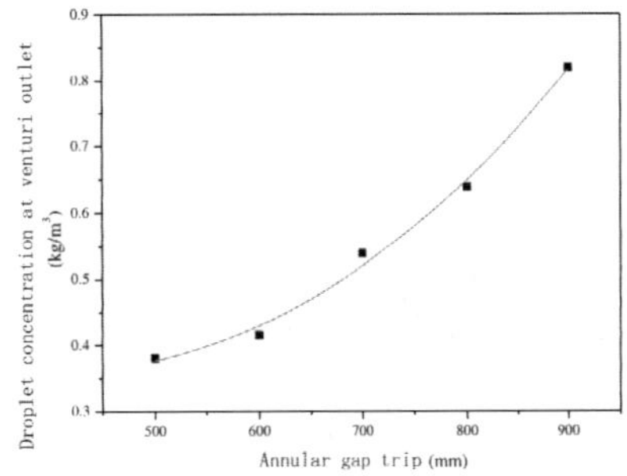

Figure 5. The variation of the droplet concentration at the venturi outlet versus the annular gap trip

It can be seen from Figure 4 that the throat openingrate decreases almost linearly with the increase of the annular gap trip, and the averagethroat velocity gradually increases with the increase of the annulargap trip. However, since the gas is compressible, the resistance or pressure of the flue gas vary with different annular gap trip,and as the annulargap trip increases, the flue gas velocity at the throat turns larger, and the impact of the gas on the water is more intense. Therefore, the flue gas velocity at the throat does not vary linearly with the annular gap trip.

Figure 5 shows the variation of the droplet concentration at the venturioutletversusthe annular gap trip when the gas flow rate and the liquid/gas ratio are constant. It can be seen from Figure 5 that, in general, the droplet concentration at the venturi outlet increases as the annular gap trip increases, and growth rate of the droplet concentration at theventuri outlet also increases as the annular stroke increases. This is because, as the annulargap trip increases, the throat gas velocity gradually increases (as shown in Figure 4). The droplet breakage mainly occurs at the throat. The larger the throat gas velocity, the better the droplet breakage effect. Therefore, the droplet concentration at the venturi outlet increasesas the annular gap trip increases. When the water spray flow rate is constant, the higher the outlet droplet concentration, the larger the specific surface area of the droplet, which helps dust particles to be encapsulated by droplets and improves dust removal efficiency of the system.

3.3. Effect of the Annular Gap Trip and Liquid/Gasratio Onpressure Drop

Figure 6 shows the variation of the total pressure drop of annular venturiversusannular gap trip under constant gas velocity and different liquid/gasratio.

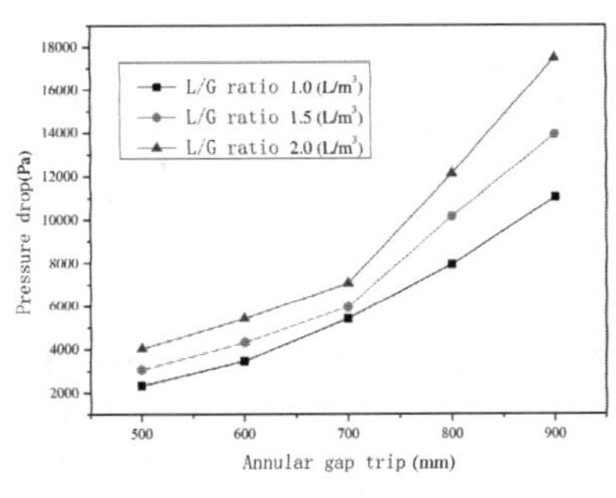

Figure 6. The variation of pressure drop versus annular gap trip

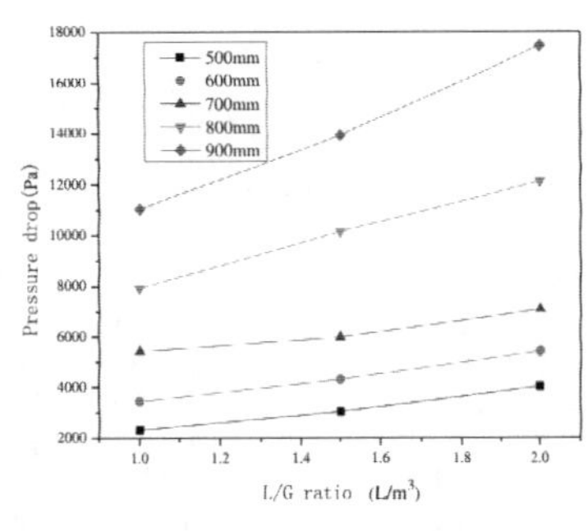

Figure 7. The variation of pressure drop versus liquid/gas ratio

Figure 6 shows that the pressure dropof the annular gap venturiincreases with the increase of the annular gap trip at the constantliquid/gas ratio. The growth rateof the pressure drop is greaterwith the annular gaptrip as the annular gap trip is larger than 700mm. This is because when the annular gap trip is larger than 700 mm, the average flow velocity at the throat increases rapidly (as shown in Figure 4), so the pressure decreases greatly at the throat, and the increase of total pressure drop with the increase of annular gas trip is also more significant. In addition, at the constantannulargap trip, the larger the liquid/gas ratio, the greater the pressure drop of the annular gapventuri. This is because with the increase of the liquid/gas ratio, the energy consumption of gas for accelerating dropletsincreases. The resistance in the annular gap increases, that is, the pressure drop of the scrubber increases.

Figure 7 shows the variation of pressure drop versus liquid/gas ratio at the constant gas flow rate and different annular gap trip. As shown in Figure 7, the larger the annular gap trip, the greater theeffect of liquid/gas ratio on pressure drop. When the annular gap trip is larger than 800mm, the effect of liquid/gas ratio on pressure drop is more significant. When the liquid/gasratio is 2.0 L/m^3, the pressure drop under the maximum annular gap trip has exceeded 17 kPa. Therefore, it is suggested that the liquid/gas ratio be controlled below 2.0 L/m^3.

4. Conclusions

Based on CFD-DPM method, the numerical simulation of gas-liquid twophase flow in annular gap venturi in new OG system was carried out. The effect of different annular gap trip and liquid/gasratio on the pressure drop of venturi and droplet concentration at venturi outlet under the constant gas flow rate was analyzed. Theconclusions are as follow:

(1)The pressure drop increases greatly at the contraction section, the pressure drop at the contraction section accounts for about 75% of the total pressure drop, and the pressure drop at the throat accounts for about 20% of it. The pressure reaches a minimum at the throat and the partially builds up during the extended section. The gas velocity is accelerated to its maximum value at the throat.As the annulargap trip is 500~900mm,the average flow velocity at the throat ranges from 106 to 192m/s.

(2) At the constant gas flow rate and annular gaptrip, the pressure drop of annular gapventuri is proportional to the liquid/gasratio. The larger the annular gap trip, the greater the effect of liquid/gas ratio on the pressure drop. Especially when the annular gap trip is larger than 800mm, the influence of liquid/gas ratio on pressure drop is more significant. Therefore, it is suggested that the liquid/gas ratio be controlled below 2.0L/m^3.

(3) Under the constant gas flow rate and liquid/gas ratio, the pressure drop of the annular gap venturi is proportional to the annular gap trip, and the pressure drop increases greatly when the annular gaptrip is larger than 700 mm; the droplet concentration at theventuri outlet is proportional to the annular gap trip, and the growth rate of the droplet concentration at the outlet also increases with the annular gap trip. When the annular gap trip is larger than 700 mm, the droplet concentration at the venture outlet increases greatly, which is more conducive to improving the dust removal efficiency of the system, however, the pressure drop of venturi increases greatly with the annular gap trip, and excessive pressure drop will reduce the service life of the system equipment.

5. Acknowledgement

This study was financially supported by Provincial Major Project of Natural Science Research for Colleges and Universities of Anhui Province (grant no. KJ2017ZD06).

6. References

[1] Pak S I and Chang K S. Performance estimation of a venturi scrubber using a computational model for capturing dust particles with liquid spray [J]. *Journal of Hazardous Materials*, 2006, **vol5 no105**, pp560-573.
[2] Lu T and Liu C G. Influence of gas rate, spraying rate, and opening rate on performance of flow and heat transfer in venturi tube for converter gas [J]. *Journal of Thermal Science and Technology*, 2007, **vol6 no4**, pp309-312.
[3] Yu X D. Research on the pressure drop characteristics of multiphase flow in the ring silt washer of converter gas [D]. Chongqing University, 2008.
[4] Guerra V G, Goncalves J A S and Coury J R. Experimental investigation on the effect of liquid

injection by multiple orifices in the formation of droplets in a venture scrubber[J]. *Hazardous Materials*, 2009, **vol161 no1**, p 351.

[5] Sharifi A and Mohebbi A. A combined CFD modeling with population balance equation to predict pressure drop in venturi scrubbers [J]. *Research on Chemical Intermediates*, 2014, **vol40 no3**, pp1021-1042.

[6] Zhang X, Wang H J, Wu S F et al. Optimization of venturi tube gas-liquid coupling numerical simulation[J]. Metallurgicall Equipment, 2014, no212, pp 54~57.

[7] Lu T and Huang C. Performance analysis of flow and heat transfer of spraying cooling for converter gas with high temperature in venturi tube [J]. *Journal of Thermal Science and Technology*, 2007, **vol6 no1**, p33.

[8] Zhang J Q. Calculation of performance of oxygen converter gas wet purification system and simulation of flow and heat transfer of a venture scrubber [D]. Beijing University of Chemical Technology: College of Mechanical Engineering, 2015.

[9] Zhang X. The converter flue gas purification and recycling system optimization and engineering practice [D]. Yanshan University: College of Mechanical Engineering, 2012.

[10] Han Z Z, Wang J and Lan X P. FLUENT: Example and Analysis of Fluid Engineering Simulation [M]. Beijing: Beijing Institute of Technology Press, 2004, pp5-44.

Distance Prediction of Slope-Foot Landslide in Southwest of China Based on GA-BP Neural Network

Hao Zhou[1], Yong Chen[1,2] and Ruoying Tian[1]

[1]School of Resources and Environmental Engineering, Wuhan University of Science and Technology, Wuhan; 430081, China.
[2]Hubei Key Laboratory for Efficient Utilization and Agglomeration of Metallurgic Mineral Resources, Wuhan; 430081, China.
*Email: Yongchen1968@163.com

Abstract. Landslide is a typical geological hazard in southwest China. In order to reasonably assess the disaster-causing range of earthquake-induced landslides in southwest China, the prediction of landslide movement distance is very important. By making principal component analysis of the factors affecting the distance of landslide movement and combining with the mechanism of landslide movement, the three principal components are named as kinetic energy factor, slope factor and resistance factor, which are used as input neurons of BP neural network. The initial weights and thresholds of BP neural network are optimized by using the global search ability of genetic algorithm, and the prediction models of maximum vertical motion distance (H) and maximum horizontal motion distance (L) of landslide are constructed based on the optimized network. At the same time, on the basis of principal component analysis, a multivariate regression prediction model is constructed, and the results of network model prediction are compared with those of multivariate regression prediction. The optimization effect of genetic algorithm for BP neural network is obvious. The BP neural network model after optimization is accurate and stable in predicting the motion distance of slope foot landslide. The prediction errors of maximum horizontal and vertical motion distances below 10% accounted for 86.67% and 93.33% respectively, which is superior to the original BP neural network and multiple regression prediction model.

1. Introduction

Landslide is one of the major geological disasters in China, which brings huge economic losses to the people every year. For landslides whose horizontal movement is not obviously hindered, the movement distance of landslides is not only affected by the volume and height difference, but also related to the change of terrain gradient. Although there may be more gradient changes in the process of landslide movement, the gradient difference between the moving and accumulating zones of landslides at the foot of the slope is at the foot of the slope, and it is also related to landslide movement. The action is very significant [1]. The average slope (α) of sliding surface, the slope difference between sliding area and accumulation area, that is the angle of slope foot (β) and the average slope (γ) of accumulation area, all have great influence on the movement distance of landslide. Slope foot landslides caused by earthquakes are common in mountainous areas of Southwest China due to topography and geological structure. Landslide movement distance is an important index for Evaluating Landslide disasters, and its reasonable prediction can provide reference and basis for local disaster prevention and mitigation work.

At present, many models and theories have been put forward for the study of landslide movement at home and abroad, such as multi-factor coupling mechanism, air lubrication mechanism and energy transfer mechanism [2]. Through qualitative and quantitative research, Huajun Meng et al. concluded that landslide volume, site conditions and geological structure are the important factors affecting landslide movement [3]. According to the movement distance of landslide, Ran Tang and Qiang Xu design simulation experiments from the dynamic point of view, deduce the theoretical calculation formula of the movement distance of landslide [4]. Based on the statistical idea, through numerical simulation and regression analysis, Jian Guo, Shanshan Hao et al. obtained the prediction methods of different types of landslide movement distances, and established the relationship between the influencing factors and the movement distances of landslides [5-6]. However, due to the complexity of the internal mechanism of landslide, the prediction accuracy of the method and model for the distance of landslide movement needs to be improved.

In this paper, principal component analysis is used to extract the principal component of factors affecting landslide movement, and the mechanism of landslide movement is analysed. Then, a network model is constructed based on BP neural network optimized by genetic algorithm. The network model is trained and tested with 75 slope-foot landslides caused by earthquakes in southwestern China as samples, and the slope-foot type caused by earthquakes in southwestern China is analysed. The prediction of landslide motion distance is expected to improve the prediction accuracy of Landslide Motion distance. At the same time, new ideas and methods are introduced for the study of this kind of problems.

2. Research Method

2.1. GA-BP Neural Network Model

Artificial neural network (ANN) is a mathematical model inspired by the principle of biological neural network perceiving the world. It has strong ability to deal with non-linear problems. It is very suitable for problems with complex internal mechanism such as landslide movement and difficult to be solved by traditional statistical methods. BP neural network is a multi-layer feedforward neural network based on error back propagation algorithm. It has the characteristics of learning and self-adapting. The whole training process includes two processes: data forward transmission and error feedback correction. In this paper, the influencing factors of landslide motion distance are studied, and the prediction model of GA-BP neural network is constructed to predict the motion distance of slope foot landslide caused by earthquake.

2.1.1. BP Basic Structure. In BP neural network, the data is transferred to the hidden layer by the input layer according to the weight, and then to the output layer through the non-linear processing. The output layer adjusts the connection weights and thresholds in the process of data transmission through error feedback. A typical BP neural network consists of an input layer, a hidden layer and an output layer. Its topological structure is shown in figure 1.

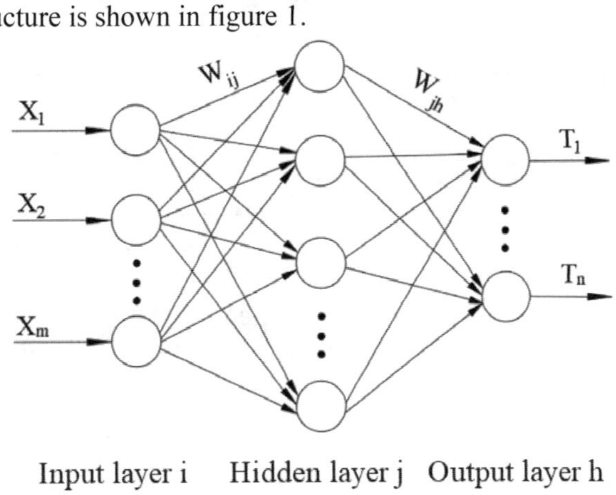

Figure 1. 3-Layer BP Neural Network

2.1.2. Setting of Network Parameters. The number of hidden layer neurons and the number of single layer neurons have a significant impact on the performance of the whole network. The method of determining the number of hidden layer neurons is not unique. The following formula (1) is selected to determine the number of hidden layer neurons preliminarily and adjust it appropriately in the actual training process. The selection of transfer function, learning algorithm, efficiency, initial threshold and weight may also have a great impact on the performance of the whole network. In the training process, the appropriate learning algorithm and efficiency are determined according to the results. The initial threshold and weight are optimized by genetic algorithm.

$$N = \sqrt{m+n} + t \tag{1}$$

Where: N is the number of neurons in the hidden layer, m is the number of neurons in the input layer, n is the number of neurons in the output layer, and it is an integer between 0 and 10.

2.1.3. Optimizing BP Neural Network by Genetic Algorithms. BP neural network has strong ability of data fitting and prediction, but the determination of initial weight and threshold has great influence on the performance of the neural network. With the help of the optimization ability of genetic algorithm, the initial weight and threshold of BP neural network are optimized, so that the neural network model can approach the most excellent solution quickly and accurately in the training process, avoiding local convergence or divergence. The optimization process of genetic algorithm includes population initialization, fitness function selection, selection operation, crossover operation and mutation operation.

2.2. Principal Component Analysis

The intrinsic mechanism of landslide is complex. In order to analyse the mechanism of landslide movement, and to reflect the intrinsic relationship among various factors more fully in the construction of network model, and to improve the accuracy of prediction model, the principal component analysis of factors affecting landslide movement is made by using MATLAB 2016, and the principal component obtained is used as the input of neural network neurons.

2.2.1. Analysis of the correlation of various factors. The original data is standardized by Min-max method (formula 2) to eliminate the influence of each factor dimension. Then the correlation coefficient matrix of each factor is obtained by the function Corrcoef. Then KMO and Bartlett tests are performed on each factor to determine whether the selected factor is suitable for principal component analysis.

$$x_i' = \frac{x_i - min(x)}{max(x) - min(x)} \tag{2}$$

Where: x_i is the first initial value of index X, x_i' is the value of x_i after standardizing, max(x) and min(x) are the maximum and minimum values of index X respectively.

2.2.2. Get the principal component expression. Using the PCA function of MATLAB 2016 to do correlation analysis, the eigenvalue, the contribution rate of principal component and the score matrix are obtained, and the principal component expression is as follows.

$$Y_i = X * W_i \tag{3}$$

Where: Y_i is the first principal component, X is the vector composed of all factors affecting landslide motion, and W_i is the score vector of the second principal component.

3. Results and Analysis

3.1. Influencing Factors of Landslide

The main factors affecting the movement of slope foot landslide are geological structure, lithology, seismic grade, water content of landslide body, volume of landslide body, slope, topographic fluctuation, elevation, normalized vegetation index and so on. The landslide samples selected in this study are slope-foot landslides caused by Wenchuan earthquake. The geological structure, lithology and water content of the landslide body in this area are relatively close. Earthquake is the starting factor of landslide, so magnitude will not affect the subsequent movement of the landslide body. These four factors are not taken into account in the construction of the model. Normalized vegetation index has a low correlation with landslide development [7], and has little influence on the distance of landslide movement. In order to improve the efficiency of model training, this factor is discarded when building the model. According to the relevant research [8-10], the following six factors are finally determined as research indicators: the length of slip surface above the foot of slope (L), the average slope degree (α), the angle of slope foot (β), the average slope gradient (γ), the volume of the landslide body (V), and the height difference of the front and rear edges of the landslide body (H).

3.2. Principal Component Analysis Results

Making principal component analysis of the determined research indicators, the correlation coefficient matrix can be obtained (table 1), it can be seen that most of the selected factors are highly correlated. KMO and Bartlett tests for each factor showed that the KMO test coefficient was 0.606, bigger than 0.5, and the Sig value of Bartlett test was 0.001, less than 0.05. In summary, we can see that there is a strong correlation between the selected factors, which is suitable for principal component analysis. Table 2 shows that the common factor variance of each factor is close to 1.0, which indicates that the extracted principal components can better reflect the information of the original data.

Eigenvalues and principal component contribution rate is shown in table 3. The cumulative variance contribution rate of the first three principal components reaches 92.87%, and the eigenvalues of principal components 1, 2 and 3 are 3.17, 1.35 and 1.03 respectively, which are all bigger than 1. These three principal components can fully reflect the influence of each factor on the movement distance of landslide. The principal component score matrix is shown in table 4, from which the expressions of the three principal components are as follows:

$$Y_1 = 0.4x'_V + 0.51x'_L + 0.5x'_H + 0.09x'_\alpha + 0.41x'_\beta - 0.39x'_\gamma \quad (4)$$

$$Y_2 = -0.34x'_V - 0.18x'_L + 0.01x'_H + 0.8x'_\alpha + 0.45x'_\beta + 0.1x'_\gamma \quad (5)$$

$$Y_3 = 0.27x'_V + 0.26x'_L + 0.34x'_H + 0.31x'_\alpha - 0.4x'_\beta + 0.69x'_\gamma \quad (6)$$

Where: x'_V, x'_L, x'_H, x'_α, x'_β, x'_γ respectively indicate the volume of landslide body, the length of sliding surface above the foot of slope, the height difference of landslide body, the average gradient of sliding surface, the angle of foot of slope and the average gradient of accumulation area after standardizing.

It can be seen that the volume of landslide body (V), the length above the foot of the slope (L) and the height difference of the landslide (H) all have higher loads on the first principal component Y_1, which reflects the energy accumulation ability of the landslide in the acceleration stage, and can be called the kinetic energy factor; the average slope of the sliding surface (α) and the angle of the slope foot (β) both have high load on the second principal component Y_2, which reflects the acceleration ability of the landslide body in the acceleration stage , and can be called acceleration factor; the angle of the slope foot (β) and the average slope of the accumulation area (γ) both have high load on the third principal component Y_3, which reflects the energy loss of the landslide body from excessive to deceleration stage, and can be called resistance factor.

Table 1. Correlation matrix

Index	V	L	H	α	β	γ
V	1.000	0.715	0.672	0.055	0.492	-0.509
L	0.715	1.000	0.941	0.066	0.442	-0.447
H	0.672	0.941	1.000	0.332	0.522	-0.330
α	0.055	0.066	0.332	1.000	0.501	0.216
β	0.492	0.442	0.522	0.501	1.000	-0.736
γ	-0.509	-0.447	-0.330	0.216	-0.736	1.000

Table 2. Common factor variance

Index	Initial	Extraction	Index	Initial	Extraction	Index	Initial	Extraction
V	1.000	0.731	H	1.000	0.930	β	1.000	0.992
L	1.000	0.935	α	1.000	0.985	γ	1.000	0.998

Table 3. Eigenvalues and principal component contribution rate

Component	Initial principal component		
	Eigenvalue	Contribution of variance/(%)	Cumulative/(%)
1	3.17	53.11	53.11
2	1.35	22.56	75.67
3	1.03	17.20	92.87
4	0.41	6.88	99.75
5	0.01	0.24	99.99
6	0.01	0.01	100.00

Table 4. Principal component scoring matrix

Index	Principal component score		
	Principal component one	Principal component two	Principal component three
V	0.40	-0.34	0.27
L	0.51	-0.18	0.26
H	0.50	0.01	0.34
α	0.09	0.80	0.31
β	0.41	0.45	-0.40
γ	-0.39	0.10	0.69

3.3. Network Model Prediction and Analysis

Finally, the genetic population size is 20, the number of evolutions is 50, the crossover probability is 0.4, the mutation probability is 0.2, the network learning efficiency is 0.1, and the target error is 0.01. Based on this, the prediction model of slope foot landslide motion distance caused by earthquake is constructed. During the experiment, 60 samples were randomly selected from the sample population as the training set of the network model, and the remaining 15 samples were used as the test set. Absolute error (AE), absolute percentage error (APE) and average absolute percentage error (MAPE) are used to evaluate the accuracy of prediction results. The calculation formulas are as follows:

$$E_{ae} = \mathrm{abs}(L_P - L_{max}) \tag{7}$$

$$E_{ape} = \frac{\mathrm{abs}(L_P - L_{max})}{L_{max}} \times 100\% \tag{8}$$

$$E_{mape} = \frac{1}{n}\sum_{i=1}^{n}\left(\frac{\mathrm{abs}(L_P - L_{max})}{L_{max}} \times 100\%\right) \tag{9}$$

Where: E_{ae} is the absolute error of prediction, E_{ape} is the relative error of prediction; E_{mape} is the average relative error of prediction; L_P is the predicted value of the model, L_{max} is the expected value of the model, n is the capacity of the test set.

In order to verify the superiority of the neural network compared with the traditional mathematical statistics method and the optimization ability of genetic algorithm to BP neural network in this kind of problems, the least squares method is used to do multiple regression and BP neural network is used to predict the distance of Landslide Motion on the basis of principal component analysis, and the maximum vertical motion distance H The regression equation of the maximum horizontal motion distance L can be found in formula (10) and (11). The prediction results of each method can be seen in figure 2 and 3 below. The absolute error can be seen in figure 4 and 5 below.

$$L = -184.15 - 0.85Y_1 + 2.42Y_2 + 4.26Y_3 \quad (10)$$

$$H = 58.83 + 0.38Y_1 - 0.39Y_2 + 1.44Y_3 \quad (11)$$

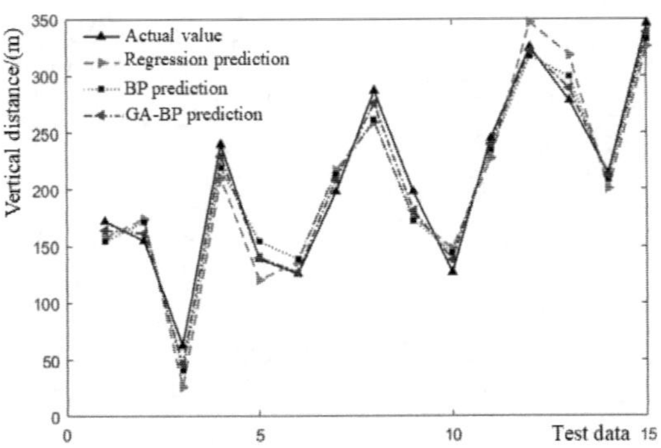

Figure 2. Prediction of maximum horizontal motion distance

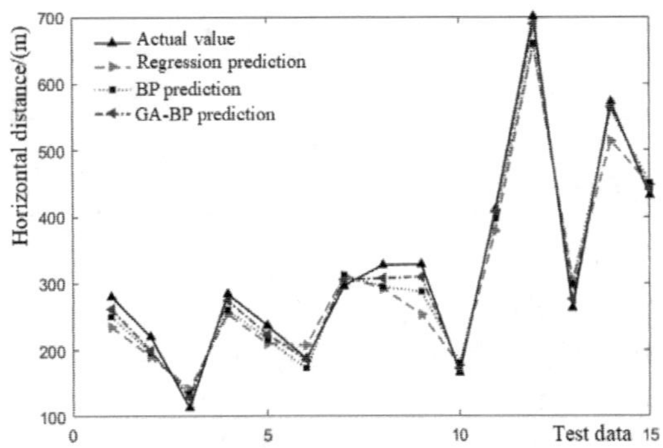

Figure 3. Prediction of maximum vertical motion distance

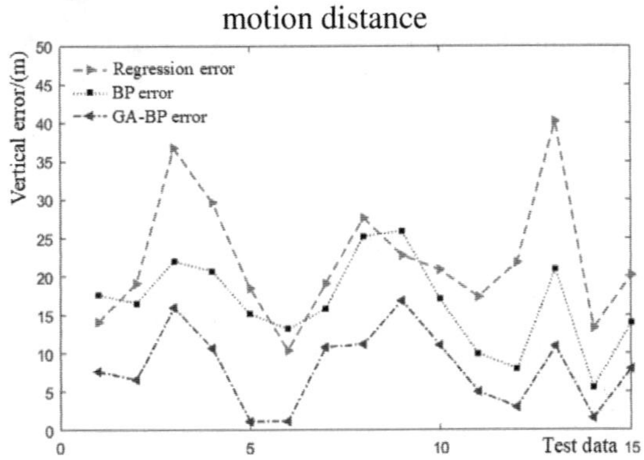

Figure 4. Absolute error of vertical motion distance

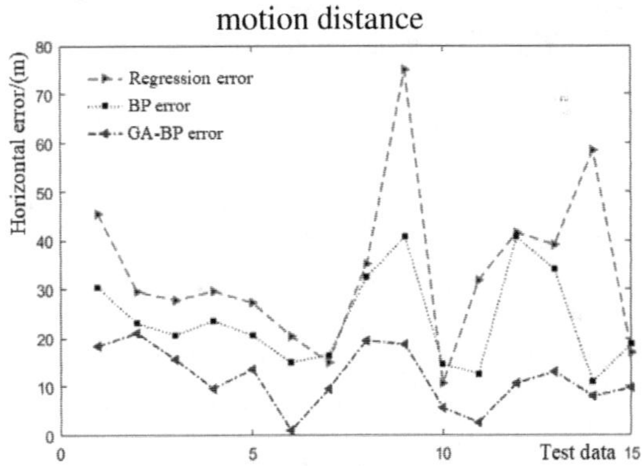

Figure 5. Absolute error of horizontal motion distance

Through the analysis of the predicted data, GA-BP neural network has the best prediction effect on the landslide movement distance. The minimum absolute error of vertical and horizontal movement distance is 1.16 meters and 1.01 meters, the minimum relative error is 0.74% and 0.54%, and the relative error is less than 10% which accounts for 93.33% and 86.67% of the test samples respectively. The standard deviations of errors are 4.87 and 5.91 respectively. The accuracy and stability of GA-BP neural network prediction are better than those of multiple regression and BP neural network prediction. The specific prediction errors are shown in table 5 and 6 below.

Table 5. Absolute error distribution

Prediction model	AE Absolute error/(m)			AE analysis	APE Average percentage error/(%)			APE analysis	
	Max	Min	AVG	Standard deviation	Max	Min	AVG	<=10%	Proportion /(%)
Regression	40.19	10.43	22.11	8.02	58.31	5.82	13.35	8	53.33
BPNN	25.86	5.59	16.52	5.69	34.91	2.45	9.99	9	60.00
GA-BPNN	16.86	1.16	8.12	4.87	25.38	0.74	5.12	14	93.33

Table 6. Relative error distribution

Prediction model	AE Absolute error/(m)			AE analysis	APE Average percentage error/(%)			APE analysis	
	Max	Min	AVG	Standard deviation	Max	Min	AVG	<=10%	Proportion/(%)
Regression	75.04	10.69	33.60	16.33	24.59	3.94	11.66	8	53.33
BPNN	40.72	10.98	23.63	9.46	18.26	1.92	8.62	10	66.67
GA-BPNN	21.04	1.01	11.77	5.91	13.83	0.54	4.57	13	86.67

4. Conclusion and Discussion

Firstly, the principal components affecting the movement distance of slope-foot landslide are extracted by principal component analysis, and the intrinsic mechanism of landslide movement is studied. Then, the prediction model of the movement distance of slope-foot landslide is constructed based on BP neural network optimized by genetic algorithm, and the movement distance of slope-foot landslide is predicted, and the results of the network model were compared with those of the multiple regression prediction. The conclusions are as follows:

(1) The GA-BP neural network model is accurate and stable in predicting the distance of landslide motion, and its strong ability to deal with non-linear problems is suitable for solving Landslide Motion Problems with complex internal mechanism.

(2) GA-BP neural network model is superior to the original BP neural network model in predicting accuracy and stability. It shows that in this problem, genetic algorithm has a great optimization on BP neural network from the training process of the network to the output of the results.

(3) For the prediction of landslide movement distance, GA-BP neural network is the best, followed by BP neural network. The deviation based on multiple regression prediction is large and unstable, and the effect is the worst. It shows that the neural network is more suitable for the study of this kind of problem than the traditional mathematical statistics method.

Using GA-BP neural network to construct the prediction model of landslide motion distance is an attempt to combine artificial intelligence with practical engineering problems. It introduces new methods and new ideas for this kind of research. At the same time, it has practical significance to predict the impact range and disaster intensity of slope foot landslides caused by earthquakes in southwestern China. Because of the complexity of influencing factors and intrinsic mechanism of landslide movement, this paper only selected the foot-slope landslide movement caused by earthquake in southwestern China as the research object. It is found that the prediction model of landslide motion distance is also affected by the scale interval of landslide, and further study is needed for this problem.

5. References

[1] FAN Xiaoyi, LENG Xiaoyu et al. The movement distance and topographic factors of slope foot type and transition earthquake landslide [J]. Geotechnical Mechanics, 2015, 36(5): 1380-1388.
[2] LI Ruikuan, LIANG Qingguo et al. Research achievements and prospects of landslide mechanism and slope stability [J]. Soil and Water Conservation in China, 2019(3): 40-42.

[3] MENG Huajun, JIANG Yuanjun, et al. Variation analysis of influencing factors of landslide movement distance in Dujiangyan mountain area before and after Wenchuan earthquake [J].Journal of Geomechanics, 2017, 23(6): 904-913.

[4] TANG Ran, XU Qiang, et al. Calculating model of moving distance of push-type landslide [J].Geotechnical Mechanics, 2018, 39(3): 1009-1020.

[5] ZHENG Guang, XU Qiang, et al. Study on the Formula for Calculating the Movement Distance of Rock Landslide-Clastic Flow [J]. Geotechnical Mechanics, 2019, 40(12): 01-10.

[6] HAO Shanshan, LI Menghua, et al. Significance analysis of soil erosion factors in loess hilly gully region [J]. Science of Soil and Water Conservation, 2019, 17(2): 77-86.

[7] ZHANG Feng. Application of PSO-RBF Neural Network in Landslide Risk Prediction in Historic Landslide Developing Areas [J].Geographic and Geographic Information Science, 2018, 34(4): 118-124.

[8] LIAO Jun, DONG Qian, et al. Influencing factors of motion parameters of high-level landslides based on UDEC [J].Science, Technology and Engineering, 2018,18(34): 29-36.

[9] WEI Changli, CHEN Liang, et al. Prediction of Causes and Stability of Landslide in Ozhai Village, Upper Reaches of Minjiang River [J]. South-to-North Water Transfer and Water Science and Technology, 2019,17(3):185-192.

[10] WANG Wenpo, HAN Aiguo, et al. Sensitivity Analysis of Environmental Factors for Landslide Disaster Pregnancy in Puge County, Sichuan Province [J]. Journal of Yangtze Academy of Sciences. 2018, 38 (9): 63-67.

IOP Conf. Series: Materials Science and Engineering **730** (2020) 012053 doi:10.1088/1757-899X/730/1/012053

Complex Environmental Characteristics of Coastal Salt Spray Area and Effect of Carbon Dioxide Content on Corrosion of Steel Bars

Jun Liu[1], Lu Hu[1] and Yingqiu Song[2*]

[1]Guangdong Provincial Key Laboratory of Durability for Marine Civil Engineering, Shenzhen University, Shenzhen, Guangdong, China.

[2]Normal university, Shenzhen University, Shenzhen, Guangdong, China.

*Email: wookee2012@163.com, *Corresponding author: Yingqiu Song

Abstract: In the coastal atmospheric environment, corrosive media exists due to natural and man-made pollution, which has a great impact on the durability of concrete structures. Taking the environment of Shenzhen coastal salt spray area as an example, the environmental parameters were observed on site, and the content of carbon dioxide in the atmosphere was measured. The influence of steel corrosion on steel was analyzed. The study found that the annual temperature change range of the coastal salt spray area is 15 °C to 37.9 °C, and the change is first rising and then decreasing. The temperature rises in the first half of the year, and the temperature gradually decreases in the second half of the year. The highest humidity in the measurement area is 74.7%. The minimum is 56%, and the overall level is maintained at a high level, which is between 55% and 75%. The carbon dioxide content of the five measuring points is consistent throughout the year, and the carbon dioxide content is high in spring and winter.

1. Introduction

In the construction of marine and coastal structures, concrete and steel are currently used as building materials. In most environments, the normal service life of concrete structures is long. However, due to the harmful medium in the environment, it will penetrate into the interior of the concrete structure, and some chemical reaction will occur in contact with the saturated solution of calcium hydroxide and cement hydration products. May cause premature failure of concrete structures [1,2]. In the coastal environment, due to natural and man-made reasons, there are many corrosion factors, which have a great influence on the durability of materials. Among them, sulfur dioxide, acid rain and chloride ions have great influence. Since concrete is a porous material, there are often defects such as capillaries, pores, and bubbles in its interior, which have a certain gas permeability. The CO2 in the air first penetrates into the pores and capillaries filled with air inside the concrete, and then the liquid phase dissolved in the capillary interacts with substances such as calcium hydroxide and hydrated calcium silicate produced during the hydration of the cement to form carbonic acid [3,4]. Calcium hydroxide is one of the main hydration products of cement. For ordinary Portland cement, the calcium hydroxide formed by hydration can reach 10% to 15%. On the one hand, calcium hydroxide maintains the high alkalinity of concrete, and on the other hand, it is one of the most unstable components in concrete, and it is easy to neutralize with acidic media in the environment, thereby carbonizing concrete. During the carbonization process, the calcium hydroxide in the concrete is consumed, which may cause damage to the passivation film of the steel bar, thereby causing corrosion of the steel bar.

Shenzhen is located in southern China and is a typical marine atmosphere. To analyze the impact of environmental factors on structural durability, it is necessary to first define the coastal environmental parameters. In this paper, the station is set up in the coastal salt fog area, and the humidity, temperature and carbon dioxide content are observed. It has certain practical significance for studying the durability of the concrete structure in the coastal salt spray area.

2. Experiment

In Shenzhen, the salt spray area environment was selected, an observation station was set up on site, and measuring points were set at 0, 100, 200, 400, 1000 meters from the coastline. The carbon dioxide concentration, temperature and humidity of the salt spray area were measured by a carbon dioxide meter. The measurement time of each measurement point is 5 minutes, and the measurement results are stable and recorded.

3. Results and Analysis

3.1. The variation Law of Temperature in Coastal Salt Spray Area

Table 1. Temperature changes in Shenzhen from 2016 to 2018

Year	Average temperature/°C	Maximum temperature/°C	Minimum temperature/°C
2016	23.2	36.7	1.7
2017	23.2	36.9	7.4
2018	23.4	35.6	4.2

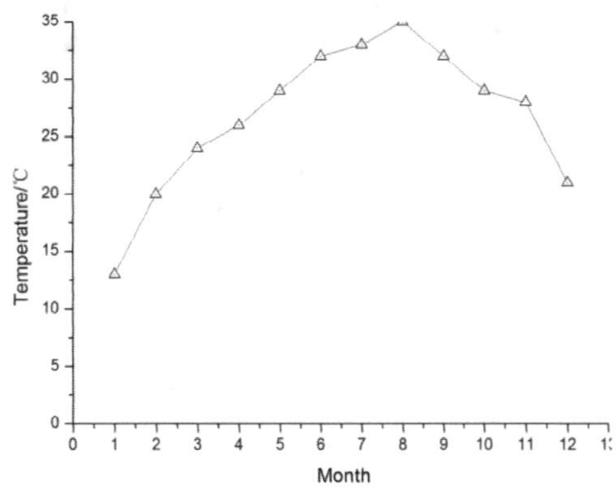

Figure 1. Average temperature of the measuring point as the month

The annual average temperature reaches 23 degrees Celsius, and the highest temperature from 2016 to 2018 is 36.9 degrees Celsius and the lowest is 1.7 degrees Celsius. In view of the high temperature characteristics of Shenzhen, it has a significant impact on the durability of concrete. The corrosion of steel bars in concrete is a dynamic electrochemical process. The corrosion rate is mainly controlled by environmental factors, besides the influence of its own materials. It mainly includes factors such as ambient temperature and relative humidity. The following is the change of temperature in the coastal salt spray zone from the temperature records of each measuring point through one year.

The temperature changes of the five measuring points in the coastal salt fog area are basically the same. The annual temperature change range is from 15 °C to 37.9 °C, and the change is first rising and then decreasing. The temperature in the first half of the year is rising continuously, and the temperature is gradually decreasing in the second half. 1,2,3,4,5,6,11,12 months, the temperature is lower, the temperature is higher in July, August and September, which is in line with the subtropical maritime climate in the coastal areas of Shenzhen, which is characterized by the cold spring and winter and the hot summer and autumn. From January to December, the temperature in the construction area of the west port area is 37.9 °C, the lowest is 13.9 °C, the lowest temperature in January, and the highest temperature in August.

3.2. Change Law of Humidity in Coastal Salt Spray Area

Table 2. Changes in rainfall and humidity in Shenzhen from 2016 to 2018

year	Total rainfall /mm	Rainy days/d	Average relative humidity
2016	2490.6	153	79
2017	1967.1	127	77
2018	1957.2	121	76

Shenzhen is the rainy season from May to September every year. The average rainfall from 2016 to 2018 is 2490.6 mm, and the annual average relative humidity is 77.3%. The environmental humidity has a great influence on the micro-environment inside the concrete, which determines the size of the pore water saturation inside the concrete, and has the resistivity of the structural concrete, the diffusion rate of oxygen inside the structure, and the supply of corrosion reaction electrochemical corrosion. Larger impact, which has an effect on the corrosion rate of the steel. Shenzhen is a high-humidity environment with an annual average relative humidity of about 77%. It is above the critical humidity for most of the year, so it provides sufficient conditions for the electrochemical corrosion of steel bars, which is highly prone to humid atmospheric corrosion.

According to annual observations and records, the highest humidity in the measurement area is 74.7%, the lowest is 56%, and the overall level is maintained at a high level, about 55% to 75%. The changes in humidity at the five measuring points were basically the same. The humidity gradually increased in the first half of the year and the humidity dropped in the second half. The higher humidity months are concentrated in summer and autumn. In the second and third quarters, rainfall is abundant, the temperature is humid and hot, and the relative humidity is between 70% and 75%. The overall temperature is high, and the month is generally characterized by high temperature and rain. In the first quarter and the fourth quarter, the rainfall is relatively small, the temperature is dry and cold, and the relative humidity is between 60% and 65%, which is low overall.

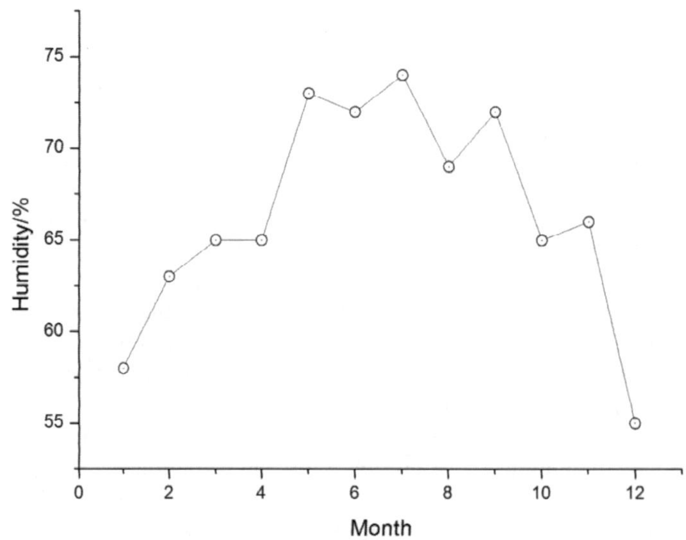

Figure 2. Average humidity as a function of the month

IOP Conf. Series: Materials Science and Engineering **730** (2020) 012053 doi:10.1088/1757-899X/730/1/012053

3.3. Factors Affecting Carbon Dioxide Content in Coastal Salt Spray Area

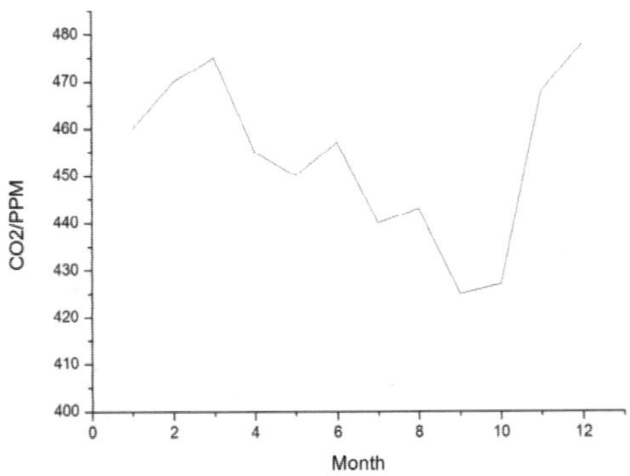

Figure 3. Carbon dioxide content of measuring point one changes with the month

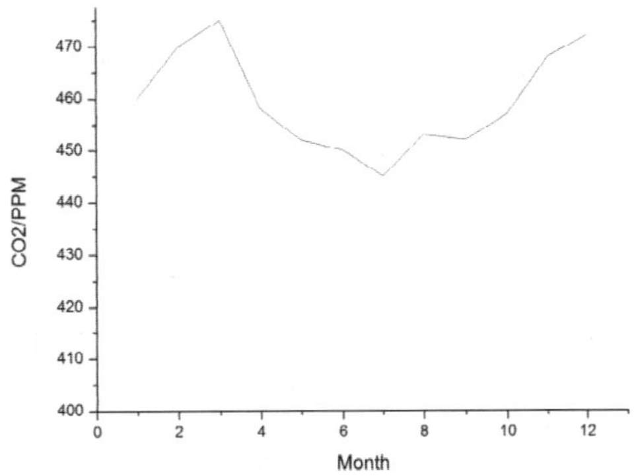

Figure 4. Carbon dioxide content of measuring point five changes with the month

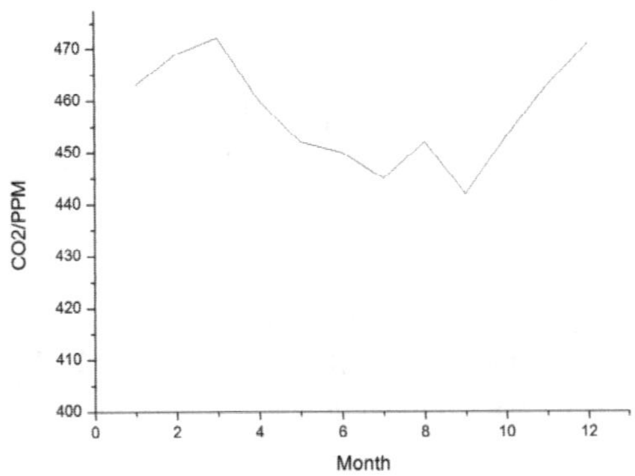

Figure 5. Average carbon dioxide content in the survey area as the month

The carbon dioxide in the coastal salt spray area will cause carbonization of concrete. The carbonization of concrete refers to the action of acid gas such as CO2 and SO2 in the air and calcium hydroxide in the liquid phase of concrete, and the neutralization process of CaCO3 and H2O [5,6]. In the coastal environment, there are many corrosion factors. It is of great significance to study the atmospheric carbon dioxide content and the influencing factors on the durability of concrete structures. The following is a measure of the variation of carbon dioxide content at each measurement point with the month and weather.

The change of carbon dioxide content in the five measuring points is consistent in the whole year, the carbon dioxide content in spring and winter is higher, and the carbon dioxide content in summer and autumn is lower; however, the change of carbon dioxide content in each measuring point is small, and each test in the same month The carbon dioxide content is similar. It can be seen that in the case of a small area, the carbon dioxide content of each place is basically at the same level. The carbon dioxide content in this area is basically between 460 (ppm) and 480 (ppm). From the average carbon dioxide year of Figure 5, it can be seen that the carbon dioxide content of spring air in Shenzhen is the highest in March, and the lowest overall content in July, August and September is between 450 and 480 (ppm); the carbon dioxide content is rising from January to March. After that, it continued to drop to September, and then continued to rise to December. During the whole year, the content of carbon dioxide is high in autumn and winter, and the land vegetation is reduced accordingly. The photosynthesis is reduced and the carbon dioxide content is high. The annual carbon dioxide content is low in spring and summer, and the land vegetation increases during the period. The effect is increased and the carbon dioxide content is low. Second, affected by the monsoon, the spring and summer winds blow from the sea to the land, bringing a lot of oxygen; the autumn and winter winds blow from the land to the sea, and the carbon dioxide content increases accordingly.

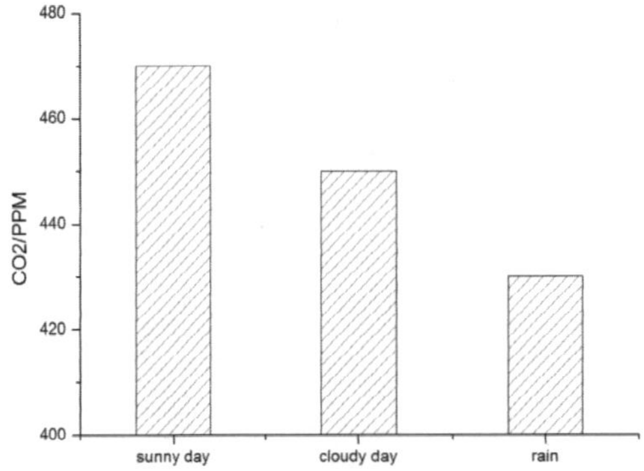

Figure 6. Effect of weather on carbon dioxide content

Figure 6 analyzes the effect of weather on carbon dioxide content, with the highest carbon dioxide content in sunny days, followed by cloudy days and rainy days. Because carbon dioxide is soluble in water, the air humidity in the rain is high and the air pressure is high, so the carbon dioxide content in the air is low. The difference between the carbon dioxide content of sunny days and cloudy days is about 20 ppm, and the difference of carbon dioxide content between sunny days and rainy days is about 40 ppm.

4. Conclusion

1. The average annual temperature from 2016 to 2018 is above 23 degrees Celsius, the maximum temperature is 36.9 degrees Celsius and the lowest is 1.7 degrees Celsius. The temperature changes of the five measuring points in the coastal salt spray area are basically the same. The temperature in the first half of the year is rising continuously, and the temperature is gradually reduced in the second half of the year.

2. The highest humidity in the measurement area is 74.7%, the lowest is 56%, and the overall temperature is maintained at a high level, about 55% to 75%. The humidity changes of the five measuring points are basically the same, and the humidity gradually rises in the first half of the year. The humidity will drop in the second half of the year.

3. The change of carbon dioxide content in the five measuring points is consistent in the whole year. The carbon dioxide content in spring and winter is high. The carbon dioxide content in summer and autumn is low. The carbon dioxide content in sunny air is the highest, followed by cloudy days and rainy days.

5. Acknowledgement

The financial supports from Natural Science Foundation of China (No. 51578340), Shenzhen city science and technology project (No. JCYJ20170302143133880, JCYJ20180305124008155), Shenzhen International Cooperation Research Project (No. GJHZ20180928155602083) and Research projects in key areas of Guangdong(2019B111107002).

6. References

[1] Jun Liu, Qiwen Qiu, Xiaochi Chen, Xiaodong Wang, Feng Xing, Ningxu Han, Yijian He. Degradation of fly ash concrete under the coupled effect of carbonation and chloride aerosol ingress [J]. Corrosion Science, 2016, 112: 364-372

[2] Jun Liu,Qiwen Qiu,Xiaochi Chen,Feng Xing,Ningxu Han,Yueshan Ma. Understanding the interacted mechanism between carbonation and chloride aerosol attack in ordinary Portland cement concrete[J].Cement and Concrete Research.2017,95:217-225

[3] Hermawana, Puti F. Marzukia, Muhamad Abduha, et al. Identification of source factors of carbon

dioxide (CO2) emissions in concreting of reinforced concrete [J]. Procedia Engineering, 2015, 125:692 – 698

[4] Anna M. Grabiec, Daniel Zawal, Jakub Szulc. Influence of type and maximum aggregate size on some properties of high-strength concrete made of pozzolana cement in respect of binder and carbon dioxide intensity indexes [J]. Construction and Building Materials, 2015, 98:17-24

[5] Phil Purnell, Leon Black. Embodied carbon dioxide in concrete: Variation with common mix design parameters [J]. Cement and Concrete Research, 2012, 42:874–877

[6] J.H.M. Visser. Influence of the carbon dioxide concentration on the resistance to carbonation of concrete [J]. Construction and Building Materials, 2014, 67:8–13

IOP Conf. Series: Materials Science and Engineering **730** (2020) 012054 doi:10.1088/1757-899X/730/1/012054

Effects of Atrazine on the Growth of *Suaeda Heteroptera*

Yujie Huo[2,3], Haifeng Wei [1,2,3*],Yumeng Zhao [2,3],Jie He [2,3] and Mingliang Zhang [2,3]

[1]Dalian Ocean University, Key Laboratory of Mariculture& Stock Enhancement in North China's Sea, Ministry of Agriculture and Rural Affairs, Dalian 116023, China.
[2]Dalian Ocean University, School of Marine Science and Environment Engineering, Dalian 116023, China.
[3]KeyLaboratory of Coastal Marine Environmental Science and Technology of Liaoning Province, Dalian 116023, China.
Email: weihaifeng@dlou.edu.cn

Abstract: To explore the ecological degradation of *Suaeda Heteroptera* in Panjin Red Beach wetland, the effects of Atrazine on the growth and physiological effects of *Suaeda Heteroptera* seedlings were studied by indoor hydroponics. Experimental results show that the leaves and roots of the *Suaeda Heteroptera* seedlings are very sensitive to Atrazine. And the LD_{50} of the number of leaves of *Suaeda Heteroptera* seedlings were 34.2 μg/L. At low concentrations (3.8-11.4 μg/L), Atrazine promoted the growth of seedling leaves; at higher concentrations (22.8-34.8 μg/L), it inhibited leavse growth and had a significant dose-effect relationship Atrazine had different effects on height and root length of *Suaeda Heteroptera* , but had no effect on seedlings stem length. The formation of chlorophyll b in leaves of *Suaeda Heteroptera* treated by Atrazine was inhibited. Also, the main factor for the decrease of total chlorophyll consent is Atrazine.

Atrazine is a triazine pesticide and is a selective herbicide. Its chemical properties are relatively stable and can only be hydrolyzed by heating in acidic or alkaline media [1]. Atrazine is one of the most easily produced triazenyl benzenes. Because Atrazine is easy to synthesize, cost-effective and effective, it has become the herbicide with the highest utilization rate among farmers [3].

The *Suaeda Heteroptera* community is a major component of wetland and a dominant species of coastal tidal flats. Since 2000, it has begun to show significant degradation, and unexplained deaths have led to a reduction in their community area. Although Chinese scholars have carried out some research on the causes of the degradation of the *Suaeda Heteroptera*, there are still many shortcomings in the current research. In the early on-site monitoring, a variety of herbicides were found in the water and soil of the growing area of *Suaeda Heteroptera*. Therefore, studying the effect of Atrazine on the growth of *Suaeda Heteroptera* has important practical significance for further revealing the cause of its degradation. [4]

1. Materials and Methods

1.1. Materials

The seeds of *Suaeda Heteropteraare* from the coastal beach of Panjin, Liaoning Province. The seawater is taken from the sand-filtered seawater in the Blackstone reef of Dalian. Atrazine, produced by Hefei Jiuyi Agricultural Development corporation, the active ingredient is 38%. Plant growth nutrient solution.

1.2. Methods

1.2.1. Test method
Before the experiment, 10 seedlings (18 groups) were selected, and the seedlings of *Suaeda Heteroptera* were vigorously tested. The data of seedling height (8-10 cm), stem length (2.5-3.5 cm), root length (2.5-3.5 cm), number of leaves (10-12), and number of roots (6-8) were recorded. According to the investigation of the use of pesticides in the Red Beach wetland. Set 5 groups of Atrazine concentration (3.8, 11.4, 19, 22.8, 26.6, 34.2 μg/L). Three replicates per treatment were performed in a 100 ml glass beaker. Place in a well-ventilated and well-lit place at room temperature to ensure that the seedlings can grow normally. The experiment time is two weeks.

1.2.2. Determination of indicators
1.2.2.1. Plant height, stem length, root length, number of leaves, number of roots
On the 3rd, 8th and 11th day, the plant height, stem length and root length of the seedlings were measured with a scale. The number of leaves and the number of roots were recorded by counting method. (plant height: full length from the bottom of the main root to the top of the blade; stem length: from the bottom of the last pair of blades to the root; root length: from the root to the bottom of the main root.)
1.2.2.2. Determination of chlorophyll content
Grind the seedlings with a grinder and put them into a 10 ml test tube, then add 6 ml of 95% alcohol solution and heat in a 75 °C water bath for 5 min to complete the extraction. After filtration, dilute to 10 ml colorimetric tube with 95% alcohol solution. Measured with a 752N UV-Vis spectrophotometer.
Calculation formula:
Chlorophyll a concentration: $Ca = 13.95A665 - 6.88A649$
Chlorophyll b concentration: $Cb = 24.96A649 - 7.32A665$
Chloroplast pigment content:

$$\text{Chloroplast pigment content}(mg/g) = \frac{(Ca + Cb) \times \text{Extract volume V} \times \text{Dilution factor}}{m}$$

1.2.3. Data Statistics and Processing
The experimental data was analyzed by SPSS 20.0 and Excel statistical software. One-way ANOVA was performed with SPSS 20.0, and the Ducan method was used for comparison. The significance level was set to 0.05.

2. Results

2.1. Effects of Atrazine on Plant Height, Stem Length and Length of Main Roots of Suaeda Heteroptera
As can be seen from Figure 1, the 11th day, the Atrazine solution with a concentration of 3.8-26.6 μg/L had no significant effect on the plant height growth of the seedlings, and the paired T test method ($P>0.05$) showed no significant difference. At the concentration of 34.2μg/L, the plant height of the seedlings was significantly different from that of the control group ($P<0.05$), indicating that the high concentration of Atrazine solution significantly inhibited the plant height growth of the seedlings. There was a significant dose-effect relationship between Atrazine and plant height, $y = -0.1252x + 11.772$, $R^2 = 0.8926$.

As can be seen from Figure 2, When the concentration of Atrazine was 3.8-34.2 μg/L, the stem length of seedlings fluctuated in a small range, and there was no significant difference ($P>0.05$).

As can be seen from Figure 3, the 11th day, when the concentration of Atrazine was 3.8-19 μg/L, the root length of the seedlings of Suaeda Heteroptera fluctuated within a small range, ($P>0.05$). At the concentration of 26.6—34.2μg/L, there was a significant difference between the root length of the seedlings and the control group, and there was a significant dose-effect relationship between Atrazine concentration and root length, $y=-0.0825x+4.2292$, R2 = 0.8368.

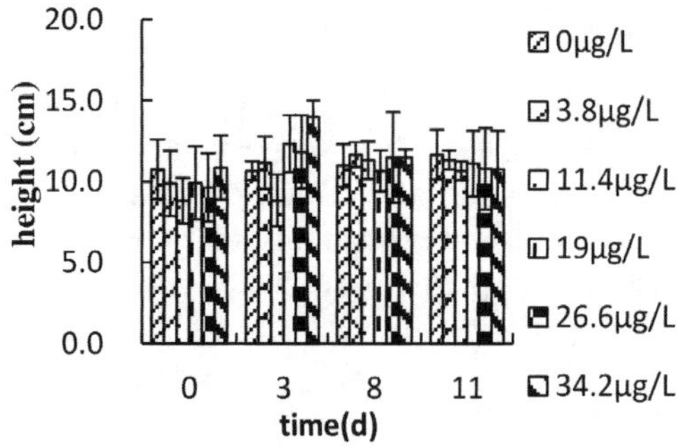

Figure 1. Effect of Atrazine on the height of *Suaeda Heteroptera*

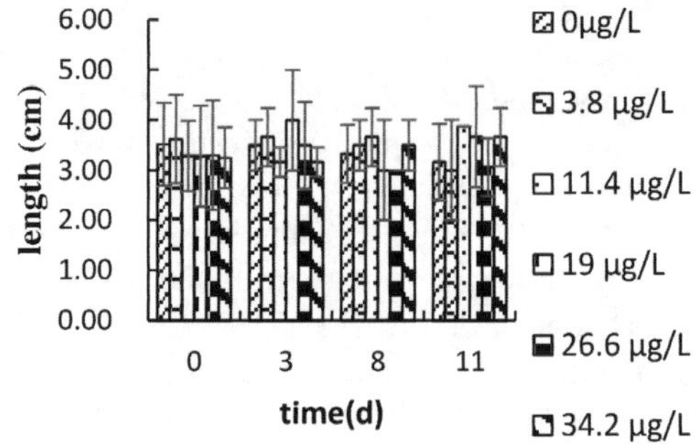

Figure 2. Effect of Atrazine on stem length of *Suaeda Heteroptera*

2.2. The Effect of Atrazine on the Number of Leaves and Roots of Suaeda Heteroptera

The effect of Atrazine on the growth of *Suaeda Heteroptera* was directly reflected in the changes in the number of leaves and roots. According to the number of leaves and roots, the extent of the effect of Atrazine on the growth of *Suaeda Heteroptera* was determined. The effects of Atrazine on the leaves and roots of *Suaeda Heteroptera* are shown in Figures 4, 5.

As can be seen from Figure 4, on the 11th day showed obvious changes. When the concentration of Atrazine was 11.4 – 19 μg/L, the number of leaves was not significantly different from that of the control group ($P>0.05$). There was a significant dose-effect relationship between Atrazine concentration and leaf number between 26.6 and 34.2 μg/L, $y= -0.7161x+16.62$, $R^2 =0.7245$. The number of leaves decreased as the concentration increased ($P<0.05$), and there was a significant change at 34.2 μg/L, and the inhibition rate was 50%.

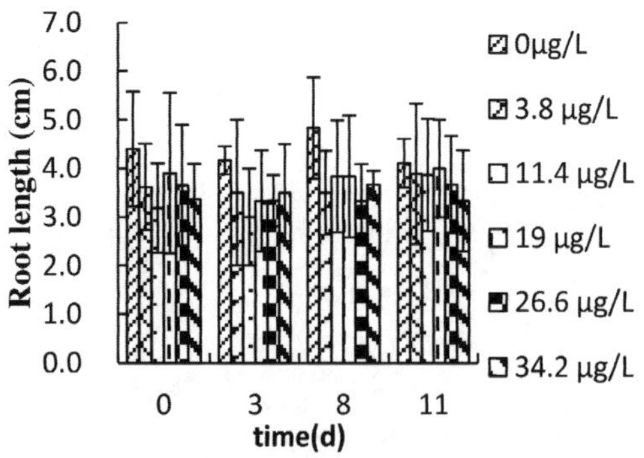

Figure 3. Effect of Atrazine on root length of *Suaeda Heteroptera*

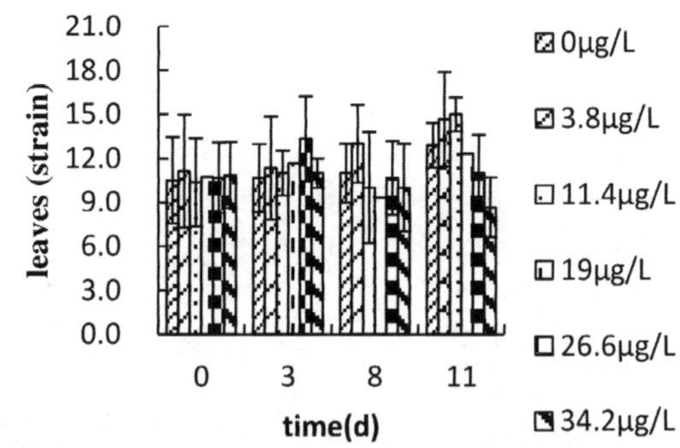

Figure 4. Effect of Atrazine on the Number of Leaves of *Suaeda Heteroptera*

Figure 5 shows that there is a clear gradient in the number of seedling roots on the 8th and 11th day, and the higher concentration of Atrazine the smaller the number of roots. At 11th day, when the concentration of Atrazine was 3.8 μg/L, the number of roots of seedlings was significantly different from that of the control group ($P<0.05$). And when the concentration of Atrazine was 11.4-38 μg/L, there was a significant change. There was a significant dose-effect relationship between the concentration of Atrazine and the number of roots. The correlation equations and coefficients were $y= -1.7758x+23$ and R2=0.937, respectively. The results showed that the concentration of Atrazine in the range of 3.8-38 μg / L had a significant inhibitory effect on the number of seedling roots, and the inhibition rate was 72%. The growth of the roots of Suaeda Heteropterais very sensitive to the concentration of Atrazine in the environment, so when the concentration of Atrazine exceeds 3.8 μg/L, it may cause slow growth or even death.

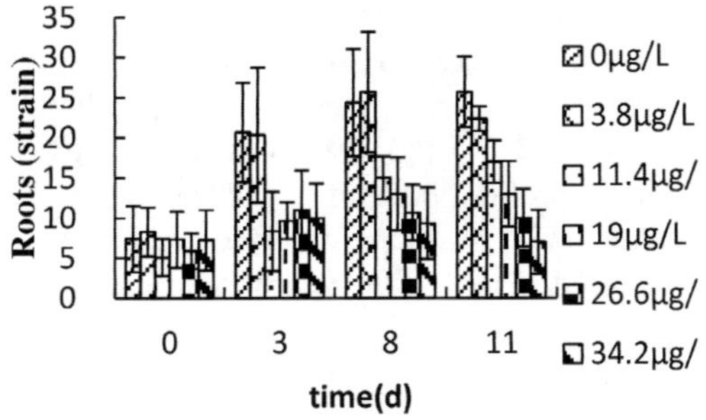

Figure 5. Effect of Atrazine on the Number of Roots of *Suaeda Heteroptera*

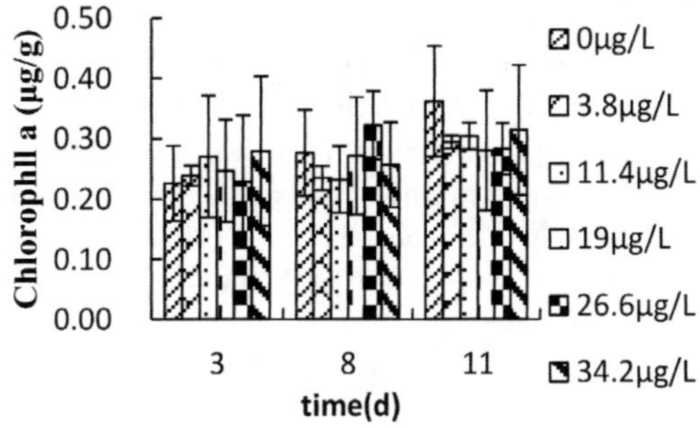

Figure 6. Effect of Atrazine on Chlorophyll a of *Suaeda Heteroptera*

2.3. Effects of Atrazine on Chlorophyll A and Chlorophyll B in Seedlings of Suaeda Heteroptera

As can be seen from Figure 6, Within the experimental concentration range (3.8-34.2 μg/L), there was no significant difference between the treatment group and the control group ($P>0.05$), indicating that Atrazine had no significant effect on the content of chlorophyll a in *Suaeda Heteroptera.*

It can be seen from the data in Figure 7, on the 11th day, the content of chlorophyll b in the seedlings showed a significant change trend. When the concentration of Atrazine was between 3.8-34.2 μg/L, the chlorophyll b content of the seedlings showed a downward trend. The chlorophyll b content was significantly different from that of the control group at 3.8-19 μg/L ($P<0.05$). At 26.6-34.2 μg/L, there was a significant difference between the two groups, and the inhibition rate was 57%. There was a significant dose-effect relationship between Atrazine concentration and chlorophyll b content, $y= -0.0315x+0.5563$, $R^2 =0.8107$, indicating that the Atrazine solution with a concentration of 3.8-34.2 μg/L had a significant effect on the content of chlorophyll b in the seedlings of S. salsa, which inhibited the synthesis of chlorophyll b.

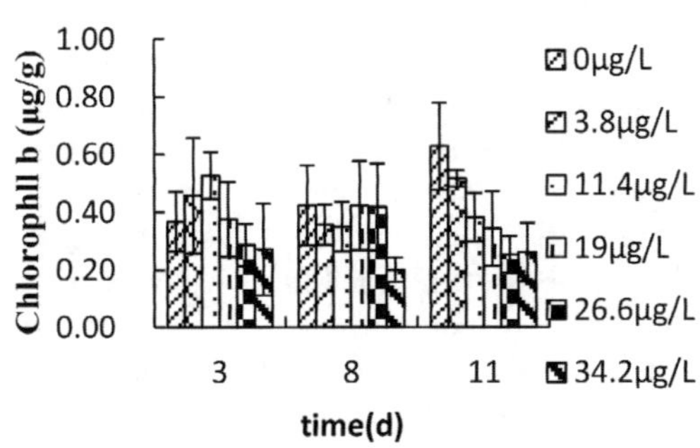

Figure 7. Effect of Atrazine on Chlorophyll b of *Suaeda Heteroptera*

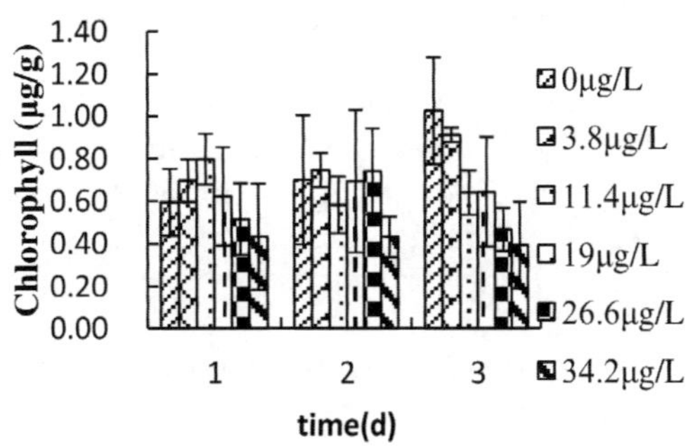

Figure 8. Effect of Atrazine on Chlorophyll of *Suaeda Heteroptera*

2.4. The Effect of Atrazine on the Total Chlorophyll Content of Suaeda Heteroptera

It can be seen from Figure 8 that there is a clear trend on the 11th day. When the concentration of Atrazine was 3.8-34.2 μg/L, the total chlorophyll content of *Suaeda Heteroptera* showed a significant downward trend. At the concentration of 3.8-19 μg/L, the total chlorophyll content of the seedlings was significantly different from the control group ($P<0.05$), and at the concentration of 26.6-34.2 μg/L, there was a significant change and the inhibition rate was 55%. There was a significant dose-effect relationship between Atrazine concentration and total chlorophyll content, $y= -0.0612x+1.0159$, $R^2 =0.9133$, indicating that the solution of Atrazine had a significant inhibitory effect on the chlorophyll synthesis of *Suaeda Heteroptera.*

From the appearance point of view, on the 14th day after treatment, when the concentration of Atrazine was 26.6-34.2 μg/L, the whole Suaeda Heteroptera showed atrophy, the leaves turned yellow, the rhizome showed bending and lodging, and gradually withered. At the concentration between 11.4 and 19 μg/L, the leaves in the lowermost layer showed atrophy. In summary, Atrazine has an inhibitory effect on the synthesis of chlorophyll in the seedlings of Suaeda Heteroptera. When the concentration of Atrazine was low, the degree of inhibition was small; and when the concentration exceeded 11.4 μg/L, a significant inhibitory effect was observed.

3. Discussion

3.1. Relationship Between Concentration of Atrazine and Growth Index of Suaeda Heteroptera

Low concentration of Atrazine had no significant effect on plant height, and high concentration inhibited the growth of *Suaeda Heteroptera*. Jinfeng Ma et al. pointed out that there was no significant difference between the experimental group and the control when the concentration of Atrazine was 1040-2080 g.a.i/hm^2. The plant height was basically consistent with the control. The low concentration of Atrazine had no significant effect on the emergence of medlar. The concentration of Atrazine reached 2460 g.a.i/hm^2 had a significant inhibitory effect on the plant height of the scorpion. This conclusion is consistent with the results of this experiment [5]. It indicated that when the plants were under the stress of Atrazine the photosynthesis electron transfer was blocked, and the formation of NADPH and ATP was reduced, which affected the carbon fixation and inhibited the normal growth of plants [6-9].

The concentration of Atrazine has a significant effect on the root growth of seedlings. Dandan Liu et al. found that when the Atrazine was 0.05 mg/L, there was no significant effect on the root growth of soybean seedlings. When the concentration of Atrazine was greater than 0.1 mg/L, the growth of radicle was significantly inhibited [10]. Under stress, the receptor protein in the plant transmits a stress signal, which produces a large amount of active oxygen, causing membrane lipid peroxidation and destroying the organelles, causing extracellular permeability of the cells, loss of electrolytes, increased conductivity, and inhibition of seedling growth.

Low concentration of Atrazine promoted the growth of seedling leaves and inhibited at high concentrations. Xianglin Liu et al. also showed that acetochlor can promote the plant height and root number of rice at low concentration. When the concentration of acetochlor reaches a certain value, it has a strong inhibitory effect on rice growth, and the root Inhibition is stronger than inhibition of stems [11], which is similar to the results of this experiment. At low concentrations, auxin promotes growth, while at high concentrations it inhibits plant growth. Higher concentrations cause plant death and can even act as herbicides [12]. In this experiment, the effect of Atrazine was similar to that of auxin.

3.2. Relationship Between Concentration of Atrazine and Physiological Indexes of Suaeda Heteroptera

Atrazine has no significant effect on chlorophyll a of *Suaeda Heteroptera*. There is a significant effect on chlorophyll b, showing a significant dose-effect relationship. The total chlorophyll content and the change trend of chlorophyll b were basically the same. The chlorophyll biosynthesis process consists of L-glutamyl-tRNA→chlorophyll a→chlorophyll b, which requires 15 steps of reaction involving 15 enzymes. The above experimental results show that the synthesis of chlorophyll a is not significantly affected, and the synthesis of chlorophyll b is inhibited, which may be caused by chlorophyll a to chlorophyll b. Atrazine affects the synthesis of certain enzymes [13]. Liu Jimin et al. showed that the chlorophyll content in cabbage leaves showed a significant downward trend compared with the control. There was a significant dose-effect relationship. There was a significant difference in the chlorophyll content between the control and the high concentration of Atrazine. There was no significant difference in the low concentration [14]. This conclusion is consistent with the results of this test.

4. Conclusion

The effects of Atrazine on different parts of the seedlings of *Suaeda Heteropterawere* quite different. From the perspective of seedling growth, there are obvious rules for the variation of plant height, root length and root number of seedlings. When the concentration of Atrazine reaches a threshold, below this concentration, there is no obvious effect on the species, but above this concentration, an inhibitory effect is produced. However, in the experiment of seedling stem length change, the law is not obvious, and the growth difference at each concentration is not significantly obvious.

From the aspect of chlorophyll content, Atrazine had a great influence on the total consent of chlorophyll b and chlorophyll in seedlings, and its synthesis was obviously inhibited, and the linear relationship showed a continuous decreasing trend. There is no obvious rule for the content of chlorophyll a, and the difference is not obvious at each concentration. The low concentration of Atrazine solution promoted the leaves growth of the seedlings, and the high concentration inhibited the growth of the seedlings, indicating that the low concentration of Atrazine promoted the growth of the plants. It may be that the different parts of the Suaeda Heteropteraare different in sensitivity to Atrazine, so the threshold is also different.

5. References

[1] Changling Liu 2002 *World Pesticide Encyclopedia, Herbicide Roll* [M] (Chemical Industry Press)

[2] Yaoxun Qu 2015 *Comprehensive understanding of Atrazine* [J]. Today's pesticides (10): 23-24

[3] Jiayun Lu 2001 *Plant pathogenic mycology* [M]. (Agriculture Press)

[4] Yi Qiu and Ningzi Xie 2008 *Research progress on the harm and non-biodegradation of pesticide atrazine* [J] Guangdong Chemical Industry 35(1):73-77

[5] Jinfeng Ma et al. 2014 *The effect of atrazine on the morphology and physiological indexes of Gardenia seedlings* [J] Crop Jr. (4): 138-142

[6] Caihong Yang, Liand Feng, Xingshan Tian 2016 *Effects of soil residues from atrazine on growth and chlorophyll fluorescence parameters of four vegetables* [J] Chinese vegetables (3): 53-59

[7] Kulikova N A and Perminova I V 2002 *Binding of Atrazine to Humic Substances from Soil: Peat and Coal Related to Their Structure* [J] Environ Sci Technol 36 (17): 3720–3724

[8] Piccolo A et al. 1998 *Atrazine Interactions with Soil Humic Substances of Different Molecular Structure* [J] J Environ Qual 27: 1324–1333

[9] Kellly W R and Wilson S D 2000 *Movement of Bromide Nitrogen-15 and Atrazine Through Flooded Soils* [J]. J Environ Qual 29 (4): 1085–1094

[10] Liu Dandan et al. 2017 *Relief of the toxic effect of degrading bacteria on atrazine in the early stage of soybean growth* [J]. Soybean Science 36(6): 938-942

[11] Liu Xianglin et al. 2014 *Study on Sensitivity of Different Varieties of Rice to Acetochlor* [J]. Chinese Agricultural Science Bulletin 30(18): 209-213

[12] Gu Xiaohong et al. 2017 *Application of Plant Growth Regulators in the Growth and Cultivation of Medicinal Plants* [J]. Modern Chinese Medicine 19(2):295-305

[13] Wang Pingrong et al. 2009 *Research progress in chlorophyll biosynthesis of higher plants* [J]. Journal of Northwest Botanical Sciences 29(3): 629-636

[14] Liu Jimin et al. 2011 *The effect of atrazine on the growth and physiological effects of Chinese cabbage seedlings* [J]. Journal of Southern Agricultural Sciences 42(7): 752-755

IOP Conf. Series: Materials Science and Engineering **730** (2020) 012055 doi:10.1088/1757-899X/730/1/012055

Distribution Features of Phytoplankton and Its Correlation with Environmental Factors of Baima Lake

Jun Wang[1], Sushu Wu[1,*], Bowen Fan[1,2] and Wenmeng Chen[1]

[1]Jiangsu Hydraulic Research Institute, 97 Nanhu Road, 210017 Nanjing, China.
[2]Hohai University, 1 Xikang Road, 210098 Nanjing, China.
*Email:wsspp8@163.com

Abstract. In order to deal with the degradation of ecological function in Baima Lake, 11 Sampling sites were set up in 2014, 2016 and 2017 in Huaian City, Jiangsu Province. Sampling was conducted monthly to monitor and analyze the temporal and spatial changes of phytoplankton community structure in the lake. The results showed that 119 species of phytoplankton were identified in 2014, 109 species in 2016 and 114 species in 2017, belonging to 8 phyla. Phytoplankton abundance got top in summer and bottom in winter. Its spatial distribution was higher in northwest and central, and lower in northeast and south. The health status of lakes in different years was evaluated by Shannon-Wiener diversity index, Pielou evenness index and SI pollution index. Baima Lake is a medium-eutrophic (moderate) water body. Pearson correlation analysis was carried out between phytoplankton and environmental factors.

1. Introduction

Phytoplankton are a tiny plant that suspend in water. Because of its small size, short life cycle and fast reproduction speed, phytoplankton are vulnerable to various environmental factors and change in a short period[1]. In water, phytoplankton can reflect the water environment changes, for example, it can better reflect the eutrophication level compared with physical and chemical indicators[2-3].

At present, the relationship between phytoplankton community structure, diversity, ecological distribution and environmental impact factors is a hot topic at home and abroad[4-7]. However, little research has been carried out on the relationship between phytoplankton and environmental factors in Baima Lake, so it is necessary to make up for this vacancy.

Baima Lake is located in the downstream of the Huaihe River Basin. It belongs to Jinhu, Hongze, Chuzhou and Baoying counties (districts), 119°02′E~119°12′E, 33°09′ N~33°19′N.The lake area is about 113.4km^2, the average annual precipitation is 956 mm and the average annual evaporation is 810 mm. The main rivers entering and leaving the lake include Caoze River, Xunhe River, Huahe River, Yongji River, Wenshan River, Xinhe River, Yunxi River, etc.

Baima Lake plays an irreplaceable role in ensuring the local residents' living water, social and economic development and ecological environment balance[8]. In recent years, a series of human activities have caused the degradation of the ecological environment of Baima lake and severely weakened the comprehensive function of the lake, such as illegal encroachment of waters, illegal sand mining and random discharge of sewage around the lake[9]. Based on the data of 2014, 2016 and 2017 phytoplankton ecological survey, the objective of this paper is to analyze the changes of phytoplankton community structure to reflect the current status of Baima Lake water environment quality, providing scientific support for the lake management.

2. Methods

According to the area, topographic profile, cultivation distribution and main rivers entering and leaving Baima Lake. 11 sampling sites were set up in the study, with sampling frequency once a month, in 2014, 2016 and 2017.

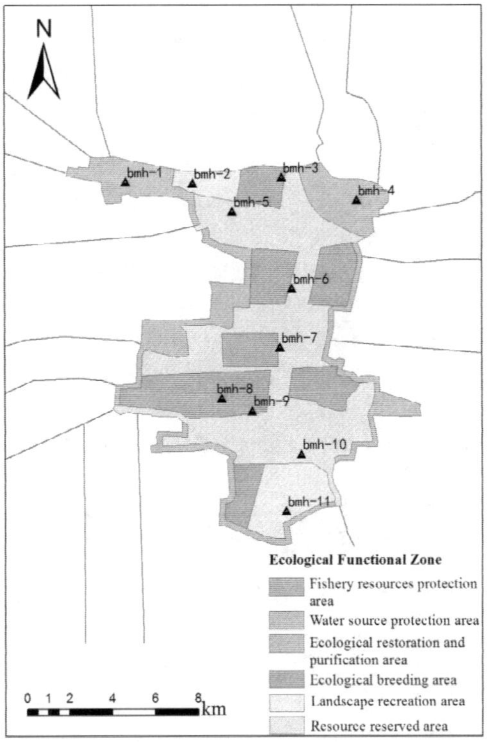

Figure 1. Phytoplankton Monitoring Points in Baima Lake

In addition to phytoplankton, other analytical indicators in water samples include water temperature, transparency, conductivity, pH, dissolved oxygen (DO), total phosphorus (TP), total nitrogen (TN), ammonia nitrogen (NH3-N), permanganate index (CODMn).The monitoring of various ecological indicators was in accordance with the local standard of Jiangsu province, "lake water ecology monitoring standard" (DB32/T3202-2017).

Phytoplankton can be used as a biological indicator to indicate water quality, because the changes in the community structure of phytoplankton are one of the direct consequences of the evolution of the water environment. In this paper, Shannon-Wiener diversity index, Pielou evenness index and SI pollution index were used to evaluate the water quality of Baima Lake[10].

Shannon-Wiener diversity:

$$H' = -\sum_{i=1}^{S} \frac{n_i}{N} \log_2 \frac{n_i}{N} \tag{1}$$

Pielou evenness index:

$$J = \frac{H'}{H_{max}} \tag{2}$$

SI pollution index:

$$SI = \frac{\sum s \times h}{\sum h} \tag{3}$$

Diminance:

$$Y = \frac{n_i}{N} \times f_i \tag{4}$$

Where n_i is the number of individuals of No.*i* kind of phytoplankton, N is the total number of individuals, and S is the total number of species in the sample. H_{max} is $\log_2 S$, which represents the maximum value of the diversity index. s is the indicator level of algae pollution, and h is the estimated quantity level of the algae. f_i is the frequency that No.*i* species appears at each station, and species with a Y value bigger than 0.02 are the dominant species.

3. Results and Discussion

3.1. Species Composition and Dominant Species

The phytoplankton identified in each year mainly belongs to *Chlorophyta, Bacillariophyta, Cyanophyta* and *Euglenophyta.* The dominant species are shown in Table 1. The dominant species in 2014 were *Chlorella sp., Cyclotella sp., Melosira granulata var. angustissima, Syndera Acus, Kephyrion ovale,* Chroomonas acuta, *Pseudanabaena catenata, and Phormidium sp*. The dominant species in 2016 is similar to that in 2017. Compared to 2014, there are another three species: *Scenedeamus quadricauda, Crypomonas erosa and Aphanizomenon gracile.*

Table 1. Dominant Species of Phytoplankton of Baima Lake in Different Years

Phytoplankton species	2014		2016		2017	
	f	Y	*f*	Y	*f*	Y
Chlorella sp.	1	0.1265	1	0.1152	1	0.0989
Scenedeamus quadricauda	-	-	1	0.0204	1	0.0288
Cyclotella sp.	1	0.1716	1	0.1293	1	0.1341
Melosira granulata var. angustissima	1	0.0332	1	0.045	1	0.0498
Synedra acus	1	0.0406	1	0.0548	1	0.0542
Kephyrion ovale	0.91	0.0225	0.91	0.0221	1	0.0364
Chroomonas acuta	1	0.0394	1	0.0392	1	0.0391
Crypomonas erosa	-	-	0.91	0.0232	1	0.0256
Aphanizomenon gracile	-	-	0.73	0.0229	0.73	0.0209
Pseudanabaena catenata	0.91	0.0236	0.91	0.0295	0.91	0.0213
Phormidium sp.	1	0.1863	1	0.1458	1	0.1483

f is the frequency that No.*i* species appear;Y indicates Phytoplankton species' diminance

3.2. Temporal and Spatial Distribution of Phytoplankton

Fig.2 reflects the monthly changes in phytoplankton abundance of Baima Lake in 2014, 2016 and 2017, and the trends during each year are basically similar. In 2014 and 2016, The average abundance at January was the lowest in that year, and the lowest in 2017 appeared in February. The average abundance of phytoplankton in the whole lake has increased significantly from June to August every year, and both peaked in August, then began to decline in September, and fell to a relatively low level in December. Among them, the abundance of phytoplankton reached a maximum of three years in August 2014, which was 47.3×10^6cells/L. In January, there was a three-year minimum of 1.70×10^6 cells/L. The lowest value per year was at the range of 1.5×10^6 cells/L ~ 3.5×10^6cells/L, but the peak value was significantly lower than the previous year.

The phytoplankton abundance in Baima Lake was the highest in summer, followed by autumn, and the abundance was the lowest in winter. This is because green algae, diatom and cyanobacteria dominate phytoplankton species in Baima Lake. The optimum temperature for green algae growth is 30°C ~ 35°C, the optimum temperature of diatom is 20°C, the optimum temperature of cyanobacteria is 40°C, and in the local temperature range, the higher the temperature, the faster the green algae and cyanobacteria reproduce, so phytoplankton abundance could reach the highest value in summer.

The interannual variations of spring and winter abundance were not much different. The phytoplankton abundance in summer 2016 was significantly lower than that in 2014, and in 2017, the abundance in summer declined remarkablely compared to 2016. This change was due to the fact that after the implementation from fish to lake in 2016, the flow of the lake accelerated, the amount of fishing bait in the lake decreased, and the problem of eutrophication of the lake was alleviated, which greatly reduced the abundance of phytoplankton in Baima Lake in summer.

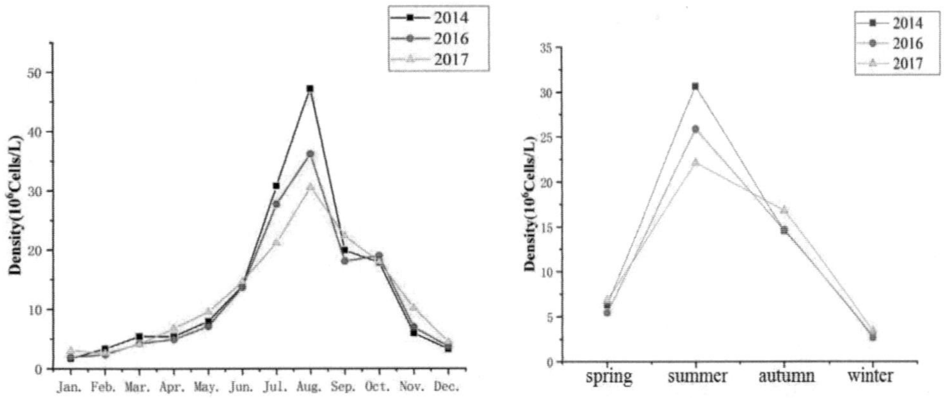

Figure 2. Monthly (left) and Seasonal (right) Variation of Phytoplankton Abundance in Baima Lake

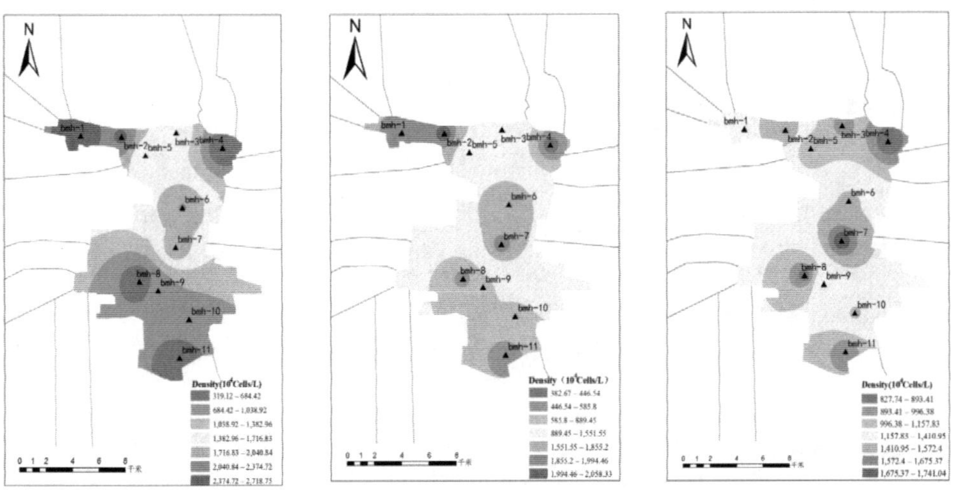

2014 **2016** **2017**

Figure 3. Spatial Distribution Difference of Phytoplankton Abundance

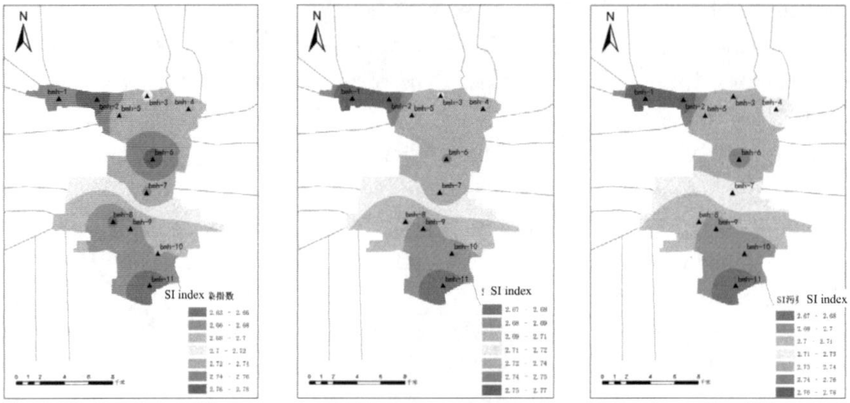

2014 **2016** **2017**

Figure 4. SI pollution index distribution difference

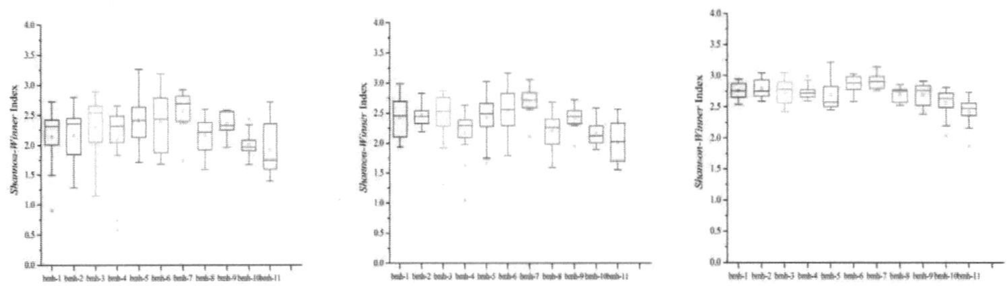

2014 **2016** **2017**

Figure 5. Shannon-Wiener diversity index distribution difference

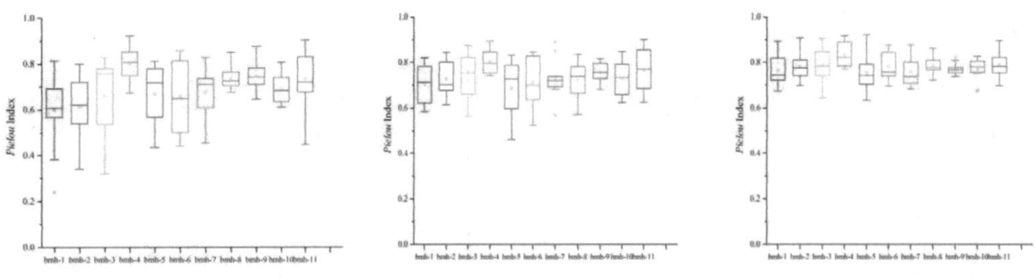

2014 **2016** **2017**

Figure 6. Pielou evenness index distribution difference

Fig.3 and Fig.4 respectively show the spatial difference of phytoplankton abundance and SI pollution index in Baima Lake for three years. Fig.5 and Fig.6 respectively show the spatial difference of Shannon-Wiener diversity index and Pielou evenness index.

Based on the spatial distribution of phytoplankton abundance, Shannon-Wiener diversity index, Pielou evenness and SI pollution index in 2014, 2016 and 2017, the following points could be summarized:

To the waters in the northwestern part of Baima Lake, especially in the landscape entertainment area, the annual abundance was the largest, and the SI index also reached the top of all measurement sites. The annual abundance of the northeastern part of Baima Lake was always relatively small, but the Pielou evenness was always the largest of all sites; The annual abundance of the waters in the central fishery resource conservation area was second only to the northwest, and the Shannon-Wiener diversity index was the largest in all three years. The average annual phytoplankton abundance in the southern part of Baima Lake was relatively small. The diversity index and the SI pollution index in the southern landscape entertainment area were the smallest over all functional areas, indicating that the water quality in the south was better.

3.3. Valuation Index

The water quality of Baima Lake was evaluated by Shannon-Wiener diversity index, Pielou evenness index and SI pollution index[11-12]. Compared with 2014 and 2016, the diversity index and Pielou evenness index in Baima Lake increased slightly in 2017, indicating that the stability of phytoplankton community structure kept increasing. The pollution index of 2017 was slightly lower than that of 2016, indicating that after the implementation of returning fishery to lake, the water body of Baima Lake is still medium-rich (medium) water, but the water quality has improved to some extent.

Table 2. Comparison of water quality evaluation of Baima Lake in different years

	Shannon-Wiener diversity index	Pielou evenness index	SI pollution index
2014	2.23(Mild pollution)	0.69(Mild pollution)	2.83(Severe pollution)
2016	2.36(Mild pollution)	0.73(Mild pollution)	2.84(Severe pollution)
2017	2.71(Mild pollution)	0.78(Mild pollution)	2.72(Severe pollution)

3.4. Correlation Analysis

Pearson correlation analysis was carried out on the dominant species abundance and water physicochemical index data of phytoplankton at 11 sites in Baima Lake in 2017 (12 months). The physical and chemical indicators were temperature, transparency, conductivity, TN, TP, etc. A total of 9 variables, the results are shown in Table 3.

In 2017, there were 11 dominant species of algae, each of them was significantly positively correlated with water temperature. Among them, the correlation coefficient of *Cyclotella sp.* with temperature was the highest, 0.775. The correlation coefficient of *Chlorella sp.*, *Scenedeamus quadricauda* and *Phormidium sp.* between temperature exceeds 0.6. This also well explains the seasonal variation of phytoplankton abundance. The higher the temperature, the more favorable it is for the growth of phytoplankton in Baima Lake, so the phytoplankton abundance was the largest in summer and the smallest in winter. There were 7 species of phytoplankton with extremely significantly negative correlation with DO, The 6 species of phytoplankton showed a extremely significantly positive correlation with the conductivity, and 5 species of phytoplankton and NH_3-N showed a extremely significantly negative correlation. *Kephyrion ovale* had a very significantly correlation with all environmental factors except conductivity.

Pearson correlation analysis showed that temperature is the main factor affecting the growth of phytoplankton, followed by DO, conductivity and pH. A large number of studies have shown that different DO, conductivity, pH and transparency of water bodies can affect the structure of phytoplankton community. For example, diatom communities need different conductivity optimum values to survive, so the water conductivity will affect the phytoplankton community structure. Transparency mainly affects the growth of phytoplankton by affecting the light absorption and scattering of water. With the reproduction of phytoplankton, the pH value of water will change. The pH value also has an important influence on the structure and distribution of phytoplankton community. pH and DO concentration can reflect the growth of algae and changes of water environment[13-14]. Water nutrients (TP, TN, NH_3-N) and permanganate index are also important factors affecting the spatial and temporal variation of phytoplankton. It is necessary to control the pH range and inflow of nitrogen and phosphorus, which is beneficial to the water quality of Baima Lake improvement.

Table 3. Correlation analysis between dominant phytoplankton species and environmental factors

	Temperatu-re	Transparency	Conductivi-ty	pH	DO	TP	TN	NH_3-N	COD_{Mn}
Chlorella sp.	0.656**	-0.008	0.469**	-0.206*	-0.378**	-0.098	-0.11	-0.347**	-0.025
Scenedeamus quadricauda	0.606**	-0.227**	0.364**	-0.101	-0.242**	-0.094	-0.095	-0.331**	-0.222*
Cyclotella sp.	0.775**	0.06	0.561**	0.018	-0.258**	-0.189*	-0.113	-0.402**	0.035
Melosira granulata var. angustissima	0.543**	-0.155	0.383**	0.044	-0.049	-0.091	0.054	-0.196*	0.037
Synedra acus	0.438**	0.409**	0.202*	-0.403**	-0.517**	-0.184*	-0.352**	-0.307**	-0.284**
Kephyrion ovale	0.324**	0.502**	0.000	-0.342**	-0.414**	-0.369**	-0.343**	-0.190*	-0.329**
Chroomonas acuta	0.493**	0.133	0.223*	-0.227**	-0.355**	-0.233**	-0.151	-0.220*	-0.176*
Crypomonas erosa	0.459**	0.026	0.219*	-0.215*	-0.306*	-0.153	-0.113	-0.200*	-0.205*
Aphanizomenon gracile	0.303**	-0.223*	0.238**	0.228**	0.127	-0.084	-0.109	-0.167	0.006
Pseudanabaena catenata	0.227**	-0.092	0.083	-0.189*	-0.328**	-0.032	-0.085	-0.188*	-0.175*
Phormidium sp.	0.632**	-0.204*	0.44**	0.121	-0.017	-0.103	0.01	-0.296**	0.007

Note:* represents a significant level of 0.05(double tailed test);**represents a very significant level of 0.01(double tailed test)

4. Conclusion

The distribution characteristics of phytoplankton in Baima Lake in 2014, 2016 and 2017 were discussed, and the correlation analysis was carried out with environmental factors. The main conclusions are as follows:

The abundance of phytoplankton in Baima Lake was the highest in summer and the lowest in winter, and the interannual variation of abundance in spring and autumn was not remarkable. The phytoplankton abundance in 2016 and 2017 was significantly lower than that in 2014, indicating that the lake water nutrition management has achieved initial results.

Baima Lake belongs to medium-eutrophic (moderate) water body. The average annual abundance of phytoplankton showed the characteristics of "larger in the northwest and middle, smaller in the northeast and south". And the Shannon-Wiener diversity index of central fishery resource protection area was the largest, the Pielou eveness index in the northeast was the smallest. The SI pollution index of the landscape entertainment area in the northwest was the largest over all sampling sites, while the SI index of the southernmost landscape entertainment area was the smallest. Though the overall water quality has improved, but the water pollution in the northwest is still serious.

Temperature is the main factor affecting the phytoplankton abundance in Baima Lake. DO, conductivity, pH, water nutrients, COD_{Mn} and transparency are also important factors affecting the spatial and temporal variation of phytoplankton.

5. Acknowledgement

This work is financially supported by Innovation capacity building project of Jiangsu provincial Department of science and technology (BM2018028) and Ministry of water resources technology demonstration project (SF-201816).

6. References

[1] LIU Yu, SHEN Jianzhong. Application of algae biological evaluation in water quality

monitoring[J]. Reservoir Fisheries, 2008(4):5-7.
[2] ZHAO Yibin, XU Wude, GUO Yuxin. Biological indication and water environment[J].Water Resources Protection, 2002(2):11-16.
[3] Salmaso N, Morabito G, Buzzi F, et al. Phytoplankton as indicator of the water quality of the deep lakes south of the Alps[J].Hydrobiologia,2006, 563:167-187.
[4] Cemeno P, Teixeira I G, Branco M, et al.Sampling the limits of species richness in marine phytoplankton communities[J]. Plankton Research, 2014, 36: 1135-1139.
[5] WANG Yibing, HOU Zeying, YE Bibi.The characteristics of spatial and temporal variations of phytoplankton in Poyang Lake and their influencing factors, Acta Scientiae Circumstantiae, 2015(5): 1310-1317.
[6] Zhao Xiu-Xia, Fang Ting, Yang Kun, et al. Community structure characteristics of phytoplankton and related environmental factors in summer in Tuohu Lake, Anhui, China.Plant Science Journal, 2018, 36(5):687-695.
[7] SHEN Rongrong. Distribution pattern of phytoplankton in the lower reaches of the Yangtze River and its correlation with environmental factors[D].Shanghai:Shanghai Normal University, 2019.
[8] JIA Shuoling. Practice and Thoughts on Aquatic Protection of Baima Lake[J].Jiangsu Water Resources, 2013[3]:28-30.
[9] YANG Wanhong, Zhou Xia, Che Jinling. Analysis on the Influence of the Retirement of the Baima Lake in Huai'an City on the Water Quality[J]. Jiangsu Water Resources,2014[8]:34-36.
[10] MA Jianxin, ZHEN Zhenhu, LI Yunping et al. Distribution characteristics of phytoplankton in Laizhou Bay[J]. Transaction of Oceanology and Limnology, 2002(4):63-67.
[11] LI Renquan, WANG Mingquan, SUN Ming et al. Analysis of Constitution Features and Species Diversity Index of Indicator Algae in Chicheng Lake, Journal of Anhui Agricultural Sciences, 2009, 37(2):773-776.
[12] ZHANG Jingping, HUANG Xiaoping , JIANG ZhiJian, et al. Assessment of the Pearl River Estuary pollution by water comprehensive pollution index and biodiversity index[J]. Journal of Tropical Oceanorgaphy, 2010, 29(1):69-76.
[13] WEN Xinyu, ZHANG Hucai, CHANG Fengqin, et al. Seasonal stratification of vertical section of Lugu Lake. Advances in Earth Science,2016, 31(8):858-869.

IOP Conf. Series: Materials Science and Engineering **730** (2020) 012056 doi:10.1088/1757-899X/730/1/012056

Study on Ecological Vulnerability of Dianchi Lake Basin Based on GIS-Based Principal Component Analysis

Xiaoliang Ma[1,2], Shuangyun Peng[1,2*], Cong Lin[1,2], Zhiqiang Lin[1,2] and Yanlian Zhou[1]

[1]College of Tourism and Geography, Yunnan Normal University, Kunming 650500, China.

[2]Yunnan Geospatial Information Technology Engineering Technology Research Center , Kunming 650500, China.

*Corresponding Author: Shuangyun Peng. First Author Email: 2485795304@qq.com.

Abstract. Ecological vulnerability is one of the key concerns of global environmental change, sustainable development, and human-land relations. This paper takes the Dianchi Basin as a research area, uses GIS principal component analysis method, selects relevant index factors to determine the weight, and finally combine evaluation index model to comprehensively analyze the ecological vulnerability of the Dianchi Basin. After research, it is found that: ①The ecological vulnerability of the Dianchi Basin is relatively high as a whole, of which 69.72% of the ecological vulnerability grades are above Grade 3. ②The ecologically fragile areas in the Dianchi Basin are generally scattered and concentrated in a small area, and the closer to the area of Dianchi, the higher the vulnerability level. ③The ecological vulnerability of the Dianchi Basin is the result of multiple factors. The research has reference significance for the ecological construction and economic development of Dianchi Basin in the future.

1. Introduction

In recent years, with the development of social economy, the increase of population and the irrational use of resources, the coordination capacity of the ecosystem itself has been declining, and the ecological environment has become more and more fragile. In this context, in order to seek harmony between man and nature, better organize human production and construction activities, and carry out related ecological restoration work, it is essential to effectively evaluate the ecological environment of a certain region. The ecological vulnerability research is an important evaluation method for regional ecological environment assessment, and it is also the research method that best reflects the correlation between various natural and human impact factors in the region.

The research on "ecological vulnerability" can be traced back to the beginning of the last century. American ecologist Clements introduced the concept of ecological transition zone into the field of ecological research [1]. Later, with the increasing attention of scholars and people, the research on "ecological vulnerability" has been flourishing, especially in the 1980s and 1990s, "ecological fragile zone" and "human-land coupled system vulnerability". Etc. began to appear more and more frequently in related research fields [2-4]. After entering the 21st century, the study of "ecological fragility" has entered a brand-new stage. It has realized the transition from the single factor study of early ecological transition to the multi-factor study of comprehensive ecological environment, from the application of simple analytical models to the new statistical model of data rate use. In China, with the development concept of "Clear waters and green mountains are mountains of gold and silver", research based on ecological environment vulnerability has become increasingly hot. Therefore, there have been some evaluation and analysis methods with high matching and accurate results in China, especially in recent years, based on GIS principal component analysis, the obtained vulnerability evaluation results are closer to reality. For example, Fu Gang et al. Evaluated the ecological vulnerability of Beijing based on the GIS spatial principal component analysis method and found that the overall ecological vulnerability of Beijing was at a medium level, and different surface coverage showed different levels of vulnerability [5]. Ma Jun et al. Evaluated the ecological vulnerability of the Three Gorges Reservoir area based on spatial principal component analysis and found that the ecological vulnerability of the Three Gorges Reservoir area is the result of the interaction between human activities and the natural environment [6].

Therefore, this paper draws on the above research results, taking Dianchi Basin as the research area, based on the GIS spatial principal component analysis method, selecting relevant index factors and assigning weights, establishing an evaluation index system, and obtaining the vulnerability distribution data of different spaces in Dianchi Basin. Comprehensively evaluate the ecological fragility index by comprehensively influencing factors to determine the ecological vulnerability of Dianchi Basin, so as to provide a reference for the socio-economic sustainable development and ecological restoration of Dianchi Basin.

2. Study Area Overview and Data Sources

2.1. Overview of the Study Area

Dianchi Basin is a collective term for the runoff area around Dianchi. The basin area is about $2920km^2$, the average elevation is 2048m, height difference up to 1094m. The basin involves nine districts and counties in Kunming, including Wuhua, Panlong, Guandu, Xishan, Chenggong, Xundian, Fumin, Songming, and Jinning. The basin concentrated 54.23% of the city's population and contributed 78.58% of the city's GDP with a land area of approximately 13.9% of the city. The highly clustered economies and highly dense population distribution make the natural environment of the region affected by human activities frequently, and the ecological environment becomes more and more fragile. At the same time, ecological and environmental problems have increasingly become an important factor restricting the economic and social development of the Dianchi Basin. Therefore, it is of great significance to study the ecological environment vulnerability of the region, which can not only guide the ecological environment restoration and reconstruction, but also provide a basis for regional sustainable development.

2.2. Data Source

This paper selects three main data, including: 1) Remote sensing image data, Landsat 8 OLI_TIRS data of Dianchi Basin in 2017, with a spatial resolution of 30m, derived from geospatial data cloud for obtaining land use type data and vegetation coverage data. 2) DEM data, ASTER GDEM data of Dianchi Basin in 2017, with a spatial resolution of 30m, derived from geospatial data cloud, used to obtain elevation data and topographic data of the study area. 3) Demographic and economic data, derived from the Yunnan Statistical Yearbook and the Kunming Statistical Yearbook, used to obtain population density data and per capita GDP data.

3. Research Methods

3.1. Construction of Ecological Vulnerability Evaluation Index Model

The construction of the evaluation index system will affect the accuracy of the final evaluation results. Based on the regional characteristics of the Dianchi Basin, this paper builds a single unit of ecological vulnerability assessment for the Dianchi Basin by selecting natural and human factors based on the four basic principles of ease of selection, scientificity, significance, and integrity. Factor evaluation level model (see Table 1).

Table 1. Single factor evaluation level model

Indicators	Potentially vulnerable areas(1)	Negligible vulnerable areas(3)	Light vulnerable areas(5)	Medium vulnerable areas (7)	Strong vulnerable areas(9)
Elevation(m)	>2400	2200-2400	2000-2200	1900-2000	<1900
Slope(°)	0-5	5-10	10-15	15-25	>25
NDVI	>0.3	0.2-0.3	0.1-0.2	0-0.1	<0
Land use type	Forest	Plow land	Grass Water	Unused land	Urban
The population density(person/km²)	200-300	300-500	500-700	700-1000	>1000
Per capita GDP(Ten thousand yuan/person)	>19	16-19	12-16	10-12	<10
Rainfall(mm)	>1000	950-1000	900-950	850-900	<850

3.2. Evaluation Index Data Processing

In order to eliminate the problem of the inaccuracy of the evaluation results due to the inconsistency of the index dimension and the order of magnitude, the selected index factors need to be standardized. Formula (1) is used when the standardized value is positively correlated with ecological vulnerability;formula (2) is used when the standardized value is negatively correlated with ecological vulnerability [7].

$$X_i' = \frac{X_i - X_{imin}}{X_{imax} - X_{imin}} \tag{1}$$

$$X_i' = 1 - \frac{X_i - X_{imin}}{X_{imax} - X_{imin}} \tag{2}$$

In formulas (1) and (2): X_i' is the standardized value of the i-th index, X_i is the initial value of the i-th evaluation index, and X_{imin} and X_{imax} are the minimum and maximum values of the i-th evaluation index in the study area. Generally, the normalized value is between (0,1). The larger the standardized value, the more vulnerable the ecosystem is, and the more vulnerable it is to external disturbances 3.

3.3. Selection of Evaluation Factors

Transform based on the GIS principal component analysis tool to obtain the eigenvalues and contribution rates of the converted principal factors (see Table 2). From Table 2, it can be seen that the contribution rate of the first six principal component factors has reached 95.9635%, with only 4.03% information loss and high credibility. Therefore, the number of main factors is determined to be 6.

Table 2. Eigenvalue and contribution rate of each main factor

Main factor	Eigenvalues	Contribution rate /%	Cumulative contribution rate /%
1	7.8645	36.5536	36.5536
2	6.1357	28.5182	65.0718
3	2.5042	11.6393	76.7111
4	2.2812	10.6029	87.314
5	1.0817	5.0277	92.3417
6	0.7792	3.6218	95.9635
7	0.6042	2.8085	98.772

3.4. Factor Weight Determination

Calculate the common factor variance of each evaluation factor according to the factor matrix [7], see formula (3) for details.

$$T_j = \sum_{j=1}^{m} \lambda_{jk}^2 (j = 1,2...7, \ k = 1,2...,6) \quad (3)$$

In the formula(3), j is the number of original indicators, k is the main component number, m is the total number of main components, and m=6.

After calculating the data, the weight coefficients of the ecological vulnerability impact factors in the Dianchi Basin can be obtained. The specific data are shown in Table 3.

Table 3. Factor weights

Factor	DEM	Slope	NDVI	Land use	Population density	Per capita GDP	Rainfall
Weights	0.1377	0.1872	0.1964	0.1826	0.1497	0.0781	0.0683

3.5. Comprehensive Evaluation Method of Ecological Vulnerability

Through the multi-factor superposition analysis, formula (4) is used to calculate the ecological vulnerability, and the ecological vulnerability of Dianchi Basin is comprehensively evaluated from the aspects of resources, environment, and economic and social development.

$$EVD_i = \sum_{j=1}^{7} P_{ij} W_j \quad (4)$$

In formula (4),EVD_i is the ecological vulnerability of the i-th evaluation unit, and the final result is a composite value, whose value is between 0-1; P_{ij} is the j-th index of the i-th unit; W_j is the weight of each index (See Table 3).

4. Results and Analysis

4.1. Analysis of Ecological Vulnerability Level in Dianchi Basin

According to the results of superposition principal component analysis, the vulnerable areas of Dianchi Lake Basin are divided into five types(see Table 4): I potentially vulnerable area, II micro-fragile zone, III mildly vulnerable area, IV moderately vulnerable area and V highly vulnerable area. On the whole, the ecological vulnerability of the Dianchi Basin is relatively high, and the vulnerable areas of mild and above account for 69.72%, of which the highly vulnerable areas account for 13.29%, the moderately vulnerable areas account for 22.06%, and the mildly vulnerable areas account for 34.37 %, potentially vulnerable areas and micro-fragile zone account for a relatively small proportion.

Table 4. Statistical table of ecological vulnerability of Dianchi Basin

Grade	Fragile type	Area(km²)	Area ratio(%)
I	Potentially vulnerable area	337.66	11.62
II	Micro-fragile zone	542.39	18.66
III	Mildly vulnerable area	999.07	34.37
IV	Moderately vulnerable area	641.22	22.06
V	Highly vulnerable area	386.39	13.29

4.2. Distribution of Ecological Vulnerability of Main Factors to Dianchi Basin

Based on the ArcGIS principal component analysis method, the distribution map of the influence of each principal factor on the ecological vulnerability of the Dianchi Basin was obtained (Figure 1). It can be found that the ecologically fragile areas of the Dianchi Basin show different distribution characteristics on each factor: 1) Except for Dianchi, the lower the NDVI value, the higher the vulnerability level, such as the main urban area of Kunming. 2) The lower the per capita GDP value, the higher the vulnerability level, such as Jinning County, a traditional agricultural area. 3) The higher the slope, the ecological risk is affected by landslides, mudslides and other geological disasters, such as upstream mountainous areas. 4) The lower the elevation, the higher the vulnerability level. For example, Dianchi has the lowest elevation, and a large amount of production and domestic sewage flows into the Dianchi along with the river, which aggravates the pollution of the Dianchi. 5) Urban land types have the highest levels of vulnerability, mainly because cities are the most densely populated areas and the most frequent areas of human activity. 6) The higher the population density, the higher the vulnerability level, such as the main urban area. The above conclusions have effectively verified the correctness of the experimental model, and the calculation results are consistent with the model preset results.

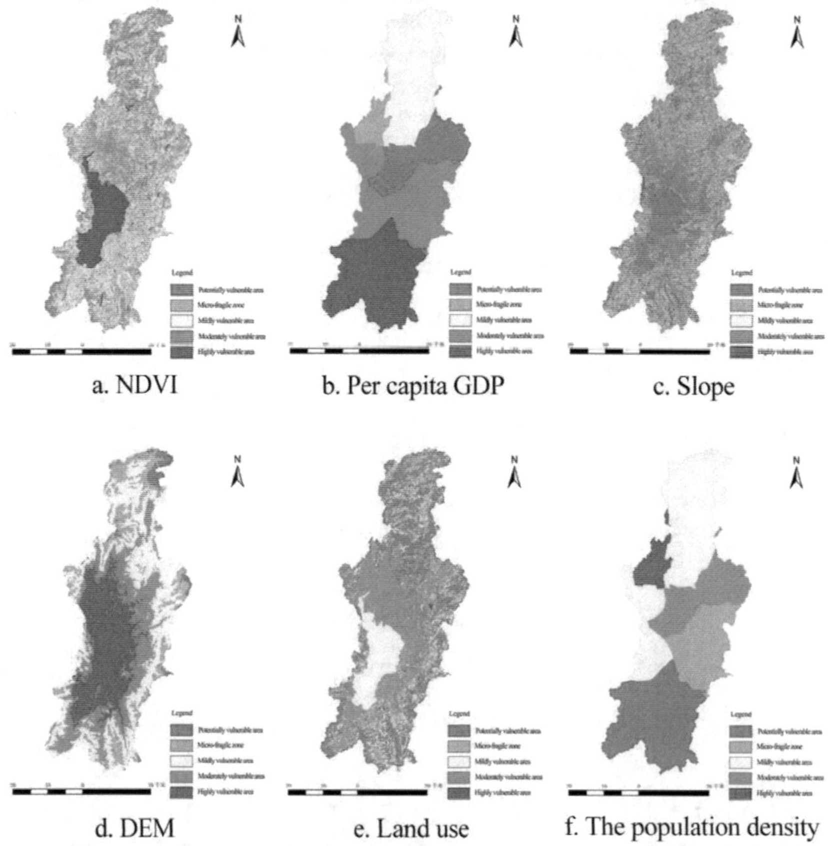

Figure 1. Distribution of main factors affecting ecological vulnerability in Dianchi Basin

4.3. Analysis of Spatial Characteristics of Ecologically Fragile Areas in Dianchi Basin

In order to better display the spatial characteristics of the ecologically fragile areas in the Dianchi Basin, a hierarchical distribution map of the ecological vulnerability in the Dianchi Basin was obtained by superposition analysis of principal factors (see Figure 2). According to the distribution map, the ecologically fragile areas in the Dianchi Basin are generally scattered and concentrated in a small area. The entire Dianchi surface is concentrated in highly vulnerable areas, and most of the remaining areas are distributed in different levels of fragile areas. And, the closer to the Dianchi area, the higher the vulnerability level. Although the ecologically fragile areas are scattered in space, the overall index factors affect each other and make different areas show the same vulnerability characteristics.

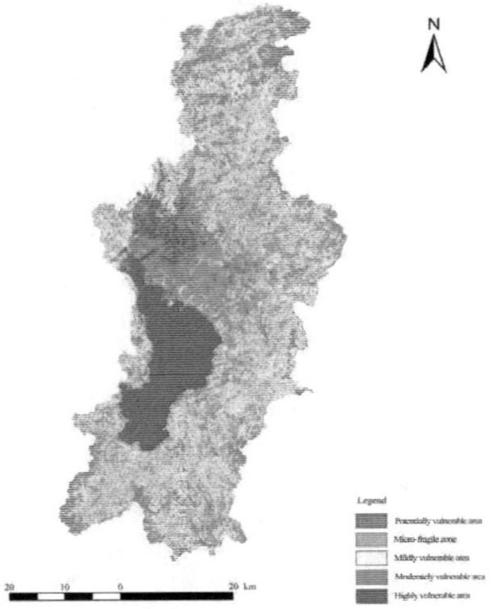

Figure 2. Dianchi Basin ecological vulnerability level distribution map

5. Conclusion

The ecological environment is constrained and affected by a variety of factors, and different factors have different weights on the regional ecological environment, resulting in different results. Therefore, this paper takes the Dianchi Basin as the research area, uses the GIS principal component analysis method, selects the relevant index factors and assigns weights based on natural and human factors, and finally establishes an evaluation index system to analyze the characteristics and causes of the ecological environment vulnerability in the river basin. The study found that the ecological vulnerability of the Dianchi Basin is generally high, and 69.72% of the regions with a vulnerability level of III or higher should be given sufficient attention. In terms of spatial distribution, the ecologically fragile areas in the Dianchi Basin are scattered and concentrated in a small area, and the closer to the area of Dianchi, the higher the vulnerability level. In addition, the ecological fragile areas of the Dianchi Basin show different distribution characteristics on each single factor, but these single factors as a whole interact and interact with each other, which makes different regions also show the same vulnerability characteristics. After researching these characteristics, it is consistent with the model presets, which shows the reliability and authenticity of the evaluation results, and further proves that the ecological vulnerability of the Dianchi Basin is the result of multiple factors acting together, thus providing ecological construction and sustainable development for the Dianchi Basin Provide important reference.

6. Acknowledgement

Fund Project: Project supported by the National Natural Science Foundation of China (41971369, 41561086, 41661082, 41461038).

7. References

[1] SHAN Yanrui. New Progress in Comprehensive Research on Natural Disasters: Vulnerability Research [J]. Regional Research and Development, 2000, 19(2): 73-77.
[2] Smithers J, Smit B. Human adaptation to climatic variability and change [J]. Global Environmental Change, 1997 (7): 129-146.
[3] Smit B, Burton I, Klein R J T, et al. The science of adaptation: A framework for assessment [J]. Mitigation and Adaptation Strategies for Global Change, 1999 (4): 199-213.
[4] XU Jun,LI Guifang,WANG Yuhong. Summary and Prospect of Research on Ecological Vulnerability at Homeand Abroad [J]. East China Economic Management, 2016, 30(04):149-162.
[5] FU Gang, BAI Jiade, QI Yue, et al. Evaluation of ecological vulnerability in Beijing based on GIS [J]. Journal of Ecology and Rural Environment. 2018, 34(9): 830-839.

[6] Ma Jun, Li Changxiao, Wei Hong, Ma Peng, Yang Yujing et al. Ecological vulnerability assessment of the Three Gorges reservoir area. Acta Ecologica Sinica, 2015, 35(21): 7117-7129.
[7] YU Kun-yong, LIU Jian, HUANG Wei-you, XU Zhang-hua, CHEN Zhi-fei. Analysis of Ecological Vulnerability of Minjiang River Basin Based on GIS Technology [J]. Journal of Jiangxi Agricultural University, 2009, 31 (03):568-573.

IOP Conf. Series: Materials Science and Engineering **730** (2020) 012057 doi:10.1088/1757-899X/730/1/012057

A Study on Environmental Cost of China's Oil Production and Utilization

Lijuan Wang[1], Xiaojun Chen [1,*], Qian Tang[1], Qiang Sun[2], Huili Wang[1] and Ling Jin[1]

[1]Chinese Academy of Environmental Planning, Beijing, China
[2]Information Center, Ministry of Ecological Environment, Beijing, China
*Email: chenxj@caep.org.cn

Abstract: This paper puts forward the environmental cost of China's oil production and utilization in the aspect of atmosphere. Every aspect of oil production and utilization from exploitation to processing, and petroleum products consumption can produce negative effects on the environment, such as environmental pollution. This paper estimates the environmental cost of China's oil production and utilization in 2015, and the results show that the environmental cost per metric ton of oil consumption is 487.9 RMB. Water pollution and water damage costs are not revealed.

1. Introduction

The production and utilization of oil caused a series of negative effects on environment in China. The environmental cost of oil exploitation and utilization is not fully reflected in the current market system. In order to promote the internalization of oil's environmental costs in China, we analyze the air pollution cost of oil, which refers to the impacts of oil-production, oil-processing, and oil-consumption on the atmosphere and climate, and the monetized environmental costs for the above three areas in China.

2. The Environmental Impact of Oil Utilization

2.1. Oil-production

The oil production process also produces volatile organic compounds (VOCs), sulfur dioxide (SO_2), nitrogen oxides (NOx), particulate matter (PM) and other atmospheric pollutants, which are the important precursors of $PM_{2.5}$ and O_3. In 2015, the emissions of VOCs, SO_2, NOx and PM in China's petroleum exploration industry were 122 thousand tons, 29 thousand tons, 28 thousand tons and 9 thousand tons respectively.

Crude oil is transported mainly via shipping and pipelines, including railways and highways. Marine oil spill accidents bring huge economic loss to fisheries, aquaculture, coastal tourism, marine transportation, and other industries, and causes abnormal changes in the marine ecosystem. China releases about 1.28 million tons of oil into the sea every year (including industrial emissions, oil spills from ships, oil spills from accidents, leakage from undersea oil fields and blowout accidents).

2.2. Oil-processing

Petroleum processing, coking, nuclear fuel processing (Industrial classification for national economic activities), and raw chemical materials and chemical product manufacturing are all also important

IOP Conf. Series: Materials Science and Engineering **730** (2020) 012057 doi:10.1088/1757-899X/730/1/012057

sources of atmospheric pollutants. In 2015, the emissions of VOCs, SO_2, NOx and PM in the petroleum processing industry were 1.7 million tons, 0.2 million tons, 0.2 million tons and 84 thousand tons respectively.

In the process of petroleum processing, there are a variety of industrial solid wastes, including waste catalysts, waste adsorbents, waste ceramic balls, etc., as well as coal ash slag and domestic garbage generated by power stations. The petroleum processing industry produces 370 million tons of general industrial solid waste, accounting for 11.8% of total industrial output. The output of hazardous waste was 9.15 million tons, accounting for 23.0% of the total output of industrial hazardous waste.

2.3. Oil-consumption

China's refined oil products (including gasoline, kerosene, and diesel) account for about 53% of total oil consumption, used mainly for motor vehicles and non-road mobile machineries. The exhaust emitted by mobile sources (Mobile sources include motor vehicles and non-road mobile machineries such as construction machinery, agricultural machinery, small general machinery, diesel generator sets, ships, railway diesel locomotives, and aircraft.) contains hundreds of different pollutants, including NOx, VOCs, PM, hydrocarbons, etc., and is an important precursor for $PM_{2.5}$ and O_3. The health effects of air pollution are characterized by the number of premature deaths caused by IHD, stroke, COPD and LC. In 2015, the emissions of NOx from mobile sources were 11.5 million tons. The emissions of NOx from mobile sources (including motor vehicles and non-road mobile machinery) accounted for 98.2%, of the emissions of air pollutants from oil-related industries, respectively.

An air quality model was used to simulate the effects of atmospheric pollutant emissions from oil-related industries on the average annual concentration of $PM_{2.5}$. The results showed that, in 2015, petroleum-related industries accounted for 10.9% of the national $PM_{2.5}$ concentration, with transportation accounting for 9.8% and the petrochemical industry accounting for 1.1%.

Plastics are one of the main products of the petroleum industry. According to relevant research reports, plastic production accounts for 6% of total oil consumption, and its production, consumption, and disposal has negative environmental impacts. Some plastics can be recycled, but most of them are incinerated, disposed of in landfills, or otherwise discarded. Incinerated plastics can result in significant pollution, landfills can destroy soil structure and hinder plant growth, and plastics discarded in the sea can harm marine ecology and affect human health.

3. The Environmental Costs of Oil

3.1. Method of environmental cost accounting

The governance cost method apply to calculate different cost. Here we use the physical and value quantity of air pollution to calculate the environmental cost of oil.

$$\boldsymbol{C_i = Physical\ quantity * Value\ quantity}$$

C is the cost of each subitem, i is air pollution.

Governance method is used to calculate the cost of air pollution. Indicators of each subitem can be found in the next section.

3.2. The Calculation of Oil Environmental Costs

In the oil-producing and processing sectors, the main environmental costs to consider are air pollutant emissions. Water consumption, water pollution, soil pollution, solid waste discharge and the environmental impacts of accidents, for example petrochemical and chemical explosions, were not considered in this article.

The accounting results show that atmosphere environmental cost of China's oil production and utilization was 487.9RMB per metric ton of oil consumption in 2015.

Table1. Atmosphere Environmental costs of oil

Section	Category	Indicator	Accounting result (yuan/ton) Note1
Oil-producing	Air pollution	Quantity of SO_2, NOx, PM and VOCs emissions , Virtual disposal cost of pollutants	2.4
Oil-processing	Air pollution	Quantity of SO_2, NOx, PM and VOCs emissions , Virtual disposal cost of pollutants	33.5
Oil-consuming	Air pollution	Quantity of SO_2, NOx, PM and VOCs emissions , Virtual disposal cost of pollutants	452.0
Subtotal			**487.9**

Note: 1. The environmental cost accounting results of petroleum in the table are amortized to the oil consumption per ton.

4. Policy Suggestions

1. Reduce the environmental impact of oil exploitation and utilization from the source

(1) Strengthen prevention and control of petrochemical and motor vehicle pollution and reduce pollutant discharge.

(2) Strengthen oil transportation safety management and reduce environmental pollution caused by oil leakage.

2. Promote the clean use of petroleum through economic measures

(1) Strengthen the monitoring and restoration of the impact of oil exploration enterprises on the ecological environment, and gradually establish a feasible ecological compensation and restoration mechanism.

(2) Promote internalization of environmental costs through tax measures.

On the production side, research and adjust the structure of tax policies, from taxing individual VOCs to taxing total VOCs emissions. Improve VOCs equivalent value to promote VOCs emission reduction.

On consumption side, improve the fuel consumption tax system to take environmental costs into consideration.

3. Promote public participation to reduce oil consumption

(1) Raise the public awareness of the importance for an oil cap in China.

(2) Promote green consumption and green transportation, lead the public to participate in plastics recycling, and strengthen plastics reduction.

4. Scientific and technical support

(1) Strengthen the development and promotion of new technologies such as green oil exploitation, pollutant emission control and ecological environment restoration.

(2) Promote the development of waste plastics recycling technology and reduce the impact of waste plastics on the environment.

5. Acknowledgement

We gratefully acknowledge the support of National key research and development program (No.2016YFC0207504).

6. References

[1] Alexeeff S, Schwartz J, Kloog I, et al. Consequences of kriging and land use regression for PM2.5 predictions in epidemiologic analyses: insights into spatial variability using high-resolution satellite data[J]. J Expo Sci Environ Epidemiol, 2015,25(2):138-144.

[2] Blastic. Toxicity of Plastics[EB/OL]. https://www.blastic.eu/knowledge-bank/impacts/toxicity-plastics/.

[3] Chai Hong-yun , Zhao Dong-feng , Lu Lei. Preliminary research on environmental impact post-assessment in the process of oil exploration [J]. Modern Chemical Industry,2017,37(02):17-19+21.

[4] China Institute of Water Resources and Hydropower Research. Study on the synergetic effect of water resources in total coal control [R]. 2015.

[5] Etkin, D.S. (2000), "Worldwide Analysis of Marine Oil Spill Cleanup Cost Factors", Arctic and Marine Oil Spill Program Technical Seminar

[6] Fang Yu , Jinnan Wang , Dong Cao , Hongqiang Jiang. Technical guidelines for environmental economic accounting in China [M].Beijing: China Environmental Science Press,2009.

[7] Jisheng Jiang ,Yong Wang ,Min Qi, et al. Geological environment problems and the compound harnessing in Daqing City [J]. Journal of Heilongjiang Hydraulic Engineering College , 2004 , 31(4):86-89

[8] Production, use, and fate of all plastics ever made, Science Advances, Jul. 2017

[9] Plastic`s Europe Market Research Group (PEMRG), 2016

[10] Plastic`s Europe Market Research Group (PEMRG), 2019

[11] Plastics Europe Market Research Group (PEMRG), 2016

[12] http://www.itopf.com/knowledge-resources/documents-guides/document/trends-in-oil-spills-from-tankers-over-the-past-ten-years/

[13] Tang X, Bai Y, Duong A, et al. Formaldehyde in China: production, consumption, exposure levels, and health effects[J]. Environ Int, 2009,35(8):1210-1224.

[14] US EPA. Toluene[EB/OL]. https://www.epa.gov/sites/production/files/2016-09/documents/toluene.pdf.

[15] UN News, Jun. 2014 , https://www.unenvironment.org/news-and-stories/press-release/plastic-waste-causes-financial-damage-us13-billion-marine-ecosystems

IOP Conf. Series: Materials Science and Engineering **730** (2020) 012058 doi:10.1088/1757-899X/730/1/012058

Enhanced Ecosystem Services by Land Use Composition Adjustment Can Reduce Weather Disasters Loss under Climatic Change

Yixiao Kuang[1], Yu Peng[1,*], Bing Sun[2], Chunqian Jiang[3] and Weiguo Sang[1]
[1] College of life and Environmental Sciences, Minzu University of China, 27 South Street, Zhongguancun, Haidian District, Beijing 100081, China.
17400195@muc.edu.cn (K.YX); swg@muc.edu.cn (S.WG)
[2]Institute of tropical forestry, China Academy of forestry 510520, Guangzhou, Guangdong, China.
[3]Institute of forestry, China Academy of Forestry Sciences, Beijing, 100091, China.
*Email: yuupeng@163.com

Abstract: How to evaluate ecosystem services has been intensively documented. However, the function of ecosystem services in alleviating the economic loss of weather disasters has been seldomly noticed. In Hunan Province of China, we investigated the loss of weather disasters: drought, flood, hail, rain storm, and heavy snow from 120 counties during 1980-2010. The land use changes and ecosystem services were also assessed. The results indicated that enhanced ecosystem services brought by land use changes have substantially reduced the intensity of weather disasters, and reduce the loss of disasters if they occurred. These potential function in mediating weather disaster should be considered in the assessment of ecosystem services. With global climate changes, we should optimize the land uses to face increasing weather extremes.

1. Introduction

Natural disasters have increasingly happened along climate change, seriously affects the survival of human life, property security and social and economic sustainable development. According to the World Bank's latest report, "Undestructible: Enhancing the Resilience of Poor People to Natural Disasters", extreme natural disasters force 26 million people into poverty every year, causing global losses equivalent to $520 billion in annual consumption [1]. To a certain extent, the occurrence of disasters destroys the harmonious development between eco-economy. It is urgent to carry out the survey of potential natural and ecological disasters, protect the ecological environment and promote the sustainable development of economy and society [2].

However, few studies have examined whether the intensity of natural disaster (indicated by economic loss) is associated with ecosystem services and how people evaluate [3]and consider the function of alleviating loss for natural disasters in ecosystem services assessment [4]. Natural disasters can cause severe destruction for livelihoods of local rural communities. It can further reduce life quality.

To clarify the relationship between ecosystem service and natural disaster is key for exploring an alternative pathway to combat increasingly extreme weathers and consequently natural disasters, promoting the sustain development of local population, economy and society. The key object of the study is to identify the contribution of ecosystem services in reducing intensity and economic loss of natural disasters. This assessment can provide valuable reference for decision-making managers in the coordinated management of regional ecosystems [5], especially where has ecological red line [6]. It is of theoretical and practical significance.

2. Materials and Methods

2.1. Study Area
Since 5000 years ago, there have been ancient human activities. Now Hunan Province has 13 provincial municipalities, 1 autonomous prefecture, with a total area of 218,000 square kilometers. Hunan is a continental subtropical monthly rheumatic and humid climate. It is cold and warm in spring, and the weather changes dramatically. It often accompanies with strong convective weather processes such as gale, hail and rainstorm. There is little rain in summer and autumn, and drought almost every year, which has a serious impact in varying degrees. In July 2017, the cultivated land area was 3.788 million hectares, the natural grassland area was 6.373 million hectares, and the forest area was 10.369 million hectares.

2.2. Methods

2.2.1. Land use change analysis methods
Landsat-TM remote sensing images are used as the basic data source. the land uses of Hunan province during 1980-2000 were identified by the method of human-computer interaction interpretation, field investigation and verification. The land uses of Hunan Province in 1980, 1990, 1995 and 2000 are demonstrated in Figure1.

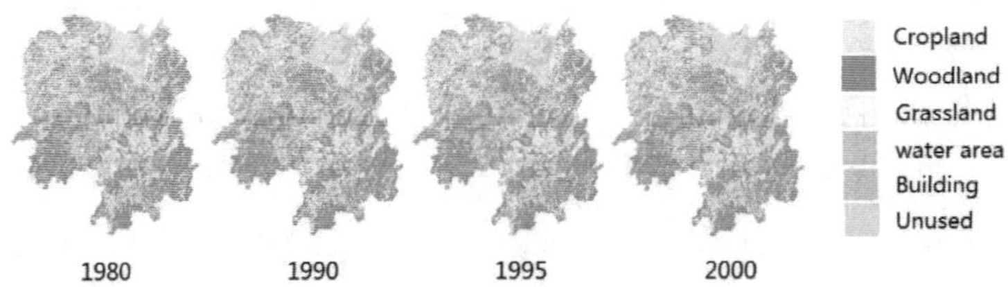

Figure 1. Land use change from 1980 to 2000 in Hunan Province produced by Landsat TM imageries

The spatial overlay is carried out by using GIS tools according to map algebra method. Based on the analysis, the transfer matrix of land use/cover types in each time period was obtained.

$$C_{ij} = A_{ij}^{k} \times 100 + A_{ij}^{k+1} \qquad \text{formula (1)}$$

Among them, C_{ij} is from Period k Land Use/Cover Change Map to Period k+1. A_{ij} denotes the change of land use/cover types from I to j in k+1 period Area of type.

2.2.2. Valuation methods of ecosystem services value
We used the method of Wang et al. [7] to evaluate the ecosystem service values (ESVs). We calculated the ecosystem service values of Hunan province at county level from 1980 to 2000 by formula (2).

$$U=\sum_{i=1}^{n} Si \times Ai \qquad (2)$$

In this formula, U is the total value of ecosystem services of a county, S_i is the ecosystem service value of one kind of ecosystem, and A_i is the area value of this ecosystem, i is the ith order, and n is the number of ecosystem types. In this study, n is 6.

2.2.3. Natural Disaster Data Source and Analysis
The data of natural disasters in this paper are derived from *Xiangxi Statistical Yearbook*, *Hunan Volume of China Meteorological Disasters* and *Statistical Yearbook of China Meteorological Disasters*. We calculated the total loss of each county in Hunan province to match the tempo-spatial scale which land uses and ecosystem service used.

Pearson correlation coefficients between ecosystem service values and economic loss of natural disasters were calculated, aiming at to reflect the degree of linear correlation between two variables.

3. Results

3.1. Land Use Change Transfer Matrix

From 1980 to 2000, land use change has summarized in Table 1.

Table 1. Land Use Transfer Matrix of Western Hunan from 1980 to 2000

1980\2000	Cropland	Woodland	Grassland	Water	building	Unused	roll-ou
Cropland	61093	256	8	338	274	5	881
Woodland	135	131820	95	110	86	2	428
Grassland	10	133	7480	9	3		155
Water	75	17	1	6582	8	5	
building					2429		
Unused	4	9	0	57	1	744	71
roll-in	224	415	104	514	372	12	

Results show: The largest transfer out of cultivated land is 881 square meters, followed by water area followed by forest land 428 square meters. Cropland was mainly converted to water area, followed by urban construction, Woodland mainly converted to cultivated land, followed water area.

3.2. Change of Ecosystem Service Values

We calculated the change of ecosystem service values from 1980 to 2000, shown in Table 2.

Table 2. Change of ecosystem service values in Hunan Province from 1980 to 2000 (RMB. ￥10^4)

Year	Cropland	Woodland	Grassland	Water	Unused	Total
1980-1990	-37391.00	2243.1408	1086.0724	153742.096	-1053.365	118626.943
1990-1995	-43167.69	41124.248	-10395.26	181695.204	-184.801	169071.687
1995-2000	11448.368	-48227.5272	1396.3788	-63892.8192	147.8408	-99127.7587

Corresponding to land use, from 1995 to 2000, he total value of ecosystem services in Hunan Province decreased by 99127.7587 million yuan, respectively.

3.3. Occurrence and Economic Loss of Meteorological Disasters

The losses caused by meteorological disasters in Table 3 basically shows:

Table 3. Natural disaster losses in Hunan, 1980-2000

Year	Farmland loss (10^4 hm^2)	food loss (10^8kg)	collapsed houses (10^4)	population casualties	livestock casualties	total loss (10^4 yuan)
1980	239.1	19.24	15.2	365	930	45808.72
1990	388.3	22.43	12.78	339	10876	299246
1995	287.6	22.63	1.63	633	978	1400000
2000	317.1	11.26	14.98	176	7600	1081796

In Farmland disaster area ,the affected area of farmland increased by 1.492 million hectares in 1990, decreased by 1.07 million hectares in 1995 and increased by 295,000 hectares in 2000;Grain loss increased by 319 million kg in 1990, by 0.2 billion kg in 1995 and by 1.137 billion kg in 2000; The direct economic losses in 1990 decreased by 253437.307 million yuan compared with 1980, increased by 1100753.973 million yuan in 1995 and decreased by 318204 million yuan in 2000 compared with 1995.The loss situation were the largest in 1990, the largest reduction in grain production in 1995 and the largest direct economic loss of 14000 million yuan in 1995.

The spatial regularity subregion is mainly in Western, southern and Northern Hunan Yueyang, among which Western Hunan is more serious, with the largest loss in 1980 being Yueyang 3227.44 million yuan, followed by Sangzhi County 12.31 million yuan; in 1990, the largest loss was Longshan County 51.26 million yuan, followed by Yongshun 30.75 million yuan; in 1995, the largest loss was Sangzhi County 104.94 million yuan, followed by Linwu. County 90.6 million yuan; the largest loss in 2000 was 113.2 million yuan in Cili County, followed by 45.486 million yuan in Yongshun County.

3.4. Service Effects of Ecosystem Services on Reducing Natural Disaster Losses

The above analysis of land use, ecosystem service value, and meteorological disasters of their respective changes in the law. According to the results of principal component analysis and R-type cluster analysis,followed by the use of polygonal maps for further analysis of the relationship between them. Divide into two groups according to the associated indicators, as shown in Figure 2:

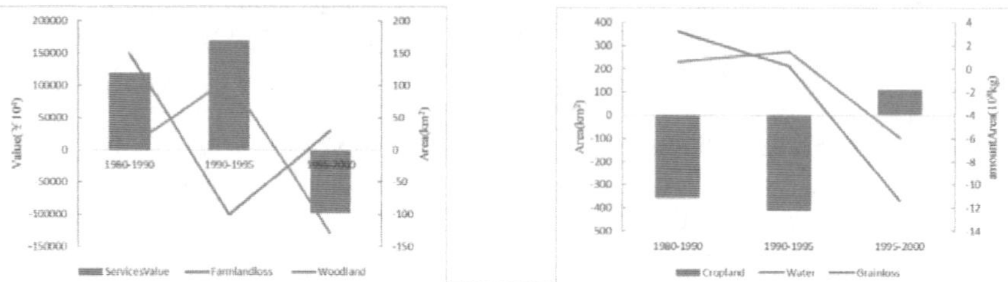

a. ServicesValue, Farmland loss and Woodland b. Cropland, Grainloss and Waterland

Figure 2. The Relation between the Change of Service Value and Shear of Meteorological Disasters

Fig.2a shows the value of ecosystem services increases first and then decreases with the increase of forest land area, and the affected area of farmland decreases first and then increases with the year. Fig.2b Reflections the loss of grain decreases with the increase of farmland area or with the decrease of water area. Therefore, ecosystem services can reduce the loss of meteorological disasters to a certain extent.

The generally show that the value of cultivated land services increased by 233.39 million yuan in 1990 compared with 1980, and the loss of natural disasters decreased by 170.536 million yuan; in 1995, the value of cultivated land services decreased by 89.035.74 million yuan compared with 1990, and the loss of natural disasters increased by 137.124.64 million yuan; in 2000, the value of cultivated land services increased by 1794.4529.26 million yuan compared with 1995, and natural disasters decreased by 318.24 million yuan. The situation of sub-counties is more obvious in Yueyang and Western Hunan, such as Cili County, Baojing County and Guzhang County, which shows that the total value of ecosystem services has increased and the loss of natural disasters has decreased.

3.5. Relationship Between Ecosystem Service and Natural Disaster Loss

The Pearson's correlation coefficients between ecosystem service value and economic loss has been calculated. Scatter plots of ecosystem services response to natural disasters in Hunan counties and districts in 1980, 1990, 1995 and 2000 were analyzed to reveal the relationship between two datasets.

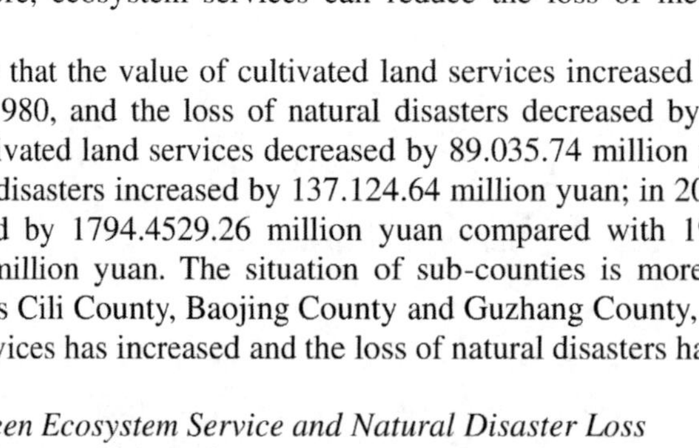

Figure 3. Scatter plots of ecosystem service value and economic loss value, respectively

Figure 3 reflects the trend of approximately linear relationship between ecosystem service value and direct economic loss in each county in each year. Pearson correlation, regression and significance test was obtained by inputting SPSS 22.0 software for correlation analysis in four years as shown in the Table 4.The integration is as follows:

Table 4. The correlation and regression analysis of ecosystem services to natural disaster losses

Year	Correlation analysis		regression analysis		
	coefficient	p-value	coefficient	F	p-value
1980	0.387	0.046	-716.488	3.603	0.016
1990	0.546	0.000	-2242.189	6.802	0.000
1995	0.519	0.000	3716.331	6.305	0.000
2000	0.479	0.055	2454.660	0.884	0.543

The results showed that the correlation coefficient between ecosystem services and meteorological disasters was 0.3-0.6 in 1980-2000, which was moderately correlated. The P value of the first three years was less than 0.05 significantly correlated, as same as meteorological disasters and the regression of six types of land use, reflecting that there was a certain correlation between ecosystem services and the spatial distribution of meteorological disasters in the counties.

The situation of sub-counties is more obvious in Yueyang and Western Hunan, such as Cili County, Baojing County and Guzhang County, which shows that the total value of ecosystem services has increased and the loss of natural disasters has decreased.

4. Discussion

In calculating the value of ecosystem services, this paper also can use Xie Gaodi et al [8]. to estimate the value of different types of ecosystem services, different methods will bring some errors to the results. No matter which method is used, the criteria for the same kind of comparison are the same, and the price can be analyzed and counted on the same benchmark. The results are objective and consistent.

According to the spatial distribution data of geological hazards in the Resource and Environment Science Data Center of the Chinese Academy of Sciences, 7 types of geological hazards in 8181 geological hazards in Hunan Province are 793 collapses, 755 collapses, 354 debris flows, 9 land subsidence, 77 ground fissures, 5337 landslides and 856 slopes. According to previous data and investigations, the main geological hazards in Hunan are concentrated in the northwest, west, South and east areas of Hunan. The main geological hazards occurred in Yuanling County, followed by Shimen County, Guidong County, Cili County, Yanling County, Anhua County, Xupu County, Xinshao County, Zhangjiajie City and Chenzhou City. A total of 54 counties and cities had geological hazards, accounting for 44.3% of the total counties and cities in the province [9].

The occurrence of natural disasters and direct economic losses in Xiangxi Prefecture show the regular pattern of periodic fluctuation in a period of six years.after the construction of the Yangtze River Gorge Reservoir in 2004, the area of water ecosystem has increased, the ecosystem services have increased, the occurrence of flood disasters has decreased, and the affected area and direct economic loss of cultivated land caused by natural disasters in Hunan Province have started from 2004. Compared with 2003, it has a large decrease [10].

On November 5, 2019, on the occasion of the 40th anniversary of the first World Climate Conference (Geneva, 1979), 11263 scientists from 153 countries warned in bioscience that the whole world is facing a climate crisis, threatening the fate of natural ecosystems and human beings [11]. The status of ecosystem services is a key factor for the occurrence or severity of natural disasters. Therefore, the key to effectively avoid and mitigate natural disasters is to restore and protect ecosystem services [12].

This paper studies the relationship between ecosystem services and natural disasters in Hunan Province. The results of correlation analysis and regression analysis of Hunan counties and districts in

2010 were compared. It can also refine the correlation analysis between farmland, forest land, farmland, grain, house collapse, livestock casualties and other indicators, and select other representative years, such as the effect of ecosystem services on meteorological disasters in 2013 or 2016. The results are different in numerical value, but the qualitative impact of ecosystem service capability on natural disaster reduction is unchanged.

5. Conclusion

This paper compares the correlation relationship between ecosystem service indicators and natural disasters by land use change in four natural years of Hunan Province and counties from 1980 to 2000, as well as the changing regularity of four years and the spatial distribution regularity of 120 counties. The results show that the coefficients of spatial-temporal correlation analysis and regression analysis of natural disasters are Medium correlation, while significance test p value is less than 0.05, which affect the ecosystem service efficiency. It should be the most remarkable. The following conclusions are drawn:

(1) On the spatial scale, the ecosystem service value of Hunan Province and its counties is related to natural disasters to a certain extent, with large land area and high ecosystem service value, complex terrain, underdeveloped economy, less exploitation and utilization, natural disasters occur frequently, but the intensity of natural disasters in counties with high ecosystem services is significantly lower than that in regions with low ecosystem services, ecosystem services play a role in reducing meteorological natural disasters to a certain extent.;

(2) On the time scale, with the increase of forest land, the ecosystem service value increases, the disaster area decreases, and the cultivated land area increases. The area of water area is less, and the loss of grain is reduced. The improvement of ecosystem service capacity has a certain effect on the reduction of natural disasters.

Usually, the ultimate goal of ecosystem services effectiveness assessment is to estimate marginal value and provide decision-making information to balance the impact of ecosystem services on human well-being before decision support [13].

6. References

[1] New World Bank report says natural disasters cause economic losses of $520 billion . *J. International Finance,* **2017** (1): 72-72.

[2] Fan Yuanyuan. Coupling relationship between ecological environment, Regional Disasters and socio-economic activities. *Shanxi University of Finance and Economics,* **2018**.

[3] Lau J D, Hicks C C, Gurney G G, et al. Disaggregating ecosystem service values and priorities by wealth, age, and education. *J. Ecosystem Services,* **2018**, 29:91-98.

[4] Deng Min. Progress in Environmental Risk Assessment of Natural Disasters. *J. Journal of Jilin Radio and Television University,* **2018**.

[5] Wang Xiaofeng. Assessment of ecosystem service effect of national ecological barrier area. *Beijing. Graduate School of Chinese Academy of Sciences.* **2014**.

[6] Zhou Tingting. Radiation effects of ecological protection red line and its ecological services in typical flood storage areas. *Inner Mongolia University*, **2016**.

[7] Wang M , Sun X . Potential impact of land use change on ecosystem services in China. *J. Environmental Monitoring and Assessment,* **2016,** 188(4):248.

[8] Xie Gaodi, Lu Chunxia, Leng Yunfa, Zheng Du, Li Shuangcheng. Valuation of ecological assets in Qinghai-Tibet Plateau. *J. Journal of Natural Resources,* **2003,** 18 (2): 189-196.21:4050.

[9] Yangyuelong, Zhou Qun, Lin Jian. The relationship between major geological hazards and topography in Hunan. *Hazardous Science,* **2007,** 22 (3): 36-40.

[10] Gong richao, Xiao Guoan, Zhou Changfeng. Study on the trend model of natural disaster loss in Hunan Province. Journal of natural disasters, 2010,19 (05): 69-74.

[11] William J Ripple, Christopher Wolf, Thomas M Newsome, Phoebe Barnard, William R Moomaw Author Notes. World Scientists' Warning of a Climate Emergency. *BioScience,*05 November **2019**,biz088, https://doi.org/10.1093/biosci/biz088

[12] Li Wenhua. Ecosystem Service Function and Natural Disaster Mitigation [A]. Paper

Compilation VIII of the 2002 Symposium on Natural Disaster Mitigation of the Chinese Association of Science and Technology [C]. Chinese Association of Science and Technology, Chinese Society of Ecology, **2002**:8.

[13] Nemec K T, Raudsepp-Hearne C. The use of geographic information systems to map and assess ecosystem services. *Biodiversity and Conservation*, **2013,** 22(1): 1–15.

IOP Conf. Series: Materials Science and Engineering **730** (2020) 012059 doi:10.1088/1757-899X/730/1/012059

Which Provinces Should Pay More Attention to CO_2 Emissions in China's Construction Industry?

Fen Li and Jianbao Chen

College of Mathematics and Informatics, Fujian Normal University, Fuzhou 350117, China.
Email: jbjy2006@126.com

Abstract. Construction industry has existed the problems of high investment and pollution. As a big country in the construction industry, China has a vast territory and the CO_2 emissions of the construction industry vary greatly among different regions. So which provinces should pay more attention to the CO_2 emissions of the construction industry is a question worthy of consideration. Based on the provincial panel data of China from 1997 to 2015, this paper extends the STIRPAT model and empirically analyses the impact of influencing factors in different quantiles by using quantile regression panel model. Research results show that: (1) The effect of population size, economic growth, energy structure and industrial scale on CO_2 of construction industry in different quantiles are positive, while the effect of energy efficiency is negative. (2) The effect of population size on CO_2 emissions in the 50th-75th quantile provinces is greater than those in the other quantile provinces; the impact of economic growth in the 10th-25th quantile provinces is higher than those in the other percentile provinces; the influence of energy structure in the 25th-50th and 50th-75th quantile provinces are stronger than those in the other quantile provinces; the influence intensity of industry scale in the upper 90th quantile provinces is the highest in all the quantile provinces; the effect of energy efficiency in the upper 90th quantile provinces is lower than those in the other percentile provinces. According to these results, we put forward some corresponding policy suggestions.

1. Introduction

Greenhouse gas emissions has become a main course of global warming and environmental pollution, and the carbon dioxide (CO_2) accounts for about 70% of the greenhouse gas. China is the largest CO_2 emitter in the world at present, its total emissions of CO_2 have reached 9232.6 million tons in 2017, 1.81 times that of the United States. With the increasing international pressure, China has become the focus of a global plan to reduce carbon emissions. According to the data provided by the World Bank, in order to achieve the goal of energy saving and emission reduction by 2030, 70% of the potential emission reduction in construction industry. Therefore, it is of great practical significance to study the main driving forces of CO_2 emissions in China's construction industry for formulating effective environmental protection and emission reduction strategies.

In recent years, scholars have conducted a large number of studies on the main driving forces of CO_2 emissions in the construction industry with people's attention to environmental pollution and energy consumption, and achieved fruitful results. We can find that the main methods to study on the influencing factors of CO_2 emissions in the construction industry include life cycle assessment method[1-3], input-output method[4-5], decomposition method[6] and so on. These traditional methods based on mean regression to study CO_2 emissions in construction industry. However, the economic variables are often non-normal distributed, and the tail of distributions hide important information. We know that the quantile regression does not need to assume zero-mean, homoscedasticity and normal distribution of stochastic error term, it can reduce the limitation of residual distribution and the quantile estimators are more effective than the OLS estimators when disturbance items are not normal[7]. Beyond that, the quantile regression can reflect the influence of the change of independent variables on the whole conditional distribution of dependent variable and estimate different parameter values at different quantile levels of dependent variable compared with the OLS method. Third, the quantile regression is monotonous and homogeneous, which can transform the unstable sample data in the regression process without affecting the estimation effect. Last, the quantile regression estimates parameters by minimizing the sum of absolute weighted residuals and it is less sensitive to outliers and more robust in parameter estimation. In order to overcome the shortcomings of the existing research methods, this paper will apply the quantile regression model to investigate the impacts of the influencing factors of CO_2 emissions in China's construction industry under different level of CO_2 emissions with a provincial panel data of China during the period of 1997-2015.

2. Model and Methodology Specification

2.1. Quantile Regression Model

The quantile regression method was first proposed by Koenker and Bassett[8]. Regression for independent variable X based on conditional quantile of dependent variable Y, and the regression models under all quantiles are obtained. The mathematical expression of the quantile regression model is as follows:

$$y_i = x_i'\beta_\theta + \mu_{\theta i}, 0 < \theta < 1 \tag{1}$$

$$Quant_\theta(y_i \mid x_i) = x_i'\beta_\theta \tag{2}$$

where y denotes the dependent variable, x is a vector of the explanatory variable, μ denotes the random error term, whose conditional quantile distribution is equal to zero. $Quant_\theta(y_i \mid x_i)$ is the θth quantile of the dependent variable.

The estimator of regression coefficients in the quantile θ is $\hat{\beta}_\theta$ which satisfies:

$$\hat{\beta}_\theta = \underset{\beta}{\arg\min} \sum_{y_i \geq x_i'\beta} \theta \left| y_i - x_i'\beta \right| + \sum_{y_i < x_i'\beta} (1-\theta) \left| y_i - x_i'\beta \right| \tag{3}$$

Different parameter estimates can be obtained by giving different θ values. We use the linear programming to solve the problem of (3), and apply the bootstrap method to estimate the standard deviation and confidence interval of the parameters in the quantile regression.

The quantile regression model is quite different from the traditional segmental regression method. The traditional segmental regression estimates parameters based on subset of samples, while the quantile regression estimates the parameters of different quantiles using all sample data[9]. Koenker[10] discussed the details of the quantile regression methods.

2.2. Construction of Empirical Model

IPAT model was first proposed by Ehrlich and Holdren[11] to reflect the impact of population on environmental stress. The mathematical expression of the model is as follows:

$$I = P \cdot A \cdot T \tag{4}$$

where I, P, A and T indicate the pollution intensity of a pollutant, the population size, the economic development level of a country and the technology development level respectively.

Although the IPAT model is concise and intuitive, it also has some limitations. First, the IPAT assumes that the elasticity coefficients of the impact of population size, economic prosperity and technological level on the environment are consistent, which conflicts with the environmental Kuznets curve hypothesis. Second, the unity of the dimensions on both sides of the equation limits other factors that may affect environmental pressure. Third, the hypothesis testing unable to proceed[12]. In order to overcome the above limitations, this paper adopts the Stochastic Impacts by Regression on Population, Affluence and Technology (STIRPAT) model, which based on the IPAT model and proposed by Dietz and Rosa[13]:

$$I_t = aP_t^b A_t^c T_t^d \xi_t \tag{5}$$

where a represents the intercept term; I, P, A and T are the same as in equation (4); b, c and d represent the elastic coefficients of environmental effects with respect to P, A and T respectively, ξ_t is the random error term; since the model is used for annual data analysis, subscript t depicts the year. In order to eliminate possible heteroscedasticity, all variables are logarithmic processed. The sample data set is the panel data, so the equation (5) after logarithmic transformation can be written as follows:

$$\ln I_{it} = \ln a + b\left(\ln P_{it}\right) + c\left(\ln A_{it}\right) + d\left(\ln T_{it}\right) + e_{it} \tag{6}$$

combined with this study and considered the actual situation of China's construction industry, we further expand the STIRPAT model to the following form:

$$\begin{aligned}\ln CO_{2it} &= \ln a + \beta_1 \ln POP_{it} + \beta_2 \ln PGDP_{it} + \beta_3 \ln ENE_{it} \\ &\quad + \beta_4 \ln ENS_{it} + \beta_5 \ln IS_{it} + e_{it}.\end{aligned} \tag{7}$$

where the CO_2 (10,000 tons) represents the CO_2 emissions of the construction industry of each province; *POP* (10,000 people) is the total population size of each province at the end of the year; *PGDP* (yuan) is economic growth and is represented by GDP per capita of each province which is converted into constant prices (1997=100) to cut the effects of inflation; *ENE* (10,000 yuan per tce) is energy efficiency, which is used to measure the technical development level of the construction industry, represents the actual output of the unit energy consumption in the construction industry; *ENS* is calculated by dividing coal consumption in the construction industry by its total energy consumption; *IS* reflects the scale effect of construction industry and is calculated by added value of construction industry divided by GDP; a and e are the intercept and disturbance terms, respectively. The panel regression model (7) can carry out mean regression on the influencing factors of CO_2 emissions in China's provincial construction industry. In the actual economic problems, the CO_2 emissions and the influencing factors may not follow the normal distribution. In order to study the impact of various factors on CO_2 emissions from China's construction industry in different quantiles, we introduce quantiles into the model (7) and obtain the econometric model for empirical analysis:

$$\begin{aligned}Q_\tau(\ln CO_{2it}) &= (\ln a)_\tau + \beta_{1\tau} \ln POP_{it} + \beta_{2\tau} \ln PGDP_{it} \\ &\quad + \beta_{3\tau} \ln ENE_{it} + \beta_{4\tau} \ln ENS_{it} + \beta_{5\tau} \ln IS_{it},\end{aligned} \tag{8}$$

where $Q(\ln CO_{2it})$ and $(\ln a)_\tau$ represent τth quantile in the dependent variable and constant term respectively. $\beta_{1\tau}$, $\beta_{2\tau}$, $\beta_{3\tau}$, $\beta_{4\tau}$ and $\beta_{5\tau}$ indicate the regression parameters of τth quantile in the explanatory variables.

3. Data Source and Description

The variables involved in model (8) are panel data of 30 provinces, municipalities and autonomous regions in China's mainland from 1997 to 2015. Due to the lack of relevant variable data in Tibet, it is not included in the sample. Provincial construction industry CO_2 emissions data is collected from CEADs; the data of coal consumption is from China Energy Statistics Yearbook (1998-2016); the other explanatory variables data sets are from China Statistics Yearbook (1998-2016) and 30 Provincial Statistical Yearbook (1998-2016). The definitions and statistical description of all the variables in this study are shown in Table 1.

Table 1. Definition and statistical description of variables in the model (8)

Variable	Definition	Units of measurement	Mean	Std.dev.	Min	Max
CO_2	Total CO_2 emission	10,000 tons	139.16	154.05	10	2220
POP	Population size	10,000 people	4258.23	2670.43	280	10849
PGDP	Per capita GDP	Yuan	19985.62	5715.89	2250	79255.90
ENE	Energy efficiency	10,000 yuan per tce	15.70	14.89	0.6924	109.30
ENS	Energy structure	Percent	20.46	19.19	0.31	99.90
IS	Industry scale	Percent	6.73	1.86	3.25	14.03

4. Empirical Results

4.1. Test of Multicollinearity

In this paper, Klein's method [14] is used to examine the multicollinearity test. From the results of likelihood ratio test (F = 25.8834, P = 0.0000) and Hausman test (χ^2 = 13.3422, P = 0.0204), we can see that the fixed effect panel regression model should be used to fit the sample data and obtains R^2=0.602. The correlation coefficient matrix of explanatory variables in this paper is shown in Table 2. As can be seen from Table 2, the absolute value of correlation coefficients of each explanatory variable is less than R^2, which shows that the correlation between explanatory variables will not cause multicollinearity problems in the model, and all variables are suitable for regression estimation.

Table 2. The correlation coefficient matrix

	ln*POP*	ln*PGDP*	ln*ENE*	ln*ENS*	ln*IS*
ln*POP*	1.00	-0.03	0.25	-0.10	-0.36
ln*PGDP*	-0.03	1.00	0.39	-0.46	-0.25
ln*ENE*	0.25	0.39	1.00	-0.25	-0.14
ln*ENS*	-0.10	-0.46	-0.25	1.00	0.10
ln*IS*	-0.36	-0.25	-0.14	0.10	1.00

4.2. Unit Root Test and Co-integration Test

We applied the Levin-Lin-Chu (LLC) to implement the panel unit root test, and the results are shown in Table 3. It can be seen that all variables are non-stationary at the 1% significance level, but their first order difference series are stationary, it indicates that the variables in this study are first order single integration. Further, we implemented panel co-integration test to examine whether there is a long-term equilibrium relationship between the CO_2 emissions of the construction industry and its main driving forces. In KAO panel test, ADF = -6.123 and P = 0.0000, which rejects the primitive hypotheses that there is no co-integration relationship. The results are indicative of a significant co-integration relationship between the dependent variable and the explanatory variables.

Table 3. Results of panel unit root tests

Variable	Test form	Test statistic	Results	First difference	Test form	Test statistic	Results
$\ln CO_2$	(0, 0)	5.85	non-stationary	$\Delta\ln CO_2$	(C, T)	-17.09***	stationary
ln*POP*	(0, 0)	11.62	non-stationary	Δln*POP*	(C, T)	-11.10***	stationary
ln*PGDP*	(C, T)	4.48	non-stationary	Δln*PGDP*	(C, T)	-2.52***	stationary
ln*ENE*	(0, 0)	4.87	non-stationary	Δln*ENE*	(C, T)	-12.65***	stationary
ln*ENS*	(C, 0)	-0.62	non-stationary	Δln*ENS*	(C, T)	-18.86***	stationary
ln*IS*	(0, 0)	4.68	non-stationary	Δln*IS*	(C, T)	-13.12***	stationary

Remark: The optimal lag is based on the Akaike Information criterion (AIC) and Schwartz information criterion (SC). ***, **, * indicates a significant at 1%, 5% and 10% significance level respectively. In the Test form, 0, C and T represent constant, intercept and trend items respectively.

4.3. Normal Distribution Tests

We tested the normality of all variables before regression analysis through the Q-Q plot, and the results are shown in Figure 1 (a-f). The observed values of all variables deviate from the straight line, indicating that all variables do not conform to the normal hypothesis. Accordingly, OLS regression can't reveal the important information contained in the tail of the data comprehensively, which further proves that it is reasonable to use quantile regression model for empirical analysis.

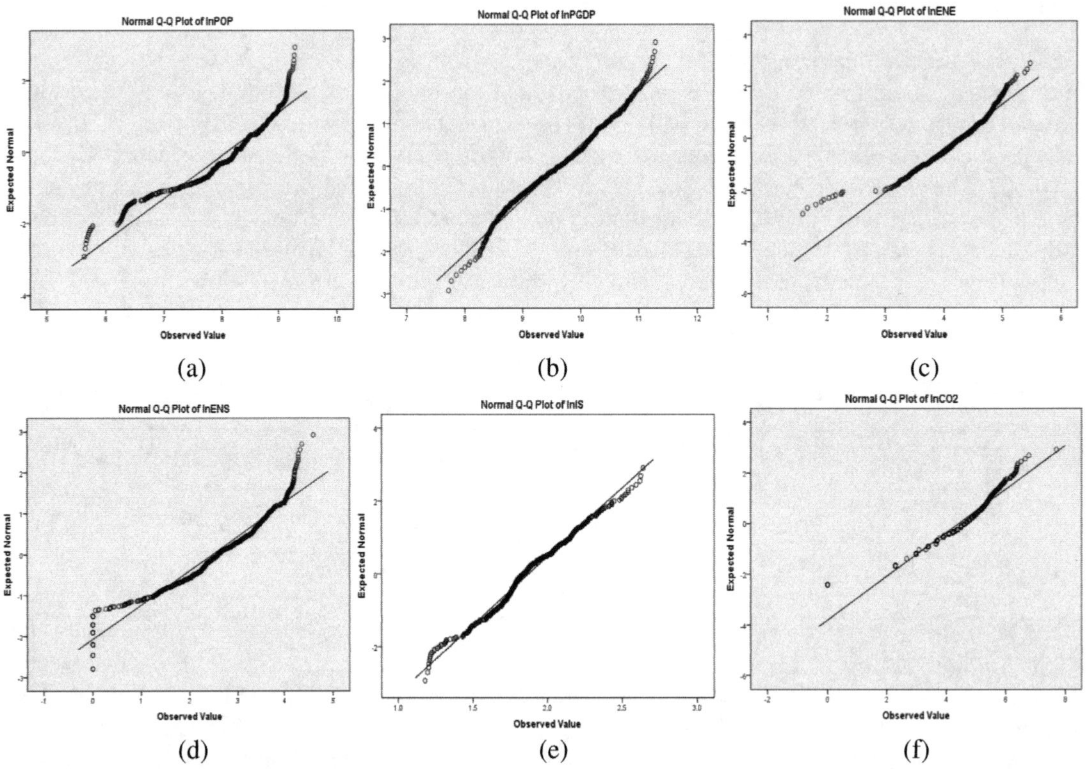

Figure 1. Normal Q-Q plots of ln*POP*, ln*GDP*, ln*ENE*, ln*ENS*, ln*IS* and lnCO_2 respectively

4.4. Quantile Regression Results Analysis

Quantile regression can directly reveal the marginal impact of explanatory variables on CO_2 emissions of different quantiles in construction industry. Different quantile functions can be obtained under different quantiles, which can reveal the influence of the main driving forces on the dependent variables of corresponding quantiles. Therefore, five representative quantiles (10, 25, 50, 75 and 90) are selected for quantile regression, and the estimation results are analyzed in depth. According to annual average CO_2 emissions of China's construction industry from 1997 to 2015, 30 provinces in China are divided into six groups (Table 4). It can be seen from table 4 that although the economic development of some provinces is quite different, the annual CO_2 emissions of the construction industry have little change.

Table 4. Provincial distribution in term of CO_2 emissions in the construction industry

Quantile	Province
The lower 10th quantile group	Heilongjiang, Guangxi, Hainan
The 10th-25th quantile group	Jiangsu, Jiangxi, Guizhou, Qinghai, Ningxia
The 25th-50th quantile group	Beijing, Hebei, Henan, Guangdong, Chongqing, Gansu, Xinjiang
The 50th-75th quantile group	Tianjin, Shanxi, Liaoning, Anhui, Fujian, Sichuan, Yunnan, Shannxi
The 75th-90th quantile group	Inner Mongolia, Jilin, Shanghai, Hunan
The upper 90th quantile group	Zhejiang, Shandong, Hube

Table 5. Estimation results: quantile regression panel model and fixed effects linear panel OLS model during the period of 1997-2015

Variables	Quantile regressions					OLS
	10th quant	25th quant	50th quant	75th quant	90th quant	
Intercept	-11.535***	-13.040***	-13.144***	-12.708***	-9.948***	-12.583***
ln*POP*	0.546***	0.634***	0.722***	0.721***	0.446***	0.587***
ln*PGDP*	1.274***	1.301***	1.216***	1.224***	1.070***	1.315***
ln*ENE*	-0.947***	-0.775***	-0.655***	-0.677***	-0.409**	-0.762***
ln*ENS*	0.295***	0.332***	0.355***	0.321***	0.310***	0.320***
ln*IS*	0.034	0.258*	0.423***	0.391***	0.799***	0.361***

Remark: ***, **, * indicates a significant at 1%, 5% and 10% significance level respectively.

The estimation results of quantile regression are shown in Table 5 and Figure 2. We give the estimation results of OLS in Table 6 for comparative analysis. From Table 5, it can be seen that most of the parameter estimation results of the influencing factors are significant at the 10% test level. The regression coefficients of ln*POP*, ln*GDP*, ln*ENS* and ln*IS* are all positive, indicating that population size, economic growth, energy consumption structure and industrial scale have a positive impact on CO_2 emissions of construction industry; the regression coefficients of ln*ENE* are estimated to be negative, indicating that technological level effectively inhibits the growth of CO_2 emissions of construction industry. In addition, the elasticity coefficient of ln*PGDP* is the largest in all quantiles, indicating that economic growth is the main factor affecting CO_2 emissions in construction industry, and population size and energy consumption structure have a greater impact. Especially, the results of parameter estimation of ln*IS* are not significant in low quantiles, indicating that industrial scale only has more influence on the provinces where CO_2 emissions in construction industry are in the middle and high quantiles. From Figure 2, we can see that the parameters trends of ln*POP* and ln*ENS* up first and then down with the increase of quantiles, ln*PGDP* down, and ln*ENE* and ln*IS* up. These show that the effects of the influencing factors of CO_2 emissions in construction industry are heterogeneous, and several interesting phenomena of quantile regression results are discussed below.

The effects of population size on CO_2 emissions in the 50th-75th quantile provinces are greater than those in the other quantile provinces, which may be caused by the different annual growth rate of population in different regions. Generally speaking, the increase of population will promote the construction demand of housing, transportation, infrastructure and so on, which will lead to the increase of energy and building materials consumption, and CO_2 from the construction industry will increase. The larger the population, the stronger the human economic activity will be, which will further lead to an increase in CO_2 emissions from the construction industry. Due to the implementation of the national policy of Family Planning, the population growth rate in all regions of China has been at a low level in recent years. According to the relevant data of China Statistical Yearbook, the annual average population growth rate during 1997-2015 period in the 50th-75th quantile provinces is 1.16%, is far larger than those in the lower 10th quantile provinces (0.46%), the 10th-25th quantile provinces (0.64%), 25th-50th quantile provinces (-0.86%), 75th-90th quantile provinces (0.92%) and upper 90th quantile provinces (0.61%). Therefore, the impact of population size on CO_2 emissions from construction industry in the 50th-75th quantile provinces is stronger than those in the other quantile provinces.

The influence of economic growth on CO_2 emissions of construction industry shows a downward trend with the increase of quantile, and the provinces with 10th-25th quantile CO_2 emissions of construction industry have the greatest impact, which may be caused by the different proportion of fixed-asset investment in construction industry in the whole society of the region. Infrastructure construction is the premise of economic development. Therefore, investment in fixed assets has always been greatly supported by central and all levels governments as one of the important strategic industries that promote economic development. However, the large-scale investment in fixed assets has driven the rapid development of the construction industry, consumed a large number of steel, cement and fossil fuels, and produced a large number of CO_2 emissions. After calculating the annual average growth rate of fixed assets investment in construction industry in 1997-2015 based on China Statistical Yearbook, we found that the annual average growth rate of fixed assets investment in the 10th-25th quartile provinces is higher than those in other quartile provinces. The rapid growth of fixed assets investment leads to the high annual growth rate of energy consumption in construction industry, and the annual growth rate of CO_2 emissions in construction industry also increases. Therefore, the impact of economic growth on CO_2 emissions in construction industry in 10th-25th quantile provinces is greater than those in other quantile provinces.

The inhibition of energy efficiency on CO_2 emissions from construction industry in the upper 90th quantile provinces are less than those in the other quantile provinces. This may be related to the difference of patent technology level, R&D investment and the number of R&D personnel in different regions. With the development of new technologies such as energy saving and emission reduction, energy efficiency plays an increasingly prominent role in improving the environment and enhancing the competitiveness of enterprises. On the one hand, patents can protect technological R&D achievements effectively, and encourage enterprises to expand investment in technological R&D. Relevant data from China Science and Technology Statistics Yearbook show that the average annual growth rate of patents granted in the upper 90th quantile provinces during 1997-2015 is lower than those of other quantile provinces. The slow growth in the number of patents has led to the slow progress of technological level in provinces with CO_2 emissions above 90th quantile compared with other quantile provinces. On the other hand, the R&D of new technology needs a lot of R&D funds and R&D personnel. Therefore, R&D funds and R&D personnel determine the level of energy efficiency. We found that annual growth rate of R&D investment and R&D personnel of R&D institutions in the upper 90th quantile provinces during 1997-2015 is much lower than those in other quantile provinces based on China Statistical Yearbook. The slow growth of R&D investment and R&D personnel is also one of the reasons for the slow technological progress in the upper 90th quantile provinces. The slow progress of energy saving and emission reduction technology results in energy efficiency has less inhibition on CO_2 emissions of provinces in the upper 90 quantile compared with other quantile provinces.

The elasticity coefficient of energy structure in the median is the largest, which has a higher impact on the construction industry in the 25th-50th and 50th-75th quantile provinces than other percentile provinces. This may be due to the difference in the proportion of coal consumption to the total energy consumption of construction industry in different provinces. China is currently the largest producer and consumer of coal in the world. Coal is rich in reserves and low in price, which has become the main source of energy consumption in China. Because coal combustion produces a large amount of CO_2, coal consumption accounts for a significant proportion of total energy consumption, which will inevitably increase the CO_2 emissions of the construction industry. We identified that the annual average proportion of coal consumption in total energy consumption of provinces in 25th-50th and 50th-75th quantile during 1997-2015 is higher than those of provinces in the other quantile based on China Energy Statistics Yearbook. Therefore, the impact of energy structure on CO_2 emissions from construction industry in 25th-50th and 50th-75th quantile provinces is greater than that of other quantile provinces.

The impact of industrial scale on CO_2 emissions of construction industry shows an upward trend with the increase of quantile. The elasticity coefficient in 90th quantile is greater than that of 10th, 25th, 50th and 75th quantile, which is mainly due to the differences of construction industry scales in different province. The construction industry is a production sector specializing in civil engineering, housing construction and equipment installation, such as factories, railways, bridges, ports, roads and public infrastructure construction. The development of the construction industry needs to consume a lot of building materials, such as steel, cement, aluminum and glass products. The large consumption of these building materials will inevitably lead to a large amount of CO_2 emissions in the construction industry. According to the China Statistical Yearbook, the average annual output value of the construction industry of the upper 90th quantile provinces in 1997-2015 is 518.294 billion yuan, and those in the lower 10th, 10th-25th, 25th-50th, 50th-75th and 75th-90th quantile provinces are 69.363 billion yuan, 22.632 billion yuan, 216.258 billion yuan, 196.976 billion yuan and 162.932 billion yuan respectively. We can see that the scale of construction industry in the upper 90th quantile provinces are much larger than those in other percentile provinces. Therefore, the impact of industrial scale on CO_2 emissions of construction industry in the upper 90th quantile provinces is stronger than those of other percentile provinces.

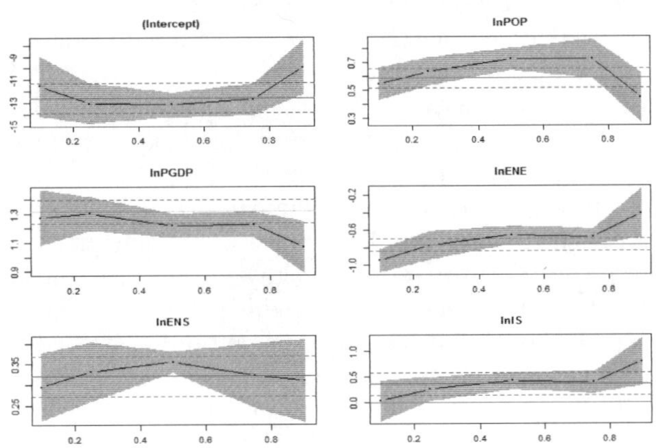

Figure 2. The effects of influencing factors on the CO_2 emissions in construction industry

Remark: Shaded areas represent 95% confidence bands for the quantile regression estimates. The horizontal dotted lines depict the 95% confidence intervals for the OLS estimates. The vertical axis represents the elasticities of explanatory variables.

5. Conclusions and Policy Implications

Based on the panel data of 30 provinces in China from 1997 to 2015, this paper extends the STIRPAT model and uses the quantile regression panel model to empirically study the main influencing factors of CO_2 emissions in China's construction industry, focusing on the different effects of influencing factors in different quantiles. The results show that: (1) Population size, economic growth, energy structure and industrial scale have positive impacts on carbon emissions of construction industry, while technological level has negative impacts. (2) The impact of population size on CO_2 emissions from construction industry in 50th-75th quantile provinces is larger than those in other percentile provinces, which may be related to the different population growth rates in different regions. The impact of economic growth on provinces in 10th-25th quantile is greater than those of provinces in other quantile, which may be caused by the difference of fixed-asset investment in construction industry in different provinces. The difference in patented technology, R&D expenditure and R&D personnel cause energy efficiency has less inhibition on CO_2 emissions of construction industry in the upper 90th quantiles provinces than those of provinces in other quantile. Energy structure has stronger impact on CO_2 emissions of provinces in the 25th-50th and 50th-75th quantiles than other percentile provinces may owe to obvious differences in coal consumption. The effects of industrial scale on CO_2 emissions of provinces in the upper 90th quantile are higher than those of provinces in other quantile because of the differences in scale of construction industry. Therefore, all levels governments and relevant enterprises should pay more attention to the heterogeneous effects of these factors on the construction industry CO_2 emissions in different quantile provinces. People-oriented, tailored to local conditions, different energy saving and emission reduction schemes should be adopted according to the characteristics and problems of different regions, rather than "one-size-fits-all".

The above research conclusions imply the following policy suggestions: (1) The 10th-25th quantile provinces need to further optimize their economic structure and reduce excessive dependence on fixed-assets investment. Local governments should establish special governance mechanism to guide and encourage social capital to develop new energy high-tech industry. (2) For the 25th-50th and 50th-75th quantile provinces of CO_2 emissions in construction industry, the local governments should encourage residents adopt green material in the decoration process and expand the construction of hydropower and nuclear power stations to reduce coal consumption for power generation during construction. The local construction companies can use energy saving composite materials to replace hollow clay bricks, which can reduce the coal consumption for the production of clay bricks. In addition, the application of 3D printing technology in the construction industry is the future development direction of the local construction industry. (3) The upper 90th quantile provinces need to further improving energy efficiency of construction industry. Firstly, the local government should encourage construction enterprises develop and apply new technologies, new equipment, new materials and new craft for green construction, make reasonable indicators of construction energy consumption. Secondly, the local governments should support the merger and acquisition of large construction enterprises with advantages, and accelerate the construction industry to use energy-saving, efficient and environment friendly construction equipments and tools in line with national and industrial standards. Thirdly, local governments should increase investment in R&D funds and encourage scientific research institutions to train technical personnel related with building energy saving and adopt flexible and effective policies to attract relevant professionals.

6. References

[1] Escamilla EZ, et al. Industrial or traditional bamboo construction? Comparative life cycle assessment (LCA) of bamboo-based buildings. *Sustainability*, 2018; **10**(9): 3096.
[2] Balasbaneh AT, Marsono AKB. New residential construction building and composite post and beam structure towards global warming mitigation. *Environmental Progress and Sustainable Energy*, 2018; **37**(4): 1394-1402.
[3] Jonggeon L, Sungho T, Rakhyun K. A study on the analysis of CO_2 emissions of apartment housing in the construction process. *Sustainability*, 2018; **10**(2).
[4] Cho S, Na S. The reduction of CO_2 emissions by application of high-strength reinforcing bars to three different structural systems in South Korea. *Sustainability*, 2017; **9**(9): 1652.
[5] Chen J, Wu Y, Yan H, et al. An analysis on the carbon emission contributors in the Chinese construction industry. *Proceedings of the 20th International Symposium on Advancement of Construction Management and Real Estate*, Springer Singapore, 2017.
[6] Lu Y, Cui P, Li D. Carbon emissions and policies in China's building and construction industry: Evidence from 1994 to 2012. *Building and Environment*, 2016; **95**: 94-103.
[7] Andreas H. Cluster-robust bootstrap inference in quantile regression models. *Journal of the American Statistical Association*, 2017; **112** (517): 446-456.
[8] Koenker R, Bassett G. Regression quantiles. *Econometrica*, 1978; **46**(1): 33-50.
[9] Cade B S, Noon B R. A gentle introduction to quantile regression for ecologists. *Frontiers in Ecology & the Environment*, 2003; **1** (8): 412-420.
[10] Koenker R. Quantile Regression. *Cambridge University Press*, New York, US, 2005.
[11] Ehrlich P R, Holden J P. Impact of population growth. *Science*, 1971; **171**: 1212-1217.

[12] York R, Rosa EA, Dietz T. STIRPAT, IPAT and ImPACT: Analytic tools for unpacking the driving forces of environmental impacts. *Ecological Economics*, 2003; **46**: 351-365.
[13] Dietz T, Rosa EA. Rethinking the environmental impacts of population, affluence and technology. *Human Ecology Review*, 1994; **2**(1): 277-300.
[14] Klein L R, Marquez J R. Economic theory and econometrics. *The Economic Journal*, 1985; **95**(380): 1096.

Wastewater Emission Reduction by China's Industrial Imports of and Its Influencing Factors

Wangsong Xie[1] and Qirong Zhu[2]

[1]Business College of Wuxi Taihu University, Wuxi, 214064, China;
Email:xiewangsong@126.com
[2]International Business College of Shandong University of Finance and Economics, Jinan, 250014, China;Email:13688617525@163.com

Abstract. By using the input-output model and the data of China's industrial segment industries, this paper measure the wastewater emission reduction intensity and the wastewater emission reduction of China's industrial imports in 2012 and 2015. At the same time, it also uses the LMDI to analyze the impact of the imports changes in scale, the structural and the intensity of the wastewater emission reduction by the imports on China's industrial wastewater emission. The results are as follows:(1) In 2012 and 2015, China's industrial product import caused its industrial wastewater discharge to decrease by 15% and 14% respectively; (2) The decreases in the scale of the imports and the intensity of wastewater emission reduction lowered the reduction of the emission but the change in the structure of the imports improved the capacity of the reduction.

1. Introduction

The 2017 China Ecological Environment Bulletin showed that the water quality of Class IV, Class V and inferior Class V accounted for total of 32% in one thousand nine hundred and forty China's evaluation and assessment sections of surface water in 2017 and among six thousand one hundred and twenty-four monitoring points for groundwater quality in China, sixty percent of them are at low levels [1]. This means that the problem of water pollution in China is serious. The main cause of the water environmental pollution is that the wastewater discharged by China's industrial production contains a large amount of harmful substances such as nitrogen oxides, ammonia nitrogen, cyanide and heavy metals. These harmful substances flowing into China's rivers and underground polluted the surface water, groundwater and soil, which endanger the safety of China's drinking water and food safety. Therefore, it is imperative to reduce the discharge of industrial wastewater in China.

Usually, there are three ways to reduce industrial wastewater pollution. Firstly, environmental protection technology is used to reduce the wastewater discharge of enterprises; secondly, the industrial structure is adjusted, namely, the productions of pollution-intensive industries are cut down. Finally, environmental tax is levied on polluting industries to internalize the cost of pollution enterprises. These three ways of reducing pollution are all from the domestic perspective. In addition, China can use imports to reduce its pollutant emissions. For example, in March 2012, the conference of 'Appropriate Expansion of Import Scale to Alleviate Resource Pressures' held by China's State Council pointed out that with the constraints on resources and environment being strengthened and the improvement of the living standards of the people, China needed to appropriately increasing its imports while keeping the growth of exports [2]. In recent years, what has been the environmental effects brought by the industrial products imports on China? Which factors affect the effects of the imports? This paper will study these issues.

The International Federation of Advanced Research Institutions (IFIAS, 1974) proposed the concept so called implicit contaminants to measure total pollutants discharge from production of upstream intermediate inputs to the processing, manufacturing and transportation of final products. After that, the concept has been widely used to analyze the environmental impact of trade by academic circles. Wyckoff and Roop (1994) found that French industrial imports had reduced its carbon emissions by 40% in 1985[3].Dietzenbacher etal (2007) found that the pollutants implied in India's imports of are twice as high as that implied its exports. They thought that India's import trade helps protect the country's ecological environment [4]. Peters (2007) studies the influencing factors of China's CO^2 emissions and found that the implied carbon emissions in China's imports had offset the growth of CO^2 emissions implied in the country's exports, and import had reduced China's CO^2 emissions [5]. Li (2013) measured the CO^2 emissions implied in Macao's imports from 2005 to 2009, and found that the CO^2 emissions implied in the imports were 3to4 times that implied exports in the Area at the same period [6] .Wang (2006) analyzed the CO_2 emissions implied in trade between China and Japan. They found that the CO_2 emissions implied in China's imports from Japan are one-sixth of that implied in China's exports to Japan which was not good for China's environment [7]. Zhang (2009) analyzed the impact of foreign trade on China's SO_2 emissions during the period 1987-2006 and found that the sulfur implied in China's imports had been increasing for many years but the growth of imports was much lower than that of its exports. He thought that the growth of China's foreign trade had led to a huge surplus of sulfur emissions. He suggested that China should further pay the role of China's import in environmental protection [8]. Xu (2010) measured the pollutants emissions implied in the imports of 42 sectors in China in 2002. His study showed that China's imports had reduced its environmental pollution [9]. Yan and Xie (2011) analyzed the impacts of China's agricultural foreign trade on the environment in the country from 1995 to 2009, and found that the agricultural imports had reduced China's environmental pollution [10]. Ni etal. (2012) measured the pollutant emissions implied in the trade between China and OECD countries in 2002 and 2007 and found that the CO_2 emissions implied in China's imports were $1.49 \times 10^8 m^3$ and1.1×10^5 m^3 in the these two years respectively [11]. Liu etal (2015) measured the CO_2 emissions implied in China's imports and exports from 2001 to 2012 and found that the CO2 emissions implied in China's imports in 2012 were 7.5 times that in 2001[12]. Zhu etal(2018) found that the emission reductions of industrial wastes by China's imports in 2002 and 2012 were equivalent to 20% and 22% of its industrial waste emissions in the same period. They thought that expansion of China's import trade were conducive to its environment [13].

The data used in the literatures are out of date, and it is difficult to show the environmental benefits brought by China's imports recently. In addition, there are few researches on the influencing factors of China's import environmental effects. This paper use China's latest input-output data to analyze the effects of industrial imports on the reduction of industrial wastewater discharge in the country and to reveal the factors affecting the effects of the imports on the wastewater reduction, and to look for method to improve the environmental benefits of China's imports.

2. Methodology and Model

2.1. Calculating Wastewater Emission Reduction Intensity of Industrial Imports

The reduction of the wastewater emission reduction by China's industrial imports is calculated based on substitution effects. Namely, the reduction is measured according to the wastewater discharge emitted from the production of similar products in China. Because China's imports of industrial products not only reduce the production of similar domestic products that directly reduce the discharge of industrial wastewater in China, but also reduce the consumption and production of domestic inputs in the production process of the product which indirectly reducing China's discharge of industrial wastewater. Therefore, the reduction of wastewater discharge by China's industrial imports can be measured by input-output model (Machado etal., 2001; Chen Ying etal., 2008; Zhu etal., 2016) [14] [15] .

When input-output model is used to calculate the wastewater emissions reduction of China's industrial imports, it is necessary to use formula (1) to calculate the direct discharge of industrial wastewater from the production sector(No.i) and then the wastewater reduction intensity of the same sectoris obtained by using input-output model,.

$$d_i = Q_i / X_i \quad (i = 1,\ 2,\ \ldots,\ n) \tag{1}$$

$$p_i = d_i\,(I - A)^{-1}\,(i = 1,\ 2,\ \ldots,\ n) \tag{2}$$

Where, d_i is the direct wastewater emissions intensity from (No.i) sector. Q_i are the reduction of wastewater emissions from (No.i) sector in China, X_i is China's output. n is the number of (No.i) import sector. $(I-A)^{-1}$ is the Leontief inverse matrix, p_i indicates the intensity of wastewater reduction by (No.i) import.

2.2. Calculating Reduction of Wastewater Emissions by Industrial Imports

Using formula (3), reductions of the wastewater emissions by China's industrial imports sectors are calculate, and then by adding the reductions, total whole of industrial wastewater emissions reductions by China's import sectors is gotten.

$$Q = \sum_{i=1}^{n} Q_i = \sum_{i=1}^{n} IM_i \times \beta_i = \sum_{i=1}^{n} IM \times \theta_i \times p_i\,(i = 1,\ 2,\ \ldots,\ n) \tag{3}$$

Where, Q is total of wastewater emission reductionby the import, β_i is the reduction intensity of importing (No.i) sector, IM_i is (No.i)sector import, θ_i is the proportion of (No.i) sector imports to China's total imports of industrial products, IM is total industrial products imports of China.

2.3. Model of Factors Affecting Wastewater Emission Reduction in China's Industrial Imports

From the base year to the reporting year, the change in the wastewater discharge reduction by China's industrial products imports can be expressed as the effects of changes in the scale of the import, the imports structure and the intensity of wastewater reduction of the imports shown in equation (6).

$$\Delta Q_{IM} = Q_{IM}^{t} - Q_{IM}^{0} = \Delta Q_{IM}^{G} + \Delta Q_{IM}^{S} + \Delta Q_{IM}^{T} \tag{4}$$

Where, ΔQ_{IM} represents the total changes in the reduction of wastewater discharge by China's industrial imports from the base year to the reporting year. Q_{IM}^{0} and Q_{IM}^{t} are the reduction by the imports in base year and reporting year respectively. ΔQ_{IM}^{G}, ΔQ_{IM}^{S} and ΔQ_{IM}^{T} are changes in the scale of China's industrial products, structure of the imports and the intensity of wastewater emissions reduction by the imports respectively

Using the method of log-average weight decomposition, the formula of the effects factors is obtained.

$$\Delta Q_{IM}^{G} = \sum_{i=1}^{n} \frac{Q_{IMi}^{t} - Q_{IMi}^{0}}{\ln Q_{IMi}^{t} - \ln Q_{IMi}^{0}} \times \ln \frac{IM_t}{IM_0} \tag{5}$$

$$\Delta Q_{IM}^{S} = \sum_{i=1}^{n} \frac{Q_{IMi}^{t} - Q_{IMi}^{0}}{\ln Q_{IMi}^{t} - \ln Q_{IMi}^{0}} \times \ln \frac{\omega_i^{t}}{\omega_i^{0}} \tag{6}$$

$$\Delta Q_{IM}^{T} = \sum_{i=1}^{n} \frac{Q_{IMi}^{t} - Q_{IMi}^{0}}{\ln Q_{IMi}^{t} - \ln Q_{IMi}^{0}} \times \ln \frac{\beta_i^{t}}{\beta_i^{0}} \tag{7}$$

Where, IM_0 and IM_t are China's industrial product imports in base year and reporting year respectively, ω_i is the proportion of (No.i) importing sector to the total industrial imports in China, β_i is the wastewater reduction intensity of the importing sector. Q_{IMi}^{0} and Q_{IMi}^{t} are the reductions wastewater discharge by the importing sector in the bass year and that in the reporting year respectively.

3. Data

At present, the latest China's Input-Output Table is the '2015 National Input-Output Table', which is an extension of the '2012 National Input-Output Table'. In order to compare China's wastewater emissions reductions by the industrial imports in these two years, this paper uses 2015 National Input-Output Table and 2012 National Input-Output Table. The data of China's output and imports of industrial sectors used in this paper also come from the database of the input-output table in these two years. The wastewater discharge of the industrial sectors comes from the China Environmental Statistics Yearbook. In order to match the industrial sectors in the input-output table with the industrial sectors in the yearbook, the industrial sectors were merged into 24 sectors shown in table1.

4. Empirical Analysis

4.1. Intensity of Wastewater Reduction by China's Industrial Imports

Using formula (1) and formula (2), the wastewater emissions reduction intensity of China's industrial importing sectors in 2012 and 2015 was calculated. Tables 1 that the reduction intensity of the importing sectors such as paper and stationery, manufacturing, electrical and mechanical and equipment manufacturing, clothing and leather and down and its products, textile materials, transportation equipment, other manufactures, chemicals and metal products is higher than the average reduction intensity of the wastewater emission by China's industrial importing sectors. These import sectors had higher capability of reducing wastewater emission in China. By comparison, the reduction intensity of import sectors such as oil and gas extraction, water production and supply, coal mining and washing, petroleum processing and coking and nuclear fuel processing, non-metallic mining and mining, wood and furniture manufacturing, metal mining, food and tobacco is significantly lower than the average reduction intensity of China's industrial imports. The reduction intensity of these importing sectors is lower.

From 2012 to 2015, the wastewater emission reduction intensity of China's importing sectors changed greatly. In 2012, the average wastewater emission reduction intensity was 6.84 m^3 /ten thousand Yuan, but in 2015, the intensity dropped to 5.26 m^3 /ten thousand Yuan. Namely, the intensity decrease 23% during the three years. In the light of changes in China's importing sectors, the reduction intensity of sectors such as coal extraction and washing, waste utilizing, gas production and supply was increased, but the reduction intensity of the rest of the importing sectors was decreased among which paper and stationery, food and tobacco, clothing and leather and down and its products, other manufacture, electricity and heat production and supply industries, the textile raw materials and the chemicals reduced significantly, all of which declined more than 30%.

Table1 shows that the simple average and the weighted average of the wastewater emission reduction intensity of China's industrial importing sectors were 6.84 m^2/ ten thousand Yuan and 6.38 m^2/ ten thousand Yuan respectively in 2012. Obviously, the former is larger than the latter. That means that the imports with high reduction intensity of wastewater emission in China is relatively small in the year (according to the relationship between the simple average and the weighted average, when the weight of the large number is small, the simple average is greater than its weighted average), which is not conducive to reducing industrial wastewater discharged in China.

In 2015, the simple average and weighted average of the reduction intensity of China's importing sectors were 5.26 m^2/ ten thousand Yuan and 5.97 m^2/ ten thousand Yuan respectively. Obviously, the former is smaller than the latter, which is conducive to reducing China's wastewater emissions in this year. By comparing the structural differences in China's industrial products import between 2012 and 2015, the result showed China's industrial products imports structure had been changing towards a good condition that is conducive to reducing China's wastewater discharge.

Table 1. Intensity of wastewater emission reduction in China's importing sectors in 2012 and 2015

Sector	Intensity of wastewater emissions reduction [cubic meter/ten thousand Yuan (RMB)]		Changes in the intensity during 2012 and2015 (%)
	2012	2015	
Coal mining and washing industry	4.38	4.45	1.60
Oil and gas extraction industry	3.20	2.76	-13.75
Metal mining industry	5.45	4.55	-16.51
Non-metallic mining industry	4.65	3.94	-15.27
Food and tobacco industry	5.07	3.18	-37.28
Textile raw material industry	12.84	8.48	-33.96
Clothing and leather and down and their products industry	12.87	8.29	-35.59
Wood and furniture manufacturing industry	5.76	4.05	-29.69
Paper and stationery manufacturing industry	15.16	7.96	-47.49
Petroleum and coke and nuclear fuel processing industry	4.43	3.92	-11.51
chemical industry	10.21	6.65	-34.87
Non-metallic mineral industry	6.76	4.98	-26.33
Metal smelting and processing industries	7.26	6.42	-11.57
metal products industry	7.05	5.59	-20.71
General and special equipment industry	6.24	5.22	-16.35
Transportation equipment industry	7.22	6.21	-13.99
Electrical and mechanical and equipment industry	7.54	5.66	-24.93
Communications equipment, computer and other electronic equipment industry	5.93	5.28	-10.96
Instrument and office machinery	6.00	4.27	-28.83
Other manufacturing industries	9.28	5.91	-36.31
Waste utilizing industries	1.86	5.32	186.02
Electricity and heat.	6.28	4.02	-35.99
Gas production and supply industry	4.47	6.26	40.04
Water production and supply	4.15	2.98	-28.19
Simple average	6.84	5.26	-23.10
weighted average	6.38	5.97	-6.43

4.2. Analysis of Emission Reduction of Wastewater by China's Industrial Imports

Using formula (3), wastewater emission reductions by China's industrial imports sectors were calculated in 2012 and 2015. Table 2 shows that the reductions in 2012 were 711.177 million cubic meters, which reduced the total discharge of industrial wastewater in China by 15% in this year. Among China's importing sectors, communication equipment, computer and other electronic equipment, chemicals, metal products, metal mining and processing, oil and natural gas, and transportation equipment all reduced the wastewater emission by more than 43 million cubic meters.

In 2015, the wastewater emission reduction decreased to 534.147 million cubic meters, which reduced the total amount of industrial wastewater discharged in China by 14%. Among China's importing sectors, communication equipment and computers and other electronic equipment, chemicals, metal smelting and rolling processing, transportation equipment manufacturing, general and special equipment manufacturing, metal mining and mining, electrical and mechanical and equipment manufacturing imports cut down more wastewater discharged than other importing sectors, each of which reduced wastewater emissions by more than 230 million cubic meters.

It can be found that the emission reduction of the wastewater caused by the import in 2015 decreased by 25% compared with that in 2012. The main reason is due to the decrease of emission reduction of the wastewater by some importing sectors such as fabricated metal products, chemicals, oil and gas, metal smelting and rolling processing, electrical and machinery and equipment, paper and stationery, communications equipment and computer and other electronic equipment, oil and coke and nuclear fuel processing, coal mining and washing, instruments and meters, and office machinery manufacturing and other 20 manufactures.

Table 2. The wastewater discharge reduction by China's importing sectors in 2012 and 2015

Sector	Wastewater reduction (Ten thousand cubic meter) 2012	2015	Changes in wastewater reductions during 2012-2015 (%)
Coal mining and washing industry	7941.16	3426.29	-56.9%
Oil and gas extraction industry	45717.85	23838.40	-47.9%
Metal mining industry	45796.22	27060.75	-40.9%
Non-metallic mining industry	1832.72	1305.12	-28.8%
Food and tobacco industry	17141.58	14200.72	-17.2%
Textile raw material industry	12066.20	9510.86	-21.2%
Clothing and leather and down and their products industry	14344.70	14808.00	3.2%
Wood and furniture manufacturing industry	2920.02	3508.96	20.2%
Paper and stationery manufacturing industry	22311.46	16137.94	-27.7%
Petroleum and coke and nuclear fuel processing industry	12765.83	8002.26	-37.3%
chemical industry	125651.47	80447.00	-36.0%
Non-metallic mineral industry	4749.77	4387.33	-7.6%
Metal smelting and processing industries	64997.59	55924.84	-14.0%
metal products industry	62962.53	4713.57	-92.5%
General and special equipment industry	31899.57	42083.04	31.9%
Transportation equipment industry	43466.92	42910.45	-1.3%
Electrical and mechanical and equipment industry	30070.64	23855.36	-20.7%
Communications equipment, computer and other electronic equipment industry	142379.37	136950.91	-3.8%
Instrument and office machinery	17423.86	13293.34	-23.7%
Other manufacturing industries	320.02	220.09	-31.2%
Waste utilizing industry	4278.18	7475.36	74.7%
Electricity and heat.	139.34	86.27	-38.1%
Gas production and supply industry	0.00	0.00	-
Water production and supply	0.00	0.00	-
Total	711177.00	534146.86	-24.9%

4.3. Factors Affecting the Reduction of Wastewater Emission by China's Industrial Imports

Using formula (4), formula (5) and formula (6), the effects of China's industrially importing scale change, the importing structure changes, and the effects of changes in wastewater emission reduction of the imports were calculated during 2012 and 2015 shown in Table 3.

Table 3 shows that during the times, wastewater emissions increased 590 million cubic meters and 1.38 billion cubic meters in China due to the decline of its industrial products import and the fall of the reduction intensity respectively, and the wastewater emissions decreased 202 million cubic meters because of structural changes in this period. Offsetting the positive and the negative effects mentioned, the reduction of the wastewater emission by the imports decreased by 1770.3million cubic meters during these three years.

From the effects of the change in imports scale of sectors, the import scale effects of wastewater reduction by China's importing sectors were negative, Among which, imports of metal products, oil and natural gas exploitation, metal mining, oil and coke and nuclear fuel processing, metal smelting and processing decreased so significantly that the reduction of wastewater emissions decreased obviously.

From the effects of changes in the intensity of wastewater emission reduction by the importing sectors, except that the effects of the changes in the intensity of waste utilizing, coal mining and washing industries, gas production and supply, water production and supply were positive, the effects of the changes in intensity of other importing sector were all negative due to the decline of the wastewater emission reduction intensity of these importing sectors in past three years. Specially, the effects of changes in chemicals, communication equipment and computers and other electronic equipments, metal, paper and cultural and stationery, food and tobacco, general and special equipments, transportation equipments, metal decreased significantly.

In terms of the effects of structural change in China's importing sectors, the effects of wastewater discharge reduction in 17 importing sectors such as communication equipment and computers and other electronic equipments, general and special equipments, transportation equipment, clothing and leather and down and its products, chemical, paper and stationery, food and tobacco, metals, electrical and mechanical and equipment were positive were positive but the structural change effects of wastewater discharge reduction in other importing sectors were negative.

Table 3. Decomposition of wastewater emission reductions in China's importing sectors from 2012 to 2015 (Unit: Ten thousand cubic meter)

Sector	Scale effects	Structural effects	Intensity effects	Net effects
Coal mining and washing industry	-522.93	-4077.11	85.16	-4514.87
Oil and gas extraction industry	-3271.18	-13638.3	-4969.99	-21879.5
Metal mining industry	-3467.04	-8841.09	-6427.34	-18735.5
Non-metallic mining industry	-151.3	-118.82	-257.48	-527.6
Food and tobacco industry	-1521.24	5868.83	-7288.45	-2940.86
Textile raw material industry	-1045.43	2944.77	-4454.67	-2555.34
Clothing and leather and down industry	-1419.02	8293.17	-6410.85	463.3
Wood and furniture manufacturing industry	-312.08	2030.06	-1129.03	588.94
Paper and stationery industry	-1855.5	7959.96	-12278	-6173.52
Petroleum and coke and nuclear fuel processing industry	-993	-2523.12	-1247.46	-4763.57
chemical industry	-9869.78	8129.93	-43464.6	-45204.5
Non-metallic mineral industry	-444.56	1477.5	-1395.38	-362.44
Metal smelting and processing industries	-5875.39	4223.08	-7420.44	-9072.75
metal products industry	-2187.83	-50846.6	-5214.54	-58249
General and special equipment industry	-3578.58	20322.44	-6560.4	10183.46
Transportation equipment industry	-4204.75	10156.47	-6508.19	-556.47
Electrical and mechanical equipment industry	-2613.43	4096.72	-7698.58	-6215.29
Communications equipment, computer and other electronic equipment industry	-13596	24380.32	-16212.8	-5428.46
Instrument and office machinery	-1486.24	2548.25	-5192.52	-4130.52
Other manufacturing industries	-25.99	46.51	-120.45	-99.93
Waste utilizing industries	-557.75	-2265.48	6020.42	3197.19
Electricity and heat.	-10.78	7.09	-49.38	-53.06
Gas production and supply industry	0	0	0	0
Water production and supply	0	0	0	0
Industrial sector change effect	-59009.76	20174.61	-138194.99	-177030.14

5. Conclusions and Policy Suggestions

Industrial imports made great contributions to protecting China's environment. In 2012, the imports reduced wastewater discharge by 711million cubic meters in China, which reduced its industrial wastewater discharge by 15% in that year. In 2015, the imports cut down China's wastewater discharge by 535 million cubic meters which reduced China's industrial wastewater discharge by 14%.

Obviously, the imports played an important role in reducing the discharge of industrial wastewater in China. Currently, water pollution is a serious problem in China. As a big importer of industrial products, the positive role of industrial import in reducing China's wastewater discharge should be paid more attention to.

The research results in this paper show that during 2012 and 2015, the decrease of China's import industrial products increased the wastewater discharge by 590.1 million cubic meters in China, which is not conducive to improving China's environment so the industrial imports should be increased in order to alleviate China's environmental protection pressures.

China's import structure should be optimized to increase the environmental benefits of industrial imports by cutting down China's tariffs of some industrial products. The research results in this paper shows that compared with the proportion of the imports with the higher wastewater emission reduction intensity in 2012, the proportion were significantly increased, which brought huge environmental benefits to China in 2015. The research results also shows that wastewater emission reduction intensity of import in paper, cultural and educational products, garment and leather, textile raw materials, chemicals electrical and machinery and equipments, transportation equipment manufacturing, metal products were higher. Therefore increase these imports reducing by cutting down their tariffs can further enhance the environmental protection benefits of China's industrial import.

6. References

[1] Ministry of the Environment of the People's Republic of China: 2017 China's Ecological Environment,Bulletin[EB/OL] http://www.mee.gov.cn/gkml/sthjbgw/qt/201805/W020180531606576563901.pdf, 2018-05-31.

[2] Office of the State Council. 2012. Appropriate expansion of imports to alleviate resource and environmental pressures [EB/OL]. Http://money. 163. Com/12/0331/03/7TT58LTC00253B0H. html.

[3] Wyckoff, A.W,J.M.Roop.1994.The implementation of Imports of Manufactured Products: Implications for International A agreements on Greenhouse Gas Emissions [J]. Energy Policy (22): 187-194.

[4] Dietzenbacher, E., Mukhopadhyay, K. An Empirical Examination of the Pollution Haven Hypothesis for India: towards a Green Leontief Paradox?[J].Environmental and Resource Economics,2007,36(4):427-449.

[5] Peters,G.P. China's Growing CO_2 Emissions:A Race between Increasing Consumption and Efficiency Gains [J]. Environmental Science and Technology, 2007,41(17): 5939-5944.

[6] Li J S, Chen G Q, Lai T M, etal. Embodied Greenhouse Gas Emission by Macao [J]. Energy Policy, 2013, (59): 819-833.

[7] Wang Wenzhong, Cheng Yongming. Global Warming and Greenhouse Gas Emissions—The Problem of CO_2 Emissions in Sino-Japanese Trade[J].Ecological Economy,2006(7):22-25.

[8] Zhang Youguo. Energy and Environmental Costs of China's Trade Growth[J].Quantitative Economics & Technology Economics Research, 2009(1): 16-30.

[9] Xu Hui. Environmental Cost Transfer of China's Import and Export Trade——Based on Input-Output Model Analysis [J]. World Economy Study, 2010(1): 51-55.

[10] Kuang Yuanpei,Xie Jie. An Empirical Analysis of the Resource Effect and Environmental Effects of China's Agricultural Products Trade[J].International Trade Issues,2011(11):138-147.

[11] Ni Hongfu,Li Shantong,He Jianwu.Study on the Measurement of Trade Implicit Pollutants and the Green Transformation of Structure[J].China Population, Resources and Environment,2012,22(5): 164-169.

[12] Liu Xiangxia, Wang Rui, Chen Xuezhong. Analysis of China's Foreign Trade Eco-environment and Green Trade Transformation—An Empirical Study Based on Implicit Carbon[J].Resource Science, 2015(2): 280-290.

[13] Zhu Qirong, Wang Yuping, Liu Xuan. Environmental Benefits and Improvement Ways of Import Trade of Industrial Products—Based on the Analysis of Reducing Industrial Waste Gas and Wastewater Emissions [J]. International Economics and Trade Research, 2018(8): 34-48.

[14] Machado, G., R. Schaeffer, E. Worrell. Energy and Carbon Embodied in the International Trade

IOP Conf. Series: Materials Science and Engineering **730** (2020) 012060 doi:10.1088/1757-899X/730/1/012060

of Brazil: An Input Output Approach [J]. Ecological Economics, 2001, (39): 409-424.
[15] Chen Ying, Pan Jiahua, Xie Laihui. The Connotation Energy and Its Policy Implications inChina's Import and Export Commodities [J]. Economic Research, 2008(7): 11-25.

IOP Conf. Series: Materials Science and Engineering **730** (2020) 012061 doi:10.1088/1757-899X/730/1/012061

Analysis of Changes in Ecological Patterns in the Upper Reaches of the Minjiang River One of Typical Disaster Prone Area of China

Guangwen Ma, Yang Yu, Guang Wang, Fenying Zhang, Haijiang Liu, Lihuan He and Yeyao Wang*

China National Environmental Monitoring Center, State Environment Protection Key Laboratory of Environmental Monitoring Quality Control, Beijing 100012, China.
First author: guangwenm@163.com
*Email: yeyaowang@163.com

Abstract. Mountain torrents and debris flows are rapid transport processes for soil and water materials in small watersheds in mountainous areas. They have a strong destructive power and often result in devastating natural disasters. Natural ecosystems can regulate natural disasters, reduce the risk of damage, increase the stability of disaster-prone environments and reduce the vulnerability of downstream areas. The upper reaches of the Minjiang River (URM) in Sichuan Province, China is a typical area prone to flood and debris flow disasters. An established scientific model was used to analyze land cover change in the URM to discover the optimum land use pattern. Based on the analysis of the pattern of temporal and spatial changes of ecosystems in the URM, the results showed that the amount of ecosystem change in the URM was small, and the overall change rate of ecosystems was only 1.7%. There was an obvious trend of landscape fragmentation in the whole URM region, and the change of ecosystems had clear regional differences. Human economic activities were the main reason for the changes in the spatial pattern of ecosystems from 1985 to 2013.

1. Introduction

Mountainous areas have steep terrain and plentiful sedimentary material. High intensity rainfall or rapid snow melt can lead to disastrous flash floods and debris flows [1]. The simultaneous occurrence of mountain floods and debris flow (MFDF) will magnify the corresponding hazards of each single process, which is of particular concern [2]. For effective disaster early warning, it is essential to identify typical areas prone to MFDF so that the distribution and pattern of temporal and spatial changes to ecosystems can be analyzed. Yang et al. developed comprehensive index values to evaluate the landscape ecological security of Maoxian County in Sichuan Province in 2000, 2007 and 2015, using changes of ecological pattern over time from multi-stage remote sensing images [3]. Based on remote sensing data from 2000, 2005, 2008 and 2013, Zhao Weiquan analyzed the evolution of landscape patterns in the Chishui River Basin [4]. Normalized difference vegetation index (NDVI) data have been used to study both long-term and large-scale trends of change in land ecosystem vegetation health and growth, and short-term changes in land distribution patterns in local areas [5-6]. Previous research has used a range of scientific theories and models to analyze land resources, improve the ecological security coefficient and explore the values of ecosystem services [7-8]. In this paper, we analyzed the distribution and pattern of ecosystems and their temporal and spatial changes in the URM as a case study of a typical disaster-prone area to optimize the land use pattern and reduce the occurrence of MFDF.

2. Materials and Methods

2.1. Study Area

The URM was selected as a typical disaster-prone area for MFDF. The URM includes six counties with an area of 25,470 km^2. It is located in the north of Sichuan Province (Figure 1). Previous research has provided information on ecological service functions, disaster risk assessment and zoning in the URM [9-11].

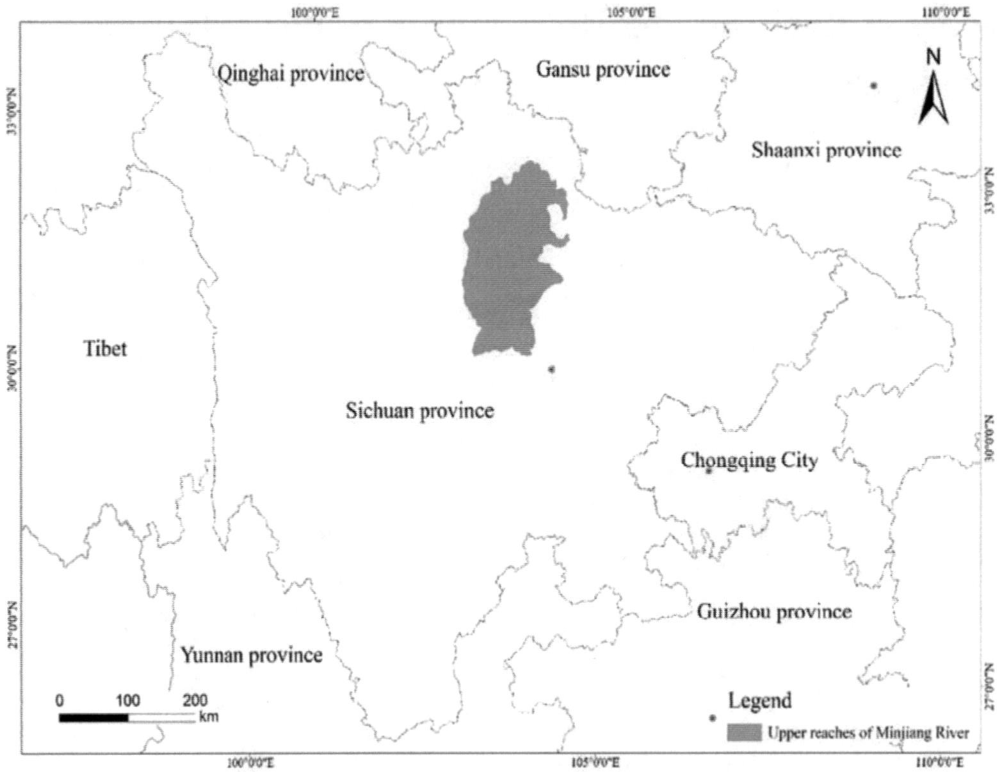

Figure 1. Location of the upper reaches of the Minjiang River Basin

2.2. Data Sources

The study used ecosystem classification data interpreted from remote sensing images for 1985, 2000 and 2013 to assess the URM ecosystem patterns. The characteristics of vegetation types were used to design the land cover classification system. The land cover classification had two levels. Level 1 had 6 categories and the level 2 classification had 38 categories. These reflected the dynamic monitoring of ecosystem types in the URM and allowed us to extract important indicators of ecosystem structure. The types of land cover met the needs of ecological environment protection and management.

2.3. Evaluation Index for Ecosystem Pattern Change

The distribution and pattern of ecosystems in the URM and their temporal and spatial changes were analyzed based on data for land use and land cover in 1985, 2000 and 2013. The types, distribution, proportion and spatial pattern of terrestrial ecosystems were evaluated, and the characteristics of transformation between various types of ecosystem were analyzed. The ecosystem pattern and change evaluation index is shown in Table 1. The index was used to examine the spatial distribution, allocation and proportion of different ecosystem types and analyze the temporal and spatial changes in ecosystems in the URM from 1985 to 2013.

Table 1. Ecosystem pattern and change evaluation index

Evaluation content	Evaluating indicator
Composition of ecosystem	Ecosystem area Proportion of ecosystem composition
Changes in ecosystem composition	Change rate of ecosystem type area
Characteristics and changes of landscape pattern in ecosystem	number of patches(NP) Mean patch size(MPS) Mean patch size of type(MPST) Edge Density (ED)(m/hm^2) contagion index(CONT) (%)

3. Results and Discussion

3.1. Changes in Ecosystem Composition

3.1.1. The range of natural ecosystem change
The change in forest and grassland was relatively small at less than 2%. The net decrease of forest area was 91.3 km^2, with a decrease rate of 0.6%, and an average annual decrease of 3.3 km^2 in the URM from 1985 to 2013. The grassland area increased by 43.5 km^2, with an average annual increase of 1.6 km^2, and an increase rate of 0.7%. In general, the areas were relatively stable. The area of other natural ecosystem types increased by 161.0 km^2, with an average annual increase of 5.8 km^2, accounting for 6.0% of the total, and the total area also remained stable. Wetland areas increased rapidly by 13.4 km^2 in area and 10.9% in proportion.

3.1.2. The range of artificial ecosystem change
Artificial surface area increased significantly by 84.5 km^2, an average of 3.0 km^2 per year, with an increase rate of 601.5%. However, there was a large net reduction in cultivated land area of 211.0 km^2, with an average annual reduction of 7.5 km^2, accounting for 21.5% of the total in the URM from 1985 to 2013. This shows that the area of cultivated land decreased significantly in the URM.

3.2. Intensity of Ecosystem Change

3.2.1. The intensity of ecosystem change
From the perspective of changes in level 1 ecosystem types, the overall change rate of ecosystems from 1985 to 2000 was 0.1%, or 23.1 km^2. The rate of conversion in the second half of the study was much larger, with 413.7 km^2 (1.6%) of level 1 classified ecosystems being converted. Over the whole period 1985–2013, the overall change rate of ecosystems was 1.7% (436.1 km^2) in the URM (Table 2).

In the level 2 classification of ecosystem types, the overall change rate of the ecosystem was 0.2% from 1985 to 2000, with 38.4 km^2 of different ecosystem types being converted. Between 2000 and 2013, the rate of change was 2.4% (614.9 km^2). Over the period 1985–2013, the overall change rate of ecosystems was 2.6% or 653.7 km^2 in the URM (Table 2).

Table 2. Comprehensive change rate and conversion intensity of ecosystems in the URM

classification	indicator	1985~2000	2000~2013	1985~2013
Level 1 classification	EC(%)	0.09	1.62	1.71
	LCCI(%)	-0.05	-1.55	-1.60
Level 2 classification	EC(%)	0.15	2.42	2.57
	LCCI(%)	-0.06	-2.01	-2.07

3.2.2. The stages of ecosystem change
The main changes occurred from 2000 to 2013 in the URM. According to the change of ecosystem types of the level 1 classification, the comprehensive change rate of the ecosystem was 1.7% from 1985 to 2013, 1.6% from 2000 to 2013, and 0.2% from 1985 to 2000. According to the change of ecosystem types of level 2 classification, the comprehensive change rate of ecosystem was 2.6% from 1985 to 2013, 2.4% from 2000 to 2013, and 0.2% from 1985 to 2000 in the URM.

3.3. Change of Ecosystem Pattern

3.3.1. Fragmentation of the regional landscape
The trend of landscape fragmentation was obvious, and the distribution of landscape types became more scattered in the URM. The number of patches increased from 42,232 to 44,447, an increase of 5.2% while the average patch area decreased from 60.3 ha to 57.3 ha, a decrease of 5.0%. The boundary density increased from 35.8 m/ha to 36.9 m/ha and the aggregation index decreased from 64.6% to 64.1% from 1985 to 2013 (Table 3). The four indicators show a clear trend of fragmentation of landscape types.

Table 3. Landscape pattern characteristics and changes in level 1 ecosystems in the URM

year	NP	MPS(ha)	ED(m/ ha)	CONT(%)
1985	42232	60.3	35.8	64.6
2000	42356	60.1	35.9	64.5
2013	44447	57.3	36.9	64.1

3.3.2. Changes in fragmentation patterns
The fragmentation of wetlands and artificial surfaces decreased, the fragmentation of cultivated land and other ecosystem types increased, and the fragmentation of forest and grassland remained stable from 1985 to 2013. The area of wetland patches in the URM increased from 6.7 ha to 7.5 ha (11.9%), while the area of artificial surface patches increased from 24.6 ha to 42.3 ha (72.0%). In contrast, the area of cultivated land patches decreased from 15.0 ha to 11.6 ha (22.7%) and the area of other types of patches decreased from 61.0 ha to 50.6 ha. The area of forest patches decreased by 1.3% from 208.3 ha to 205.6 ha; the area of grassland patches decreased by 2.6% from 27.4 ha to 26.7 ha.

3.4. Regional Differences of Ecosystem Changes

3.4.1. Extension along the main rivers and valleys
The overall rate of ecosystem change in the URM was mainly characterized by extension along both sides of the large river valley from 1985 to 2013. The change of ecosystem types was most obvious along the main stream of the Minjiang River and its tributaries, such as the Xiaoxinggou and Weilong Rivers, especially in some reaches of the main stream of the Minjiang River.

3.4.2. The change of ecosystem types in the vicinity of main urban settlements

The ecosystem types of the URM changed most rapidly around the main towns and surrounding areas. In the small watersheds around Dujiangyan, Wenchuan, Maoxian, Heishui and Songpan, the overall change rate of ecosystems was significantly higher than that of the surrounding areas from 1985 to 2013.

4. Conclusion

The intensity of ecosystem change in the URM was relatively small, and the overall rate of ecosystem change was only 1.7% from 1985 to 2013. Changes in ecosystems showed clear stages: the major changes occurred between 2000 and 2013, with conversion of forest to other types, and the conversion of cultivated land to forest land and artificial surfaces. The overall landscape fragmentation trend of the URM was obvious and the distribution of landscape types was more scattered. The fragmentation of wetlands and artificial surfaces landscapes was reduced, the fragmentation of arable land and other ecosystem types increased, and the fragmentation of forest and grassland was generally stable. Human economic activities in the URM were the main reason for the change of ecosystem spatial patterns. Around 60.4% of ecosystem conversion was related to the changes in the cultivated land and artificial surface ecosystems from 1985 to 2013.

5. Acknowledgement

This work was supported by the National Natural Science Foundation of China (No.31400449).

6. References

[1] Ugur Ozturk, Dadiyorto Wendi, Irene Crisologo, et al. Rare flash floods and debris flows in southern Germany [J]. Science of the Total Environment, 2018, 626(6): 941-952.

[2] Elisa Destro, William Amponsah, Efthymios I. Nikolopoulos, et al. Coupled prediction of flash flood response and debris flow occurrence: Application on an alpine extreme flood event [J]. Journal of Hydrology, 2018, 558 (3): 225-237.

[3] YANG Bin, LI Maojiao, CHENG Lu, et al. Landscape ecological security assessment based on multi-temporal remote sensing data in Maoxian County of Sichuan[J]. Engineering of Surveying and Mapping, 2018, 27(4): 41-48. (in Chinese)

[4] ZHAO Weiquan, YANG Zhenhua, SU Weici, et al. Ecological risk assessment and management of watershed based on landscape pattern change —a case study of the chishui river basin in guizhou[J]. Resources and Environment in the Yangtze Basin, 2017, 26(8):1218-1227. (in Chinese)

[5] ZHANG Xueyan, Hu Yunfeng, Zhuang Dafang, et al. NDVI spatial pattern and its differentiation on the Mongolian Plateau[J]. Journal of Geographical Sciences, 2009, 19(4): 403-415.

[6] Terry L Sohl, Peter R Claggett. Clarity versus complexity: Land-use modeling as a practical tool for decision maker [J]. Journal of Environmental Management, 2013, 41(16): 235-243.

[7] ZHOU Hao, LEI Guoping, ZHAO Yuhui, et al. Simulation of Dynamics of Land Use in Naoli River Valley Based on CA-Markov Model [J]. Journal of Ecology and Rural Environment, 2016, 32 (2): 252-258. (in Chinese)

[8] LU Wentao, DAI Chao, GUO Huaicheng. Land use scenario design and simulation based on Dyna-CLUE model in Dianchi Lake Watershed [J]. Geographical research, 2015,34(9) : 1619-1629. (in Chinese)

[9] WANG Shan, PENG Peihao, LIU Qin, et al. Primary analyse of reducing flood effect of different vegetation cover in small watershed of disaster prone area based on GIS: A case study in upper Minjiang river Basin [J]. Journal of Catastrophology, 2016, 31(4): 210 -214. (in Chinese)

[10] WANG Jun, YU Yan, OU Guo-qiang, et al. Flash Flood risk Zoning of Areas Hit by Wenchuan Earthquake in the Upper Reach of Minjiang river [J]. Journal of Yangtze River Scientific Research Institute, 2017, 34(1): 54-60. (in Chinese)

[11] DU Jun, DING Wenfeng, REN Hongyu. Relationships between different types of flash flood disasters and their main impact factors in the Sichuan province [J]. Journal of Yangtze River

Scientific Research Institute, 2015, 24(11), 1977-1983. (in Chinese)

IOP Conf. Series: Materials Science and Engineering **730** (2020) 012062 doi:10.1088/1757-899X/730/1/012062

Inherent Analysis on Feature and Mechanism of Renewable Energy System in ST-3E System

Chunyan Diao[1*], Bojun Cui[1], Yunjie Wu[1] and Jianfeng Li[2]

[1]College of Eco-environmental Engineering, Guizhou Minzu University, Guiyang 550025, China.
[2]Beijing Municipal Institute of Labor Protections, Beijing, People's Republic of China.
[*]Email: diaoxuanji@126.com

Abstract. Development of renewable energy is significant for energy security and also an important solution to cope with the climate change. Nowadays, due to its unique energy endowment, China's energy security and sustainable development becomes a big issue. In this paper, in the ST-3E renewable energy system, the five features and five inherent mechanism are analyzed. In the area of sustainable development, this work can further promote the ST-3E renewable energy system model to help governments and international agencies to make energy strategies and policies.

1. Introduction
For tackling energy security and climate change, the research and analyses for development and policy of renewable energy system have been done. Taichen and Jin-Li (2008) conducted an in-depth research on the stimulation of policy in realizing sustainable development [1]. Joanna (2007) studied the relationship between market-driven competition and renewable energy from the perspectives of China's electricity market reform and renewable energy policy changes [2]. Chunyan etc (2013) ST-3E system of renewable energy is such a system in which energy, economy and environment can sustainably develop with renewable energy-an effective component of the harmonious society and driven by technology progress [3]. This system comprised of S (Society), T (Technology), E (Energy), E(Economy) and E (Environment). This paper analyzed the five features and five inherent mechanism in the ST-3E renewable energy system, which further improved the connotation of ST-3E system. It further lays the foundation for energy strategies and policies making.

2. Features of ST-3E Renewable Energy System

2.1. Integrity
The five elements in the renewable energy system are closely related as integrity. The development of one system element is severely affected by the development of other system elements, supported and restricted by the others; The development of a system element depends on its former development path. It is accumulative and path-dependent, and derives the executive factors which directly affect the renewable energy system. Therefore, the integrity of the renewable energy system determines that the simple superposition of the optimal development mode of all subsystem elements cannot guarantee the optimal development of the whole renewable energy system. As a result, it is essential to study the development of the renewable energy system from the perspective of system theory.

2.2. Hierarchy
The renewable energy system is an ordered, multi-level-structured entity. Its diversity and unity is reflected through levels of hierarchy. Not only does the renewable energy system include five system elements of society, technology, energy, economy and environment, but there are many different levels of hierarchies in all system elements, which interrelate and interact with each other. The system element layer and the executive element layer can only abstractly describe the different levels of the renewable energy system. The sub levels could be further divided.

2.3. Dynamics
The renewable energy system and all their inner layers continue to improve with new emerging features all the time. Quantitative and qualitative changes reflect the dynamics of development of the renewable energy system. When the system is at a stable state, progressive quantitative changes make dynamic characteristics of the system's development predictable. However, when quantitative changes accumulate to a threshold, the original balance will be broken, and the dynamic characteristics are often uncertain. The most revolutionary element is technology. Technology shows great uncertainty during quantitative and qualitative change process of the renewable energy system and has the most direct impetus on human energy system. The changes of renewable energy system are an integral part of energy system changes.

2.4. Openness
The renewable energy system is a complex system with high openness. Its development is closely related to external environment. The national emission reduction responsibilities under the background of climate change, requires further development of the renewable energy system. Economic elements after the financial crisis changed the speed of renewable energy development. Energy resource endowments in different countries make renewable energy development strategies differ dramatically, however, in order to deal with the issue of climate change, different countries have to cooperate with each other in the development of renewable energy. Technology innovation of renewable energy and international energy cooperation affect the capacity of renewable energy development, greenhouse gas emission reduction conditions, knowledge innovation and property rights protection in different countries.

2.5. Complexity
The system elements and executive elements in the renewable energy system are all interdependent and interactive. And those relationships are in different forms, they can be linear or nonlinear, unidirectional or multi-directional, stable or unstable, etc. The complexity of the system structure and their interrelationships create the complexity of the system development process, making the system development in disorder, and inevitability in occasionally.

3. Inherent Mechanism of ST-3E Renewable Energy System

3.1. The Sustainable Development of Renewable Energy Aims at Benefiting Social System
With harmonious social development as its ultimate goal, the sustainable development of renewable energy shows a human-oriented thought since society consists of human beings. Various actions as taken by human beings aims to make living standards higher, environment better, energy supply more continuous, and relationships among human beings as well as between human beings and nature more harmonious.

One conflict between the development of renewable energy and social objectives is the cost of renewable energy. The cost of renewable energy is higher than conventional energy, and whether extra part of cost can be accepted by society is the decisive factor to its sustainable development. Using green price as the baseline to judge the level of social environmental awareness determines whether renewable energy can be developed on a large scale.

In the report delivered at the Seventeenth National Congress of the CPC, the Communist Party made a political commitment publicly to the world: in terms of environmental protection, China will make efforts together with other countries to help each other, move forward in concert, and jointly take good care of the earth that human beings rely on. From the domestic point of view, the report proposed the idea and development goals of building an ecological civilization the first time. The report pointed out: "we will build an ecological civilization, basically form an industry structure, a growth pattern and a consuming model that save energy and resources and protect the environment. Meanwhile circular economy will achieve a considerable scale, and the proportion of renewable energy should be increased significantly. The emissions of main pollutants will be effectively under control, and the environment will be greatly improved. The concept of an ecological civilization should be firmly established in the society." This is the first time that China developed the social development philosophy of human living in harmony with nature from a height of national strategy. The construction of an ecological civilization is an important cultural aspect of the scientific development concept put forward by the Communist Party of China. It is also an important ideological guarantee to achieve economic model transition and sustainable development, displaying the scientific development concept of "people-orientation". This is the first time environment and human development are considered together, sublimating and expanding Taoists' development concept of unity of mankind and nature proposed 2,500 years ago.

With China's rapid economic development, people's standard of living and way of life have undergone fundamental changes. In the course of China's economic take-off and building a moderately prosperous society, it has experienced the impact of environment deterioration on people's lives and the constraints resulting from economic and social progress. Among those impacts, the most fundamental are human health cost caused by environment deterioration, and the intensification of conflicts among social groups as well as social instability resulting from resource depletion. Environmental problem has become a huge challenge for sustainable development of economy. Various natural disasters caused by environmental problems gave profound lessons to the human society. Meanwhile, intensified group incidents have greatly affected social stability and national security.

In the new environmental situation, we should promptly adjust our development policies and environmental protection strategies. We should adopt an integrated policy to set up a sustainable social and economic system. Meanwhile, we should keep improving the governance structure and encourage participation through various cooperative partnerships and incentive policies. We should give more support to the civil environmental protection organizations and strengthen our environmental cooperation internationally to promote the sustainable development process.

A green and sustainable social development mode is the inevitable choice for China to create a harmonious society and a harmonious world. Renewable energy, as the primary means of optimizing energy structure, plays an important role in achieving this goal.

3.2. Economic System: the core of the sustainable development of renewable energy

The normal operation and development of economic system, the core of the development of the entire renewable energy system, provides support with other key system factors. Only the continuous and uninterrupted support can ensure more funds and materials being put into renewable energy and environmental protection to push the progress of the entire society. The harmonious development of economic system can be seen in the growth of economic aggregate, in the improvement of economic structure, and in its decrease of the enormous pressure on energy and environment system imposed by economic growth.

The inseparability between economy and society leads to the integration of economy and social ideology. The social responsibility of economic steering and social publicity is to form psychological preference over renewable energy, so that investment in renewable energy can be stimulated in economic system. The development speed of renewable energy and the industrialization of technologies can be achieved only if the investment is enlarged; and meanwhile, the state should support its development through finance and tax policies and attract sustainable investment to it by stabilizing electricity price.

As for the importance of the interdependence of China's economic development and energy supply, economic security and energy security have become the two most important issues of national security.

In the "Eleventh Five-Year" period, China's unit GDP energy consumption increased instead of decreasing, making energy-saving situation more pressing. The reason of the persistent rise of energy intensity in recent years is not the increase of unit product energy consumption, but the rapid expansion of energy-intensive heavy industry. In fact, the new round of high economic growth is driven by the accelerated development of heavy industry, which coincides with the middle stage of industrialization in developed countries, whose energy density also showed the "inverted U-type" curve of left-side effect.

China has proposed its national energy policy to ensure economic security. Recently, China started changing system factors such as adjusting energy price policies, taxation policies and energy government in order to inhibit the disordered development of high-energy-consumption industries as well as promoting adoption of advanced and applicable technology. In the medium and long term, China will adopt technology leapfrogging strategy, achieving technological innovation in major fields and improving the contribution rate of technical energy saving; meanwhile, the economic structure will be significantly optimized. Therefore, sustainable development of renewable energy is an inevitable choice to achieve the short-term and long-term national goals.

3.3. Environment System Motivates the Sustainable Development of Renewable Energy

The quality of environmental system affects directly the living conditions, the level of resource stock, economic development foundation and the speed of technology progress. The key to its healthy development is to make social and economic development compatible to the carrying capacity of environment, which requires society and economy develop at a speed not exceeding the carrying capacity of environment. Renewable energy is duty-bound to put energy supply in line with environmental constraint and to bring sustainable development into reality.

The development of renewable energy and environmental protection has been promoted to the height of national strategy, so the definition to the cost of renewable energy cannot be limited to the cost within the system. The comprehensive cost of environmental that combines externality and internality price must be considered. Consideration of the comprehensive cost of renewable energy and its development under corresponding policies can result in industrial structure optimization indirectly.

Climate change has become a common challenge for human beings. Apart from tackling climate change, China also has to face serious environmental pollution. Climate change and environmental issues will accompany China's society and economy transformation. The resource constraint and environmental pollution during the transformation period have appeared in the initial stage of economic development, and have endangered the foundation of the country's sustainable development. Energy production and consumption is the main source of China's environmental pollutants, approximately 90% emissions of sulfur dioxide and nitrogen oxides in waste gases are caused by energy production and consumption activities. From international experience of reducing greenhouse gases, enhancing energy efficiency and developing renewable energy are the two main measures.

Developing clean renewable energy can not only reduce local environmental pollution and economic loss caused by using fossil energy, but also decrease greenhouse gas emissions. This makes a contribution to mitigating global climate change as well as the greenhouse gas emission reduction pressure that China might face in the future. More importantly, determination and actions to develop renewable energy is one of the main symbols representing that China is a responsible country for reducing greenhouse gas emissions.

3.4. Energy System: foundation for sustainable development of renewable energy

Energy has become a kind of strategic material for the economic progress of modern society. Many aspects of human activities require energy, especially for modern economy. The development of modern economy has strong relevance and dependence on energy consumption, and only safe and secure energy supply can spur the consistent progress of economy. The harmonious development of energy system is represented in its quantity, but the consumption structure of energy also has great influence on economy and environment. The sustainable development of renewable energy can turn the sustainable supply and optimal structure of energy system into reality, decreasing the emission of greenhouse gas and pollutants.

Renewable energy and renewable resource are closely connected. As a kind of energy, the large-scale application of renewable resource is an inevitable choice for human beings to develop sustainably. When reaching certain scale, it can reduce the consumption of non-renewable resources, giving dual protection to environment and resources. And also, renewable resource affects the output of renewable energy directly. It turns energy system from non-renewable to renewable through technical means for a sustainable development.

China is in a period of rapid economic growth. Nowadays, China is the world's largest coal producer and consumer, the second largest energy producer and consumer, and the largest oil importer after OECD. The dependence of China's economy on energy will continue to increase, and energy security has become a bottleneck restricting its economic and social development. The resource reserve of traditional energy that China's development relies on, its production capacity and the level of protection are in a serious situation. Estimated by the relevant state agencies, there will be a large gap between the conventional energy supply and demand, reaching 18%, 20% and 30% in 2020, 2030 and 2050 respectively, according to the constraints of production capacity that can be actually increased, the environmental capacity as well as the effect after the implementation of energy-saving policies. In that case, renewable energy will historically shoulder the responsibilities of filling energy supply gap, optimizing energy structure, and protecting the environment.

3.5. Technical System: backbone for the sustainable development of renewable energy

Renewable energy system directs the energy revolution. It serves as a substitution for conventional energy consisting of wood, coal and petroleum. And its sustainable development shall gradually make it from supplement to strategic energy. By means of ever-going technical progress, the use of renewable energy can improve the utilization efficiency and change mankind's energy consumption mode. To sum up, technological system shall affect the direction and speed of energy development of the entire world.

Technology is the most revolutionary factor. Its immediate action is to enhance the transformation efficiency of renewable energy. The enhancement of transformation efficiency not only shows that the level of renewable energy makes the large-scale application of renewable energy possible but also predicts that the completion of energy transformation will raise renewable energy to a strategic significance at the middle of this century.

Science and technology are primary productive forces, so the level of renewable energy technologies is often used to measure a country's overall strength. Major developed countries all devote lots of money in R&D, resulting in fast development of renewable energy technologies.

Therefore, when discussing sustainable economic development these days, we should make the level of renewable energy technologies the main indicator of China's energy structure adjustment. Improvements of technology can effectively decrease the cost of renewable energy and make it more economical. Meanwhile, the consensus of the world about the global climate change will generate huge demands for new energy technology. It will also greatly stimulate China's economy. Moreover, the development of renewable energy and the country's emission reduction commitments will benefit each other to win the initiative in the international political game of climate change.

4. Conclusion

The ST-3E renewable energy system is an integral, hierarchical, dynamic, open and complex system. Therefore, the study of the development and evolution pattern of the ST-3E system must be based on systematic and comprehensive study of its structure and their correlations. Further discussion of the inherent mechanism among all the ST-3E system elements, all executive elements and different levels in the system is crucial for the development of renewable energy, and also for the sustainable development.

The sustainable development of ST-3E renewable energy system, as the link and center of the five key factors of the system and the five key factors of operation, is directly related to the healthy development of system factors and operation factors. It is not a simple addition. It should be upgraded to the height of national policy and strategy for optimization and also requires different stimulation policies and institutional guarantees in line with specific development situation.

5. Acknowledgement

The work was financially supported by Guizhou Education National Science Foundation(Grant No. Guizhou Education Contract of KY [2012] 049), Youth Science and technology talent growth project of Guizhou Provincial Education Department (Guizhou Education Cooperation KY Word [2017]130) and National Natural Science Foundation (41365009).

6. References

[1] Taichen Chien & Jin-Li Hu. (2008). Renewable energy: An efficient mechanism to improve GDP. Energy Policy, Vol.36, pp 3045-3052

[2] Joanna I. Lewis & Ryan H. Wiser. (2007). Fostering a renewable energy technology industry: An international comparison of wind industry policy support mechanisms. Energy Policy, Vol.35, pp 1844-1857

[3] Chunyan Diao, Jianfeng Li, Research on Sustainable Development Mechanism of Renewable Energy using System Theory, Advanced Material Research, 2013, Volume 779-780, pp 1320-1323

Study on Water Quality Characteristics and Production Coefficients of Urban Domestic Sewage in Shanghai

Mina Song[1], Chunyan Diao[1,*], Lei Jia[2], Lei Yang[2] and Bojun Cui[1]

[1]College of Eco-environmental Engineering, Guizhou Minzu University, Guiyang 550025, China.
[2]Shanghai Environmental Protection Co., Ltd., Shanghai 200233, China.
[*] Email: diaoxuanji@126.com

Abstract. The existing coefficient of urban domestic sewage production cannot reflect the differences among regions. It is difficult to meet the needs of environmental protection work in Shanghai. This study divided Shanghai into three tiers. According to the 2017 on-line monitoring data of urban sewage treatment plants in main urban areas, Xincheng district and Xinshi town, the concentration of pollutants of each tier was calculated. And the production coefficients of urban domestic source pollutants at all tiers in Shanghai was obtained by combining with the urban per capita comprehensive domestic water consumption. The results showed that the COD_{cr} concentration changes in Xincheng district were relatively large, and that changes in the main urban area and Xinshi town were relatively small. The pollutant production coefficients of the main urban area, Xincheng district and Xinshi town were COD_{cr}: 75.6 g/d.cap, 66.95 g/d.cap, 28.02 g/d.cap, NH_3-N: 7.76 g/d.cap, 5.32 g/d.cap, 2.04 g/d.cap.

1. Introduction

With the rapid growth of population and social economy, the discharge of wastewater from domestic sources in Shanghai accounted for 85% of the total amount of wastewater in 2017[1], the proportion of living source pollutants in the total urban pollutants was increased, which has became one of the main reasons affecting the quality of urban water environment. Therefore, ascertaining the situation of urban living source pollution is very important for water pollution reduction and environmental management decisions.

The existing domestic sewage production coefficient research in China was mainly divided into three tiers, individual [2-3], residential quarters [4-7], sewage treatment plants [8-9], in addition, there are scholars on campus [10] as a study area. Individuals and residential communities lack research on wastewater from service industries such as catering and accommodation. The campus environment was relatively closed, the staff was fixed, and it was not representative because it was special. The resulting per capita pollutant production coefficient may differ from the actual pollution situation.

The object of this study was wastewater from urban living sources, which mainly includes wastewater from residents' household life and tertiary industry, schools, and office activities. By collected Shanghai different development tiers of sewage treatment plant influent water quality monitoring data, combined with the urban water supply, formulate pollutant production coefficients for urban living sources in Shanghai.

2. Research Methods

2.1. Statement of Problem
Considering the city's districts economic, social and industrial development level and the status quo, the "Shanghai Urban Master Plan" (2017-2035) and "Shanghai Water Supply Plan" (2017-2035 years) as the basis, the city divided into major urban areas, Xincheng district, and Xinshi town, and determine the production coefficients of COD_{cr} and NH_3-N at different tiers.

The urban domestic sewage is collect by the sewage pipe network and then discharged to the sewage treatment plant, when the rainwater and industrial wastewater mixed in the sewage pipe network are relatively small, it can be considered that the influent concentration of the sewage treatment plant is equivalent to the urban sewage produced concentration [8]. Collecting different tiers of 2017 sewage treatment plant influent water quality on-line monitoring data, analysis of water quality characteristics of urban life source. In order to reduce the impact of rainfall on the concentration of urban sewage, through screening of meteorological factors (the rainfall 3 consecutive days did not exceed 1 mm), obtaining the import concentration of urban sewage treatment plants at different tiers, and calculating the production coefficient of urban living source water pollutants in combination with per capita comprehensive domestic water consumption at different tiers.

2.2. Methods
The estimation of pollutant generation per capita per day (GCPD) values was conducted according to the following equations (1) [11]:

$$F=Q\cdot k\cdot C/1000 \quad (1)$$

In the formula:

F——Single factor production coefficient of comprehensive domestic sewage in a certain level of town, g/d.cap;

Q——Urban per capita comprehensive domestic water consumption, L/d·cap;

k——Comprehensive urban drainage coefficient, dimensionless;

C——Average concentration of urban comprehensive domestic sewage, *mg/L*.

The value of k is select as follows: when $Q\leq150$ L//d·cap, k is 0.8; when $Q\geq300$ L/d·cap, k is 0.9; when 150 L/d·cap $<Q<$300 L/d·cap, k is determined by interpolation.

2.3. Key Parameter
Comprehensive urban water consumption is includes domestic water consumption and public service water consumption. According to the "Shanghai System Plan" (2017-2035), the per capita residential water consumption index in the main urban area is generally 120-160 L/d·cap, and Xincheng district is generally 120-150 L/d·cap, Xinshi town is generally 80-100 L/d·cap. Water indicators for public services can be characterized by tertiary industry coefficients [12]. According to the residents' water consumption and tertiary industry coefficients the average value of urban comprehensive living water consumption in the main urban area is 310 L/d·cap, in Xincheng district is 240 L/d·cap, in Xinshi town is 117 L/d·cap.

2.4. Data Processing and Quality Control
Sewage treatment plant data uses SPSS 25 software, using descriptive statistics in statistical analysis to check whether the data conform to the normal distribution. According to the Dixon test, the data of the main pollutant concentration values outside the standard deviation of $\pm$ 3 times are outliers, and the outliers are excluded. Using ORIGIN 2017 for data analysis and graphing.

3. Results and Discussion

3.1. Concentration of Domestic Sewage
In order to avoid the influx of river water and groundwater and the leakage of drainage network, using the ratio of the dry season to the annual average (that is, the correction coefficient) is shown in Table 1, the contaminant concentration was corrected to achieve the closest to the actual concentration of the

object, the corrected concentration of domestic sewage is shown in Table 2.

Table 1. Calculation results of pollutant the correction coefficient

Paramater	Main urban area	Xincheng district	Xinshi town
COD_{cr}	1.06	0.96	1.07
NH_3-N	1.11	0.92	1.03

Table 2. Concentration of domestic sewage at different tiers (mg/L)

Paramater	Main urban area	Xincheng district	Xinshi town
COD_{cr}	270.98	324.39	299.40
NH_3-N	27.80	25.79	21.76

It can be seen from Table 2 that the comparison of COD_{cr} concentration was Xincheng district > Xincheng district > main urban area, and the comparison of NH_3-N concentration was main urban area > Xincheng district > Xinshi town.

3.2. Analysis of the Distribution Characteristics of Sewage Water Quality

3.2.1. Variation of pollutant concentration. As shown in Figure 1(a), the COD_{cr} concentration tiers of three tiers variation having certain similarity, in 2017 the overall upward trend of January to March, April to August decreased, the concentration of October to December increased, small peak in September, and the main urban area of the highest values of the year in September. The minimum COD_{cr} concentration in all three tiers was appeared in August. The COD_{cr} concentration changes in Xincheng district were relatively large, and the COD_{cr} concentration changes in the main urban area and Xinshi town were relatively small.

As shown in Figure 1(b), the three tiers of NH_3-N concentration in whole was January 2017 to March rise, from April to September decreased, and fluctuations occur. The main urban and Xinshi town concentrations were rised of October to December, Xincheng district concentration was increased in November, but decreased in December concentration. Three tiers of NH_3-N concentration of relatively stable changes.

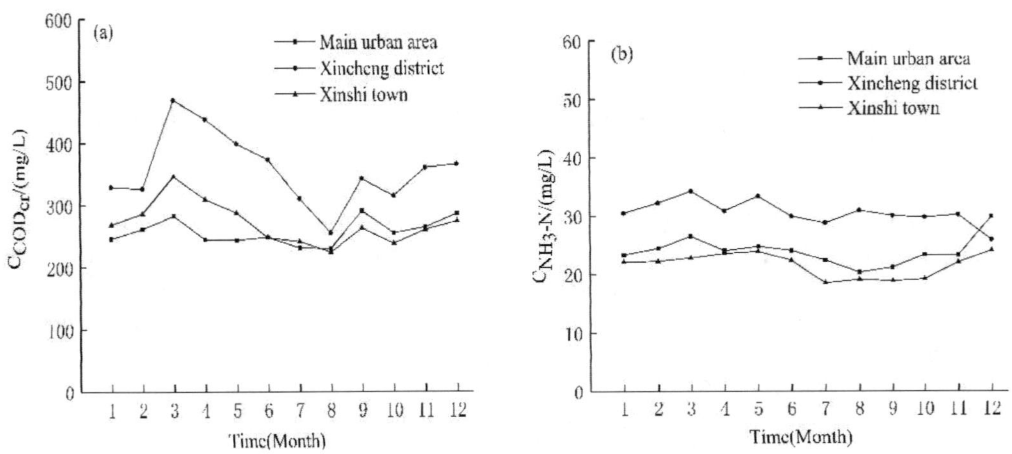

Figure 1. Annual variation curve of pollutant concentration

3.2.2. Accumulation probability analysis of sewage water quality. As shown in Figure 2(a), the concentration range of the higher probability distribution of COD_{cr} in the main urban area was 150-350 mg/L, and its cumulative probability was 86.2%. As shown in Figure 2(b), the concentration range of Xincheng district COD_{cr} with a high probability distribution was 150-450 mg/L, and its cumulative probability was 78.4%. As shown in Figure 2(c), the concentration range of the higher probability distribution of COD_{cr} in Xinshi town was 100-450 mg/L, and its cumulative probability was 87.7%.

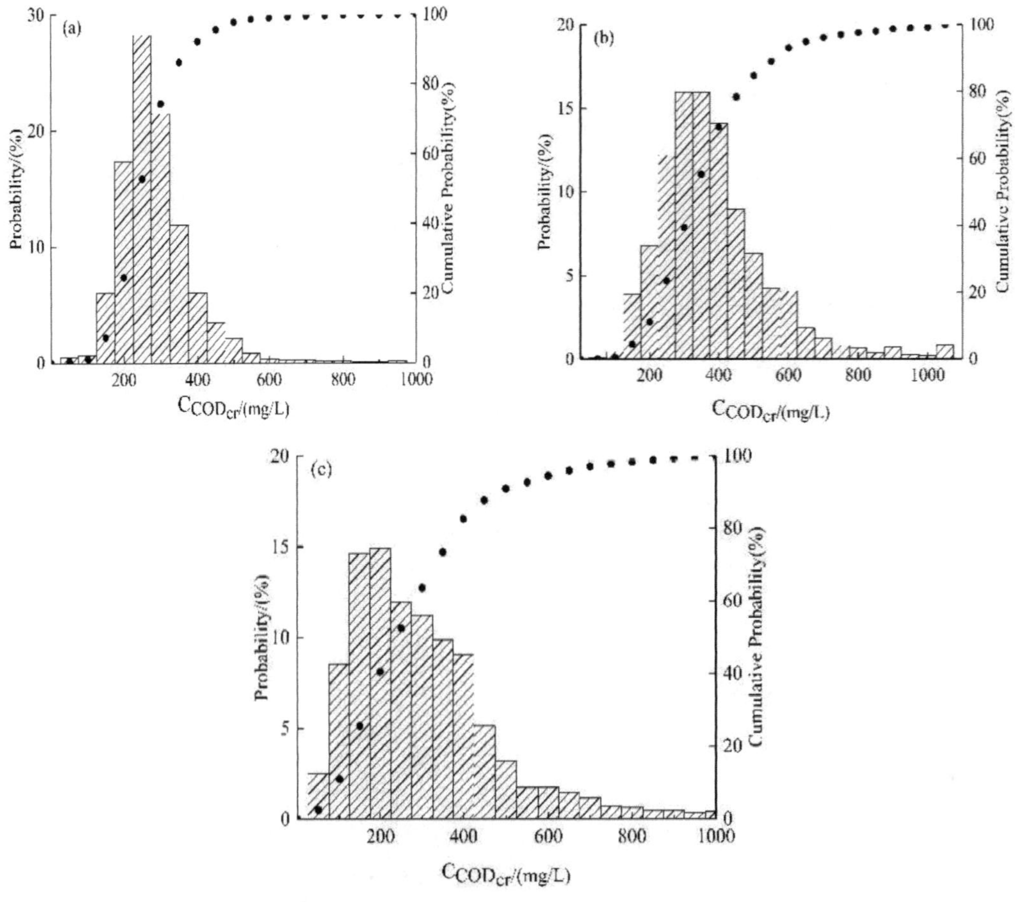

Figure 2. COD_{cr} cumulative probability curve

As shown in Figure 3(a), the concentration range of NH_3-N with a high probability distribution in the main urban area ranges from 20-45 mg/L, and its cumulative probability was 98.9%. As shown in Figure 3(b), the concentration range of Xincheng district NH_3-N with a high probability distribution was 20-50 mg/L, and its cumulative probability was 93.2%. As shown in Figure 3(c), the concentration range of NH_3-N in Xinshi town with a high probability distribution was 10-40 mg/L, and its cumulative probability was 93.0%.

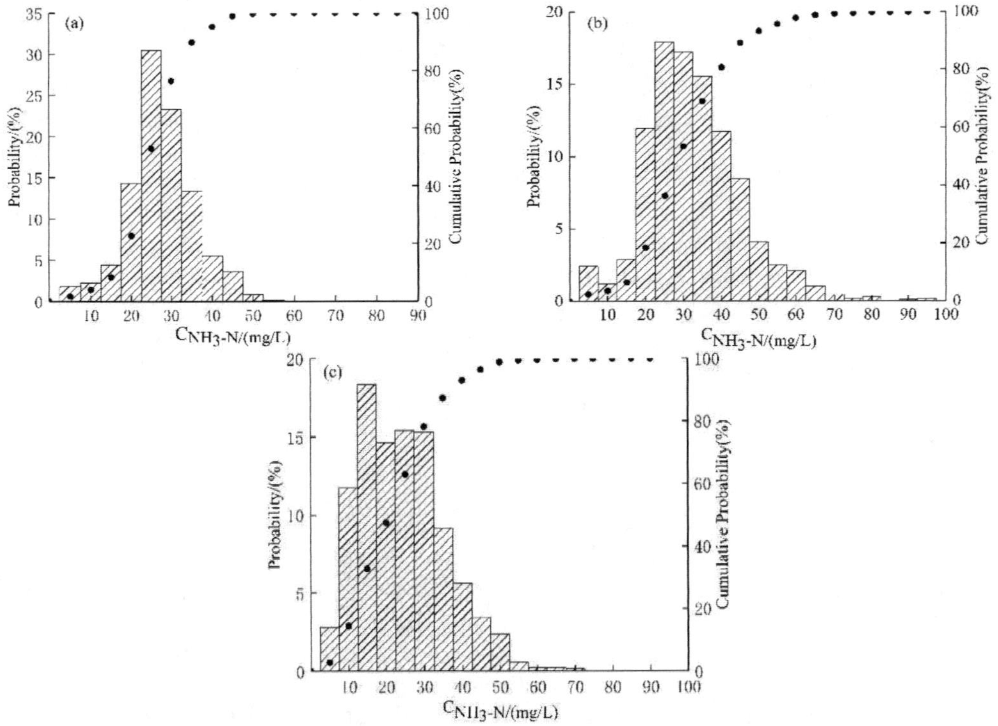

Figure 3. NH_3-N cumulative probability curve

3.3. Determine the Production Coefficient of Urban Living Source Pollutants

According to formula (1), the production coefficients of COD_{cr} and NH_3-N in different tiers are obtained, as shown in Table 3.

Table 3. Pollutant production coefficients [g/d.cap]

Paramater	Main urban area	Xincheng district	Xinshi town
COD_{cr}	75.60	66.95	28.02
NH_3-N	7.76	5.32	2.04

According to the literature [2-10], the production coefficients of different pollutants in different city were different. Based on the research results of urban sewage treatment plants [8-9], the COD_{cr} production coefficient was between 27.9 g/d.cap to 69.9 g/d.cap, with an average was 51.1 g/d.cap. The NH_3-N production coefficient was between 2.6 g/d.cap to 7.9 g/d.cap, with an average was 5.4 g/d.cap. In this study, in addition to Xinshi town COD_{cr}, both the main urban area and Xincheng district were higher than the earlier studies average. NH_3-N in addition to the main urban area, Xincheng district and Xinshi town were below the average in the previous study.

Huan Zhu [5] calculated that the results of Shanghai were COD_{cr} 54.5 g/d.cap, NH_3-N 6.4 g/d.cap. The average production coefficient of each pollutant in the main urban area of this study was higher 29.98% than the result of Zhu Huan. The COD_{cr} production coefficient of Xincheng district was higher 22.84% than that of Zhu Huan, and the NH_3-N production coefficient was lower 16.88% than that of Zhu Huan. The average coefficient of each pollutant calculated by Zhuhuan was higher 58.36% than Xinshi town.

4. Conclusion

The results show that the production coefficients of COD_{cr} and NH_3-N were both in the main urban area> Xincheng district > Xinshi town. The main urban area and Xincheng district were close to each other in terms of the coefficient of pollutant production, but it was significantly different in Xinshi town. Therefore, carry out a stratified study was necessary in Shanghai. The layer-by-layer measurement of the pollutant production coefficient can more accurately reflect the pollution status of domestic pollution sources in Shanghai, and understanding the characteristics and rules of pollution production at different tiers. The coefficients measured in this study can provide a basis for environmental management departments to correctly judge the current status of the environment, especially for the total control of COD_{cr} and NH_3-N in Shanghai's 13th Five-Year Plan. This study used online monitoring data, only COD_{cr} and NH_3-N pollutants. In the next step, actual measurement methods can be use to monitor other water pollutants and calculating the production coefficient.

5. Acknowledgement

The work was financially supported by Special Project of Shanghai Environmental Protection Bureau (H418505), National Natural Science Foundation (41365009) and the innovative talents of the 2014 Guizhou “thousand” level.

6. References

[1] Shanghai statistical yearbook 2018 *Shanghai municipal statistics bureau.*
[2] Zhencheng Xu, Junneng Wang, Xiaocheng Zhuang, Xiaochun Peng 2009 Study on nitrogen emission and model of human excreta *Journal of Anhui Agricultural Sciences* Vol.37 No 33 p 16510.
[3] Xiuli Xiang 2009 Study on generation processess of nitrogen, phosphor in domestic sewage of five representative families in Guangzhou *Jinan University.*
[4] Shuang Hu 2008 Generation coefficient and dischargr coefficient of domestic pollution sourcrs in Chongqing *Chongqing University.*
[5] Huan Zhu, Huaizheng Li, Jianfeng Ye, Wei Fu 2010 Coefficients of major pollutants in domestic sewage in Shanghai *China Environmental Science* Vol.30 No.1 p.37.
[6] Yuanyuan Wei 2012 Pollution load study of domestic sewage in Zhenjing city *Jiangsu University.*
[7] Hongkui Shi 2013 The study of the pollution generation and discharge coefficient of different types of domestic pollution source in Hohhot *Inner Mongolia University.*
[8] Jun Qian, Shanshan Yang, Jie Xiao, et al 2009 Study on pollutant production rate in sichuan province *Sichuan Environment*, Vol.28 No.1 p.27.
[9] Aiping Liu, Xiaowen Liu, Zhongyin Chen, Kaiming Li 2011 Investigation and accounting of the municipal domestic sources pollution load in Pearl River Estuary *China Environmental Science* Vol 31 No S1 p 53.
[10] Songzhu Jiang, Junyan Wang, Xiuhong Liu, et al 2017 Study on the production coefficient of sewage pollutants in urban living sources *Water & Wastewater Engineering* No S1 p 87.
[11] Department of planning and finance 2001 State environmental protection administration introduction to environmental satistics China Environmental Science Press.
[12] Shiqing Zhu 1998 Forecast of Shanghai's planned sewage volume in 2020 *Shanghai municipal engineering* No 4 p 17.

IOP Conf. Series: Materials Science and Engineering **730** (2020) 012064 doi:10.1088/1757-899X/730/1/012064

Drainage and Total Nitrogen Loss of Cropland under Drip Irrigation and Subsurface Pipe Drainage

Shan Zhang[1], Qian Tan[1,2], Hang Zhao[1] and Tong Zhang[1]

[1]College of Water Resources and Civil Engineering, China Agricultural University, Beijing 100083, China.
[2]Institute of Environmental & Ecological Engineering, Guangdong University of Technology, Guangdong 510006, China.
Email: qian_tan@cau.edu.cn

Abstract. Water saving irrigation as an effective measure to improve the utilization rate of water resources and alleviate water deficit has been widely practiced in arid and semi-arid regions. In this study, an integrated drip irrigation and subsurface pipe drainage system was developed to study the total nitrogen emission concentration of sunflowers during the whole growth period. The test and results show that total nitrogen concentration of farmland drainage in the spring irrigation period exceeded the standard V of "Surface Water Quality Standard". In addition, prolonging the leaching time or reducing the irrigation quota can effectively reduce the concentration of total nitrogen in farmland drainage.

1. Introduction

Water scarcity and non-point source pollution have been a serious issue for crop production in arid and semi-arid regions in China (*1, 2*). Poor natural water endowment and mismanagement of water resources are the main contributor to water shortage as well as water pollution (*3, 4*). Water saving irrigation, especially drip irrigation, as an effective water resources management measure, greatly improved the utilization rate of water resources and effectively alleviated water deficit. It has been widely practiced in arid and semi-arid regions in China. Therefore, it is significant to research the influence of drip irrigation system on pollutant emission concentrations.

Nitrogen is not only an important nutrient in crop growth but also one of the main components of eutrophication. Over the last a few decades, there have been a number of studies have considered the change rule of TN concentration based on drip irrigation system. For instance, Peng et al. (*5*) developed an integrated water-saving irrigation and controlled drainage system to monitor the TN and TP load in South China and confirmed the system's desirable function of water saving and pollutant reduction. Sepaskhah and Tafteh (*6*) revealed the effect of different N application rates on rapeseed yield, drainage water and N leaching. Ye et al. (*7*) found that the main water pollution including TN, TP and COD of paddy filed could decrease at least 22% by using water saving irrigation. Many years researches have shown that water saving irrigation is one of the effective tools to achieve good results of water saving and pollution decreasing.

However, major challenges still remained for controlling non-point source pollution by water saving irrigation technology. Because the farmland drainage is too small in arid region, most of the previous work focus on paddy field, few studies consider the change of pollutant concentration in farmland drainage letting alone incorporation into water saving irrigation. Therefore, this study will take Hetao irrigation region of Inner Mongolia province as an example, and combine drip irrigation with subsurface pipe drainage to monitor the TN emission concentration of sunflower, a local characteristic crop, during the whole growth period. The objective of this research is to reveal the change rule of TN emission concentration of crops in arid region under water saving irrigation and provide reasonable suggestions for local water saving and pollution control according to the research results.

2. Materials and Methods

Experiments were conducted in the single sunflower-growing region (May to August) in 2019 at Heji Experiment Station (40°44′N-107°16′E), Hetao Irrigation District (Figure 1), one of the main grain production regions in the west of the Inner Mongolia Autonomous Region, China. The experimental site has an area of approximate 7.84 ha. The study site has a typical continental desert climate with a mean annual precipitation of 130mm to 250mm and evaporation of 2,100mm to 2,300mm. Therefore, irrigation is very important for crop growth in Hetao Irrigation District, which will consume [4.35, 5.19] billion m^3 water resources from Yellow River every year. In order to reduce soil salinity, salt-leaching was fixed to May and October each year and consumed [1.34, 1.84] billion m^3 of the irrigation water. Eutrophication of Lake Wuliang Suhai has become severer and severer because of the release of various pollutants of farmland wastewater.

This experiment makes sunflowers as research object where the plant spacing was 60 cm and planting density was 37,000 plant/hm^2. This experiment used the ARIES TM drip irrigation tape produced by Netafim. The dripper spacing is 30cm and the dripper discharge is 1.0L / h. The drainage pipe consists of 8 suction pipes and 1 collector pipe. The row spacing of drip irrigation tape is 40cm. The suction pipe is PE bellows with diameter 4 inches, distance 30m, buried depth 1.5m and slope 0.5%. The collector pipe is PE bellows with diameter 6 inches. One collecting well is installed at the end of the intersection of the suction pipe and the collecting pipe, which is a concrete culvert with diameter 2m and depth 4.5m. The water of collecting well was discharged into the main drain by pump. In addition, the collecting well could monitor drainage through electromagnetic flowmeter.

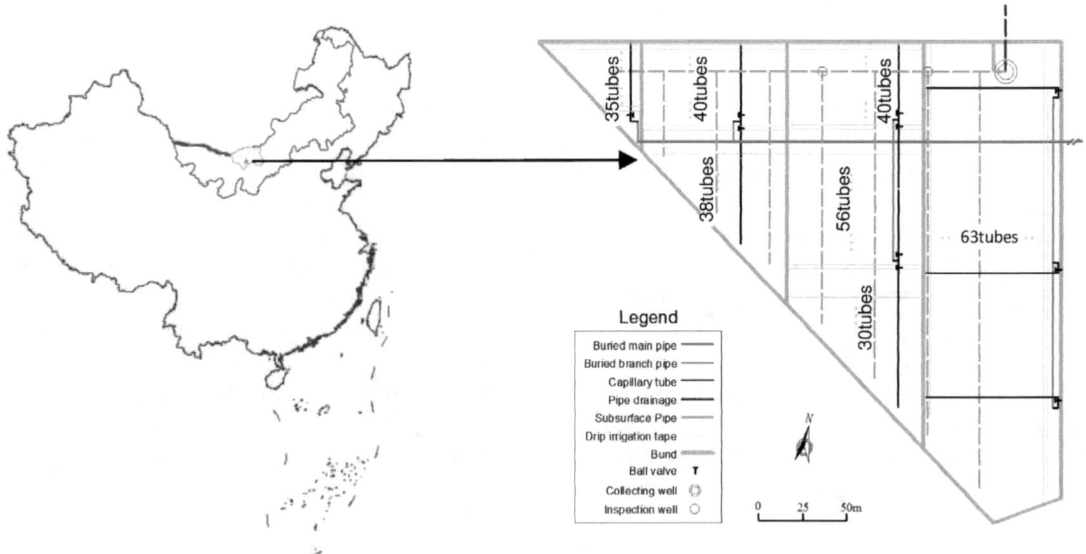

Figure 1. Location of experiment site and layout of pipe layout

3. Results and Discussion

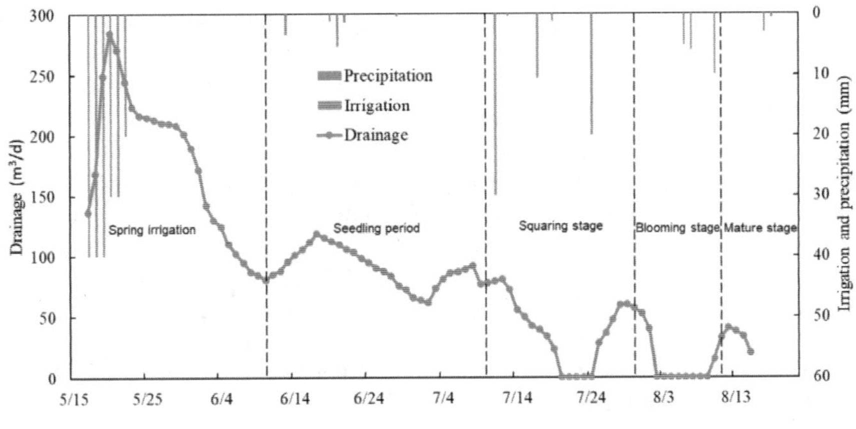

Figure 2. Drainage, irrigation and precipitation during the experimental period

There are four times of irrigation, one is before sowing (spring irrigation period), and the remaining three times are distributed in the growth period. The irrigation before sowing lasts a long time and the irrigation amount is large. But the irrigation amount in the growth period is relatively small and the duration is short. Most of the rainfall is from June to August, and most of the rainfall is less than 10mm. in the aspect of water output, the daily water drainage first increased rapidly and then decreases tortuously. In detail, the discharge was 284.5m^3 on May 20, which is the maximum for the whole year and more than double of the drainage, 135m^3, on May 17. Then, the drainage continued to drop from May 21 until sowing was completed. After entering the seedling period, there was a short-term recovery of drainage. This may be because the Yellow River had entered into the high flow period, raising the groundwater table of Hetao Irrigation district. But as crops grow, the demanded for water increases, so the drainage decreased again.

Figure 2 also shows that irrigation is usually arranged when drainage is zero. This means that the natural water resources had been unable to fully meet the growth needs of crops, after entering the squaring stage. Irrigation is necessary for crops growth. However, it is worth noting that the drainage has increased significantly a few days before irrigation. Through field investigation and analysis, it can be inferred that this may be because the experimental field is close to the canal, and the water level of canal will rise significantly before irrigation, which will lead to the rise of groundwater table in the nearby field. In addition, there was no significant change in the drainage before and after precipitation, which means that precipitation would not supply groundwater in Hetao Irrigation district.

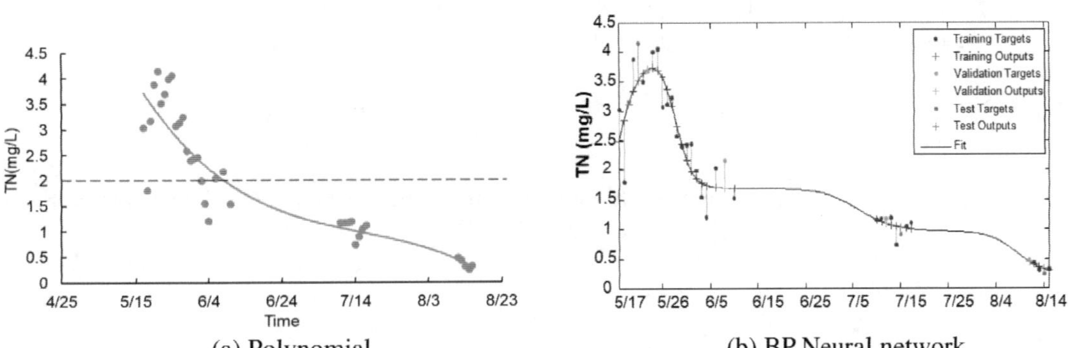

(a) Polynomial (b) BP Neural network

Figure 3. Simulation for concentration of TN in agricultural drain

Figure 3 shows the relationship between time and emission concentrations of TN based on polynomial and BP neural network. In particular, the structure of BP neural network is designed as three-layer feed forward neural networks, including 3 neurons in input layer, 5 neurons in hidden layer and 1 neuron in output layer. Among the samples, 80% points would be selected as training set, 10% points would be singled out for validation set and the rest 10% points would be chosen as test set. The Determination Coefficients R^2 are 0.8040 and 0.9146, respectively, that is to say, as time passed by, the TN emission concentrations would be significantly reduced, which agreed well with the finding of earlier related studies(*8-10*). K-Folder-Cross Validation Further (K-CV, where K=10) would be chosen to calculate the Predicted Determination Coefficient (predR^2) of fitting curve of polynomial and BP neural network, which are 0.7670 and 0.8014, respectively. Because of the non-linear mapping ability and adaptation and other such characteristics BP neural network, it is feasible for TN emission concentrations fitting, with higher precision than polynomial fitting.

The fitting result of BP neural network indicated that TN emission concentrations would increase firstly and then decrease. The concentration of TN of agriculture drain changed dramatically during the spring irrigation stage, rising from 2mg/L to 4mg/L in ten days, and then decreasing rapidly to about 2mg/L. This may be related to the application of nitrogen before sowing. In addition, the change of TN concentration was relatively slow and the concentration stable at 1.5-2mg/L during seeding stage. In mature stage the TN concentration of drainage was lower than 0.5mg/L, and the possibility via runoff was very little.

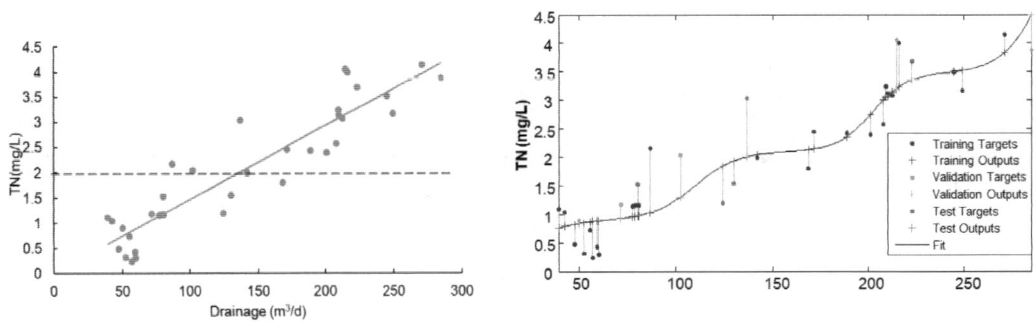

(a) Polynomial (b) BP Neural network

Figure 4. The relationship between drainage and the concentration of TN in agricultural drain

The relationship between drainage and emission concentration of TN based on polynomial and BP neural network as indicated in Figure 4. The statistical parameters were R^2 and PredR^2 for polynomial (0.8494 and 0.8342) and BP neural network (0.9126 and 0.7041) respectively. It can be seen that the fitting precision of BP neural network is better than using polynomial, but the latter one shows more stable. That is to say that there is a positive linear relationship between the agricultural drainage and the emission concentration of TN. Combined with Figure 3, a large amount of drainage during the spring irrigation period is also the main factor leading to the rapid loss of nitrogen.

4. Conclusion

This study take Hetao irrigation region of Inner Mongolia province as an example and combine drip irrigation with subsurface pipe drainage to monitor the TN emission concentration of sunflower during the whole growth period. The test and results show that the largest amount of drainage was produced in spring irrigation stage, while irrigation and rainfall did not promote the drainage in crops' growth period. Therefore, further study could pay more attention to the water movement before sowing. In addition, the experimental results also showed that TN concentration of farmland drainage exceeded the standard V of "Surface Water Quality Standard" during the spring irrigation period. This means that the salt leaching would easily lead to nitrogen loss and water pollution at this time. However, after a growth period, most of the nitrogen in the soil has been absorbed by the crops. Salt leaching after harvest can effectively prevent nitrogen loss and water pollution. It is well known that the purpose of spring irrigation and autumn irrigation is to leach excess salt from the soil. Therefore, the management department can alleviate the local water pollution by regulating the leaching scheme, such as reducing the water cost of autumn irrigation to encourage local farmers to discharge salt in autumn. Moreover, due to the direct ratio between the drainage and the loss of nitrogen, prolonging the leaching time or reducing the irrigation quota can also effectively reduce the concentration of TN in farmland drainage.

5. Acknowledgement

This research was supported by the National Natural Science Foundation of China (No. 51822905; 51779255; 51639009; and 51621061).

6. References

[1] Q. Tan, S. Zhang, R. Li, Optimal Use of Agricultural Water and Land Resources through Reconfiguring Crop Planting Structure under Socioeconomic and Ecological Objectives. *Water-Sui* **9**, (2017); published online EpubJul (Artn 48810.3390/W9070488).

[2] N. Zarrineh, K. Abbaspour, A. van Griensven, B. Jeangros, A. Holzkämper, Model-Based Evaluation of Land Management Strategies with Regard to Multiple Ecosystem Services. *Sustainability* **10**, 3844 (2018)10.3390/su10113844).

[3] C. Dai, Q. Tan, W. T. Lu, Y. Liu, H. C. Guo, Identification of optimal water transfer schemes for restoration of a eutrophic lake: An integrated simulation-optimization method. *Ecol Eng* **95**, 409-421 (2016); published online EpubOct (10.1016/j.ecoleng.2016.06.080).

[4] Y. P. Cai, Q. Q. Rong, Z. F. Yang, W. C. Yue, Q. Tan, An export coefficient based inexact fuzzy bi-level multi-objective programming model for the management of agricultural nonpoint source pollution under uncertainty. *Journal of Hydrology* **557**, 713-725 (2018); published online EpubFeb (10.1016/j.jhydrol.2017.12.067).

[5] S. Peng, Y. Luo, J. Xu, S. Khan, X. Jiao, W. Wang, Integrated Irrigation and Drainage Practices to Enhance Water Productivity and Reduce Pollution in a Rice Production System. *Irrigation and Drainage* **61**, 285-293 (2012)10.1002/ird.684).

[6] A. R. Sepaskhah, A. Tafteh, Yield and nitrogen leaching in rapeseed field under different nitrogen rates and water saving irrigation. *Agricultural Water Management* **112**, 55-62 (2012)10.1016/j.agwat.2012.06.005).

[7] S. G. Ye, S. Z. Zheng, X. Chen, *Study on pollution deduction and environmental impact of water-saving irrigation for paddy fields*. C. K. Zhang, H. W. Tang, Eds., Advances in Water Resources and Hydraulic Engineering, Vols 1-6 (Tsinghua Univ Press, Beijing, 2009), pp. 539-544.

[8] L. Zhang, Z. Wang, J. Chai, Y. Fu, C. Wei, Y. Wang, Temporal and Spatial Changes of Non-Point Source N and P and Its Decoupling from Agricultural Development in Water Source Area of Middle Route of the South-to-North Water Diversion Project. *Sustainability* **11**, 895 (2019)10.3390/su11030895).

[9] Q. Li, Y. Hu, Q. Jia, C. Song, Establishment and application of the estimation model for pollutant concentrfation in agriculture drain. *IOP Conference Series: Earth and Environmental Science* **121**, 032046 (2018)10.1088/1755-1315/121/3/032046).

[10] S. Li, L. Zhang, Y. Du, H. Liu, Y. Zhuang, S. Liu, Evaluating Phosphorus Loss for Watershed Management: Integrating a Weighting Scheme of Watershed Heterogeneity into Export Coefficient Model. *Environmental Modeling & Assessment* **21**, 657-668 (2016)10.1007/s10666-016-9499-1).

IOP Conf. Series: Materials Science and Engineering **730** (2020) 012065 doi:10.1088/1757-899X/730/1/012065

Analysis of Disturbance Factors in Zhalong Wetland Landscape Dynamics

Yuqing Yang[1], Adu Gong[1,3], Yuhong Zhang[2], Yanling Chen [1] and Tingting Zeng[1]

[1] Faculty of Geographical Science, Beijing Normal University, No.19, XinJieKouWai St., HaiDian District, Beijing 100875, P. R. China.

[2] Faculty of Geographical Science, Harbin Normal University, No.1, ShiDa Road, LiMin Economic Development Zone, Harbin 150025, P. R. China.

[3,*]Email: gad@bnu.edu.cn

Abstract. Based on remote sensing image data and meteorological data, this paper analyzes dynamic landscape disturbance factors of zhalong wetland through landscape pattern index and other methods, the results show that: (1) In recent 40 years, zhalong wetland landscape type pattern space, landscape fragmentation phenomenon is more serious.(2) The total area of swamp continues to decrease, and the bare land and cultivated land also show a decreasing trend, but the grassland and water area increase.(3) The temperature fluctuation, precipitation change and the government's policy of building railways and ditches have a profound impact on the landscape pattern of wetland.(4) At the early stage of the study, zhalong wetland landscape dynamic influencing factors are mainly natural factors such as temperature and precipitation;Since the 1990s, human activities such as government building of railways, highways, ditches and other factors have gradually increased the impact on wetland landscape dynamics.

1. Introduction

The evolution of landscape dynamic spatial pattern and the analysis of disturbance factors are the hot issues in the field of geography and ecology. As early as 1979, zhalong wetland was rated as a provincial nature reserve, designated as a national one in 1989, and listed in the "world important wetland list" in 1992.In recent decades, with the rapid progress of social economy and the continuous improvement of urbanization, the quantity and quality of wetland area in zhalong area have been greatly reduced, and the biodiversity and sustainable development of wetland have been seriously affected. Therefore, it is urgent to analyze wetland changes and interfering factors.

Taking zhalong region as the research object, based on remote sensing and GIS technology, with 1979, 1989, 2016, the third phase of remote sensing image data and meteorological data from 1979 to 2016 on the basis of data source, and statistical methods to analyze the landscape pattern index of landscape dynamic change and rainfall, air temperature, the relationship between the factors such as roads, canals, discusses the influence factors of zhalong wetland landscape changes in recent years (relevant academic research shows that 1989 is the abrupt change point of the zhalong wetland landscape dynamic process [1]).

2. Research Area and Data

Zhalong wetland is located in qiqihar city, heilongjiang province, bordering lindian, fufu and tailai counties. Located in the lower reaches of wuyuer river and the eastern songnen plain, the longitude and latitude are respectively 123°46'~124°36' and 46°51'~47°13'(Figure 1). Zhalong wetland is located in the mid-latitude area, which is the most primitive and complete wetland ecosystem preserved at the same latitude in northern China. It covers an area of 2,100 square kilometers, of which 1,240 square kilometers are wetlands.

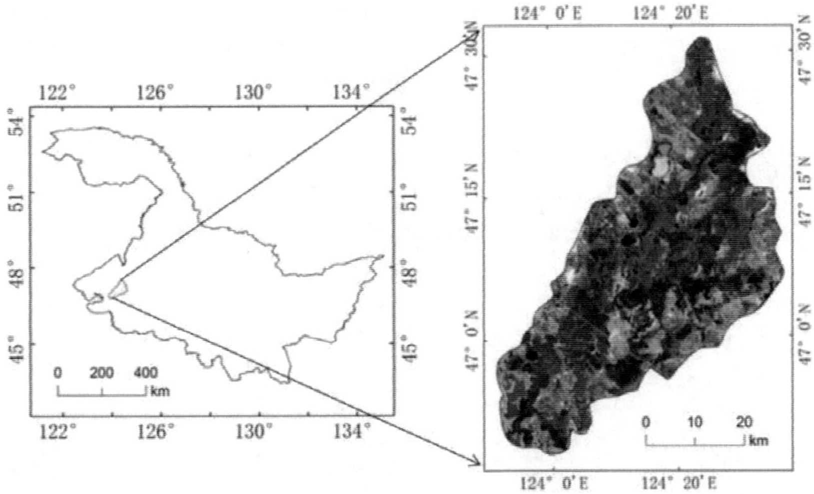

Figure 1. Overview of the research area and remote sensing images

Remote sensing image data used in this study are shown in Table 1, which are downloaded from the website of teaching and research office and geospatial data cloud (http://www.gscloud.cn/).Meteorological data mainly in qiqihaer city in heilongjiang province meteorological sites in 1979-2016 data set value of the month, from the China meteorological network (http://cdc.cma.gov.cn/home.do).

Table 1.Zhalong wetland remote sensing image parameters

Number	Date Acquired	Sensor Type	Spatial Resolution
1	1979.08.24	Landsat3 MSS	30m
2	1989.08.09	Landsat5 TM	30m
3	2016.04.16	Landsat8 OLI	30m

3. Research Method

3.1. Image Classification

According to the accuracy requirements of this study and the prior knowledge of staff, this study selects the maximum likelihood method in supervised classification for classification. It first obtains the probability of each pixel belonging to each category, and then assigns the pixel to the group with the largest corresponding probability value. Then, the results of classification are manually identified, categories combined and other operations. According to the research purpose of this paper, images are divided into 6 categories: swamp, water, cultivated land, village, grassland and bare land.

3.2. Landscape Pattern Index

Landscape pattern index is to divide the study area into various landscape patch units according to different ecological structures, and then analyze each characteristic index of spatial landscape pattern in the study area quantitatively. Its main indicators are: landscape separation, interference intensity naturalness, landscape diversity, dominance, fractal dimension and so on. Based on the research purpose, the landscape pattern parameters calculated in this paper include patch type area, landscape type percentage, landscape separation degree and landscape diversity [2].

3.3. Statistical Method

In this paper, mann-kendall(m-k) test is used to test the abrupt change of temperature and precipitation data in zhalong area from 1979 to 2016.Manmann Kendall test is one of the main research methods on the trend of change and mutation of time series. It can determine the exact year of mutation, which is less affected by human activities, and can reach a high quantitative level. Moreover, it has a wide range of detection, which is very suitable for the analysis and inspection of hydrological and meteorological data.

4. Result Analysis

Three phases of remote sensing images were classified by the maximum likelihood method in supervised classification.The classification results are shown in Figure 2.

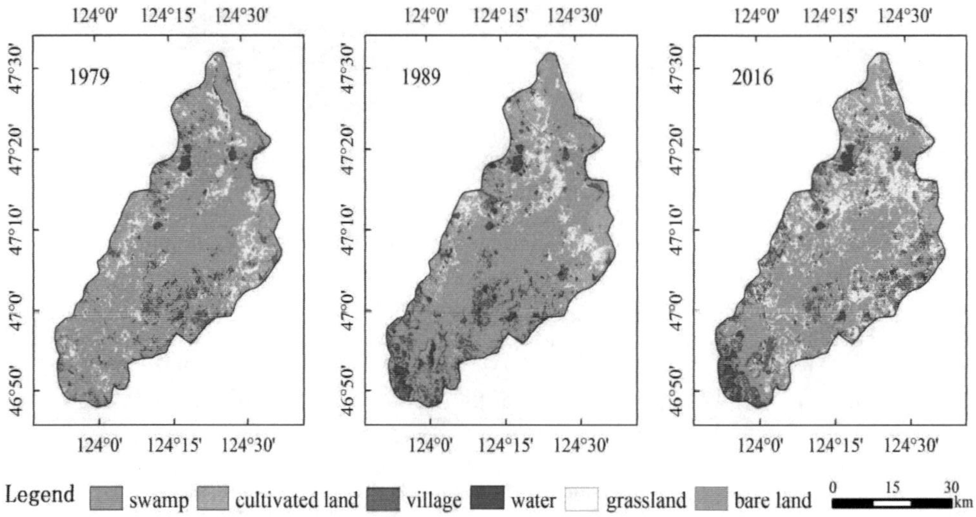

Figure 2. Classification results of zhalong district supervision

4.1. Analysis of Dynamic Changes of Landscape

Fragstats4.2 software is used to calculate the landscape pattern index of classified images, and the results are as follows:

4.1.1. Wetland landscape type conversion analysis. As can be seen from Table 2, the area of swamp in 1979 was 109128.05hm2, accounting for about 48% of the total area of zalong area; in 1989 and 2016, the area of swamp showed a decreasing trend; in 2016, the area of swamp was 83591.41 hm2, accounting for only 37% of the total area. The change of grassland and water is obvious, among which the grassland has increased from 21165.29hm2 in 1979 to 53439.82hm2 in 2016, and the water has increased from 5416.91hm2 in 1979 to 8945.69hm2 in 2016, which is related to the local policies of building ditches and widening water sources. Since 1989, zhalong wetland as a national nature reserve, the area of villages and towns has not changed greatly, accounting for about 11%.The area of bare land decreased from 13878hm2(6%) in 1979 to 10512hm2(5%) in 2016. Some bare land was transformed into grassland under the combined action of external forces such as climate change and human factors [3].

Table 2. Landscape type changes of zhalong wetland in different years

Landscape Index	Patch Type Area (hm^2)			Percentage of Landscape Type		
Year	1979	1989	2016	1979	1989	2016
Swamp	109128.05	95304.64	83591.41	48.48	42.34	37.14
Grassland	21165.29	41748.68	53439.82	9.40	18.55	23.74
Cultivated Land	49700.79	45679.15	42429.50	22.08	20.29	18.85
Villages	25802.67	22900.41	26156.82	11.46	10.17	11.62
Bare Land	13878.12	10691.09	10512.04	6.17	4.75	4.67
Water	5416.91	8754.95	8945.69	2.41	3.89	3.97

4.1.2. Analysis of spatial variation of wetland landscape. According to the classification results supervised by Figure 2, it can be seen that at the early stage of the study, the swamp in the landscape types of zhalong area were concentrated, Other land types were mostly distributed in the surrounding area of zhalong area. From the images of 1989 in the middle of the study, it can be seen that the water content in the south of zhalong area increased significantly, while the grassland area in the north increased significantly. In the later period of 2016, landscape fragmentation was serious, especially in the northern and central zhalong areas. According to relevant information, since 1990, under the decision of water conservancy departments at all levels and local governments in zhalong district, various water conservancy projects and national roads and railways at all levels have been built inside the wetland, which seriously damaged the integrity of zhalong wetland environment.

Table 3.Landscape index list of zhalong in different years

Year \ Index	DIVISION	SHDI
1979	0.85	1.42
1989	0.82	1.42
2016	0.92	1.55

The larger the degree of separation index, the higher the degree of separation, that is, the landscape is more fragmented.It can be seen from the table that, compared with 1979 and 1989, the degree of separation index increased significantly in 2016, and the degree of separation of wetland landscape was obvious.The greater the diversity index is, the higher the degree of fragmentation will be, and the greater the uncertainty information will be.

4.2. Annual Average Temperature, Precipitation Changes

From 1979 to 2016, the overall temperature in zhalong area showed an upward trend. Figure 3 shows that the average annual temperature in zhalong area has increased by 0.024°C in recent years. From the distribution trend, in the early 1980s, the annual temperature change in zhalong area was relatively large, and in the late 1980s to the early 21st century, the annual temperature change in zhalong area was relatively small, the temperature during this period may be in the warm period of the whole research period.

Zhalong area belongs to semi-arid area and the distribution of precipitation is unbalanced. In this paper, the precipitation anomaly percentage in zhalong region from 1979 to 2016 was calculated. When the precipitation anomaly percentage is positive, the precipitation reaches the standard. When it is negative, the precipitation is not up to the standard. From 1979 to 2016, the precipitation in zhalong area was seriously substandard around 1979 and 1996, but the temperature basically showed an upward trend, which may lead to an increase in evapotranspiration and damage the wetland ecosystem (Figure 3).

IOP Conf. Series: Materials Science and Engineering **730** (2020) 012065 doi:10.1088/1757-899X/730/1/012065

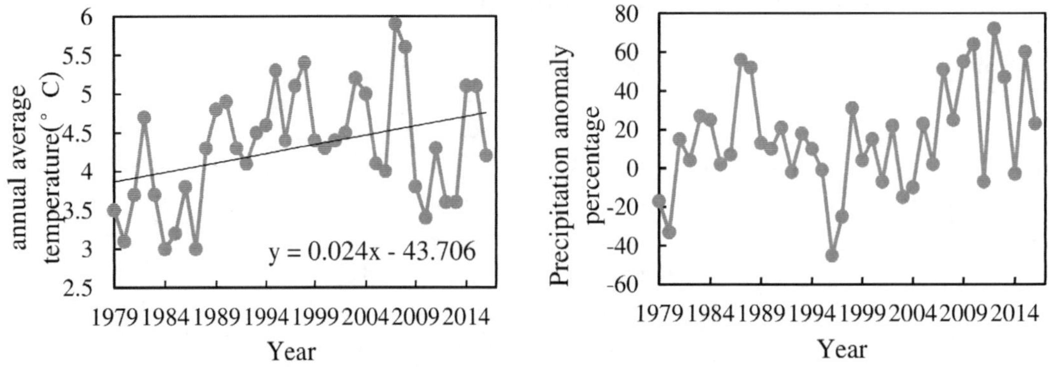

Figure 3. Schematic diagram of annual temperature change and precipitation anomaly in zhalong region from 1979 to 2016

4.3. Abrupt Change Detection of Temperature and Precipitation

Mann-kandall method was used to test the meteorological data of zhalong wetland.

The results of the temperature trend test (Figure 4) show that: the average annual temperature of zalong wetland is in a process of increasing, and has a significant increase trend since 1987. It passed the 0.05 significance test (U0.05=1.96) from the mid-1990s to 2012, and even passed the 0.001 significance test (U0.001=2.56).According to the position of the intersection of UF (k) and UB (k) curves, it was determined that the annual average temperature increase in zhalong wetland was an abrupt behavior, mainly starting from 1988[4].

From 1985 to 2008, the precipitation of zhalong wetland showed a decreasing trend, but this trend was not significant and did not pass the 0.05 significance test. From 2008 to 2016, it showed an increasing trend, and the trend was not significant.According to the intersection of UF (k) and UB (k) curves, the annual precipitation of zhalong wetland showed abrupt changes in 1979, 1989, 1991 and 2009(Figure4).

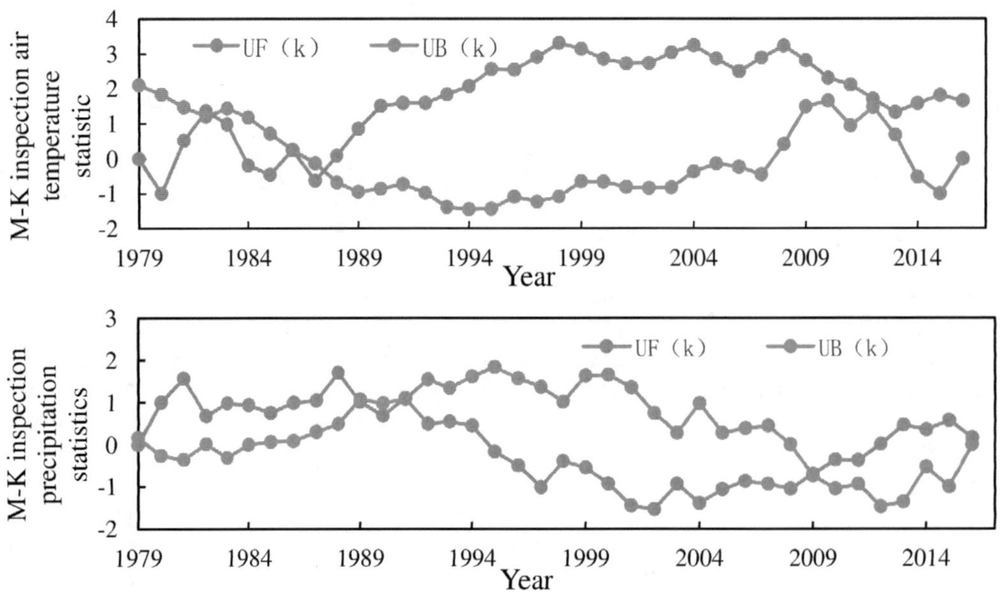

Figure 4. Average temperature and precipitation curve of zhalong wetland from 1979 to 2016

4.4. Human Factor Analysis

With the development of social economy and urbanization, the area around zhalong is more seriously affected by human beings. In 1980, zhalong wetland inside only 301 national road from the north through; In 1986, tele highway bridge was completed and qidu highway was completed. In 1992, national highway 301 was changed. From 1990 to 2000, the construction of roads in and around the wetlands reached a peak, including village-tun roads in several core areas. Moreover, various levels of water conservancy departments and local areas in zhalong area to build a variety of major water conservancy facilities. These roads and ditches are being built and put into use, which seriously damage the environment around the wetlands and affect the normal growth of wetlands.

5. Summary

According to the analysis of disturbance factors of zhalong wetland landscape dynamics in this paper, it can be seen that: 1. From 1979 to 2016, the wetland landscape of zhalong wetland changed greatly in terms of quantity and spatial distribution. Wetland area always accounts for the highest proportion, with the passage of time, the number of decline trend, the spatial distribution to the southwest shift. The area of grassland and water bodies increased significantly, the area of cultivated land and bare land decreased slightly, and the area of villages and towns remained basically unchanged.2. In recent 40 years, the average temperature in zhalong area has shown an upward trend. Since 1980, the dual effects of continuous temperature rise and precipitation decrease in zhalong area have changed the climate environment in zhalong area and affected the landscape dynamics of wetlands.3. Since the 1990s, railways, highways, ditches and large-scale water conservancy projects have been built in and around zhalong wetland, seriously damaging the natural hydrological model and landscape spatial structure of the wetland, and increasing the fragmentation and separation degree of the wetland landscape.

6. References

[1] Zhang Yuhong, Su Liying, Yu Wanhui, Zhang Hongyun. Dynamic change characteristics of zhalong wetland landscape [J]. Acta geographica sinica.2015, 70 (1) :131-142.

[2] Liu Di, Cao Chunxiang. Remote sensing diagnosis of wetland changes -- a case study of ordos gulls national nature reserve in Inner Mongolia [M]. Master's thesis of the university of Chinese academy of sciences.2017.

[3] Khaled H Hamed. Trend detection in hydrologic data: The mann-kendall Trend test under The scaling hypothesis [J].Journal of Hydrology.2008, 349(3):350-363.

[4] Zhen Shuo, Dong Liqin, Zheng Rudin, Yao Pengju. Landscape pattern and changes of ruoergai plateau swamp in 2007 and 2016 [J]. Wetland science.2017, 15(4):522-525.

IOP Conf. Series: Materials Science and Engineering **730** (2020) 012066 doi:10.1088/1757-899X/730/1/012066

Comparative Study on Several Determination Methods of Chlorophyll Content in Plants

Jing Li, Xiaoyan Zhou, Jiaping Zhou, Ran Shang, Yu Wang and Pei Jing

School of Environmental Science and Engineering, Shandong Agriculture and Engineering University, Jinan Shandong, 250100, China.
Email: ripplelj@126.com

Abstract. Chlorophyll was determined in the N,N-dimethylformamide (DMF) and the mixture solution ($V_{alcohol}/V_{acetone}$=1:1) respectively, and the extracted effects were compared. The results showed that the extracted effects of chlorophyll content in the mixture solution method is better than that of DMF method when the chlorophyll content was low, while DMF method is better when the chlorophyll content was high. DMF extraction method with long extraction time was better to the coniferous forest vegetation and *Ginkgo biloba* Linn., because the plant's well-developed cuticle on the blade surface could hinder the dissolution of chlorophyll in the extract in a short time. The SPAD values were measured when leaves were determined by above two methods meanwhile, and were linearly fitted with the actual chlorophyll content values to obtain the linear correction equation. It found that SPAD values had a good linear relationship with chlorophyll content (with mg/dm^2 fresh weight, R^2>0.96). Linearity range of SPAD value is 25~60 in DMF method, and 26~50 in the mixture solution method.

1. Introduction

Chlorophyll is an important photosynthetic pigment in plants. The chlorophyll content of plants can reflect the photosynthetic energy efficiency, so it can be used as a basic physiological index to study plant growth and development.[1]

There are many methods for the determination of chlorophyll content of plants by spectrophotometer [2]. Among them, Arnon method [3,4], DMF (N,N-dimethylformamide) method[5,6] and mixture solution ($V_{alcohol}/V_{acetone}$=1:1) method[7] were widely used. Arnon method was put forward by Mackinney[3], then was explained and deduced by Arnon[4]. DMF method was put forword by Moran [5,6]. The mixture solution method was put forward by Xianzheng Zhang in 1985[7]. Arnon method of chlorophyll extraction was cumbersome steps and heavy workload, and influencing accuracy easily by unstable the extraction [8]. DMF method was simple and efficient because DMF has the same effect of inhibiting chlorophyll enzyme activity as acetone but poisonous. The mixture solution method was simple, too. There are many improvements and comparative studies on this method at present [1,8,9].

In the 1960s, Japan began the study in nondestructive chlorophyll analyzer [10]. The nondestructive chlorophyll analyzer can measure chlorophyll content without damage to plants, which makes up for the shortcoming that the above methods can only measure chlorophyll in vitro. Therefore this type of analyzer developed rapidly, and the research about on nondestructive chlorophyll determination technology and the correlation between the results of nondestructive chlorophyll analyzer and actual chlorophyll content appeared extensively [10-14]. At present, SPAD value which represented nondestructive testing data was widely used.

In this study, DMF method and the mixture solution method ($V_{alcohol}/V_{acetone}$=1:1) were selected to compare the extracting effection of chlorophyll content used seven plant species. SPAD values were determined at the same time to establish correction equation in order to provide reference for rapid extraction and determination of chlorophyll in batch samples.

2. Materials and Methods

2.1. Plant Material

Seven species of plants were selected as plant materials for this study, including *Fraxinus chinensis* Roxb., *Ginkgo biloba* Linn., *Magnolia denudata* Desr., *Platycladus orientalis* (L.) Franco, *Pinus tabulaeformis*, maize and soybean. The experimental plants of *P. alba* var. were two years-old seedlings, maize and soybean were seedling which potted in early April, and the other experimental plants were adult trees.

2.2. Chlorophyll Content Measurements

In April and June, fresh leaves of *Fraxinus chinensis* Roxb., *Ginkgo biloba* Linn., *Magnolia denudata* Desr. were picked for chlorophyll content determination in weight (mg/g) and in area (mg/dm^2).

In May, fresh leaves of *Platycladus orientalis*(L.) Franco, *Pinus tabulaeformis*, maize and soybean were picked for chlorophyll content determination in weight (mg/g) and in area (mg/dm^2).

The tested leaves were washed with steam water and the surface water of leaves was absorbed with filter paper.

The leaves of *Fraxinus chinensis* Roxb., *Ginkgo biloba* Linn. (in June), *Magnolia denudata* Desr. were sampled with a hole punch (disc, 1cm in diameter) on one side of the main vein, then 4 pieces in each group (area was π) were added DMF. The other side of the main vein of leaves was sampled as above (disc, 1cm in diameter, area is π) and added the mixture solution ($V_{alcohol}/V_{acetone}$=1:1).

Four leaves of maize and soybean were picked respectively. Two discs were sampled with a hole punch (1 cm in diameter), each one disc was picked to mix in a group with four leaves and tested. Because the leaves of *Ginkgo biloba* Linn. (in April) were small, the chlorophyll content is determined in weight (mg/g). The leaves of *Platycladus orientalis*(L.) Franco and *Pinus tabulaeformis* were needles, therefore the chlorophyll content is determined in weight (mg/g).

Samples were cut into small pieces about 1 mm, and then put into test tube.

2.2.1. DMF method. The extract (DMF, 5 ml) was added into each test tube, and was kept away from light for 24 hours at 4 °C, shaking occasionally. The sample supernatant was measured at 647nm and 664.5nm by 722 visible spectrophotometer. The chlorophyll content was calculated by formula (1-3)[5,6].

$$Chla=12.70A664.5-2.79A647 \quad (1)$$

$$Chlb=20.70A647-4.62A664.5 \quad (2)$$

$$Total\ Chl=17.90A647+8.08A664.5 \quad (3)$$

2.2.2. The mixture solution method. The absorption spectra of chlorophyll solution extracted by the mixture solution method were consistent with those of chlorophyll extraction solution by Arnon method[4,7]. Therefore, the formula of Arnon method was used to calculate the chlorophyll content here.

The extract (the mixture solution, $V_{alcohol}/V_{acetone}$=1:1, 10ml) was added into each test tube, and was kept away from light for 2 hours at 50°C [7,8]. The sample supernatant was measured at 645nm and 663nm by 722 visible spectrophotometer. The chlorophyll content was calculated by formula (4-6).

$$Chla=12.70A663-2.69A645 \quad (4)$$

$$Chlb=22.90A645-4.68A663 \quad (5)$$

$$Total\ Chl=20.20A645+8.02A663 \quad (6)$$

2.2.3. SPAD-502 plus analyzer. The SPAD values were measured by SPAD-502 plus (Konica Minolta) at the same time when leaves were sampled as the above method.

3. Result

3.1. Comparison of Two Extraction Methods

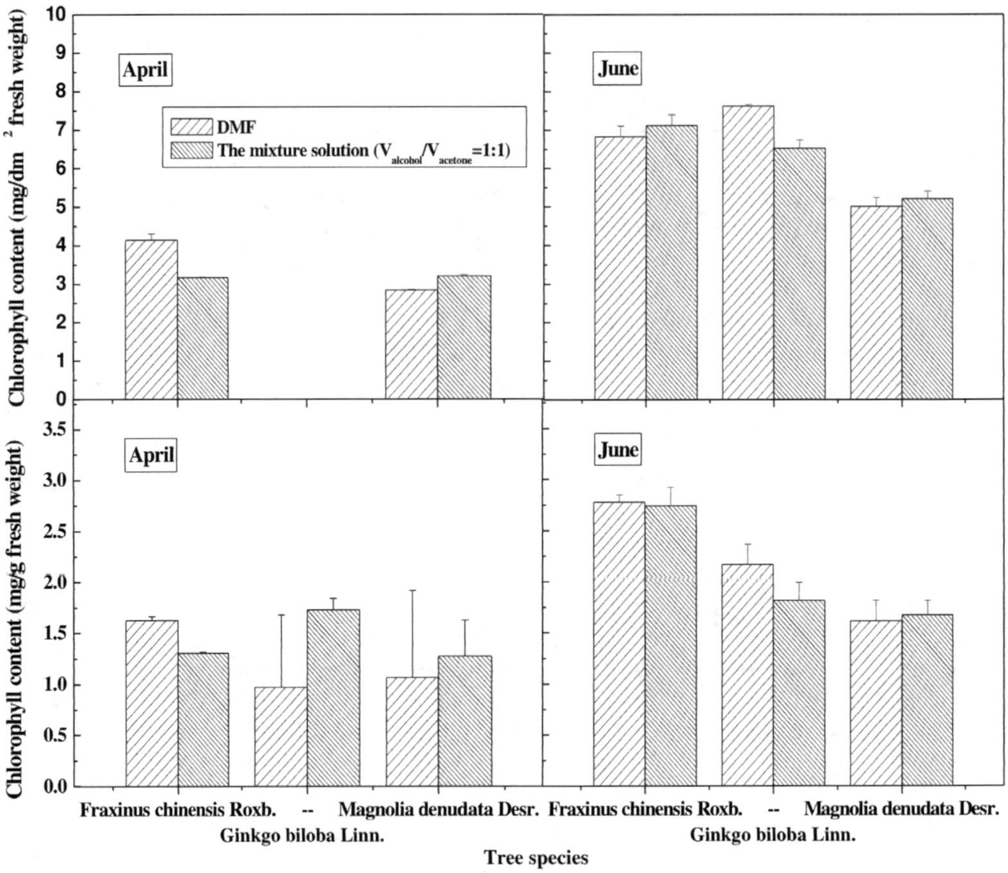

Figure 1. Chlorophyll content (mg/g fresh weight, mg/dm^2 fresh weight) of *Fraxinus chinensis* Roxb., *Ginkgo biloba* Linn. and *Magnolia denudata* Desr. with DMF method and the mixture solution method ($V_{alcohol}/V_{acetone}$=1:1) in April and June.

Figure 1 and figure 2 show the results of chlorophyll content determination of seven plants by two different extraction methods in different months. The mixture solution method has higher measured values in *Magnolia denudata* Desr., maize and soybean. The DMF method has higher measured values in *Platycladus orientalis*(L.) Franco, *Pinus tabulaeformis.* The measurements are similar with two extraction methods in *Fraxinus chinensis* Roxb..The mixture solution method has higher measured values in *Ginkgo biloba* Linn. in April, but it is higher in June with the DMF method.

3.2. Changeing of Chlorophyll Content

Fig. 1 shows the changes of chlorophyll content in April and June in three plants (*Fraxinus chinensis* Roxb., *Ginkgo biloba* Linn., *Magnolia denudata* Desr.). In April, leaves of plants are mainly tender leaves which is with light leaf color and low chlorophyll content. In June, the leaves are mostly mature leaves when plants grow vigorously. The color of leaves in June is greener than that in April, therefore all the three plants show the measured values are higher in June than in April.

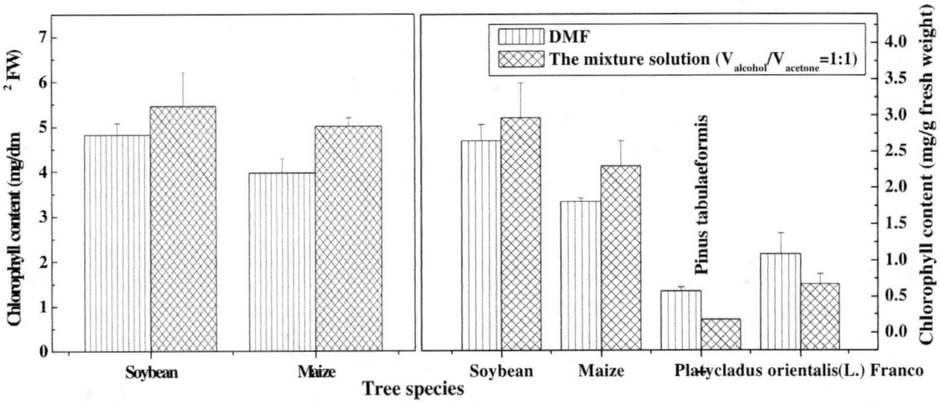

Figure 2. Chlorophyll content (mg/g fresh weight, mg/dm^2 fresh weight) of soybean, maize, *Pinus tabulaeformis* and *Platycladus orientalis* (L.) Franco, with DMF method and the mixture solution method ($V_{alcohol}/V_{acetone}$=1:1) in May.

3.3. SPAD Linear Correction Equation

The SPAD values were measured when leaves were determined by DMF method and the mixture solution method meanwhile. The SPAD values were linearly fitted with the actual chlorophyll content values to obtain the linear correction equation. The independent variable was SPAD, and the dependent variable was chlorophyll content. Linear correction equation and R^2 are shown in Table 1.

Table 1. SPAD linear correction equation

Determined methods	Linearity range of SPAD value	Unit of Chlorophyll content	Linear correction equation	R^2
DMF Method	25~60	mg/dm^2 FW	y = 0.186x - 1.933	0.9859
		mg/g FW	y = 0.1305x - 2.4186	0.7252
The mixture solution method ($V_{alcohol}/V_{acetone}$=1:1)	26~50	mg/dm^2 FW	y= 0.1879x - 1.8667	0.9602
		mg/g FW	y= 0.1219 - 1.9835	0.8797
	26~60	mg/dm^2 FW	y= 0.1787x - 1.5844	0.984
		mg/g FW	y= 0.0339x + 0.7437	0.2221

Table 1 shows that the linear relationship between chlorophyll content expressed in area and SPAD value is better. Linearity range of SPAD value is 25~60 in DMF method. Linearity range of SPAD value is 26~50 in the mixture solution method. When the measurement exceeds 50, the R^2 of linear correction equation appears serious deviation (y = 0.0339x + 0.7437, R^2 = 0.2221, mg/g FW).

4. Discussion

Chlorophyll is a kind of magnesium porphyrins, which is unstable. It can be decomposed in the presence of light, acid, alkali, oxygen and oxidant etc.

DMF can inhibit the activity of chlorophyll enzymes effectively, and the extraction method is easy to operate. Therefore DMF method is widely used at present. However, DMF method has a long extraction time (24h) and DMF is high toxicity. Therefore, the mixture solution method (acetone/ethanol) was proposed by domestic scholars and several improvements about the method were made. The mixture solution method is easy to operate and is widely used at present too, because of the short extraction time (2h) and fast determination. But the mixture solution method needs higher temperature and needs to avoid light when extracting, while the improper operation is easy to affect the extracted effects of the chlorophyll content.

Three plant species were selected to compare the above two extraction methods in April and June. The results show that the extracted effects of chlorophyll content in the mixture solution method is better than that of DMF method when the chlorophyll content was low in April, while DMF method is better when the chlorophyll content was high in June. That may be related to the interannual variation of chlorophyll content and the extracted time of the two methods. The extracted time of the mixture solution method (2h) is shorter than that of DMF method (24 hours). When the chlorophyll content is low, the extraction effect of the two methods is similar. When the chlorophyll content is high, the short extracted time would lead to the incomplete extraction of chlorophyll. Therefore, it is suggested to use DMF method when the plant species with high level of chlorophyll, and to use the less area or weight of leaves with the mixture solution method.

The extracted effect of the DMF method is obviously better than with the mixture solution method in *Platycladus orientalis*(L.) Franco, *Pinus tabulaeformis and Ginkgo biloba* Linn. (in June). *Platycladus orientalis*(L.) Franco and *Pinus tabulaeformis* are coniferous forest vegetation which has well-developed cuticle on the blade surface[15]. *Ginkgo biloba* Linn. is deciduous tree specie, while its leaf is similar to that of *Platycladus orientalis* (L.) Franco and *Pinus tabulaeformis* with well-developed cuticle on the blade surface. Insoluble proteins in cuticle may hinder the dissolution of chlorophyll in the extract in a short time. Therefore, DMF method is recommended for the extraction of chlorophyll from leaves with thicker cuticle. DMF method could be more completely destroyed the protection of cuticle to leaves with a long extracted time.

At present, the main methods for determining chlorophyll content are spectrophotometer method and the nondestructive chlorophyll analyzer. SPAD value is relative chlorophyll content which is measured by the nondestructive chlorophyll analyzer, therefore the spectrophotometer method that can determine absolute chlorophyll content is more widely used [1]. The spectrophotometer method has a variety of determination methods because of using the different solvents. Most of the determination methods have heavy workload, complicated steps, long extracted time or strict extraction conditions. Therefore, improper operation can lead to errors easily, and the determination methods are not suitable for the extraction and determination of a large number of samples [8]. SPAD-502 plus chlorophyll meter could be used to measure the chlorophyll content rapidly and in real time. After recording the SPAD values, the actual chlorophyll content could be obtained through the linear correction equation. Therefore, SPAD value and chlorophyll content could be carried out simultaneously at the beginning of the experiment in order to obtain the linear correction equation and the linearity range of SPAD value. That could save time for chlorophyll content determination for the subsequent experiment.

A number of studies have used the SPAD value to express chlorophyll content directly at present [16-20]. However, table 1 show that the SPAD values which could indicate the chlorophyll content accurately has a certain linearity range. If the SPAD values exceed the linearity range, the quantitative conversion and analysis of chlorophyll content could not be carried out. Therefore, in order to obtain more reliable data, it is recommended to obtain the the linear correction equation and the linearity range of SPAD value first according to the different extraction methods of chlorophyll content with the characteristics of plant leaves. At the same time, the different units of chlorophyll content need to be considered when the linear correction equation and the linearity range of SPAD value are obtained.

5. Acknowledgement

This work was supported by Young Teachers' Scientific Research Fund of Shandong Agriculture and Engineering University (QNKJY201803), Shandong Education Department Fund (J18KB061) and Poverty Alleviation Program of Shandong Agriculture and Engineering University (FPKJ201809).

6. References

[1] Fen X, Wei L, Hong-Yan L, Ling-Yun Z, Ding D, Zhen Z 2016 Comparisonon methods of chlorophyll extraction in camellia sinensis. *Journal of Tea Communication* 43(04):37-40

[2] Zheng-Shu S and Xian-Zheng Z 1989 Comparison of several methods for determining plant chlorophyll content *Plant Physiology Communications* 5:77-78

[3] Mackinney G 1941 Absorption of Light by Chlorophyll Solution *J. B iol. Chem.* 140: 315-322

[4] Arnon D I 1949 Copper Enzymes in Isolated Chloroplasts Phenoloxidases in Beta Vulgaris *Plant Physiol.* 24:1-15

[5] Rami M and Dan P 1980 Chlorophyll Determination in Intact Tissues Using N,N-Dimethylformamide *Plant Physiol.* 65:478-479

[6] Rami M 1982 Formulae for Determination of Chlorophyllous Pigments Extracted with N,N-Dimethylformamide *Plant Physiol.* 69: 1376-1381

[7] Xian-Zheng Z 1986 Determination of Plant Chlorophyll Content by Acetone-Ethanol Mixture Method *Liaoning Agricultural Science* 3:26-28

[8] De-Xiao L, Wan-Wei H, Hai-Yan Y 2006 Fast-soaking methods of chlorophyll from maize leaf *Jour. of Northwest Sci-Tech Univ. of Agri. and For. (Nat. Sci. Ed.)* 34(11):65-67

[9] Fa-Shui H, Zheng-Gui W, Gui-Wen Z 2001 Soaking and synergistic extracting reaction of chlorophyll from spinach *Chinese Journal of Applied Chemistry* 18(7):532-535

[10] Zhang-Qing Y, Jian-Dong H, Tie-Cheng D, Shun W, Jian-Zhong W, Wei-Bo S, Yu-Min Z 2019 Research Progress of Nondestructive Diagnostic Technique of Chlorophyll in Plants *Chinese Agricultural Science Bulletin* 35(07):139-144

[11] Uddling J, Gelang-Alfredsson J, Piikki K, Pleijel H 2007 Evaluating the relationship between leaf chlorophyll concentration and SPAD-502chlorophyll meter readings *Photosynthesis research* 91(1):37-46

[12] Cate T.M. Perkins T.D. 2003 Chlorophyll content monitoring in sugar maple (acer saccharum) *Tree Physiology* 23(15):1077-1079

[13] Xiao-Sheng Z, Yue-Qin Z, Lei P, Chang-Jun J, Ye-Yun L 2012 Estimate of leaf chlorophyll and nitrogen content in camellia sinensis by ccm-200 *Journal of Anhui Agricultural University* 39(01):150-153

[14] Pingzeng L, Guansheng Z, Zhenmin D, Wenshan W, Yucun W, Qingkuan M, Shusheng B 2012 Development of a low-cost crop chlorophyll detector *Computer Measurement & Control* 20(4):1155-1158

[15] Chen-Hai H and Jian L 1993 Analysis of the Anatomical Structure of Tree Leaves from Western Liaoning *Journal of Shenyang Agricultural University* 24(4):269-273

[16] Guo-Xiang S,Xiao-Chan W, Ye S, Yong-Qian D, Wei L 2019 Measurement Method Based on Multispectral Three-Dimensional Imaging for the Chlorophyll Contents of Greenhouse Tomato Plants *Sensors (Basel, Switzerland)* 19(15)

[17] Li-Ping R, Jia-Qi Z, Dong-Hui C, Ji-Sheng L, Yan Z, Yu-Fu G 2019 Effects of simulated acid rain on physiology and chlorophyll fluorescence parameters in Acer truncatum seedlings *Non-wood Forest Research* 3:44-51

[18] Rui L,Jun-Hui C,Yong-Lin Q,Ming-Shou F 2019 Possibility of using a SPAD chlorophyll meter to establish a normalized threshold index of nitrogen status in different potato cultivars *Journal of Plant Nutrition* 42(8)

[19] Swiader J M and Moore A 2002 Spad-chlorophyll response to nitrogen fertilization and evaluation of nitrogen status in dryland and irrigated pumpkins *Journal of Plant Nutrition* 25(5):1089-1100

[20] Yu-Jie H, Lin-Nan W, Jian-Rong F, Baimaduoji, Fu-Xia W, Xin-Min R 2016 Effects of NaCl stress on photosynthetic characteristics of leaves of the two grape cultivars *Xinjiang Agricultural Sciences* 53(10):1794-1800

IOP Conf. Series: Materials Science and Engineering **730** (2020) 012067 doi:10.1088/1757-899X/730/1/012067

Three-Dimensional Hydrodynamic and Water Quality Integrative Simulation System Based on GIS of Donghu in Wuhan

Xiaojuan Li, Mutao Huang* and Ronghui Wang

School of Hydropower Engineering & Information Engineering, Huazhong University of Science & Technology, Wuhan 430074, China.

*Email: M201873778@hust.edu.cn

Abstract. The three-dimensional hydrodynamic and water quality integrative simulation system is established based on the coupling of three-dimensional hydrodynamic model and WASP water quality model and GIS control technology in this paper. It integrates the functions of unstructured grid generation, three-dimensional hydrodynamic and water quality simulation and three-dimensional visual display. When the basic lake data is imported, the system can simulate and display the three-dimensional dynamic change process of lake hydrodynamic parameters and water quality parameters in a one-stop way. Numerical simulation was carried out on its hydrodynamic and water quality, taking Donghu in Wuhan as the study area. The system was tested with topographic data, water quality monitoring data, hydrological data and meteorological data of Donghu in 2018. Compared with the measured results, the error is small, which shows that the system could be used for hydrodynamic and water quality simulation of inland lakes.

1. Introduction

Under the both of human activities and climate change factors, Lake Eutrophication is becoming increasingly prominent, as a common environmental problem in the global water ecosystem. A common phenomenon of eutrophication is planktonic algae mass-propagation and frequent algal bloom. To further study the mechanism of algal bloom formation in eutrophic lakes and the relationship between various environmental factors and bloom, it is necessary to develop the prediction technology of algal bloom. It can provide scientific theoretical basis and decision-making support technology for Lake Eutrophication management, ecological restoration and watershed comprehensive management, and help to improve the ability of environmental management departments to reduce the ecological hazards and health risks caused by algae blooms, which is of great ecological and environmental significance.

At present, the main research methods include field observation, remote sensing inversion, physical model experiment and numerical simulation. However neither site observation nor remote sensing observation can fully and continuously express the spatio-temporal information of lakes. Physical models are too complex to get the data it need. Numerical simulation has the advantage of low operating cost and controllable process, which can not only reshape the historical dynamic evolution of water environment, but also predict its future trend. It has been widely used in the modeling and simulation of water ecosystems such as rivers, estuaries, lakes and oceans.

The establishment of a numerical model of lake ecological dynamics through the coupled hydrodynamic and water quality model can simulate the continuous evolution of water quality ecological variables in time and space and provide an efficient platform for data integration and process understanding of the interaction between lake hydrodynamic water quality and water ecology under complex spatio-temporal background.

For example, Fu et al. coupled WASP with EFDC hydrodynamic model and integrated USEPA WASP model in a GIS platform [1]; Ng et al. integrated a GIS with a three-dimensional hydrodynamic sediment and heavy metal transport numerical model and applied to the Pearl River Estuary in China [2]. Therefore, many achievements have been made in establishing hydrodynamic water quality models based on GIS. However, there are several weaknesses associated with the existing systems, especially for lake modeling [3]. The formation mechanism of algal bloom is very complex, which limits the application scope of the model. The ecological dynamics model needs to be parameterized and discretized to describe the process of water state change and introduce various errors. The calibration of multiple parameters in mathematical model needs to be improved. The natural flow trend of lakes is relatively weak, and the existing one-dimensional and two-dimensional water environment models cannot accurately reflect the three-dimensional structure of water flow, so it is difficult to correctly simulate the flow of lakes. The limitations of existing research methods have brought great challenges to researchers to accurately understand the evolution mechanism of lake blooms, grasp the development trend of blooms, and take feasible measures to control and prevent blooms.

In this paper, we present the lake three dimensional hydrodynamic and water quality integrative simulation system based on GIS. This system integrates some key technologies, such as ELCOM, WASP water quality model, GIS and component development. The functions of it include multi-source remote sensing image management, electronic map management, unstructured grid automatic generation, 3d hydrodynamic water quality coupling simulation, spatial data and attribute data comprehensive query, water quality thematic map making, etc. The applicability of the system is illustrated by a case study in Lake Donghu in Wuhan. Relying on Landsat8 image, domestic high-resolution image, DEM data, GIS data and other data in Donghu, it organically integrates remote sensing, GIS, coupled modeling, big data technology and other disciplines, to realize the integrated management of lake property, vector, image and elevation data.

2. Model

The system has been embedded with a self-compiled 3D hydrodynamic model.

Based on the Navier-Stokes three dimensional water flow equation, the model adopts vertical static pressure assumption, unstructured triangular grid in horizontal direction and σ coordinate transformation in vertical direction. The calculation formula of sigma coordinates is defined as:

$$\sigma = \frac{z-\xi}{h+\xi} = \frac{z-\xi}{H} \tag{1}$$

Where σ is the transformed vertical coordinate, $\sigma \in (0,-1)$, where σ=0 corresponds to the static water surface, where σ=-1 corresponds to lake bottom; z is the water level in the actual physical domain, z=0 is the reference static water surface; ξ is the water level of the water surface relative to z=0; h is the water depth from the base still water surface to the bottom of the lake; $\mathrm{H} = \mathrm{h} + \xi$ is the total water depth. The hydrodynamic control equation in the coordinate system is defined as

$$\frac{\partial \xi}{\partial t} + \frac{\partial Hu}{\partial x} + \frac{\partial Hv}{\partial y} + \frac{\partial \omega}{\partial \sigma} = 0 \tag{2}$$

Momentum conservation equation is defined as:

$$\frac{\partial Hu}{\partial t} + \frac{\partial Hu^2}{\partial x} + \frac{\partial Huv}{\partial y} + \frac{\partial uw}{\partial \sigma} = -gH\frac{\partial \xi}{\partial x} + \mathrm{BV3}_x + \frac{1}{H}\frac{\partial}{\partial \sigma}\left(\mu_v \frac{\partial u}{\partial \sigma}\right) + F_x + fvH \tag{3}$$

$$\frac{\partial Hv}{\partial t} + \frac{\partial Hv^2}{\partial x} + \frac{\partial Hv^2}{\partial y} + \frac{\partial vw'}{\partial \sigma} = -gH\frac{\partial \xi}{\partial y} + \mathrm{BV3}_y + \frac{1}{H}\frac{\partial}{\partial \sigma}\left(\mu_v \frac{\partial v}{\partial \sigma}\right) + F_y - fuH \tag{4}$$

$$\mathrm{BV3}_x = -\frac{gH}{\rho_0}\left[\frac{\partial}{\partial x}\int_\sigma^0 (\rho-\rho_0)d\sigma + (\rho-\rho_0)\sigma\frac{\partial H}{\partial x}\right] \tag{5}$$

$$\mathrm{BV3}_y = -\frac{gH}{\rho_0}\left[\frac{\partial}{\partial y}\int_\sigma^0 (\rho-\rho_0)d\sigma + (\rho-\rho_0)\sigma\frac{\partial H}{\partial y}\right] \tag{6}$$

$$F_x = \frac{\partial}{\partial x}\left(\mu_h H \frac{\partial u}{\partial x}\right) + \frac{\partial}{\partial y}\left(\mu_h H \frac{\partial u}{\partial y}\right) \tag{7}$$

$$F_y = \frac{\partial}{\partial y}\left(\mu_h H \frac{\partial v}{\partial y}\right) + \frac{\partial}{\partial y}\left(\mu_h H \frac{\partial v}{\partial y}\right) \tag{8}$$

Where u, v and w are the x, y, z velocity components, respectively; p is the hydrodynamic pressure; ρ_0 is reference density of water body; $\mu_h$is turbulent viscosity coefficients in horizontal; μ_v is vertical directions; f is coriolis force coefficient, $f = 2\Omega \sin\theta, \Omega = 2\pi/86184$, θ is the rotation frequency of the earth, g is the acceleration of gravity; $BV3_x$ and $BV3_y$ are the x, y baroclinic gradient term, F_x and F_y are the x, y horizontal diffusion term of momentum [4].

The system uses the Water Analysis Simulation Program (WASP) as water quality model. The model includes eight water quality indicators, including inorganic phosphorus (OPO4), organic phosphorus (OP), ammonia nitrogen (NH)-N), nitrate nitrogen (NO3--N), organic nitrogen (ON), chlorophyll a (chl-a), dissolved oxygen (DO) and carbon biochemical oxygen demand (CBOD) [5].

Integrating the three-dimensional hydrodynamic model, WASP and GIS controls, the system establishes the three-dimensional hydrodynamic water quality model, which can simulate the eutrophication status of the lake and display it in the form of 3D.

3. Introduction to System

The overall framework of the system is shown in figure 1. Using C# language, basing on ArcGIS Engine, it provides unstructured grid automatic generation function, three-dimensional hydrodynamic water quality numerical simulation function and three-dimensional visualization function.

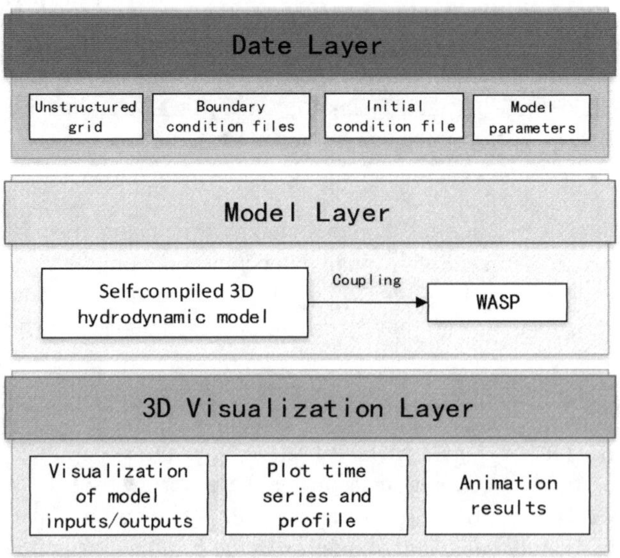

Figure 1. Frame diagram of system

3.1. Unstructured Grid Automatic Generation Function

In view of the problem that the existing structure grid can't deal with the complex flow field boundary and variable bottom elevation, this system uses Delaunay triangulation algorithm control to provide fast and automatic generation of unstructured grid based on GIS. This system provides simple mesh and fine mesh and terrain interpolation tools. After grid generation is completed, the system provides statistics and histogram drawing functions to facilitate users to analyze the quality of generated grid [6].

First, import the shapefile file of Lake Boundary and Lake Topography document. The next step is to set the grid generation parameters. Then the unstructured grid is generated automatically. Figure 2 presents mesh generator.

3.2. Three-dimensional Hydrodynamic Water Quality Numerical Simulation Function

Coupled three-dimensional hydrodynamic model and WASP water quality model, the hydrodynamic water quality model was established to simulate eutrophication of lakes.

Firstly, set the initial conditions in the initial conditions interface. Initial conditions include initial water temperature, initial concentration of water quality variable (chlorophyll water nitrogen and phosphorus concentration water temperature dissolved oxygen suspended matter transparency lake mud nitrogen and phosphorus) initial wind speed and so on. Figure 3 presents initial water quality variable concentration parameter setting interface.

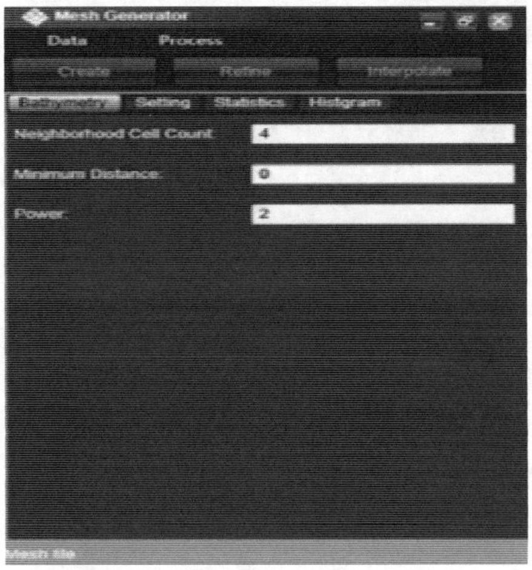

Figure 2. Mesh generator

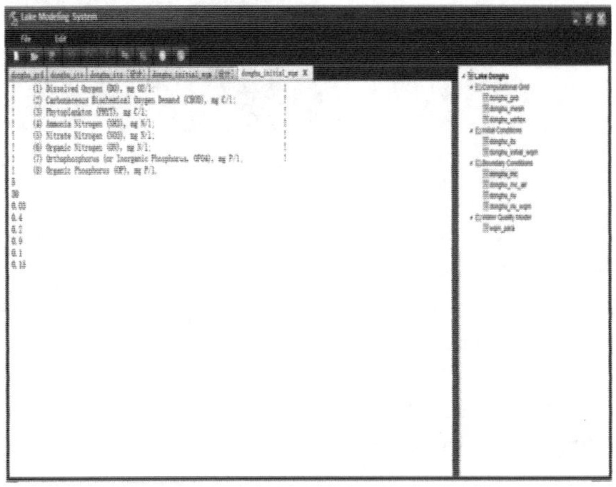

Figure 3. Initial water quality variable concentration parameter setting interface

Then set the boundary conditions in the boundary conditions interface, including atmospheric drive data, inflow and outflow boundary discharge, concentration of pollutants in the inflow and outflow and so on. Figure 4 presents setting interface of pollutant concentration data in Inflow and Outflow Rivers.

Finally, set the model parameters, including model time step, depth stratification coefficient, bottom roughness, turbulence model parameters, degradation parameters of pollutants, temperature coefficient and atmospheric reoxygenation parameters. Figure 5 presents water quality model parameter setting interface.

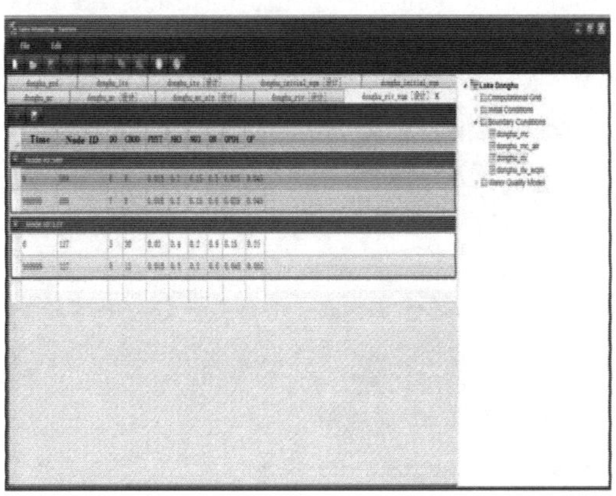

Figure 4. Setting interface of pollutant concentration data in Inflow and Outflow. Rivers

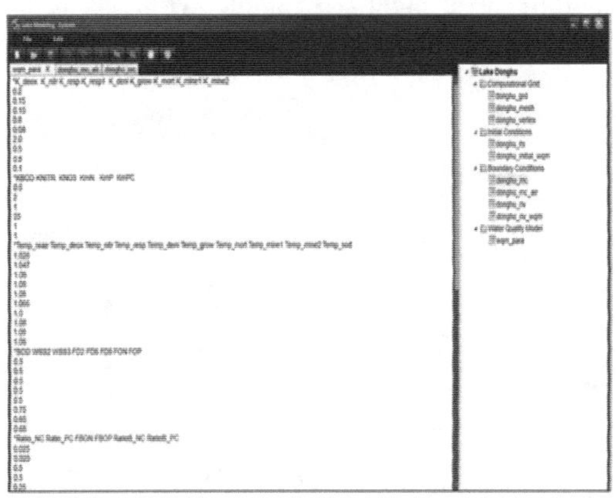

Figure 5. Water quality model parameter setting interface

After completing all settings, click run. The 3D-hydrodynamic water quality numerical model began to operate. The numerical simulation results are stored in the spatio-temporal database. Figure 6 presents model operation interface.

Figure 6. Model operation interface

3.3. *Three-dimensional Visualization Function*

Three-dimensional Visualization Function module shows the result of hydrodynamic water quality simulation of lakes, using digital elevation model and real projective processing, modeling and visualization of large-scale terrain simulation, dynamic 3D scene interaction, all kinds of control chart. Through the selection and editing of three-dimensional simulation elements, the three-dimensional velocity field, temperature field and water quality concentration field of lakes under different hydrological conditions are displayed.

4. Case

4.1. *Study Area*

Wuhan Donghu, located in the central city of Wuhan, covers an area of 33 square kilometers. It is a typical shallow inland lake, which is seriously eutrophication due to sewage generated by human activities around the lake, aquaculture, etc. This section takes Wuhan Donghu as the study area, and runs this system to simulate the water quality of the lake.

4.2. *Model Run*

First according to the remote sensing image of December 2018, extract the water body boundary of the Donghu applies water body extraction technology. According to the DEM data of Donghu, extract and obtain the terrain elevation data of lake. Then import the boundary file and elevation data file of Donghu into the system, adjust the parameters, and automatically generate the unstructured grid. Figure 7 is complex mesh of Donghu.

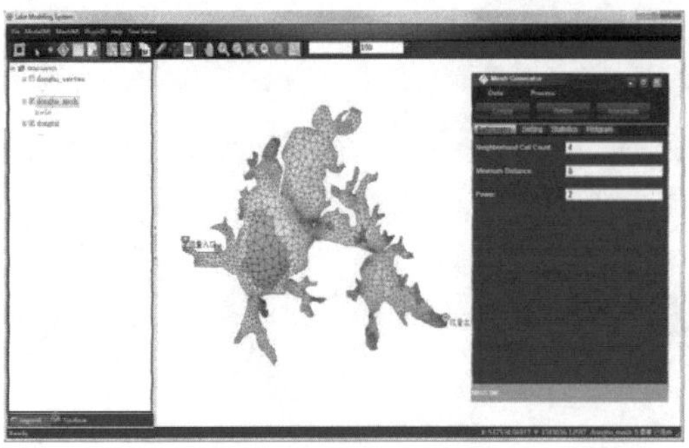

Figure 7. Complex grid

Then import the Donghu initial condition file and boundary condition file to determine the model parameters. When all the Settings were completed, the model was run to obtain dynamic variation characteristics of hydrodynamic parameters and water quality parameters (dissolved oxygen, ammonia, nitrogen, nitrate, nitrogen, organic nitrogen, inorganic phosphorus, organic phosphorus, carbon, biochemical oxygen demand, and chlorophyll a). Figure 8 is 3D visualization simulation results of CBOD of Donghu.

Figure 8. 3D visualization simulation results of CBOD of Donghu

5. Conclusion

In this paper, a three-dimensional hydrodynamic water quality simulation system is built based on GIS. Its main functions include (1) Providing the functions of pre-processing and post-processing of three-dimensional hydrodynamic water quality numerical calculation based on GIS control, and realizes the integrated management and comprehensive query of multi-source and heterogeneous data; (2) Providing a technique for rapid generation of unstructured grids; (3) Providing lake 3D visualization simulation tool based on 3D GIS, realizing the fine simulation and trend prediction of the dynamic process of the evolution of lake hydrodynamic water quality under different hydrological conditions.

The case study shows that the system can well simulate and display the dynamic process of water quality evolution.

6. Acknowledgement

This work was financially supported by the National Natural Science Foundation of China under Grant No. 51579108, the National Key R&D Program of China under Grant No.2017YFC0405900, the Fundamental Research Funds for the Central Universities under Grant No. 2017KFYXJJ203.

7. References

[1] Peng S, Fu G Y Z and Zhao X H *J. Zhejiang Univ. Sci.* Integration of USEPA WASP model in a GIS platformA, 11 (12) (2010), pp. 1015-1024

[2] Ng S M Y, Wai O W H, Li Y S, Li Z L and Jiang Y W 2009 *J Adv. Eng. Softw.* Integration of a GIS and a complex three-dimensional hydrodynamic, sediment and heavy metal transport numerical model, 40, 391–401.

[3] Huang MT and Tian Y 2019 J. ISPRS International Journal of Geo-Information, An Integrated Graphic Modeling System for Three-Dimensional Hydrodynamic and Water Quality Simulation in Lakes, 8(1), 18.

[4] Huang MT and Tian Y 2014 *J. CHINESE JOURNAL OF HYDRODYNAMICS(A)*, Three-dimensional lake hydrodynamic numerical modeling and its application to Lake Donghu ,29(01):114-124.

[5] Ambrose J R B, Wool T A, Connolly J P and Schanz R W. A Hydrodynamic and Water Quality Model: Model Theory, User's Manual and Programmer's Guide; U.S. Environmental Protection Agency: Athens, GA, USA, 1988; pp. 280–297.

[6] Shewchuk J R. Delaunay refinement algorithms for triangular mesh generation. Comput. Geom. 2002, 22, 21–74.

IOP Conf. Series: Materials Science and Engineering **730** (2020) 012068 doi:10.1088/1757-899X/730/1/012068

A Survey on Multi-Product Sustainable Economic Production Quantity Model Considering Wastewater Emission Costs

Yu-Hua Chiang Li[1,3], L L Sy[2,4], K C Lomibao[2,5] and J D German[2,6] H M Wee[1,3*]

[1]Department of Industrial and Systems Engineering, Chung Yuan Christian University, Chungli 3202, Taiwan, R.O.C.
[2]School of Industrial Engineering and Engineering Management, Mapúa University, Manila, Philippines.
[3]g10602409@cycu.edu.tw
[4]llsy@mymail.mapua.edu.ph & jdgerman@mapua.edu.ph
[5]kclomibao@mymail.mapua.edu.ph
*Email: weehm@cycu.edu.tw

Abstract. Implementing sustainability in production processes is the goal of many companies. A traditional economic production quantity (EPQ) model is used to determine the optimal number of products to produce in one period. Few researchers have considered carbon emission costs. In this study, we made a survey on multi-product sustainable economic production quantity model considering wastewater emission costs. The objective of this study is investigate the possible research to determine the economic production quantity to minimize the total costs per production cycle and waste emissions. The survey consider researches under three scenarios: (1) sustainable EPQ model that considers all CO2 emissions from inventory holding and production; (2) sustainable EPQ with shortage backorder; and (3) multi-product sustainable EPQ model with shortage.

1. Introduction

Wastewater emission and demand for fresh water continues to increase with industrialization. Wastewater emission is a known global problem, where 80% of water effluents from various industries is discharged into rivers, lakes, and oceans that are left untreated [1], eventually leading to eutrophication, human health risks and water scarcity. Efforts to address these problems are already done by various researchers, utilizing different process integration methods to reduce water emission in industrial plants. These include linear models [2], nonlinear models [3], pinch analysis [4], genetic algorithms [5], management strategies [6], benchmarking [7], and wastewater reuse, recovery, and treatment methods [8].

Harmful gaseous releases are also produced from manufacturing industries, including carbon dioxide (CO2), sulphur oxides (SOx), nitrogen oxides (NOx), and methane (CH4) emissions. These gases are emitted through fossil fuel combustion, production, transportation, and other procedures. A 2016 study [9] revealed that annual CO2 emission can go up to 11 million tons and is continually increasing while SOx and NOx emissions have increased during the years 1990 and 2016 respectively due to continuous fuel combustion [10]. In contrast, CH4 emission associated from landfills, coal mining, and distribution of natural gas and petroleum products have decreased. CH4, however, still impacts the environment on a global scale as a greenhouse gas (GHG) due to its high global warming potential compared to other GHGs. Similarly, numerous researches have been conducted to address

GHG emission problems, focusing on inventory management [11], fuel saving strategies [12], and optimizing transhipment routes [13].

The problems regarding emission of pollutants are still not solved although information and procedures on how to integrate processes considering environmental factors are readily available. This may be attributed to the substantial costs and expenses for the implementation of sustainable processes, a main barrier to achieve a cleaner production system [14]. Other researchers linked the problem to the Green Attitude-Behaviour Gap of consumers, a term defined by [15] as the comparison between favourable attitude of customers and their actual purchasing activities for green products. It noted that consumers are more concerned with product price than green product attributes. This means that implementing sustainability is a challenge to industries.

Numerous researchers have already done the extension of the EPQ model. Some studies considered different types of demand [16], incorporated shortage situations [17], and included deteriorating rates [18]. Researches on sustainable inventory models integrating environmental costs through direct accounting approach are also available since there is a fervent need for environmental sustainability management. These include the work of Taleizadeh et. al. [17], where an EPQ model is integrated to have carbon emission costs through direct accounting approach considering different shortage situations, and Zadfajar and Gholamian [2] who developed an inventory model that considers the emission of SOx, NOx, BOD, and COD from the manufacturing process of pulp and paper mills and its water effluents. The model was developed using exact non-linear constrained solution method. Another research [19] created a sustainable EPQ model considering carbon emission from production, warehousing, and waste disposal activities then extended it to a full backordering situation.

This paper aimed to provide a more realistic EPQ model that can be used to determine the right quantity to produce while minimizing the total costs per production cycle, taking into account of sustainable issues. Studying the EPQ model is more logical since water effluents and GHG emissions are more apparent during production processes. The sustainable EPQ model extends the study by [17] to consider a full backordering and multi-product situations. The model applies optimization methods to consider CO2 emission cost, wastewater costs and BOD/COD emission. Finally, the CH4 emission was incorporated from the waste disposal procedure.

As the world's need for environmental sustainability management grows, tax laws and emission costs are continuously imposed on various industries. However, companies still tend to exclude sustainability management on their system because of the costs and expenses that comes with its implementation and maintenance. The costs may be compromised by increasing the product price, but the consumers would be less willing to purchase the product. For these reasons, the manufacturers must reduce cost by optimizing an EPQ model. Production processes are the main sources of harmful emissions for a manufacturing system. In order to become more globally competitive, manufacturers would want to consider sustainability in their production system without affecting or minimizing their total cost.

2. Finding the Gap: Literature Review

2.1. Sustainable Inventory Models

The EOQ model, developed by [20] in 1913 help companies determine the optimal quantity to order. Two years after, the same author extended his work by presenting a model for finding the EPQ that involves only holding and set-up costs. Since then the importance of environmental issues, sustainable development, and concept of sustainability receives more attention than before [21]. As newer sustainable machines and treatment processes are invented for different types of industries, both producers and buyers are still not accustomed to applying sustainability in the production system because of the implementation costs and expenses [14]. By then, numerous researchers studied and developed sustainable inventory models to respond to environmental needs of the corporate world. In a recent study by [17], four (4) sustainable EPQ models were analysed and applied to different shortage situations. Using direct accounting approach, the researchers developed the sustainable EPQ (SEPQ) with carbon emission costs from holding, obsolescence, and production under different backordering situations. The study revealed that although these did not generate the least cost; an

SEPQ with partial model is the ideal and realistic model to use. Similarly, [19] developed an SEPQ model through optimization methods considering carbon emission costs from production, warehousing, and waste disposal activities for a full backordering situation. A broader perspective has been adopted by [22], who implemented a study to a single-vendor single-buyer situation accounting for carbon emission from transporting, warehousing, and inventory holding of deteriorating items. This provided insights on determining the optimal production quantity while minimizing total inventory and carbon emission costs. The work of [2] on the other hand, presented a revised EOQ model by adding income from waste sales as well as sulphur, nitrogen, BOD, and COD emission costs. The investigators compared all the developed models to the real data from a pulp and paper mill in Iran and found that through environmental ergonomics, both inventory costs and emissions can be minimized. Together these studies provided important insights to both the manufacturers and buyers in implementing system sustainability. This study will attempt to fill the gaps and add new considerations in the models developed from past researches.

2.2. Pulp and Paper Mill Industry

Numerous literatures on sustainability to various types of industries including the pulp and paper mill industry is evident over the years. This industry has been receiving widespread attention due to its high-energy consumption [23]. The pulp and paper mill industry generate different pollutants that depend on the type of pulping process [8]. Paper production includes three (3) major activities: pulp making, paper formation, and finishing. During the pulp making and paper formation process, coatings and fillers are added to the pulp that converts it into paper. This process requires an intensive resource and energy that results to huge emission of GHGs, water effluents, and other pollutants [24]. Various researches on sustainability methods include generating environment-friendly management strategies [6], treatment processes [8], and integration of inventory models [2].

2.3. Emissions during the Pulp and Paper Production

The emissions of pulp and paper processes are presented in Table 1. Wood handling, pulping, and chemical recovery units emit specific amounts of SOx in each process. It can also be observed that NOx emissions are present only in pulping and chemical recovery. Wastewater emission can be found in all units except in packaging while wastewater treatment unit produces refined water (BOD and COD) and dry sludge. The dry sludge is then disposed to landfills that produce methane. Evidently, air and water pollutions are present in the pulp and paper mill industry.

Table 1. Source of Emissions from a Pulp and Paper Mill Process

Considered Processes	NO_x	SO_x	BOD	COD	CO_2	CH_4
Wood Handling	•	•	•	•		
Pulping		•	•	•		
Chemical Recovery	•	•				
Pulp Washing			•	•		
Pulp Bleaching			•	•		
Paper Making			•	•		
Finished Paper Storage						
Wastewater Treatment			•	•		
Disposal of Solid Waste (Dry Sludge)						•
Whole Production Process (Electricity)					•	

3. Conclusion

This study focuses on literature review for sustainable Economic Production Quantity (EPQ) models. At proposal to extend the sustainable EPQ model is then is suggested to consider shortage and multi-product production system with warehouse capacity constraint. The main goal is to provide insights for researcher to minimize GH emission and the total cost per production cycle. If a manufacturer applies the proposed models to their current system, the company will become more sustainable and cost effective. The proposed sustainable EPQ models may provide significant insights to various manufacturing industries. Future research can focus on other objectives such as decreasing emission of hazardous wastes during production and broadening the models to consider simultaneously the perspectives of the manufacturer and the buyer.

4. References

[1] International Water Association, I. W. A. (2018). "The Reuse Opportunity." Wastewater Report

[2] Zadjafar, M. A. and Gholamian, M. R. (2018). "A sustainable inventory model by considering environmental ergonomics and environmental pollution, case study: Pulp and paper mills." Journal of Cleaner Production 199: 444-458.

[3] Oliveira-Esquerre, K. P., Seborg, D. E., Mori, M., & Bruns, R. E. (2004). "Application of steady-state and dynamic modeling for the prediction of the BOD of an aerated lagoon at a pulp and paper mill: Part II. Nonlinear approaches." Chemical Engineering Journal 105(1-2): 61-69.

[4] Jacob, J., Kaipe, H., Couderc, F., & Paris, J. (2002). "Water network analysis in pulp and paper processes by pinch and linear programming techniques." Chemical Engineering Communications 189(2): 184-206.

[5] Shafiei, S., Domenech, S., Koteles, R., & Paris, J. (2004). "System closure in pulp and paper mills: network analysis by genetic algorithm." Journal of Cleaner Production 12(2): 131-135.

[6] Helin, K., Käki, A., Zakeri, B., Lahdelma, R., & Syri, S. (2017). "Economic potential of industrial demand side management in pulp and paper industry." Energy 141: 1681-1694.

[7] Rogers, J. G., Cooper, S. J., & Norman, J. B. (2018). "Uses of industrial energy benchmarking with reference to the pulp and paper industries." Renewable and Sustainable Energy Reviews 95: 23-37.

[8] Pokhrel, D. and T. Viraraghavan (2004). "Treatment of pulp and paper mill wastewater—a review." Science of The Total Environment 333(1): 37-58.

[9] Ritchie, H. R., M. (2018) "CO_2 and other Greenhouse Gas Emissions."

[10] United States Environmental Protection Agency, E. P. A. (2018). "Inventory of U.S. Greenhouse Gas Emissions and Sinks"

[11] Tang, S., Wang, W., Cho, S., & Yan, H. (2018). "Reducing emissions in transportation and inventory management:(R, Q) Policy with considerations of carbon reduction." European Journal of Operational Research 269(1): 327-340.

[12] Seddiek, I. S. and M. M. Elgohary (2014). "Eco-friendly selection of ship emissions reduction strategies with emphasis on SOx and NOx emissions." International Journal of Naval Architecture and Ocean Engineering 6(3): 737-748.

[13] Mirzapour Al-e-hashem, S. M., Rekik, Y., & Hoseinhajlou, E. M. (2017). "A hybrid L-shaped method to solve a bi-objective stochastic transshipment-enabled inventory routing problem." International Journal of Production Economics.

[14] Shi, H., Peng, S. Z., Liu, Y., & Zhong, P. (2008)." Barriers to the implementation of cleaner production in Chinese SMEs: government, industry and expert stakeholders' perspectives." Journal of cleaner production 16(7): 842-852.

[15] Joshi, Y., and Rahman, Z. (2015). "Factors affecting green purchase behaviour and future research directions." International Strategic management review 3(1-2): 128-143.

[16] Pal, B., Sana, S. S., & Chaudhuri, K. (2013). "A mathematical model on EPQ for stochastic demand in an imperfect production system." Journal of Manufacturing Systems 32(1): 260-270.

[17] Taleizadeh, A. A., Soleymanfar, V. R., & Govindan, K. (2018). "Sustainable economic production quantity models for inventory systems with shortage." Journal of Cleaner Production 174: 1011-1020.

[18] Wee, H.-M. and Wang, W.-T. (2012). "A supplement to the EPQ with partial backordering and phase-dependent backordering rate." Omega 40(3): 264-266.

[19] Daryanto, Y. and Wee, H.-M. (2018). "Sustainable Economic Production Quantity Models: An Approach toward a Cleaner Production." ICAMM. Tokyo.

[20] Harris, F.W. (1915). What Quantity to Make at Once: In: The Library of Factory Management, Vol. V. Operation and Costs. Chicago: A. W. Shaw Company, 47-52.
[21] Shankar, K. M., Kannan, D., & Kumar, P. U. (2017). "Analyzing sustainable manufacturing practices–A case study in Indian context." Journal of Cleaner Production 164: 1332-1343.
[22] Tiwari, S., Daryanto, Y., & Wee, H. M. (2018). "Sustainable inventory management with deteriorating and imperfect quality items considering carbon emission." Journal of Cleaner Production 192: 281-292.
[23] Shan, Y., Liu, Z., & Guan, D. (2016). "CO2 emissions from China's lime industry." Applied energy 166: 245-252.
[24] Sun, M., Wang, Y., Shi, L., & Klemeš, J. J. (2018). "Uncovering energy use, carbon emissions and environmental burdens of pulp and paper industry: A systematic review and meta-analysis." Renewable and Sustainable Energy Reviews 92: 823-833.
[25] Cárdenas-Barrón, L. E. (2001). "The economic production quantity (EPQ) with shortage derived algebraically." International Journal of Production Economics 70(3): 289-292.
[26] Wee, H. M., Huang, Y. D., Wang, W. T., & Cheng, Y. L. (2014). "An EPQ model with partial backorders considering two backordering costs." Applied Mathematics and Computation 232: 898-907.
[27] Chiu, S. W., Tseng, C. T., Wu, M. F., & Sung, P. C. (2014). "Multi-item EPQ model with scrap, rework and multi-delivery using common cycle policy." Journal of applied research and technology 12(3): 615-622.
[28] Nobil, A. H., Sedigh, A. H. A., & Cárdenas-Barrón, L. E. (2016). "A multi-machine multi-product EPQ problem for an imperfect manufacturing system considering utilization and allocation decisions." Expert Systems with Applications 56: 310-319.
[29] GAPS Guidelines. (2015). Pulp and Paper Manufacturing.
[30] Taft, E. W. (1918). "The most economical production lot." Iron Age 101: 1410-1412.
[31] Pasandideh, S. H. R., Niaki, S. T. A., Nobil, A. H., & Cárdenas-Barrón, L. E. (2015). "A multiproduct single machine economic production quantity model for an imperfect production system under warehouse construction cost." International Journal of Production Economics 169: 203-214.
[32] Kuhn, H. and Tucker, A. (1951). "Nonlinear Programming." Proceedings of the Berkeley Symposium on Mathematical Statistics and Probability. 481-492.
[33] Ghosh, S.K., Sarkar, T., and Chaudhuri, K. (2015). "A Multi-Item Inventory Model for Deteriorating Items in Limited Storage Space with Stock-Dependent Demand". American Journal of Mathematical and Management Sciences 34: 147–161.

IOP Conf. Series: Materials Science and Engineering 730 (2020) 012069 doi:10.1088/1757-899X/730/1/012069

Jason-1 Observation of Typhoon Storm Surge and Analysis Based on Numerical Simulation

Zhou Qiu[1], Gang Zheng[1], Jingsong Yang[1], Guoqi Han[2], Xiaohui Li[3], Chuanzheng Zhang[1], Lizhang Zhou[1], and Yan Wang[1]

[1]State Key Laboratory of Satellite Ocean Environment Dynamics, Second Institute of Oceanography, Ministry of Natural Resources, Hangzhou 310012, China.
[2]Fisheries and Oceans Canada, Northwest Atlantic Fisheries Centre, St. John's, NL A1C 5X1, Canada.
[3]School of Electronic and Information Engineering, Beihang University, Beijing 100191, China.
*Email: zhou_qiu@outlook.com

Abstract. Storm surge is one of the most disastrous marine disasters to coastal regions. It is greatly essential to research on storm surge and mitigate this damage. In our paper, Typhoon hit the Japan Coast on 31 July, 2004. Jason-1 and two tide-gauge stations all captured the obvious sea level anomalies. The magnitude of this storm surge was approximately 0.4 m. We further used the Unstructured Grid Finite-Volume Community Ocean Model (FVCOM) to simulate this storm surge. The method based on the relationships between the maximum wind speed of best track and that of CFSR data was adopted to reconstruct wind field. This method reduced the root mean squared error (RMSE) from 0.11 m to 0.07 m at Owase station. The accuracy was improved by 35%. Similar to Owase station, the RMSE was reduced from 0.085 m to 0.06 m at Cape Muroto station. The accuracy was improved by 30%.

1. Introduction

Satellite altimetry, a technique for measurement of sea surface height, has developed more than 40 years. The spatial resolution has revolutionarily improved from initial meter-level to centimeter-level accuracy today. Much more progress has been made in the ocean application. And satellite altimetry opens up some new possibilities in the study of mesoscale climate variability, coastal currents, ocean tide and terrestrial water storage variations.

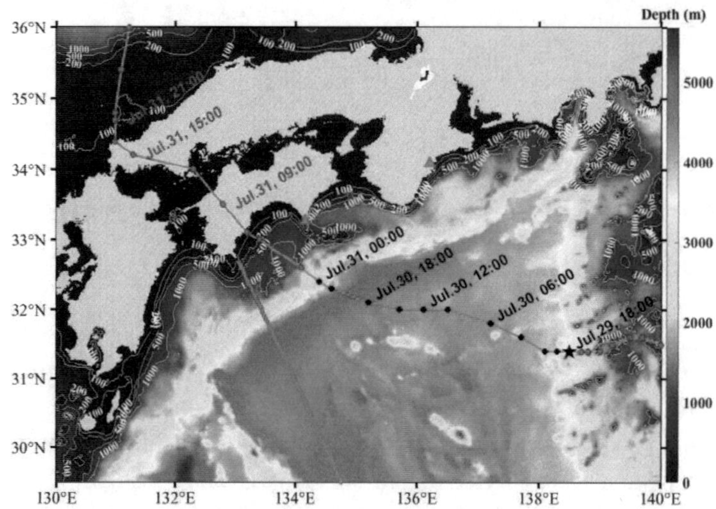

Figure 1. Map showing the study area in Japan with bathymetric data in meters. The green line with black dots represents the typhoon track at specific time, and the pink triangles denote the locations of tide-gauge stations. The pink line is the ground track of the Jason-1 altimeter.

However, the application of the altimeter on the storm surge is still in its infancy. In the past decade, some scientific researches show the ability of storm surge observation based on satellite altimetry. On 28 August 2005, the hurricane-induced storm surge was captured by the Geosat Follow-on. The results unveiled the prelude to the altimeter observation of the storm surge [1]. Lillibridge et al. pointed out that HY-2A satellite captured storm surges caused by hurricane Sandy off the coast of New York, and could provide the valuable cross-shelf structure of the storm surge, but the deeper analysis was not conducted [2]. Chen et al. combined the HY-2A satellite and tide-gauge station to further analyze the storm surge induced by hurricane Sandy. The results showed that the storm surge observed by the altimeter was similar to that observed by the tide station. Besides, the cross-shelf exponential decaying scale and the propagation speed were analyzed and discussed. The results indicated that altimeter could reveal some characteristics of the storm surge well to supplement the deficiency of tide-gauge station [3]. Moreover, the storm surge off China Coast was captured by TOPEX/Poseidon altimeter. The accuracy of the model was corrected by the mean of the altimeter and achieved good results unexpectedly. The results showed that the altimeter data had capacity to improve the accuracy of numerical model [4].

In this study, Typhoon Namtheun hit southwestern Japan on 31 July, 2004 and flooded homes and businesses. Before the landfall of Typhoon Namtheun, the Jason-1 flew over the Japan Coast at 17:00 UTC on 29 July (Figure 1) and captured an obvious sea level anomaly (SLA). The rest of this paper is organized as follows. In Section 2, altimeter data, tide-gauge data and numerical model setup are described. In section 3, the SLAs based on altimeter data, tide-gauge data and numerical simulation are further discussed. A conclusion is presented in Section 4.

2. Data and Methods

2.1. Jason-1 Data

The Jason-1 satellite was launched on 7 December, 2001. This satellite is mainly used for measurement of sea level height, significant wave height and wind speed. The cycle is about 10 days. To improve the accuracy of coastal altimeter data, the Center for Topographic studies of the Ocean and Hydrosphere (CTOH) in France reprocessed the Geophysical Data Record products by using the X-TRACK processor. Thus, Jason-1 track 112 SLAs of cycle 94 in this paper were obtained by CTOH [5].

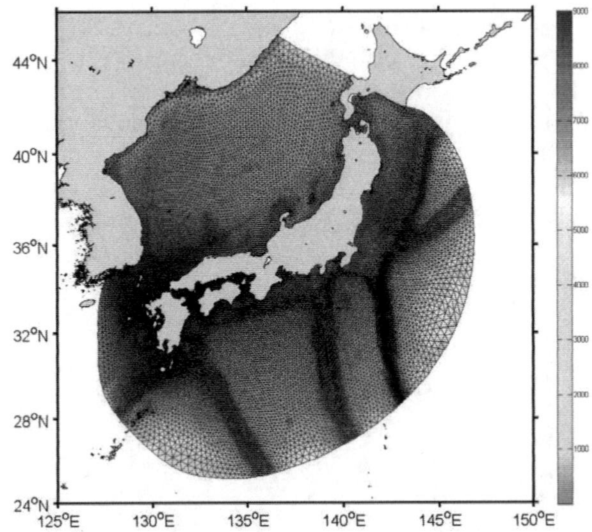

Figure 2. The research area was discretized with 52036 grids and 26829 nodes. The water depth distribution was from the Etopo1 dataset.

2.2. Tide-Gauge Data

Hourly tide-gauge data at Cape Muroto station and Owase station were obtained from the Japan Meteorological Agency (JMA, http://www.data.jma.go.jp/gmd/kaiyou/db/tide/genbo/index.php). The data period is from 28 July to 06 August.

2.3. Numerical Model Setup

In our study, the Unstructured Grid Finite-Volume Community Ocean Model (FVCOM) was used to simulate the storm surge caused by Typhoon Namtheun. The details of FVCOM can be found in [6]. Figure 2. shows that the grid of the research area, and the grid was generated by the Surface-water Modeling Solution (SMS). The maximum grid size is about 20 km, and the minimum grid size near Shikoku, Japan is about 1 km.

At the open boundary, the model was driven by the tide height data generated by the Oregon State University (OSU) Tidal Inversion Software (OTIS). At the sea surface, the model was forced by the sea surface wind and pressure data obtained from the National Centers for Environmental Prediction (NCEP) Climate Forecast System Reanalysis (CFSR) 6-hourly products. During the Typhoon Namtheun, two numerical experiments were conducted. One still was forced by the CFSR wind at the sea surface, another was forced by the reconstructed wind field that we discussed in Section 2.3.

2.4. Wind Field Reconstruction

The accuracy of the wind force is essential for the numerical simulation. Generally, the CFSR wind data underestimate the maximum wind speed of the typhoon. Based on the relationship between the maximum wind speed of best track and that of CFSR, the approach in [4] was adopted as follows:

$$V_w = \left(\frac{r}{R} \times ntimes + \frac{R-r}{R}\right) \times V_{ncep}, \qquad 0 < r \leq R \#(1)$$

$$V_w = \left(\frac{4 \times R - r}{3 \times R} \times ntimes + \frac{r-R}{3 \times R}\right) \times V_{ncep}, \qquad R < r \leq 4 \times R \#(2)$$

Where V_w is the reconstructed wind speed, r is radial distance from the typhoon center, R is the radius of the maximum wind, *ntimes* is a parameter, and V_{ncep} is background wind speed.

3. Results and Discussion

During the Typhoon Namtheum event, Jason-1 captured the obvious SLAs. We can see that the SLAs had the significant increase from 29.7° toward the coast. In order to explore the coastal SLAs induced by Typhoon Namtheun, the tide-gauge results were represented by blue lines in Figure 4. Obviously, the SLAs around 31 July had the significant increase at Owase station and Cape Muroto station. The maximum magnitude of SLAs recorded by Owase station was approximately 0.41 m, and the maximum magnitude of SLAs recorded by Cape Muroto was approximately 0.34 m. Note that the SLAs of two stations both have residual oscillation after Typhoon Namtheun passing.

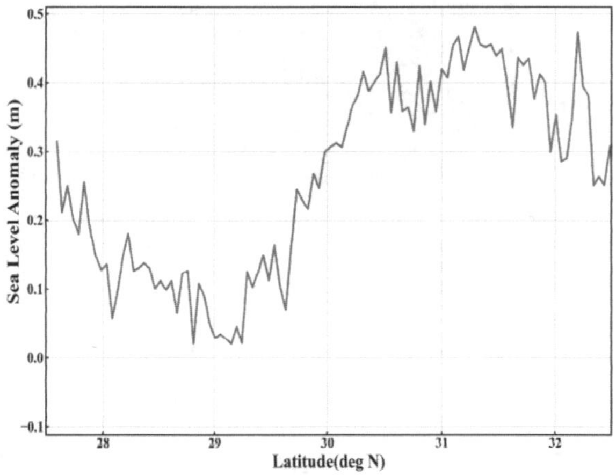

Figure 3. Map showing the sea level anomalies of Jason-1 track 112 SLAs of cycle 94.

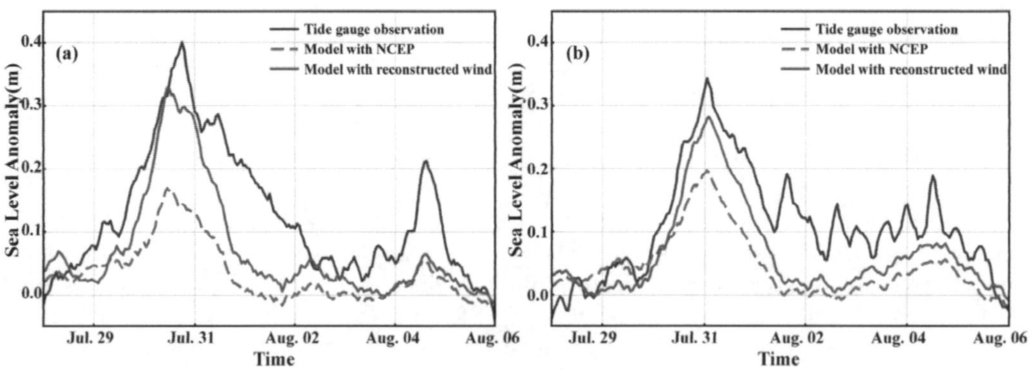

Figure 4. Comparison of simulated and observed SLAs. (a) Owase station, (b) Cape Muroto station.

This storm surge was simulated with above CFSR data and reconstructed wind data, respectively. The simulation results are shown in Figure 4. At Owase station, we can see that both peak times were well simulated. However, the peak value of the simulation forced by CFSR wind data was not ideal, the root mean squared error (RMSE) was 0.11 m. The simulation results forced by reconstructed wind data were better, and the RMSE was 0.07 m. The latter accuracy was improved by about 35%. At Cape Muroto station, the simulated SLAs had the similar results. The RMSE of the simulation forced by CFSR wind data was approximately 0.085 m and that of the simulation forced by reconstructed wind data was approximately 0.06 m. The accuracy was improved by about 30%. However, the simulated residual oscillations of these two approaches were both not ideal. Further research about simulation of residual oscillation is needed in the further. In conclusion, the method to reconstruct the wind field in our study is effective.

4. Conclusion

During the Typhoon Namtheun event, the obvious surges were captured by Jason-1 and tide-gauge stations. The magnitude was approximately 0.4 m. This result suggests that altimeter products can be applied to monitor storm surge phenomena. The two ocean dynamical numerical simulations were further conducted to analyzed this storm surge. The results show that the simulated SLAs forced by reconstructed wind (based on the relationships between the maximum wind speed of best track and that of CFSR data) were better than that forced by CFSR wind data. The RMSE was reduced from 0.11 m to 0.07 m at Owase station, and the RMSE was reduced from 0.085 m to 0.06 m at Cape Muroto station. The accuracy can be improved by 30%-35%.

5. Acknowledgement

This work was supported by the Project of State Key Laboratory of Satellite Ocean Environment Dynamics, Second Institute of Oceanography (No. SOEDZZ1804), the National Key R&D Program of China (No. 2016YFC1401003), the National Natural Science Foundation of China (No. 41676167 and No. 41776183), and the Scientific Research Fund of the Second Institute of Oceanography, Ministry of Natural Resources (No. JT1703). The altimeter data were provided by the Center for Topographic studies of the Ocean and Hydrosphere, France. The tide-gauge data were provided by Japan Meteorological Agency. The depth data were from the National Centers for Environmental Information, NOAA, USA. The typhoon best track data were from the China Meteorological Administration.

6. References

[1] R. Scharroo, W. H. F. Smith, and J. L. Lillibridge, "Satellite altimetry and the intensification of Hurricane Katrina," *Eos Trans. Am. Geophys. Union,* vol. 86, no. 40, pp. 366-366, 2013.

[2] J. Lillibridge, M. Lin, and C. Shum, "Hurricane Sandy storm surge measured by satellite altimetry," *Oceanography,* vol. 26, no. 2, pp. 8-9, 2013.

[3] N. Chen, G. Han, J. Yang, and D. Chen, "Hurricane Sandy storm surges observed by HY-2A satellite altimetry and tide gauges," *J. Geophys. Res. Oceans,* vol. 119, no. 7, pp. 4542-4548, 2014.

[4] X. Li, G. Han, J. Yang, D. Chen, G. Zheng, and N. Chen, "Using Satellite Altimetry to Calibrate the Simulation of Typhoon Seth Storm Surge off Southeast China," *Remote Sens.,* vol. 10, no. 4, 2018.

[5] F. Birol *et al.*, "Coastal applications from nadir altimetry: Example of the X-TRACK regional products," *Adv. Space Res.,* vol. 59, no. 4, pp. 936-953, 2017.

[6] C. Chen, R. Beardsley, and G. Cowles, "An Unstructured Grid, Finite-Volume Coastal Ocean Model (FVCOM) System," *Oceanography,* vol. 19, no. 1, pp. 78–89, 2006.

Drought Crisis: A Path-analysis in Beijing-Tianjin-Hebei Region

Wenhui Qiao [a], Yanru Pu [a], Zhuo Wei [a], Peng Wang [a,*]

[a]Faculty of Civil Engineering and Mechanics, Jiangsu University, Zhenjiang 212013, China.

*Email: upeswp@ujs.edu.cn

Abstract. Urban drought has seriously hindered the sustainable development. Reducing urban drought vulnerability (UDV) entails more effective and systematic strategies. We discussed the ways to reduce UDV for government in the Beijing-Tianjin-Hebei (BTH) region from building vulnerability assessment model suitable for urban system, correlation analysis and path analysis. Results indicated that (1) the drought vulnerability of 13 cities in the BTH region fluctuated continuously from 1987 to 2016; (2) the factors affecting UDV are systemic and have the regional characteristic; and (3) the path analysis provide the specific directions for government. This study provides a theoretical basis for the government to manage drought risk.

1. Introduction

Drought is a serious natural disaster characterized by water deficit, which hinders sustainable urban development. Global economic losses caused by droughts are estimated at US$ 6–8 billion per year, which is significantly more than that from other meteorological disasters (Desbureaux and Rodella, 2019; Smith and Katz, 2013). As the shelter of population and various economic activities, cities are especially vulnerable to drought disasters. Urban drought has become an extremely aggressive natural phenomenon that has direct and indirect impacts on a city's wellbeing, including the lives of residents and the development of economic activities (Zhang et al., 2019).

Many scholars had noticed the urgency of the drought and tied to mitigate urban drought. Studies in urban drought focused on urban water management and water scarcity. For "water management", these studies are qualitatively described based on interviews and workshops with the inhabitants using social practices (Sletto et al., 2019; Workalemahu et al., 2019). Herslund and Mguni (2019) examined the water management practices of households based on interviews and workshops with the inhabitants at two case sites. For "water scarcity", the modeling, prediction, and factor analysis of water scarcity have been extensively discussed (Clercq et al., 2018; Mashhadi Ali et al., 2017). Sharvelle et al. (2017) developed and demonstrated an Integrated Urban Water Model (IUWM) with the input data: land cover, imperviousness, population, households, daily climate and so on. However, previous studies were limited in system and comprehensiveness for research perspective.

The concept of vulnerability would give us a new perspective on drought risk reduction. Current studies have emphasized the transition from "crisis management" to "risk management" for disasters mitigation (Sharvelle et al., 2017). For drought vulnerability, many studies had provided various study frameworks and indicators. However, they are mostly concentrated in agriculture drought vulnerability (Hannaford, 2018). Rainfall variability, terrain, agricultural share of GDP, illiteracy rate and infrastructure are widely used in agricultural drought vulnerability (Li et al., 2016; Shen et al., 2019). It must be admitted that drought vulnerability studies at the urban scale are more referential and accurate. And based on literature review, we propose that urban drought vulnerability refers to the extent to which the balance of urban water demand and supply is disrupted by drought and the city is unable to cope with. Due to the lack of relevant studies, most cities are poor in the basic knowledge of mitigating the adverse effects and risks of drought, and even lack the adaptive capacity to urban drought.

Therefore, this study is to provide the government with specific knowledge to manage the drought and reduce the adverse of urban drought. Based on this purpose, this study discusses urban drought vulnerability from the following aspects: 1) calculation of urban drought vulnerability through the research framework suitable for urban systems, 2) correlation analysis between indicators, and 3) path analysis based on structural equation model. This study would provide specific theoretical basis to understand how to reduce urban drought vulnerability and ease negative effects of urban drought.

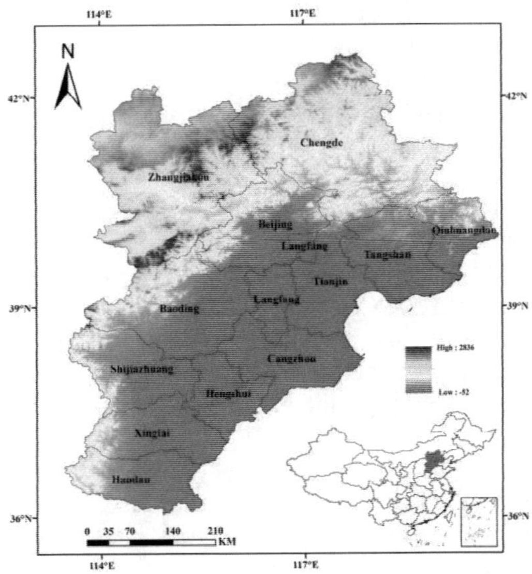

Figure 1. Hypsometric map of the Beijing-Tianjin-Hebei (BTH) region

2. Study Area

The Beijing-Tianjin-Hebei (BTH) region (36°01'N - 42°37'N, 113°04'E - 119°53'E, Fig.1) is formed by two municipalities, Beijing and Tianjin, and 11 cities of the Hebei Province. The BTH region is the driving force of China's economic development and policy implementation. The BTH region has an area of 218,000 m^2. In 2018, the GDP of the BTH region reached 8.5 trillion yuan. Since the implementation of Beijing-Tianjin-Hebei integration policy in 2014, the economy of BTH has developed rapidly and life services have become more convenient.

However, the BTH region is faced with the serious problem of drought. BTH region are more severe drought event happened in 1992, 1993, 1997, 1999, 2000, 2014. In addition, the water resources have been overexploited due to the increasing demand for water by economic and population growth. For example, the utilization of shallow groundwater is 130%.

3. Methodology

3.1 Urban Drought Vulnerability Model

3.1.1. Index selection

According to IPCC, we considered vulnerability as a function of sensitivity, adaptive capacity and exposure. In order to select indicators legitimately, we reviewed lots of studies related to water management, water shortage and agricultural drought vulnerability. Based on literature review and regional characteristics, we selected 16 indicators suitable for the urban system, shown in Table X. Particularly, we calculated drought severity (DS) using SPI index. We chose a 12-month SPI index, which is usually tied to streamflows, reservoir levels, and even groundwater levels at longer timescales.

Table 1. Indicators of urban drought vulnerability.

Indicator	Category
Population density	Exposure
Built-up area	Exposure
Drought severity	Exposure
High temperature days	Exposure
Per GDP water consumption growth rate	Sensitivity
Unemployment rate	Sensitivity
Area of paved roads	Sensitivity
Gross of domestic product of secondary and tertiary industries	Sensitivity
Fixed investments	Sensitivity
Passenger traffic	Sensitivity
Coverage rate of afforestation in developed area	Adaptive capacity
Number of beds in health care institutions	Adaptive capacity
Production capacity of urban tap water	Adaptive capacity
Per capita disposable income	Adaptive capacity
Number of students enrolled in institutions of higher education	Adaptive capacity
Public Budget Expenditure	Adaptive capacity

3.1.2. Calculation of weighting

The weight of indicators can be divided into objective and subjective weighting. Subjective weighting depends on subjective factors, and the amount of qualitative data is larger than that of quantitative data. This study selected principal component analysis (PCA) to define weighting.

Weight is defined according to the rotated factor load matrix: first, square factor load is calculated; Then, the weighted factor internal load is calculated by dividing the square factor load by the variance proportion of each factor interpretation. Then, the weighted load between factors is obtained by dividing the weighted load within factors by the proportion of variance of each factor interpretation to the total accumulated variance of each factor. Finally, determine the final weight.

Exposure, sensitivity and adaptive capacity were calculated using the following formula:

$$E \ or \ S \ or \ A = \sum_{i=1}^{n} y_{ij} w_j \quad (1)$$

Following methods from IPCC, urban drought vulnerability (UDV) was calculated using the following formula:

$$UDV = E + S - A \quad (2)$$

3.2. Analysis of Urban Drought Vulnerability

Correlation analysis (based on weighting), and path analysis (based on SEM) was conductive for basic understanding and subsequent in-depth analysis.

The Pearson correlation coefficient algorithm is chosen to realize feature selection through dimensionality reduction, which has many advantages, such as high efficiency, accurate calculation, strong practicability (Peng et al., 2018). The correlation coefficient ris used to measure the correlation between variable x and y, and the value range of r is$[-1,1]$; we define:

$$r = \frac{\sum_{i=1}^{n}(x_i - \bar{x})(y_i - \bar{y})}{\sqrt{\sum_{i=1}^{n}(x_i - \bar{x})^2}\sqrt{\sum_{i=1}^{n}(y_i - \bar{y})^2}} \quad (3)$$

IOP Conf. Series: Materials Science and Engineering **730** (2020) 012070 doi:10.1088/1757-899X/730/1/012070

Where $\bar{x}$ and $\bar{y}$ represented the mean of data set x and y, respectively. When $r > 0$, it indicates that variable x and variable y are positively correlated, and when $r < 0$, it indicates that variable x and variable y are negatively correlated. The larger $|r|$ is, the more significant relationship between x and y. Generally, the variable correlation degrees can be measured as in Table 2.

Table 2. Correlation coefficients

Ranges	Descriptions
$0.8 < r \leq 1$	Extremely correlated
$0.6 < r \leq 0.8$	Strong correlation
$0.4 < r \leq 0.6$	Medium correlation
$0.2 < r \leq 0.4$	Weak correlation
$0 < r \leq 0.2$	Hardly correlated

4. Results and Discussion

4.1. UDV Scores of 13 Cities

The results of weighting of exposure, sensitivity and adaptive capacity were shown in Fig.2. Sensitivity and adaptive capacity are the dominant factors contributing to UDV in BTH region. Both dimensions accounted for more than 75% in all cities. And seven cities (Tianjin, Shijiazhuang, etc.) account for more than 80% of drought vulnerability. Among them, Shijiazhuang was the largest, at 84.33%. Thus, the decrease of sensitivity is as important as the improvement of adaptive capacity. This implied that decrease sensitivity and improving adaptive capacity play important roles in reducing urban drought vulnerability.

Based on principal component analysis, we calculated the UDV scores of 13 cities from 1887 to 2016 (Fig.3). The drought vulnerability of 13 cities was constantly fluctuating and irregular. This will present a great challenge for the government to manage drought risk.

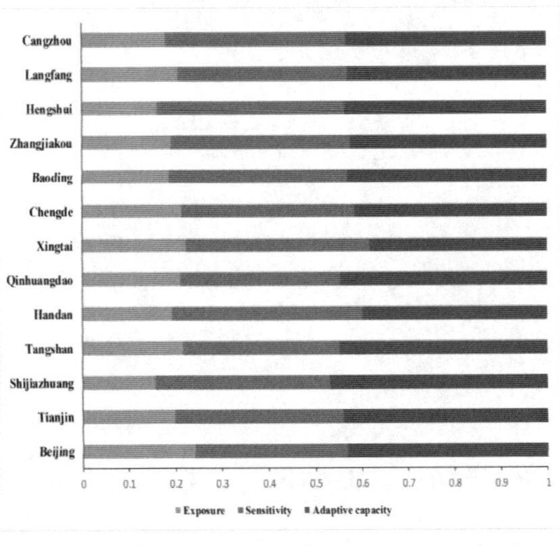

Figure 2. Contribution of three dimensions to UDV

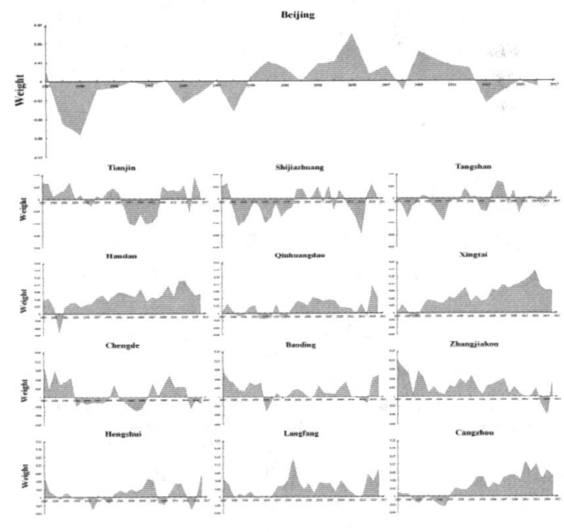

Figure 3. UDV trend graph for 13 cities

4.2. Correlation Analysis

Correlation analysis is an important method to explore how to reduce urban drought vulnerability. Based on the correlation analysis of 16 indicators in each city, we found that the indicators were significantly correlated in generally. This indicates that the factors affecting UDV are systematic. It is understandable that cities are complex systems, which consist of many sub-systems, such as socio-economic sub-systems, infrastructure sub-systems, and so on. This is where the difficulty and the key to reducing vulnerability to urban drought. What' more, the indicators of 13 cities in semi-arid areas showed different correlation results in detail. This suggests the regionalism characteristics of UDV. Thus, it is necessary to reduce drought vulnerability from a system perspective. The formulation of policies should be based on the characteristics of each city.

Based on the finding, we further analyzed the correlation between policy indicators (public budget expenditure) and other indicators, as shown in Fig 4. There is a significant correlation between policy indicators and most indicators. This indicates the relationship between policy indicators and other indicators. Revealing their internal links would give governments a clear direction in managing drought risk.

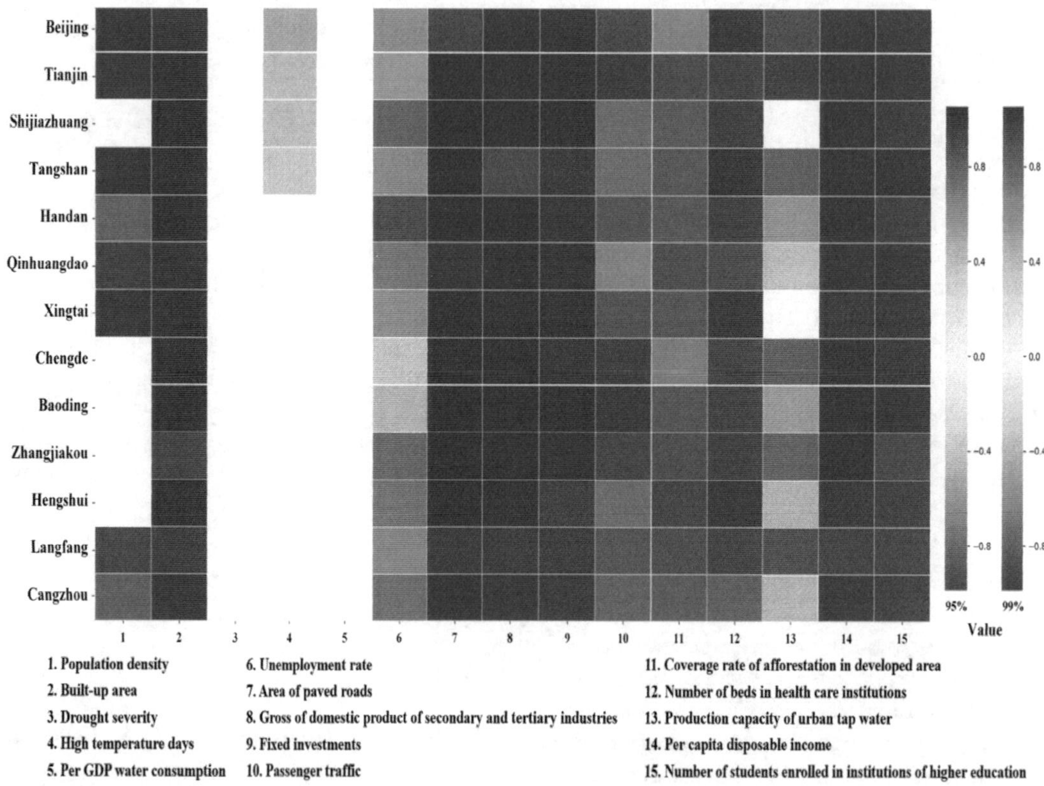

Figure 4. Correlation between "public budget expenditure" and the other 15 indicators in 13 cities

4.3. Path Analysis

Structural Equation Modeling, which is widely used in travel behavior analysis, can reflect the relationship analysis among variables. Due to the large number of cities, we selected Beijing as the main research object. Beijing is the core of China's socio-economic development. Climate change, rapid population growth and vigorous economic development have led a great pressure of urban drought to Beijing. From the above analysis results, it can be found that the drought vulnerability of Beijing is extremely unstable during 1987-2016. And high sensitivity is dominant reason contributed to urban drought vulnerability. Based on the correlation analysis, we explored the relationship between public budget expenditure and 13 other indicators. According to China statistical methods and dimensions, public budget expenditure indicator involves multifarious elements: general public service, education, science, technology, culture, media, social security, employment, healthcare, transportation, and so on. Fig 5 showed the constructed SEM for urban drought vulnerability, in which "indicator" in this figure come from the 13 indicators in Table 3. And Table 3 listed the results of effects of 13 indicators on drought vulnerability reduction when policies were proposed and effectively implemented. More clearly, other things being equal, for every unit increase in the policy related to per capita disposable, the UDV will be reduced by 0.95 units. From the results, we found that the policies are most effective in reducing "gross of domestic product of secondary and tertiary industries". However, it is admitted the difficulty to achieve in the process of urbanization. Therefore, for some indicators that are prone to change, such as medical infrastructure (0.9) and higher education (0.77), the government should formulate targeted policies for them.

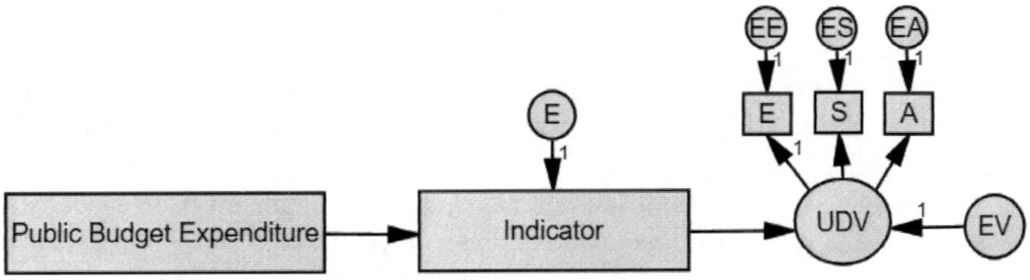

Figure 5. SEM of UDV reduction

Table 3. Results of paths and effects

Path	Effects
Public Budget Expenditure - >Population density - > UDV	-0.91
Public Budget Expenditure - >Built-up area - > UDV	-0.76
Public Budget Expenditure - >Area of paved roads - > UDV	-0.64
Public Budget Expenditure - >Gross of domestic product of secondary and tertiary industries - > UDV	-0.99
Public Budget Expenditure - >Fixed investments - > UDV	-0.95
Public Budget Expenditure - >Passenger traffic - > UDV	-0.65
Public Budget Expenditure - >Coverage rate of afforestation in developed area - > UDV	-0.57
Public Budget Expenditure - >Number of beds in health care institutions - > UDV	-0.90
Public Budget Expenditure - >Production capacity of urban tap water - > UDV	-0.77
Public Budget Expenditure - >Per capita disposable income - > UDV	-0.95
Public Budget Expenditure - >Number of students enrolled in institutions of higher education - > UDV	-0.77

5. Conclusions

This study assessed urban drought vulnerability and explored the pathways that reduce urban drought vulnerability from a system perspective.

The weighting results showed that 13 cities in BTH region have the characteristics of high sensitivity and high adaptability. The UDV score indicates that the vulnerability of 13 cities in the BTH region to drought fluctuated continuously from 1987 to 2016. This will present a great challenge for the government to manage the risk of drought.

Correlation analysis gave us a new perspective to reduce UDV. The 16 indicators of 13 cities are significantly correlated in generally, but the correlation varies among cities. We observed that the factors affecting UDV are systematic interactions and have regional characteristics.

For providing the government with a clear direction to reduce drought vulnerability, based on correlation results and path analysis, we found that Beijing should focus on the adjustable factors. Among them, the improvement of per capita disposable income and the improvement of medical infrastructure are relatively important directions.

6. Acknowledgement

This work was supported by the National Natural Science Foundation of China Grant No. 51908249], the Natural Science Foundation of the Jiangsu Higher Education Institutions of China Grant No. 19KIB560012] and the High-level Scientific Research Foundation for the introduction of talent for Jiangsu University [Grant No. 18JDG038].

7. Reference

[1] Clercq, D. De, Smith, K., Chou, B., Gonzalez, A., Kothapalle, R., Li, C., Dong, X., Liu, S., Wen, Z., 2018. Identi fi cation of urban drinking water supply patterns across 627 cities in China based on supervised and unsupervised statistical learning. J. Environ. Manage. 223, 658–667. https://doi.org/10.1016/j.jenvman.2018.06.073

[2] Desbureaux, S., Rodella, A.S., 2019. Drought in the city: The economic impact of water scarcity in Latin American metropolitan areas. World Dev. 114, 13–27. https://doi.org/10.1016/j.worlddev.2018.09.026

[3] Hannaford, M.J., 2018. Long-term drivers of vulnerability and resilience to drought in the Zambezi- Save area of southern Africa , 1505 – 1830. Glob. Planet. Change 166, 94–106. https://doi.org/10.1016/j.gloplacha.2018.05.001

[4] Herslund, L., Mguni, P., 2019. Examining urban water management practices – Challenges and possibilities for transitions to sustainable urban water management in Sub-Saharan cities. Sustain. Cities Soc. 48, 101573. https://doi.org/10.1016/j.scs.2019.101573

[5] Li, X., Philp, J., Cremades, R., Roberts, A., He, L., 2016. Agricultural vulnerability over the Chinese Loess Plateau in response to climate change : Exposure , sensitivity , and adaptive capacity. Ambio 45, 350–360. https://doi.org/10.1007/s13280-015-0727-8

[6] Mashhadi Ali, A., Shafiee, M.E., Berglund, E.Z., 2017. Agent-based modeling to simulate the dynamics of urban water supply: Climate, population growth, and water shortages. Sustain. Cities Soc. 28, 420–434. https://doi.org/10.1016/j.scs.2016.10.001

[7] Sharvelle, S., Dozier, A., Arabi, M., Reichel, B., 2017. A geospatially-enabled web tool for urban water demand forecasting and assessment of alternative urban water management strategies. Environ. Model. Softw. 97, 213–228. https://doi.org/10.1016/j.envsoft.2017.08.009

[8] Shen, Z., Zhang, Q., Singh, V.P., Sun, P., Song, C., Yu, H., 2019. Agricultural drought monitoring across Inner Mongolia , China : Model development , spatiotemporal patterns and impacts. J. Hydrol. 571, 793–804. https://doi.org/10.1016/j.jhydrol.2019.02.028

[9] Sletto, B., Tabory, S., Strickler, K., 2019. Sustainable urban water management and integrated development in informal settlements : The contested politics of co-production in Santo Domingo , Dominican Republic. Glob. Environ. Chang. 54, 195–202. https://doi.org/10.1016/j.gloenvcha.2018.12.004

[10] Smith, A.B., Katz, R.W., 2013. US billion-dollar weather and climate disasters: Data sources, trends, accuracy and biases. Nat. Hazards 67, 387–410. https://doi.org/10.1007/s11069-013-0566-5

[11] Workalemahu, L., Byskov, L., Mguni, P., 2019. What makes a champion for landscape-based storm water management in Addis Ababa? Sustain. Cities Soc. 46, 101378. https://doi.org/10.1016/j.scs.2018.12.006

[12] Zhang, X., Chen, N., Sheng, H., Ip, C., Yang, L., Chen, Y., Sang, Z., Tadesse, T., Lim, T.P.Y., Rajabifard, A., Bueti, C., Zeng, L., Wardlow, B., Wang, S., Tang, S., Xiong, Z., Li, D., Niyogi, D., 2019. Urban drought challenge to 2030 sustainable development goals. Sci. Total Environ. 693, 133536. https://doi.org/10.1016/j.scitotenv.2019.07.342

Overview of Deep Convolutional Neural Network Approaches for Satellite Remote Sensing Ship Monitoring Technology

Xiunan Li, Peng Chen and Kaiguo Fan

State Key Laboratory of Satellite Ocean Environment Dynamics, Second Institute of Oceanography, Ministry of Natural Resources, Hangzhou, Zhejiang Province, China.
*Email: chenpeng@sio.org.cn

Abstract. Remote sensing (RS) monitoring of ships has important significance in both military and civilian fields. The RS ship detection aims to locate the position of the ship in the remote sensing image and extract its characteristics. Traditional ship target detection algorithms cannot meet the demands for speed and precision of SAR remote sensing and optical remote sensing data. With the development of artificial intelligence technology, the target detection technology such as deep learning algorithms has made significant progress in RS ship detection. Deep learning has become a heated topic in research. This paper has analyzed and summarized previous researches on the application of deep learning algorithms in ship detection technologies based on SAR and optical remote sensing images in recent years and has provided suggestions for future studies. In the future, deep learning-based technologies for RS ship detection will use more data, such as data from multiple sensors in multiple channels. Deep neural networks will also include more rules and specialized knowledge. Its structure will become more complicated and eventually develop into a neural network like the human brain.

1. Introduction

With the rapid development of remote sensing (RS) information science, RS technology has been widely used in military and civilian fields as a comprehensive technology and has played an important role in terrestrial resource survey, ocean exploration, military investigation, strike analysis, and assessment [1]. Satellite RS ship monitoring can be applied to many areas including management of marine fisheries, water traffic management and supervision, and surveillance of sewage from ships. Synthetic Aperture Radar (SAR) is not affected by clouds or rain, so it can be used 24 hours per day[2]. It is a main research direction of RS ship detection. While a visible light (VL) image is susceptible to factors such as light and clouds, it has high resolution and clear texture, and it can well distinguish a target in the image. On clear and sunny days, the image has rich information and shows clear features of the target structure. VL images have incomparable advantages in the field of maritime reconnaissance, especially ship identification. VL images compliment SAR images for marine target monitoring [3, 4].

Both optical and SAR remote sensing images have seen development bottlenecks using traditional ship detection algorithms. For example, the terrain effect in SAR images and effects of thin clouds in optical images still exist. Thanks to the development of artificial intelligence, the Convolutional Neural Networks (CNN) technology has made significant progress in the fields of image classification, detection, and segmentation. The progress of deep learning-based CNNs on ordinary images has drawn attention to many scholars on introducing deep learning into RS ship detection. This paper will summarize, analyze, and discuss recent years' researches on the use of deep learning technology in RS ship detection, and will discuss the future development of RS ship detection.

2. Development of CNN in Target Detection

This section describes the normal structure of a manuscript and how each part should be handled. LeCun et al. have established LeNet5[5], a modern neural network architecture of CNN, mainly for handwritten character recognition (Figure 1). The structure of a CNN has three features: local connections, weight sharing, and spatial or temporal sub-sampling. These features allow a CNN to be invariant to a certain degree of translations, scaling, and distortions [6]. AlexNet[7] proposed by Hinton et al. in 2012 was the winner of the ImageNet competition. Following AlexNet, a larger number of deeper CNN structures have been established, such as VGG[8], GoogleNet[9] and ResNet[10]. The record of the ImageNet competition has been constantly refreshed.

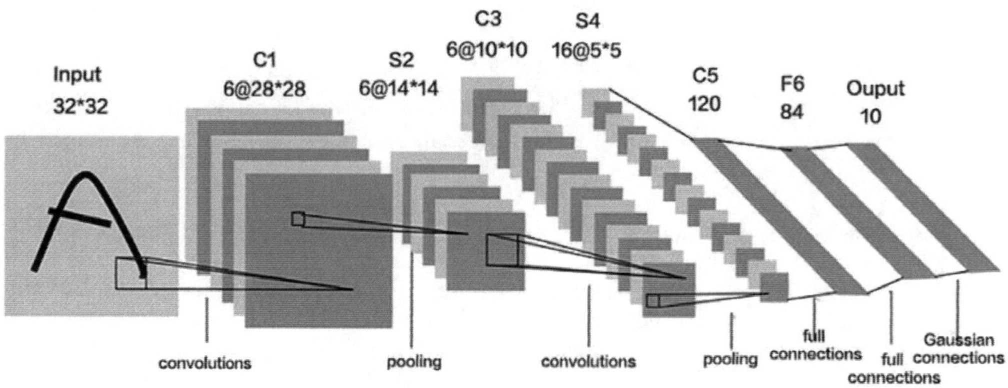

Figure 1. Schematic Diagram of CNN Structure [5]

Image classification determines the type of subject in an image, for example, whether the subject is a cat or a person in an image, while target detection determines the type of object in the image and also locates the subject, combining image classification and positioning. Given the great success of CNN in image classification, researchers have begun to improve it and introduce it into the field of target detection, in which it has also made significant progress. Several target detection frameworks have been born and are commonly used nowadays, such as the region proposal-based Faster R-CNN[11], regression-based YOLO[12], and SSD[13] algorithm.

3. Application of CNN to Ship Detection in SAR Imagery

Ship detection in SAR imagery has the following two problems: missed detections of small targets and the false alarms occurred with noises in blurred land, small islands, and azimuthal directions. CNN can be used for classification and it be used as a network for feature extraction of the target detection framework. There are two common approaches to solve false alarms and missed detections with CNNs. One is to combine traditional detection algorithms and deep learning algorithms; the other is to use the target detection framework with deep learning for ship detection.

3.1. Ship Detection Based on CFAR and Deep Learning Algorithm

Constant false alarm rate (CFAR)[14,15] is the most in-depth and widely-used algorithm among traditional target detection algorithms. Its key concept is modelling the distribution of sea clutter with regional sea clutter statistics. The threshold for dividing target pixels is obtained based on the density curves of the model and the false alarm rate to obtain, which is then used for detecting targets with high grayscale values in an SAR image[16]. A large number of studies and experiments have shown that even at complicated sea state, CFAR detectors can still produce quite satisfactory detection results [17]. It is generally believed that the sea clutter has a Rayleigh distribution with a calm sea and a K distribution at most sea states. Considering that, most of the SAR algorithms for ship detection use the CFAR detection with K distribution[18, 19]. In the face of various sea states, if a single distribution model is used, the detection accuracy may not be satisfactory in some cases. In 2000, Jiang et al.[20]

modelled a probability distribution of the radar backscattering from a sea surface using the probabilistic neural network (PNN). The algorithm can select the sea clutter model adaptively, but it has low accuracy in detecting images with high sea surface background noises. Chen et al.[21] have improved the algorithm which maintains the accuracy with high background noises and is faster than Jiang et al.'s algorithm.

CFAR detects the false alarms occurred with noises in blurred land, small islands, and azimuthal directions. An et al.[22] have found that CNN classification techniques can be used to remove false alarms and improve detection accuracy. Its technical flow chart is shown in Figure 2. Firstly, separate the sea and land by the fully convolutional network (FCN) with 18 layers (ten convolutional layers, five maxpool layers, three deconvolutional layers). Secondly, calculate the distribution of sea clutter and compare it with the K distribution, Gamma distribution and Rayleigh distribution. The results show that in terms of high scores, the Rayleigh distribution is close to the K distribution and is better than the Gamma distribution, and that the Rayleigh distribution is the most efficient in calculation. Thirdly, use the CFAR algorithm for ship detection; finally, classify the ship slices generated from the CFAR detection with a 7-layer deep CNN (two convolutional layers, two maxpool layers, three fully-connected layers) and remove false alarms. The experimental results of UFS images suggest that this method has effectively reduced the false alarm rate.

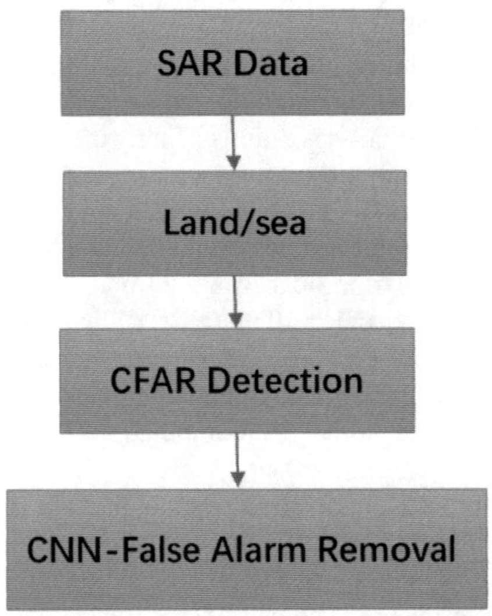

Figure 2. CFAR+CNN Flow Chart

Faster Region-CNN (R-CNN) is directly applied to ship target detection. Its missed detection rate is very high for ships, especially small targets. Kang et al. [23] believe that the missed detection is due to the low confidence score of the test frame generated from the Faster R-CNN of small ships. The deep learning-based detection algorithms are based on features. Small targets have few features, so the their detection should use the pixel-based CFAR detection. The specific detection process includes two steps (see Figure 3). The first step is to perform the Faster R-CNN detection with VGG16 as the network for feature extraction, and then generate regions through the Region Proposal Network (RPN), and finally obtain classification scores and coordinates in object classification. The second step is to determine targets through the CFAR detection of objects with the classification scores between 0.3 and 0.8. If an object has a score above 0.8, it is directly confirmed as a target. If its score is below 0.3, then it is considered as the background. The results of experiments on the Sentinue-1 data suggest that the detection rate of the experiment using is 16% higher than using Faster R-CNN alone, but the false alarm rate has increased by 3.6%. This is because the CFAR algorithm inevitably introduces a small number of false alarms while increasing the detection rate.

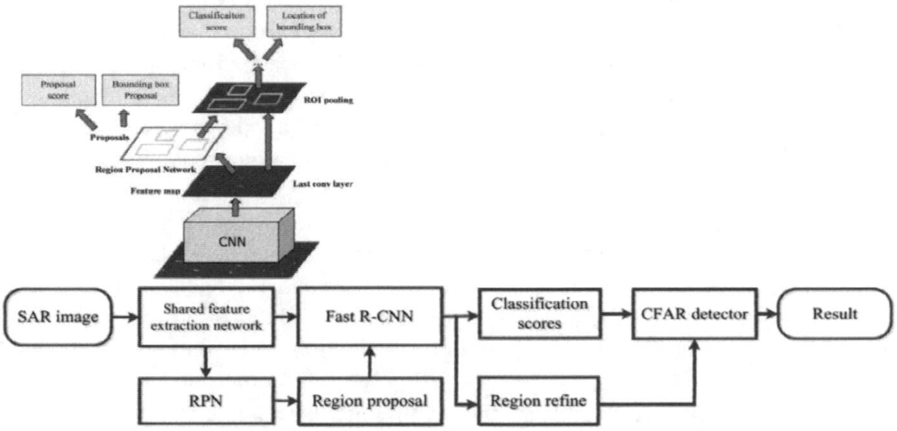

Figure 3. Faster R-CNN+CFAR Detection [23]

The above two schemes both adopt a multi-model strategy, combining the CFAR and CNN for ship target detection. The algorithm is mainly based on the traditional CFAR algorithm, while the CNN is only used for false alarm removal. The combination of Faster R-CNN and CFAR is mainly based on the target detection framework with deep learning, while the CFAR is used for tackling missed detections. Multi-model approaches usually run slower than single-model approaches. The calculation of the parameters of the CFAR algorithm also require a large amount of data.

3.2. Ship Detection Based on Target Detection Framework with Deep Learning

It is not satisfactory to apply deep learning-based CNN models to SAR images. This is because a general CNN is only proposed in the region of the last layer of feature maps. A CNN has higher resolution in shallow layers and higher semantic information in deep layers. Small-scale targets have little information on deep network feature maps. Considering the above, many scholars have improved the feature extraction network and explored the combinations of information at different levels in order to solve this problem.

It is generally believed that the deeper the CNN is, the better the performance is. However, problems such as gradient disappearance or gradient explosion arise with the increase of network depth, making the network difficult to be trained. ResNet can make the training easier while increasing the depth of the network. However, when it comes to small targets, it is difficult to retain information about small targets in feature maps generated by deep networks. Therefore, Jiao J. et al.[24] merged the feature maps generated from four residual blocks of ResNet101. The specific combining process is: first, make the feature map in the last three stages of rennet the same as the one in the first stage by sampling grid sizes of its nearest neighbor. Second, make the number of channels in the last three stages the same as that in the first stage through a 1x1 convolution. Third, add them up for combining. Finally, obtain the feature map we need through a 3×3 convolution (see Figure 4). According the tests on RadarSat2, TerraSAR-X, and Sentinue-1, the accuracy rate was up to 92.8% and the F1 score was 0.879.

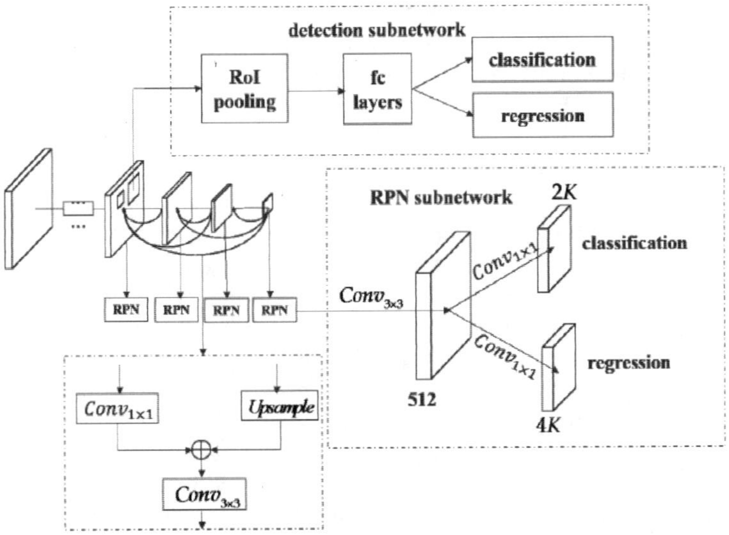

Figure 4. Target Detection Process [24]

Contextual information helps to increase confidence in decision making. Kang et al.[25]use not only multilayer fusion but also contextual information to improve the accuracy of detection. They have used VGG16 as the feature extraction network. In VGG16, conv1_2, conv2_2, conv3_3, conv4_3, and conv5_3 are called conv1, conv2, conv3, conv4, and conv5, respectively. Conv1 was reduced by max pooling; con3 was made in the size as conv2 through sampling; and the three feature maps were merged by L2 regularization. The borders of three layers were obtained from ROI pooling. After extracting context features, three ROI features and the three context features were combined and input into two fully-connected layers. The classification and the borders were obtained by fusing the fully-connected layers (see Figure 5). The F1 score was 0.873 with the Sentinue-1 data.

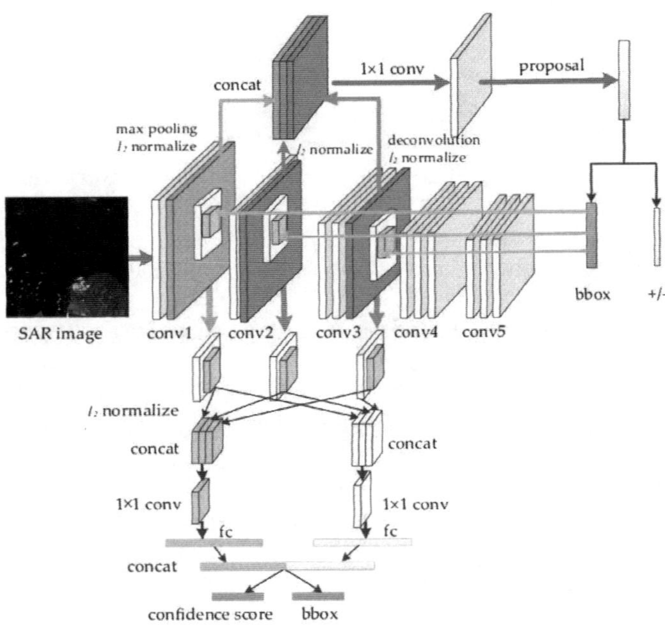

Figure 5. Fused-layer CNN with Contextual Information [25]

Single deep learning detection frameworks can meet the requirements of ship detection accuracy. The key point is the fusion of different feature layers. Feature fusion aims to provide solutions to multi-scale ship detection problems, especially the information loss in deep CNNs for small targets. There are no answers to the problems of false alarms with some land and islands and blurred azimuthal directions.

4. Optical Image for Ship Inspection

The difficult issues in the field of target detection in optical remote sensing imagery remain to be the detection of small targets, the impact of complicated near-shore sea states, and the false alarms caused by cloud coverage and broken clouds. Traditional optical RS target detection algorithms, such as the algorithm based on image gray values, are suitable for cases where the sea is relatively calm with well-distributed texture and low gray-level waterbody[26, 27, 28]. Edge-based algorithms are influenced by a lot of interference from the edges of large wave and light wave blocks[29]. In terms of algorithms based on fractal theory and fuzzy logic, if an image contains disturbance of clouds, the self-similarity of background will decrease, and the fitting error of fractal model will be quite large [30]. All the above traditional algorithms will have a lower detection accuracy at complicated sea states.

Deep learning applications for ship detection in optical imagery are mainly based on three major detection frameworks, namely Faster R-CNN, SSD, and YOLO, together with the improved versions of these frameworks.

Unlike SAR images, optical remote sensing images are more similar to natural images and match human vision more. As deep CNNs have slow convergence speed and small data set, overfitting is easy to occur in training. Therefore, people often train deep CNNs using migration learning. Nie et al. [31] have used the SSD detection framework, VGG-16 for feature extraction and migration learning to train the network, as shown in Figure 6. They tested on the Google Earth data and discovered the average precision (AP) of the SSD framework can reach 87.9% using migration learning. This has proven the effectiveness of migration learning for detecting ship targets in optical remote sensing images.

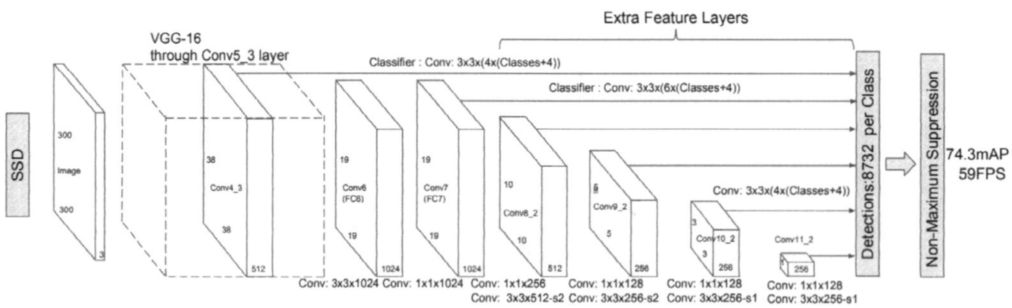

Figure 6. Flow Chart of SSD Detection [13]

Ships are multi-scale on remote sensing images. Large ships may have hundreds of thousands of pixels, while small ships may have only a dozen pixels. Li et al.[32] believe that information about shapes may be more important than local details regarding the identification of multi-scale ships. Therefore, they have proposed the Hierarchical Selective Filtering-Net (HSF-Net) based on the Faster R-CNN target detection algorithm (see Figure 7). With respects to HSF-Net, three different sizes of kernels were used for different proportions of anchor frames. The receptive fields from the smallest to the largest are 1×1, 3×3, and 5×5. 1×1 kernel was used for detecting small-scale targets, while 5×5 kernel was for detecting large-scale targets. Tests were conducted with Google Earth, GaoFen-2, and UAV data, and the AP achieved 89.32%.

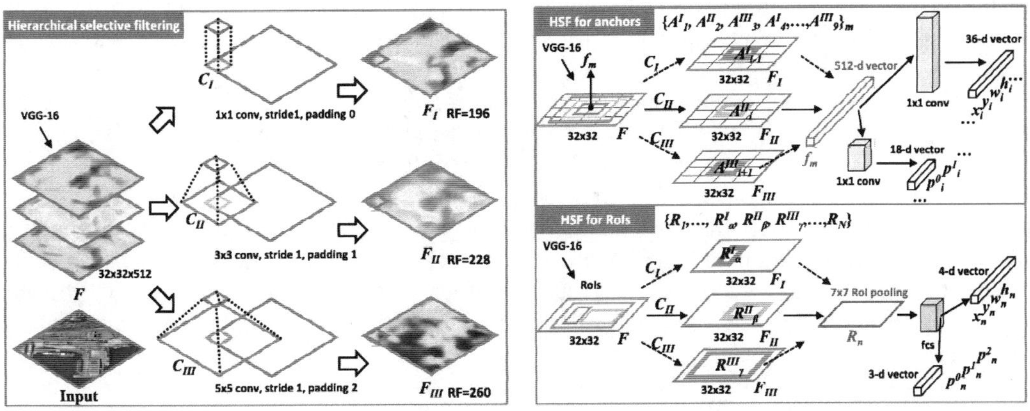

Figure 7. Schematic diagram of HSF-Net [32]

The detection of congested small-scale ships has always been a daunting task in the field of ship detection. Congested ships confuse the algorithm, so the algorithm takes multiple ships as one ship. Then, ships are often marked by only one test frame, causing missed detection of ships. Van Etten et al. have proposed a You Only Look Twice (YOLT) [33] framework based on the YOLO detection framework to tackle the detection of congested small-scale ships. In an original YOLO network, a 416×416 output picture is converted into a 13×13 feature map through feature extraction by 32 time-down sampling. If the target is less than 32 pixels, it will be difficult to be detected. Therefore, the author used the 16-time down sampling to obtain a 26×26 feature map. This method has enriched the information of small-scale ships and has also connected the information in shallow and deep layers with the passthrough layer. These strategies all contribute to the detection of small targets. The F1 score of the Digital Globe data for congested small-scale ships was 0.82.

Optical remote sensing images can accelerate the training of the network through migration learning. On the last layer of feature maps, using larger feature maps and kernels in different sizes can help solve the problems related to multi-scale ships and congested ships. However, the above studies did not consider the multi-channel characteristics of optical remote sensing images, and they did not analyze the remote sensing images with complicated sea states and cloudy sky.

5. Conclusion and Discussion

Remote sensing ship detection with deep CNNs has made great progress so far. A few studies have provided solutions to the problems occurred in traditional SAR and optical RS algorithms for ship detection. For example, concerning the difficulty small targets posed on deep learning target detection, it is generally believed that using deep learning directly does not have satisfactory detection results. Different strategies and methods are required to integrate shallow and deep layers to highlight the information of small targets. The combination of CFAR and CNN has borrowed the classification capacity of CNN. The algorithm is mainly based on the CFAR algorithm, while the CNN plays a role of reducing false alarms. The combination of Faster R-CNN and CFAR is mainly based on the deep learning-based target detection framework, while the CFAR is used for tackling missed detections. Both the above two methods adopt the multi-model strategy. As two models are used instead of one, a multi-model algorithm usually run more slowly than a single-model algorithm, and it requires a large amount of data to estimate some parameters of the CFAR algorithm. In terms of the influence of cloud on optical remote sensing, deep learning algorithms have good performance if the data is large enough. Our own experiments also show that its performance surpasses traditional algorithms, which is a very big progress. All the studies mentioned previously have failed to fully analyze and discuss the false alarms caused by islands, land margins, and other factors, which will also be a difficult task in follow-up studies.

In the future, deep learning-based RS ship detection will develop in the direction of business application. Based on the analysis of current issues, we believe that the technology can be further improved in the following aspects:

(1) With respects to date, multi-channel remote sensing data should be used to their full potential. SAR data have multiple polarization modes, and optical data have multiple channels. The above studies did not consider the influence of SAR data polarization on ship detection. In terms of optical data, only traditional true color images were used for ship detection. Future research should use more data from channels, such as four channels of high-resolution optical data, in order to improve detection performance. Where the training samples are small, data enhancement strategies such as rotation, flipping, and random addition of noise can be used to improve the generalization ability of the model. Training samples should be large enough to cover as many types of false alarms as possible including small islands.

(2) With regard to the network, future research should focus on how to effectively add some priori or specialized domain rules or knowledge to make it reflected in the network. An example of that would be using multiple classification networks. Regarding convolution implementation of sliding windows, if the scale of the ship changes greatly, different window sizes will be used. This will increase the calculation amount. If the ships are congested, and the results can be quite poor, hierarchical classification can be introduced to the structure design.

In summary, deep learning convolutional neural networks have significantly promoted the development of remote sensing technology for ship detection. Deep learning algorithms are essentially data-driven. Remote sensing data will see an explosive growth due to the development of sensing technology. In the future, the remote sensing technology for ship detection may no longer be required to distinguish optical data and SAR data, or to consider different imaging modes with different resolutions. Building and training an end-to-end model could suffice. Let us wait and see.

6. Acknowledgement

This work was supported by the National Natural Science Foundation of China (No.41476088).

7. References

[1] Chen L, Wang Z R, Han Z, Wang G Q, Zhou H T, Shi H, Hu C and Long T 2017 A review of ship detection and recognition based on optical remote sensing image. *science and technology review*, **35**(20): 77-85.

[2] Nieto-Hidalgo M, Gallego A J, Gil P and Pertusa A 2018 Two-Stage convolutional neural network for ship and spill detection using SLAR images. *IEEE Transactions on Geoscience and Remote Sensing*, **56**(9):5217-30.

[3] Wang Y Q, Ma L and Tian Y 2011 State-of-the-art of ship detection and recognition in optical remotely sensed imagery. *Acta Automatica Sinica*, **37**(9):1029-39.

[4] Liu F, Shen T S and Ma X X 2017 Convolutional neural network based multi-band ship target recognition with feature fusion. *Acta Optica Sinica* ,**37**(10): 1015002.

[5] Lecun Y, Bottou L, Bengio Y, and Haffner P 1998 Gradient-based learning applied to document recognition. *Proceedings of the IEEE*, **86**(11): 2278-2324.

[6] Zhou X Y, Wang K, Li and L Y 2017 Review of object detection based on deep learning. *Electronic Measurement Technology*, **40**(11):89-93.

[7] Krizhevsky A, Sutskever I and Hinton G E 2012 ImageNet classification with deep convolutional neural networks. *Proceedings of the 25th International Conference on Neural Information Processing Systems. Lake Tahoe, Nevada: Curran Associates Inc.*:1097–1105.

[8] Simonyan K and Zisserman A 2014 Very deep convolutional networks for large-scale image recognition. (preprint arXiv:1409.1556).

[9] Szegedy C, Liu W, Jia Y Q, Serment P, Reeds Scott, Anguelov D, Erhan D, Vanhoucke V and Rabinovick A 2015 Going deeper with convolutions. *Proceedings of the IEEE conference on computer vision and pattern recognition*, 1-9.

[10] He K M, Zhang X Y, Ren S Q and Sun J 2016 Deep residual learning for image recognition. *Proceedings of the IEEE conference on computer vision and pattern recognition*,770-778.

[11] Ren S Q, He K M, Girshick R and Sun J 2017 Faster R-CNN:towards real-time object detection with region proposal networks. *IEEE Transactions on Pattern Analysis and Machine Intelligence*,**39**(6):1137-1146.

[12] Redmon J, Divvala S, Girshick R and Farhadi A 2016 You only look once: unified, real-time object detection. *Proceedings of the IEEE conference on computer vision and pattern recognition*, 779-788.

[13] Liu W, Anguelov D, Erhan D, Szegedy C, Reed S, Fu C Y, and Berg A C 2016 SSD:single shot multibox detector. *European conference on computer vision, Springer, Cham,* 21-37.

[14] Eldhuset K 1996 An automatic ship and ship wake detection system for spaceborne SAR images in coastal regions. *IEEE transactions on Geoscience and Remote Sensing*, **34**(4): 1010-19.

[15] Wackerman C C, Friedman K S, Pichel W G, Clement-colon P and Li X 2001 Automatic detection of ships in RADARSAT-1 SAR imagery. *Canadian Journal of Remote Sensing*, **27**(5): 568-577.

[16] Gao G, Liu L, Zhao L J, Shi Gong T and Kuang G Y 2009 An adaptive and fast CFAR algorithm based on automatic censoring for target detection in high-resolution SAR images. *IEEE transactions on geoscience and remote sensing*, **47**(6): 1685-1697.

[17] Tang M N, Lin T Q and Wen G J 2011 Overview of ship detection methods in remote sensing image. *Application Research of Computers*, **28**(1):29-36.

[18] Vachon P W, Thomas S J, Cranton J,Edel H R and Henschel M D 2000 Validation of ship detection by the RADARSAT synthetic aperture radar and the ocean monitoring workstation. *Canadian Journal of Remote Sensing*,**26**(3): 200-212.

[19] Vachon P W, Adlakha P, Edel H, Henschel M, Bruce Ramsay B, Flett D, Rey M, Staples G and Thoms S 2000 Canadian progress toward marine and coastal applications of synthetic aperture radar. *Johns Hopkins APL Technical Digest*, **21**(1): 33-40.

[20] Jiang Q, Aitnouri E, Wang S and Zhou D 2000 Automatic detection for ship target in SAR imagery using PNN-model. *Canadian Journal of Remote Sensing*, **26**(4): 297-305.

[21] Chen P, Huang W G, Fu B and Shi A Q 2005 An improved CFAR algorithm for ship detection in SAR imagery. *Journal of Remote Sensing*, **9**(3):260-264.

[22] An Q Z, Pan Z X, You H J 2018 Ship detection in Gaofen-3 SAR images based on sea clutter distribution analysis and deep convolutional neural network. *Sensors*,**18**(2): 334.

[23] Kang M, Leng X G, Lin Z and Ji K F 2017 A modified faster R-CNN based on CFAR algorithm for SAR ship detection. *2017 International Workshop on Remote Sensing with Intelligent Processing (RSIP). IEEE*,1-4.

[24] Jiao J, Zhang Y, Sun H, Yang X, Gao X, Hong W, Fu K and Sun X 2018 A densely connected end-to-end neural network for multiscale and multiscene SAR ship detection. *IEEE Access*, **6**: 20881-20892.

[25] Kang M, Ji K F, Leng X G and Lin Z 2017 Contextual region-based convolutional neural network with multilayer fusion for SAR ship detection. *Remote Sensing*, **9**(8): 860.

[26] Wang M, Luo J C and Ming D 2005 Extract ship targets from high spatial resolution remote sensed imagery with shape feature. *Geomatics and Information Science of Wuhan University*, **30**(8): 685-688.

[27] You X J, Xu S S and Hou L 2015 A new method for ship detection based on feature fusion in optical image. *Computer Engineering and Applications*, **41**(19):199-202.

[28] Xiao L P, Cao J and Gao X Y 2007 Detection for ship targets in complicated background of sea and land. *Opto-Electronic Engineering*, **34**(6):6-10.

[29] Zhu C G, Zhou H, Wang R S and Guo J 2010 A novel hierarchical method of ship detection from spaceborne optical image based on shape and texture features. *IEEE Transactions on geoscience and remote sensing*, **48**(9): 3446-3456.

[30] Zhang D X, He S H and Yang S Q 2009 Ship targets detection method based on multi-scale fractal feature. *Laser and Infrared*, **39**(3):315-318.

[31] Nie G H, Zhang P, Niu X, Dou Y and Xia F 2017 Ship detection using transfer learned single shot multi box detector. *ITM Web of Conferences. EDP Sciences*, **12**: 01006.

[32] Li Q P, Mou L C, Liu Q J,Wang Y H and Zhu X X 2018 HSF-Net: Multiscale deep feature embedding for ship detection in optical remote sensing imagery. *IEEE Transactions on Geoscience and Remote Sensing*, (99): 1-15.

[33] Van Etten A 2018 You only look twice: Rapid multi-scale object detection in satellite imagery. (preprint arXiv:1805.09512).

AUTHOR INDEX